Emerging Markets

Performance, Analysis and Innovation

CHAPMAN & HALL/CRC FINANCE SERIES

Series Editor

Michael K. Ong

Stuart School of Business
Illinois Institute of Technology
Chicago, Illinois, U. S. A.

Aims and Scopes

As the vast field of finance continues to rapidly expand, it becomes increasingly important to present the latest research and applications to academics, practitioners, and students in the field.

An active and timely forum for both traditional and modern developments in the financial sector, this finance series aims to promote the whole spectrum of traditional and classic disciplines in banking and money, general finance and investments (economics, econometrics, corporate finance and valuation, treasury management, and asset and liability management), mergers and acquisitions, insurance, tax and accounting, and compliance and regulatory issues. The series also captures new and modern developments in risk management (market risk, credit risk, operational risk, capital attribution, and liquidity risk), behavioral finance, trading and financial markets innovations, financial engineering, alternative investments and the hedge funds industry, and financial crisis management.

The series will consider a broad range of textbooks, reference works, and handbooks that appeal to academics, practitioners, and students. The inclusion of numerical code and concrete real-world case studies is highly encouraged.

Published Titles

Decision Options®: The Art and Science of Making Decisions, **Gill Eapen**

Emerging Markets: Performance, Analysis, and Innovation, **Greg N. Gregoriou**

Introduction to Financial Models for Management and Planning, **James R. Morris and John P. Daley**

Stock Market Volatility, **Greg N. Gregoriou**

Forthcoming Titles

Portfolio Optimization, **Michael J. Best**

Proposals for the series should be submitted to the series editor above or directly to:
CRC Press, Taylor & Francis Group
4th, Floor, Albert House
1-4 Singer Street
London EC2A 4BQ
UK

CHAPMAN & HALL/CRC FINANCE SERIES

Emerging Markets

Performance, Analysis and Innovation

Edited by

Greg N. Gregoriou

State University of New York (SUNY)

Plattsburgh, U. S. A.

CRC Press
Taylor & Francis Group
Boca Raton London New York

CRC Press is an imprint of the
Taylor & Francis Group, an **informa** business
A CHAPMAN & HALL BOOK

Neither the editor nor the publisher is responsible for the accuracy of each chapter.

CRC Press
Taylor & Francis Group
6000 Broken Sound Parkway NW, Suite 300
Boca Raton, FL 33487-2742

© 2010 by Taylor and Francis Group, LLC
CRC Press is an imprint of Taylor & Francis Group, an Informa business

Printed in the United States of America on acid-free paper
10 9 8 7 6 5 4 3 2 1

International Standard Book Number: 978-1-4398-0448-3 (Hardback)

Library of Congress Cataloging-in-Publication Data

Emerging markets : performance, analysis and innovation / editor, Greg N. Gregoriou.
     p. cm.
   Includes bibliographical references and index.
   ISBN 978-1-4398-0448-3 (hardcover : alk. paper)
   1. Investments--Developing countries. 2. Stock exchanges--Developing countries.
3. Structural adjustment (Economic policy)--Developing countries. 4. International economic integration. I. Gregoriou, Greg N., 1956- II. Title.

HG5993.E5634 2009
332.64'2091724--dc22   2009018855

Visit the Taylor & Francis Web site at
http://www.taylorandfrancis.com

and the CRC Press Web site at
http://www.crcpress.com

Contents

Preface

Emerging markets experienced dramatic growth in the early 1990s when investors worldwide opted to complement traditional portfolios, due to their low positive correlation, with developed markets. Although the ride has been bumpy, investors should not neglect this area. Because emerging markets rely heavily on the U.S. economy, their long-term prospects appear promising even though they are decoupling from developed markets. I hope this collection of exclusive chapters sheds some light on what lies ahead for emerging markets with the latest research from both academics and practitioners.

Acknowledgments

I would like to thank the handful of anonymous referees for reading and selecting the final chapters for this book. I also thank the finance editor, Dr. Sunil Nair, at Chapman-Hall/CRC Press/Taylor & Francis Group in London for his valuable comments and suggestions as well as Sarah Morris and Jessica Vakili for their support. I also extend my thanks to Suganthi Thirunavukarasu, project manager, for her assistance.

Greg N. Gregoriou

Editor

Greg N. Gregoriou is a professor of finance in the School of Business and Economics at the State University of New York (Plattsburgh). A native of Montreal, Professor Gregoriou obtained his joint PhD in finance at the University of Quebec at Montreal, which merges the resources of Montreal's four major universities: McGill, Concordia, Quebec and HEC. Professor Gregoriou's interests focus on hedge funds and managed futures. In addition to his university studies, Greg has completed several specialized courses from the Canadian Securities Institute. Greg has published over 50 academic articles in over a dozen peer-reviewed journals, such as the *Journal of Portfolio Management*, the *Journal of Futures Markets*, the *European Journal of Operational Research*, the *Annals of Operations Research*, the *Computers and Operations Research*, as well as 20 book chapters. Greg is a hedge fund editor and an editorial board member for the *Journal of Derivatives and Hedge Funds*, a London-based academic journal, and also an editorial board member of the *Journal of Wealth Management* and the *Journal of Risk and Financial Institutions*. He has published 31 books with Chapman-Hall/CRC Press, John Wiley & Sons, McGraw-Hill, Elsevier Butterworth-Heinemann, Palgrave-MacMillan, and Risk books.

Contributor Bios

David E. Allen is a professor of finance at Edith Cowan University, Perth, Western Australia. He is the author of three monographs and over 70 refereed publications on a diverse range of topics covering corporate financial policy decisions, asset pricing, business economics, funds management and performance bench marking, volatility modeling and hedging, and market microstructure and liquidity.

Mohamed El Hedi Arouri is currently an associate professor of finance at the University of Orleans, France, and a researcher at the EDHEC Business School, France. He received his master's degree in economics and his PhD in finance from the University of Paris X Nanterre. His research focuses on the cost of capital, stock market integration, and international portfolio choice. He has published articles in refereed journals such as the *International Journal of Business and Finance Research, Frontiers of Finance and Economics, Annals of Economics and Statistics, Finance,* and *Economics Bulletin*.

Michael F. Bleaney is a professor of economics at the University of Nottingham, where he has taught since 1978. He received his BA and PhD from the University of Cambridge. He is an editor of the *Journal of African Economies* and *Economics Bulletin*, and has been a visiting scholar at the International Monetary Fund Research Department on several occasions. He has published over 80 papers in refereed journals, mainly on macroeconomics and development economics.

Paul Brockman is an associate professor and Matteson Professor of Financial Services at the University of Missouri-Columbia. Paul received his BA degree from Ohio State University (*Summa Cum Laude*), his MBA from Nova University, and his PhD (Finance) from Louisiana State University. Prior to his current position, Paul taught at the Hong Kong

Polytechnic University and the University of Manitoba. His private sector experience includes several years working as a CPA, cash manager, and commodity trader. Paul has published in a number of journals including the *Journal of Finance*, the *Journal of Financial Economics*, the *Journal of Financial and Quantitative Analysis*, the *Journal of Corporate Finance*, the *Journal of Empirical Finance*, the *Journal of Banking and Finance*, the *Journal of Financial Research*, *Financial Review*, and *Review of Quantitative Finance and Accounting*.

Robert D. Brooks is a professor in the Department of Econometrics and Business Statistics at Monash University. He has published a number of papers on empirical finance including papers in the *Journal of Banking and Finance*, the *Journal of International Money Finance*, and *Emerging Markets Review*.

Silvio John Camilleri is a lecturer in the banking and finance department at the University of Malta. He completed his PhD program at Loughborough University, focusing on the microstructure of emerging securities markets. Dr. Camilleri has published papers in journals and scholarly collections, including *Economics of Emerging Markets* (Lado Beridze, ed., Nova Science Publishers, 2008).

Thomas C. Chiang is the Marshall M. Austin Professor of Finance at Drexel University. His recent research interests are financial contagion, international finance, and financial econometrics. He is the author of numerous articles in refereed journals and two books. His articles have been published in the *Journal of International Money and Finance*; *Quantitative Finance*; the *Journal of Money, Credit and Banking*; and the *Pacific-Basin Finance Journal*, among others. Dr. Chiang received his PhD from the Pennsylvania State University, with a concentration in financial economics.

Joseline Chimhini is a former MBus student at Edith Cowan University, Perth, Western Australia, and is employed at the Commercial Bank of Zimbabwe.

Imed Chkir is an associate professor of finance at Telfer School of Management, University of Ottawa. His research interests include dividend policy, capital structure, and cross-listing. He has published in journals such as the *Journal of Financial Research* and the *Journal of Multinational Financial Management*. His most recent book is *Fondement de la Finance d'Entreprise*.

Lamia Chourou is an assistant professor of finance at the Faculty of Law, Economics & Political Sciences, University of Sousse. Her research interests include executive compensation and corporate governance. Dr. Chourou has participated in many finance conferences such as FMA, EFA, SFA, and EFMA. She has published in refereed journals such as the *Journal of Multinational Financial Management* and *Canadian Investment Review.*

Mehmet A. Civelek received his bachelor's degree from the Faculty of Economics, Istanbul University (1965); his MBA from Graduate School of Business, New York University (1971), and his PhD from the Graduate Faculty, New School for Social Research (1976). His long-term teaching assignments include Ege University (1976–1982), Yarmouk University (1982–1992), Eastern Mediterranean University (1992–1995), and Dokuz Eylül University (1995–1999). His research interests include stock market efficiency, banks' portfolio behavior, and empirical issues in monetary policy. Professor Civelek has published articles in national and international journals and books.

Carolyn V. Currie is a member of the Association of Certified Practising Accountants, the Chartered Secretaries Association, and a fellow of Finsia, a merger of the Australian Institute of Banking and Finance and the Securities Institute. Her experience represents almost four decades in the public and private sectors, as a merchant banker, regulator, internal auditor, and financial trainer. For the last 15 years, she has been a senior lecturer in financial services at the University of Technology, Sydney, as well as the managing director of her own consulting company and several private investment companies.

A. Fatih Dalkılıç graduated with a bachelor's and master's degree in business administration from Dokuz Eylül University. Dalkılıç spent his sophomore year at Saxion University in the Netherlands as an exchange student. His master's thesis was titled "Earnings management and its role on financial reporting." Since 2002, he has been working as a research assistant and continues his PhD studies on "Professional judgment in IFRS." Dalkılıç spent the 2008 spring term at the University of Wisconsin, River Falls, as an exchange academic staff.

Mathijs A. van Dijk is an associate professor of finance at the Rotterdam School of Management (Erasmus University). He was a visiting scholar at the Fisher College of Business (Ohio State University) during the academic year 2005–2006 and at the Fuqua School of Business (Duke University) in the period January–April 2008. His research focuses on international finance. His work on financial economics has been published in various journals, including the *Financial Analyst Journal*, the *Journal of International Money and Finance*, and the *Review of Finance*. He has presented his work at numerous international conferences as well as seminars at, among others, Dartmouth, Harvard, and INSEAD. In 2008, he received a large grant from the Dutch National Science Foundation for a 5-year research program on liquidity black holes.

George D. Dounias is an associate professor in the Department of Financial and Management Engineering at the University of the Aegean, Chios, Greece, and the director of the Management and Decision Engineering Laboratory. Dr. Dounias was also recently appointed the Head of the Department of Financial and Management Engineering, University of the Aegean. He received his diploma and his PhD in production and management engineering from the Technical University of Crete. His interests lie in the areas of artificial intelligence, decision making, and complex dynamic systems. Since 1997, Dr. Dounias has served as an EU project evaluator/reviewer for the DG-INFSO, Administrations Sector, and for the period 2002–2003 served as the vice-chairman of the Human Medical & Healthcare Committee for the European Research Network EUNITE. Since 2005, Dr. Dounias has represented the University of the Aegean in the European Research Network NISIS. He has published several research papers and has served as an editorial board member, a guest editor, a reviewer of international journals, and in organizing/program committees of international conferences, summer schools, and special sessions.

M. Banu Durukan received her bachelor's degree from the Faculty of Business, Dokuz Eylül University (1993). She obtained her MBA from the Graduate School of Management, Boston University (1995), and received her doctorate degree from the Graduate School of Social Sciences, Dokuz Eylül University (1997). Her research interests include the stock market, investments, behavioral finance, and capital structure. Professor Durukan is a member of the Dokuz Eylül University Faculty of Business and the head of the Division of Accounting and Finance. She has published articles in national and international journals and books. She is currently teaching and carrying out her research activities at the Faculty of Economics, University of Ljubljana.

Cumhur Ekinci is an assistant professor at Istanbul Technical University (ITU). He received his BA in economics from Bogazici University, his MA in finance from the University of Paris I Pantheon-Sorbonne, and his PhD in finance from the University of Aix-Marseille III. Dr. Ekinci worked in the trading room at CNAM in Paris and has given courses in financial markets, accounting, and investment at CNAM, the University of Aix-Marseille II, and ENPC in addition to ITU. His research interests include market microstructure; high-frequency data; and competition among market venues, hedge funds business, and algorithmic trading.

Craig Ellis is an associate professor of finance at the University of Western Sydney, Australia. His primary research interests include topics relating financial asset return distributions and the statistical and economic implications of nonrandom behavior for financial asset pricing. Craig has published and refereed numerous articles in journals including *Chaos Solitons and Fractals, Economics Letters, International Review of Financial Analysis*, and *Physica A*.

Dean Fantazzini is a lecturer in econometrics and finance at the Moscow School of Economics, Moscow State University. He graduated with honors from the Department of Economics at the University of Bologna (Italy) in 1999. He obtained his master's in financial and insurance investments at the Department of Statistics, University of Bologna (Italy) in 2000 and his PhD in economics in 2006 at the Department of Economics and Quantitative Methods, University of Pavia (Italy). Before joining the Moscow School of Economics, he was a research fellow at the Chair for Economics and Econometrics, University of Konstanz (Germany), and at the Department of Statistics and Applied Economics, University of Pavia (Italy). Dean is a specialist in time series analysis, financial econometrics, and multivariate dependence in finance and economics, with more than 20 publications.

Viviana Fernandez received her BA and master's in economics from the Catholic University of Chile, and her PhD in economics from the University of California at Berkeley. She is currently an associate professor in the Department of Industrial Engineering of the University of Chile, and an external research associate of the INFINITI Group, Trinity College Dublin. She has published in *The Review of Economics and Statistics, Studies of Nonlinear Dynamics and Econometrics, Energy Economics*, the *Journal* *of Financial Intermediation*, the *Journal of Futures Markets*, and *Physica A*, among others. She is currently an associate editor of the *International Review of Financial Analysis* (Elsevier).

Thomas J. Flavin is a lecturer in financial economics at the National University of Ireland, Maynooth. Thomas received his PhD in finance from the University of York (UK). He has been a visiting scholar at the Federal Reserve Bank of Atlanta, the University of Cambridge, and the University of York. He has published his research in leading peer-reviewed journals such as the *Journal of International Money and Finance, International Review of Economics and Finance, Applied Financial Economics*, the *Journal of Financial Markets, and Institutions and Money*, among others.

Don U.A. Galagedera is a senior lecturer in the Department of Econometrics and Business Statistics at Monash University. He has published a number of papers on empirical finance including papers in *Emerging Market Review*, the *Journal of Multinational Financial Management*, and *Quantitative Finance*.

Aaron Garay received his bachelor's degree in economics from Universidad del Pacifico (Peru) and has collaborated as a research assistant at Universidad del Pacifico Research Center (CIUP).

Mark D. Griffiths is the Jack Anderson Professor of Finance at the Farmer School of Business at Miami University, where he oversees the Student Investment Fund and teaches a wide variety of finance courses. He has authored numerous journal articles and several books, and conducts research on issues related to transactional impediments to entrepreneurship as well as issues in the United States and international money markets. He received his PhD from The University of Western Ontario.

Massimo Guidolin received his PhD from the University of California in 2000. He is a Chair Professor of Finance at Manchester Business School. He has also served as an assistant vice-president and senior policy consultant (financial markets) within the U.S. Federal Reserve system (St. Louis FED), where he still acts as an advisor. Since December 2007 he has been a codirector of the Center for Analysis of Investment Risk at Manchester Business School. His research focuses on predictability and nonlinear dynamics in financial returns, with applications to portfolio management, and sources and dynamics of volatility and higher-order moments in equilibrium asset pricing models. His research has been published in the *American Economic Review*, the *Journal of Financial Economics*, *Review of Financial Studies*, the *Journal of Business*, and the *Journal of Econometrics*, among others.

Javed Iqbal is a PhD student in the Department of Econometrics and Business Statistics at Monash University and a lecturer in the Department of Statistics at Karachi University. He has published a number of papers on empirical finance including papers in the *Journal of Multinational Financial Management* and the *International Journal of Business*.

Vassilis N. Karavas is the managing director at Credit Agricole Asset Management Alternative Investments. He has extensive experience in the alternative investments industry, and several years of experience in the area of information systems. He received his PhD in management science from the Isenberg School of Management at the University of Massachusetts, and his MSc and BSc in industrial engineering from the Technical University of Crete, Greece. He has presented his research at numerous professional and academic conferences worldwide. He is a coeditor of two books on hedge funds and CTAs, and has published in various academic and professional journals in the areas of quantitative methods, hedge fund performance, and asset allocation.

Dimitris Kenourgios is a lecturer in the Department of Economics at the University of Athens. He studied economics at the University of Athens (BSc, 1995) and banking and finance at the University of Birmingham, UK (MSc, 1996). He also holds a PhD in finance from the University of Athens, Faculty of Economics (2000). His main research interests are analyses of international financial markets and financial risk management.

Jill R. Kickul is the director of the Stuart Satter Program in Social Entrepreneurship in the Berkley Center for Entrepreneurship at New York University's Stern School of Business. She has authored numerous journal articles and received several best paper awards on entrepreneurship education development and curriculum design. Her current research interests include innovation strategies for new and emerging business as well as the evaluation and measurement of the impact of social ventures. She received her PhD from Northern Illinois University.

Berna Kirkulak is an assistant professor of finance at Dokuz Eylül University in Turkey. She received her PhD in economics from Hokkaido

University in Japan. The majority of her research centers around initial public offerings, venture capital, mergers and acquisitions, dividend policy, and corporate governance. She has published articles in Turkish, English, and Japanese. Her research studies were funded by grants from organizations such as the Japanese Ministry of Education (MONBUSHO), the Turkish Scientific and Technical Research Institution (TUBİTAK), the Dutch Ministry of Education (NUFFIC), and the Ministry of Education of the People's Republic of China. Recently, she was appointed as a visiting scholar at Southeast University in China.

Bülent Köksal received his PhD in economics from Indiana University, Bloomington, in 2005 by completing a market microstructure dissertation, which examined the strategies of the NYSE specialists. Currently, he is in the economics department of Fatih University in Istanbul and teaches courses in econometrics and statistics. His research is primarily concerned with issues related to the connection between aggregate economic activity and the stock market, the effect of political events on the stock market, and the linkages between market microstructure variables and asset pricing models.

Thomas Lagoarde-Segot received his BA and MSc degrees in economics from the Université de la Méditerranée (France) and his PhD in finance from Trinity College Dublin (Ireland). He is an assistant professor of finance at Euromed Marseille, Ecole de Management, where he teaches corporate finance, emerging markets finance, and research methods. His academic research focuses on emerging markets finance and development economics.

R. McFall Lamm, Jr. is the chief investment officer for Stelac Advisory Services in New York. He was previously the chief strategist for Deutsche Bank's global hedge fund business and investment management group in London. Dr. Lamm is a frequent speaker at conferences and events around the globe. He also writes market commentary that is disseminated worldwide and is often quoted in the news media. In addition, Dr. Lamm is an energetic writer, having published numerous book chapters and professional articles in publications such as the *Journal of Portfolio Management*, the *Journal of Economic Dynamics and Control*, the *Journal of Alternative Investments*, and many others.

Laurence Le Poder received her master's in economics and her PhD in finance from the Université Paul Cézanne (France). She is an associate professor of economics at Euromed Marseille, Ecole de Management. She teaches economics, banking economics, monetary economics, and international economic environment and financial systems. Her academic research is on financial institutions and development economics.

Peter B. Lerner received his undergraduate and graduate education in physics at the Moscow Institute for Physics and Technology and the Lebedev Institute for Physical Sciences. He conducted research at the Los Alamos National Laboratory and the Pennsylvania State University. During this time, he authored more than 50 papers and book contributions in optics, atomic physics, materials science, and arms control. In 1998, Peter graduated from the Katz School of Business (University of Pittsburgh) with an MBA and worked for 2 years as a risk quant in energy trading. He received his PhD in finance from Syracuse University in 2006.

Kian-Ping Lim is a senior lecturer at the Labuan School of International Business and Finance, Universiti Malaysia Sabah. He is currently on study leave to pursue his PhD at Monash University.

Stephen J. Lubben is the Daniel J. Moore Professor of Law at Seton Hall University School of Law. Professor Lubben joined Seton Hall after several years in practice with Skadden, Arps, Slate, Meagher & Flom in New York and Los Angeles, where he represented parties in Chapter 11 cases throughout the country. He received his bachelor's degree from the University of California, Irvine; his JD, magna cum laude, from Boston University School of Law, where he was an editor of the *Boston University Law Review*; and his LLM from Harvard Law School, where he was a teaching fellow. Following graduation from Boston University, he clerked for the now Chief Justice John T. Broderick, Jr. of the New Hampshire Supreme Court. Professor Lubben was the principal investigator under a $345,000 grant from the American Bankruptcy Institute that funded the 2007 ABI Chapter 11 Fee Study, the leading empirical study of professional fees in Chapter 11 bankruptcy cases. His recent research has focused on professionals in Chapter 11 and the effect of credit default swaps on Chapter 11 reorganizations. He is a frequent speaker at distressed investing and corporate reorganization conferences throughout the world.

Gary McCormick is a visiting assistant professor at North Texas University. Gary received his BA degree from Lock Haven State University, his MS from Villanova University, his MBA from Pennsylvania State University-Great Valley, and his PhD (Finance) from the University of Missouri. Previously, Gary has taught at Washington State University. His private sector experience includes an extensive career in information technology, which includes a decade as an independent consultant for the financial services sector.

Robert W. McGee is the director of the Center for Accounting, Auditing, and Tax Studies at Florida International University in Miami. He has published more than 50 books and more than 480 scholarly papers in the fields of accounting, taxation, economics, law, and philosophy. He recently published two books on corporate governance, titled *Corporate Governance in Transition Economies* and *Corporate Governance in Developing Economies*, both published by Springer.

Samuel Mongrut received his doctoral degree in financial economics from Universidad de Barcelona (Spain), his master's in economics from Maastricht University (the Netherlands), and his bachelor degree in business administration from Universidad del Pacifico (Peru). He also studied quantitative economics at the Netherlands Network of Quantitative Economics (NAKE) and he has participated in research projects at the National Opinion Research Center (NORC) at the University of Chicago. Currently he is a professor of finance at the Graduate School of Business and Leadership (EGADE, Zona Centro) of the Instituto Tecnologico y de Estudios Superiores de Monterrey, Campus Queretaro (Mexico). He is also a visiting professor at the Toulouse Business School (France), the Barcelona Business School (Spain), the International Business School of Vilnius University (Lithuania), Universidad Federico Santa Maria (Chile), Universidad EAFIT (Colombia), and Centrum Business School (Peru), among others.

Duc Khuong Nguyen is a professor of finance and the head of the Department of Economics, Finance and Law at ISC Paris School of Management (France). He received his MSc and PhD in finance from the University of Grenoble II (France). His principal research areas concern emerging markets finance, market efficiency, volatility modeling, and risk management in international capital markets. His most recent articles have been published in refereed journals such as *Review of Accounting and Finance,* the *American Journal of Finance and Accounting, Economics Bulletin,* and *Bank and Markets.*

Thuy Thu Nguyen is a lecturer at the Faculty of Business Administration, Foreign Trade University (Hanoi). She received her PhD in finance from the Rotterdam School of Management, Erasmus University, the Netherlands. Her research focuses on empirical corporate finance, including international capital structure, product market competition, and governance. She has published in the *Journal of Finance and Banking.* Her work has been presented at various international conferences and seminars, including the annual meetings of the European Financial Association, the European Financial Management Association, and the Financial Management Association.

Begoña Torre Olmo is a professor of banking and finance at the University of Cantabria. She is assistant to the vice-chancellor of teaching staff at the University of Cantabria and a member of the Board of the Public Project Finance Company of the Cantabria Regional Government. She is the coordinator of doctorate programs at important Mexican universities. Her research interests include the mutual funds industry and corporate finance distress, and she has published several papers in academic journals.

Terry J. O'Neill has been the head of School of Finance and Applied Statistics at the Australian National University (ANU). He is an applied statistician with international recognition for his involvement in experiment-based research and modern computer-intensive statistical techniques. His research on classification methods has appeared in top tier journals for 25 years.

Mehmet Orhan is an associate professor in the economics department of Fatih University, Istanbul, and is also the director of the Social Sciences Institute, which is responsible for the coordination of graduate programs. He obtained his PhD from Bilkent University, Ankara, and has a graduate degree from the industrial engineering department of the same university. His main interests include theoretical and applied econometrics. He has published articles in *Economics Letters,* the *International Journal of Business, Applied Economics,* and the *Journal of Economic and Social Research.* His theoretical research interests include HCCME estimation, robust estimation techniques, and Bayesian inference. He is currently working on IPO performance, hedge fund returns, tax revenue estimation, and international economic cooperation as part of his applied research studies.

Elisa Ossola received her MA from the University of Insubria, Italy, in 2007. She is a PhD student in economics and finance within the Pro*Docs project at USI (University of Italian Switzerland) in Lugano. Her research focuses on the econometrics of jump process and their applications in finance.

Serdar Özkan is an associate professor of accounting at Izmir University of Economics. He received his master's degree in accounting and PhD in business in 1995 and 2000, respectively, both from Dokuz Eylül University, Turkey. He teaches accounting courses at graduate and undergraduate levels. His research interests include theory and application of international accounting standards, accounting education, capital markets research in accounting, corporate governance, and accounting information systems. He serves as a reviewer and works on the editorial boards of national journals.

Ekaterini Panopoulou is a lecturer in the Department of Statistics and Insurance Science at the University of Piraeus, Greece, and is a research associate at the Institute for International Integration Studies (IIS) in Dublin. She completed her PhD in econometrics at the University of Piraeus before working as a postdoctoral fellow at the National University of Ireland, Maynooth. Ekaterini's work has been published in leading academic journals such as *The Econometrics Journal*, the *Journal of International Money and Finance*, the *Journal of Applied Econometrics*, and the *International Journal of Forecasting*, among others.

Jack Penm is currently an Academic Level D at the ANU. He has an excellent research record in the two disciplines in which he earned his PhDs: one in electrical engineering from the University of Pittsburgh and the other in finance from ANU. He is an author/coauthor of more than 80 papers published in various international journals.

Anastasia Petraki was initially trained at the University of Athens, Greece, and at the University of Mainz, Germany. In 2007–2008, she worked on the Leverhulme Trust's project on a change of share-ownership around the world and on a Glasshouse project on short-termism. She is currently doing her PhD at the School of Management, University of Bath, United Kingdom, pursuing research on the risk of private equity for investments in pension funds.

Peter Roosenboom is a professor of entrepreneurial finance and private equity at RSM Erasmus University and a member of Erasmus Research Institute of Management (ERIM). He holds a PhD in finance from Tilburg University. His research interests include corporate governance, venture capital, and initial public offerings. His work has been published in the *Journal of Corporate Finance*, *Contemporary Accounting Research*, the *European Financial Management* the *Journal*, *Applied Economics*, *International Review of Financial Analysis*, the *Pacific-Basin Finance Journal*, the *Journal of Accounting & Public Policy*, *International Journal of Accounting*, and the *Journal of Management & Governance*. He is a coeditor of the book *The Rise and Fall of Europe's New Stock Markets* that has appeared in the series "Advances in Financial Economics." He has also contributed book chapters on the subjects of initial public offerings, mergers and acquisitions, venture capital, and corporate governance.

Samir Saadi is a research associate and part-time instructor of finance at the Telfer School of Management, University of Ottawa. He is currently a PhD candidate at Queen's School of Business. His research interests include executive compensation and international finance. His work has been published in refereed journals such as the *Journal of Multinational Financial Management* and the *Journal of International Financial Markets, Institutions and Money*.

Aristeidis Samitas is an assistant professor in the Department of Business Administration at the University of the Aegean. He studied economics at the University of Athens (BSc, 1995) and banking and finance at the University of Birmingham (MSc, 1996). He also holds a PhD in economics and finance from the University of Athens (2000). His main research interests are financial risk management and portfolio management.

Marcelo Sánchez holds a PhD in economics (University of California at Berkeley, 2000) and currently works in the Euro Area Macroeconomic Developments Division at the European Central Bank. His areas of expertise are international macroeconomics (with a focus on emerging market economies and currency unions) and the impact of oil shocks on industrial economies.

Willem Schramade is an assistant professor of corporate finance at the School of Economics at Erasmus University. He holds a PhD in finance from RSM Erasmus University. His research interests include corporate governance, valuation, and corporate finance. His work has been published in the *Journal of Corporate Finance* and the *Pacific-Basin Finance Journal.*

R. Todd Smith is a professor in the Department of Economics at the University of Alberta. He holds a BA in economics from the University of Saskatchewan and a PhD in economics and finance from Queen's University. He served as an economist in the capital markets and financial studies division of the research department of the International Monetary Fund (IMF) during 1994–1997 and as a senior economist in the financial stability division and the emerging markets division of the international capital markets department of the IMF during 2000–2002. Smith's research and teaching interests are in financial economics, monetary economics, and macroeconomics. He has published extensively in these areas.

Ebru Guven Solakoglu is an assistant professor in the banking and finance department at Bilkent University in Ankara, Turkey. She is an applied economist, and focuses her teaching and research on cross-section time series econometrics and its applications to microeconomics. She has previously worked as an assistant professor and a vice-chair in the Department of Economics at Fatih University. Prior to this, she worked for American Express as a manager economist in the Platform Decision Science Group and as an econometrician in the Fraud Modeling Group at International Risk Management. Even before this, she was an instructor in the Department of Agricultural and Resource Economics at North Carolina State University. She obtained her PhD from North Carolina State University from the Agricultural and Resource Economics in March 2001. She obtained her PhD in March 2001 from North Carolina State University (Department of Agricultural and Resource Economics).

M. Nihat Solakoglu is an assistant professor in the banking and finance department at Bilkent University in Ankara, Turkey. He was previously an assistant professor in the Department of Management at Fatih University. Prior to this, he worked for American Express in the United States in the international risk management, international information management, information and analysis, and fee services marketing departments. He received his PhD in economics and his master's degree in statistics from North Carolina State University. His main interests are applied finance and international finance. His papers have been published in *Applied Economics, Applied Economics Letters*, the *Journal of International*

Financial Markets, Institutions & Money, and the *Journal of Economic and Social Research,* among others.

Spyros Spyrou obtained his PhD in finance from Brunel University (UK) in 1997. He is currently at Athens University of Economics & Business (Department of Accounting & Finance) and serves on the managing committee for postgraduate courses in accounting and finance. Prior to this he was a lecturer and postgraduate admissions tutor at the Department of Economics & Finance, University of Durham (UK); a lecturer and MA program leader at Middlesex University Business School (School of Economics); and an analyst in a brokerage firm (Greece). His research interests are in asset pricing, investor behavior, and emerging markets. He has published many research articles in international refereed academic journals, in professional journals, and in newspapers. He is also the author of the book *Money & Capital Markets.*

Maike Sundmacher is a lecturer in finance at the School of Economics & Finance, University of Western Sydney. She teaches corporate finance, bank management, and credit risk management. Currently, she is pursuing her PhD at the Macquarie Graduate School of Management and researches in the areas of capital markets and risk management in financial institutions.

Lin Tan is an assistant professor of finance at California State Polytechnic University, Pomona. Her research interests are in Chinese capital markets, international finance, and corporate finance. Her work has been published in *Quantitative Finance,* the *Pacific-Basin Finance Journal,* and the *International Review of Economics and Finance.* Dr. Tan received her PhD in finance from Drexel University.

Oktay Taş is an associate professor and the chair of accounting and finance at Istanbul Technical University. After a BA and an MA in accounting and finance at Marmara University, he pursued his PhD at the Technical University of Berlin. Professor Tas teaches financial management, portfolio management, and investment. His areas of interest are corporate finance, financial analysis, and auditing and financial derivatives.

Dermot Tennyson is a control engineer and financial manager at the University of Anáhuac in Mexico. He is currently pursuing his PhD at the University of Cantabria (Spain) working in security price anomalies in the Mexican financial markets.

R.D. Terrell is a financial econometrician and an officer in the general division of the Order of Australia. He served as the vice-chancellor of the ANU from 1994 to 2000. He has also held visiting appointments at the London School of Economics, the Wharton School, the University of Pennsylvania, and the Econometrics Program (Princeton University). He has published several books and research monographs and around 80 research papers in leading journals.

Nikos S. Thomaidis holds a PhD in financial engineering with artificial intelligence from the University of the Aegean, Greece; an MSc in mathematics and finance from Imperial College, University of London; and a BSc in industrial engineering from the Technical University of Crete, Greece. He currently works as a vice director of research & development with Kepler Asset Management on Wall Street in New York. His research interests focus on the application of computational intelligent methods (artificial neural networks, genetic algorithms, particle swarm optimization) to statistical arbitrage, portfolio optimization, volatility forecasting, and nonlinear econometric models.

Kaya Tokmakçıoğlu is a teaching and research assistant at Istanbul Technical University. He has a BS in textile engineering and an MA in management from the same university. He is pursuing his PhD at the Department of Management Engineering and works in the field of financial econometrics.

Sirimon Treepongkaruna has been appointed a senior lecturer in finance at Monash University. She has previously worked as a lecturer at the Australian National University and Lincoln University. She graduated from Chulalongkorn University with a BSc in statistics (Hons) in 1992; an MBA from George Washington University, Washington, D.C., in 1995; and a PhD in finance from the University of Queensland in 2001. Prior to this, Sirimon served as a foreign exchange derivatives dealer at Bangkok Bank.

Alonso Valenzuela holds a bachelor's degree in economics from Universidad del Pacifico (Peru) and has collaborated as a research assistant at Universidad del Pacifico Research Center (CIUP).

Eline van Niekerk holds a master's degree in finance and investments from RSM Erasmus University. She wrote her master's thesis on "Bonding through Cross-Listing: An Analysis of Emerging Markets." She initially worked for the equity capital markets department of JP Morgan in London and is currently at OC&C Strategy Consultants in the Netherlands.

Anson L.K. Wong is the head of Credit Risk Research of CT Risk Solutions coordinating Basel II credit risk projects and the development of credit risk analytics. She is an experienced business intelligence and data mining analyst, having worked for HSBC and Experian. Anson received her PhD from the City University of Hong Kong; MPhil and BSSc from Lingnan University of Hong Kong; and MEcon from the University of Hong Kong.

Michael C.S. Wong is an associate professor of finance at City University of Hong Kong, Hong Kong; director of Global Association of Risk Professionals, and the president of CT Risk Solutions. He is a well-known risk management expert, serving as a consultant of more than 20 banks and enterprises. Michael was awarded his PhD by the Chinese University of Hong Kong, MPhil by the University of Cambridge, and MA by the University of Essex.

Chunchi Wu received his PhD from the University of Illinois-Urbana in 1982. He was an assistant, associate, and full professor at Syracuse University from 1983 to 2005, serving as chairman of the finance department (1987–1990) and as head of the PhD program of its School of Management. In 2004–2006, he was head of financial direction of Singapore Management University. In 2006, he joined the faculty of the University of Missouri as J.E. Smith Professor of Finance at the Robert J. Trulaske College of Business.

Eliza Wu is a senior lecturer in the School of Banking and Finance at the University of New South Wales, from where she received her PhD in finance in 2005. She specializes in emerging financial market research and has worked as a research fellow at the Bank for International Settlement's Asia-Pacific office. Her works have been published in the *Journal of Banking and Finance*, the *Journal of Fixed Income, Emerging Markets Review*, the *International Journal of Finance and Economics*, and the *Journal of International Financial Markets Institutions and Money*.

Anna Zalewska is a professor of finance at the School of Management, University of Bath. Having both a mathematical and economic background (in 1998 she received her PhD in mathematics at the Polish Academy of Sciences, Warsaw, and her PhD in economics at the London Business School), her research interests cover a broad range of subjects, mostly in financial economics. Her publications include papers in the *Journal of Financial Economics, European Economic Review*, the *Journal of Empirical Finance, Economics Letters*, and the *European Journal of Finance*. She has also contributed to several books. Professor Zalewska works and publishes mainly in the following areas: (1) privatization, and market risk and regulation; (2) governance and managerial incentives; (3) emerging markets; and (4) impact of pension reforms on the development of financial markets. She has also advised government bodies and leading international companies on financial issues.

Contributors

David E. Allen
School of Finance and Business
  Economics
Edith Cowan University
Joondalup, Western Australia,
  Australia

Mohamed El Hedi Arouri
Laboratoire d'Économie
  d'Orléans
Université d'Orléans
Orléans, France

Michael F. Bleaney
School of Economics
University of Nottingham
Nottingham, United Kingdom

Paul Brockman
Department of Finance
University of Missouri
Columbia, Missouri

Robert D. Brooks
Department of Econometrics
  and Business Statistics
Monash University
Clayton, Victoria, Australia

Silvio John Camilleri
Department of Banking
  and Finance
University of Malta
Msida, Malta

Thomas C. Chiang
Department of Finance
Drexel University
Philadelphia, Pennsylvania

Joseline Chimhini
School of Finance and Business
  Economics
Edith Cowan University
Joondalup, Western Australia,
  Australia

Imed Chkir
Telfer School of Management
University of Ottawa
Ottawa, Ontario, Canada

Lamia Chourou
Faculty of Law
Economics and Political Sciences
University of Sousse
Sousse, Tunisia

Mehmet A. Civelek
Department of Economics
Dokuz Eylul University
Izmir, Turkey

Carolyn V. Currie
Public Private Sector Partnership
    Pvt Ltd.
Sydney, Australia

A. Fatih Dalkılıç
Faculty of Business
Dokuz Eylül University
Izmir, Turkey

Mathijs A. van Dijk
Rotterdam School of
    Management
Erasmus University
    Rotterdam
Rotterdam, the Netherlands

George D. Dounias
Department of Financial
    Engineering & Management
University of the Aegean
Chios, Greece

M. Banu Durukan
Faculty of Economics
Ljubljana University
Ljubljana, Slovenia

Cumhur Ekinci
Department of Management
    Engineering
Istanbul Technical
    University
Istanbul, Turkey

Craig Ellis
School of Economics and Finance
University of Western Sydney
Rydalmere, New South Wales,
    Australia

Dean Fantazzini
Moscow School of Economics
Moscow State University
Moscow, Russia

Viviana Fernandez
Department of Industrial
    Engineering
University of Chile
Santiago, Chile

Thomas J. Flavin
Department of Economics,
    Finance & Accounting
National University of Ireland,
    Maynooth
Maynooth, Ireland

Don U.A. Galagedera
Department of Econometrics
    and Business Statistics
Monash University
Caulfield East, Victoria, Australia

Aaron Garay
Universidad del Pacifico
Research Center
Lima, Peru

Mark D. Griffiths
Department of Finance
Miami University
Oxford, Ohio

Massimo Guidolin
Manchester Business School
University of Manchester
Manchester, United Kingdom

Javed Iqbal
Department of Statistics
Karachi University
Karachi, Pakistan

Vassilis N. Karavas
Managing Director
Credit Agricole Asset
  Management Alternative
  Investments
Chicago, Illinois

Dimitris Kenourgios
Department of Economics
National and Kapodistrian
  University of Athens
Athens, Greece

Jill R. Kickul
Berkley Center for
  Entrepreneurial Studies
New York University Stern School
  of Business
New York, New York

Berna Kirkulak
Department of Business
  Administration
Dokuz Eylül Universitesi
Izmir, Turkey

Bülent Köksal
Department of Economics
Fatih University
Istanbul, Turkey

Thomas Lagoarde-Segot
School of Management and
  Business of Marseille
Euromed Marseille Ecole de
  Management
Marseille, France

R. McFall Lamm, Jr.
Chief Investment Officer
Stelac Advisory Services
New York, New York

Laurence Le Poder
School of Management and
  Business of Marseille
Euromed Marseille Ecole de
  Management
Marseille, France

Peter B. Lerner (Retired)
Whitman School
  of Management
Syracuse University
Syracuse, New York

Kian-Ping Lim
Labuan School of International
  Business and Finance
Universiti Malaysia Sabah
Labuan, Malaysia

Stephen J. Lubben
School of Law
Seton Hall University
Newark, New Jersey

Gary McCormick
Department of Finance
University of Missouri
Columbia, Missouri

Robert W. Mcgee
School of Accounting
Florida International University
Miami, Florida

Samuel Mongrut
Escuela de Graduados en
  Administracion y Direccion
  de Empresas Zona Centro
Instituto Tecnológico y
  de Estudios Superiores de
  Monterrey–Campus Queretaro
Queretaro, Mexico

Duc Khuong Nguyen
Paris School of Management
Institut Supérieur du Commerce
  de Paris
Paris, France

Thuy Thu Nguyen
Faculty of Business Administration
Foreign Trade University
Hanoi, Vietnam

Begoña Torre Olmo
Department of Economics
University of Cantabria
Cantabria, Spain

Terry J. O'Neill
School of Finance and Applied
  Statistics
The Australian National University
Canberra, Acton, Australia

Mehmet Orhan
Department of Economics
Fatih University
Istanbul, Turkey

Elisa Ossola
Institute of Finance
University of Lugano
Lugano, Switzerland

Serdar Özkan
Department of Business
  Administration
Izmir University of Economics
Izmir, Turkey

Ekaterini Panopoulou
Department of Statistics and
  Insurance Science
University of Piraeus
Piraeus, Greece

Jack Penm
School of Finance and Applied
  Statistics
Australian National University
Canberra, Acton,
  Australia

Anastasia Petraki
School of Management
University of Bath
Bath, United Kingdom

Peter Roosenboom
Rotterdam School of
  Management
Erasmus University
  Rotterdam
Rotterdam, the Netherlands

Samir Saadi
Queen's School of Business
Queen's University
Kingston, Ontario, Canada

Aristeidis Samitas
Department of Business
  Administration
University of the Aegean
Chios, Greece

Marcelo Sánchez
European Central Bank
Frankfurt am Main, Germany

Willem Schramade
Erasmus School of Economics
Erasmus University Rotterdam
Rotterdam, the Netherlands

R. Todd Smith
Department of Economics
University of Alberta
Edmonton, Alberta, Canada

Ebru Guven Solakoglu
Department of Banking and Finance
Bilkent University
Ankara, Turkey

M. Nihat Solakoglu
Department of Banking and Finance
Bilkent University
Ankara, Turkey

Spyros Spyrou
Department of Accounting & Finance
Athens University of Economics &
  Business
Athens, Greece

Maike Sundmacher
School of Economics and Finance
University of Western Sydney
Rydalmere, New South Wales,
  Australia

Lin Tan
Department of Finance, Real
  Estate and Law
California State Polytechnic
  University
Pomona, California

Oktay Taş
Department of Management
  Engineering
Istanbul Technical University
Istanbul, Turkey

Dermott Tennyson
Financial Manager
Universidad Anáhuac
Huixquilucan, Mexico

R.D. Terrell
National Graduate School
  of Management
Australian National University
Canberra, Acton, Australia

Nikos S. Thomaidis
Department of Financial
  Engineering & Management
University of the Aegean
Chios, Greece

Kaya Tokmakçıoğlu
Department of Management
  Engineering
Istanbul Technical University
Istanbul, Turkey

Sirimon Treepongkaruna
Department of Accounting
  and Finance
Monash University
Caulfield East, Victoria, Australia

Alonso Valenzuela
Universidad del Pacifico
Research Center
Lima, Peru

Eline van Niekerk
Finance & Investments
RSM Erasmus University
Rotterdam, the Netherlands

Anson L.K. Wong
Department of Economics
 and Finance
City University of Hong Kong
Kowloon, Hong Kong

Michael C.S. Wong
Department of Economics
 and Finance
City University of Hong Kong
Kowloon, Hong Kong

Chunchi Wu
Robert J. Trulaske, Sr. College
 of Business
University of Missouri
Columbia, Missouri

Eliza Wu
School of Banking and
 Finance
The University of New South
 Wales
Sydney, New South Wales,
 Australia

Anna Zalewska
School of Management
University of Bath
Bath, United Kingdom

CHAPTER **1**

Growth Prospects of New and Old Emerging Markets

Anastasia Petraki and Anna Zalewska

CONTENTS

1.1 INTRODUCTION

One of the most impressive technical developments in the last 20 years has been the stunning reduction in the time it takes to diffuse information around the world and the huge growth in the associated market for information. Whether we think of natural phenomena (e.g., hurricanes and earthquakes) or human and market activities (e.g., stock market crashes, government coups, wars, etc.), news is perpetually flashing around the world 24 h a day, 7 days a week. The effect has been particularly significant for financial markets, which in the last couple of decades have seen major new technologies introduced (electronic trading and transfer systems), new financial instruments developed (e.g., enhanced capital advanced preferred security [ECAPS]), and new markets created across the globe. For many countries, the creation of a new stock market is the biggest event or at least one of most important events in shaping and/or reforming the financial structure of the country. Hence, it could be argued that the burst of new stock markets is the big global financial innovation of modern history!

When one compares the current situation with that of the early 1980s, the dramatic change in the number, geographic coverage and growth rates of stock markets stands out:

- The number of countries with stock markets is almost two and half times the 1985 figure (142 compared to 58). Put in another way, a staggering number of countries (84) have opened their first stock exchange in the last two decades.* Almost by definition, all these new markets are classed as "emerging"; hence, there are currently nearly five times as many countries with emerging stock markets (118) than countries with developed stock markets (24). So in terms of pure numbers, the emerging markets dominate, whereas in the mid-1980s the position was more equally balanced (34 countries with emerging stock markets compared to 24 countries with developed stock markets).†

- Turning to growth, the capitalization of emerging stock exchanges has grown almost 11,000% and the volume of share trading by 19,000% since the early 1980s. This is an astonishing statistics especially when

---

* In some cases (e.g., Russia and China), more than one exchange have been opened in a country.
† As a country, we consider all the 192 countries recognized by the UN, plus Hong Kong and Taiwan.

stacked up against the corresponding growth rates for the developed stock exchanges over the same period of time, which are around 1900% and 6500%, respectively.

This trend to create new stock markets is not accidental or driven purely by technological progress. Numerous efforts and programs have been put in place to stimulate financial liberalization and the evolution of capital markets (e.g., privatization programs and pension reforms), not only by individual governments but also by international bodies (e.g., International Monetary Fund [IMF], World Bank [WB], and World Federation of Stock Exchanges [WFSE]), that have been engaged in the process of stimulating the stock market development. Why is this happening?

First, although there is still a lot of research that needs to be done, the existing studies clearly document a strong causality between stock market development and economic growth (e.g., Demirguc-Kunt and Levine, 1995; Calderon and Liu, 2003; Beck and Levine, 2004; Claessens et al., 2006). Therefore, developing efficient capital markets, and financial markets more generally, is often perceived as an important step in economic development and reduction of poverty. Second, the liberalization of capital flows among the developed markets has resulted in a decrease of diversification benefits, which in turn has forced big (mostly institutional) investors to search for alternative investment opportunities. Hence, the development of emerging markets has, in part, been driven by the demand from international investors. Third, domestic institutional investors, both in developed and in developing countries, can be a driving force for change. Pension reforms, and in particular, the creation of vast compulsory and voluntary saving schemes, are expected to stimulate investment and economic growth, but to do so they require diverse investment opportunities, and hence the development of local stock markets (e.g., Impavido et al., 2003; Zalewska, 2006). Finally, the desire to provide incentives, stimulate innovation, and improve efficiency has resulted in numerous privatization programs. Naturally, stock markets were needed to absorb the assets of privatized enterprises and to provide financing opportunities for them and for governments (government bond markets were created in parallel to many equity markets and sometimes were created prior to the equity markets).

The opening of many new stock markets has been strongly linked to the change in the shareownership of individuals around the world. Although the numbers of individual investors have increased significantly in many developed countries (e.g., in Canada, the number of shareholders increased from about 2.4 million in 1983 to 12.4 million in 2004; in Japan, it increased from 19.3 million in 1980 to 39.3 million in 2006; in the United Kingdom, the number of shareholders tripled between 1980 and 2005 from 3 to 9.1 million; and in the United States, traditionally known for the broad ownership, the number of shareowners increased from 42.4 million in 1983 to 62.9 million in 2005), it is the emerging markets again where the changes are most spectacular. For example, although as a percentage of the total population the numbers are still low, there are 80 million shareholders in China alone. Thus, China contributes more to the world's aggregate statistics than the United States and Canada together. The lack of comprehensive data for emerging markets makes general comparisons difficult but, as Grout et al. (2008) report, there are at least 138 million shareholders in the emerging markets.* Although it is less than in the developed countries (around 173 million), this is a very high figure particularly if one takes into account the short history of the majority of the emerging markets and the much lower disposable income of people living in developing economies. It is, however, interesting that the newcomers contribute nearly 100 million of the 138 million shareholders.

In this chapter, we document the growth of emerging markets with special focus on the "new emerging stock markets," i.e., those markets that started to operate after 1985 in contrast to the "old emerging markets" that operated before 1985. Although the past enormous growth of emerging markets was driven by the newcomers, we argue that the future growth of emerging markets as a group will lie in the "organic" growth of the existing exchanges. The trend to open new stock markets is over and if the emerging markets wish to strengthen their position on the world financial scene, they will have to work hard to achieve it. For example, we show that the perception of corruption in the emerging markets has not improved since the 1980s, and in cross-sectional comparisons, emerging markets lag far behind their developed rivals. If the emerging markets fail to attract international investors and assure investors that they provide good investment opportunities, the emerging markets' development may slow down significantly as they will be unable to generate enough domestic capital required to maintain high growth rates.

---

* The number of shareholders in the emerging markets is underestimated, because data for only 18 developing countries were available. However, the emerging markets that are accounted for are those biggest in the sense of a stock market development and are located in most populated countries.

This chapter is organized as follows: Section 1.2 documents the growth of emerging stock markets; Section 1.3 argues why the future emerging markets' growth must be "organic;" Section 1.4 compares and contrasts the quality of governance in countries with developed, emerging, and no markets; and this chapter concludes with Section 1.5.

1.2 EMERGING MARKETS BOOM

The growth of capital markets around the world has been faster over the last couple of decades than ever before. The capitalization of the world stock markets exceeded US$66 trillion in December 2007 implying that they have grown by about 2500% since the early 1980s. There are several factors driving it, but the growth of emerging stock markets is one of the most significant factors given that they have grown nearly 11,000% over that period of time. In nominal terms, capitalization of the emerging markets reaching US$8.3 trillion by the end of 2007, accounting for over 27% of the world stock market capitalization. It is not just that the capitalization of the emerging markets that has grown, they have become more active as well. The volume of share trading of the emerging markets has also increased dramatically in nominal and relative terms. The 2007 figure of over US$7.6 trillion translates into a growth rate of over 19,000% since the early 1980s. Even though, this accounts for only 13% of the world volume of share trading, it is a great achievement given that the emerging market's share was only around 2.5%–5% in the early 1980s. Both in size and liquidity, the emerging markets are now becoming an important investment partner of the developed markets.

Figure 1.1 shows the US$-denominated market capitalization of the developed and of the emerging stock markets over the period 1981–2007 (left-hand-side axis). It is clear that the emerging markets' share started to be "noticeable" by late 1980s, and over time it has continued to grow to its current two digit figure. Figure 1.1 also shows how many of the countries that did not have stock markets before 1985 have opened at least one stock market in each calendar year (right-hand-side axis). The trend to open stock markets was particularly strong in the 1990s and now appears to have slowed down. Indeed, since 1986, there have only been 2 years, 2004 and 2006, when no stock markets were opened by at least one country.

A consequence of the massive creation of stock markets is that currently there are more countries with stock markets that are no more than 20 years old, than there were countries with stock markets (both developed and emerging) in the early 1980s (84 against 58). This means that the

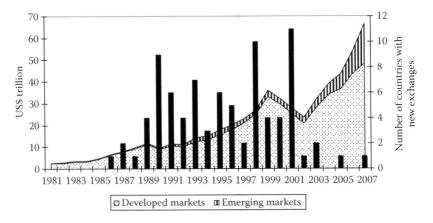

FIGURE 1.1 Market capitalization of the developed and emerging stock markets (left-hand-side axis), and the number of countries that opened stock exchanges (right-hand-side axis); end of year figures. (Data from Individual stock exchanges.)

number of countries with emerging stock markets has increased more than threefold from 34 in the early 1980s to 118 in 2007. So currently, there are nearly five times as many countries with emerging stock markets than countries with developed stock markets (24).

Figure 1.2 shows the distribution of countries that have stock markets in 2007, separated into eight disjoint geographical regions. Despite the common assumption that if one is talking about an emerging market, then it must concern an Asian or Latin American one, Africa and Europe are the continents with the highest number of countries with emerging stock markets. Indeed, Africa has 29 countries with emerging stock markets and of these 22 have opened in the last 20 years. Africa is also the continent with the largest number of countries that do not have a stock market. Europe, on the other hand, appeared to be a place of developed markets once the Portuguese and the Greek stock exchanges officially became classified as developed in the mid-1980s. However, since the collapse of communism, 26 countries opened stock markets pushing Europe to the top of the league of geographical regions with the highest number of countries with emerging stock exchanges.

This increase in the number of exchanges also has an interesting demographical aspect. The number of people living in countries that offer stock market investment opportunities has increased dramatically.

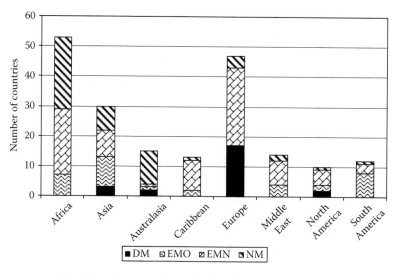

FIGURE 1.2 Number of countries with developed (DM), emerging (old and new, denoted EMO and EMN, respectively), and no stock markets (NM) by geographical region; end of 2007 statistics.

In the early 1980, about 55% of the world's population lived in countries with stock markets, including less than 40% living in countries with emerging stock markets. Currently, nearly 80% of the world's population lives in countries with emerging stock markets, including nearly 40% living in countries that have opened stock markets after 1985. When we add to this the 14% (approximately) of the world's population that live in countries with developed stock markets, we can easily say that nearly everyone lives in a country with a stock market. Figure 1.3 shows, for 1985 (Panel A) and 2007 (Panel B), the percentage of the world population living in countries with developed stock markets, emerging stock markets, and countries that do not have a stock market.

In the light of these statistics it can be argued that although most of the wealth is still in the hands of the developed markets, population wise, the emerging markets dominate. However, this "dominance" is highly driven by two countries—India and China. Each of these two countries contributes about 50% to the population of the group they belong to, i.e., India to the population of the old emerging markets and China to the population of the new emerging markets. This fact can have far-reaching consequences. Although China is still in the group of middle-low and

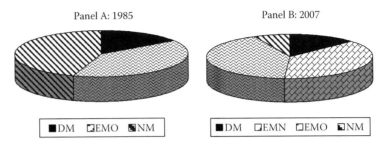

FIGURE 1.3 The percentage of the world population living in countries with developed (DM), emerging (old and new, denoted EMO and EMN, respectively), and no stock markets (NM); 1985 (Panel A) and 2007 (Panel B).

India in the group of low-income countries, they are fast growing economies with a great appetite to develop their financial sector. Given this position, if shareholder ownership grows over time, then in the near future, China and India together may have more shareholders than the rest of the world.

To close the discussion of the growth of the emerging markets, Table 1.1 shows results of seven cross-sectional regressions that seek to assess the contribution of the new emerging markets to the aggregate growth of stock markets. The regressions cover the period 1985–2006. We start the regressions in 1985 as this is the earliest year for which comprehensive data across exchanges for the control variables are available. This is also the last year without new emerging stock markets. The analysis is performed for (1) the whole period 1985–2006 for which there are data for 47 countries out of 58 that had stock markets, (2) 1990–2006 for which there are 57 observations out of 75,* (3) 1995–2006 with 79 observations out of 101 countries with stock markets, and (4) 2000–2006 with 94 observations out of 119 countries with stock markets.

To control for the change in the economic situation of each country, we introduce the percentage change in GDP PPP per capita (in US$), and the percentage change in corruption. The latter is measured as the change in the International Country Risk Guide (ICRG) corruption index. The growth of stock markets is measured by the percentage change in market capitalization of the corresponding exchange or exchanges if in a particular

---

* In this regression, the annualized change of GDP PPP per capita, market capitalization, and corruption for Ghana, Hungary, Iran, and Saudi Arabia cover the 1991–2006 period. This is because no data for these countries could be found for 1990.

TABLE 1.1 Regression Results

	1985–2006	1990–2006		1995–2006		2000–2006	
Intercept	0.559	0.448	0.336	−0.324	−0.254	−0.209	−0.142
	(0.948)	(1.071)	(0.787)	(−0.909)	(−0.756)	(−0.834)	(−0.559)
ΔGDP PPP per capita in US$	0.419534**	−0.054	0.039	0.977**	0.807**	7.639**	6.328*
	(2.226)	(−0.164)	(0.115)	(2.505)	(2.181)	(2.442)	(1.961)
ΔICRG corruption index	33.793	−7.435	−9.678	1.265	−2.288	2.161	1.979
	(1.196)	(−0.877)	(−1.123)	(0.205)	(−0.387)	(1.158)	(1.065)
Emerging market dummy	1.819**	0.817***		0.589*		0.565**	
	(2.572)	(2.733)		(1.924)		(2.241)	
Old emerging market dummy			0.689**		0.106		0.396
			(2.191)		(0.328)		(1.442)
New emerging market dummy			1.347**		1.113***		0.754***
			(2.601)		(3.390)		(2.689)
Observations	47	57	57	79	79	94	94
R^2	0.249	0.160	0.185	0.127	0.240	0.169	0.189
Adjusted R^2	0.197	0.113	0.122	0.092	0.199	0.141	0.153

***, **, and * indicate significance at 5%, 10%, and 15%, respectively.

country there is more than one stock exchange. To make the comparison across the periods possible, we annualize the three variables. In addition, dummies are used in each regression to indicate emerging markets. To test how significant new emerging markets are, we separate the emerging market dummy into two dummies, one for the old and one for the new emerging markets, for all the periods but 1985–2006.

The 2006 figures are used as the end of the period because it is the most recent year for which the ICRG corruption index is available. The ICRG corruption index is used since it goes back to the 1980s and hence offers the longest time series among available corruption measures. The ICRG corruption index aims to measure the corruption within the political system that is a threat to foreign investment by distorting the economic and financial environment, reducing the efficiency of government and business. It varies between 1 (most corrupt) and 6 (least corrupt). We control for corruption since numerous studies discuss its significance for economic and financial sector development (e.g., La Porta et al., 1997).

The results of Table 1.1 are clear and consistent across periods—the emerging markets have been growing faster than the developed markets, and it has been the new emerging markets that have been the drivers of this growth. The estimated coefficients for the emerging market dummy are significant for the four periods in question, although in the 1995–2006 regression the significance is only at 10% level. When the dummy is separated into two, the new emerging market dummy is the one that is highly significant. The estimated coefficient for the new emerging markets is also much higher than the one for the old emerging markets. The old emerging market dummy is significant in the 1990–2006 period only. This may result from the fact that very few new emerging markets are included in the regression (only 5 out of 43 that already existed). The lack of data for the markets created in between 1985 and 1990 is the reason of this poor representation.

The change in the corruption index does not carry significant explanatory power. We have also tried to use the level of the corruption index rather than its change (because the corruption index does not change much), but this regression delivers results that are consistent with the ones described earlier, and therefore to save space we do not present them.

Consistent with previous research, we find that economic growth is positively related to the growth of stock markets (only in the 1990–2006 regression is the estimated coefficient negative, but this is not statistically significant).

In the light of this evidence, a natural question is to ask whether emerging markets and, in particular, new emerging markets can retain their reputation for fast growth or at least continue to grow faster than the developed and the old emerging exchanges over the next decade or so.

Simple logic suggests that, since the economic growth of developing countries is on average higher than the economic growth of the developed countries, it is not at all implausible that emerging stock markets, as a group, will grow faster than the developed stock exchanges. However, it would be naïve to believe that the emerging markets will outpace the developed markets as easily as they have done in the last couple of decades. There are several reasons to support this view and be at least a bit pessimistic about the growth opportunities of the emerging markets. Here, we discuss a few reasons in support of this view.

1.3 WHAT NEXT?

The growth figures of the emerging markets as a group have two components: the growth of the old emerging markets and the emergence and subsequent growth of the new exchanges. The new emerging markets already contribute more than 55% to the total capitalization of the emerging markets with their US$10 trillion capitalization accumulated over the last couple of decades. This money has been raised in the process of stock market creation (from scratch) and often massive privatization programs (e.g., so far the biggest IPO offerings in the world are a result of privatization programs). However, is it sensible to believe that the development of the emerging markets can be significantly supported by the emergence of new stock markets in the future? First, these new stock markets would either have to be in addition to the existing newly opened exchanges or created in countries that do not have stock markets so far.

However, opening additional new exchanges in countries where exchanges already exist is not a process that can bring any significant change. Most of the existing emerging stock markets are small and illiquid enough to discourage local authorities from opening more exchanges. In general, local authorities realize the trouble early enough and restrict the number of new exchanges (they focus on one or two). When multiple exchanges pop up, it is usually a sign of a lack of control and forthcoming financial and economic distress. Bulgaria and Russia are good examples of such problems. Bulgaria had 15 stock exchanges in the early 1990s, with just 21 stocks listed on them. The 1995–1996 economic crisis resulted in collapse of all of them. Similarly, 56 exchanges of Russia that sprung off after mass privatization of the early 1990s did not survive the economic collapse of 1998. The few successful stories concern old emerging markets. For instance, the opening of the National Stock Exchange (NSE) in India in 1993 (equity trading started in 1994) seems to nicely complement the Bombay Stock Exchange (BSE) that has been in operation since 1875. The newly opened exchange is nearly as big as the old one (end of 2007 market capitalization of the NSE was US$1.66 billion and of the BSE was US$1.82 billion), and already more active with the value of share trading of US$751.4 billion against US$343.8 billion of the BSE.

Another possibility is that those 52 countries that still do not have stock markets will open them in coming years. Although the creation of new stock markets has been a dominant factor of the emerging markets' growth, one cannot expect this trend to prevail. The "no-exchange countries" are small in population and often economically weak. Figure 1.3 shows that currently less than 7% of the world's population lives in countries that do not have a stock market. Figure 1.4 confirms that the countries without stock markets are often very poor. Indeed, according to the WB classification, 26 countries without a stock market, i.e., half of the group, are low-income countries. The only rich countries without stock markets are

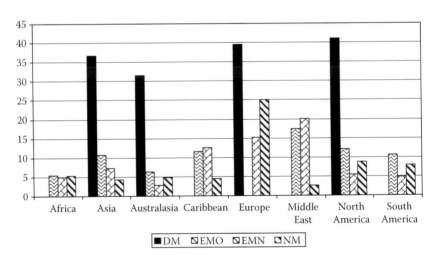

FIGURE 1.4 Regional GDP PPP per capita with separation for countries with developed (DM), emerging old (EMO), emerging new (EMN) stock markets, and without stock markets (NM). All figures are expressed in US$ for the end of 2007.

located in Europe. These are Andorra, Lichtenstein, Monaco, and San Marino. Obviously, there is no immediate need for these countries to open stock exchanges, and even if they decided to do so, it is not reasonable to expect that these new exchanges could have a dramatic impact on the emerging markets as a group. In addition, it is hard to believe that citizens of, say, Monaco feel deprived of possibility of investing in stock markets and would need one in Monaco to invest in equity.

Therefore, the growth of the emerging markets cannot be further stimulated by new countries "joining the club" to the degree it has happened over the last two decades. The days when the growth figures were inflated by newcomers with high growth potentials are over. Consequently, if the emerging markets are to grow, it must be the "organic" growth of the existing exchanges.

1.4 GROWTH PROSPECTS OF THE EXISTING EMERGING STOCK MARKETS

The path to become a developed market can be long and is not always straightforward. It is not just the regulation of the exchange or sophistication of its electronic trading system that matters but a broad range of factors that have roots in the economic and political system of a country in which the exchange operates. In the mid-1980s, there was only a few exchanges that were upgraded to the status of a developed market, and there have been none since. Indeed, currently the emerging markets group seems more diverse than it was a few decades ago. Next to very big exchanges trading thousands of companies and having the capitalization of trillions of US$ (e.g., China) there are minute exchanges trading just a few shares with the capitalization less than a billion of US$ (e.g., Armenia). Figure 1.5 shows that BRIC, as Brazil, Russia, India, and China, are commonly referred to contribute toward 60% of the emerging markets capitalization, with China contributing more (35%) than the other three markets together.

With such strong concentration of capital, these few big markets can drive the averages up or down depending on their performance; however, it does not change the fact that when we want the emerging markets to develop, we have in mind a broad range of countries (usually small) rather than a few big ones. In the absence of development of the small markets, the whole experiment of the creation of stock markets will be a wasted time and money. Unfortunately, the developing countries are not rich

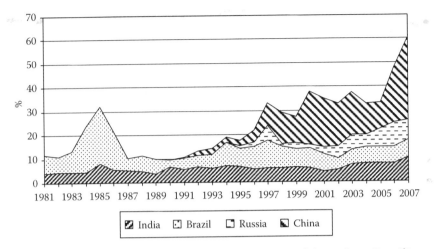

FIGURE 1.5 The time-path of market capitalization of the Indian, Brazilian, Russian, and Chinese stock markets as a percent of market capitalization of the emerging stock markets.

enough to afford to misallocate vast sum of money. Therefore, doing it right is of vital importance.

Stock markets do not operate in a vacuum but are strongly interlinked with the legal structure of the country in which they operate. The long-run growth of markets is strongly correlated with the development of the letter of the law and governance and a country's ability to implement them. Monitoring, regulation, and shareholder protection are all vital for securing efficiency of stock market operations and trading, and, as a result, to secure lower investment risk. Emerging markets are particularly weak in this regard. There is a strong negative correlation between the level of economic development, i.e., its wealth, and its lawlessness. Countries that suffer from corruption, poor efficiency of governmental institutions and officials, political instability, etc., are those that remain poor. In consequence, the development of stock markets in such countries is hampered and may not be possible at all.

It is difficult to find an objective measure of a country's lawlessness, as there are many aspects of it, and these are themselves often hard to measure. Therefore, it may be informative to look at such country indicators as bureaucracy, corruption, effectiveness of government, voice and accountability, political stability and regulatory quality to get some feel for what the characteristics of emerging markets are, and how much they differ from the developed countries.

To provide a better understanding of the environmental differences among countries with developed, emerging or no exchanges, Figure 1.6 presents averages of various WB governance indicators for 2007 for the four market groups discussed in this chapter separated into geographical regions in which they operate. The averages are equally weighted to provide a general picture for a region and group of markets rather than concentrate on a few dominant markets. Panel A presents the averages for the four regions with developed stock markets (Australasia, Asia, Europe and North America), and Panel B shows the averages for the four regions without developed stock markets (Africa, Caribbean, Middle East, and South America).* Each of the six governance indicators is measured in units ranging from about −2.5 to 2.5, with higher values corresponding to better governance outcomes.

Figure 1.6 gives a strong and clear message. There is a huge difference between the developed and emerging markets. Across all the four regions that have developed stock markets, the average scores of the developed markets in these regions are positive. In contrast, the emerging markets, both old and new, are negative with the exception of Europe and Caribbean. The figure also confirms our earlier concern about countries without stock markets. The statistics for Africa, Asia, and Australasia are lowest out of presented ones.

Unfortunately, the WB indicators presented earlier have been calculated since 1997 and for selected years only. This does not help when trying to identify a long-term trend, especially that there is nearly no variability in the indicators across years they are calculated for. Therefore, to have an assessment of changes over time, Figure 1.7 shows the 1981–2006 time-path of the ICRG corruption index averaged (equally weighted) for each of the group of markets discussed in this chapter.

It is interesting to note that (1) the corruption index decreases over time for the four groups in question; (2) the countries with old emerging markets, with new emerging markets, and without exchanges manifest a very similar level of corruption with the new emerging markets being marginally less corrupt, and countries without exchanges being most corrupt (though this is not statistically significant); and finally (3) the gap between

---

* The World Bank's Aggregate Governance Data Set. The presented statistics are averages of governance indicators that reflect the statistical compilation of responses on the quality of governance given by a large number of enterprise, citizen, and expert survey respondents in industrial and developing countries, as reported by a number of survey institutes, think tanks, non-governmental organizations, and international organizations.

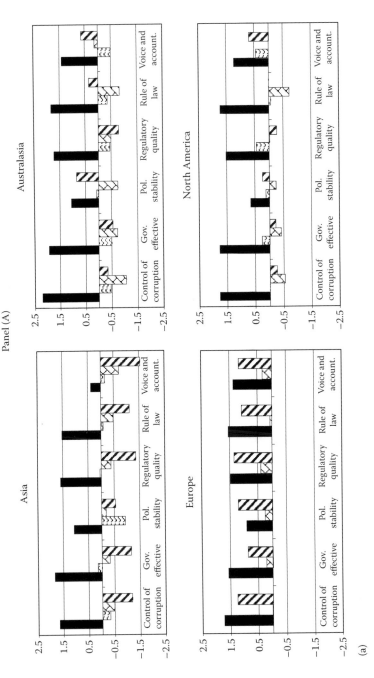

FIGURE 1.6   Regional averages of statistical compilation of responses on the quality of governance as reported by the World Bank's Aggregate Governance Data Set. *Note:* Individual responses vary between −2.5 (bad) and +2.5 (good). The six indicators are (1) voice and accountability, measures various aspects of the political process, civil liberties, and political and human rights, measures the extent to which citizens of a country are able to participate in the selection of governments; (2) political stability, measures perceptions of the likelihood that the government in power will be destabilized or overthrown by possibly unconstitutional and/or violent means, including domestic violence and terrorism; (3) government effectiveness, quality of public service provision, the quality of the bureaucracy, the competence of civil servants, the independence of the civil service from political pressures, and the credibility of the government's commitment to policies;

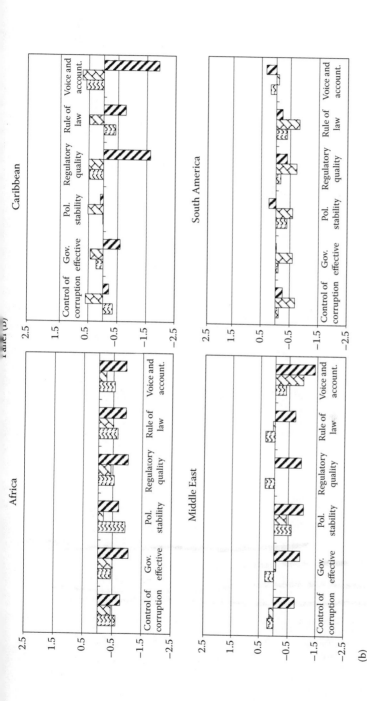

FIGURE 1.6 (continued) *Note:* (4) regulatory quality, measures of the incidence of market-unfriendly policies (e.g., price controls and inadequate bank supervision), perceptions of the burdens imposed by excessive regulation in areas such as foreign trade and business development; (5) rule of law, measures the extent to which agents have confidence in and abide by the rules of society (perceptions of the incidence of crime, the effectiveness and predictability of the judiciary, and the enforceability of contracts); and (6) control of corruption: measures the extent of corruption (exercise of public power for private gain). The black bars refer to the average for countries with developed stock exchanges, the wavy pattern bars refer to the average for countries with old emerging stock markets, the angular brick pattern bars refer to the average for the countries with new emerging stock markets, and the stripy pattern bars show the average for the countries without stock markets.

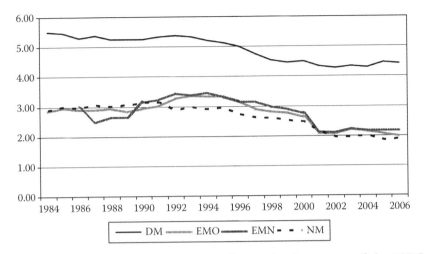

FIGURE 1.7 The time-path of the equally weighted averages of the ICRG corruption index for countries with developed stock markets, emerging stock markets for the period 1981–2007. At a country level, the index can vary between 1 (most corrupt) and 6 (least corrupt).

the developed markets and the other markets remains more or less constant over time. Since the index is based on investors' perception rather than some hard-core fundamentals, the decrease in corruption observed for the nondeveloped markets in the first half of the 1990s may be the result of investors' general enthusiasm for broadening investment opportunities and/or a lack of understanding of the investment climate of the emerging markets. As soon as the Asian Crisis arose, the nondeveloped countries corruption index starts declining with the biggest drop observed around 1999–2000 when investors, hurt by the burst of the e-commerce bubble, projected their pessimism on their assessment of the emerging markets.

All this leads to the conclusion that the growth prospects of emerging markets may be weaker than one would wish for. It is not only that they score low at all available indicators of governance and corruption, but the gap between them and the developed markets does not narrow down over time. On the contrary, it seems to expand in relative terms. It is interesting that despite numerous programs to stimulate the development of emerging markets, investors are still not convinced about the emerging market environment. Whether this is fair or not (it is beyond the scope of this chapter to debate) it does not change the fact, however, that it is investors who shape these markets to a high degree. Therefore, if the international investors do not find the emerging markets a good place to invest in, the emerging markets may find it difficult to grow fast using their own capital only.

1.5 CONCLUSION

This chapter discusses the growth of the emerging markets with a special focus on the exchanges that have been created since 1985 in countries that did not have any exchanges prior to the date. We provide evidence that the high growth of the emerging markets has been driven by these newcomers. However, we argue that this trend is over because opening additional exchanges in countries with markets that already struggle for liquidity or opening additional exchanges in countries currently without a stock exchange will not help to maintain the high growth figures observed in the last couple of decades. Hence, we believe that the future growth of the emerging markets must be "organic." Building up the inner-strength, rather than a simple physical expansion, is the only way forward. To achieve this, however, the emerging markets must improve their governance and legal framework. We show that in terms of governance and legal framework the emerging markets lag far behind developed markets and no improvement over the last 20 years or so can be detected. This conclusion has important policy implications but it is somewhat depressing, given the numerous programs and initiatives that have been put in place to stimulate the growth and development of capital markets in developing countries.

ACKNOWLEDGMENT

The authors would like to thank The Leverhulme Trust for financial support (grant F/00 351/T).

REFERENCES

Beck, T. and Levine, R. (2004) Stock markets, banks, and growth: Panel evidence. *Journal of Banking and Finance*, 28(3): 423–442.

Calderon, C. and Liu, L. (2003) The direction of causality between financial development and economic growth. *Journal of Development Economics*, 72(1): 321–334.

Catalan, M., Impavido, G., and Musalem, A.R. (2000) Contractual savings or stock markets development: Which leads? World Bank Policy Research Working Paper No. 2421.

Claessens, S., Klingebiel, D., and Schmukler, S.L. (2006) Stock market development and internationalization: Do economic fundamentals spur both similarly? *Journal of Empirical Finance*, 13(3): 316–350.

Demirguc-Kunt, A. and Levine, R. (1995) Stock market development and financial intermediaries: Stylized facts. World Bank Policy Research Working Paper No. 1462.

Grout, P.A., Megginson, W., and Zalewska, A. (2008) *The New Shareholders: Twenty-five Years On*. Mimeo, University of Bath, Bath, U.K.

Impavido, G., Musalem, A.R., and Tressel, T. (2003) The impact of contractual savings institutions on securities markets. World Bank Policy Research Working Paper 2948.

La Porta, R., Lopez-de-Silanes, F., and Shleifer, A. (1997) Legal determinants of external finance. *Journal of Finance*, 52(3): 1131–1150.

Zalewska, A. (2006) Is locking domestic funds into the local market beneficial? Evidence from the polish pension reforms. *Emerging Markets Review*, 7(4): 339–360.

CHAPTER 2

Are Emerging Stock Markets Less Efficient? A Survey of Empirical Literature

Kian-Ping Lim and Robert D. Brooks

CONTENTS

2.1 INTRODUCTION

Market efficiency has been defined in many different ways, and the literature has not come to terms with a standard definition. However, the definition given by Fama (1970) is the most widely used in the academic literature and quoted in most standard finance textbooks. Specifically, the efficient markets hypothesis (EMH) defines a market as efficient when security prices fully reflect all available information, implying that new information is quickly reflected in prices. So, after nearly four decades, what is the current state of the literature on EMH? Lo (2008) notes that even after thousands of published articles, there is still no consensus among economists on whether financial markets are efficient. One of the reasons given by the author for this state of affairs is that the EMH is not a well-defined and empirically refutable hypothesis due to the existence of the joint hypothesis problem, i.e., market efficiency is determined within the context of a particular asset pricing model.

A more pertinent problem highlighted by Lo (2008) is that statistical tests of the EMH may not be the most informative means for gauging the efficiency of a given market as their focus is on testing the all-or-nothing notion of absolute market efficiency. Campbell et al. (1997), Lo (1997, 2008), and Lo and MacKinlay (1999) have repeatedly argued that perfect efficiency is an unrealistic benchmark that is unlikely to hold in practice. Instead, these authors offer the notion of relative efficiency, that is the efficiency of one market measured against another, for example, the New York Stock Exchange versus the Paris Bourse, futures market versus spot market, or auction markets versus dealer markets. In other words, it is more useful to know the degree of efficiency rather than proclaiming whether a market is or is not efficient. Hence, it is not surprising to learn that after decades of empirical investigation, little is known about the differences in the degree of efficiency across markets and what characteristics are associated with greater levels of informational efficiency.

However, there has been a shift in research focus in recent years in measuring the relative efficiency of stock markets based on (1) how much private-firm-specific information is incorporated into stock prices using the market model R-square statistic of Morck et al. (2000) and the private information trading measure of Llorente et al. (2002); (2) how quickly market-wide information is capitalized into stock prices using the price delay measure of Hou and Moskowitz (2005); (3) how closely stock prices follow a random walk using conventional autocorrelation-based measures; and (4) how persistent stock prices deviate from a random walk using rolling test statistics. The growing importance of this literature has motivated the World Bank Financial Sector Development Indicators (FSDI) project, as part of its ongoing effort to measure various aspects of financial sector development and facilitate direct cross-country comparisons, to construct a composite indicator for assessing the relative informational efficiency of stock markets around the world.

2.2 HOW TO MEASURE THE DEGREE OF INFORMATIONAL EFFICIENCY?

2.2.1 Market Model R-Square Statistic

The stock price synchronicity measure proposed by Morck et al. (2000), in particular, their average market model R-square statistic, has inspired extensive studies on stock market efficiency. Briefly, the stock price synchronicity is measured by the percent of total firm-level return variation explained by local and U.S. market indexes in ordinary least squares regression as follows:

$$r_{i,t} = \alpha_i + \beta_{1,i} r_{m,jt} + \beta_{2,i} \left[r_{U.S.,t} + e_{j,t} \right] + \varepsilon_{i,t} \tag{2.1}$$

where

$r_{i,t}$ is the return on stock i

$r_{m,jt}$ is a domestic market index that represents the whole market of country j

$r_{U.S.,t}$ is the U.S. market index return that proxies for the global market

t is a 2 week period time index

The rate of change in the exchange rate per U.S. dollar is $e_{j,t}$.

The coefficient of determination from the estimation of Equation 2.1, R_{ij}^2, measures the percent of the variation in the biweekly return of stock i in country j explained by variations in country j's market return and the U.S. market return. Country-level R-squares are aggregated across stocks as follows

$$R_j^2 = \frac{\sum_i R_{ij}^2 \times \mathrm{SST}_{ij}}{\sum_i \mathrm{SST}_{ij}} \qquad (2.2)$$

where SST_{ij} is the sum of squared total variations. A higher R_j^2 indicates that stock prices in country j frequently move together.

Morck et al. (2000) argue that their synchronicity measure is inversely related to the amount of firm-specific information impounded into individual stock prices, with more firm-specific information being associated with a lower market model R^2. The intuition is that when a firm's stock return is strongly correlated with the market return, then its stock price is less likely to convey firm-specific information. Subsequent study by Durnev et al. (2003) using U.S. stocks provides evidence supporting the aforementioned information-efficiency interpretation, establishing the empirical link between stock price synchronicity and stock price informativeness.* Specifically, Durnev et al. (2003) find that firms with a lower R^2 value observe a higher association between current return and future earnings, which they regard as evidence suggesting that lower R^2 statistic signals more informative stock price and hence more efficient stock market. Since then, many empirical studies utilize the market model R^2 as an inverse measure of informational efficiency. Despite its popularity, the validity of the information-efficiency interpretation of R^2 does not go unchallenged (see Ashbaugh-Skaife et al., 2006; Chan and Hameed, 2006; Hou et al., 2006; Teoh et al., 2006; Griffin et al., 2007; Kelly, 2007; Saffi and Sigurdsson, 2007).

2.2.2 Private Information Trading Measure

Llorente et al. (2002) examine the volume–return dynamics using a model in which investors trade both to share risk (hedging trade) and speculate

---

* The authors define stock price informativeness as the amount of information about future earnings that is reflected in stock prices.

on their private information (speculative trade). In their theoretical model, if hedging-motivated trades dominate, return accompanied by high volume tends to reverse in the subsequent period. In contrast, if speculative trading in a stock is significant, conditioned on high volume, return becomes less likely to reverse and can even exhibit persistence in the subsequent period. Hence, the difference in the degree of informed trading gives rise to the cross-sectional variation in the relation between volume and return autocorrelation. This private information trading measure has been adopted by Durnev and Nain (2007), Ferreira and Laux (2007), and Fernandes and Ferreira (2008a,b) as a proxy of informational efficiency. To construct the measure for each firm year, the following time-series regression is estimated:

$$r_{i,t} = \alpha_i + \beta_{1,i} r_{i,t-1} + \beta_{2,i} r_{i,t-1} V_{i,t-1} + \varepsilon_{i,t} \qquad (2.3)$$

where

$r_{i,t}$ is the return on stock i in week t

$V_{i,t}$ is the weekly trading volume

The regression coefficient, $\beta_{2,i}$, on the interaction term indicates whether stocks are dominated by hedging trades or trades generated by private information. As noted earlier, their model prediction is that hedging trades generate negatively autocorrelated returns while speculative trades tend to exhibit positive return autocorrelation. Hence, a statistically significant positive coefficient for $\beta_{2,i}$ suggests the dominance of informational trades while for stock with predominantly risk-sharing trades, the coefficient should be significantly negative. However, when neither dominates, $\beta_{2,i}$ is insignificantly different from zero. The magnitude of the coefficient captures the degree of information-based trading with higher value indicating more informed trading.

2.2.3 Price Delay Measure

Hou and Moskowitz (2005) developed a parsimonious measure that captures the average delay with which a firm's stock price responds to information.[*] Specifically, the authors perform a regression of each stock's weekly

---

[*] However, the price delay measure is not new to the literature. It can be traced back to Brennan et al. (1993) and Mech (1993).

return on contemporaneous market returns and 4 weeks of lagged market returns that is employed as the relevant news to which stock responds. The unrestricted model is

$$r_{i,t} = \alpha_i + \beta_i r_{m,t} + \sum_{n=1}^{4} \delta_i^{(-n)} r_{m,t-n} + \varepsilon_{i,t} \qquad (2.4)$$

where

$r_{i,t}$ is the return on stock i

$r_{m,t}$ is the local market index return in week t

If the stock responds immediately to market news, then β_i will be significantly different from zero, but none of the $\delta_i^{(-n)}$ will differ from zero, indicating that there would be no improvement in the R^2 by adding the lagged market returns to the regression. If, however, stock i's price responds with a lag, then some of the $\delta_i^{(-n)}$ will differ significantly from zero.

The restricted model constrains the coefficients of the lagged market returns to zero:

$$r_{i,t} = \alpha_i + \beta_i r_{m,t} + \varepsilon_{i,t} \qquad (2.5)$$

The R-squares from Equations 2.4 and 2.5 are used to calculate the commonly used delay measure:

$$\text{Delay} = 1 - \frac{R^2_{\text{restricted}}}{R^2_{\text{unrestricted}}} \qquad (2.6)$$

The larger the value of the delay measure, the more the variation in stock return captured by lagged market returns, indicating greater delay in the response of stock price to market-wide information. The ability of the delay measure to capture the speed of information processing has been utilized by a number of recent studies as an inverse measure of informational efficiency (Griffin et al., 2007; Saffi and Sigurdsson, 2007; Wu, 2007; Bae et al., 2008). Elsewhere, Bris et al. (2007) and Saffi and Sigurdsson (2007) also used shorter lag via the cross-autocorrelation measure defined as the correlation between contemporaneous stock return and lagged 1 week market return, i.e., $\rho = \text{Corr}(r_{i,t}, r_{m,t-1})$.

2.2.4 Autocorrelation-Based Measures

According to Malkiel (2003), Fama's (1970) definition of efficiency is associated with the view that stock prices would move unpredictably. This underlying logic is that price changes occur only in response to genuinely new information, which by definition is unpredictable, hence the resulting price changes must be random. In fact, the origin of the EMH is generally traced back to the landmark work of Samuelson (1965) who has been widely credited for giving academic respectability for the random walk hypothesis. Specifically, Samuelson (1965) demonstrates that in an informationally efficient market, price changes must be unforecastable if they fully incorporate the expectations and information of all market participants. Hence, the evidence of return autocorrelation is widely interpreted as reflecting a delay adjustment to information. For instance, Theobald and Yallup (2004) developed formal speed of adjustment estimators that are functions of autocorrelations to gauge the speed with which new information is reflected in stock prices (see also references cited therein). Their research is motivated by extant behavioral models that show both price under- and overreactions would induce particular autocorrelation patterns into the return series.

In the context of relative efficiency, the magnitude of the autocorrelation coefficient can be used to gauge how closely stock price follows a random walk. Motivated by the appealing statistical property of the variance ratio statistic, a number of recent studies employ the absolute deviation from one of the variance ratios to measure the degree of informational efficiency (see Gu and Finnerty, 2002; Griffin et al., 2007; Boehmer and Kelley, 2008; Chordia et al., 2008). This is because the variance ratio is one plus a weighted sum of the autocorrelation coefficients for stock return with positive and declining weights, and both positive and negative autocorrelations represent departures from market efficiency. Briefly, the variance ratio should be equal to one for all holding periods under the null hypothesis that stock price follows a random walk with uncorrelated increments. If variance ratios are significantly greater (less) than one, returns are positively (negatively) serially correlated.

2.2.5 Rolling Test Statistics

The financial econometrics literature has witnessed the growth of advanced statistical tests designed to uncover other forms of stock price deviations from random walk, such as the long-term correlations of price changes and nonlinear serial dependence in return series. However, due to the portmanteau nature of these statistical tests, the issue of interest is only on whether the market under study is or is not efficient. This is because the magnitude of the test statistics or their corresponding *p*-values does not indicate the degree of market efficiency. In recent years, there emerged a strand of literature that computes these test statistics in rolling subsamples in order to capture the evolution of market efficiency over time, which can be rationalized within the adaptive markets hypothesis of Lo (2004, 2005). Since the rolling subsamples show how often the random walk hypothesis is rejected by the test statistic over the full sample period, it is possible to assess the relative efficiency of those sampled stock markets by comparing the percentages of subsamples with significant test statistic. In other words, a higher percentage indicates more persistent deviations from random walk over the sample period, and hence a lower degree of informational efficiency (see Cajueiro and Tabak, 2004; Lim, 2007; Lim and Brooks, 2008).

2.3 ARE EMERGING STOCK MARKETS LESS EFFICIENT?

According to conventional wisdom, developed stock markets are expected to be more efficient in incorporating information into prices than emerging ones. This section assesses whether the degree of country-level market efficiency reported by earlier studies conforms to this expectation. When assigning market status of a country, we follow Standard & Poor's *Global Stock Markets Factbook* 2006 that classifies "developed" and "emerging" markets based on (1) the income level of the country (gross national income per capita) as defined by World Bank and (2) the existence of investment restrictions such as currency repatriation restrictions, capital controls, and foreign share ownership limitations.[*]

2.3.1 Morck et al. (2000)

Morck et al. (2000) propose two country-specific measures of stock price synchronicity: (1) the fraction of stocks that move in the same direction and (2) the market model R-square statistic. Our discussion focuses on the latter since it is the one that receives the maximum attention in empirical literature. Panel C of Table 2 in Morck et al. (2000) presents the average R^2s of firm-level regressions of biweekly stock returns on local and U.S. market indexes for 40 countries in 1995, denominated

---

[*] Other major index providers such as FTSE, Russell, MSCI, and Dow Jones have different criteria, but the country classifications are in general consistent with those provided by S&P.

in local currency units. The results show that high-income countries have lower R^2s while low-income economies account for the high ratios of market-wide return variation to total return variation. When we classify these countries according to their market status, the means for developed and emerging markets are 0.116 and 0.273, respectively. Our t-test of the equality of means between these two groups confirms that the difference is statistically significant at the 1% level. The results indicate that developed markets in general have lower R^2s and hence higher degree of informational efficiency than emerging markets. However, on a country-by-country basis, some emerging markets are found to be more efficient than their developed counterparts. For instance, stock market in Indonesia is more efficient than those in the developed category—Finland, Sweden, Belgium, Hong Kong, Italy, Singapore, Greece, Spain, and Japan. Notably, the latter five developed markets have higher R^2s than the emerging markets in Brazil, the Philippines, South Korea, and Pakistan.

2.3.2 Jin and Myers (2006)

Jin and Myers (2006) expand Equation 2.1 to include two lead and lag terms for the local and U.S. market indexes in order to correct for the nonsynchronous trading. Using weekly data for stocks in 30 countries, from 1990 to 2001, and 10 more countries for part of the period, the authors find that the average R^2s vary across countries and have been declining over time internationally. The market model, R^2s, for individual stocks are averaged for each country and year using variance weights as in Morck et al. (2000) and equal weights. These computed values for each country are then averaged across time and reported in Table 2 of Jin and Myers (2006). On the basis of equal-weighted country-average R^2s, the means for developed and emerging markets are 0.285 and 0.326, respectively, indicating that developed markets are more efficient than their emerging counterparts. Our t-test of the equality of means between these two groups confirms that the difference is statistically significant at the 1% level. Individually, the emerging markets in Columbia and Russia only trail behind Canada and Denmark, indicating that they are as efficient as the developed markets in terms of the amount of firm-specific information being incorporated into stock prices. The R^2s for Chile, Czech Republic, Peru, and South Africa are lower than the mean for developed markets. Using variance-weighted R^2s do not alter much the general conclusions.

2.3.3 Fernandes and Ferreira (2008a)

Fernandes and Ferreira (2008a) employ firm-specific stock return variation for each country as their primary measure of informational efficiency. Specifically, the authors compute firm-specific return variation relative to total return variation for each country as follows:

$$\frac{\sigma_{\varepsilon j}^2}{\sigma_j^2} = \frac{\sigma_{\varepsilon j}^2}{\sigma_{mj}^2 + \sigma_{\varepsilon j}^2} \tag{2.7}$$

where
$\sigma_{\varepsilon j}^2$ is the average firm-specific return variation in country j
σ_j^2 denotes total variation in stock return for country j
σ_{mj}^2 is the average variation in country j's stock return that is explained by market factors

$\sigma_{\varepsilon j}^2 / \sigma_j^2$ is precisely one minus the R_j^2 of Equation 2.1, and hence it is a direct measure of informational efficiency where a higher value of $\sigma_{\varepsilon j}^2 / \sigma_j^2$ indicates that more firm-specific information is impounded into individual stock prices.

The sample period in Fernandes and Ferreira (2008a) spans from 1980 to 2003, covering 48 countries with developed and emerging markets equally divided. Relative firm-specific stock return variation is the median across all firms for each country in each year estimated using the two-factor international model and monthly excess returns denominated in U.S. dollars. Table 1 of Fernandes and Ferreira (2008a) reports the time-series averages of median relative firm-specific return variation for each country. In comparison, the annual country-level relative firm-specific return variation is 5.47% higher in developed markets than emerging markets, conforming to the expectation that developed markets are relatively more efficient. We conduct the t-test of the equality of means between these two groups and find that the difference is statistically significant at the 5% level. Once again, some emerging markets outperform their developed counterparts in terms of informational efficiency. For instance, based on the overall ranking of countries by $\sigma_\varepsilon^2 / \sigma^2$, Czech Republic occupies the second spot behind the United States. Other emerging markets that make it to the top-10 list are Peru (fifth), Turkey (eighth), and China (ninth).

2.3.4 Fernandes and Ferreira (2008b)

Fernandes and Ferreira (2008b) also employ the median relative firm-specific stock return variation as a direct measure of informational efficiency. The countries and sample period covered in their sample is similar to Fernandes and Ferreira (2008a). The differences in terms of research design are (1) firms from U.S. are excluded; (2) weekly returns instead of monthly returns are utilized; and (3) the regressions are performed at the firm level and not country level. Nevertheless, our main interest is on the median relative firm-specific return variation across all firms for each country as reported in the first column of their Table 1.

The country-level median relative firm-specific return variation varies widely, from a minimum of 0.569 in Venezuela to a maximum of 0.926 in Peru. In fact, the values of $\sigma_\varepsilon^2/\sigma^2$ in Fernandes and Ferreira (2008b) are higher than those reported by Fernandes and Ferreira (2008a) for all countries where the averages across 47 countries are 0.782 for the former study and 0.550 for the latter. It is unclear whether this is due to the frequency of returns (monthly versus weekly) or the difference in the number of firms selected by both studies. Based on these country-level $\sigma_\varepsilon^2/\sigma^2$ values, there is a 6.25% difference between developed and emerging markets. Our t-test of the equality of means between these two groups confirms the difference is statistically significant at the 5% level, indicating that the stock prices in developed markets convey more firm-specific information than those in emerging markets. Ranking all countries by $\sigma_\varepsilon^2/\sigma^2$ reveals that the most efficient is the emerging market in Peru, pushing Canada to second place. Other emerging markets that make it to the top-10 list are China (5th), Turkey (9th), and South Africa (10th).

2.3.5 Griffin et al. (2007)

The most comprehensive study so far is the one conducted by Griffin et al. (2007) who compare the degree of informational efficiency over the period 1994–2005 in 56 stock markets around the world, comprising 33 emerging and 23 developed markets. To capture different aspects of market efficiency, the authors employ four empirical measures, namely, the price delay measure, absolute deviation of variance ratio minus one, abnormal returns associated with postearnings announcement, and the average market model R-square. The construction of five size portfolios based on the U.S. market breakpoints allows comparison between large and small capitalization stocks not only in each country but also across countries for different size-ranked portfolios.

Their first set of analysis reveals that price delay is unexpectedly larger in the more developed stock markets, suggesting that emerging markets are on average more efficient than their developed counterparts. Two of the largest stock markets in the world by market capitalization, the United States and the United Kingdom, exhibit the largest delay with over 10% of return variation that can be explained by past returns for the bottom two size quintiles. Secondly, the variance ratio statistics are computed at the individual stock level, and then averaged across stocks using the absolute deviation of variance ratio minus one to obtain within-country equally weighted portfolio level averages. Across countries, stock prices in developed markets do not seem to follow more closely a random walk than emerging markets. Their t-test result confirms that the difference between the averages for both groups is statistically insignificant. Individually, some emerging markets appear to be more efficient than their developed counterparts. For instance, stock markets in the United Kingdom, Canada, and the United States exhibit absolute variance ratios that are larger than Venezuela, Israel, China, and Turkey, especially for stocks in the bottom two size quintiles.

Finally, Griffin et al. (2007) estimate the country-level R^2 in similar fashion as Morck et al. (2000) by averaging across all firms in a country using variance weights. Across countries, developed markets have lower R^2s than their emerging counterparts in all size portfolios, but the differences of means between these two groups are statistically significant only in the largest three size quintiles. Again, some emerging markets like China, Czech Republic, Romania, and Bulgaria exhibit higher pricing efficiency than many developed markets. The authors single out the discrepancy between Morck et al. (2000) and their finding for China. While this market records the second highest R^2 in the former study, Griffin et al. (2007) find that China is not only the most efficient emerging market but also ranks second overall behind the United States using stock data over 1994–2005.

2.3.6 Lim and Brooks (2008)

Lim and Brooks (2008) compute the bicorrelation statistic in rolling subsamples in order to compare the persistence of stock price deviations from a random walk for the stock markets of 50 countries over the period 1995–2005. Specifically, a higher percentage of subsamples with significant bicorrelation statistic indicates more persistent nonrandom walk price movements over time, and hence a lower degree of informational efficiency. Consistent with other studies, developed markets as a group, on average, are still more efficient than their emerging counterparts. The percentage of subsamples with significant bicorrelation statistic for developed markets is 8.65% lower than emerging markets, and the subsequent t-test of the equality of means between these two groups confirms that the difference is statistically significant at the 1% level. On a country-by-country basis, some emerging markets appear to be more efficient than the developed ones, at least in incorporating information contained in past returns into stock prices. In fact, sorting the results from the lowest percentage of significant rolling subsamples to the highest, four emerging markets occupy the top five in the chart, with Hong Kong being the only representative from developed countries. Specifically, Thailand (17.45%) is the most efficient market, followed by Jordan (17.75%), Hong Kong (18.35%), South Korea (24.12%), and Malaysia (25.84%).

2.3.7 World Bank FSDI Project

The World Bank FSDI project constructs a composite efficiency index based on the average of the following three indicators: (1) the stock price synchronicity measure of Morck et al. (2000); (2) the private information trading measure of Llorente et al. (2002); and (3) equity transaction costs inferred from daily stock price movements using the approach outlined by Lesmond et al. (1999). A higher value for the composite efficiency index indicates a more informational efficient stock market. Figure 2.1 presents the 2004 index constructed by FSDI for 58 stock markets around the world, comprising 25 developed and 33 emerging markets. Again, developed markets are still more efficient than their emerging counterparts, with the former obtaining an average index of 5.39 while the latter scoring 4.42. Our subsequent t-test of the equality of means between these two groups confirms that the difference is statistically significant at the 1% level. On a country-by-country basis, it is clear from Figure 2.1 that emerging markets occupy the tail end of the ranking table. For instance, the most inefficient market is Nigeria, followed by Jordan, China, Sri Lanka, Indonesia, the Philippines, and Venezuela. Though there are seven developed markets in the top-10 list, the most efficient market is Mexico, pushing the United States to second place. Hungary and Peru are other two emerging markets that have been given high score for their performance in information processing.

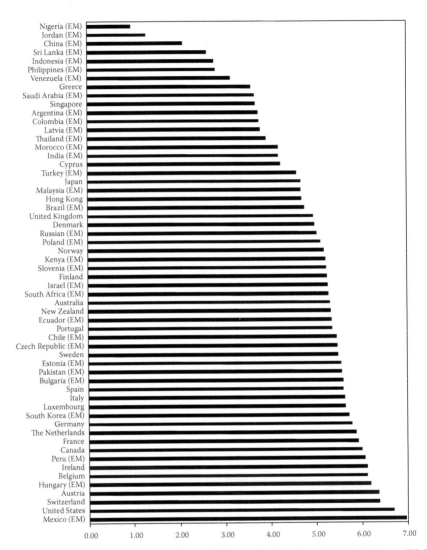

FIGURE 2.1 FSDI equity market efficiency index for 2004. *Notes:* EM denotes emerging markets; a higher value for the efficiency index indicates more efficient stock market. (From http://www.fsdi.org/. With permission.)

2.4 CONCLUSION

Our literature survey indeed finds that emerging markets, in general, are less efficient than their developed markets when informational efficiency is measured in terms of how much private-firm-specific information is incorporated into stock prices and how persistent stock prices deviate from a random walk. Nevertheless, there are some emerging markets that perform exceptionally well in their informational role. Of course, the literature does not just stop there. Morck et al. (2000) find that the cross-country differences in property rights protection can explain why stock prices move together more in poor economies (emerging markets) than rich economies (developed markets). The justification is that poor property rights protection might cause arbitrageurs to shun the stock markets of these economies, leaving them to noise traders. The model of De Long et al. (1990) predicts that when the proportion of noise traders in the market is above a critical level, this effect causes noise trading to grow in importance relative to informed trading, and they eventually dominate the market. Lim and Brooks (2008) also find that the persistent stock price deviations from random walk in emerging markets can be largely attributed to low-income economies providing poor protection of private property rights. This is because the dominance of noise traders, who are prone to sentiment not fully justified by information, and their correlated trading can cause stock prices to deviate from their random walk benchmarks for persistent periods of time. The policy implication from the aforementioned two studies for emerging markets is unambiguous.

Other important factors found to explain the cross-country differences in market model R-square include public investor protection (Morck et al., 2000), stock market liberalization (Li et al., 2004), corporate transparency (Jin and Myers, 2006), securities laws (Daouk et al., 2006), short sales restrictions (Bris et al., 2007), market liquidity (Bris et al., 2007), market volatility (Griffin et al., 2007), and insider trading laws (Beny, 2007; Fernandes and Ferreira, 2008a). It is worth highlighting that though securities laws exist in most countries, their effects on informational efficiency have only been examined in recent years using measures of relative efficiency. Such an investigation provides useful feedback and helps market regulators to improve the functioning of stock markets. These measures of relative efficiency also allow researchers to address empirically the role of stock market in the real economy (for theoretical work, see, for example, Dow and Gorton, 1997; Subrahmanyam and Titman, 2001; Dow and Rahi, 2003; Goldstein and Guembel, 2008). For instance, using the market model R-square, it has been shown that the stock market is not an economic sideshow, instead efficient market prices enhance the efficiency of capital allocation (Wurgler, 2000), induce higher productivity and faster economic growth (Durnev et al., 2004a), facilitate more efficient corporate capital investments (Durnev et al., 2004b), exert a strong positive effect on firms' real investments

(Chen et al., 2007), and lead to a less demanding corporate board structure with a lesser degree of independence (Ferreira et al., 2007).

REFERENCES

Ashbaugh-Skaife, H., Gassen, J., and LaFond, R. (2006) Does stock price synchronicity represent firm-specific information? The International Evidence. MIT Sloan Research Paper No. 4551-05, available at http://ssrn.com/abstract=768024.

Bae, K.H., Ozoguz, A., and Tan, H. (2008) Do foreigners facilitate information transmission? SSRN Working Paper, available at http://ssrn.com/abstract=1160063.

Beny, L.N. (2007) Insider trading laws and stock markets around the world: An empirical contribution to the theoretical law and economics debate. *Journal of Corporation Law*, 32(2): 237–300.

Boehmer, E. and Kelley, E. (2008) Institutional investors and the informational efficiency of prices. *Review of Financial Studies*, forthcoming.

Brennan, M.J., Jegadeesh, N., and Swaminathan, B. (1993) Investment analysis and the adjustment of stock prices to common information. *Review of Financial Studies*, 6(4): 799–824.

Bris, A., Goetzmann, W.N., and Zhu, N. (2007) Efficiency and the bear: Short sales and markets around the world. *Journal of Finance*, 62(3): 1029–1079.

Cajueiro, D.O. and Tabak, B.M. (2004) Ranking efficiency for emerging markets. *Chaos, Solitons and Fractals*, 22(2): 349–352.

Campbell, J.Y., Lo, A.W., and MacKinlay, A.C. (1997) *The Econometrics of Financial Markets*. Princeton University Press, Princeton, NJ.

Chan, K. and Hameed, A. (2006) Stock price synchronicity and analyst coverage in emerging markets. *Journal of Financial Economics*, 80(1): 115–147.

Chen, Q., Goldstein, I., and Jiang, W. (2007) Price informativeness and investment sensitivity to stock price. *Review of Financial Studies*, 20(3): 619–650.

Chordia, T., Roll, R., and Subrahmanyam, A. (2008) Liquidity and market efficiency. *Journal of Financial Economics*, 87(2): 249–268.

Daouk, H., Lee, C.M.C., and Ng, D. (2006) Capital market governance: How do security laws affect market performance? *Journal of Corporate Finance*, 12(3): 560–593.

De Long, J.B., Shleifer, A., Summers, L.H., and Waldmann, R.J. (1990) Noise trader risk in financial markets. *Journal of Political Economy*, 98(4): 703–738.

Dow, J. and Gorton, G. (1997) Stock market efficiency and economic efficiency: Is there a connection? *Journal of Finance*, 52(3): 1087–1129.

Dow, J. and Rahi, R. (2003) Informed trading, investment, and welfare. *Journal of Business*, 76(3): 439–454.

Durnev, A., Li, K., Morck, R., and Yeung, B. (2004a) Capital markets and capital allocation: Implications for economies in transition. *Economics of Transition*, 12(4): 593–634.

Durnev, A., Morck, R., and Yeung, B. (2004b) Value-enhancing capital budgeting and firm-specific stock return variation. *Journal of Finance*, 59(1): 65–105.

Durnev, A., Morck, R., Yeung, B., and Zarowin, P. (2003) Does greater firm-specific return variation mean more or less informed stock pricing? *Journal of Accounting Research*, 41(5): 797–836.

Durnev, A. and Nain, A.S. (2007) Does insider trading regulation deter private information trading? International evidence. *Pacific-Basin Finance Journal*, 15(5): 409–433.

Fama, E.F. (1970) Efficient capital markets: A review of theory and empirical work. *Journal of Finance*, 25(2): 383–417.

Fernandes, N. and Ferreira, M.A. (2008a) Insider trading laws and stock price informativeness. *Review of Financial Studies*, in press.

Fernandes, N. and Ferreira, M.A. (2008b) Does international cross-listing improve the information environment? *Journal of Financial Economics*, 88(2): 216–244.

Ferreira, D., Ferreira, M.A., and Raposo, C.C. (2007) Board structure and price informativeness. ECGI-Finance Working Paper No. 160/2007, available at http://ssrn.com/abstract=983524.

Ferreira, M.A. and Laux, P.A. (2007) Corporate governance, idiosyncratic risk, and information flow. *Journal of Finance*, 62(2): 951–989.

Goldstein, I. and Guembel, A. (2008) Manipulation and the allocational role of prices. *Review of Economic Studies*, 75(1): 133–164.

Griffin, J.M., Kelly, P.J., and Nardari, F. (2007) Measuring short-term international stock market efficiency. SSRN Working Paper, available at http://ssrn.com/abstract=959006.

Gu, A.Y. and Finnerty, J. (2002) The evolution of market efficiency: 103 years daily data of the dow. *Review of Quantitative Finance and Accounting*, 18(3): 219–237.

Hou, K. and Moskowitz, T.J. (2005) Market frictions, price delay, and the cross-section of expected returns. *Review of Financial Studies*, 18(3): 981–1020.

Hou, K., Peng, L., and Xiong, W. (2006) R^2 and Price inefficiency. Fisher College of Business Working Paper No. 2006-03-007, available at http://ssrn.com/abstract=954559.

Jin, L. and Myers, S.C. (2006) R^2 around the world: New theory and new tests. *Journal of Financial Economics*, 79(2): 257–292.

Kelly, P.J. (2007) Information efficiency and firm-specific return variation. SSRN Working Paper, available at http://ssrn.com/abstract=972775.

Lesmond, D.A., Ogden, J.P., and Trzcinka, C.A. (1999) A new estimate of transaction costs. *Review of Financial Studies*, 12(5): 1113–1141.

Li, K., Morck, R., Yang, F., and Yeung, B. (2004) Firm-specific variation and openness in emerging markets. *Review of Economics and Statistics*, 86(3): 658–669.

Lim, K.P. (2007) Ranking of efficiency for stock markets: A nonlinear perspective. *Physica A*, 376: 445–454.

Lim, K.P. and Brooks, R.D. (2008) Why do emerging stock markets experience more persistent price deviations from a random walk over time? A country-level analysis. SSRN Working paper, available at http://ssrn.com/abstract=1194562.

Llorente, G., Michaely, R., Saar, G., and Wang, J. (2002) Dynamic volume–return relation of individual stocks. *Review of Financial Studies*, 15(4): 1005–1047.

Lo, A.W. (1997) Introduction. In: A.W. Lo (Ed.), *Market Efficiency: Stock Market Behaviour in Theory and Practice*. Volumes I and II. Edward Elgar, Cheltenham, U.K.

Lo, A.W. (2004) The adaptive markets hypothesis: Market efficiency from an evolutionary perspective. *Journal of Portfolio Management*, 30: 15–29.

Lo, A.W. (2005) Reconciling efficient markets with behavioral finance: The adaptive markets hypothesis. *Journal of Investment Consulting*, 7(2): 21–44.

Lo, A.W. (2008) Efficient markets hypothesis. In: S.N. Durlauf and L.E. Blume (Eds), *The New Palgrave Dictionary of Economics Online*, 2nd edn. Palgrave Macmillan, Basingstoke, U.K., doi:10.1057/9780230226203.0454.

Lo, A.W. and MacKinlay, A.C. (1999) *A Non-Random Walk Down Wall Street*. Princeton University Press, Princeton, NJ.

Malkiel, B.G. (2003) The efficient market hypothesis and its critics. *Journal of Economic Perspectives*, 17(1): 59–82.

Mech, T.S. (1993) Portfolio return autocorrelation. *Journal of Financial Economics*, 34(3): 307–344.

Morck, R., Yeung, B., and Yu, W. (2000) The information content of stock markets: Why do emerging markets have synchronous stock price movements? *Journal of Financial Economics*, 58(1–2): 215–260.

Saffi, P.A.C. and Sigurdsson, K. (2007) Price efficiency and short-selling. SSRN Working Paper, available at http://ssrn.com/abstract=949027.

Samuelson, P. (1965) Proof that properly anticipated prices fluctuate randomly. *Industrial Management Review*, 6(2): 41–49.

Subrahmanyam, A. and Titman, S. (2001) Feedback from stock prices to cash flows. *Journal of Finance*, 56(6): 2389–2413.

Teoh, S.H., Yang, Y.G., and Zhang, Y. (2006) *R*-Square: Noise or firm-specific information? SSRN Working Paper, available at http://ssrn.com/abstract=926948.

Theobald, M. and Yallup, P. (2004) Determining security speed of adjustment coefficients. *Journal of Financial Markets*, 7(1): 75–96.

Wu, J. (2007) Short selling and the informational efficiency of prices. SSRN Working Paper, available at http://ssrn.com/abstract=1002652.

Wurgler, J. (2000) Financial markets and the allocation of capital. *Journal of Financial Economics*, 58(1–2): 187–214.

CHAPTER 3

How "Normal" Are Emerging Market Returns?

Craig Ellis and Maike Sundmacher

CONTENTS

3.1 INTRODUCTION

The assumption that consecutive price changes are independent and identically distributed (IID) has particular implications when working with financial asset returns. For investors, knowledge of the distribution is vital to measure the underlying risk of loss associated with a particular asset or trading strategy. In empirical studies of asset returns, the nature of the distribution also provides information about the return-generating process. Statistical tools for testing model fit and the significance of model parameters are also conditioned on particular types of distribution (Fama, 1965). Finally, and arguably most important, general forms of asset pricing models such as the Capital Asset Pricing Model (CAPM) and the Black–Scholes Option Pricing Model explicitly require Normality of the underlying distribution of asset returns. More general implications for financial asset returns of a stationary Normal distribution include the likelihood of a given percentage variation in price remaining constant over time and the assumption that the probability of price change in either direction (positive or negative) is equal in each period.[*]

This chapter considers the distributional and time-series characteristics of 10 emerging markets from the Asian-Pacific region. The "emerging markets" classification adopted in this chapter follows directly from that employed by Standard and Poor's (S&P) for the construction of the *S&P Emerging Stock Market Indices*. The relevant conditions to be classified as an emerging market according to the S&P standard are that the market meets at least one of the two following general criteria: (1) it is located in a low- or middle-income economy as defined by the World Bank and (2) the investable market capitalization is low relative to GDP (Standard and Poor's, 2007). The emerging markets examined herein comprise China (Shanghai A 30), India (S&P CNX 500), Indonesia (Jakarta SE Composite), Korea (KOSPI), Malaysia (KLCI Composite), Pakistan (Karachi SE 100), the Philippines (Philippines SE Composite), Sri Lanka (Colombo SE All), Taiwan (Taiwan SE Weighted), and Thailand (Bangkok SET). The sample data comprise interday returns over a period from January 1, 1980 to August 29, 2008.[†]

The objective of research in this chapter is to examine the distributional characteristics of the aforementioned emerging markets and to test specifically the overall goodness of fit of the Normal distribution to interday returns in these markets. Our research shows that the assumption

---

[*] Given specific values of the moments of the distribution (μ and σ), the stability of the distribution and independence of successive price changes, the returns generating process may for instance be defined as either a Weiner process, strict random walk, or standard Brownian motion.

[†] Owing to data availability, returns data for all the emerging markets are not available for the entire sample period. Specific beginning-of-sample dates and number of observations for each market are provided in Table 3.2.

of Normality of the distribution grossly underestimates the incidence of large returns (both above and below the mean) yet closely approximates the incidence of small returns. The general class of power-law distributed models is presented as an alternative to the Normal distribution for Asian-Pacific emerging market returns.

The remainder of this chapter is structured as follows: Section 3.2 discusses the Normal distribution within the context of financial asset returns and discusses evidence of Normality in returns for different classes of financial assets. Power-law distributed models as an alternative candidate distribution for financial asset returns are also discussed. Section 3.3 describes the employed data set, research methodology, and summaries of the results. Specific evidence of the goodness of fit of the Normal distribution is provided, by way of example, via detailed numeric and graphical evidence for particular markets. Section 3.4 provides some concluding remarks.

3.2 DISTRIBUTION OF FINANCIAL ASSET RETURNS

The Normal distribution as a candidate distributional model for financial asset returns was first proposed by Bachelier (1900), whose pioneering research into transactions in the French Exchange formed the basis of the modern Random Walk Hypothesis of financial asset returns.* A continuous random variable, X, has a Normal distribution if the density curve of X is a Normal curve defined by

$$f(x) = \frac{1}{\sigma\sqrt{2\pi}} \exp\left(-\frac{1}{2}\frac{(x-\mu)^2}{\sigma^2}\right) \qquad (3.1)$$

where the variable parameters of the distribution $E(X) = \mu$ and $Var(X) = \sigma^2$ are the mean and variance of the distribution, respectively. That no higher moments (e.g., skewness and kurtosis) exist for the Normal distribution first implies that positive and negative price changes have an equal probability of occurrence, and second that distribution of price changes is approximately bell shaped, which means that the probability of a particular

---

* Fama (1965) notes that while modern random walk theory does not explicitly demand a Normal distribution—rather only a distribution whose parameters are stationary over time, and one which is consistent with independent consecutive returns—the assertion of Bachelier (1900) that consecutive prices changes are IID and have a finite variance leads their distribution to be Normal.

price change occurring varies proportionately to the magnitude of the price change. Simply put, small price changes are more likely than large.

When modeling asset returns according to the assumption of Normality, the magnitude of price changes may not be known a priori. In most practical applications, the solution to this problem is to assume price changes are standard Normally distributed. Standard Normal distributions specifically have zero mean and unit variance. A well-known feature of these processes is that approximately 95% of all the price changes are observed to be within ±2 standard deviations of the mean change in price.

Normal distributions as defined by Equation 3.1 are both strictly stable and infinitely divisible. Although the term "stable" is often—rather loosely—used in finance with inference to the stationarity of the distribution's mean and variance over time, its real meaning is with respect to the fact that any linear combination of random variables drawn from a given distribution are also a part of that distribution. Assuming then that individual asset returns are identically Normally distributed, stability implies that returns to portfolio combinations of these assets will also be Normally distributed. By the implication of being infinitely divisible, the Normal probability distribution function is fully continuous. Neither transactions time nor money is infinitely divisible however.*

The near Normality of financial asset returns is regarded as an expected outcome of the application of Central Limit Theorem to independent summands. The mean–variance portfolio theory (Markowitz, 1952), the CAPM (Sharpe, 1964), and the Black–Scholes Option Pricing Model (Black and Scholes, 1973) are but some substantial models in modern finance that are proposed on the basis of the Normality asset returns.

Challenging the IID model proposed by Bachelier (1900), Mandelbrot (1963) observed that speculative prices contradicted the IID hypothesis by the fact that returns for many assets exhibited high-peaked and long-tailed (leptokurtic) distributions and appeared neither independent nor stationary. The observation led him to propose a stable (Paretian) power-law distribution for asset returns of the form $\Pr\{U > u\} = u^{-\alpha}$. Characterized by the four parameters, α, β, δ, and γ, Pareto distributions belong to the class of Levy skew alpha-stable continuous distribution. Moments of the Pareto distribution exist up to $|\alpha|$. The parameter α is

---

* The term "transactions time" is used herein to describe the points in calendar time when asset prices (values) are physically recorded by an exchange, commonly in discrete units of 1s.

a tail exponent and determines the probability contained in the tails of the distribution. The parameter β determines the skewness of the distribution, and the parameters δ and γ determine the location and scale of the distribution respectively. Given $\alpha > 1$, the location parameter δ is therefore equal to the mean of the distribution. Variance is likewise only known and finite given $\alpha = 2$. Given the parameter values $\alpha = 2$ and $\beta = 0$, the Normal distribution can therefore be described as a special form of the stable Pareto distribution. The significance of the stable Paretian distribution is that it allows for positive and negative price changes, the magnitude of which is, well outside of the boundaries of the Normal curve.

Testing Mandelbrot's proposal for a number of different U.S. common stocks, Fama (1965) likewise found little support for the Normal distribution. Further evidence of non-Normality in the distribution of common stocks is extensive and includes Longin (1996), Pagan (1996), Peiró (1999), Jondeau and Rockinger (2003), and references therein. The general failure of the standard Normality assumption can be seen to arise from two sources: the instability of financial asset probability distributions (Tucker and Scott, 1987; Kearns and Pagan, 1997) and the overall poor fit of the Normal curve against observed changes in financial asset prices (Mandelbrot, 1963; Fama, 1965; Pagan, 1996).

3.3 EMERGING MARKET RETURN DISTRIBUTIONS

Figures pertaining to the domestic market capitalization, number of listed companies, and average daily value of share trading for Asian-Pacific stock markets in Table 3.1 illustrate the significant differences in size and value between emerging markets in the region. The delineation of Asian-Pacific stock markets by size and value is of interest as it allows for the potential impact of market size and share value on the distributional and time-series characteristics of the markets to be examined.

Ranked by market capitalization and daily share value, we expect that the distribution of returns for the comparatively larger markets of China and Korea should be more closely Normal than those of Sri Lanka for instance.

The time-series properties of the natural logarithm of interday returns $\ln(P_t/P_{t-1})$, for emerging Asian-Pacific stock markets are presented in Table 3.2. The sample means of interday returns for all markets reported in Table 3.2 are slightly positive, but not significantly different from zero. Indicative of the long-run growth in these markets, the sum of the daily ln

TABLE 3.1   Market Size and Value (in US$ Millions) as in July 2008

	Domestic Market Capitalization	**Number of Listed Companies**	**Average Daily Value of Share Trading**
China (Shanghai)	2,143,473.1	863	9699.3
India	1,117,352.1	4917	1272.0
Indonesia	198,067.2	397	446.8
Korea	875,699.5	1787	5956.0
Malaysia	262,195.2	985	296.0
Pakistan	46,258.2	653	3157.7
Philippines	73,250.9	246	81.4
Sri Lanka	7,447.8	234	5.7
Taiwan	582,693.7	716	3263.1
Thailand	159,789.3	529	371.2

Source: Data from World Federation of Exchanges (2008) Focus. No. 186 August. Available at: http://www.world-exchanges.org, except for data for Pakistan sourced from the Karachi Stock Exchange (http://www.kse.com.pk/kse4/phps/mktglance01.php).

returns, and index points gain (loss) is positive for all series. The Karachi SE 100 (Pakistan) recorded the highest index point gain over the sample period (8621.4 points), and the Bangkok S.E.T. (Thailand) the lowest gain (535.0 points). The sum of ln returns for a standard Brownian motion process should be approximately zero since gains tend to equal losses. The sum of ln returns for all the emerging markets in Table 3.2 is positive; however, implying that returns in these markets may be better described by a submartingale process with mean expected returns greater than zero.

To determine the degree of randomness in interday ln returns each series is tested for the presence of autocorrelation up to 10 lags, with the results for lags 1, 2, 5, and 10 presented in Table 3.2. Without exception, there are no significant autocorrelations at any lag. A nonparametric runs test is also conducted. The runs test *p*-value determines for α levels above the critical value if there is sufficient evidence to conclude that the interday returns are not random. Consistent with the finding for autocorrelation, this test accepts randomness of interday returns for all series at the 0.01 level of confidence.

Interday returns in all emerging markets are non-Gaussian on the basis of the higher moments of the distribution. Small degrees of negative skewness, indicative of the small probability of large losses being offset by a higher probability of small gains, are evident for the Indian, Korean,

TABLE 3.2 Emerging Markets Descriptive Statistics

	China	India	Indonesia	Korea	Malaysia	Pakistan	Philippines	Sri Lanka	Taiwan	Thailand
Start of period	2/01/1991	2/01/1991	4/04/1983	2/01/1980	2/01/1980	2/01/1989	2/01/1986	2/01/1985	2/01/1980	2/01/1980
Count	4612	4607	6629	7478	7477	5129	5911	6172	7478	7478
Mean	4.60E-04	4.97E-04	4.61E-04	3.37E-04	2.24E-04	5.37E-04	5.11E-04	5.22E-04	3.43E-04	2.04E-04
Standard deviation	0.0233	0.0163	0.0153	0.0162	0.0145	0.0153	0.0171	0.0107	0.0178	0.0147
Skewness	1.2694	-0.4639	3.1630	-0.2317	-0.3884	-0.2653	0.3107	1.0317	0.0377	-0.0396
Kurtosis	20.7077	7.5779	99.0236	6.7804	35.2565	6.8806	11.5682	38.1346	9.1086	9.2276
Maximum	0.2886	0.1248	0.4031	0.1002	0.2082	0.1276	0.1618	0.1829	0.1991	0.1135
Minimum	-0.1791	-0.1432	-0.2253	-0.1737	-0.2415	-0.1321	-0.1579	-0.1390	-0.1966	-0.1606
Sum of ln returns	2.1196	2.2899	3.0557	2.5170	1.6730	2.7530	3.0189	3.2215	2.5637	1.5220
Index points										
Start of period	128.8	353.4	102.0	119.0	206.5	586.9	131.3	96.1	542.7	149.4
End of period	2202.4	3489.1	2165.9	1474.2	1100.5	9208.3	2688.1	2408.6	7046.1	684.4
Gain (loss)	2073.6	3135.7	2063.9	1355.2	894.0	8621.4	2556.8	2312.5	6503.4	535.0
Autocorrelation										
Lag 1	0.0406	0.1241	0.2631	0.0537	0.1019	0.0933	0.1569	0.2779	0.0306	0.1249
p-Value	0.0052	0.0000	0.0000	0.0000	0.0000	0.0000	0.0000	0.0000	0.0052	0.0000
Lag 2	0.0382	0.0064	0.0844	-0.0090	0.0288	0.0407	0.0032	0.0401	0.0547	0.0427
p-Value	0.0000	0.0000	0.0000	0.0000	0.0000	0.0000	0.0000	0.0000	0.0000	0.0000
Lag 5	0.0220	0.0079	-0.0221	-0.0302	0.0435	0.0203	-0.0105	0.0613	0.0091	0.0164
p-Value	0.0000	0.0000	0.0000	0.0000	0.0000	0.0000	0.0000	0.0000	0.0000	0.0000

(continued)

TABLE 3.2 (continued) Emerging Markets Descriptive Statistics

	China	India	Indonesia	Korea	Malaysia	Pakistan	Philippines	Sri Lanka	Taiwan	Thailand
Lag 10	-0.0071	0.0730	0.0432	0.0171	0.0226	0.0514	-0.0050	-0.0008	0.0352	0.0384
p-Value	0.0000	0.0000	0.0000	0.0000	0.0000	0.0000	0.0000	0.0000	0.0000	0.0000
Runs test (p-value)	0.0000	0.0000	0.0000	0.0000	0.0000	0.0000	0.0000	0.0000	0.0020	0.0000
Anderson -Darling (p-value)	0.0000	0.0000	0.0000	0.0000	0.0000	0.0000	0.0000	0.0000	0.0000	0.0000
Ryan–Joiner (p-value)	<0.0100	<0.0100	<0.0100	<0.0100	<0.0100	<0.0100	<0.0100	<0.0100	<0.0100	<0.0100

Malaysian, Pakistani, and Thai markets. An intuitive explanation of the observed negative skewness concerns the asymmetric price response to good versus bad news and indicates that participants in these markets may tend to overvalue bad news and undervalue good news. Larger relative degrees of positive skewness in the remaining markets, particularly those of China and Indonesia imply the opposite, i.e., participants are biased to good news. Significant levels of excess kurtosis in all the emerging markets (Indonesia in particular) are also noted. Kurtosis generally implies that a higher level of the observed variance in these markets is due to infrequent extreme deviations from the mean, as opposed to frequent smaller deviations. Indicative of the findings for all emerging markets, the distribution of Jarkarta SE interday returns in Figure 3.1, however, demonstrates the relative impact of a significantly greater than expected number of small deviations relative to a small number of larger than expected deviations (as much as ±8 s.d. above and below the mean) on the overall variance of returns provided in Table 3.2.

Formal testing for the Normality of interday returns is also conducted using the Anderson–Darling and Ryan–Joiner tests. A graphical example of the Anderson–Darling test output for the Malaysian KCLI Composite Index is provided in Figure 3.2. For series conforming to a Normal distribution, the Anderson–Darling plot of observed versus Normal probabilities will fall along the diagonal line at all levels. Deviations above or below

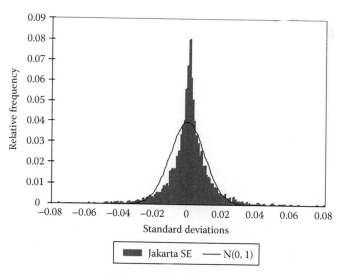

FIGURE 3.1   Distribution of Jakarta SE interday returns.

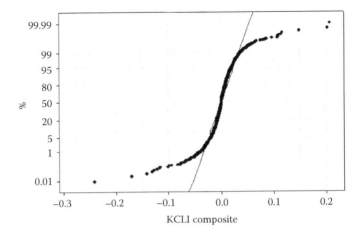

FIGURE 3.2 KCLI Composite Index normal probability plot.

the line indicate that the series under observation does not conform to a Normal distribution. In the case of the KCLI Composite Index, the fact that the observed deviations are above the line at low probabilities and below the line at high probabilities is evidence of a heavy-tailed distribution for Malaysian interday stock market returns.

As for KCLI interday returns, the Anderson–Darling and Ryan–Joiner p-values reject the null hypothesis of normality for the return distributions in all remaining markets. This result is not surprising since leptokurtic, non-Normal distributions are common in financial time-series (Pagan, 1996). Excluding testing for time-variation in either the skewness or kurtosis of the distributions of Asian-Pacific stock market returns, the findings presented in Table 3.2 suggest while the null hypothesis that developed market interday returns are independent cannot be conclusively rejected, there is overwhelming evidence that the distribution of returns in all the emerging markets is highly non-Gaussian.

Examining the tails of the distribution of KCLI interday returns in more detail, Table 3.3 shows the observed and predicted number of Normalized percentage returns $(P_t - P_{t-1})/P_{t-1}$, at each level.

The predicted number of returns at each level is calculated first using the observed mean (μ) and standard deviation (σ) of interday returns, in column 3 of Table 3.3. Column 4 of Table 3.3 then predicts the expected number of returns at each level assuming a standard Normal distribution with zero mean and unit standard deviation. The last column of Table 3.3 gives the ratio of the observed number of returns at each level to the

TABLE 3.3   KCLI Composite Index Observed versus Predicted Returns by Magnitude

± (%)	Observed	Gaussian N(μ,σ)	Gaussian N(0,1)	Observed/ Gaussian N(μ,σ)
1	5305	5587.73	5104.47	0.95
2	1461	1233.29	2031.37	1.18
3	415	502.73	319.98	0.83
4	166	129.74	19.71	1.28
5	57	21.18	0.47	3
6	25	2.18	0.0042712628	11
7	18	0.14	0.0000147323	127
8	7	0.01	0.0000000191	1, 203
9	2	0.0001500526	0.0000000000	13, 329
10	4	0.0000024309	0.0000000000	1, 645, 462
11	1	0.0000000247	0.0000000000	40, 455, 943
12	6	0.0000000002	0.0000000000	38, 133, 351, 252
13	4	0.0000000000	0.0000000000	IND
>13	5	0.0000000000	0.0000000000	IND

Note: IND: Gaussian prediction = 0, therefore cannot be determined.

corresponding predicted number of returns in column 3. Where the predicted number of returns is equal to zero at 30 decimal places, the ratio of observed returns to predicted returns in Table 3.3 is indeterminate, and is denoted *IND*.

Ratio values <1 in Table 3.3 indicate that the observed number of returns is less than the predicted number. Ratios >1 conversely indicate that the actual number of returns exceeds the predicted number.

It can be seen from Table 3.3 that both the Normal distribution N(μ,σ) and the standard Normal distribution N(0,1) are reasonable approximations for the distribution of percentage price changes in the KCLI Composite Index up to ±3% as the ratio of the observed to Gaussian is near 1. The poor fit of the Normal distribution is apparent, however, by the fact that over 3.9% of all percentage price changes (295 out of 7476 price changes) are greater than ±3%. The Normal distribution for the KCLI Composite Index predicts that approximately 2.0% of all price changes are higher than ±3% and the standard Normal distribution 0.3%.

The ratio of observed to predicted percentage returns for all Asian-Pacific emerging markets is provided in Table 3.4, using the same method as described for the last column of Table 3.3. Consistent with findings already described for the KCLI Composite Index, ratios in Table 3.4 demonstrate the overall poor fit of the Normal distribution to interday price

TABLE 3.4 Comparative Fit of the Gaussian Normal Distribution, N(μ, σ)

±	1%	2%	3%	4%	5%	6%	7%	8%	9%	10%	>10%
China	0.92	1.73	1.03	0.58	0.57	0.63	0.76	1.14	2.17	5.43	271
India	0.84	1.47	1.12	1.17	2.34	6.64	28.34	248	1,245	11,066	3,879,629,343
Indonesia	1.03	0.95	0.67	0.71	1.96	5.53	26.95	373	1,901	82,922	16,500,131,652
Korea	0.86	1.38	1.09	1.32	2.74	7.49	33.01	229	2,593	—	3,037,468,922
Malaysia	0.95	1.18	0.83	1.28	2.69	11.45	126.73	1202	13328	1645461	38173807195
Pakistan	0.89	1.24	1.05	1.62	3.74	6.67	45.72	823	14,354	171,308	157,078,547,019
Philippines	0.86	1.44	1.02	1.03	1.86	4.07	16.55	88	756	7,295	197,826,162
Sri Lanka	1.02	0.78	0.97	3.27	30.98	739.88	83,603.72	16,747,353	1,584,310,702	IND	IND
Taiwan	0.84	1.43	1.12	1.15	1.84	4.60	21.72	20	234	2,600	652,922
Thailand	0.91	1.21	1.03	1.50	3.26	14.63	109.70	1,835	59,933	705,457	5,572,243,378

changes of Asian-Pacific emerging markets. That all the markets show significant deviations from Normality due to heavy left and/or right tails suggests that the distribution of interday returns may in fact be better described by a power-law distribution.

The application of power-law models to financial time-series is motivated by their ability to capture more successfully the "outlier" behavior of financial returns described in the aforementioned figures and tables. Numerous recent studies including Lux (1996), Gopikrishnan et al. (1999), Weron and Weron (2000), Podobnik et al. (2002), Storer and Gunner (2002), Kaizoji and Kaizoji (2004), and Matosa et al. (2004) find strong evidence of power-law distributions in a variety of different developed market indices and individual equities. The value of the tail exponent ($|\alpha| = 2$ for Normal distribution) is found to be $\alpha \simeq 3$ for both tails of developed market index and equity returns universally (Matia et al., 2004; Yan et al., 2005).

Evidence of power-law distributions in emerging market index and equity returns is less known. Applying different power-law models to equities listed on the Korea Stock Exchange and Korea Securities Dealers Automated Quotations, Kim et al. (2004) find that the rank distribution, cumulative probability, and probability density of returns in these markets is consistent with a power-law distribution. Coronel-Brizio et al. (2003) derive similar results for daily values of the Mexican IPC stock index and Matia et al. (2004) for equities listed on the National Stock Exchange of India. Investigating individual equities on the Shanghai Stock Exchange and Shenzhen Stock Exchange, Yan et al. (2005) find similar evidence of power-law behavior in daily returns. Consistent with Matia et al. (2004), Yan et al. (2005) also find significant asymmetries in the value of the tail exponent when the left and right tails of the distribution are tested separately.

The comparative fit of the power-law model to the Gaussian Normal distribution is shown in Figure 3.3 for the Taiwan SE Weighted Index, where the Y-axis values are the log of the observed number of returns at each level and the X-axis values are the log of the percentage price changes. The poor relative fit of the Normal distribution in the figure is most evident for returns higher than ±3% (approximately −1.5 on the log scale). The exponent value $|\alpha| = 4.8511$ for the tails of interday Taiwan SE returns is outside the range required for a stable Levy process ($1 \leq \alpha \leq 2$), yet is consistent with the general class of hyperbolic nonstable distribution (Zhaoxia and Gençay, 2003).

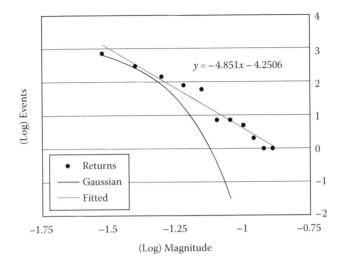

FIGURE 3.3   Taiwan SE, Gaussian versus power-law fit.

Power-law tail exponents $|\alpha|$ for all emerging markets under consideration are provided in Table 3.5. Panel (a) of Table 3.5 shows the exponent value for the left and right tails together, panel (b) the left tail (negative) only, and panel (c) the right tail (positive) only. The mean tail exponent for all markets is also provided. Consistent with earlier discussed findings, tail exponent values for all emerging markets indicate the distribution of returns to be highly non-Gaussian. Indicative of the microstructure of these exchanges, significant differences are, however, evident across the markets. The range of α values for all markets when both the left and right tails are taken together is $|\alpha| = 3.0961$ (China) to $|\alpha| = 4.8511$ (Thailand). Comparing the left versus right tails across markets reveals even greater divergences in the exponent values, with a values ranging from $|\alpha| = 2.7208$ (China) to $|\alpha| = 4.4837$ (Taiwan) for the left tail only and $|\alpha| = 2.9189$ (China) to $|\alpha| = 4.5786$ (India) for the right tail only. The greater relative difference across markets when positive and negative returns are tested individually is consistent with the fit of the power-law specification being optimized by modeling the left and right tails of the distribution separately.

Relative to previous findings by Matia et al. (2004) and Yan et al. (2005) for equities listed on the Indian and Chinese exchanges, respectively, we find significant evidence of asymmetry in the distribution of returns

TABLE 3.5 Power-law Tail Exponents, $|\alpha|$

	China	India	Indonesia	Korea	Malaysia	Pakistan	Philippines	Sri Lanka	Taiwan	Thailand
				(a) Left and right tails						
Tail exponent, $|\alpha|$	3.0961	4.4226	3.7030	4.6987	3.6898	4.4559	4.3698	3.6544	4.8511	4.3176
p-Value (difference to mean)	0.0000	0.0192	0.0154	0.0126	0.1484	0.0000	0.0000	0.0086	0.0000	0.0264
Mean tail exponent, $|\alpha|$ = 4.1259										
				(b) Left tail						
Tail exponent, $|\alpha|$	2.7208	4.0436	3.9400	4.1016	4.4044	4.0564	3.8352	3.7523	4.4837	3.6000
p-Value (difference to mean)	0.0000	0.5184	0.2764	0.1909	0.3033	0.6710	0.0000	0.0154	0.3037	0.0000
p-Value (difference to left/right)	0.0000	0.0000	0.0000	0.0047	0.0000	0.0000	0.0000	0.0001	0.3037	0.0000
Mean tail exponent, $|\alpha|$ = 3.8938										
				(c) Right tail						
Tail exponent, $|\alpha|$	2.9189	4.5786	3.9871	4.4086	4.5426	4.1949	4.0269	3.8851	4.2848	3.8026
p-Value (difference to mean)	0.0000	0.0518	0.6092	0.2842	0.0000	0.7169	0.0000	0.1330	0.6155	0.0386
p-Value (difference to left/right)	0.3090	0.1411	0.0034	0.0000	0.0001	0.0244	0.0000	0.0000	0.0001	0.0000
p-Value (difference to left)	0.0804	0.0000	0.6402	0.2108	0.4707	0.0000	0.0000	0.0000	0.3434	0.0325
Mean tail exponent, $|\alpha|$ = 4.0630										

using the power-law fit for India only. The power-law tail exponent for the Shanghai A 30 in panel (a) by contrast is not significantly different from the mean for all markets, nor do exponent values for the left versus right tails significantly differ. Indicative of a relatively greater number of large gains versus large loses, significant differences in the left versus right tails are also observed for the emerging markets of Pakistan, the Philippines, and Sri Lanka. These results are consistent with our prior conjecture that the distribution of returns for the comparatively larger emerging markets should be more closely Normal than those of the smaller markets when ranked by capitalization and number of listed companies.

3.4 CONCLUSION

Knowledge of the underlying distribution of financial asset returns is vital to measure the underlying risk of loss associated with investment in a particular market. Under the common assumption of Normality, Value-at-Risk estimates for instance are relatively simple given that the statistical properties of the Normal distribution are well known. On this basis, general forms of asset pricing models such as the CAPM and the Black–Scholes Option Pricing Model also explicitly require Normality of the underlying distribution of asset returns. "Risk management tools such as volatility, covariance and Value-at-Risk, all so critical to how we deal with risk, become meaningless if the return pattern does not match the famous bell curve" (Mauldin, 2007).

This chapter tests for Normality in interday returns for 10 emerging stock markets in the Asia-Pacific region: China, India, Indonesia, Korea, Malaysia, Pakistan, the Philippines, Sri Lanka, Taiwan, and Thailand. The markets exhibit significant differences in size—as measured by capitalization, numbers of listed companies, and value of share trading—and ultimately differences in their distributional characteristics, in particular, the higher order moments.

Summarizing some of the commonly observed behaviors in statistical finance, the following "stylized facts" about financial time-series data are noted by Bouchaud (2002): return distributions are highly non-Gaussian (Normal) and are characterized by power-law, or heavy tails. Furthermore, the tails of emerging market return distributions are relatively heavier than those corresponding to developed markets. Not surprisingly, all these "facts" appear consistent with the observed behavior of Asian-Pacific emerging stock market returns considered in this chapter.

While the Gaussian Normal distribution has been shown to provide a reasonable approximation of the observed distribution of emerging market interday returns in the center region (around ±3% above and below the mean), we find consistent evidence of fat tails using a variety of different techniques. In response to this finding, we fit a general class of power-law distributed model to capture the tail behavior of emerging market returns. Consistent with prior evidence for individual equities listed on emerging market exchanges, we find that the emerging market returns exhibit comparatively larger tails than found in developed markets. Our findings serve to question the applicability of standardized models of financial asset pricing and risk analysis in emerging markets.

REFERENCES

Bachelier, L. (1900) Theorie de la Speculation. *Annales de l'Ecole Normale Superieure*, 3(17): 21–86.

Black, F. and Scholes, M. (1973) The pricing of options and corporate liabilities. *Journal of Political Economy*, 81(3): 637–659.

Bouchaud, J. (2002) An Introduction to statistical finance. *Physica A*, 313(1): 238–251.

Coronel-Brizio, H.F., de la Cruz-Laso, C.R., and Hernandez-Montoya, A.R. (2003) Fitting the power-law distribution to the Mexican stock market index data. Available at: http://arxiv.org/abs/cond-mat/0303568

Fama, E.F. (1965) The behavior of stock market prices. *Journal of Business*, 38(1): 34–105.

Gopikrishnan, P., Plerou, V., Amaral, L.A.N., Meyer, M., and Stanley, H.E. (1999) Scaling of the distribution of fluctuations of financial market indices. *Physical Review E*, 60(5): 5305–5316.

Jondeau, E. and Rockinger, M. (2003) Testing for differences in the tails of stock-market returns. *Journal of Empirical Finance*, 10(5): 559–581.

Kaizoji, T. and Kaizoji, M. (2004) Power law for ensembles of stock prices. *Physica A*, 344(1–2): 240–243.

Kearns, P. and Pagan, A. (1997) Estimating the density tail index for financial time series. *Review of Economics and Statistics*, 79(2): 171–175.

Kim, K., Yoon, S.M., and Chang, K.H. (2004) Power law distributions for stock prices in financial markets. Available at: http://arxiv.org/abs/cond-mat/0412014v1

Longin, F. (1996) The asymptotic distribution of extreme stock market returns. *Journal of Business*, 69(3): 383–408.

Lux, T. (1996) The stable Paretian hypothesis and the frequency of large returns: An examination of major German stocks. *Applied Financial Economics*, 6(6): 463–475.

Mandelbrot, B.B. (1963) The variation of certain speculative prices. *Journal of Business*, 36(4): 392–417.

Markowitz, H.M. (1952) Portfolio selection. *Journal of Finance*, 7(1): 77–91.

Matia, K., Pal, M., Saluunkay, H., and Stanley, H.E. (2004) Scale-dependent price fluctuations for the Indian stock market. *Europhysics Letters*, 66(6): 909–914.

Matosa, J.A.O., Gama, S.M.A., Ruskind, H.J., and Duarte, A.M.E. (2004) An econophysics approach to the Portuguese stock index PSI-20. *Physica A*, 342(3–4): 665–676.

Mauldin, J. (2007) Financial crisis and why risk valuation tools in practical portfolio selection are meaningless. Available at: http://www.marketoracle.co.uk/article2466.html

Pagan, A. (1996) The econometrics of financial markets. *Journal of Empirical Finance*, 3(1): 15–102.

Peiró, A. (1999) Skewness in financial returns. *Journal of Banking and Finance*, 23(6): 847–862.

Podobnik, B., Grosse, I., and Stanley, H.E. (2002) Stochastic processes with power law stability and a crossover in power law correlations. *Physica A*, 316(1): 153–159.

Sharpe, W.F. (1964) Capital asset prices: A theory of market equilibrium under conditions of risk. *Journal of Finance*, 19(3): 425–442.

Standard and Poor's (2007) The S&P emerging market indices. Methodology, definitions, and practices. Available at: http://www2.standardandpoors.com/spf/pdf/index/

Storer, R. and Gunner, S. (2002) Statistical properties of the Australian 'All Ordinaries' index. *International Journal of Modern Physics*, 13(7): 893–897.

Tucker, A.L. and Scott, E. (1987) A study of diffusion processes for foreign exchange rates. *Journal of International Money and Finance*, 6(4): 465–478.

Weron, A. and Weron, R. (2000) Fractal market hypothesis and two power-laws. *Chaos, Solitons and Fractals*, 11(1): 289–296.

World Federation of Exchanges (2008) Focus. no 186 August. Available at: http://www.world-exchanges.org

Yan, C., Zhang, J.W., Zhang, Y., and Tang, Y.N. (2005) Power-law properties of Chinese stock market. *Physica A*, 353(1): 425–432.

Zhaoxia, X. and Gençay, R. (2003) Scaling, self-similarity and multifractality in FX markets. *Physica A*, 323(1): 578–590.

CHAPTER 4

Emerging Markets Exposure: Equities or Hedge Funds?

Vassilis N. Karavas, Nikos S. Thomaidis, and George D. Dounias

CONTENTS

4.1 INTRODUCTION

The two most recent decades have been characterized by the revolution in various types of investments among which are investments in emerging markets via hedge fund and the more traditional asset classes. As the number of funds dedicated to emerging markets grew so were the opportunities for hedge funds in these regions. Hedge funds have grown rather rapidly over the past years and have exhibited increased assets under management (AUM) and increased number of funds. More specifically the Center for International Securities and Derivatives Markets (CISDM)* reports that the AUM in the hedge fund industry grew almost 55-fold from $39 billion in 1990 to $2165 billion at the end of 2007. The growth in the number of funds experienced an increase, albeit less dramatic, by almost 16-fold from 590 funds in 1990 to slightly over 9700 by the end of 2007. A caveat for the industry was the fact that outside the inner hedge fund circle, it really became synonym to speculation, excessive leverage, high risk, and sometimes hubris and/or fraud. Although we do not share this view, it is true that a few high-profile hedge funds were characterized by one or more of these attributes starting with Long-Term Capital Management, Bayou, Beacon Hill, Lipper & Co, Bear Stearns High-Grade Structured Credit Strategies. The current difficulties in the credit markets and the global economies as a whole let us believe that the impact may be significant in those investors that the number of hedge funds that will be liquidated is going to exhibit a significant increase relative to the historical norms.

More sensitive in such markets tend to be funds that have primary exposure to the various emerging markets, which historically tend to be more volatile and sometimes affected by the overall economic circumstances in the developed economies. Our interest in this study is in traditional and hedge fund investments dedicated to emerging markets around the globe, the number of which has increased drastically over the past few years. According to HFR† AUM in the hedge fund industry dedicated to emerging markets grew from a little over $100 million in 1990 to over $100 billion in the beginning of 2008. Further development of the BRIC‡ economies as well as of other emerging markets such as Mexico and Korea attracted a number of investment managers to implement various hedge fund strategies for their investors. The number of hedge funds grew

---

* CISDM is a financial research center at the University of Massachusetts, Amherst.
† Hedge fund research.
‡ Brazil, Russia, India, and China.

significantly. Performance has been outstanding for the most part of the previous years. However, all investors including hedge funds are impacted by the regional and global economic conditions and sometimes are subject to criticism as whether they are the cause or the victims of such conditions (Fung et al., 2001).

Following the collapse of the Russian market in 1998, it took several years before investors actively returned to the emerging markets and increased infrastructure and the overall investments to these markets. The year 2007 was extremely profitable for many emerging economies, more notably China and Russia. However, despite the continued interest in these markets, investors are getting more wary about the short-term growth pace and performance, while the expectations for the long-term growth have not changed for the moment. It still remains to observe how these markets will be affected by the overall credit crunch and whether the decoupling will take place.

In this chapter, we explore the benefits, of investing in emerging markets via various asset classes with an emphasis in hedge funds in the context of institutional portfolios as well as endowment simulated allocations. In Section 4.2, we describe the data used in the analysis and the various portfolio constituents as well as assumptions used in our analysis. Thereafter, we present the performance and risk characteristics of each asset class separately as well as for each of the benchmark portfolios created. In Section 4.4, we use a Markowitz-based optimization to allocate across the various asset classes in the endowment portfolio and explore the benefits of investing in hedge funds with exposure to emerging markets. Subsequently, we explore the behavior of the various asset classes during periods in which a simulated U.S. Institutional Portfolio had its worst, average, and best performance.

4.2 DATA AND METHODOLOGY

In this chapter, we utilize traditional equity and fixed income asset classes alongside the liquid and less liquid alternative investments as proxies for the institutional- and the endowment-type portfolios described later in this section. For most asset classes used in this chapter there exist multiple providers therefore we deemed necessary in Appendix 4.A.1 to present the details, the source, and the type of index used for each case. For the hedge fund investment benchmarks there have been numerous research papers discussing potential biases in the performance proxies used (Fung and Hsieh, 2001; Karavas and Siokos, 2003). Similar approaches in the use of hedge fund benchmarks as well as for studying various institutional portfolios have been used in previous research by Karavas (2000) and Schneeweis and Martin (2000). Returns for all data series are expressed as monthly holding period returns starting in January 1999 through September 2008. The time frame was chosen as such to explore the increased interest in emerging market investing following the Russian default. This is not to undermine the danger of such potential significant drawdowns, but to investigate the period that characterized the enormous growth in these markets following the Russian default.

For the purposes of studying the effect of including emerging markets exposure in portfolio, we create three benchmark portfolios: (1) U.S. Institutional Portfolio, which is comprised of U.S. Stocks, Bonds, and alternative investments; (2) Global Institutional Portfolio, which is comprised of global stocks, bonds, and alternative investments; and (3) Endowment Portfolio, which simulates the allocations of prominent endowment portfolios, which are diversified across multiple asset classes, and are characterized by unique risk, return, or liquidity profiles. For each of the three benchmark portfolios we create a similar portfolio in which certain allocations have been replaced with the corresponding asset class exposure in emerging markets.

The U.S. and Global Institutional portfolios are represented by allocations to equities, bonds, illiquid (private equity, venture capital, and leveraged buyouts [LBOs]), and liquid alternative investments (commodities and hedge funds). The Endowment Portfolio is a more diversified portfolio across the various geographic regions and asset classes. More specifically the portfolios analyzed are as follows:

Portfolio I—U.S. Institutional: 40% S&P 500, 40% Lehman U.S. Agg., 5% illiquid alternative investments, 5% commodities, and 10% hedge funds.

Portfolio II—U.S. Institutional w. Emerging Markets: 35% S&P 500, 40% Lehman U.S. Agg., 5% illiquid alternative investments, 5% commodities, and 5% hedge funds, 5% Morgan Stanley Capital International (MSCI) Emerging Markets, and 5% hedge funds—emerging markets.

Portfolio III—Global Institutional: 40% MSCI World, 40% Lehman Global Agg., 5% illiquid alternative investments, 5% commodities, and 10% hedge funds.

Portfolio IV—Global Institutional w. Emerging Markets: 35% MSCI World, 40% Lehman Global Agg., 5% illiquid alternative investments, 5% commodities, and 5% hedge funds, 5% MSCI emerging markets, and 5% hedge funds—emerging markets.

Portfolio V—Endowment: 10% S&P 500, 8% Russell 2000, 8% DJ Euro Stoxx, 8% NIKKEI, 5% Lehman U.S. Agg., 5% Lehman Euro Agg., 3% Lehman U.S. High Yield, 3% Lehman Global High Yield, 10% FTSE NAREIT, 7% LPX Listed Private Equity Venture, 3% LPX Buyout, 10% commodities, and 20% hedge funds.

Portfolio VI—Endowment w. Emerging Markets: 8% S&P 500, 6% Russell 2000, 6% DJ Euro Stoxx, 6% NIKKEI, 5% Lehman U.S. Agg., 5% Lehman Euro Agg., 2% Lehman U.S. High Yield, 2% Lehman Global High Yield, 2% Lehman Emerging Markets High Yield, 10% FTSE NAREIT, 7% LPX Private Equity Venture, 3% LPX Buyout, 10% commodities, and 15% hedge funds, 5% hedge funds—emerging markets.

4.3 DESCRIPTIVE STATISTICS

In this section, we analyze the performance characteristics of the various asset classes by region on a standalone basis to better understand the inherent characteristics of these investments and the respective contribution when these are part of a portfolio. Following this analysis, we explore the statistical attributes of the benchmark portfolios for the same time period.

As Table 4.1 shows, the previous 10 years were rather flat for the equity markets in the developed countries. To the contrary, the performance of the emerging equity markets was rather strong, but so was the underlying volatility. On the fixed-income side, performance was moderate to flat for the developed countries, while on the high-yield side the emerging markets exhibited strong performance with the corresponding sector globally remaining on the moderate side. On the alternative investments side, while the Real Estate sector was strong, the remaining more traditional illiquid sectors—private equity and buyout—were below their long-term average historical performance. Commodity investments were profitable as were most hedge funds, especially those with focus on emerging markets.

Volatility for the various asset classes focused on emerging markets was at high levels ranging from 20% to almost 60% for specific regional equity

TABLE 4.1   Descriptive Statistics of Index Performance: January 1999–September 2008

	Ann. Rate of Return %	Ann. Std. Dev. %	Monthly MIN %	Monthly MAX %	Sharpe Ratio	Maximum Drawdown %
Traditional Assets						
Equity indices						
S&P 500	1.1	14.1	−10.9	9.8	−0.18	−44.7
Russell 2000	6.4	19.0	−15.1	16.5	0.14	−35.1
Dow Jones STOXX 50	3.1	19.4	−18.0	17.4	−0.03	−54.9
NIKKEI 225	−1.3	19.7	−16.1	11.8	−0.25	−66.9
MSCI World Index	1.9	14.1	−11.9	8.9	−0.13	−46.8
MSCI Emerging Markets	12.9	22.4	−17.5	13.8	0.41	−48.1
MSCI Brazil	17.5	39.8	−30.3	36.5	0.35	−68.9
MSCI India	15.7	30.3	−19.7	18.7	0.40	−60.5
MSCI China	11.4	20.5	−17.7	20.1	0.38	−48.5
MSCI Russia	29.7	44.5	−29.7	61.1	0.59	−49.4
MSCI Turkey	13.0	58.7	−41.2	72.3	0.16	−83.5
Bond indices						
Lehman U.S. Agg.	5.3	3.5	−3.4	2.6	0.46	−3.6
Lehman Euro Agg.	0.8	2.8	−3.0	2.5	−1.03	−7.4
Lehman Global Agg.	4.8	5.5	−3.7	4.8	0.20	−7.4
Lehman U.S. Corp High Yield	4.3	7.9	−8.0	7.5	0.08	−12.5
Lehman Pan-Eur High Yield	5.0	13.7	−12.0	13.2	0.10	−35.9
Lehman EM High Yield	11.6	10.2	−7.5	8.4	0.78	−14.5
Lehman Global High Yield	7.0	7.9	−8.3	6.7	0.42	−12.2
LIBOR (yield)						
Bloomberg Generic USD Libor 1 M	3.7	1.8	1.1	6.8		
Alternative Investments						
Traditional alternative investments						
FTSE NAREIT US	12.1	15.0	−15.3	9.7	0.56	−28.1
LPX Listed Private Equity Venture	−1.6	30.1	−19.5	32.9	−0.18	−84.2
LPX Buyout	10.8	15.1	−16.4	7.5	0.47	−43.5
Commodities						
DJ AIG Commodity	11.8	15.8	−11.9	12.3	0.51	−27.7

TABLE 4.1 (continued)   Descriptive Statistics of Index Performance:
January 1999–September 2008

	Ann. Rate of Return %	Ann. Std. Dev. %	Monthly MIN %	Monthly MAX %	Sharpe Ratio	Maximum Drawdown %
Hedge funds						
HFR Fund Weighted	8.8	6.8	−4.7	7.7	0.75	−10.9
HFR Emerging Markets	14.8	12.7	−9.0	14.8	0.88	−22.6
HFR EM Russia Eastern Europe	28.4	20.6	−15.5	30.1	1.20	−31.0
HFR EM Asia ex-Japan	11.5	13.5	−8.4	12.4	0.58	−31.0
HFR EM Latin America	11.2	16.2	−10.5	19.2	0.46	−32.0

markets and from 10% to 20% for bond and hedge fund investments in these regions. In Table 4.1, we also present the maximum drawdown* for the past almost 10 years and it is evident that maximum loss of wealth can take place when investing in emerging markets as well as in certain types of the traditional illiquid alternative investments. On the "safer" side were the bond markets and hedge funds.

When comparing the asset classes in the emerging markets space, we observe that although the equity asset class exhibits strong performance in most cases the corresponding hedge fund indices offer similar returns. However, in all cases the volatility of the hedge fund indices is significantly lower, which results in better Sharpe ratios for these investments. This can be explained by the more diverse nature of the hedge fund investments and the fact that the corresponding indices are combination of multiple funds.

In Table 4.2, we observe the same statistics for the various benchmark portfolios with and without Emerging Market exposure. As expected, the exposure to emerging markets in various asset classes has three major consequences for the underlying portfolios: (1) it improves the performance in terms of return, (2) it increases the risk in terms of volatility, and (3) it poses a limit on the *downside* risk, as measured by the maximum drawdown. Overall, the risk-adjusted return is higher and exposed portfolios tend to be more efficient. Looking at Table 4.2, the Endowment Portfolio is the one with the highest volatility, best Sharpe ratio, but the largest maximum

---

* The maximum drawdown measures the maximum loss possible in a given period as this is represented by the peak to trough performance.

TABLE 4.2 Descriptive Statistics of Portfolio Performance:
January 1999– September 2008

	Avg. Annual RTN %	Std. Dev. %	Monthly MIN %	Monthly MAX %	Sharpe Ratio	Maximum Drawdown %
Portfolio I: U.S. Institutional	4.6	7.0	−5.7	4.4	0.14	−16.2
Portfolio II: U.S. Institutional w. EM	5.6	7.5	−6.4	4.8	0.25	−15.4
Portfolio III: Global Institutional	4.7	7.7	−7.3	5.0	0.14	−19.1
Portfolio IV: Global Institutional w. EM	5.6	8.3	−7.8	5.6	0.24	−19.1
Portfolio V: Endowment	6.4	9.8	−7.4	7.3	0.28	−22.4
Portfolio VI: Endowment w. EM	7.7	10.5	−8.1	8.2	0.38	−22.1

drawdown. We should emphasize here that these portfolios have not been optimized but rather have been put together to represent the typical allocations of large institutions and/or endowment funds.

In Figure 4.1a, we observe the risk–return profile of each investment on a standalone basis and how this profile is depicted relative to the other investment under consideration. Investments to the right of the graph are more volatile than those to the left side, and investments on the upper part are better performing than those on the lower part. Emerging markets dominate the volatile and better performing sections, while fixed income is on the lower side of volatility and performance. Hedge fund investments maintain a balance with low volatility and moderate to high performance.

Similarly, in Figure 4.1b, we see the risk–return profile of the various benchmark portfolios discussed earlier. The endowment portfolios have the highest return but also the highest volatility. Generally, portfolios with various degrees of exposure to emerging markets dominate the northeast corner in the risk–return graph, indicating higher returns and higher standard deviation than the corresponding portfolios without exposure to emerging markets. The correlation of the various assets classes on a standalone basis are shown in Table 4.3.

From the tables discussed in this section, we conclude that investing in emerging markets can be beneficial from a risk–reward point of view

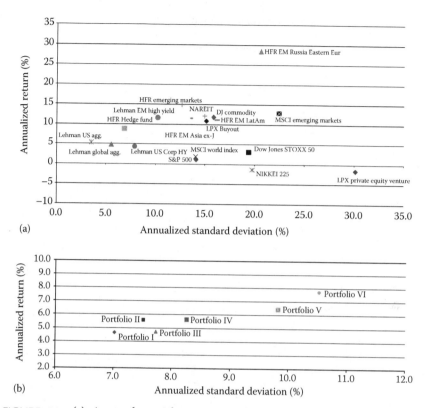

FIGURE 4.1 (a) Asset class risk–return trade-off: January 1999–September 2008 and (b) Portfolios risk–return trade-off: January 1999–September 2008.

without taking into account other idiosyncratic risks for the markets in consideration. It appears however to be rather beneficial in the context of diversified portfolios across various asset classes and regions. The benefits are shown to be on performance, Sharpe ratio, and maximum drawdown point of view. In Section 4.4, we explore various ways of looking at the portfolio construction process and means of identifying optimal portfolios when the volatility and performance are the primary considerations.

4.4 PORTFOLIO OPTIMIZATION

Markowitz-based efficient frontier determination is conducted using historical performance to evaluate the potential positioning of emerging markets investments in a portfolio context. Initially, the portfolios will be optimized with no allocation constraints. The Markowitz approach for

TABLE 4.3 Index Correlation Matrix: January 1999–September 2008

	S&P 500	Russell 2000	Dow Jones STOXX 50	NIKKEI 225	MSCI World Index	MSCI Emerging Markets	Lehman U.S. Agg.	Lehman Euro Agg.	Lehman Global Agg.
S&P 500	1.00								
Russell 2000	0.72	1.00							
Dow Jones STOXX 50	0.80	0.72	1.00						
NIKKEI 225	0.53	0.47	0.41	1.00					
MSCI World Index	0.96	0.75	0.89	0.62	1.00				
MSCI Emerging Markets	0.73	0.70	0.72	0.61	0.83	1.00			
Lehman U.S. Agg.	−0.17	−0.15	−0.15	0.01	−0.15	−0.14	1.00		
Lehman Euro Agg.	−0.04	−0.06	0.00	−0.06	−0.03	−0.14	0.40	1.00	
Lehman Global Agg.	−0.09	−0.09	0.09	0.13	0.03	0.02	0.70	0.47	1.00
Lehman U.S. Corp High Yield	0.55	0.58	0.48	0.35	0.58	0.59	0.13	−0.02	0.12
Lehman EM High Yield	0.52	0.50	0.49	0.38	0.56	0.59	0.26	0.12	0.19
Lehman Global High Yield	0.60	0.61	0.57	0.41	0.65	0.67	0.21	0.05	0.23
FTSE NAREIT US	0.36	0.47	0.30	0.13	0.35	0.30	0.05	−0.01	0.12
LPX Listed PE Venture	0.64	0.69	0.68	0.61	0.72	0.77	−0.13	−0.17	0.00
LPX Buyout	0.61	0.65	0.68	0.44	0.70	0.68	0.00	0.00	0.18
DJ AIG Commodity	0.10	0.11	0.19	0.35	0.23	0.35	0.05	0.00	0.25
HFR Fund Weighted	0.69	0.82	0.70	0.60	0.78	0.84	−0.06	−0.11	0.02
HFR Emerging Markets	0.66	0.67	0.64	0.61	0.76	0.91	−0.06	−0.14	0.03

	Lehman U.S. Corp High Yield	Lehman EM High Yield	Lehman Global High Yield	FTSE NAREIT US	LPX Listed Private Equity Venture	LPX Buyout	DJ AIG Commodity	HFR Fund Weighted	HFR Emerging Markets
S&P 500									
Russell 2000									
Dow Jones STOXX 50									
NIKKEI 225									
MSCI World Index									
MSCI Emerging Markets									
Lehman U.S. Agg.									
Lehman Euro Agg.									
Lehman Global Agg.									
Lehman U.S. Corp High Yield	1.00								
Lehman EM High Yield	0.59	1.00							
Lehman Global High Yield	0.93	0.82	1.00						
FTSE NAREIT US	0.37	0.32	0.40	1.00					
LPX Listed PE Venture	0.50	0.47	0.56	0.20	1.00				
LPX Buyout	0.60	0.51	0.66	0.54	0.56	1.00			
DJ AIG Commodity	0.15	0.19	0.22	-0.08	0.22	0.24	1.00		
HFR Fund Weighted	0.56	0.57	0.65	0.25	0.82	0.65	0.36	1.00	
HFR Emerging Markets	0.54	0.63	0.66	0.22	0.78	0.59	0.33	0.90	1.00

the portfolio optimization attempts to find the optimal allocation to the various asset classes in a way that the resulting portfolio performance is maximized while the volatility is restricted to a maximum level.

However, as reported in previous studies, such portfolios are sensitive to return forecasts (Chopra and Ziemba, [1993] Black and Litterman, [1992]) and the accompanying portfolio constraints. Our analysis is exploring the hypothetical ex-post optimal asset allocations to investigate the asset classes that are preferred at an optimization level. We fully understand that from a practical point of view optimizing using a rear view mirror brings inherent limitations to the resulting portfolio. It, however, provides an insight of how the various asset classes are assembled based on more stable statistical properties such as covariance (if the period is long enough as we have chosen here). Any attempt to forecast the first moment for the underlying asset classes can be unreliable especially in the current market environment, where unquantifiable factors can be more determining for the performance of the assets, e.g., liquidity and credit squeeze. To conclude the word of warning about the limitations of this approach, we believe that historical attributes can be useful to portfolio construction as long as short-term expectations and market dislocations are taken into account.

In the Markowitz-style portfolios, we have examined a number of potential risk profiles for optimal portfolios. These risk profiles are used for both instances of portfolio optimization, without and with allocation constraints. These risk profiles are based on the acceptable annualized volatility that a portfolio manager may have. The various levels of risk in the analysis intend to capture a variety of risk preferences on the institutional side, while they provide useful information regarding the allocation changes as the risk aversion is relaxed. In Table 4.4, we present the results for the portfolio optimization without allocation constraints. The objective function is to maximize performance in the period under consideration without allowing for leverage i.e., weights cannot exceed 100% of the capital available. As per the results, in the most risk-averse scenario (5% volatility) the maximum allocation is on Lehman U.S. Aggregate, while HFR Emerging Market receives the majority of the remaining allocation. This is justified by the slightly negative correlation between the two asset classes, which contributes to the reduction in volatility and the contribution in performance because of the hedge fund Emerging Market exposure.

As the risk tolerance grows, we observe a gradual increase in the hedge fund emerging markets exposure. In the case of 15% volatility we have more than 90% of the portfolio in hedge funds (emerging markets).

TABLE 4.4   Optimal Weights Using Historical Risk, Return, and Correlation Data: January 1999–September 2008

Ann. Std. Dev. %	5.00	7.50	10.00	12.50	15.00
Ann. return %	9.27	11.61	13.82	15.70	15.75
Sharpe ratio	1.12	1.06	1.01	0.96	0.80
Maximum drawdown %	−7.66	−12.04	−16.00	−22.45	−24.43
S&P 500					
Russell 2000					
Dow Jones STOXX 50					
NIKKEI 225					
MSCI World Index					
MSCI Emerging Mkts. %					9.72
Lehman U.S. Agg. %	60.03	34.21	9.79		
Lehman Euro Agg.					
Lehman Global Agg.					
Lehman U.S. High Yield					
Lehman EM High Yield					
Lehman Global High Yield					
FTSE NAREIT US %	5.70	14.52	23.66	2.00	
LPX Private Equity & Venture					
LPX Buyout					0.00
DJ AIG Commodity %		5.44	11.84		
HFR Fund Weighted					
HFR Emerging Markets %	34.27	45.83	54.72	98.00	90.28

Performance for the underlying portfolios varies from 9.27% in the case of 5% volatility to 15.75% in the case of 15% volatility. As a consequence of the increase in the risk tolerance, the maximum drawdown is also gradually increased. One observation is that the Sharpe ratio for the portfolio drops as we allow risk to greater levels, which implies that we are not rewarded at the same rate in performance terms per unit of risk we undertake. In the case of portfolio optimization with no allocation constraints it is evident that we may have optimal portfolios with low risk and Sharpe ratio of over 1, but with very limited asset classes being represented in the portfolios. Such concentrated portfolios are not the institutional or endowment type portfolios. For this purpose, the second set of optimizations is based on the same premises but with allocation constraints to ensure a minimum diversification to asset classes that are most common to institutional

TABLE 4.5 Optimal Weights Using Historical Risk, Return, and Correlation Data—Minimum Allocation Constraint to S&P 500 and Lehman Aggregate of 20% and HFR Hedge Fund Index of 10%: January 1999–September 2008

Ann. Std. Dev. %	5.00	7.50	10.00	12.50	15.00
Ann. return %	6.33	8.51	9.86	10.14	10.15
Sharpe ratio	0.53	0.64	0.62	0.52	0.43
Maximum drawdown %	−7.22	−10.75	−15.27	−18.79	−24.71
S&P 500 %	20.00	20.00	20.00	20.00	20.00
Russell 2000					
Dow Jones STOXX 50					
NIKKEI 225					
MSCI World Index					
MSCI Emerging Mkts. %				10.08	33.03
Lehman U.S. Agg. %	53.49	25.49	20.00	20.00	20.00
Lehman Euro Agg.					
Lehman Global Agg.					
Lehman U.S. High Yield					
Lehman EM High Yield					
Lehman Global High Yield					
FTSE NAREIT US %	5.52	20.09	6.59		
LPX Private Equity & Venture					
LPX Buyout					
DJ AIG Commodity %	11.00	24.42	5.19		
HFR Fund Weighted %	10.00	10.00	10.00	10.00	10.00
HFR Emerging Markets %			38.22	39.92	16.97

portfolios (Table 4.5). We use as an example the minimum allocations for a U.S. Institutional Portfolio. These constraints require that we have a minimum allocation to U.S. Equities of 20%, U.S. Bonds (predominantly investment grade) of 20%, and hedge funds of 10%. Subsequently, we let the optimization to choose the optimal allocation for the remaining 50% of the portfolio.

Results indicate similar trends as before. The Lehman U.S. Aggregate exposure is increased to over 50% for the case of the lowest risk tolerance examined (5% volatility), with the remaining allocations to FTSE NAREIT, and Commodities. As the portfolio allows for greater volatility the exposure to emerging markets is increased via equity and hedge fund allocations. More specifically, in the case of 15% volatility except for the allocation dictated by the constraints, the remaining allocation is split by approximately 2/3 (namely, 33% of the overall portfolio) in equities in emerging markets (MSCI emerging markets) and 1/3 (namely, 17% of the portfolio) in HFR emerging markets. Similar to the observations we had in the previous set of optimizations, we see that the Sharpe ratio of the riskier* portfolios is lower than that of the less risky ones.

It is evident that in the context of portfolio selection, exposure to emerging markets offers unique benefits to a portfolio with or without allocation constraints. When an optimization model is used, the results show a preference toward hedge fund exposure to emerging markets rather than the outright equity exposure. This is explained by the better risk–return profile of the hedge fund investments—higher Sharpe ratio. While historically, and over a period of time the benefits are evident in this context, it is important to take into consideration the underlying risks of such exposures alongside the higher volatility. Proper diversification and portfolio construction is necessary to provide the adequate cushion for potential drawdowns so that the performance benefit can be realized by the overall portfolio allocation.

4.5 ASSET CLASSES IN PERIODS OF EXTREME RETURN MOVEMENTS

In this section, we focus on the individual asset classes under various extreme return movements of the U.S. Institutional Portfolio. We attempt to explore further the potential contribution of the emerging market asset classes during periods of (1) market stress, (2) average market behavior, and (3) strong markets. For the purposes of this analysis, we have used as a proxy for the market behavior the performance of the U.S. Institutional Portfolio. Previous research (Schneeweis et al., 1996) has shown that the risk and return characteristics of various alternative investment strategies relative to the risk and return pattern of a typical stock/bond portfolio are conditional on the return environment of the stock/bond portfolio. For the purposes of identifying these periods, we ranked the performance of

---

* Portfolio risk in this analysis is proxied by the volatility consistent with the Markowitz approach.

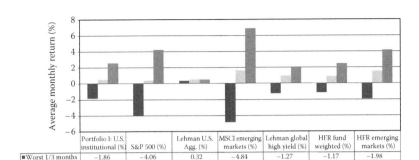

	Portfolio I: U.S. institutional (%)	S&P 500 (%)	Lehman U.S. Agg. (%)	MSCI emerging markets (%)	Lehman global high yield (%)	HFR fund weighted (%)	HFR emerging markets (%)
■Worst 1/3 months	−1.86	−4.06	0.32	−4.84	−1.27	−1.17	−1.98
Mid 1/3 months	0.51	0.36	0.51	1.61	0.97	0.87	1.52
■Best 1/3 months	2.55	4.23	0.47	6.91	2.08	2.47	4.14

(a)

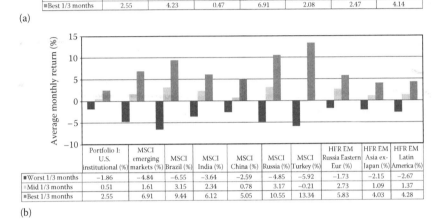

	Portfolio I: U.S. institutional (%)	MSCI emerging markets (%)	MSCI Brazil (%)	MSCI India (%)	MSCI China (%)	MSCI Russia (%)	MSCI Turkey (%)	HFR EM Russia Eastern Eur (%)	HFR EM Asia ex-Japan (%)	HFR EM Latin America (%)
■Worst 1/3 months	−1.86	−4.84	−6.55	−3.64	−2.59	−4.85	−5.92	−1.73	−2.15	−2.67
Mid 1/3 months	0.51	1.61	3.15	2.34	0.78	3.17	−0.21	2.73	1.09	1.37
■Best 1/3 months	2.55	6.91	9.44	6.12	5.05	10.55	13.34	5.83	4.03	4.28

(b)

FIGURE 4.2   (a) Average monthly return of major asset classes in 39-month periods with worst, mid, and best return for the U.S. Institutional Portfolio January 1999–September 2008; and (b) average monthly return of emerging markets asset classes in 39-month periods with worst, mid, and best return for the U.S. Institutional Portfolio January 1999–September 2008.

all asset classes by the performance of the U.S. Institutional Portfolio. Then we examined the average performance and correlation during the worst 39 months (out of 117) for the U.S. Institutional Portfolio, the middle 39 months, and the best performing 39 months. The results show the relative contribution of the asset classes to the overall portfolio during periods of stress, average performance, and exceptional performance (Figure 4.2).

In Table 4.6, we observe that the best period for the U.S. Institutional Portfolio coincides with the best period for almost all asset classes with the exception of the bond portfolios (U.S., Global, and Euro Aggregate). Similar are the results for the middle and worst set of periods. At the same time the emerging markets asset classes' performance numbers cover a wider range of losses or gains, indicating that there is indeed a comovement but with higher volatility.

TABLE 4.6   Average Monthly Return in 39-Month Periods with Worst, Mid, and Best
Return for the U.S. Institutional Portfolio: January 1999–September 2008

	All (%)	Bottom (%)	Middle (%)	Top (%)
Portfolio I: U.S. Institutional	0.40	−1.86	0.51	2.55
S&P 500	0.18	−4.06	0.36	4.23
Russell 2000	0.66	−4.27	1.09	5.17
Dow Jones STOXX 50	0.41	−4.47	0.82	4.89
NIKKEI 225	0.05	−4.09	0.64	3.61
MSCI World Index	0.24	−4.08	0.59	4.21
MSCI Emerging Markets	1.23	−4.84	1.61	6.91
MSCI Brazil	2.01	−6.55	3.15	9.44
MSCI India	1.61	−3.64	2.34	6.12
MSCI China	1.08	−2.59	0.78	5.05
MSCI Russia	2.96	−4.85	3.17	10.55
MSCI Turkey	2.40	−5.92	−0.21	13.34
Lehman U.S. Agg.	0.44	0.32	0.51	0.47
Lehman Euro Agg.	0.07	0.11	0.03	0.08
Lehman Global Agg.	0.40	0.23	0.43	0.56
Lehman U.S. Corp High Yield	0.38	−1.41	0.85	1.69
Lehman EM High Yield	0.97	−1.00	1.33	2.57
Lehman Global High Yield	0.59	−1.27	0.97	2.08
FTSE NAREIT US	1.05	−1.12	1.50	2.76
LPX Listed Private Equity Venture	0.24	−6.86	−0.05	7.61
LPX Buyout	0.96	−2.39	1.35	3.91
DJ AIG Commodity	1.04	−0.30	0.80	2.62
HFR Fund Weighted	0.72	−1.17	0.87	2.47
HFR Emerging Markets	1.23	−1.98	1.52	4.14
HFR EM Russia Eastern Europe	2.28	−1.73	2.73	5.83
HFR EM Asia ex-Japan	0.99	−2.15	1.09	4.03
HFR EM Latin America	0.99	−2.67	1.37	4.28

In order to further explore the effect of these investments to a U.S.
Institutional Portfolio, we performed the same exercise but with correlations in Table 4.7, Figure 4.3a and b. It is interesting to observe that in the
vast majority of asset classes the correlation increases during the negative months. This implies that such investments would not have been a
significant break to the downdraft during periods of stress for an institutional portfolio. Such result brings more emphasis to the diversification
and appropriate portfolio construction so that the performance benefit is
maintained even following months of market stress.

TABLE 4.7 Correlation of Asset Classes with the U.S. Institutional Portfolio
in 39-Month Periods with Worst, Mid, and Best Return for the U.S. Institutional
Portfolio: January 1999–September 2008

	All	Bottom	Middle	Top
S&P 500	0.95	0.86	0.57	0.87
Russell 2000	0.75	0.59	0.31	0.07
Dow Jones STOXX 50	0.81	0.80	0.57	0.32
NIKKEI 225	0.61	0.42	0.20	0.24
MSCI World Index	0.96	0.91	0.71	0.77
MSCI Emerging Markets	0.80	0.71	0.28	0.04
MSCI Brazil	0.66	0.70	0.24	0.12
MSCI India	0.50	0.41	0.37	−0.13
MSCI China	0.63	0.58	0.36	0.17
MSCI Russia	0.54	0.22	−0.03	0.47
MSCI Turkey	0.53	0.35	0.21	0.29
Lehman U.S. Agg.	0.05	−0.05	0.09	0.00
Lehman Euro Agg.	0.02	0.08	0.08	0.07
Lehman Global Agg.	0.11	0.06	0.20	0.03
Lehman U.S. Corp High Yield	0.62	0.51	0.38	−0.04
Lehman EM High Yield	0.62	0.55	0.24	0.43
Lehman Global High Yield	0.71	0.58	0.46	0.20
FTSE NAREIT US	0.36	0.17	0.17	−0.18
LPX Listed Private Equity Venture	0.73	0.56	0.45	0.22
LPX Buyout	0.70	0.64	0.53	−0.06
DJ AIG Commodity	0.28	0.15	0.05	0.09
HFR Fund Weighted	0.81	0.63	0.31	0.29
HFR Emerging Markets	0.76	0.61	0.27	0.28
HFR EM Russia Eastern Europe	0.60	0.39	0.12	0.39
HFR EM Asia ex-Japan	0.70	0.57	0.22	0.08
HFR EM Latin America	0.71	0.69	0.30	0.32

This analysis was performed only for the U.S. Institutional Portfolio,
however conclusions for the Global Institutional and Endowment portfolios are similar and therefore we refrain from presenting them here.

4.6 CONCLUSION

We created three benchmark portfolios to represent the typical allocations for a U.S. Institution, a Global Institution and an Endowment and
presented the descriptive statistics for the underlying constituents on a
standalone and on a portfolio basis over the past 10 years. For all three

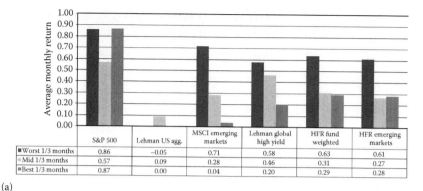

(a)

	S&P 500	Lehman US agg.	MSCI emerging markets	Lehman global high yield	HFR fund weighted	HFR emerging markets
■ Worst 1/3 months	0.86	−0.05	0.71	0.58	0.63	0.61
▨ Mid 1/3 months	0.57	0.09	0.28	0.46	0.31	0.27
■ Best 1/3 months	0.87	0.00	0.04	0.20	0.29	0.28

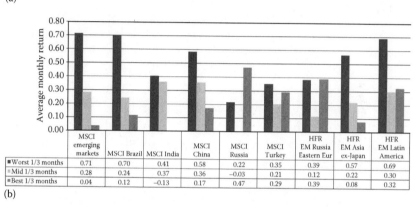

	MSCI emerging markets	MSCI Brazil	MSCI India	MSCI China	MSCI Russia	MSCI Turkey	HFR EM Russia Eastern Eur	HFR EM Asia ex-Japan	HFR EM Latin America
■ Worst 1/3 months	0.71	0.70	0.41	0.58	0.22	0.35	0.39	0.57	0.69
▨ Mid 1/3 months	0.28	0.24	0.37	0.36	−0.03	0.21	0.12	0.22	0.30
■ Best 1/3 months	0.04	0.12	−0.13	0.17	0.47	0.29	0.39	0.08	0.32

(b)

FIGURE 4.3   (a) Correlation of major asset classes with the U.S. Institutional Portfolio in 39 month periods with worst, mid, and best return for the U.S. Institutional Portfolio January 1999–September 2008; and (b) correlation of emerging markets asset classes with the U.S. Institutional Portfolio in 39 month periods with worst, mid, and best return for the U.S. Institutional Portfolio January 1999–September 2008.

benchmark portfolios we examined the effect of including small allocations to emerging markets, and concluded that a small allocation would be beneficial from a performance and Sharpe ratio point of view.

We used optimization without allocation constraints to explore the optimal portfolios under the Markowitz selection context. Results showed that allocation to emerging markets Hedge Funds was favored and these allocations could be even higher as the risk tolerance increased. Such optimization, however, produced suboptimal portfolios in terms of concentration because no diversification constraints had been incorporated in the model. The next step was to impose certain minimum allocation constraints to equities, bonds, and hedge funds, and allow 50% of the portfolio to be selected according to a Markowitz (mean–variance) approach. The results were similar, in that exposure to emerging markets was favored as the risk tolerance increased. However, at the low risk portfolios there was no allocation to emerging markets, while the optimization allocated to hedge funds (emerging markets) as we increased the risk tolerance. At the very high levels of risk tolerance, allocation to emerging markets was represented predominantly by outright equity exposure rather than hedge funds.

At the final step, we explored the behavior of the underlying components of these portfolios and the various asset classes during three sets of periods for the institutional portfolio: (1) market stress, (2) average market behavior, and (3) strong markets. We observed that the asset classes for the emerging markets exhibited amplified performance characteristics i.e., greater losses and greater gains when the benchmark portfolio exhibited losses and gains, respectively. The analysis was carried one step further to examine the correlation of the various asset classes with the U.S. Institutional Portfolio during the same periods under consideration. The results revealed, especially for the emerging markets asset classes, that correlation increased to high levels during the worst months for the U.S. Institutional Portfolio. The conclusion from these observations was that the performance benefits may be limited or may disappear if portfolio selection is not carefully designed to take into account the significant maximum drawdown that may be observed in these markets.

APPENDIX 4.A.1   Indices

Index Name	Asset Class	Geographical Focus
Equity indices		
S&P 500	Equity	United States
Russel 2000	Equity	United States
Dow Jones STOXX 50	Equity	Europe
NIKKEI 225	Equity	Japan
MSCI World Index	Equity	Global
MSCI Emerging Markets	Equity	Emerging markets
MSCI Brazil	Equity	Brazil
MSCI India	Equity	India

(continued)

APPENDIX 4.A.1 (continued)   Indices

Index Name	Asset Class	Geographical Focus
Equity indices		
MSCI China	Equity	China
MSCI Russia	Equity	Russia
MSCI Turkey	Equity	Turkey
Bond indices		
Lehman U.S. Agg.	Bond	United States
Lehman Euro Agg.	Bond	Europe
Lehman Global Agg.	Bond	Global
Lehman U.S. Corp High Yield	Bond/high yield	United States
Lehman Pan-Eur High Yield	Bond/high yield	Europe
Lehman EM High Yield	Bond/high yield	Emerging markets
Lehman Global High Yield	Bond/high yield	Global
LIBOR(Yield)		
Bloomberg Generic USD Libor 1 m	LIBOR	United States
Traditional alternative investments		
FTSE NAREIT US	Real estate	North America
LPX Listed Private Equity Venture	Private equity venture	Global
LPX Buyout	Buyout	Global
Commodities		
DJ AIG Commodities		
Hedge funds		
HFR fund weighted	Hedge funds	
HFR Emerging Markets	Hedge funds	Eastern markets
HFR EM Russia Eastern Europe	Hedge funds	Eastern Europe
HFR EM Asia ex-Japan	Hedge funds	Asia ex-Japan
HFR EM Latin America	Hedge funds	Latin America

REFERENCES

Black, F. and R. Litterman. (1992) Global portfolio optimization, *Financial Analyst Journal*, 48(5): 28–43.

Chopra, V. and W. Ziemba. (1993) The effect of error in means, variances, and covariances on optimal portfolio choice, *Journal of Portfolio Management*, 19(2): 6–11.

Fung, W. and D. A. Hsieh. (2001) Benchmarks of hedge fund performance: Information content and measurement biases, Available at http://papers.ssrn.com/sol3/papers.cfm?abstract_id = 278744.

Fung, W., D. A. Hsieh, and K. Tsatsaronis. (2001) Do hedge funds disrupt emerging markets?, Brookings–Wharton Papers on Financial Services, Washington.

Karavas, V. (2000) Alternative investments in the institutional portfolio, *Journal of Alternative Investments*, 3(3): 11–26.

Karavas, V. and S. Siokos. (2003) The hedge fund indices universe. In: G. Gregoriou, V. Karavas, and F. Rouah (Eds.), *Hedge Funds: Strategies, Risk Assessment, and Returns*. Beard Books, Washington DC.

Schneeweis, T. and G. Martin. (2000) The benefits of hedge funds: Asset allocation for the institutional investor, Lehman Brothers, New York.

Schneeweis, T., R. Spurgin, and M. Potter. (1996) Managed futures and hedge fund investment for downside equity risk management, *Derivatives Quarterly*, (Fall): 62–72.

CHAPTER 5

Equity Returns in Emerging Markets: Prospects for the BRICs

R. McFall Lamm, Jr.

CONTENTS

5.1 INTRODUCTION

Few investments rival the extraordinary performance of emerging market stocks over the past decade. Many optimists expect this pattern to continue in coming years as the ongoing transition to free markets and open economies continues to yield rewards. That said, navigating the ups and downs of emerging markets is not without its perils—as evidenced by the 2008 experience. Risks include unexpected policy changes, currency system manipulation, new capital controls, and capricious regulation in a revisionist backlash that reverses the market-opening process. Further, as is increasingly evident, emerging markets are not immune to events in developed markets such as bank crises and recession.

Within emerging markets, arguably the BRICs—Brazil, Russia, India, and China—offer the greatest potential for future growth. These four countries contain half the earth's population and stand to benefit the most from scale effects as large corporate champions evolve. Their wherewithal to dramatically reshape the global status quo is evident since the beginning of this millennium as the BRIC economies grew from a third to half the size of the U.S. economy. Equity market performance and capitalization exploded in unison. An obvious question for investors is: Can the superior returns posted by the BRICs continue and, if so, by how much as their economies and capital markets continue to evolve?

This chapter examines this issue using a somewhat unusual approach where basic models of BRIC equity market behavior are constructed and presumed to converge to developed market standards at the end of a 10 year adjustment period. Section 5.2 provides a brief review of BRIC market performance. This is followed by an explanation of the analytical approach and its rationale. Statistical estimates of the model are then presented and their efficacy examined. Projections of future BRIC equity market returns are then made and the implications assessed. The key conclusion is that even if one believes that growth rates slow and market behavior evolves to resemble that of developed markets over the next decade, BRIC equity markets should outperform.

5.2 BACKGROUND

Emerging markets are of major interest because their stock markets have exhibited extraordinary outperformance versus developed countries over lengthy periods of time. Unfortunately, most periods of strong market gains and exuberance have been followed by cataclysm and acute underperformance. For example, emerging market equities outperformed developed markets by double-digits from the late 1980s until the peso crisis in 1994 when prices collapsed. This was superseded by several years of robust returns—until the Asian currency crisis in 1997 and the Russian debt default in 1998 precipitated another crash. Then, after a hiatus that lasted until 2002, came another round of extraordinary market performance during the global expansion that culminated at the market peak in 2007 (Figure 5.1).

The emerging markets today are not what they were in the past, however. Indeed, the old emerging market regime of vicarious policy error has been replaced with a new paradigm where stable macro foundations have been built—especially in the BRICs. Chronic fiscal and trade deficits have largely been eliminated and replaced by surpluses. Foreign exchange shortages have been supplanted by massive surpluses that dwarf anything in modern history. The BRIC governments have also universally accepted the role of markets as a resource allocation mechanism now that capitalist enterprise has unambiguously demonstrated its capability to generate great wealth and rapid economic growth.

Despite this, some vestiges of the old emerging markets order remain. For example, China maintains a predatory currency fix, regulates capital flows, and limits foreign ownership of domestic companies. Its citizens face restrictions on owning assets outside of China and must choose between bank deposits at painfully low rates set by the government or investing in the domestic equity market or property. In addition, energy prices are controlled and now are considerably below the world price level.

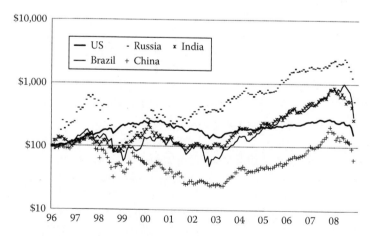

FIGURE 5.1 BRIC equity market returns versus the United States.

Conditions in the other BRICs are somewhat more liberal but each country retains its own special constraints and nuances.

While the BRICs remain linked with developed markets—they are still reliant on trade and investment flows as evidenced again in 2008—each possesses its own unique history and this has influenced market structure. Brazil experienced a catastrophic sovereign debt default in the 1980s followed by hyperinflation in the 1990s. It has struggled paying down a large foreign debt ever since. Like Brazil, Russia experienced hyperinflation and sovereign debt default. However, Russia's experience was concentrated in a shorter time frame with out-of-control inflation in the early 1990s—after the break-up of the Soviet Union—culminating in the 1998 debt default that had repercussions around the world.

India possesses a more stable and long-lived democratic government with institutions in place for nearly half a century. Its stock market also has a long history. However, India has been handicapped by an excessively socialist and antifree market tilt with frequent government intercession. This has come with substantial bureaucratic bloat, import/export restrictions, price controls, and capital flow limits that have hindered development.

The BRIC economies differ dramatically in the extent of resource specialization. China imports raw materials and exports manufactured items. Brazil is resource rich where exports of iron ore, soybeans, sugar, and other commodities represent a key driver of growth in recent years. Russia's commodity export basket differs from Brazil in that it is dominated by petroleum-based products and natural gas. Metals exports are also important but, unlike Brazil, agricultural exports are immaterial. India is more service oriented with labor as its chief resource, especially a well-educated professional class.

Despite these differences, BRIC equities generally have performed well as a market-oriented integration with the rest of the world ensued over the past decade (Table 5.1).* At the same time, the BRICs experienced incredible economic growth with cumulative gross domestic product (GDP) rising from a third of that for the United States to nearly half over the past 5 years (Figure 5.2). This was accompanied by an extraordinary surge in available information where not only do companies in the BRICs report earnings on a timely basis, understand cost control, and are sensitive to shareholder

---

* Note that annual returns reported in the table include the sharp stock market sell-off in September and October 2008. Therefore, average returns are somewhat lower than if other time periods are selected. Also, we begin with 1996 due to data availability and since it represents a relatively neutral mid-cycle starting point.

TABLE 5.1   BRIC Equity Market Performance versus the World—January 1996–October 2008

Metric	World	Emerging Markets	United States	Russia	India	Brazil	China
Return (annualized)	2.9%	2.8%	4.0%	14.1%	7.3%	10.4%	−2.4%
Volatility	16.0%	26.3%	16.6%	59.6%	32.1%	32.1%	40.0%
Skew	−1.66	−1.54	−1.27	−0.13	−0.56	−0.73	0.50
Kurt	6.68	5.34	4.67	2.66	0.96	2.29	2.63
JB test	357.26	244.10	181.49	45.99	13.93	47.17	51.00
Correlation							
World	1.00	0.83	0.96	0.57	0.51	0.72	0.52
Emerging		1.00	0.75	0.68	0.65	0.83	0.68
United States			1.00	0.53	0.44	0.65	0.51
Russia				1.00	0.33	0.60	0.36
India					1.00	0.48	0.39
Brazil						1.00	0.53

Note: Underlying equity return data cover 154 months and are from MSCI via Bloomberg.

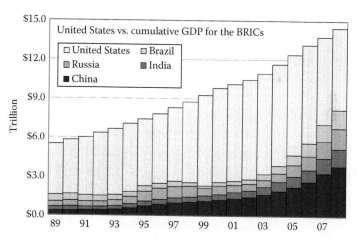

FIGURE 5.2   BRICs versus U.S. GDP.

interests, but government statistics and surveys have improved radically. All this is reflected in equity markets where trading today is increasingly driven by fundamentals and more complete information than ever before—an incredible turnaround when compared with the past.

5.3 ANALYTICAL APPROACH

What then is the future of BRIC equity markets? Is the recent outperformance poised to continue or is it a passing moment to be followed by an interlude of subpar returns? To properly address this question one needs to identify the process that generated the returns. Once this is accomplished, a more thoughtful conjecture about the future is possible.

Significant research has focused on quantifying the fundamental determinants of emerging market equity returns. However, much of this work is overtly concerned with specific structural problems that are less relevant today. For example, early studies analyzed the effects of market liberalization, exchange rate controls, the extent of return predictability in emerging markets, and the impact of sovereign debt defaults (Bekaert and Urias, 1999; Hwang and Satchell, 1999; Bourguignon and Sequier, 2000; Calvo and Mendoza, 2000; Henry, 2002; Harvey, 1995). More recently, effort has been directed to understanding the effect of capital inflows (Bekaert, et al., 2002; Miniane, 2004) and the role of speculation (Bekaert, et al., 2007). The specific case of the BRICs has been largely neglected. One exception is Gay (2008) who evaluates oil prices and exchange rates as predictors of BRIC equity market returns. Obviously, other factors than these two variables are also important.

My approach is somewhat different from the standard statistical search for fit in that some structure is posited where demand and supply determinants reflect the role of asset complementarity and substitution. Furthermore, log models are employed to represent BRIC market behavior rather than typical return-based regressions. While this adds a little complexity to deriving and testing for return forecasting capability, it has the advantage of avoiding the collinearity plague in a different way. It also embeds a more direct representation of the components of common discount models in the structural specification.

In this regard, equity prices in each BRIC are viewed as a component of the local system $y_t = A(L)y_{t-1} + B(L)x_t + C(L)z_t + e_t$ where the y_t vector includes stock prices and volume, the x_t vector includes variables for local market conditions, and the z_t vector represents global exogenous influences that affect BRIC stock prices. Because equity market behavior is my primary concern, the focus here is only on the price equations from the system. Of major interest is the special case where the lag operator $L = 0$, such that lagged values are not important in the price determination process. In this instance, the model reduces to a contemporaneous version where expectations formed off past information play no direct role—that is, prices are determined instantaneously. The adequacy of this structure, of course, can be tested directly.

The local market factors in the model are interest rates, earnings, and dividends—the standard ingredients for equity price prediction model. As for exogenous variables, global commodity prices, emerging market stock prices, and exchange rates are added to reflect complement/substitute investment possibilities.* To establish a benchmark basis for comparison, the same model is estimated for the United States. This establishes a proxy reference for equity price behavior in a developed market.

5.4 ESTIMATION RESULTS

For empirical implementation, available monthly data up to March 2008 have been used. The starting points for each BRIC are dependent on when reliable data is available—September 2001 for Brazil and Russia, June 2002 for India, and June 1997 for China. While this sample is small, it does cover at least one full global cycle—with equity prices generally falling until the market trough in late 2002, rising to the peak in October 2007, and then moving downward since that time. In addition, the sample covers a period of significant fluctuation in values with earnings, dividend yields, currencies, interest rates, and commodity prices all exhibiting substantial volatility over a wide range of extremes.

The one shortcoming in data availability is interest rates where specific allowances are made for each BRIC. For example, central bank rates are used for Brazil and Russia as proxy measures due to the absence of fluid local markets and the preponderance of offshore currency borrowing and lending. For India, the term structure is defined as the 10 year sovereign yield minus the central bank rate. And for China, the term structure is the central bank less the 3 month Treasury yield due to the fact that short rates were consistently held artificially low by the government over the

---

* Note that macro variables such as industrial production, inflation, and similar measures are excluded since these in theory are reflected in the realized corporate earnings stream. This is a key simplification compared with some broad-based approaches and avoids the statistical estimation haze of double-counting when too many correlated factors are included. In addition, the inclusion of emerging market equity prices might initially appear egregious in that it includes the BRICs. However, the BRIC weights as a percentage of the emerging market universe were very small over much of the sample where Taiwan, Korea, and South Africa equity market capitalizations are much larger.

entire sample. Otherwise, all model variables are defined in exactly the same way. BRIC and U.S. equity prices, earnings, and dividends in dollars are measured by the relevant MSCI series, as are emerging market equity prices; BRIC currencies are each country's own rate versus the dollar, while the DXY index is used for the United States; and commodity prices are measured by the CRB index.*

For the United States, where much more information is readily available, a slightly longer but similar time period from January 1996 to March 2008 is used for estimation. The term structure is defined as the 10 year less 3 month Treasury rate differential. For the U.S. model, as well as all the BRICs the use of data through March 2008 allows the April to October 2008 period to be reserved for an out-of-sample validation test that is particularly stringent because of the sharp decline in equity prices during these months.

5.4.1 Qualitative Assessment

Basic parameter estimation results are reported in Table 5.2. Because log transformations are used, the coefficients in the table represent elasticities, that is, they show the impact on stock prices in percentages for an underlying 1% change in each variable. Also, notice that in the case of the BRICs, no coefficients are reported for lagged variables. These are found to have no or extremely limited influence on equity prices for each country—as should be evident from the relatively high reported R^2 values. In contrast, last period's own price for the United States is found to play an important role—as evidenced by a highly significant coefficient and verified in additional experiments where it is omitted from the model. This indicates the past has predictive power in the case of the United States, that is, expectations formation is important.[†]

As for the estimated BRIC coefficients, higher corporate earnings are found to have a statistically significant positive influence on equity prices with the exception of India and China. Increases in dividends are found to have an insignificant impact on stock prices in all cases—no doubt due to the fact that dividends are extraordinarily low over the sample horizon for these countries and do not contribute meaningfully to stock returns.

---

[*] The use of equity prices denominated in dollars as the pertinent measure reflects our interest in globally investable stock markets and has the particular advantage of avoiding the situation where, for example, in the case of China local markets are closed to outside investors.

[†] In the interest of space limitations, detailed test procedures and results are not reported. In addition, emerging market stock prices from the U.S. model are excluded due to the small size of the asset class over most of the sample period.

TABLE 5.2   Estimates of BRIC Equity Price Determinant Models

Metric	Brazil	Russia	India	China	United States
Earnings	.099*	.180*	.021	−.033	.129*
	(.033)	(.070)	(.097)	(.091)	(.075)
Dividends	.037	.097	−.048	.128	.612*
	(.040)	(.055)	(.083)	(.082)	(.186)
Policy rate	−.062	−.492*	—	—	—
	(.055)	(.167)	—	—	—
Term structure	—	—	−1.266*	−20.06*	−2.29
	—	—	(.655)	(.170)	(1.51)
Currency	.891*	−1.812*	−.725	1.097	−.534*
	(.118)	(.491)	(.273)	(.844)	(.224)
Commodity prices	.286*	.250	−.013	−1.202*	−.308*
	(.110)	(.184)	(.149)	(.170)	(.146)
Emerging markets	.833*	.801*	1.328*	1.344*	—
	(.081)	(.134)	(.089)	(.116)	—
Own price lag	—	—	—	—	.700*
	—	—	—	—	(.062)
Intercept	−.051	−5.721*	−3.490	7.244*	6.076*
	(.425)	(2.133)	(1.248)	(2.076)	(.080)
Sample begins	Sep. 2001	Sep. 2001	June 2002	June 1997	June 1996
Adjusted R^2	.996	.974	.989	.874	.946

Note: Standard errors are presented in parentheses. The BRIC estimates are OLS while the U.S. model is estimated via nonlinear least squares.
*, denotes statistical confidence of 90% or greater.

In contrast, both earnings and dividends are statistically significant and contribute positively to stock market performance in the United States, as one would expect.

With respect to interest rates, the coefficients measure slightly different phenomena as already noted. Nonetheless, one would expect that higher long rates are a negative for stock prices since investors normally rotate from stocks to fixed income as interest rates rise. This is exactly what is found where all estimated coefficients are negative, though some are not highly statistically significant. Somewhat surprising is the absence of a high degree of statistical significance for the term structure in the United States. However, one needs to remember that a sizeable portion of the sample includes the late 1990s bubble. As such, the term structure may play a weaker role than usual as a measure of bond for stock substitution.

As for currency, appreciation in the Brazilian *real* is found to have a statistically significant positive effect on Brazilian equity prices; ruble appreciation is found to have a statistically significant negative effect on Russian stock prices; and currency appreciation has no statistically significant effect for India and China (where in the latter instance, the government completely controls rates). The sign disparity for Brazil and Russia is likely a consequence of the fact that a strengthening *real* induces repatriation as local investors bring home capital to fund local investment, whereas in Russia the opposite occurs as local investors exploit currency strength to shift investment abroad. For the United States, dollar appreciation has a negative effect on stock prices most likely because it signals future earnings weakness via offshore currency translation effects.

An acceleration in commodity prices is found to push up stock prices for resource exporters such as Brazil and Russia, but has a negative effect for commodity importers such as China and the United States (the coefficient for India is not statistically different from zero). While one might initially believe commodity price effects should be embedded in the earnings coefficient, there is a lag before higher prices today generate earnings tomorrow for exporters and vice versa for commodity importers. This expectational aspect to commodity prices is not captured in concurrent earnings—which reflect past commodity price realizations.

Last, the effects of higher emerging market stock prices are found to be positive for local equity prices for all the BRICs. This indicates that generic basket flows that drive up prices broadly in emerging markets spillover directly to BRICs. This amount is somewhat less than one-for-one in Brazil and Russia, and substantially more for India and China.

5.4.2 Predictive Capability

While forecasting is not a primary objective of the approach taken here—the goal is simply to secure a rudimentary representation of the price determination process—it nonetheless is important from a model validation perspective. As a measure of predictive accuracy, the focus here is on the extent to which the models correctly forecast market turning points. This requires that log prices be translated into forecast changes and the number of correct predictions counted. Table 5.3 presents the results, both within sample and over 7 months out-of-sample from March through October of 2008. As indicated, the BRIC models capture turning points fairly well within sample with the appropriate direction of stock price movements rightly predicted about 75% of the time.

TABLE 5.3   BRIC Equity Price Determinant Models—In and Out of Sample
Predictive Ability

Metric	Brazil	Russia	India	China	United States
In sample					
Months with positive returns	51	51	47	68	86
Correctly predicted	94%	90%	85%	72%	57%
Months with losses	27	27	22	61	59
Correctly predicted	67%	59%	68%	69%	41%
Total direction correctly predicted	85%	79%	80%	71%	50%
Out of sample					
Months with positive returns	2	2	3	2	3
Correctly predicted	100%	100%	33%	50%	67%
Months with losses	5	5	4	5	4
Correctly predicted	100%	100%	75%	80%	50%
Total direction correctly predicted	100%	100%	57%	71%	57%

Out-of-sample—a period where the 2008 sell-off intensified and markets were very volatile—the models are even better at forecasting stock turning points as falling earnings, a commodity price reversal, and other factors signaled generally declining stock prices. For Brazil and Russia, predictive accuracy is 100%, while for India and China market direction is correctly predicted in 57% and 71% of sample months, respectively. The within sample predictive accuracy of the U.S. model is not as good as that of the BRICs—it correctly forecasts market direction in only about half the cases. However, the model is partially vindicated in the out-of-sample test where direction is correctly predicted 57% of the time.

5.4.3 Structural Stability

Another question of interest is the extent to which the model parameters are sensitive to the sample—that is, whether the estimated coefficients change significantly when alternate periods are used for estimation. This may appear to be an esoteric issue—the estimates in fact summarize market behavior for a discrete period and one would expect them to evolve over time as the BRICs mature. However, if the parameters are stable over subsamples, this is usually regarded as a verification of model integrity.

To arrive at a rough assessment of model stability, each sample is split in half and reestimate the coefficients. In no case was predictive ability as measured by R^2s severely dampened. Furthermore, the sizes of estimated coefficients were generally quite similar. In addition, formal Chow tests for model stability on the split sample revealed that in only one of the five cases (China) was model stability rejected with a high level of statistical confidence and this was perhaps appropriate since a longer estimation period was used.

In summary, the overall model estimates appear fairly reasonable. Each has a high degree of explanatory power and most capture in and out of sample turning points fairly well. While imperfect in some ways, the resulting representations provide a basic approximation of equity market behavior for the BRICs and the United States.

5.5 FUTURE OF BRIC EQUITY RETURNS

Turning to the question of what these models imply for future BRIC stock prices, the estimated BRIC models are presumed to be transitory representations in an evolutionary process where they gradually morph to a developed market standard over the next decade, that is, as embodied in the U.S. model where earnings, dividends, and expectations are more important drivers of equity prices. Obviously, this is a hypothetical exercise since we would never expect markets in different countries to all behave identically. Even so, it is a provocative query—what might BRIC equity returns be as hyper-growth slows and equity markets behave like those in a developed market such as the United States?[*]

The BRIC parameter evolution to the U.S. model is taken as a simple linear process extending over 10 years. That is, the parameters in each BRIC model move 1/120 toward the U.S. model parameters each month, until convergence is complete at the end of the horizon. Implicit is the belief that capital markets open, price controls erode, and currency controls end (in the case of China) such that structural conditions in the BRICs come to approximate those of the United States after a decade. While this may appear draconian, it is actually a quite conservative stance since some

---

[*] This approach—based on an abstract but feasible market metamorphosis—avoids the fallacy of using small sample BRIC models to make long-term projections. It is also a more intriguing than accounting-based extrapolation or simplistic projections off logistic growth curves.

analysts believe that BRICs will grow faster longer and retain the residual visages of emerging markets for many years.

It must be presumed that for the trajectories of the local market and exogenous variables, that earnings, it must be presumed, parallel economic growth—which is taken as current nominal BRIC rates declining proportionately to a terminal 3% rate in real terms. Also, dividends are allowed to increase to payout rates now in evidence in the United States; U.S. growth and interest rates are permitted to move to long term norms (3% and 2% in real terms, respectively, plus 2.5% inflation) linearly over the next 5 years; and the dollar is presumed to depreciate 3% annually (to eliminate the large U.S. trade deficit). BRIC currency rates start at current levels and increase (decrease) as a function of inflation differences versus the United States, while initial inflation in each BRIC is taken as the current level moving to the U.S. rate at the end of 2010.

Table 5.4 outlines the key assumptions made and shows the resulting implied annual equity market gains for each BRICs versus the United States, while Figure 5.3 shows the equity price time paths. Each month's return is generated from that month's model (which is progressively 1/120th closer to the U.S. structure) and the month's local and exogenous variable values (which are 1/120th closer to their terminal levels). The equity return calculated for the United States over the period is 6.9% annually, which appears reasonable under the expected market conditions. The corresponding annual equity returns for Brazil, Russia, India, and China are

TABLE 5.4   BRIC Convergence Assumptions and Implied Equity Gains

Metric	Brazil	Russia	India	China	United States
		Starting Values %			Avg. %
Real GDP	5	5	8	11	3
Inflation rate	6	15	9	5	2.5
Three month interest rate	13.8	11.0	7.5	3.0	3.5
Ten year interest rate	11.0	14.0	8.6	3.7	4.5
Central bank target rate	13.8	11.0	9.0	8.0	3.5
Currency appreciation (depreciation)	−1.8	−6.6	−3.2	1.0	−3.0
Equity gain (avg. annual geometric)	13.0	17.2	15.2	9.0	6.9

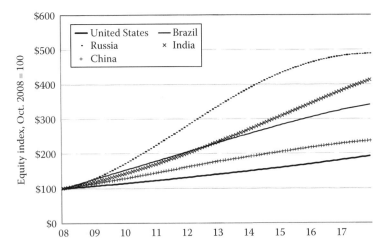

FIGURE 5.3 BRIC equity market returns under 10 year convergence to developed market behavior.

13.0%, 17.2%, 15.2%, and 9.0%, respectively. Indications are therefore that the BRIC equity markets outperform the United States over the 10 year adjustment horizon. This comes in large part from more rapid earnings and dividend growth vis-à-vis the United States but is also a function of other factors such as exchange rates and commodity prices.

To examine equity return sensitivity to alternative scenarios, a number of trials were conducted where the underlying growth assumptions are varied across a range of values. A sample of the findings is shown in Table 5.5. Generally, the outcomes are intuitive—based on the signs and magnitudes of the model coefficients previously presented. For example, weaker U.S. growth produces higher relative equity price gains for the BRICs; a higher trajectory for commodity price inflation yields greater equity market gains for Russia and Brazil versus India, China, and the United States; and a more rapid dollar depreciation produces higher U.S. equity returns (no doubt as terms-of-trade advantages shift to the United States).

Interestingly, the fundamental conclusion—that BRIC equity markets should outperform the United States—is remarkably resilient. Of course, this result does not hold up if one postulates divergence rather than convergence in macro growth paths—such as if BRIC economic growth falls below U.S. levels or inflation differentials widen rather than narrow. Nor does it hold if BRIC market structure does not evolve to match that of the United States. For example, if one merely extends the initial BRIC structural market

TABLE 5.5   BRIC Equity Market Gains under Alternative Scenarios

Scenario	Brazil	Russia	India	China	United States
Base scenario	13.1	17.2	15.2	9.0	6.9
Commodity prices + 2%	11.7	15.4	13.4	7.2	5.0
Commodity prices – 2%	14.4	19.1	17.1	10.9	8.7
U.S. inflation – 1%	13.6	17.8	15.8	9.3	5.8
U.S. inflation + 1%	12.5	16.6	14.6	8.7	8.0
U.S. real GDP – 1%	13.1	17.3	15.2	9.0	5.8
U.S. real GDP + 1%	13.0	17.1	15.2	9.0	8.0
U.S. real 10 year rate = 3%	13.1	17.2	15.2	9.0	6.8

estimates, indications are that equity price gains in Russia, Brazil, and India would be substantially lower. This indicates that a prime driver of future BRIC equity market gains will be ongoing market-based liberalization.

5.6 CONCLUSION

The analysis presented here is straightforward. Fairly simple models explaining the BRIC equity price determination process are estimated and are presumed to morph to a developed market structure—proxied by the United States—over the next decade. At the same time, BRIC growth is presumed to decay linearly to developed market rates over 10 years, while other factors such as exchange rates and commodity prices follow plausible trajectories. The results indicate that BRIC equities should continue to outperform and this finding holds up under a range of alternatives.

Obviously, there are a number of bold assumptions underlying this conclusion. The models are taken as valid representations. Convergence is arbitrarily fixed at 10 years in the future. The falloff in higher BRIC growth is presumed rapid—perhaps too much so—while the decline in inflation and interest rates may be too fast, especially for Brazil and Russia. Furthermore, cyclicality is not taken into account except in that the United States reverts to normality after a 5 year adjustment from its current low growth and interest rates. Nonetheless, the approach is fairly objective without preconceptions or conditions. It is also more sophisticated than subjective efforts one often sees where analysts assert that BRIC equity returns must be superior to developed markets because faster earnings growth alone will translate into superior equity gains. Other factors are also important and need to be considered—an effort explicitly made here.

REFERENCES

Bekaert, G., C.R. Harvey, and R.L. Lumsdaine (2002) The dynamics of emerging market equity flows, *Journal of International Money and Finance*, 21(3), 295–350.

Bekaert, G., C.R. Harvey, and C.T. Lundblad (2007) Liquidity and expected returns: Lessons from emerging markets, *Review of Financial Studies*, 20(6), 1783–1831.

Bekaert, G. and M. Urias (1999) Is there a free lunch in emerging market equities?, *Journal of Portfolio Management*, 25(2), 83–95.

Bourguignon, F. and P. Sequier (2000) Regularity and reason in emerging equity markets, *Emerging Markets Quarterly*, 4(1), 25–36.

Calvo, G.A. and E. Mendoza (2000) Capital markets crisis and economic collapse in emerging markets: An informational friction approach, *American Economic Review*, 90(2), 59–64.

Gay, R.D. (2008) Effect of macroeconomic variables on stock market returns for four emerging economies: Brazil, Russia, India, and China, *International Business & Economics Research Journal*, 7(3), 1–8.

Harvey, C.R. (1995) Predictable risk and returns in emerging markets, *Review of Financial Studies*, 8(3), 773–816.

Henry, P.B. (2002) Stock market liberalization, economic reform, and emerging market equity prices, *Journal of Finance*, 55(2), 529–564.

Hwang, S. and S.E. Satchell (1999) Empirical identification of common factors in emerging equity returns, *Emerging Markets Quarterly*, 3(4), 7–27.

Miniane, J. (2004) Foreign speculators and emerging market equities, *Journal of Finance*, 55(2), 565–613.

CHAPTER 6

Indices and Price Book, Price Earnings, and Dividend Yield Ratios in Emerging Financial Markets

Oktay Taş, Cumhur Ekinci,
and Kaya Tokmakçıoğlu

CONTENTS

6.1 INTRODUCTION

There has been an ongoing effort to predict future prices in capital markets for the purpose of either hedging or speculation. Much work has been done for mature financial markets, yet there is room for reliable research on the so-called emerging markets. A common characteristic of emerging markets is that, being influenced deeply by international political and macroeconomic conditions, they are quite volatile. It is not very unusual to see double-digit daily variations in these markets as opposed to major developed markets. Moreover, as these financial markets are still "developing," they suffer from liquidity shortage. Finally, they constitute a challenge to researchers since historical data are missing or not long enough for statistical analysis.

The main purpose of this chapter is to determine the role of some financial ratios such as price book (P/B), price earnings (P/E), and dividend yield on the general level of securities prices depicted by national market index. We use multivariate and univariate regressions to see how the variations in these ratios can be associated with the variations in index. To the best of our knowledge, this is the first study on emerging markets to integrate all the P/B, P/E, and dividend yield ratios in the same equation. Besides, we make comparisons between groups of countries. For example, it is interesting to determine if different levels of financial ratios provide additional explanation for index performance.

The main findings include a positive relation between the changes in P/B ratio (and to some extent P/E ratio) and the changes in index. Moreover, this relation is more significant in middle- and lower-ratio countries.

This chapter is organized as follows: Section 6.2 reviews the literature, Section 6.3 explains the data and the methodology, Section 6.4 presents the empirical findings, and Section 6.5 gives the concluding remarks.

6.2 LITERATURE REVIEW

The relation between security prices and financial ratios out of the accounting records of the underlying companies has been widely studied. A general conclusion is that future average returns can be predicted for long time intervals. As an example, Fama and French (1989) studied the effect of dividend yield ratio on index return. They used R^2 as a measure of the power of the relation and found the value of 0.25 with data at 4 year time intervals. The most striking result of this research is that the power of the statistical interpretation declines as the time interval for data decreases. The authors found that R^2 falls below 0.05 for monthly, quarterly, or annual time series. Campbell and Shiller (1988) obtained similar results as to the relative success of larger time intervals in predicting stock market index through dividend ratios. Additionally, they noted that price earnings (P/E) ratio is more powerful than dividend yield ratio in predicting stock market index. Lee (1996, 1998) investigated the comovement of logarithmic earnings, dividends, and stock prices by testing for the number of common stochastic trends among these series and found out that the cointegration between log book-to-market and log dividend yield ratios is consistent with the comovement of earnings, dividends, and stock prices. Due to the assumption of stable dividend policy, which is often supposed to be true for various reasons, Vuolteenaho (2002) introduced a new model in which current book-to-market ratio depends on expected future profitability, excess stock returns, and interest rates. His loglinear book-to-market model does not rely on possibly unstable dividend policy. Kothari and Shanken (1997) revealed that both dividend yield and book-to-market ratio tracked the variation in expected stock returns over time. Furthermore, Chen and Wu (1999) examined the dynamic relation between corporate dividends, earnings, and prices as well as the implications of these relations for dividend signaling and smoothing. Their results showed that dynamic relations did exist among dividends, earnings, and prices. Accordingly, current dividend volatility provides information about earnings volatility of the next fifth or sixth quarter. More recently, Jiang and Lee (2007) developed a loglinear cointegration model that explains future profitability and excess stock returns by a linear combination of log book-to-market ratio and log dividend yield. They obtained better results in terms of forecasting performance than those with simply log dividend yield or log book-to-market model.

Cole et al. (1996), by modifying the financial accounting standard FAS106,* claimed that the dividend yield and market-to-book ratio are no longer valid indicators to predict the stock market value and reached the conclusion that the new ratios increased the predictability of index returns calculated with long time intervals. On the other hand, using a cross-sectional analysis with dividend yield ratio, P/E ratio, and P/B ratio, Fama and French (1992) found statistically significant results in spite of a market risk control. More specifically, for the 1963–1990 period, size and book-to-market equity capture the cross-sectional variation in average stock returns controlled for size, E/P, book-to-market equity, and leverage. Similar results were obtained by Ferson and Harvey (1997) and Fama and French (1997) for developed markets and by Claessens et al. (1998), Patel (1998), and Rouwenhorst (1999) for emerging markets.

From a more global perspective, Achour et al. (1998) found evidence that country selection was as much important as stock selection. Likewise, Erb et al. (1995) stated that for emerging markets the choice of a portfolio based on country risk provided higher returns than the one based on conventional valuation ratios. Erb et al. (1996), tried to predict the effects of political, economic, credit-related, and financial risks on P/E, P/B, and dividend yield ratios for the 1984–1995 period. Based on a multiple regression analysis, they concluded that when used as the dependent variable, P/B ratio, and dividend yield ratio offered better results than returns or P/E ratio. Moreover, there was a high correlation between the risk variables. Furthermore, among all the explanatory variables, only political risk could have an effect on dividend yield ratio. As opposed to the general view, Harvey (1995) interestingly found that the patterns in emerging markets were more predictable than in developed markets. In their turn, Ferson and Harvey (1997) suggested that main market indicators such as P/B ratio, P/E ratio, and dividend yield ratio had an explanatory power on market risk.

Bleiberg (1994) grouped 19 emerging markets according to their simulated quarterly, semiannual, and annual returns in order to develop a market timing and asset allocation strategy. Then, on a panel data study for the 1986–1999 period, he found that P/E and P/B ratios had the power to predict returns and as previously stated, the evidence is higher with less frequent data.

Generally speaking, previous research indicates P/B and P/E ratios vary markedly across countries and this may be a signal for the varying return

---

* This statement establishes the accounting standards about postretirement benefits other than pensions in the United States.

performance. Although the literature is abundant for developed markets, very few papers handle the relation between financial ratios and return performance in emerging markets. In what follows, this chapter seeks the factors that affect stock returns in these markets.

6.3 DATA AND METHODOLOGY

The dataset, obtained from Compustat database and corrected for errors[*] and updates from each market, contains monthly index, P/B ratio, P/E ratio, and dividend yield ratio data at national market level for 14 countries for the period January 1997–October 2007. For the ease of comparison between countries, the index data of each country have been normalized by taking January 1997 as the base month (with a value of 100). Table 6.1 gives some descriptive statistics of these data while Table 6.2 ranks countries according to their average index value, P/B, P/E, and dividend yield ratios.

Figures 6.1 through 6.4 plots index, P/B ratio, P/E ratio, and dividend yield ratio of these 14 countries. All the charts contain normalized data. These charts provide with a 10 year trend of these variables. For instance, one can easily observe the rising trend in indices (Figure 6.1) and to some extent in P/B ratios (Figure 6.2). In its turn, Figure 6.4 reveals dividend yields are more heterogeneous across countries in the last 5 years.

Preliminary analysis with augmented Dickey–Fuller test shows almost all the variables are integrated of order 1. For, we take the first difference of the series to satisfy the stationary condition.

Our method consists of a classical linear regression model defined as follows:

$$DIND = a + b_1\,DPB + b_2\,DPE + b_3\,DDIV + ERR$$

where
  DIND is the variation of national stock market index
  DPB is the variation of price book ratio
  DPE is the variation of price earnings ratio
  DDIV is the variation of dividend yield ratio
  ERR is the error term

---

[*] We had to convert manually several numbers given as 0 in the database to N/A. We also had to omit some figures. For example, Russian index went down from 8603 to 342 in January 2001 and went up from 1,884 to 10,097 in April 2006. These very large variations significantly change the nature of the data.

TABLE 6.1   Descriptive Market Statistics and Financial Ratios

Country	Code	Avg. MCap.	No. of Obs.	Avg. Norm. Index	No. of Obs.	Avg. P/B	No. of Obs.	Avg. P/E	No. of Obs.	Avg. Div.	No. of Obs.
Argentina	ar	78140	130	203	130	1.63	112	**20.50**	110	*0.06*	106
Chile	cl	89622	130	133	130	1.54	127	**19.63**	128	*0.32*	130
Czech Rep	cz	21723	130	138	130	1.62	130	15.63	130	**2.97**	130
Egypt	eg	40730	130	183	130	**4.16**	108	13.67	126	**5.38**	128
India	in	302390	130	379	130	2.72	130	17.10	130	1.79	130
Mexico	mx	167483	130	282	130	2.41	128	15.72	128	*0.17*	128
Morocco	ma	19913	130	163	129	2.56	126	16.71	130	**2.88**	125
Philippines	ph	43887	130	56	130	1.60	124	15.65	127	1.47	123
Poland	pl	52269	130	148	130	1.83	130	**21.58**	107	1.51	130
Russia	ru	279970	130	451	130	*1.15*	129	*11.65*	129	1.34	129
South Korea	kr	360066	130	857	130	**3.68**	129	14.82	129	1.75	129
Thailand	th	76051	130	65	130	1.53	130	*8.81*	130	**2.78**	130
Tunisia	tn	2789	130	205	130	*1.24*	130	*11.91*	130	**4.35**	118
Turkey	tr	84067	130	1090	130	**3.57**	130	16.91	130	1.79	130

Notes: The table contains data of average market capitalization in local currency, average (normalized) main index of the country, average price/book ratio, average price/earnings ratio, and average dividend yields ratio with their numbers of observations from Jan. 1997 to Oct. 2007, respectively on their right. Dividend, P/E, and P/B ratios are ranked and some higher and lower values are cut off ad hoc. The bold (italic) numbers show the higher (lower) class for each variable.

TABLE 6.2 Average Values of (Normalized) Index, P/B, P/E, and Dividend Ratios

Country	Norm. Ind.	Country	P/B	Country	P/E	Country	Div.
Turkey	**1089.89**	**Egypt**	**4.16**	**Poland**	**21.58**	**Egypt**	**5.38**
Russia	**451.38**	**South Korea**	**3.68**	**Argentina**	**20.50**	**Tunisia**	**4.35**
India	**378.63**	**Turkey**	**3.57**	**Chile**	**19.63**	**Czech Rep**	**2.97**
Mexico	**281.80**	India	2.72	India	17.10	**Morocco**	**2.88**
Tunisia	204.56	Morocco	2.56	Turkey	16.91	**Thailand**	**2.78**
Argentina	202.89	Mexico	2.41	Morocco	16.71	India	1.79
Egypt	183.07	Poland	1.83	Mexico	15.72	Turkey	1.79
Morocco	161.65	Argentina	1.63	Philippines	15.65	South Korea	1.75
Poland	148.35	Czech Rep	1.62	Poland	15.63	Poland	1.51
Czech Republic	138.27	Philippines	1.60	Czech Rep	14.82	Philippines	1.42
Chile	132.59	Chile	1.54	South Korea	13.67	Russia	1.34
South Korea	124.96	Thailand	1.53	Egypt	11.91	*Chile*	*0.32*
Thailand	*65.11*	*Tunisia*	*1.24*	*Tunisia*	*11.65*	*Mexico*	*0.17*
Philippines	*55.80*	*Russia*	*1.15*	*Russia*	*8.81*	*Argentina*	*0.06*

Notes: Statistics are sorted in descending order and cut off ad hoc for high, middle, and low values. Bold (italic) figures show the group of high (low) values in each group.

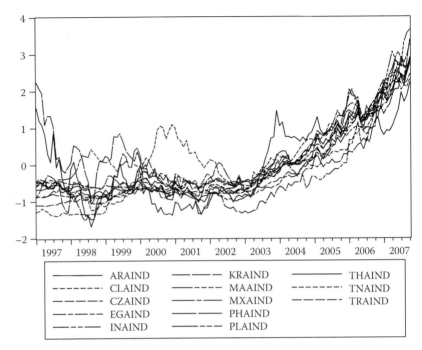

FIGURE 6.1   Normalized indices.

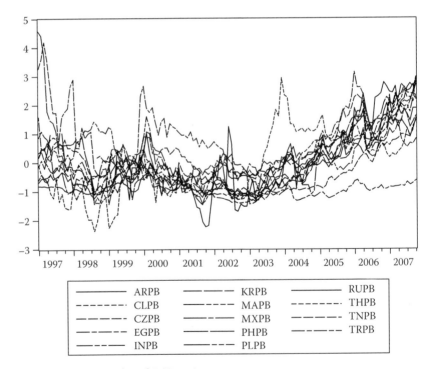

FIGURE 6.2   Normalized P/B ratios.

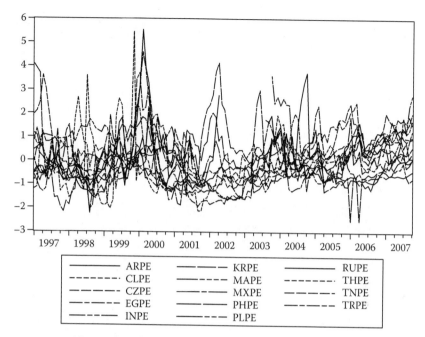

FIGURE 6.3 Normalized P/E ratios.

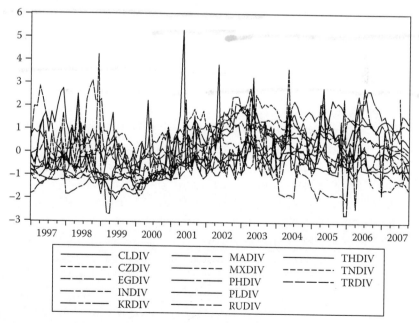

FIGURE 6.4 Normalized dividend yield ratios.

All the IND, PB, PE, and DIV series are normalized by taking their average as 100 for the overall period. This normalization does not alter at all the regression statistics, but makes the interpretation of the coefficients across countries easy.

As a second step, we run univariate regressions of DPB, DPE, and DDIV on DIND to see which of the three variables is most associated with index returns and to see the sensitivity of index returns to each variable.

6.4 EMPIRICAL FINDINGS

This section presents the empirical findings in three subsections. First, we give the results of the multivariate regression; second, we present the results of the univariate regressions; and third, we make comparisons between the countries with high, middle, and low values of P/B, P/E, and dividend yield ratios.

6.4.1 Results of the Multivariate Regression

Table 6.3 explores the regression results for the multivariate model defined earlier. According to the table, b_1 is positive and significant at 1% for all the countries except Turkey. In words, changes in P/B ratio have a significantly positive relation with changes in index. On the other hand, b_2 is significantly positive at 1% for Czech Republic, Egypt, and Turkey and at 5% for Tunisia, showing the positive relation between changes in P/E ratio and changes in index in these countries. If b_3 is negative for 8 of the 14 countries, it is significantly negative at 1% only for South Korea and at 5% for Thailand, indicating a negative relation between changes in dividend yield ratio and changes in index only in these countries. Besides, the intercept term is significantly positive at 1% for 9 of the 14 countries, at 5% for 1 country and at 10% for another. This signals there is a given level of variation in index, free from the three explanatory variables.

The explanatory power of the regressions is pretty high, as the adjusted R^2 is 0.50 on average, with a maximum of 0.86 for Thailand and a minimum of 0.24 for Russia. Durbin–Watson (D–W) statistics mostly are above the upper limit of the test statistic. This denies a serial correlation in error terms, with the exceptions of Tunisia for which D–W is below the lower limit, revealing a positive autocorrelation, and Egypt and India for which D–W is between the upper and lower limits, remaining inconclusive about autocorrelation. By the same token, Breusch–Godfrey (B–G) statistics fail to reject the null hypothesis of no serial correlation up to lag order 2 for all the countries except Tunisia and Turkey.

TABLE 6.3 Multivariate Regression Results

Country	a	b_1	b_2	b_3	Adj. R^2	D-W	B-G	DoF
Argentina	**2.15** (0.008)	**0.67** (0.000)	-0.01 (0.762)	0.00 (0.659)	0.47	**2.27**	0.26	80
Chile	**1.00** (0.002)	**0.39** (0.000)	0.02 (0.217)	0.00 (0.657)	0.34	**1.94**	0.28	119
Czech Republic	**0.95** (0.001)	**0.45** (0.000)	**0.11** (0.000)	-0.00 (0.691)	0.74	**1.77**	0.34	129
Egypt	**2.22** (0.001)	**0.41** (0.000)	**0.19** (0.003)	-0.06 (0.345)	0.45	**1.70**	0.30	102
India	**2.19** (0.002)	**0.88** (0.000)	-0.12 (0.341)	0.15 (0.145)	0.43	**1.67**	0.43	129
Mexico	**1.76** (0.000)	**0.40** (0.000)	0.08 (0.063)	0.00 (0.339)	0.42	**1.80**	0.32	125
Morocco	0.90 (0.094)	**0.51** (0.000)	0.08 (0.227)	0.02 (0.647)	0.34	**1.82**	0.59	113
Philippines	0.89 (0.097)	**0.57** (0.000)	0.06 (0.388)	-0.01 (0.104)	0.67	**1.77**	0.34	111
Poland	**1.46** (0.002)	**0.63** (0.000)	0.03 (0.356)	-0.05 (0.209)	0.63	**1.91**	0.55	105
Russia	**2.17** (0.033)	0.39 (0.000)	**0.07** (0.038)	-0.00 (0.995)	0.24	**1.86**	0.21	125

(continued)

TABLE 6.3 (continued) Multivariate Regression Results

Country	a	b_1	b_2	b_3	Adj. R^2	D–W	B–G	DoF
South Korea	0.86	**0.17**	0.03	**-0.33**	0.47	**1.88**	0.62	127
	(0.104)	(0.000)	(0.250)	(0.000)				
Thailand	-0.10	**0.82**	0.01	**-0.07**	0.86	**1.80**	0.36	129
	(0.741)	(0.000)	(0.777)	(0.023)				
Tunisia	**1.24**	**0.70**	**0.20**	-0.13	0.54	1.48	0.03	116
	(0.001)	(0.000)	(0.047)	(0.216)				
Turkey	**2.75**	-0.06	**0.25**	**-0.18**	0.39	**1.98**	0.01	109
	(0.002)	(0.363)	(0.000)	(0.000)				
Average	1.460	0.495	0.071	-0.047	0.50	1.83	0.33	116
	(0.078)	(0.026)	(0.248)	(0.359)				
Minimum	-0.100	-0.060	-0.120	-0.330	0.24	1.48	0.01	80
	(0.000)	(0.000)	(0.000)	(0.000)				
Maximum	2.750	0.880	0.250	0.150	0.86	2.27	0.62	129
	(0.741)	(0.363)	(0.777)	(0.995)				
Median	1.350	0.480	0.065	-0.005	0.46	1.81	0.33	118
	(0.002)	(0.000)	(0.222)	(0.278)				

Notes: The statistics are for the regression DIND = $a + b_1$ DPB + b_2 DPE + b_3 DDIV + ERR. D–W is the Durbin–Watson statistic, B–G is the Breusch–Godfrey statistic, and DoF is the degrees of freedom. *p*-Values of the parameter estimates are given in parentheses. Coefficients significant at 1% (5%) are typed in bold (bold and italic). D–W values above the upper bound (between the upper bound and the lower bound) at 5% are typed in bold (bold and italic). The lower part gives the average, minimum, maximum, and median values of coefficients, *p*-values (in parentheses), adjusted R^2, D–W, B–G, and DoF.

6.4.2 Results of the Univariate Regressions

In this subsection, we run univariate regressions of the changes in P/B, P/E, and dividend yield ratios on the changes in normalized index to see the sensitivity of the index to these ratios. Tables 6.4 through 6.6 give the results of these univariate regressions. A first remark is that in all the three tables, the intercept coefficient a is positive and significant at 1% for most countries. According to Table 6.4, changes in index have a higher (lower) sensitivity to changes in P/B ratio in Thailand, Poland, India, and the Philippines (in South Korea and Turkey), since the coefficient b is higher (lower) in these countries. Besides, b is 0.51 on average. This means, on average, a 1 point increase in P/B ratio is associated with a 0.51 point increase in the normalized index. Regression results are pretty well in view of the fact that all the p-values are below 0.001; average adjusted R^2 is 0.43; D–W statistics are above the upper limit (denying autocorrelation) except for Tunisia and India; and all the B–G statistics fail to reject the null hypothesis of no serial correlation in error terms.

As shown in Table 6.5, results are qualitatively similar for the relation between the changes in P/E ratio and in index. In South Korea, India, and

TABLE 6.4 Regression Results for the Relation between the Variations in Index and P/B Ratio

Country	a	b	Adj. R^2	D–W	B–G	DoF
Argentina	2.31	0.38	0.31	1.99	0.61	108
	(0.002)	(0.000)				
Chile	1.02	0.40	0.33	1.99	0.48	123
	(0.001)	(0.000)				
Czech Republic	0.97	0.51	0.70	1.80	0.50	129
	(0.001)	(0.000)				
Egypt	2.49	0.55	0.40	1.81	0.58	105
	(0.004)	(0.000)				
India	2.27	0.66	0.42	1.63	0.37	129
	(0.001)	(0.000)				
Mexico	1.74	0.43	0.41	1.84	0.31	125
	(0.000)	(0.000)				
Morocco	0.92	0.58	0.39	2.15	0.55	121
	(0.078)	(0.000)				
Philippines	*0.84*	0.63	0.69	1.77	0.34	118
	(0.048)	(0.000)				

(continued)

TABLE 6.4 (continued)   Regression Results for the Relation between the Variations
in Index and P/B Ratio

Country	a	b	Adj. R^2	D–W	B–G	DoF
Poland	**1.25**	**0.70**	0.64	**1.91**	0.47	129
	(0.001)	(0.000)				
Russia	*2.14*	0.44	0.22	**1.75**	0.15	125
	(0.037)	(0.000)				
South Korea	0.75	**0.22**	0.20	**1.82**	0.73	127
	(0.249)	(0.000)				
Thailand	−0.11	**0.89**	0.85	**1.96**	0.71	129
	(0.712)	(0.000)				
Tunisia	**1.35**	**0.47**	0.30	1.69	0.22	129
	(0.001)	(0.000)				
Turkey	**2.67**	**0.22**	0.14	**2.15**	0.29	129
	(0.005)	(0.000)				
Average	1.47	0.51	0.43	1.88	0.45	123
	(0.081)	(0.000)				
Minimum	−0.11	0.22	0.14	1.63	0.15	105
	(0.000)	(0.000)				
Maximum	2.67	0.89	0.85	2.15	0.73	129
	(0.712)	(0.000)				
Median	1.30	0.49	0.40	1.83	0.48	126
	(0.003)	(0.000)				

Notes: The statistics are for the regression DIND = a + b DPB + ERR. D–W is the
Durbin–Watson statistic, B–G is the Breusch–Godfrey statistic, and DoF is
the degrees of freedom. p-Values of the parameter estimates are given in parentheses. Coefficients significant at 1% (5%) are typed in bold (bold and italic).
D–W values above the upper bound (between the upper bound and the lower
bound) at 5% are typed in bold (bold and italic). The lower part gives the average,
minimum, maximum, and median values of coefficients, p-values (in parentheses),
adjusted R^2, D–W, B–G, and DoF.

TABLE 6.5   Regression Results for the Relation between the Variations
in Index and P/E Ratio

Country	a	b	Adj. R^2	D–W	B–G	DoF
Argentina	*2.19*	*0.07*	0.04	**1.91**	0.97	105
	(0.016)	(0.027)				
Chile	**1.11**	0.03	0.01	1.58	0.04	125
	(0.003)	(0.120)				
Czech	**1.17**	**0.26**	0.29	**1.92**	0.76	129
Republic	(0.009)	(0.000)				

TABLE 6.5 (continued)   Regression Results for the Relation between the Variations
in Index and P/E Ratio

Country	*a*	*b*	Adj. R^2	D–W	B–G	DoF
Egypt	*1.84*	0.39	0.28	1.49	0.03	122
	(0.022)	(0.000)				
India	**2.56**	**0.50**	0.26	1.49	0.11	129
	(0.001)	(0.000)				
Mexico	**2.16**	**0.23**	0.15	1.61	0.07	125
	(0.000)	(0.000)				
Morocco	*1.44*	0.29	0.13	**2.06**	0.32	127
	(0.015)	(0.000)				
Philippines	0.65	**0.46**	0.43	**1.75**	0.39	124
	(0.237)	(0.000)				
Poland	**1.89**	**0.17**	0.12	**2.11**	0.55	105
	(0.008)	(0.000)				
Russia	*2.39*	0.13	0.09	**1.98**	0.25	125
	(0.031)	(0.000)				
South Korea	1.55	**0.78**	0.06	**1.83**	0.67	129
	(0.073)	(0.004)				
Thailand	0.14	**0.27**	0.26	**2.22**	0.19	129
	(0.839)	(0.000)				
Tunisia	**1.23**	**0.37**	0.21	1.68	0.22	129
	(0.006)	(0.000)				
Turkey	**2.90**	**0.31**	0.31	**2.13**	0.00	109
	(0.002)	(0.000)				
Average	1.66	0.30	0.19	1.84	0.33	122
	(0.090)	(0.011)				
Minimum	0.14	0.03	0.01	1.49	0.00	105
	(0.000)	(0.000)				
Maximum	2.90	0.78	0.43	2.22	0.97	129
	(0.839)	(0.120)				
Median	1.70	0.28	0.18	1.87	0.24	125
	(0.012)	(0.000)				

Notes: The statistics are for the regression DIND = $a + b$ DPE + ERR. D–W is the
Durbin–Watson statistic, B–G is the Breusch–Godfrey statistic, and DoF is the
degrees of freedom. *p*-Values of the parameter estimates are given in parentheses. Coefficients significant at 1% (5%) are typed in bold (bold and italic).
D–W values above the upper bound (between the upper bound and the lower
bound) at 5% are typed in bold (bold and italic). The lower part gives the average, minimum, maximum, and median values of coefficients, *p*-values (in parentheses), adjusted R^2, D–W, B–G, and DoF.

TABLE 6.6   Regression Results for the Relation between the Variations in Index and Dividend Yield Ratio

Country	*a*	*b*	Adj. R^2	D–W	B–G	DoF
Argentina	*2.10*	−0.00	−0.01	**2.16**	0.59	94
	(0.026)	(0.483)				
Chile	**1.13**	−0.00	−0.01	1.64	0.01	129
	(0.002)	(0.957)				
Czech Republic	**1.36**	**−0.06**	0.06	**1.77**	0.50	129
	(0.009)	(0.003)				
Egypt	*2.14*	*−0.13*	0.03	1.54	0.04	125
	(0.022)	(0.029)				
India	**2.66**	**−0.40**	0.17	1.55	0.17	129
	(0.001)	(0.000)				
Mexico	**2.21**	0.00	−0.01	*1.70*	0.09	125
	(0.000)	(0.722)				
Morocco	1.62	−0.13	0.06	1.64	0.44	119
	(0.225)	(0.531)				
Philippines	0.36	*−0.13*	0.05	**1.75**	0.35	117
	(0.615)	(0.011)				
Poland	**1.69**	**−0.26**	0.22	**2.00**	0.84	129
	(0.003)	(0.000)				
Russia	*2.45*	−0.10	0.02	**1.76**	0.41	125
	(0.034)	(0.086)				
South Korea	*1.25*	**−0.40**	0.36	**1.73**	0.54	127
	(0.030)	(0.000)				
Thailand	0.12	**−0.46**	0.44	1.64	0.01	129
	(0.839)	(0.000)				
Tunisia	*1.19*	**−0.63**	0.26	1.56	0.37	116
	(0.012)	(0.000)				
Turkey	**2.44**	**−0.28**	0.31	**1.99**	0.37	129
	(0.004)	(0.000)				
Average	1.62	−0.21	0.14	1.75	0.34	123
	(0.130)	(0.202)				
Minimum	0.12	−0.63	−0.01	1.54	0.01	94
	(0.000)	(0.000)				
Maximum	2.66	0.00	0.44	2.16	0.84	129
	(0.839)	(0.957)				
Median	1.66	−0.13	0.06	1.72	0.37	126
	(0.017)	(0.007)				

Notes: The statistics are for the regression DIND = $a + b$ DDIV + ERR. D–W is the Durbin–Watson statistic, B–G is the Breusch–Godfrey statistic, and DoF is the degrees of freedom. p-Values of the parameter estimates are given in parentheses. Coefficients significant at 1% (5%) are typed in bold (bold and italic). D–W values above the upper bound (between the upper bound and the lower bound) at 5% are typed in bold (bold and italic). The lower part gives the average, minimum, maximum, and median values of coefficients, p-values (in parentheses), adjusted R^2, D–W, B–G, and DoF.

the Philippines (in Chile, Argentina, Russia, and Poland), index returns seem to be highly (lowly) sensitive to changes in P/E ratio. However, b is 0.30 on average, much weaker than the coefficient for P/B. Thus, we can assert a 1 point increase in P/E ratio is associated with a 0.30 point increase in normalized index. If all the p-values for b are below 0.01 except for Chile and Argentina, the average adjusted R^2 is 0.19, far below the one for P/B. Moreover, D–W (B–G) statistic is below the lower limit (below 0.05) for 5 (3) of the 14 countries. All these outputs show the power of the tests as well as the parameter significance is poorer in these regressions.

Finally, the results about the relation between the changes in dividend yield ratio and in index are given in Table 6.6. Accordingly, all the b coefficients are negative if not null. In Tunisia, Thailand, India, and South Korea they are particularly low. The coefficient b is 0.21 on average. This result signifies a 1 point increase in dividend yield ratio is associated with a 0.21 point decrease in the normalized index and this is more pronounced for the aforementioned countries. However, the evidence is particularly weak since only 7 of the 14 countries have a p-value below 0.01 and two others below 0.05; adjusted R^2 is 0.14 on average; D–W is below the lower limit for 6 countries and in inconclusive area for another; and B–G is below 0.05 for 3 countries.

6.4.3 Comparison for Different Levels of P/B, P/E, and Dividend Yield Ratios

In this subsection, we perform a separate analysis by grouping the countries as high-, middle-, and low-level according to their financial ratios. More specifically, we take the averages of the univariate regression estimates to compare the group of countries.

Table 6.7 depicts the regression results for each group. For those countries that have a higher (lower) P/B ratio, b is 0.33 (0.46) on average. Interestingly, b is highest (0.58) for the countries with middle levels of P/B ratio. Similar results hold for the regressions with P/E ratio. High-, middle-, and low-level groups have b-coefficients of 0.09, 0.40, and 0.26, respectively. Finally, for dividend yield ratio, high-level group has the highest b-coefficient on average (0.28) while middle- and low-level groups have average b-coefficients of 0.26 and 0.00, respectively. Nonetheless, the statistical evidence is quite weak since the only significant average b-coefficient is for the middle-level group and this is at 5%. Notice that adjusted R^2 is particularly low for P/E and dividend yield ratio regressions, and average D–W statistic of 1.63 indicates a positive autocorrelation in

TABLE 6.7 Regression Results according to High-, Middle- and Low-Level Groups

Explan. Variable	Group	a	b	Adj. R^2	D-W	B-G	DoF	No. of Countries Within the Group
P/B	High-level	1.97 (0.086)	**0.33** (0.000)	0.25	**1.93**	0.53	120	3
	Middle-level	1.25 (0.094)	**0.58** (0.000)	0.53	**1.89**	0.48	123	9
	Low-level	*1.75* (0.019)	0.46 (0.000)	0.26	*1.72*	0.19	127	2
P/E	High-level	**1.73** (0.009)	*0.09* (0.049)	0.06	**1.87**	0.52	112	3
	Middle-level	*1.78* (0.045)	0.40 (0.001)	0.24	**1.79**	0.29	124	8
	Low-level	1.25 (0.292)	**0.26** (0.000)	0.19	**1.96**	0.22	128	3
Dividend yield	High-level	1.29 (0.221)	-0.28 (0.113)	0.17	1.63	0.27	124	5
	Middle-level	1.81 (0.115)	*-0.26* (0.016)	0.19	**1.80**	0.45	126	6
	Low-level	**1.81** (0.009)	0.00 (0.721)	-0.01	**1.83**	0.23	116	3

Notes: The statistics are for the regressions $DIND = a + b X + ERR$ where X is DPB, DPE, or DDIV. D-W is the Durbin–Watson statistic, B-G is the Breusch-Godfrey statistic, and DoF is the degrees of freedom. *p*-Values of the parameter estimates are given in parentheses. Coefficients significant at 1% (5%) are typed in bold (bold and italic). D-W values above the upper bound (between the upper bound and the lower bound) at 5% are typed in bold (bold and italic). All the regression residuals for the regression with dividend yield ratio. These results should be interpreted with caution since the number of observations in each group (shown in the last column of Table 6.7) is quite small.

6.5 CONCLUSION

This chapter attempts to find a link between the variations in national stock market indices and such market-level financial ratios as P/B, P/E, and dividend yield, by using monthly data from 14 emerging markets. The analysis consists of three parts. First, we run a multivariate regression to see how the variations in these three ratios are associated with the variations in index. The results show a positive and very significant relation between the changes in P/B ratio and the changes in index. According to these results, a 1 point deviation of P/B ratio from its average is associated with a ½ point deviation (in the same direction) of the index from its average. We also detect a positive (negative) relation between the changes in P/E ratio (dividend yield ratio) and the changes in index for some countries but the evidence remains weak.

Second, we run univariate regressions of the variations in each ratio on the variations in index. These yield results that are similar to the multivariate case. In both multivariate and univariate regressions, the intercept term mostly is positive and significant. This illustrates the bullish trend witnessed in most world markets in the last 10 years prior to the crash in 2008.

Among all the three relations, the one between dividend yield and index is the least obvious. On the one hand, by giving out cash dividends, companies give up some of their equity capital. This should decrease their overall market capitalization and stock price. On the other hand, offering high dividend yields is a sign for the profitability and healthiness of a firm. Hence, this may imply a simultaneous increase in its market value. When these effects are considered marketwide, national index would decrease (increase) if the first (second) hypothesis holds. Our results weakly support the first hypothesis since the regression coefficients mostly are negative.

The last part of the study compares countries for the same effects. After having regrouped countries with higher, middle, and lower P/B, P/E, and dividend yield ratios, we try to detect whether there is a difference between these groups as for the relations described in univariate regressions. The results show that in higher P/B (P/E) countries, market index is less sensitive to variations in P/B (P/E). This may be explained by diminishing marginal returns to scale. For instance, all else being equal, we would expect the index of a low-P/B (or low-P/E) country to increase more than a high-P/B (or high-P/E) country. However, this should also hold for decreases, which is not obvious. Then, an interesting research topic would be to determine if this high sensitivity of lower-ratio countries is the same in upward and downward index movements.

Obviously, the fact that there are only a few countries in each group prevents from drawing a robust statistical inference in the sense of a panel data analysis. Nevertheless, these results give a rough idea about the sensitivity of index to each financial ratio in countries grouped according to these financial ratios.

ACKNOWLEDGMENTS

We thank İnci Gümüş from Sabancı University and Abdullah Kunt from Global Securities Co. for data facilities.

REFERENCES

Achour, D., C. R. Harvey, G. Hopkins, and C. Lang. (1998) Stock selection in emerging markets: Portfolio strategies for Malaysia, Mexico, and South Africa, *Emerging Markets Quarterly*, 2(4): 38–91.

Bleiberg, S. (1994) Price-earnings ratios as a valuation tool. In: S. Lofthouse (Ed.), *Readings in Investments*. Wiley and Sons: Chichestser, England.

Campbell, J. Y. and R. J. Shiller. (1988) Stock prices, earnings, and expected dividends, *Journal of Finance*, 43(3): 661–676.

Chen, C. and C. Wu. (1999) The dynamics of dividends, earnings and prices: Evidence and implications for dividend smoothing and signaling, *Journal of Empirical Finance*, 6(1): 29–58.

Claessens, S., S. Dasgupta, and J. Glen. (1998) The cross-section of stock returns: Evidence from the emerging markets, *Emerging Markets Quarterly*, 2(4): 4–13.

Cole, M., J. Helwege, and D. Laster. (1996) Stock market valuation indicators: Is this time different? *Financial Analysts Journal*, 52(3): 56–64.

Erb, C. B., C. R. Harvey, and T. E. Viskanta. (1995) Country risk and global equity selection, *Journal of Portfolio Management*, 21(2): 74–81.

Erb, C., C. Harvey, and T. Viskanta. (1996) Political risk, economic risk and financial risk, *Financial Analysts Journal*, 52(6): 29–46.

Fama, E. and K. R. French. (1989) Business conditions and the expected returns on stocks and bonds, *Journal of Financial Economics*, 25(1): 23–49.

Fama, E. F. and K. R. French. (1992) The cross-section of expected stock returns, *Journal of Finance*, 47(2): 427–465.

Fama, E. F. and K. R. French. (1997) Industry costs of equity, *Journal of Financial Economics*, 43(2): 153–193.

Ferson, W. E. and C. R. Harvey. (1997) Fundamental determinants of national equity market returns: A perspective on conditional asset pricing, *Journal of Banking and Finance*, 21(11–12): 1625–1665.

Harvey, C. R. (1995) Predictable risk and returns in emerging markets, *Review of Financial Studies*, 8(3): 773–816.

Jiang, X. and B. S. Lee. (2007) Stock returns, dividend yield, and book-to-market ratio, *Journal of Banking and Finance*, 31(2): 455–475.

Kothari, S. P. and J. Shanken. (1997) Book-to-market, dividend yield, and expected market returns: A time series analysis, *Journal of Financial Economics*, 44(2): 169–203.

Lee, B. S. (1996) Comovements of earnings, dividends, and stock prices, *Journal of Empirical Finance*, 3(4): 327–346.

Lee, B. S. (1998) Permanent, temporary, and nonfundamental components of stock prices, *Journal of Financial and Quantitative Analysis*, 33(1): 1–32.

Patel, S. A. (1998) Cross-sectional variation in emerging markets equity returns; January 1988–March 1997, *Emerging Markets Quarterly*, 2(1): 57–70.

Rouwenhorst, K. G. (1999) Local return factors and turnover in emerging stock markets, *Journal of Finance*, 54(4): 1439–1464.

Vuolteenaho, V. (2002) What drives firm-level stock returns? *Journal of Finance*, 57(1): 233–264.

CHAPTER 7

World Price of Covariance Risk with Respect to Emerging Markets

David E. Allen and Joseline Chimhini

CONTENTS

7.1 INTRODUCTION

Given the increasing integration of financial markets around the world, why is it that countries have different average stock returns? We analyze this in the context of emerging markets using a framework first developed by Harvey (1991). Asset pricing theory suggests that cross-sectional differences in countries risk exposures should explain cross-sectional differences in expected returns. This chapter tests whether conditional versions of the Sharpe (1964) and Lintner (1965) asset pricing model are consistent with the behavior of returns in 20 emerging markets and four developed ones. A country's risk is defined as the conditional sensitivity or covariance of its market return with a world stock return. This risk is permitted to vary through time. If the asset pricing model holds and there is only one source of risk that is priced, then the conditional covariances calculated for each emerging market should explain the differences in national stock index performance. We test this proposition.

The results for our sample of emerging markets are varied. Those markets with high capitalizations and high correlations with world markets appear to be priced more in terms of world factors. Other markets are much more influenced by local factors. The results suggest that time-varying covariances capture some but not all of the dynamic behavior of country returns. This could be because of a variety of factors such as incomplete market integration, model miss-specification, or more than one factor being priced in practice.

This chapter is organized into four sections: a brief literature review in Section 7.2, the research method is presented in Section 7.3, the results in Section 7.4, and a brief conclusion in Section 7.5.

7.2 LITERATURE REVIEW

7.2.1 Conditional CAPM

The capital asset pricing model (CAPM) is a full information equilibrium-based model which assumes that stock return distributions are time invariant and that investors have homogenous expectations. Bollerslev' et al. (1988) and Schwert and Seguin (1990) report evidence of time-varying return distributions suggesting that the moments of return distributions behave like random variables rather than constants as assumed by the CAPM. The time-varying nature of moments can be incorporated into CAPM and the modified version, a conditional CAPM can be applied to explain stock return variation. Harvey's (1989, 1991) and Harvey and Bekaert (1997) work recognizes this. Early studies of asset pricing models by Blume (1971), Gonedes (1973), and Officer (1973) also concluded that portfolio risk changes over time. Fama and French (1989) found that expected returns are high during depressions and low when business is booming. Francis and Fabozzi (1979) suggest that the instability of the beta is related to business conditions as do Jagganathan and Wang (1996). These observations have motivated the study of the conditional CAPM in which beta changes depending on the relevant information set. Howton and Peterson (1999) test a dual cross-sectional model of expected security returns. In this model, risk is allowed to vary through time according to predictions of market and economic states.

7.3 METHODOLOGY

7.3.1 Econometric Model

A conditional version of the CAPM is applied to the returns of emerging markets and the results are compared with those obtained in developed markets. Conditional refers to the use of conditioning information, some information set Z_{t-1} to calculate expected moments and to test properly the intertemporal capital asset pricing model (ICAPM) as a relation between expected returns and ex-ante risk. The conditional version of the Sharpe (1964) and Lintner (1965) asset-pricing model restricts the conditionally expected return on an asset to be proportional to its covariance with the market portfolio. The proportionality factor is the price of covariance risk which is the expected compensation that the investor receives for taking on a unit of covariance risk. The model is given as

$$E\left[r_{jt}\middle|\Omega_{t-1}\right] = \frac{E[r_{mt}|\Omega_{t-1}]}{\text{Var}[r_{mt}|\Omega_{t-1}]}\text{Cov}\left[r_{jt}r_{mt}\middle|\Omega_{t-1}\right] \qquad (7.1)$$

where

r_{jt} is the return on a portfolio of country j equity from time $t-1$ to t in excess of the risk return

r_{mt} is the excess return on the world market portfolio

Ω_{t-1} is the information set that investors use to set prices

The ratio of the conditionally expected return on the market index $E[r_{mt}|\Omega_{t-1}]$ to the conditional of the market index $\text{Var}[r_{mt}|\Omega_{t-1}]$ is the world price of covariance.* Harvey first specified a model of the conditional first moment and assumed that investors process information using a linear filter:

$$R_{jt} - r_{t-1} = Z_{t-1}\delta_j + u_{jt} \qquad (7.2)$$

$$E\left(u_{jt} | Z_{t-1}\right) = 0 \qquad (7.3)$$

where

u_{jt} is the investor's error for the return on assets j

Z_{t-1} is a row of vector of predetermined instrumental variables, which are known to the investor

δ_j is a column vector of time invariant forecast coefficients

Given the assumption on the conditional first moment, Equation 7.1 can be rewritten as

$$Z_{t-1}\delta_j = \frac{Z_{t-1}\delta_m}{E\left[u_{mt}^2|z_{t-1}\right]}E\left[u_{jt}u_{mt}|z_{t-1}\right] \qquad (7.4)$$

where

u_{mt} is the investor's forecast error on the world market portfolio

$E\left[u_{mt}^2|Z_{t-1}\right]$ is the conditional variance

$E\left[u_{jt}u_{mt}|Z_{t-1}\right]$ is the conditional

---

* For the Sharpe–Lintner model to hold in an international framework some auxiliary assumption must be made. The assumption that world market portfolios are perfectly correlated with world consumption is made. The model is viewed as testing the mean variance efficiency of the world portfolios.

Next, multiply both sides of the Equation 7.4 by the conditional variance

$$E\left[u_{mt}^2 Z_{t-1}\delta_j \middle| Z_{t-1}\right] = E\left[u_{jt}u_{mt}Z_{t-1}\delta_m Z_{t-1}\right] \tag{7.5}$$

Notice that the conditionally expected returns on the market and country portfolio are moved inside the expectation operators. This can be done because they are known conditional on the information Z_{t-1}. The deviation from the expectation is

$$h_{jt} = u_{mt}^2 Z_{t-1}\delta_j - u_{jt}u_{mt}Z_{t-1}\delta_m \tag{7.6}$$

where h_{jt} is the disturbance that should be unrelated to the information under the null hypothesis that the model is true. h_{jt} is a pricing error which implies that the model is overpriced when h_{jt} is negative and under priced when h_{jt} is positive.

The econometric model to test the asset pricing restrictions is formed by combining Equations 7.2 and 7.6

$$\varepsilon = \left(u_t u_{mt} h_t\right) = \begin{pmatrix} \left[r_t - z_{t-1}\delta\right]' \\ \left[r_{mt}z_{t-1}\delta_m\right]' \\ \left[u_{mt}^2 z_{t-1}\delta - u_{mt}u_t z_{t-1}\delta_m\right]' \end{pmatrix}' \tag{7.7}$$

where u is a $1 \times n$ (number of countries) vector of innovations in the conditional means of the country returns. The model implies that $E[\varepsilon_t|z_{t-1}] = 0$. With n countries, there are $n + 1$ columns of innovations in the conditional means (u and u_m) and n columns in h.

Hansen's (1982) generalized method of moments (GMM) is used to estimate the parameters in Equation 7.7. The GMM forms a vector of the orthogonality conditions $\mathbf{g} = \text{vec}(\varepsilon'Z)$ where ε is the matrix of forecast errors for T observations and $2n + 1$ equations and Z is a $T \times l$ matrix of observations on the predetermined instrumental variables. The parameter vector δ is chosen as to make the orthogonality conditions as close to zero as possible by minimizing the quadratic form $\mathbf{g'wg}$ where the $\mathbf{w}$ symmetric weighing matrix that defines the metric used to make $\mathbf{g}$ close to zero. The consistent estimate of $\mathbf{w}$ is formed by

$$\left[\sum_{t=1}^{T} \left(\varepsilon_t \otimes z_{t-1} \right)' \left(\varepsilon_t \otimes z_{t-1} \right) \right]^{-1} \tag{7.8}$$

The ε depends on the parameters. As a result, the estimation proceeds in stages. An initial estimate of the parameters is obtained by using an identity matrix for **w**. These parameters are used to calculate ε and a new weighing matrix. The estimation procedure is repeated with this new weighing matrix. Hansen (1982) provides the conditions that guarantee that the estimates are consistent and asymptotically normal.

The minimized value of this quadratic form is distributed χ^2 under the null hypothesis with degrees of freedom equal to the number of orthogonality conditions minus the number of parameters. This χ^2 statistic which is known as the test of the over identifying restrictions will provides a goodness of fit test for the model. A high χ^2 statistic means that the disturbances are correlated with the instrumental variables. This is a symptom of model miss-specification.

Equation 7.8 is estimated for each country. Equation 7.8 provides a test of the model's restriction that the conditionally expected excess return on a country portfolio is proportional to its conditional variance with the world return.

7.3.2 Price of Covariance Risk in Emerging Markets

The general framework provided by Equation 7.8 permits all the conditional moments; means, variance, and covariances to wander through time. In circumstances where some of these moments are constant more powerful tests can be constructed using this structure. Typically, tests of asset pricing models have assumed that expected returns are proportional to the expected return on a benchmark portfolio. A restriction of this kind can be tested:

$$\mathbf{K}_t = \mathbf{r}_t - r_{mt} \beta \tag{7.9}$$

where β is an n-vector of coefficients which could represent the ratios of conditional covariances of the country excess return to the conditional variance of the benchmark return.

7.3.3 Instrumental Variables

We use instrumental variables to predict returns in the multicountry setting. The intention is to condition the model on variables which capture the current state of the global economy. Fama and Shwert (1977), Rozeff (1984), Keim and Stambaugh (1986), Fama and French (1988), and Campbell and Shiller (1988) have shown that stock returns are predictable on the basis of the dividend yield, short-term interest rate, the spread between long- and short-term bond yields (the term structure premium), the spread between corporate and government bonds (a default risk spread), stock market returns and exchange returns as well as dummy variables for the January effect and days of the week effect. Fama and French (1989) find that the above variables are closely associated with economic conditions.

We include dividends in the test of the CAPM. Bar-Yosef and Kolodny (1976), Crockett and Friend (1988), and Christie (1990) report a positive relation between dividends and stock returns. Chen et al. (1986) find that the betas on economic variables may add to the explanatory power of the cross section of expected returns.

7.3.4 Description of Data Sources

A sample of 20 emerging markets and four developed markets is used. The emerging markets picked from markets across the globe include Argentina, Brazil, Chile, the Czech Republic, Greece, Hungary, India, Indonesia, Jordan, Korea, Mexico, Malaysia, Nigeria, the Philippines, Poland, South Africa, Thailand, Turkey, Venezuela, and Zimbabwe. The aforementioned markets are available on the emerging market database (EMDB) obtained from Datastream. The data is comparable across national boundaries and we utilized monthly indexes and data series from January 1976. The database categorizes markets into global, investable, and frontier markets.

We consider January 1985 to June 2000 using monthly data for equity indices. The countries are grouped according to data availability and 14 of the countries have data available from February 1985 while 20 countries have data from February 1994.

In order to determine how correlated the markets are with the rest of the world indexes the IFC index, the Morgan Stanley Capital International (MSCI) EAFE index, the U.S. index, and the Japan index were used. The S&P indexes are treated as a stock portfolio. The MSCI indices are composed of stocks that broadly represent stock market compositions in the different countries.

To test the CAPM relationship between each country index and the world market, the country index and a set of instrumental variables were used. The total number of observations (T) in this study range from 75 to 185.

The assets are country index monthly returns measured in excess of the 3 month U.S. Treasury bill rate. N is the number of assets and there are 24 assets, that is, the 20 emerging markets, the world index, and proxed as the market portfolio, EAFE, the United States, and Japan. The number of instruments (L) ranges between 5 and 8. These are used as conditioning information available at time $t-1$, denoted by Z_{t-1} and include the constant, the dividend yield, the default rate, the spread between U.S. Moody's Baa and Moody's Aaa, and the world returns index. One of the objectives is to find out if emerging markets are integrated or segmented and hence local instrumental variables are included in the conditioning information. The local instruments include the country-specific dividend yield, the rate of change of the exchange rate, the local market index, and a country-specific short-term interest rate. All the monthly returns were calculated in U.S. dollars to make them comparable. The dividend yield and interest rates were scaled down by a factor of 12 to derive the monthly dividend yield and interest rates.

7.3.4.1 Measure of the Return Index
Continuously compounded returns were calculated for all the variables used and were calculated as: $R_{it} = \ln\left(\dfrac{\text{New price}}{\text{Old price}}\right)$, where ln is natural log.

The excess returns were calculated for all the markets in the sample by subtracting the returns on risk free asset from the total country returns and the world market returns. Summary statistics for our basic series are shown in Tables 7.1 and 7.2. Regressions of country returns using world factors were used to predict returns. A linear regression was used to forecast country returns. The regression equation is

$$R_{j,t} = \delta_{j,0} + \delta_{j,1}\text{Wlss}_{t-1} + \delta_{j,2}\text{Spread}_{t-1} + \delta_{j,3}\text{Ustb90}_{t-1} + \delta_{j,4}\text{Wldy}_{t-1} + \varepsilon_{j,t}$$

$$(7.10)$$

where
$R_{j,t}$ is the country's conditional expected return
Wlss is the excess return on the world index
Spread is the yield spread between the U.S. Baa and Aaa rated bonds
Ustb90 is the U.S. 90 day Treasury bill rate
Wldy is the monthly dividend yield on the Standard and Poor 500 stock
index
ε is the regression error

TABLE 7.1 Summary Statistics for the Country Excess Returns and Instrumental Variables

Panel A

Variable	Mean Return	Standard Deviation	Autocorrelation					
Equity Returns			$\rho 1$	$\rho 2$	$\rho 3$	$\rho 4$	$\rho 12$	$\rho 24$
Argentina	0.0127	0.2038	0.01	0.01	0.04	-0.06	-0.10	-0.01
Brazil	0.0059	0.1798	0.02	-0.05	-0.06	0.05	0.02	0.00
Chile	0.0175	0.0790	0.17	0.20	-0.01	0.02	0.08	0.03
Czech Republic	-0.0120	0.0926	0.17	-0.15	-0.23	-0.23	-0.16	-0.07
EAFE	0.0072	0.0511	0.01	-0.08	0.00	0.03	0.05	0.06
Hungary	0.0045	0.1201	-0.08	-0.11	0.01	-0.08	-0.14	-0.07
India	0.0040	0.0925	0.07	0.02	-0.06	-0.08	-0.09	0.00
Indonesia	-0.0137	0.1490	0.22	-0.11	-0.01	0.21	-0.11	0.05
Japan	0.0038	0.0725	0.05	-0.07	0.08	0.05	0.06	0.00
Jordan	0.0005	0.0446	0.01	0.00	0.13	-0.03	0.03	0.02
Malaysia	-0.0003	0.1036	0.13	0.20	-0.12	-0.06	-0.11	0.07
Mexico	0.0137	0.1367	0.24	-0.05	-0.05	-0.02	-0.04	0.02
Nigeria	-0.0011	0.1521	-0.01	-0.06	-0.09	-0.07	0.02	-0.02
Philippines	0.0102	0.1086	0.33	0.06	0.03	0.05	0.08	-0.01
Poland	-0.0062	0.1408	-0.10	-0.09	0.00	-0.20	-0.19	-0.06
South Africa	-0.0021	0.0880	0.01	-0.04	0.01	-0.23	-0.08	-0.19
Thailand	0.0010	0.1025	0.13	0.14	-0.06	-0.13	0.04	-0.12
Turkey	0.0005	0.1208	0.13	0.07	0.08	0.07	-0.13	0.00

(continued)

TABLE 7.1 (continued) Summary Statistics for the Country Excess Returns and Instrumental Variables

Panel A

Variable	Mean Return	Standard Deviation	Autocorrelation					
Equity Returns			$\rho 1$	$\rho 2$	$\rho 3$	$\rho 4$	$\rho 12$	$\rho 24$
United States	0.0095	0.0441	-0.02	-0.06	-0.05	-0.06	0.02	0.07
Venezuela	0.0035	0.1447	-0.01	0.16	-0.01	0.00	0.02	-0.04
World	0.0079	0.0424	0.01	-0.08	-0.03	-0.05	0.03	0.10
Zimbabwe	0.0080	0.1084	0.21	0.14	0.23	0.16	-0.03	-0.03

Panel B

Variable	Mean Return	Standard Deviation	Autocorrelation					
Dividend Yield			$\rho 1$	$\rho 2$	$\rho 3$	$\rho 4$	$\rho 12$	$\rho 24$
Argentina	0.0103	0.0013	0.822	0.707	0.612	0.538	0.428	-0.038
Brazil	0.0029	0.0026	0.867	0.726	0.632	0.523	-0.054	-0.054
Chile	0.0041	0.0019	0.968	0.921	0.884	0.860	0.649	0.314
Czech Republic	0.0016	0.0006	0.845	0.750	0.693	0.594	0.155	-0.532
Greece	0.0042	0.0027	0.910	0.829	0.767	0.706	0.260	0.004
Hungary	0.0013	0.0006	0.870	0.775	0.702	0.619	0.392	0.056
India	0.0015	0.0006	0.913	0.867	0.831	0.797	0.519	0.364
Indonesia	0.0010	0.0006	0.906	0.801	0.732	0.672	-0.035	-0.205
Jordan	0.0028	0.0016	0.920	0.858	0.795	0.727	0.243	0.066
Korea	0.0016	0.0012	0.926	0.851	0.780	0.719	0.343	0.669

	Mean Return	Standard Deviation	ρ1	ρ2	ρ3	ρ4	ρ12	ρ24
Malaysia	0.0017	0.0006	0.909	0.800	0.680	0.572	-0.077	-0.311
Mexico	0.0020	0.0013	0.941	0.868	0.807	0.774	0.391	0.164
Nigeria	0.0058	0.0019	0.921	0.835	0.747	0.648	0.670	0.010
Philippines	0.0012	0.0012	0.882	0.765	0.648	0.601	0.213	0.282
Poland	0.0010	0.0005	0.857	0.743	0.605	0.479	-0.064	0.037
South Africa	0.0021	0.0003	0.802	0.677	0.501	0.372	0.022	0.789
Thailand	0.0030	0.0022	0.942	0.880	0.814	0.756	0.515	0.366
Turkey	0.0034	0.0018	0.826	0.695	0.589	0.519	0.188	0.123
Venezuela	0.0018	0.0014	0.974	0.947	0.932	0.923	0.606	-0.427
World	0.0023	0.0007	0.997	0.994	0.993	0.993	0.968	0.987
Zimbabwe	0.0544	0.0264	0.908	0.852	0.817	0.781	0.482	0.383

Variable	Mean Return	Standard Deviation	Autocorrelation					
			ρ1	ρ2	ρ3	ρ4	ρ12	ρ24
SPREAD	0.0063	0.0014	0.9554	0.9053	0.8590	0.8303	0.6878	0.5184
USTB90	0.0046	0.0012	0.9757	0.9432	0.9432	0.8822	0.5953	0.1140
World dividend yield	0.0023	0.0007	0.9900	0.9809	0.9722	0.9616	0.9184	0.9728

Notes: The statistics are based on monthly data from February 1976 to 2000:7. The country returns are calculated in U.S. dollars in excess of the holding period returns on the 90 day Treasury bill rate. The dividend yields are the average (over the past year) monthly dividend divided by the current month price level. The returns and dividend yields are from Standard and Poor Emerging Market Price Index. The instrumental variables are the return for holding a 90 day U.S. Treasury bill, the yield on Moody's Baa rated bonds less the yield on Moody's Aaa rated bonds (spread), and the dividend yield on the Standard and Poor's 500 stock index less the return 90 day bill.

TABLE 7.2 Correlations of Country Returns and Instrumental Variables

A. Equity Returns

Portfolio	Arg	Bra	Chi	Cze	Eafe	Gre	Hun	Indo	Ind	Jap	Jor	Kor	Mal	Mex	Nig	Phi	Pol	Saf	Tha	Tur	US	Ven	Wld	Zim
Argentina	1.00																							
Brazil	0.60	1.00																						
Chile	0.61	0.61	1.00																					
Czech-Republic	0.31	0.37	0.25	1.00																				
EAFE	0.51	0.48	0.47	0.25	1.00																			
Greece	0.30	0.25	0.37	0.36	0.36	1.00																		
Hungary	0.52	0.50	0.43	0.54	0.45	0.44	1.00																	
Indonesia	0.36	0.39	0.51	0.21	0.44	0.24	0.36	1.00																
India	0.13	0.24	0.39	0.38	0.20	0.32	0.32	0.20	1.00															
Japan	0.35	0.32	0.35	0.13	0.80	0.10	0.15	0.33	0.16	1.00														
Jordan	-0.01	0.10	0.14	-0.02	0.09	0.27	0.04	0.03	0.15	-0.09	1.00													
Korea	0.21	0.22	0.32	0.16	0.48	0.24	0.17	0.42	0.15	0.59	0.01	1.00												
Malaysia	0.42	0.29	0.53	0.35	0.44	0.25	0.41	0.56	0.33	0.29	-0.06	0.36	1.00											
Mexico	0.71	0.63	0.51	0.28	0.51	0.29	0.59	0.28	0.28	0.34	0.07	0.19	0.31	1.00										
Nigeria	-0.14	0.18	0.08	0.08	-0.13	-0.07	0.06	0.13	0.01	-0.16	-0.03	-0.06	0.08	0.04	1.00									
Philippines	0.48	0.40	0.56	0.22	0.54	0.24	0.43	0.66	0.13	0.39	-0.08	0.37	0.68	0.39	0.10	1.00								
Poland	0.46	0.42	0.38	0.58	0.39	0.28	0.63	0.20	0.26	0.23	0.07	0.24	0.38	0.54	0.05	0.33	1.00							
South Africa	0.56	0.44	0.55	0.24	0.58	0.41	0.45	0.35	0.19	0.46	-0.09	0.42	0.51	0.53	-0.06	0.59	0.41	1.00						
Thailand	0.41	0.34	0.45	0.16	0.52	0.16	0.29	0.62	0.13	0.44	-0.02	0.62	0.69	0.32	0.08	0.74	0.34	0.62	1.00					
Turkey	0.28	0.43	0.38	0.15	0.28	0.25	0.41	0.21	0.19	0.14	0.13	0.08	0.08	0.36	-0.15	0.20	0.30	0.29	0.11	1.00				

United States	0.50	0.43	0.46	0.16	0.69	0.29	0.51	0.46	0.12	0.43	0.08	0.32	0.44	0.48	0.02	0.54	0.43	0.49	0.55	0.25	1.00			
Venezuela	0.34	0.34	0.35	0.40	0.19	0.20	0.27	0.17	0.27	0.07	0.19	0.10	0.32	0.23	0.06	0.27	0.25	0.22	0.13	0.15	0.22	1.00		
World	0.55	0.49	0.51	0.23	0.92	0.36	0.52	0.48	0.18	0.68	0.09	0.43	0.48	0.55	−0.07	0.59	0.44	0.59	0.58	0.29	0.91	0.22	1.00	
Zimbabwe	0.27	0.33	0.41	0.25	0.22	0.12	0.23	0.31	0.35	0.27	0.04	0.42	0.29	0.23	0.10	0.28	0.30	0.26	0.28	0.13	0.25	0.25	1.00	
Ave. Cor.	0.41	0.41	0.44	0.29	0.44	0.29	0.40	0.37	0.25	0.32	0.09	0.31	0.39	0.40	0.05	0.42	0.37	0.42	0.40	0.25	0.42	0.26	0.47	0.29

B. Dividend Yields

Country	Arg	Bra	Chi	Cze	Gre	Hun	Indo	Ind	Jor	Kor	Mal	Mex	Nig	Phi	Pol	SAf	Tha	Tur	Ven	Wld	Zim
Argentina	1.00																				
Brazil	0.40	1.00																			
Chile	−0.09	0.46	1.00																		
Czech-Republic	−0.04	0.64	0.43	1.00																	
Greece	−0.15	−0.55	0.05	−0.45	1.00																
Hungary	0.47	−0.26	−0.44	−0.32	0.39	1.00															
Indonesia	−0.31	−0.27	0.21	0.13	0.35	0.02	1.00														
India	−0.07	0.60	0.66	0.47	−0.16	−0.44	0.23	1.00													
Jordan	−0.27	−0.06	0.10	−0.16	−0.02	−0.37	−0.25	0.05	1.00												
Korea	−0.48	−0.24	0.36	−0.09	0.50	−0.24	0.53	0.31	0.00	1.00											
Malaysia	0.10	0.62	0.45	0.60	−0.31	−0.22	0.36	0.66	−0.31	0.18	1.00										
Mexico	0.32	0.58	0.30	0.59	−0.31	0.03	0.07	0.27	−0.36	−0.10	0.64	1.00									
Nigeria	0.30	−0.17	−0.73	−0.26	−0.28	0.35	−0.18	−0.51	−0.27	−0.45	−0.13	−0.06	1.00								
Philippines	−0.06	0.52	0.43	0.48	−0.55	−0.49	0.24	0.61	0.04	0.17	0.71	0.33	−0.04	1.00							
Poland	0.26	0.29	0.36	0.19	0.27	0.38	0.34	0.42	−0.39	0.15	0.40	0.14	−0.30	0.12	1.00						

(continued)

TABLE 7.2 (continued) Correlations of Country Returns and Instrumental Variables

B. Dividend Yields

Country	Arg	Bra	Chi	Cze	Gre	Hun	Indo	Ind	Jor	Kor	Mal	Mex	Nig	Phi	Pol	SAf	Tha	Tur	Ven	Wld	Zim
South Africa	0.30	0.65	0.24	0.41	-0.41	0.01	-0.35	0.37	-0.03	-0.09	0.35	0.60	-0.02	0.41	0.08	1.00					
Thailand	-0.55	-0.38	0.34	0.07	0.40	-0.20	0.81	0.24	-0.14	0.72	0.26	-0.08	-0.34	0.25	0.22	-0.33	1.00				
Turkey	0.20	0.21	0.34	-0.08	0.53	0.34	-0.03	0.12	0.01	0.13	0.04	0.21	-0.37	-0.31	0.41	0.11	-0.10	1.00			
Venezuela	0.56	0.56	-0.30	0.15	-0.62	0.13	-0.67	-0.07	0.07	-0.66	0.03	0.19	0.46	0.15	-0.12	0.53	-0.84	-0.09	1.00		
World	0.02	-0.60	-0.24	-0.51	0.85	0.61	0.28	-0.43	-0.29	-0.30	-0.16	0.07	-0.68	0.16	-0.40	0.25	0.45	-0.48		1.00	
Zimbabwe	0.46	-0.06	-0.30	-0.40	0.48	0.67	-0.03	-0.15	-0.33	-0.07	0.00	-0.02	0.30	-0.42	0.34	-0.08	-0.22	0.51	0.11	0.67	1.00

Other Instrumental Variables

	Spread	UStb 90	Wldy	Wlss
Spread	1			
UStb90	0.48	1		
Wldy	0.68	0.52	1	
Wlss	0.09	-0	-0.04	1

Notes: The correlations are based on monthly data from 1985:2 to 2000:7. The country returns are calculated in U.S. dollars in excess of the holding period return on the Treasury bill rate that is closest to 90 days to maturity. The dividend yields are the average (over the past year) monthly dividends divided by the current monthly price level. The returns and dividends yields are from Standard and Poor Emerging Market Price Indices. The instrumental variables are the excess returns on the world index (Wlss), the U.S. 90 day treasury bill rate, the yield on Moody's Baa rated bonds less the yield on Moody's Aaa rated bonds (spread), and the dividend yield on the Standard and Poor's stock index less the return on the 90 day treasury bill rate.

The conditioning information variables are available at time $t-1$ and are used to predict the next period returns for time t. Table 7.3 reports the results for these regressions.

Table 7.4 is constructed in the same way as Table 7.3. However, local variables are also used. Column Z_{t-1} (1) is the same as the regression in Table 7.3 and Column Z_{t-1} (8) uses local variables only. Columns Z_{t-1} (2) to Z_{t-1} (7) use mixed variables, which include both common and local variables.

7.3.4.2 Conditional CAPM

The GMM procedure was used to estimate the parameters of the conditional CAPM set in Equation 7.1. The GMM estimator selects parameter estimates so that the correlations between the instruments and disturbances are as close to zero as possible as defined by the criteria function. There are ℓ information variables and $\ell \times (2n + 1)$ orthogonality conditions and there are $\ell \times (n + 1)$ parameters to estimate leaving $\ell \times n$ overidentifying conditions to estimate. For each country there are $5(2 + 1) = 15$ orthogonality conditions and $5(1 + 1)$ parameters to estimate leaving 5 overidentifying conditions. In a situation where there are more moment conditions than parameters, not all moment restrictions will be satisfied. The weighting matrix determines the relative importance of the various moment conditions.

$$\left[\sum_{t=1}^{T} \left(\varepsilon_t \otimes z_{t-1} \right)' \left(\varepsilon_t \otimes z_{t-1} \right) \right]^{-1}$$

Cliff (2000) points that an "optimal" weighting matrix requires an estimation of the parameter vector, yet at the same time estimating the parameters requires a weighting matrix. To solve this dependency, common practice is to set the initial weighting matrix to the identity then calculate the parameter estimate. A new weighting matrix is calculated with the last parameter estimates, then new estimates with updated weighting matrix. The iterating procedure continues until the change in the objective function is sufficiently small. The system of equations was used to estimate the parameters in Equation 7.7. From the results u_t and u_{mt}, country and world error terms as defined in Equations 7.3 and 7.4 were determined. The average conditional covariance was obtained by $u_t \times u_{mt}$ multiplied by 1000 based on a single country estimation with common instrument set.[*]

---

[*] See Harvey (1991) Table V.

TABLE 7.3 Regressions of Country Excess Returns on the Common Instrumental Variables

Portfolio	δ_0 Constant	δ_1 Wlss (−1)	δ_2 Spread (−1)	δ_3 Ustb90 (−1)	δ_4 Wldy (−1)	R^2
Argentina	0.053 [1.0750]	0.278 [0.8316]	−6.833 [−0.6756]	−8.546 [−0.9512]	18.554 [0.9613]	0.012
Brazil	0.022 [0.6422]	0.270 [1.1343]	−4.663 [−0.6466]	1.595 [0.2489]	0.834 [0.0606]	0.006
Chile	0.021 [1.3035]	0.146 [1.3286]	4.421 [−0.4045]	3.926 [−2.5602]**	8.435 [2.2078]**	0.040
Czech-Republic	0.260 [1.3561]	0.245 [0.9215]	−14.183 [−0.4902]	−38.987 [−1.5199]	−25.551 [−0.9133]	0.056
EAFE	−0.023 [0.2973]	0.009 [0.2025]	9.010 [1.3570]	−5.822 [−3.1711]	−0.240 [1.2981]	0.046
Greece	0.046 [2.2397]**	0.069 [0.4928]	−1.141 [−0.2689]	−2.490 [−0.6605]	−8.104 [−1.0007]	0.028
Hungary	0.244 [1.0019]	0.365 [0.9863]	−14.147 [−0.5595]	−33.369 [−1.0509]	−20.087 [−0.6304]	0.034
India	0.001 [0.0558]	0.167 [1.4264]	−1.434 [−0.4049]	0.239 [0.0758]	3.824 [0.5661]	0.008
Indonesia	−0.076 [−0.8961]	0.393 [1.1972]*	25.370 [1.5733]	−25.065 [−1.8104]*	9.925 [0.4211]	0.057
Japan	−0.042 [−0.9656]	0.018 [0.2005]	15.923 [2.3455]**	−10.219 [−2.8744]**	−3.531 [0.6821]	0.066
Jordan	0.004 [0.4251]	0.052 [0.7273]	−4.401 [−1.9605]*	−0.692 [−0.3356]	11.277 [2.5036]**	0.028
Korea	−0.018 [−0.7926]	0.007 [0.0446]	9.934 [2.1054]**	−9.839 [−2.3478]**	1.891 [0.2100]	0.024
Malaysia	−0.021 [−0.5494] [1.2723]	−0.092 [−0.5066] [2.6650]**	12.252 [1.6036] [−0.4368]	−16.841 [−2.2706]** [−1.5553]	8.831 [0.5869] [0.9643]	0.039
Mexico	0.036 [1.272325]	0.522 [2.665031]	−2.592 [−0.436826]	−8.198 [−1.555314]	10.920 [0.964291]	0.044
Nigeria	0.033 [0.5842]	0.134 [0.4957]	−9.089 [−0.7978]	2.183 [0.1974]	5.582 [0.2488]	0.005

Country	$\delta_{j,0}$	$\delta_{j,1}$	$\delta_{j,2}$	$\delta_{j,3}$	$\delta_{j,4}$	R^2
Philippines	−0.085 [−2.2575]**	0.214 [1.1852]*	22.648 [2.9773]**	−22.070 [−2.9886]**	22.303 [1.4887]	0.129
Poland	0.047 [0.2093]	0.290 [0.6601]	−43.652 [−0.2740]	−0.263 [−0.0070]	−18.702 [−0.5302]	0.010
South Africa	0.077 [0.5353]	−0.393 [−1.4572]	−3.738 [−0.0382]	−24.958 [−1.0184]	17.925 [0.8279]	0.053
Thailand	−0.036 [−1.6397]	−0.041 [−0.2733]	10.866 [2.4112]**	−15.736 [−3.9326]	16.437 [1.9121]*	0.058
Turkey	0.014 [0.1712]	0.524 [1.5399]	−8.747 [−0.5460]	19.115 [1.3995]	−16.340 [−0.5922]	0.028
United States	0.009 [0.9845]	−0.063 [−1.0013]	2.389 [1.2558]	−3.917 [0.9084]	−2.318 [0.2502]	0.022
Venezuela	−0.069 [−1.2960]	−0.209 [−0.8148]	11.563 [1.0691]	5.612 [0.5345]	−10.691 [−0.5019]	0.014
World	0.005 [0.5364]	−0.026 [−0.4366]	3.052 [1.7152]*	−5.271 [−3.3349]**	3.079 [0.9068]	0.038
Zimbabwe	0.020 [0.8454]	0.200 [1.2445]	−3.965 [−0.8142]	−3.628 [−0.8387]	9.374 [1.0088]	0.015

Notes: The regressions are based on monthly data from 1985:2 to 2000:7. The county returns are calculated in U.S. dollars in excess of the holding period return on the treasury bill that is closest to 90 days to maturity. The equity data are from Standard and Poor Emerging Markets Price indices. The *t*-statistics are in brackets. The model estimated is

$$r_{j,t} = \delta_{j,0} + \delta_{j,1} wlss_{t-1} + \delta_{j,2} spread_{t-1} + \delta_{j,3} ustb90_{t-1} + \delta_{j,4} wldy_{t-1} + \varepsilon_{j,t}$$

The instrumental variables are a constant, the excess return on the world index (Wlss), U.S. 90 day treasury bill rate (ustb90), the yield on Moody's Baa rated bonds less the yield on Aaa rated bonds (spread), and the dividend yield on the Standard and Poor's 500 stock index less the return on a 90 day bill.

* Indicates significant at 5%.
** Indicates significant at 1%.

TABLE 7.4 International Evidence on the Predictability of Equity Returns Using Common and Country-Specific Instrumental Variables

	Z_{t-1} (1)	Z_{t-1} (2)	Z_{t-1} (3)	Z_{t-1} (4)	Z_{t-1} (5)	Z_{t-1} (6)	Z_{t-1} (7)	Z_{t-1} (8)
	C, Wlss (t–1), Spread (t–1), Ustb90 (t–1), Wldy (t–1)	C, Rjt (t–1), Spread (t–1), Ustb90 (t–1), Wldy (t–1)	C, Rjt (t–1), Spread (t–1), Ustb90 (t–1), Ldy (t–1)	C Rjt (t–1), Spread (t–1), Ustb90 (t–1), Ldy (t–1), Wldy (t–1)	C Rjt (t–1), Spread (t–1), Ustb90 (t–1), Ldy (t–1), Ex (t–1)	C Rjt (t–1), Spread (t–1), Ustb90 (t–1), Ldy (t–1), Lir (t–1)	C Rjt (t–1), Spread (t–1), Lir (t–1), Ldy (t–1)	C, Rjt (t–1), Ldy (t–1), Lir (t–1), Ex (t–1)
Argentina	−0.018	−0.011	−0.012	−0.016	0.041	—	—	—
Brazil	0.004	−0.012	−0.018	−0.018	−0.035	−0.091	−0.074	−0.067
Chile	0.055	0.072	0.080	0.064	0.078	−0.026	−0.035	−0.038
Czech-Republic	0.005	0.014	−0.009	−0.025	−0.021	0.007	−0.050	−0.055
Greece	0.009	0.011	−0.005	0.013	−0.004	−0.018	−0.010	−0.017
Hungary	−0.018	−0.032	−0.045	−0.045	−0.050	−0.060	−0.045	−0.035
India	−0.014	−0.007	0.007	0.004	0.006	0.025	−0.003	0.003
Indonesia	0.011	0.054	0.060	0.051	0.065	0.055	0.035	0.014
Jordan	−0.004	−0.004	−0.013	−0.008	0.007	—	—	—
Korea	0.011	0.041	0.043	0.038	0.050	0.023	0.030	−0.001
Malaysia	0.018	0.025	0.037	0.036	0.042	0.065	0.024	0.103

Mexico	0.034	0.082	0.101	0.096	-0.015	0.066	0.028	0.030
Nigeria	-0.017	-0.019	-0.019	-0.025	-0.050	—	—	—
Philippines	0.110	0.160	0.176	0.179	0.165	0.171	0.153	0.140
Poland	-0.044	-0.019	-0.011	-0.018	-0.024	-0.012	0.000	-0.013
South Africa	0.001	-0.009	0.101	0.197	0.092	0.289	0.300	0.290
Thailand	0.044	0.052	0.044	0.049	0.055	0.050	0.040	0.025
Turkey	0.0017	0.003	0.039	0.044	0.044	—	—	—
Venezuela	-0.007	-0.011	-0.012	-0.018	0.016	0.023	0.027	0.021
Zimbabwe	0.010	0.085	0.089	0.084	0.080	0.097	0.102	0.099

Notes: The country returns are calculated in U.S. dollars in excess of the holding period on the U.S. Treasury bill that is closest to 90 day maturity. The equity data are from Standard and Poor's Emerging Market Price Indices. The regressions are estimated with eight different sets of conditioning information. The instrumental variables are excess world returns ($Wlss_{t-1}$), U.S. 90 day treasury bill rate ($Ustb90_{t-1}$), the yield on the Moody's rated bonds less the yield on Aaa rated bonds ($Spread_{t-1}$), the U.S. dividend yield in excess of the 90 day treasury bill rate ($Wldy_{t-1}$), the equity return for each country (Rjt_{t-1}), the dividend yield for each country (Ldy_{t-1}), the return on the U.S. exchange rate for each country (Ex_{t-1}), and level of short-term interest rates in each country (Lir_{t-1}).

7.4 RESULTS

7.4.1 Statistical Description

Table 7.1 represents summary statistics for asset excess returns and the instrumental variables over the period 1985:2–2000:6. All returns are calculated in U.S. dollars. Seven of the markets including the United States have mean excess returns higher than the world index. The world portfolio has the lowest standard deviations when all countries are considered. Harvey (1991) found the same results for developed markets. In addition, evidence shows that emerging markets are more volatile than the three developed markets in the sample except for Jordan with a standard deviation of 0.0446 that is lower than that of Japan and the EAFE. Japan has a relatively high standard deviation when compared with other developed markets. Nigeria has the lowest mean excess return (−0.0011) and a high standard deviation.

The highest mean excess return over the sample period is from Chile and Argentina has the highest standard deviation. Chile, Greece, Malaysia, Mexico, the Philippines, Thailand, and Zimbabwe display significant first-order autocorrelation. Autocorrelations measure the persistence or predictability of the market returns based on past market returns. Further, Zimbabwe continues to show significant autocorrelations up to the fourth lag.

7.4.2 What Are the Return and Risk Characteristics of These Markets?

The analysis suggests that emerging markets offer diversification benefits to investors given the risk–return characteristics of these markets but it does not follow that all markets offer the same benefits. The Asian markets had recently experienced financial crises and the Philippines is the only Asian country offering high returns for a "reasonable" level of risk. The Latin American countries offer high returns at varying levels of risk. The results of the African and European countries are suspect for various reasons. Five of the markets, South Africa, Poland, Hungary, the Czech Republic, and Turkey, have shorter observation periods and during this time they were affected by both the Asian and the Russian crises. Furthermore, some transition economies have just moved from closed economies to open economies and this shift is not an easy move. South Africa recently gained its independence and its variance is comparable to Japan's but its returns are negative.

The means, standard deviations, and autocorrelations of the countries' dividend yields are shown in panel B of Table 7.1. Zimbabwe, Argentina, Greece, and Thailand have both high means and standard deviations of dividend yields. Studies have shown that dividend yields help to forecast future returns. Autocorrelation coefficients are significant for all the countries up to lags of 4 months. They remain significant for the world market up to 24 month lags. Panel C of Table 7.1 provides summary statistics for the common instrumental variables. All the instrumental variables show high autocorrelations up to 24 month lags.

7.4.3 How Do These Markets Correlate with Each Other and the Major World Markets?

The cross-country correlations are presented in Table 7.2. The sample period is February 1985 to June 2000 (185 observations) for 14 developing markets and the EAFE, Japan, the United States, and the world market. Six countries, the Czech Republic, Hungary, Indonesia, Poland, South Africa, and Turkey have 77 observations. The correlations between the world market index and the developed markets are very high ranging from 68% to 92%. This suggests that these markets are integrated with the world market. Correlations between the world market and emerging markets vary with seven countries exhibiting a correlation of 25% or less. South Africa has the highest correlation of 59% and Nigeria has the lowest at −0.07%. South Africa also exhibits a high correlation with the United States and Japan. In addition, its correlation with other emerging markets is relatively high with an average of 42%. Nigeria also has a negative correlation with Argentina, the EAFE, Greece, Japan, Jordan, Korea, South Africa, and Turkey. Both Jordan and Nigeria are isolated from their regional economic groupings and this may explain their negative low correlations with other countries in the sample. Their low correlations may also indicate lack of integration with world markets. Their average correlations are very low, Jordan with 0.09% and Nigeria with 0.05%. The European markets; the Czech Republic, Greece, Hungary, and Poland have correlations above 25%. Another striking feature was the low correlations between countries in the same regional groupings. Buckberg (1995) reports similar results for the period 1985–1991.

During this period, only seven countries had correlations above 25% with the world. In our analysis, 13 countries have correlations above 25% with the world. This suggests that emerging markets are moving toward integration with developed markets. The average correlation between developed markets in the sample is 80.4% and for all the countries in the sample is 34%. These results suggest there are still investment opportunities in developing markets. The four countries with the highest market capitalizations, Brazil, Greece, Korea, and South Africa, have high correlations with the developed markets. Markets that were partially open during the period in question also have low correlations with world markets, for example, India, Jordan, Nigeria, Venezuela, and Zimbabwe. These countries have the lowest market capitalizations.

The correlations are based on monthly data from 1985: month 2 to 2000: month 6. The country returns are calculated in U.S. dollars in excess of the holding period return on the Treasury bill rate that is closest to 90 days to maturity. The dividend yields are the average (over the past year) monthly dividends divided by the current monthly price level. The returns and dividends yields are from Standard and Poor Emerging Market Price Indices. The instrumental variables are: the excess returns on the world index (Wlss), the U.S. 90 day Treasury bill rate, the yield on Moody's Baa rated bonds less the yield on Moody's Aaa rated bonds (spread) and the dividend yield on the Standard and Poor's stock index less the return on the 90% with both the world market and the United States. However, their correlation with Japan is low, less than 25% for all the four countries. The Latin Americas have relatively high correlations with the world, all above 49% except for Venezuela with a 22% correlation. This may suggest that these markets are integrated with the world markets. Furthermore, the Latin Americas group also shows a high level of correlation among themselves, for example, Argentina has a correlation of 72% with Mexico, 61% with Brazil, and 60% with Chile. The Latin Americas and the Asian markets are more correlated with the United States than any other region in the sample. The Asian markets are also highly correlated among themselves; yet India exhibits very low correlations with both developed and emerging markets. The average correlation for India is 25%. In Sub-Saharan Africa, the correlations between the three markets in the sample are very low, and Nigeria and South Africa have a negative correlation.

Zimbabwe has an average correlation of 29% with the markets in the sample. Table 7.2 Panel A provides the dividend yield correlations. Most of the countries exhibit negative correlation with each other.

Figure 7.1 plots a "traditional" graph of mean excess returns against standard deviation in the 24 markets. Fourteen markets have 185 observations, Turkey has 162, Indonesia has 126, and Czech, Hungary, Poland,

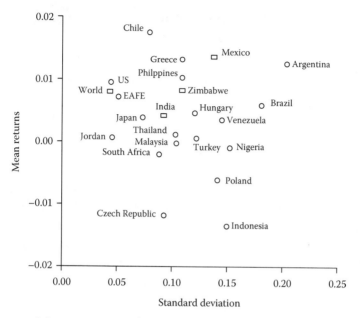

FIGURE 7.1 Mean returns and variances for emerging markets and developed markets 1985–2000.

and South Africa have 78 observations. The developed markets are clustered together with low standard deviations and low returns. The Latin American markets have high excess returns and in the case of Argentina and Brazil, high volatility. Greece, the Philippines, and Zimbabwe have similar standard deviations but the return patterns differ. The Asian markets—India, Thailand, Korea, and Malaysia—are clustered together with low excess returns and relatively high variances. Chile has the highest excess return in the sample and its variance is slightly above that of developed markets. Jordan on the other hand has low excess returns and low volatility about the same as that of developed markets. Of the five markets with a short observation period, Hungary is the only country with a positive reward/risk ratio. South Africa and the Czech Republic have negative excess returns and low risk while Poland and Indonesia offer the similar returns and their risk is high.

7.4.4 Predictability of Expected Returns Using World Factors

In this section, we report the results of a test of the Null hypothesis that the security price behavior of emerging markets is consistent with those of world markets. The assumption is that the ICAPM validly describe the return behavior of mature markets. Are market returns in emerging markets consistent with the model's predictions? The results of regressions of country returns on a world common set of instrumental variables are shown in Table 7.3. The regressions are based on monthly data from 1985:2 to 2000:6 for 18 markets and for the six countries with a shorter observation period, the data is from 1994:2 to 2000:6. The results show a pattern of predictable variation in the emerging market returns. The results of regressions of the emerging market returns on the four information variables are also detailed. The world dividend yield is insignificant for all the countries except for Greece. The constant is significant at a 5% level in the Philippines only. The world market portfolio betas have little influence on the expected returns in the emerging markets. The adjusted R^2 for Argentina, India, Jordan, Nigeria, the Unites States, and Venezuela are negative. The Philippines has the highest adjusted R^2 of 11%. The world excess return beta is significant in explaining the returns for Mexico only and the rest of the sample show insignificant effects in terms of the impact of the world excess return. The yield between the Moody's Baa and Aaa bond rates (spread) beta is statistically significant in seven markets, and for Chile and the world market at a 10% level of significance, and for EAFE, Japan, Korea, the Philippines, and Thailand at a 5% level of significance. The U.S. 90 day Treasury bill rate beta is significant at a 5% level for Japan, Korea, and Malaysia,

7.4.5 Predictability of Returns Using Common and Local Instrumental Variables

The hypothesis that emerging markets are now integrated into world markets is tested using world factor variables and local information variables. Table 7.4 presents the results of international evidence on the predictability of returns using common and country-specific instrumental variables. Z_{t-1} (1) comprises the common instrument set, that is, world excess returns, U.S. 90 day treasury bill rate, the spread between the Moody's Baa and Aaa bond rates, and the world dividend yield. In Z_{t-1} (2) the world excess return is replaced by the country excess return. Z_{t-1} (3) uses the same variables as in (2) but substitutes the world dividend yield with the country dividend yield. Z_{t-1} (4) uses the same variables as in (3) plus the world dividend yield. Z_{t-1} (5) uses the same variables as in (3) plus returns on the U.S. exchange rate for each country. Z_{t-1} (6) comprises the country excess return index, local dividend yield, local interest rate, and the U.S. 90 day Treasury bill rate. Z_{t-1} (7) uses the country excess return, local short-term interest rate and the spread as in Z_{t-1} (1). Z_{t-1} (8) uses all local instruments, excess returns, local short-term interest rates, and the local dividend yields. Columns Z_{t-1} (2) to Z_{t-1} (7) have mixed variables for common and local variables.

Argentina, Jordan, Nigeria, and Turkey did not have a local interest variable available in the data and therefore use the return on the exchange rate local dividend yield and the country excess return as the local information set. The adjusted coefficients of determination, in columns Z_{t-1} (2)–Z_{t-1} (4) do not show any statistical significance for Argentina. However, column Z_{t-1} (5) contains three local variables for Argentina, and there is a modest improvement on the adjusted R^2. For Brazil, Chile, the Czech Republic, Greece, Korea, and Thailand the world common variables regression capture the bulk of the predictable variation in country returns. Column Z_{t-1} (8) with local instrumental variable regressions shows decreasing R^2, which are negative for all the aforementioned countries except for Thailand. The R^2 for Thailand decreased from 4.4% to 1.3%.

The impact of using mixed variables differs across the sample. When the local excess returns is used in regression Z_{t-1} (2) the explanatory power is increased for Chile, the Czech Republic, Greece, Indonesia, Korea, Malaysia, Mexico, the Philippines, Thailand, Turkey, and Zimbabwe. The explanatory power increases for Argentina, India, and Poland but it remains negative. For Brazil, Hungary, Nigeria, South Africa, Turkey, and Venezuela the explanatory power decreases. The local excess returns and the local dividend yield increase the explanatory power of the regressions in 11 of the countries in the sample. For Jordan and Poland the R^2 are still negative. The R^2 decreased for seven of the markets in the sample and in six of the markets the R^2 are negative and insignificant. In column Z_{t-1} (4) two local instrumental variables, that is, the local excess return, the local dividend yield and are used and three common variables. The R^2 decreased further, in 13 of the regressions when compared with the previous column which used two local variables and two common variables.

The effect of the explanatory power increase in 14 of the regressions when column Z_{t-1} (1) and Z_{t-1} (5) with three local information variables and two common instruments are compared. In the other six markets, the R^2 decreases substantially showing no explanatory power at all. Column Z_{t-1} (6) shows the effect of the local short-term interest rates. In 12 of the countries, the R^2 is better than the regressions of the first column. However, if it is compared to column Z_{t-1} (5) it appears that the local interest rate has more explanatory power than the exchange rate variable. For example, the

Czech Republic, India, Malaysia, Mexico, the Philippines, South Africa, and Zimbabwe the R^2 increased substantially.

In column Z_{t-1} (7), nine of the regressions show an explanatory power which is better than that of the common instrument regressions. The last column contains regressions of local variables only and some interesting results follow. The markets that are more integrated with world markets have a low and negative R^2, for example; Brazil, Mexico, Chile, Greece, Korea, and Thailand. Those markets that are lackluster have a higher R^2 indicating the effect of local variables.

7.4.6 Conditional Asset Pricing with Time-Varying Moments

Table 7.5 presents the results of estimating Equation 7.7, which allows for time variation in expected returns, conditional covariance, and conditional variance. Tests of the asset pricing restrictions are provided for individual countries. The hypothesis that the world market portfolio is conditionally mean variance efficient is tested. The conditional variances are presented in column 2 of Table 7.5. Panel B presents the results for the six countries with a shorter observation period. The average conditional variance is high for all the developed markets and for seven of the developing markets. In panel B, five countries exhibit high average conditional covariances. The average pricing error and the absolute pricing errors are provided in columns four and five. The γ^2 provides a test of the model's restrictions: a high chi-squared statistics indicates poor fit and a low one indicates a good fit. Three sets of instruments variables are used, the common set; the local instrument A, comprising the common instrument set plus the local dividend yields; and local instruments B include the local excess return in place of the world excess return. The model is rejected by Malaysia using the common instrument set and local instrument set A. Argentina, Greece, Korea, and Thailand reject the model at 5% level of significance when local instrument set B is used. Indonesia rejects the model using the common instrument set. The adjusted coefficients of determination (R^2) which are the result of regressing the model's errors on the common information variables are shown in column six of Table 7.5. A small R^2 indicates that the model fits well. The R^2 are low for 15 countries in the samples. This suggests that the ratio of mean to variance is time varying. Indonesia and Malaysia have high adjusted R^2 and they reject the model in the overidentifying restrictions test using a common instrument set.

TABLE 7.5 Estimation of a Conditional CAPM with Time-Varying Expected Returns Conditional Covariances and Conditional Variances

Portfolio	Average Returns	Average Conditional Covariance	Average Error	Average Absolute Error	R^2	Common Instruments χ^2 [p-Value]	Local Instruments: A χ^2 [p-Value]	Local Instruments: B χ^2 [p-Value]
Argentina	0.0127	0.466426	0.004461	0.125911	−0.016	6.09 [0.2977]	2.21 [0.8990]	13.09 [0.0416]
Brazil	0.0059	1.965704	0.007607	0.130755	−0.005	6.14 [0.2921]	6.58 [0.3619]	5.32 [0.5038]
Chile	0.0175	0.859373	−0.00163	0.063683	0.051	7.75 [0.1708]	0.49 [0.9979]	9.52 [0.1462]
EAFE[a]	0.0072	1.9426	0.0005	0.0390	0.0035	5.73 [0.3330]		
Greece	0.0132	1.0644	0.0053	0.0754	0.0066	6.67 [0.2460]	7.24 [0.2580]	14.29 [0.0270]
India	0.0040	−0.1412	0.0019	0.0738	−0.0156	2.53 [0.7720]	7.68 [0.2628]	4.34 [0.6311]
Japan[a]	0.0038	2.2136	−0.0039	0.0555	−0.0056	4.88 [0.4304]		
Jordan	0.0005	0.2168	0.0005	0.0332	−0.0098	3.89 [0.5659]	5.92 [0.43234]	4.63 [0.5915]
Korea	0.0038	1.4794	−0.0107	0.0806	0.0045	6.23 [0.2794]	7.16 [0.3064]	10.14 [0.1188]
Malaysia	−0.0003	1.7963	−0.0004	0.0736	0.0166	19.37 [0.0016]	18.56 [0.0049]	5.28 [0.5080]
Mexico	0.0137	2.0859	−0.0024	0.0894	0.0306	6.06 [0.3004]	4.52 [0.6064]	3.97 [0.6810]
Nigeria	−0.0011	0.4198	−0.0015	0.0714	−0.0197	2.48 [0.7802]	0.73 [0.9938]	0.70 [0.9943]
Thailand	0.0010	1.9739	0.0280	0.0923	0.0523	5.75 [0.3306]	4.39 [0.6237]	10.57 [0.1026]
Venezuela	0.0035	0.2081	0.0019	0.0999	−0.0054	3.67 [0.6132]	3.39 [0.6407]	2.48 [0.8504]
Zimbabwe	0.0080	0.6294	−0.0134	0.0737	0.0044	3.42 [0.6348]	4.14 [0.6581]	8.26 [0.2200]

(continued)

TABLE 7.5 (continued) Estimation of a Conditional CAPM with Time-Varying Expected Returns Conditional Covariances and Conditional Variances

Panel B

Czech-Republic	-0.0170	0.6979	-0.0182	0.0628	-0.0028	6.7 [0.2440]	5.36 [0.4987]	4.36 [1.628]
Hungary	-0.0001	2.3549	0.0020	0.0810	-0.0271	3.62 [0.2440]	3.22 [0.7809]	4.92 [0.5536]
Indonesia	-0.0232	3.3671	-0.0241	0.1279	0.1054	17.67 [0.0033]	8.58 [0.1989]	5.36 [0.4981]
Poland	-0.0101	2.2269	-0.0124	0.0994	-0.0046	2.87 [.07206]	3.95 [0.6831]	2.83 [0.8302]
South Africa	-0.0021	1.8924	-0.0025	0.0596	0.0141	5.09 [0.4047]	3.78 [0.7063]	3.58 [0.7332]
Turkey	0.0100	2.0844	0.0162	0.1220	-0.0002	5.80 [0.3257]	5.09 [0.5322]	3.20 [0.7837]

Notes: The following system of equations is estimated with the GMM

$$\varepsilon_t = (u_t \, u_{mt} \, h_t) = \left(\begin{array}{c} [r_t - z_{t-1}\delta]' \\ [r_{mt} - z_{t-1}\delta_m]' \\ [u_{mt}^2 z_{t-1}\delta - u_{mt} u_t z_{t-1}\delta_m]' \end{array} \right)'$$

where

r_m is the excess return on the world portfolio

δ represents the coefficients associated with the instrumental variables

u is the forecast error for the country returns

u_m is the forecast error for the world market return

h represents the deviation of the country returns from the model's expected returns

There are three instrumental variable sets Z that are used in the estimation. The common set instrumental variables are a constant, the excess returns on the world index, U.S. 90 day Treasury bill rate, the yield on the Moody's Baa rated bond less the yield on the Aaa rated bonds, and the dividend yield on the Standard and Poor's 500 stock index less the return on the 90 day bill. The local instrument set A is the common set instruments plus the country-specific dividend yield. Instrument set B is the same as set A but the world excess return is replaced by the country-specific excess returns.

7.5 CONCLUSION

Our results show that the asset return behavior in emerging markets is changing even though it differs from country to country. Ostensibly, some markets became more integrated with the world markets while others become more segmented. In most markets, the correlations with developed markets increased. Yet investors still find benefits from investing in these markets because returns remain higher than those available in developed markets. The increase in integration has possibly been due to rising capital flows from developed markets. In markets where integration has increased over the years, the countries are making concerted effort to catch up with the developed world in terms of technology, innovations, and risk management. Further, they are working hard to become part of the economic regional groups. In some of the emerging markets such as Zimbabwe, Jordan, Venezuela, and Nigeria, the markets have become more segmented and isolated from the world markets. This may be because of various market imperfections such as inadequate and inaccurate information, illiquidity, and government intervention. Investing in these markets is risky and the benefits of diversification are minimal.

The predictability of returns using common instruments suggests evidence of time-varying expected returns. Further, the hypothesis that conditional mean returns are constant is rejected. Markets with high capitalizations and high correlations with world markets are affected more by world factors than local factors. This leads us to the conclusion that most markets in Asia, Latin America, and Europe are integrated with the world economy. However, South Africa, with a high market capitalization showed that it is influenced more by local factor than world factors.

REFERENCES

Bar-Yosef, S. and Kolodny, R. (1976) Dividend policy and capital market theory. *Review of Economics and Statistics*, 58(2): 181–190.

Blume, M.E. (1971) On the assessment of risk. *Journal of Finance*, 26(1): 1–10.

Bollerslev, T., Engle, R.F., and Woodridge, J.M. (1988) Capital asset pricing model with time varying covariance. *Journal of Political Economy*, 96(1): 116–131.

Buckberg, E. (1995) Emerging stock markets and international asset pricing. *The World Bank Economic Review*, 9(1): 51–74.

Campbell, J.Y. and Shiller, R. (1988) The dividend price ratio and expectation of future dividend and discount factors. *Review of Financial Studies*, 1(3): 195–228.

Chen, N., Roll, R., and Ross, S. (1986) Economic forces and the stock market. *Journal of Business*, 59(3): 383–403.

Christie, W.G. (1990) Dividend yield and expected returns: The zero-dividend puzzle. *Journal of Financial Economics*, 28(1–2): 95–126.

Cliff, T.M. (2000) *GMM and MINZ Program Libraries for MATLAB*. Krannert graduate School of Management, Purdue University, West Lafayette, IN.

Crockett, J.A. and Friend, I. (1988) Dividend policy in prospective: Can theory explain behaviour. *Review of Economics and Statistics*, 70(4): 603–613.

Fama, E. and French, K. (1988) Dividend yields and expected stock returns. *Journal of Financial Economics*, 22(1): 3–26.

Fama, E.F. and French, K.R. (1989) Business conditions and expected returns on stocks and bonds. *Journal of Financial Economics*, 25(3): 23–49.

Fama, E. and Shwert, G.W. (1977) Asset returns and inflation. *Journal of Financial Economics*, 5(2): 115–146.

Francis, J. and Fabozzi, F. (1979) The effects of changing macroeconomic conditions on the parameters of the single index market model. *Journal of Financial and Quantitative Analysis*, 14(2): 351–360.

Gonedes, N.J. (1973) Evidence on the information contents of accounting numbers: Accounting based and market based estimates of systematic risk. *Journal of Financial and Quantitative Analysis*, 8(4): 407–444.

Hansen, L.P. (1982) Large sample properties of generalized methods moments estimators. *Econometrica*, 50(4): 1029–1054.

Harvey, R.C. (1989) Time-varying conditional in the test of asset pricing models. *Journal of Financial Economics*, 24(2): 289–317.

Harvey, R.C. (1991) The world price of covariance risk. *Journal of Finance*, 46(1): 111–157.

Harvey, R.C. and Bekaert, G. (1997) Emerging equity market volatility. *Journal of Financial Economics*, 43(1): 29–77.

Howton, W.S. and Peterson, R.D. (1999) A Cross–sectional empirical test of a dual state multifactor pricing model. *Financial Review*, 34(3): 47–64.

Jagganathan, R. and Wang, Z. (1996) The conditional CAPM and the cross-section of expected returns. *Journal of Finance*, 51(1): 3–53.

Keim, D. and Stambaugh, R.F. (1986) Predicting returns in the stock markets and bond markets. *Journal of Financial Economics*, 17(2): 357–390.

Lintner, J. (1965) The valuation of risk assets and the selection of risky investments in portfolios and capital budgets. *Review of Economics and Statistics*, 47(1): 13–37.

Officer, R. (1973) The variability of the market factor of the New York stock exchange. *Journal of Business*, 46(3): 434–453.

Rozeff, M. (1984) Dividend yields are equity risk premium. *Journal of Portfolio Management*, Fall: 68–75.

Sharpe, W. (1964) Capital asset prices: A theory of market equilibrium under condition of risk. *Journal of Finance*, 19(3): 425–442.

Schwert, G.W. and Seguin, P.J. (1990) Heteroskedasticity in stock returns. *Journal of Finance*, 45(4): 1129–1156.

CHAPTER 8

Do Jumps Matter in Emerging Market Portfolio Strategies?

Massimo Guidolin and Elisa Ossola

CONTENTS

8.1 INTRODUCTION

Are emerging stock markets "special" and—if so—why are they different from the stock markets of developed countries/regions? This chapter uses a simple asset allocation framework to show that the main difference between developed and emerging stock markets lies in the higher "exposure" (or vulnerability) of emerging stock markets to systemic jumps, i.e., events able to affect worldwide stock markets causing sudden and large simultaneous price changes. Equivalently, one may say that emerging stock markets are characterized by strong departures from the standard multivariate independent and identically distributed (IID) Gaussian benchmark implied by the classical benchmark of a pure-diffusion model, in which stock prices are driven by a simple geometric Brownian motion (GBM) with constant means and second moments. These departures would be deeper and stronger than the departures—sometimes simply called "nonnormalities"—which are typical of stock returns in developed markets. In particular, this chapter focuses on three kinds of departures: asymmetries (as measured by skewness), coasymmetries (as measured by coskewness), and thick tails, i.e., tails which are "fatter" than the benchmark Gaussian tails, with the implication that higher probability mass is attached to extreme returns (both large and small). This chapter shows that a peculiar class of continuous-time models—multivariate jump-diffusions—may be used to fit the main statistical features of international stock returns (including both developed and emerging markets) and imply optimal diversification decisions that differ in important ways from the portfolio choices one would derive under the classical multivariate IID Gaussian framework.

    There are two logical threads that run through this chapter. First, we present two alternative statistical frameworks—multivariate pure- and jump-diffusion models—and proceed to show that a benchmark multivariate IID Gaussian framework is clearly inadequate to capture the main properties of most (if not all) of the international stock return series we study in this chapter. A simple pure-diffusion, implicitly mean–variance, framework is at odds with the basic statistical properties of the data as the empirical evidence suggests that both marginal and joint (unconditional) distributions for stock returns display nonzero skewness and coskewness and excessive kurtosis. The marginal unconditional distributions for most stock returns tend to concentrate more mass around the mean than a Gaussian does (we say that these distributions are leptokurtic) and at the same time they are often characterized by thicker tails than a Gaussian. One explanation for these features is given by the fact that the standard pure-diffusion benchmark ignores the presence of systemic jumps that affect most if not all markets at the same time. These jumps tend to occur at the same time across markets, implying that correlations between returns tend to change over time and to be higher in periods of high market volatility. The repeated, worldwide stock market crashes of September–October 2008 are an obvious example of systemic events with large correlated downward price movements. Adding jumps to the standard multivariate IID Gaussian model represents a relatively simple but effective way to capture the main dynamic properties of international stock returns.*

Second, this chapter develops a number of results concerning optimal international equity portfolio diversification. It is well known that one of the most important decisions most agents (individuals, households and, at times, the treasury departments of firms) face is the choice of a portfolio of assets. This means that agents decide how much to invest (in the case of households, save) and in which assets. In particular, a huge body of literature exists that shows that equity portfolio diversification should be pursed by aggressively diversifying an agent's portfolio among stocks issued in many (better, all) countries. In this chapter, we write and solve by dynamic programming methods a typical asset allocation problem in which an expected power (constant relative risk aversion) utility maximizer proceeds to select portfolio shares under the two alternative statistical frameworks—pure- and jump-diffusions—entertained in our

---

* One famous example of systemic market crash (included in our sample) is October 1997, when the drop was 3% in North America, 5.7% in continental Europe, 4.1% in the United Kingdom, 9.8% in Japan, 24.1% in the Pacific markets, 21.1% in the Latin American emerging markets, 21% in the Asian emerging markets, and 6.5% in the European and Middle Eastern emerging markets. Other events with large correlated price drops include September 11, 2001, October 1987, and the explosion of the Russian crisis in August 1998.

analysis. The risk of event-related jumps in security prices and volatility, changes the standard dynamic portfolio choice problem in several important ways. In the classical pure-jump problem, security prices are continuous, and instantaneous returns have infinitesimal standard deviations; as a result, an investor considers only small local changes in security prices when selecting the optimal portfolio. In the presence of jumps, the investor must also consider the effects of potentially large changes in security prices when selecting a dynamic portfolio strategy. Since the portfolio that is optimal for large returns need not be the same as the portfolio that is optimal for small returns, this creates a possible trade-off between considering continuous, "small" local changes that always occur in any (as small as one may consider) interval of time, and "large" price changes that do not always occur and in fact are unlikely to occur as the interval of time considered shrinks.

In this chapter, we model systemic events by constraining the jumps to simultaneously occur in all markets. However, conditional on a jump occurring, the size of the jumps is not constrained in any way and in fact jump realizations come from a random vector of independent lognormal distributions, which is a rather flexible framework. We find two main results. First, we find that jump-diffusion models are considerably more adequate than simpler, pure-diffusion based frameworks to describe the main (unconditional) properties of most stock indices under considerations, in particular those describing the behavior of emerging market returns. Second, differently from Das and Uppal (2004), where the effects of systemic jumps is shown to be rather modest, we show that systemic jumps may have a relatively large effect on optimal portfolio choices. In particular, our simulations show that even though the presence of jumps substantially changes the composition of the optimal international equity portfolio of a diversified investor, jumps may also lead to increase the overall percentage of one's portfolio that should be invested in equity indices.

This chapter contributes to a new but quickly expanding literature on the effects of nonnormalities (in some cases, of higher-order moments) on optimal portfolio choices. A few papers have explicitly connected international portfolio diversification to the effects of (co-)skewness and (co-)kurtosis on the expected utility of investors, such as Guidolin and Timmermann (2008) and Jondeau and Rockinger (2006). Differently from these papers, we do not make the role of higher-order moments explicit by taking Taylor series expansions of standard utility functions, but simply maximize expected utility using dynamic programming methods.

Guidolin and Nicodano (2009) use numerical methods in a discrete-time framework and show that the dynamics in higher-order (co-)moments generated by Markov switching models may have a significant impact on optimal international diversification.

Studying the impact of jumps on portfolio choice has a long history, going back to Merton (1971), who first studied a continuous-time consumption-portfolio problem. Many papers have considered the portfolio problem in the presence of jumps. Early papers on asset allocation in continuous-time models with discontinuous returns include, e.g., Ball and Torous (1983). More recently, Kallsen (2000) studied a continuous-time utility maximization problem in a market where risky security prices follow Lévy processes, and provided a solution for power, logarithmic and exponential utility using the duality or martingale approach. Liu et al. (2003) studied the implications of jumps in both prices and volatility on investment strategies, when a riskless asset and a stochastic-volatility jump-diffusion stock defining time-varying investment opportunities are available in the asset menu. Cvitanić et al. (2008) proposed a model where the asset returns are driven by a pure jump process having time-varying higher moments, and studied the sensitivity of the investment in the risky asset to dynamics in higher moments, as well as the resulting utility losses form ignoring the presence of higher moments. Leippold and Trojani (2008) introduced a new general class of matrix-valued jump diffusions that are convenient for modeling multivariate risk factors and managed to retain a high degree of analytical tractability for multivariate factor structures within both single-asset and multiasset settings.

Finally, a closely related paper is by Das and Uppal (2004) who evaluated the effects on portfolio choice of systemic risk, defined as the risk from infrequent events that are highly correlated across a large number of assets. In fact, in this chapter we borrow heavily—at least in terms of jump-diffusion framework and technical methods of solution—from Das and Uppal (2004). However, our focus is mostly on the implications of systemic jumps on international portfolio diversification between developed and emerging markets. Additionally, and differently from Das and Uppal's paper we find—using different/longer time series—that jumps may actually affect optimal portfolio choice to a nonnegligible degree, which is an important result to explain how rational investors may approach investing in emerging market stocks.

The rest of this chapter is organized in the following way. Section 8.2 introduces pure- and jump-diffusion models and illustrates how portfolio choice problems should be solved in the two frameworks. Section 8.3 describes the main properties of a number of Morgan Stanley Capital International (MSCI) stock returns series and documents the empirical failure of simple IID Gaussian benchmarks. Section 8.4 presents method-of-moments estimation and discusses our main estimation results. Section 8.5 presents the main asset allocation results, both using full-sample simulations that document the qualitative features of optimal portfolio weights and recursive out-of-sample experiments. Section 8.6 concludes this chapter.

8.2 OPTIMAL PORTFOLIO CHOICE UNDER UNDIVERSIFIABLE JUMPS

8.2.1 Benchmark: Pure-diffusion Processes

A workhorse of classical finance theory is represented by the simple pure-diffusion model in which stock (index) prices follow a GBM:

$$S_t^i = S_0^i \exp(a_i t + \sigma_i Z_t^i) \quad t \geq 0, \tag{8.1}$$

where

S_t^i is the price of stock (index) i, $i = 1, \ldots, n$, at time t
a_i and σ_i are the constant parameters specific to stock $i*$

Additionally, Itô's lemma implies that S_t^i is an Itô process with

$$dS_t^i = (a_i - \sigma_i^2/2)S_t^i dt + \sigma_i S_t^i dz_t = \alpha_i S_t^i dt + \sigma_i S_t^i dz_t^i, \tag{8.2}$$

and dz_t^i is the standard Brownian motion. Therefore at any time t, S_t^i has a conditional expected rate of change $\alpha_i S_t^i$ and a conditional standard deviation of $\sigma_i S_t^i$, so that one may think of α_i as the "instantaneous" expected rate of change, and of σ_i as the "instantaneous" standard deviation of the stock price change:

$$E\left[\frac{dS_t^i}{S_t^i}\right] = \alpha_i dt, \quad E\left[\left(\frac{dS_t^i}{S_t^i}\right)\left(\frac{dS_t^i}{S_t^i}\right)\right] = \sigma_i^2 dt. \tag{8.3}$$

As a result, the continuously compounded (cum-dividend) stock returns are IID normally distributed:

---

* If the stock pays any dividend, S_t^i represents the cumulant of stock prices and cash dividends, and its rate of change dS_t^i/S_t^i is equal to the instantaneous total return.

$$r_t \equiv dS_t^i / S_t^i \sim \text{IID } N(\alpha_i dt, \sigma_i^2 \, dt). \tag{8.4}$$

Equation 8.4 implies that stock returns should exhibit no skewness (i.e., their conditional and unconditional empirical distributions should be symmetric), and have tails that imply a kurtosis of 3 (or, equivalently, no excess kurtosis).

Even though Equation 8.2 is at the roots of much financial research (e.g., Black and Scholes' option pricing results), now there is strong empirical evidence that the unconditional distribution of asset returns, particularly stock market returns, is not Gaussian. In particular, as we shall see in Section 8.3, there is a vast literature (see e.g., Pagan, 1996) that has shown that stock returns at daily, weekly, and monthly frequencies typically exhibit statistically significant asymmetries (i.e., nonzero skewness) and fat tails (i.e., excess kurtosis). The unconditional skewness of a random variable r_t measures the degree to which positive deviations of a random variable from its mean are larger than the negative ones. Typically, a distribution characterized by large, negative outliers will be characterized by a negative asymmetry and hence a negative skewness. The unconditional kurtosis of a random variable r_t measures how "thick" are the tails of the distribution. The kurtosis of the normal distribution is 3. If the kurtosis exceeds 3, the distribution is said to be peaked (leptokurtic) when compared to a Gaussian; if the kurtosis is less than 3, the distribution is flat (platykurtic).

Equation 8.2 can be easily generalized to the multivariate case, when the object of interest is the vector of stock prices $\mathbf{S}_t \equiv \left[S_t^1 \ S_t^2 \ \cdots \ S_t^n \right]$:

$$d\mathbf{S}_t = \hat{\alpha}\mathbf{S}_t dt + \hat{\Sigma}\mathbf{S}_t d\mathbf{z}_t, \tag{8.5}$$

where $\mathbf{z}_t \equiv \left[z_t^1 \ z_t^2 \ ... \ z_t^n \right]'$ is an n-dimensional standard Brownian motion, and the correlations between each pair dz_l and dz_m are denoted as $\hat{\rho}_{lm} dt = E(dz_l \cdot dz_m)$, with $l, m = 1, ..., n$. Similarly, as before, it is straightforward to show that $E[d\mathbf{S}_t / \mathbf{S}_t] = \hat{\alpha} dt \ E[(d\mathbf{S}_t / \mathbf{S}_t) \times (d\mathbf{S}_t / \mathbf{S}_t)] = \hat{\Sigma} dt$ (where "/" and "×" are the element-by-element division and product operators, respectively). As a result, the continuously compounded (cum-dividend) stock returns follow a joint IID multivariate (n-dimensional) Gaussian distribution, $d\mathbf{S}_t / \mathbf{S}_t \sim \text{IID} \sim N(\hat{\alpha} dt, \hat{\Sigma} dt)$. Therefore also in this case, the individual stock returns will display zero skewness and excess kurtosis; moreover, also the coskewness and cokurtosis coefficients (to be defined in Section 8.4) will match their multivariate normal counterparts (zero in the case of the coskewness coefficients). Finally, the IID nature of the multivariate Gaussian distribution in Equation 8.5 has led researchers to stress that in this model investment, opportunities are constant over time.

8.2.2 Jump-Diffusion Processes

One simple and popular way to model the presence of asymmetries, fat tails, and nonzero cohigher moments (i.e., coskewness and cokurtosis) is to assume that stock prices follow mixed, jump-diffusion processes. In particular, one may always decompose the total change in a stock price into a normal and a nonnormal, jump-induced component. The pure-diffusion, GBM-type change in price may be due to variations in capitalization rates, temporary imbalances between market supply and demand, or the receipt of any other information which causes marginal, continuous price changes. This component may be accurately modeled as the pure-diffusion process in Equation 8.2. The "nonnormal" change may be due to the receipt of any information which causes a more than marginal change in the price of the stock, and is usually modeled as a Poisson process. When such an event occurs, we assume that there is an instantaneous jump in the ith stock price of size J_i $(i = 1, \ldots, n)$. J_i is a positive random variable that has a lognormal distribution:

$$\ln J_i \sim N(\mu_i, v_i^2), \tag{8.6}$$

where
$\quad \mu_i$ is the jump size
$\quad v_i$ is the volatility of the jump

J_i is assumed to affect stock prices through a Poisson process (dQ_t) that depends on the probability, λ, that a jump occurs, in the sense that only when a Poisson event has a realization then J_i impacts on the stock price S_t^i. Formally, we can write this model as a simple extension of Equation 8.2:

$$dS_t^i = \alpha_i S_t^i dt + \sigma_i S_t^i dz_t^i + S_t^i dQ_t, \tag{8.7}$$

where
$\quad Q_t$ is a Poisson process, assumed independent of the Brownian motion z_t^i
$\quad \lambda$ is the intensity of the Poisson process, i.e., the number of "abnormal" information arrivals per unit time

As a result, the continuously compounded stock return now has a distribution characterized as

$$r_t = \begin{cases} r \sim \text{IID } N(\alpha_i, \sigma_i^2) & \text{if there are no jumps } (L = 0); \\ r + J^{(1)} + \cdots + J^{(L)} & \text{if there are } L \geq 1 \text{ jumps,} \end{cases} \qquad (8.8)$$

where

$$\Pr\{L = l\} = \frac{\exp(-\lambda)\lambda^l}{l!}, \qquad (8.9)$$

from the standard expression for the probability function of a Poisson (λ) distribution. Since jumps are governed by a Poisson process, their intensity, λ, is assumed to be constant over time. Therefore, Equations 8.8 and 8.9 imply that the day after a jump—say, a stock market crash—another crash is equally likely as on the previous day. It is possible to show that Equations 8.8 and 8.9 imply rich patterns of skewness and fat tails (see Section 8.4). This is quite intuitive: jumps capture the arrival of discrete events producing a major impact on stock prices and as such these large (in absolute value) outliers have the effect of inflating the tails of the distribution of returns and—at least in principle—may produce arbitrary large asymmetries in distribution.

Also in this case, Equation 8.9 is easily extended to the multivariate, n-dimensional case:

$$d\mathbf{S}_t = \alpha \mathbf{S}_t dt + \Sigma \mathbf{S}_t d\mathbf{z}_t + (\mathbf{J}_t - 1)\mathbf{S}_t dQ_t(\lambda), \qquad (8.10)$$

where Q_t is a Poisson process with constant intensity $\lambda \in (0,1]$, and $(\mathbf{J}_t - 1)$, with $\mathbf{J}_t \equiv [J_t^1 \, J_t^2 \, \ldots \, J_t^n]'$, is the random vector of jump amplitudes that determine the change in each stock prices in the case in which the Poisson event occurs. Each of them has a lognormal distribution with mean μ_i and variance v_i^2, so that $E[\ln \mathbf{J}_t] = \mu$ and the covariance matrix is $\text{Var}[\ln \mathbf{J}_t] = \mathbf{v}'\mathbf{v}$. Notice that Equation 8.10 imposes that the arrival of jumps is contemporaneous across all stock market/indices with identical jump intensity λ so that $dQ_t^l(\lambda_l) = dQ_t^m(\lambda_m) = dQ_t(\lambda)$, $\forall l, m = 1, \ldots, n$. Clearly when $\lambda = 0$, then $dQ_t(\lambda) = 0$ $\forall t \geq 0$ so that Equation 8.10 simplifies to Equation 8.5; when $\lambda = 1$, Equation 8.10 becomes a pure-jump process, i.e., jumps are so frequent that they overwhelm the effect of Brownian motion shocks on stock prices. Finally, all the random variables in $\mathbf{J}_t$ are assumed to be independent, i.e., given that a system jump occurs in all markets, the realization of the jump size in each market is independent of all other markets.

The assumption that jumps simultaneously hit all markets is similar to Das and Uppal's (2004) framework in which two restrictions are imposed. First, the jump is assumed to arrive at the same time across all risky assets, i.e., jumps represent a systemic risk source. Second, conditional on a jump, the jump sign is assumed to be identical across assets, i.e., the value of all the assets jumps in the same direction. Thus, Equation 8.10 should capture large changes in asset prices and a high degree of correlation across these changes. However, notice that our model does not impose restrictions on the size or the volatility of asset-specific jumps.

8.2.3 Portfolio Problem

We consider a standard portfolio problem in which an investor selects optimal weights to maximize the expected utility from the value of her final wealth. As it is typically done in the literature, for tractability we assume that our investor has perfect knowledge (or at least, confidence) in the parameters of models (Equations 8.8 and 8.9), i.e., α, Σ, λ, μ, and ν (when $\lambda > 0$). The investor faces an asset menu composed of n risky stocks (or stock indices) and one riskless asset (indexed as asset 0). The asset prices follow the joint process:

$$\begin{aligned} dS_0 &= rS_0 dt, \\ dS_t &= \alpha S_t dt + \Sigma S_t dz_t + (J_t - 1)S_t dQ_t(\lambda). \end{aligned} \qquad (8.11)$$

For simplicity, we assume a constant interest rate $r \geq 0$. Notice that for this joint process, the total expected return vector and covariance matrix (both for stocks only) have two components: one part comes from the diffusion process of the returns, respectively α and Σ, and the other, denoted α^J and Σ^J, comes from the jump process:

$$E[r_t] = \alpha dt + \alpha^J dt, \qquad (8.12)$$

$$\text{Var}[r_t] = \Sigma dt + \Sigma^J dt. \qquad (8.13)$$

Let w_0 denote the percentage of wealth invested in the riskless asset and $\mathbf{w} \equiv [w_1\ w_2\ \dots\ w_n]'$ denote the $n \times 1$ vector of portfolio weights in each of the n risky assets. The portfolio weights must satisfy

$$w_0 + \mathbf{w}'\mathbf{1} = 1. \qquad (8.14)$$

In the absence of any income apart from dividends and capital gains from the portfolio holdings, the investor's wealth, W, starting from a positive, unit level $W_0 = 1$, will follow the dynamic law

$$\frac{dW_t}{W_t} = [r + \mathbf{w}'\mathbf{x}]dt + \mathbf{w}'(\boldsymbol{\sigma} \times d\mathbf{z}_t) + \mathbf{w}'\mathbf{J}_t dQ(\lambda) \tag{8.15}$$

where

$\mathbf{x} \equiv [\alpha_1 - r \ \alpha_2 - r \ \cdots \ \alpha_n - r]'$ is the $n \times 1$ vector of excess-returns vector
$\boldsymbol{\sigma}$ is the vector of volatilities such that $\boldsymbol{\sigma}\boldsymbol{\sigma}' = \Sigma$
$d\mathbf{z}$ is the vector of diffusion shocks, with the dot product $\times$ denoting element-by-element multiplication of σ_i and dz_t^i $(i = 1, 2, \ldots, n)$.

For concreteness and following the bulk of the literature, we consider an investor with power utility:

$$U(W_T) = \frac{W_T^{1-\gamma}}{1-\gamma} \quad \gamma > 0, \ \gamma \neq 1. \tag{8.16}$$

As it is well known, in this case γ is the constant coefficient of relative risk aversion, while $\gamma = 1$ corresponds to the case of logarithmic utility $U(W_T) =$ ln W_T. The investor's problem at time t is then to pick a policy rule for the selection of portfolio weights $\{\mathbf{w}\}$ which maximize his/her expected utility:

$$V(W_t, t) \equiv \max_{\{\mathbf{w}\}} E\left[\frac{W_T^{1-\gamma}}{1-\gamma}\right], \tag{8.17}$$

subject to the dynamics of his/her wealth in Equation 8.15.

Given that financial markets are incomplete because of the presence of jumps, we determine the optimal portfolio weights using stochastic dynamic programming. The Bellman principle and the appropriate form of Itô's lemma for jump-diffusion processes imply the following Hamilton–Jacobi–Bellman equation[*]:

$$0 = \max_{\{\mathbf{w}\}} \left\{ \frac{\partial V(W_t, t)}{\partial t} + \frac{\partial V(W_t, t)}{\partial W} W_t[\mathbf{w}'\mathbf{x} + r] + \frac{1}{2}\frac{\partial^2 V(W_t, t)}{\partial W^2} W_t^2 \mathbf{w}'\Sigma\mathbf{w} \right.$$
$$\left. + \lambda E\Big[V(W_t + W_t\mathbf{w}'\mathbf{J}_t, t) - V(W_t, t)\Big] \right\}. \tag{8.18}$$

---

[*] For further details and a complete derivation, see Ossola (2007). A classical textbook such as Ingersoll (1987) gives an introduction to dynamic programming-based solution methods in the presence of jumps.

At this point, we conjecture that the value function takes the separable form:

$$V(W_t, t) = A(t) \frac{W_T^{1-\gamma}}{1-\gamma}, \tag{8.19}$$

where $A(t)$ is a deterministic function that depends only on time. Therefore, substituting Equation 8.19 into Equation 8.18, we obtain

$$0 = \max_{\{w\}} \left\{ \frac{1}{A(t)} \frac{dA(t)}{dt} + (1-\gamma)W_t [\mathbf{w'x} + r] + \frac{(1-\gamma)y}{2} \mathbf{w'\Sigma w} \right. $$
$$\left. + \lambda E[(1+\mathbf{w'J}_t)^{-\gamma} - 1] \right\}. \tag{8.20}$$

Imposing first-order conditions from the previous equation, one gets the following system of n nonlinear equations:

$$0 = \mathbf{x} - \gamma \mathbf{\Sigma \tilde{w}} + \lambda E\left[\mathbf{J}_t \left(1 + \mathbf{\tilde{w}'J}_t\right)^{-\gamma} \right]. \tag{8.21}$$

By (numerically) solving this system of equations and taking $\mathbf{x}$, γ, $\mathbf{\Sigma}$, λ, $\mathbf{\mu}$, and $\mathbf{v}$ as given ($\mathbf{\mu}$ and $\mathbf{v}$ enter the computation of $E[\mathbf{J}_t(1 + \mathbf{w'J}_t)^{-\gamma}]$), one can determine the vector of optimal systemic weights $\mathbf{\tilde{w}}$.*

In the pure-diffusion case of $\lambda = 0$, the Hamilton–Jacobi–Bellman equation for a pure-diffusion process simplifies to

$$0 = \max_{\{w\}} \left\{ \frac{\partial V(W_t, t)}{\partial t} + \frac{\partial V(W_t, t)}{\partial W} W_t [\mathbf{w'x} + r] + \frac{1}{2} \frac{\partial^2 V(W_t, t)}{\partial W^2} W_t^2 \mathbf{w'\Sigma w} \right\}. $$
$$\tag{8.22}$$

Substituting the value function $V(W_t, t)$ and taking the first-order conditions, we obtain the familiar mean–variance solution that derives from Equation 8.17 when the set of investment opportunities is constant:

$$\mathbf{\hat{w}} = \frac{1}{\gamma} \mathbf{\Sigma^{-1} x}. \tag{8.23}$$

---

* The solution is complete only after verifying that the solution obtained satisfies the value function conjectured in Equation 8.21. See Ossola (2007) for details.

Comparing Equations 8.23 and 8.21, we note that the two equations are the same when there are no jumps ($\lambda = 0$), as in this case, Equation 8.21 simplifies to $0 = x - \gamma \Sigma \hat{w}$ which obviously has Equation 8.23 as its solution. Therefore, all differences between Equations 8.23 and 8.21 must be caused by the fact that $\lambda > 0$, i.e., jumps are possible. Equivalently, using Equation 8.23 when the process (Equation 8.11) holds would be incorrect because this would ignore the term $\lambda E[J_t(1 + \tilde{w}'J_t)^{-\gamma}]$ when solving the Equation 8.21.

In what follows, we shall be interested in comparing the portfolio of an investor who perceives security returns as following the pure-diffusion process in Equation 8.5, with the portfolio computed by an investor who takes into account systemic risk by using the jump-diffusion process in Equation 8.11. To make meaningful comparisons possible, we proceed to "neutralize" the effect of the first two moments of the distribution of returns in the following way: The first two moments for the process given in Equations 8.12 and 8.13 are set to match exactly the first two moments of the pure-diffusion process:

$$\alpha dt + \alpha^J dt = \hat{\alpha} dt \Rightarrow \alpha = \hat{\alpha} - \alpha^J, \tag{8.24}$$

$$\Sigma dt + \Sigma^J dt = \hat{\Sigma} dt \Rightarrow \Sigma = \hat{\Sigma} - \Sigma^J \tag{8.25}$$

As a result, for each stock/index, the investor using the jump-diffusion returns process takes the total expected return on the asset, $\hat{\alpha}_i$, and the covariance, $\hat{\alpha}_{lm}$, and subtracts from them α_i^J and σ_{lm}^J ($i, l, m = 1, ..., n$), respectively, with the understanding that this will be added back through the jump term, $(J_i - 1)dQ(\lambda)$. The logic of this approach is to bring out the (co-)skewness and (co-)kurtosis driven effects on optimal portfolio choice. Indeed, even though the unconditional expected returns and covariances under the compensated jump-diffusion process will match those from the pure-diffusion process, the two processes will not lead to identical portfolios: this is because jumps introduce (co-)skewness and (co-)kurtosis into the joint process for stock returns.

8.3 DATA

We use monthly MSCI equity total return index data for the period 1988: 01–2008:06, for a total of 246 observations per series. The total stock return data series (i.e., adjusted for dividends, splits, and other cash distributions) are collected for $n = 8$ countries/regions: North America (the United States and Canada), the United Kingdom, Europe ex-United Kingdom, Japan, Pacific developed ex-Japan, Emerging Markets (EM) Latin America, EM Asia, and EM Europe and Middle East (ME).* Clearly, this means that five return index series concern the developed countries and three emerging markets. All indices are free float-adjusted market capitalization weighted indices. As typical in the applied international finance literature (see e.g., De Santis and Gerard, 1997; Guidolin and Timmermann, 2007), returns are measured in U.S. dollars, and the portfolio exercises are all performed from the perspective of a U.S. investor. For this reason, the riskless rate is identified with the sample average of the 1 month U.S. Treasury bill yield (4.34%, in annualized terms).

Table 8.1 presents the summary statistics for the stock return series under investigation. Means and volatilities match the common beliefs about the rates of returns in developed vs. emerging international stock markets. In general, EM tend to display means that are not lower than those for the developed markets, although their volatilities are higher. The only surprise may be represented by the Japanese stocks that gave an average return of −0.3% per year. However, the Sharpe ratio statistics are much less in line with common expectations: even though the highest Sharpe ratio is given by EM Latin America (0.46 in annualized terms), it is interesting to notice that North American and continental European stocks yielded Sharpe ratios in the range 0.27–0.28 that exceed those for EM Asia and EM Europe and ME, both in the range 0.10–0.15. Using quantile-related statistics—median and interquartile range—to identify location and scale statistics returns have a similar picture even though median returns are systematically higher than means (the only exception occurs for Japan) and the interquartile range systematically exceeds volatility. All stock return indices show nonpositive skewness coefficients (Japan and United Kingdom essentially display zero skewness) and positive excess kurtosis. In particular, EM indices are characterized by substantially negative skewness and high excess kurtosis.

---

* As of June 2007, EM Latin America collects stock return data from Argentina, Brazil, Chile, Colombia, Mexico, and Peru. EM Asia covers equity index returns from the People's Republic of China, India, Indonesia, South Korea, Malaysia, Pakistan, the Philippines, Taiwan, and Thailand. EM Europe and Middle East collect stock returns data from the Czech Republic, Hungary, Poland, Russia, Turkey, Egypt, Israel, Jordan, and Morocco. Additionally, the Pacific developed ex-Japan index covers Australia, New Zealand, Hong Kong, and Singapore.

TABLE 8.1 Summary Statistics for International Stock Returns

	North America	Europe ex-United Kingdom	Japan	United Kingdom	Pacific Dev. ex-Japan	EM Latin America	EM Asia	EM Europe and ME
Summary statistics								
Mean	0.0823	0.0882	-0.0033	0.0602	0.0760	0.1883	0.0672	0.0857
Volatility	0.1368	0.1669	0.2192	0.1524	0.1943	0.3165	0.2483	0.2760
Sharpe ratio	0.2841	0.2683	-0.2131	0.1101	0.1674	0.4579	0.0958	0.1531
Median	0.1347	0.1369	-0.0273	0.0813	0.1131	0.3281	0.1468	0.2946
Interquartile range	0.1844	0.2077	0.3062	0.2039	0.2140	0.3451	0.3082	0.3197
Ratio median/range	0.4950	0.4501	-0.2310	0.1860	0.3257	0.8248	0.3354	0.7856
Skewness	-0.6000	-0.7177	0.0892	0.0087	-0.3698	-1.1178	-0.4792	-0.3134
Excess kurtosis	1.0158	1.5377	0.6826	0.1838	2.0269	3.3096	0.8981	2.6944
Correlation matrix								
North America	1							
Europe ex-United Kingdom	0.6990	1						
Japan	0.3690	0.4605	1					
United Kingdom	0.6637	0.7713	0.4756	1				
Pacific Dev. ex-Japan	0.6106	0.5983	0.4322	0.6184	1			
EM Latin America	0.5070	0.4451	0.3279	0.3870	0.5498	1		
EM Asia	0.5612	0.5314	0.4109	0.4453	0.7963	0.4919	1	
EM Europe and ME	0.4146	0.5338	0.5338	0.4088	0.4536	0.4621	0.4730	1

(continued)

TABLE 8.1 (continued) Summary Statistics for International Stock Returns

	North America	Europe ex-United Kingdom	Japan	United Kingdom	Pacific Dev. ex-Japan	EM Latin America	EM Asia	EM Europe and ME
Coskewness matrix								
North America	-0.6000	-0.6629	-0.2792	-0.3160	-0.4612	-0.6092	-0.5101	-0.5278
Europe ex-United Kingdom	-0.6452	-0.7177	-0.4197	-0.4476	-0.4940	-0.5123	-0.6051	-0.4520
Japan	-0.0904	-0.0884	0.0892	0.0527	-0.1590	-0.3504	-0.1848	-0.1635
United Kingdom	-0.1150	-0.2395	-0.1011	0.0087	-0.0727	-0.1367	-0.1449	-0.2010
Pacific Dev. ex-Japan	-0.3647	-0.3354	-0.2218	-0.1858	-0.3698	-0.4716	-0.3872	-0.3910
EM Latin America	-0.5588	-0.4102	-0.5804	-0.3739	-0.5855	-1.1178	-0.4265	-0.5724
EM Asia	-0.4047	-0.5020	-0.3549	-0.3024	-0.4482	-0.4218	-0.4792	-0.4243
EM Europe and ME	-0.5280	-0.3418	-0.2508	-0.3196	-0.3712	-0.4354	-0.3864	-0.3134

Interestingly, EM Latin America—the index with the highest Sharpe and median-to-interquartile range ratios—also yields the smallest skewness coefficient (−1.12) and the highest excess kurtosis coefficient (3.31). Even though we have not formally reported them, the reader may be easily persuaded that for all stock return indices (including Japan, but with a p-value between 0.01 and 0.05) a simple Jarque–Bera test of normality leads to strong (i.e., with p-values close to 0) rejection of the null hypothesis of a Gaussian distribution.

The second panel of Table 8.1 reports pairwise correlation coefficients. In general, the markets/regions under consideration tend to be substantially correlated, with an average, across-market correlation of 0.51. Moreover, all correlations coefficients are highly statistically significant (with most p-values below 0.01). One may also notice that, while in general, the developed market block tends to display very high correlations (the average for the five developed markets is 0.57, but there are peaks in excess of 0.60 for a few pairs), the correlations within the EM block and between the developed and EM blocks are lower, between 0.40 and 0.45. The last panel of Table 8.1 reports skewness and coskewness coefficients. The coskewness coefficient between the return indices l and m ($l,m = 1,2,...,n$) is defined as $\mathrm{Coskew}\left[r_t^l, r_t^m\right] \equiv \mathrm{Cov}\left[(r_t^l)^2, r_t^m\right]/\sigma_l^2\sigma_m$. As a result, coskewness may be interpreted as a scaled measure of covariance between the variance of the return on asset l and the level of returns on asset m. Clearly, high and positive coskewness levels are preferred by any risk-averse, nonsatiated portfolio optimizer because $\mathrm{Cov}\left[(r_t^l)^2, r_t^m\right] > 0$ means that higher variance from asset l is at least met (in the data) with higher mean returns on asset m or, vice-versa, that lower mean returns on asset m are "compensated" (in expected utility terms) by lower variance on asset l (see Guidolin and Nicodano, 2008, for a discussion on related concepts in an expected utility maximizing framework).* While under the benchmark of a multivariate IID Gaussian distribution of stock returns, all skewness and coskewness coefficients in Table 8.1 ought to be zero, the table clearly reveals that this is not the case in our data. Apart from the low and negative skewness

---

* Given the definition of coskewness, we have that $\mathrm{Coskew}\left[r_t^l, r_t^m\right] \neq \mathrm{Coskew}\left[r_t^m, r_t^l\right]$ and that $\mathrm{Coskew}\left[r_t^l, r_t^l\right] = \mathrm{Skew}\left[r_t^l\right]$. The first property explains why the coskewness matrix in Table 8.1 fails to be symmetric (differently from a correlation matrix); the second property justifies the fact that the coskewness coefficients on the main diagonal of the third panel of the table do coincide with the skewness coefficients reported in the first panel.

coefficients for most indices, the table also shows a few small negative (and statistically significant) coskewness coefficients, such as those involving the pairs North America–Europe ex-United Kingdom, North America–EM Latin America, Europe ex-United Kingdom–North America, Europe ex-United Kingdom–EM Asia, EM Latin America–Japan, and EM Latin America–Pacific ex-Japan. This is very important. For instance, with reference to EM Latin American returns, the finding that $\text{Coskew}\left[r_t^{\text{NA}}, r_t^{\text{EMLA}}\right]$, $\text{Coskew}\left[r_t^{\text{EMLA}}, r_t^{\text{Japan}}\right]$, and $\text{Coskew}\left[r_t^{\text{EMLA}}, r_t^{\text{Pac}}\right] \ll 0$ means that EM Latin America fails to provide a good hedge for volatility bursts in the North American markets, and that Japan and Pacific ex-Japan fail to provide a good hedge for volatility bursts in Latin American emerging markets. Of course, these properties do contribute to make all these stock markets less attractive than they would otherwise be under the benchmark of zero coskewness (i.e., under Gaussianity). Finally, notice that only one pair of markets is characterized by nonnegative coskewness (essentially, only Japan–United Kingdom and United Kingdom–Japan), with coefficients that are not statistically significant.

Overall, Table 8.1 reveals that a standard multivariate IID Gaussian model for international stock returns such as the one implied by Equation 8.5 would be largely inadequate to capture the salient properties of the returns, as it would counterfactually imply zero coskewness and excess kurtosis coefficients. The challenge then consists in finding other/better models describing the dynamics of stock returns to model and forecast stock returns and to support sound financial (portfolio) decisions. Section 8.4 proposes to estimate Equation 8.11 as a way to tackle these problems of the benchmark in Equation 8.5.

8.4 METHOD OF MOMENT ESTIMATES

8.4.1 Methodology

Although more efficient estimation approaches can in principle be found (see the discussion in Singleton, 2006), in this chapter we follow a simple method of moment strategy. The underlying idea is fairly simple: derive and use a number of moments from the assumed joint process for stock returns in Equation 8.11 equal to the number of parameters to be estimated (say, k), and proceed to solve the k (possibly nonlinear) equations to find such parameters. Starting from the benchmark case of the pure-diffusion process in Equation 8.5, the parameters to be estimated are $\hat{\alpha}$ and $\hat{\Sigma}$, with the moment conditions available being those identified in

$k = n + 0.5n(n + 1)$, Equations 8.24 and 8.25. Obviously, in this case $\hat{\alpha}$ and $\hat{\Sigma}$ can be estimated directly from the sample means, variances, and the covariances of the return series at hand. This also follows from the fact that it easy to obtain a number of moments conditions identical to the number of parameters to be estimated, k.

For the jump-diffusion process in Equation 8.11, the parameters to be estimated are α, Σ, λ, μ, and v, where μ and v characterize the random jump amplitude. Das and Uppal (2004) derive the unconditional moments of the returns processes in Equation 8.11 using the characteristic function that they have computed by exploiting its relation with the Kolmogorov backward equation. Differentiating the characteristic function then gives the moments of the returns process. The expressions for the moments of the continuously compounded returns are $(i, l, m = 1, 2, \ldots, n)$

$$E\left[r_t^i\right] = t(\alpha_i + \lambda\mu_i), \tag{8.26}$$

$$\text{Cov}\left[r_t^l, r_t^m\right] = t\left[\sigma_{lm} + \lambda(\mu_l\mu_m + v_l v_m)\right], \tag{8.27}$$

$$\text{Coskew}\left[r_t^l, r_t^m\right] \equiv \frac{\text{Cov}\left[\left(r_t^l\right)^2, r_t^m\right]}{\sigma_l^2\sigma_m} = \frac{t\lambda\left[2\mu_l v_l v_m + \mu_m\left(\mu_l^2 + v_l^2\right)\right]}{\sigma_l^2\sigma_m}, \tag{8.28}$$

$$\text{Exkurt}\left[r_t^l\right] \equiv \frac{E\left[\left(r_t^l\right)^4\right]}{\sigma_l^4} - 3 = \frac{t\lambda\left(3v_l^4 + 6v_l^2\mu_l^2 + \mu_l^2\right)}{\sigma_l^4}, \tag{8.29}$$

where σ_l and σ_m indicate the standard deviations for assets l and m, respectively. In order to obtain the equalities in Equations 8.24 and 8.25, we proceed to compensate the mean and the covariance for the jump-diffusion processes considered earlier with those for the pure-diffusion processes $(\lambda = 0)$:

$$\alpha^J = \lambda\mu, \tag{8.30}$$

$$\Sigma^J = \lambda(\mu\mu' + vv'). \tag{8.31}$$

As a result, the first two sets of moment conditions no longer depend on α and Σ, and as such only the parameters λ, μ, and v have to be estimated for a total of $k = 2n + 1$ parameters. In particular, we minimize the square of the differences between the moment conditions (Equations 8.30 and 8.31) and the corresponding (co-)moments implied by the data in the number of n sample excess kurtosis coefficients, n skewness coefficients, and $n(n - 1)$ coskewness coefficients (i.e., Equation 8.28, when $l \neq m$). Notice that with $n \geq 2$ assets, we have $k = 2n + 1 < 2n + n(n - 1)$ so that there are more moment conditions than parameters to be estimated. Therefore, we choose the k parameters to minimize the sum of squared deviations of the $2n + n(n - 1)$ moment conditions from their sample counterparts implied by the data, weighted by the squared ratio between the average skewness and excess kurtosis implied by the data (called κ):

$$\min_{\{\lambda,\mu,v\}}\left\{\sum_{l=1}^{n}\sum_{m=1}^{n}\left[\widehat{\text{Coskew}}\left[r_t^l, r_t^m\right] - \text{Coskew}\left[r_t^l, r_t^m; \lambda, \mu_l, \mu_m, v_l, v_m\right]\right]^2\right.$$
$$\left. + \kappa\sum_{l=1}^{n}\left[\widehat{\text{Exkurt}}\left[r_t^l\right] - \text{Exkurt}\left[r_t^l; \lambda, \mu_l, v_l\right]\right]^2\right\}, \tag{8.32}$$

where the κ weighting adjusts for the different scale of coskewness and kurtosis coefficients, and the notations $\text{Coskew}\left[r_t^l, r_t^m; \lambda, \mu_l, \mu_m, v_l, v_m\right]$ and $\text{Exkurt}\left[r_t^l; \lambda, \mu_l, v_l\right]$ stress the dependence of the (co-)moments from the unknown parameters to be estimated. We solve Equation 8.32 by applying numerical optimization methods, starting from a range of alternative initial conditions. Additionally, we approach the minimization problem after imposing natural bounds on the values of λ and v that are derived from the fact that λ is a probability, while v is a standard deviation, (1) $\lambda \in (0, 1]$, (2) $v > 0$; $\lambda = 0$ is ruled out because in this case, Equation 8.32 stops depending on μ and v, which are therefore not identifiable.

8.4.2 Estimation Results

Table 8.2 reports method-of-moment parameter estimates. The unique jump intensity parameter λ is estimated at 0.047 at monthly frequencies, which means that one would expect to see one jump every 21–22 months. This seems a plausible estimate also in the light of the negative estimates of the vector μ in the table—which illustrate that jumps can be interpreted as outbursts of systemic risk and contagious market crashes. The estimates

TABLE 8.2   Method of Moments Estimates of Jump-Diffusion Parameters

	North America	Europe ex-United Kingdom	Japan	United Kingdom	Pacific Dev. ex-Japan	EM Latin America	EM Asia	EM Europe
Common Poisson jump intensity								
λ	0.04648							
Country-specific jump diffusion mean								
μ	−0.08082	−0.11322	−0.06008	−0.05447	−0.05040	−0.13883	−0.13180	−0.05050
Country-specific jump diffusion volatility								
ν	0.02856	0.02153	0.04847	0.01154	0.09315	0.15618	0.05100	0.15600

for the μ vector corroborate this interpretation, as the mean jump levels are all negative and vary from −13.9% a month for EM Latin America to −5.1% for EM Europe and ME; the average estimate for the μ coefficients is in fact −8.5%. This is consistent with the notion that we are facing/estimating actual "nonnormal" jumps, since under an average (across-markets) monthly volatility of 1.8% (and a mean of 0.19% per month, as in our data), a return of −8.5% in a single month ought to be observed with a probability of 0.0000016, i.e., every 52,083 years!

Table 8.2 also shows estimates for the vector of ν parameters, obtaining values that range from 1.15% for the United Kingdom (this means that the magnitude/direction of jumps is subject to little uncertainty) to 15.6% for EM Latin America and EM Europe and ME. The average for the estimated values of ν is in fact a nonnegligible 7.1%. Given that conditionally on one systemic jump occurrence, the mean jump size is −8.5% and its volatility 7.1%, the fact that jump-induced log-changes (returns) in prices have a normal distribution, implies that on average 95% of all jumps will fall between −22.7 and +5.7%, i.e., some small chances of seeing positive jumps exists in the end.*

Table 8.2 shows that there are remarkable differences between the estimated jump process (conditional on the jump occurrence) in the case of EM markets when compared to developed markets. EM markets tend to display lower estimates of μ_i (−13.9% for EM Latin America and −13.2% for EM Asia vs. an average −8.5%) and higher estimates of ν_i (15.6% for both EM Latin American and EM Europe and ME vs. an average volatility of 7.1%). This fits the common perception that the emerging markets may display high Sharpe ratios (this is the case of EM Latin America) but also be characterized by degrees of deviations from the Gaussian benchmark—as either measured by skewness and kurtosis in Table 8.1 or by the estimated jump coefficients μ_i and ν_i in Table 8.2—that substantially differ from those typical of developed stock markets. Additionally, in the light of the results on coskewness in Table 8.1, it is not really surprising

---

* The 95% confidence intervals for each of the markets are North America [−13.7, −2.4], Europe ex-United Kingdom [−15.6, −7], Japan [−15.7, 3.7], the United Kingdom [−7.8, −3.1], Pacific ex-Japan [−23.7, 13.6], EM Latin America [−45.1, 17.4], EM Asia [−23.4, −3], EM Europe and ME [−36.3, 26.2]. Therefore, realized jumps may actually imply positive returns with a nonnegligible probability in the case of Japan, Pacific ex-Japan, EM Latin America, and EM Europe and ME.

that all the estimates of μ_i in Table 8.2 turn out to be negative: in fact $\mu_i < 0$ is necessary and sufficient for skewness coefficients to be negative, while $\mu_l, \mu_m < 0$ makes it more likely for coskewness coefficients to be negative. For instance, given the definition in Equation 8.28, it is easy to show that setting $l = m$, the sign of $2\mu_l v_l^2 + \mu_l(\mu_l^2 + v_l^2) = \mu_l(3v_l^2 + \mu_l^2)$ depends on the sign of μ_l only ($l = 1, 2, ..., n$).

In Table 8.3 we report a few statistics concerning the quality of the fit provided by the jump-diffusion model. Clearly, the fit provided is qualitatively good in the sense that the model replicates asymmetries and excess kurtosis wherever present in the data, although quantitatively the model provides an excellent fit (with an overall mean squared error for skewness coefficients of 0.28 only) for skewness coefficients, and a decent fit in terms of excess kurtosis (but with a much higher MSE of 8.99). In particular, the skewness fit is less-than-perfect in the case of Japan, EM Latin America, and EM Asia; the excess kurtosis fit is rather poor in the case of Europe ex-United Kingdom, Pacific ex-Japan, and EM Europe. The bottom panel of Table 8.3 reports similar measures of fit for the coskewness coefficients. Here, the fit provided is generally impressive with rather small MSE measures. Out of 56 off-diagonal coskewness coefficients, only 11 or 12 present some problems in terms of the fit provided by the jump-diffusion model; in fact, in only 10 cases there is a sign difference between sample and fitted coskewness coefficients (and in some of these cases the actual differences are negligible; for instance, $\widehat{\text{Cov}}[(r_t^{UK})^2, r_t^{Pac}] = -0.07$ while the jump-diffusion implied $\text{Cov}[(r_t^{UK})^2, r_t^{Pac}] = 0.01$). All in all, it remains rather remarkable that with 72 (co-)moments to fit and only 17 parameters, one is able to produce the qualitative fit illustrated in Table 8.3.

8.5 OPTIMAL ASSET ALLOCATION

We now use the estimated parameters in Tables 8.1 and 8.2 to compute two sets of optimal portfolio weights. The first set reflects the presence (as shown by Tables 8.2 and 8.3) of jumps in the joint process of international stock returns and is obtained by defining

$$f(\mathbf{w}; \lambda, \boldsymbol{\mu}, \mathbf{v}, \gamma) \equiv \left[\mathbf{x} - \gamma \Sigma \mathbf{w} + \lambda E[\mathbf{J}_t (1 + \mathbf{w}' \mathbf{J}_t)^{-\gamma}] \right]^2, \qquad (8.33)$$

and numerically solving this system of n nonlinear equations parameterized by $\lambda, \boldsymbol{\mu}, \mathbf{v}, \gamma$ to find the vector of systemic weights $\tilde{\mathbf{w}}$. Applying quasi-Newton

TABLE 8.3 Sample and Fitted Moments from Jump-Diffusion Model

		North America	Europe ex-United Kingdom	Japan	United Kingdom	Pacific Dev. ex-Japan	EM Latin America	EM Asia	EM Europe	MSE
Skewness	Data	-0.6000	-0.7177	0.0892	0.0087	-0.3698	-1.1178	-0.4792	-0.3134	
	Fitted	-0.4221	-0.9345	-0.3272	-0.1296	-0.3215	-0.1838	-1.5077	-0.0481	
	Squared diff.	0.0316	0.0470	0.1734	0.0191	0.0023	0.8723	1.0577	0.0704	0.2842
Excess kurtosis	Data	1.0158	1.5377	0.6826	0.1838	2.0269	3.3096	0.8981	2.6944	
	Fitted	0.9138	3.1178	1.2399	0.2196	1.2829	0.3121	8.1607	0.0768	
	Squared diff.	0.0104	2.4966	0.3106	0.0013	0.5536	8.9849	52.7441	6.8518	8.9942
Coskewness matrix										
North America	Data	-0.6000	-0.6629	-0.2792	-0.3160	-0.4612	-0.6092	-0.5101	-0.5278	
	Fitted	-0.4221	-0.5219	-0.2701	-0.2729	-0.0944	-0.3189	-0.4857	-0.1612	
	Squared diff.	0.0316	0.0199	0.0001	0.0019	0.1346	0.0843	0.0006	0.1344	0.0509
Europe ex-United Kingdom	Data	-0.6452	-0.7177	-0.4197	-0.4476	-0.4940	-0.5123	-0.6051	-0.4520	
	Fitted	-0.6851	-0.9345	-0.6122	-0.4836	0.0603	-0.5301	-1.0376	-0.3332	
	Squared diff.	0.0016	0.0470	0.0371	0.0013	0.3072	0.0003	0.1870	0.0141	0.0745
Japan	Data	-0.0904	-0.0884	0.0892	0.0527	-0.1590	-0.3504	-0.1848	-0.1635	
	Fitted	-0.3299	-0.4663	-0.3272	-0.2404	0.0688	-0.2577	-0.5461	-0.1738	
	Squared diff.	0.0574	0.1428	0.1734	0.0859	0.0519	0.0086	0.1305	0.0001	0.0813

United Kingdom	Data	-0.1150	-0.2395	-0.1011	0.0087	-0.0727	-0.1367	-0.1449	-0.2010
	Fitted	-0.1850	-0.2503	-0.1612	-0.1296	0.0112	-0.1429	-0.2743	-0.0883
	Squared diff.	0.0049	0.0001	0.0036	0.0191	0.0070	0.0000	0.0167	0.0127	0.0080
Pacific Dev. ex-Japan	Data	-0.3647	-0.3354	-0.2218	-0.1858	-0.3698	-0.4716	-0.3872	-0.3910
	Fitted	-0.2583	-0.2115	0.0494	-0.1169	-0.3215	-0.1796	0.0106	-0.0104
	Squared diff.	0.0113	0.0153	0.0735	0.0048	0.0023	0.0852	0.1582	0.1448	0.0619
EM Latin America	Data	-0.5588	-0.4102	-0.5804	-0.3739	-0.5855	-1.1178	-0.4265	-0.5724
	Fitted	-0.2417	-0.3073	-0.1714	-0.1602	-0.0335	-0.1838	-0.3021	-0.0992
	Squared diff.	0.1006	0.0106	0.1673	0.0457	0.3047	0.8723	0.0155	0.2240	0.2176
EM Asia	Data	-0.4047	-0.5020	-0.3549	-0.3024	-0.4482	-0.4218	-0.4792	-0.4243
	Fitted	-0.8819	-1.2633	-0.9080	-0.6505	0.2251	-0.6911	-1.5077	-0.4782
	Squared diff.	0.2278	0.5796	0.3060	0.1212	0.4533	0.0726	1.0577	0.0029	0.3526
EM Europe	Data	-0.5280	-0.3418	-0.2508	-0.3196	-0.3712	-0.4354	-0.3864	-0.3134
	Fitted	-0.0881	-0.1266	-0.0914	-0.0652	0.0233	-0.0691	-0.1516	-0.0481
	Squared diff.	0.1934	0.0463	0.0254	0.0647	0.1556	0.1341	0.0551	0.0704	0.0931

methods and choosing the initial guess to match the diffusion weights, $\tilde{\mathbf{w}}$, the function f is guaranteed to decline to lower values at each iteration step. The second set of weights ignores systemic risk and consists of the standard pure-diffusion, GBM weights $\hat{\mathbf{w}} = \gamma^{-1}\boldsymbol{\Sigma}^{-1}\mathbf{x}$. Finally, in addition to $\tilde{\mathbf{w}}$ and $\hat{\mathbf{w}}$, we also report the composition of the portfolio consisting of only risky assets, which can be obtained by dividing each individual weight by the total investment in risky assets. These weights are given by $\tilde{\mathbf{w}}/(\tilde{\mathbf{w}}'\mathbf{1})$ for the systemic jump-diffusion case and by $\hat{\mathbf{w}}/(\hat{\mathbf{w}}'\mathbf{1})$ in the pure-diffusion case.

8.5.1 Simulation Results: Do Jumps Matter?

Table 8.4 reports the results obtained using the estimates from the full-sample (1988–2008) described in Section 8.4. We compute optimal portfolio weights for a variety of risk aversion coefficients, $\gamma = 2, 3, 5, 10$, and 20. To save space, we report only the results for $\gamma = 2, 5$, and 10 in Table 8.4. However the qualitative results are not sensitive to the selected level for γ (apart from the obvious fact that an increasing percentage of the overall portfolio is invested in the riskless asset as γ increases). Therefore, in this subsection we focus our comments on the case of $\gamma = 5$, which is a typical risk aversion coefficient in the literature.

While under a pure-diffusion process, an investor is ought to invest only 19% of his total wealth in equities—and this seems a rather modest allocation to stocks, also in the light of the Sharpe ratios in Table 8.1—when jumps are taken into account, this percentage increases to 44%, which is more reasonable. This is at first counter-intuitive, as it would seem that computing optimal portfolio weights in a jump-diffusion framework corresponds to an operation by which additional risk factors (such as skewness, kurtosis, and coskewness, see Guidolin and Timmermann, 2007) are taken into account. However, one should remember that by Equation 8.17, our investor cares for the expected utility of final wealth and not for the properties (such as moments and comoments) of individual asset returns; additionally, Equations 8.24 and 8.25 have matched moments from pure- and jump-diffusion processes in such a way that their considerable differences cannot translate into any effect on either risk premia, variances, or covariances. Therefore, any difference in portfolio weights between pure- and jump-diffusion must solely derive from the effect of higher-order (co-)moments induced by the presence of jumps on the expected utility of wealth. In this regard, it is clear that by adequately choosing

TABLE 8.4 Optimal Portfolio Weights under Pure-Diffusion vs. Jump-Diffusion Models

	Total Portfolio			Risky Portfolio Only		
	Pure-Diffusion Weight	Jump-Diffusion Systemic Weight	Δ	Pure-Diffusion Weight	Jump-Diffusion Systemic Weight	Δ
Constant relative risk aversion coefficient $\gamma = 2$						
North America	0.4729	0.5315	0.0586	1.0169	0.4114	-0.6055
Europe ex-United Kingdom	1.3029	1.3728	0.0699	2.8019	1.0626	-1.7393
Japan	-1.0475	-0.9640	0.0835	-2.2526	-0.7462	1.5064
United Kingdom	-0.6798	-0.6024	0.0774	-1.4620	-0.4663	0.9957
Pacific Dev. ex-Japan	0.1396	0.2558	0.1162	0.3003	0.1980	-0.1023
EM Latin America	0.8601	1.0266	0.1665	1.8496	0.7946	-1.0550
EM Asia	-0.3641	-0.2351	0.1290	-0.7829	-0.1820	0.6010
EM Europe and ME	-0.2191	-0.0933	0.1258	-0.4711	-0.0722	0.3989
Riskless (1 month U.S. T-bills)	0.5350	-0.2919	-0.8269	0	0	
Total in risky assets	0.4650	1.2919	0.8269	1	1	
Constant relative risk aversion coefficient $\gamma = 5$						
North America	0.1891	0.2078	0.0187	1.0169	0.4725	-0.5444
Europe ex-United Kingdom	0.5212	0.5442	0.0230	2.8019	1.2374	-1.5645
Japan	-0.4190	-0.3937	0.0253	-2.2526	-0.8952	1.3574
United Kingdom	-0.2719	-0.2486	0.0233	-1.4620	-0.5653	0.8967
Pacific Dev. ex-Japan	0.0559	0.0900	0.0341	0.3003	0.2046	-0.0956
EM Latin America	0.3440	0.3949	0.0509	1.8496	0.8979	-0.9517

(continued)

TABLE 8.4 (continued) Optimal Portfolio Weights under Pure-Diffusion vs. Jump-Diffusion Models

	Total Portfolio			Risky Portfolio Only		
	Pure-Diffusion Weight	Jump-Diffusion Systemic Weight	Δ	Pure-Diffusion Weight	Jump-Diffusion Systemic Weight	Δ
EM Asia	−0.1456	−0.1055	0.0401	−0.7829	−0.2399	0.5431
EM Europe and ME	−0.0876	−0.0493	0.0383	−0.4711	−0.1121	0.3590
Riskless (1 month U.S. T-bills)	0.8140	0.5602	−0.2538	0	0	
Total in risky assets	0.1860	0.4398	0.2538	1	1	
Constant relative risk aversion coefficient γ=10						
North America	0.0946	0.1031	0.0085	1.0169	0.4954	−0.5214
Europe ex-United Kingdom	0.2606	0.2712	0.0106	2.8019	1.3032	−1.4987
Japan	−0.2095	−0.1981	0.0114	−2.2526	−0.9519	1.3006
United Kingdom	−0.1360	−0.1255	0.0105	−1.4620	−0.6031	0.8589
Pacific Dev. ex-Japan	0.0279	0.0432	0.0153	0.3003	0.2076	−0.0927
EM Latin America	0.1720	0.1951	0.0231	1.8496	0.9375	−0.9121
EM Asia	−0.0728	−0.0545	0.0183	−0.7829	−0.2619	0.5210
EM Europe and ME	−0.0438	−0.0264	0.0174	−0.4711	−0.1269	0.3443
Riskless (1 month U.S. T-bills)	0.9070	0.7919	−0.1151	0	0	
Total in risky assets	0.0930	0.2081	0.1151	1	1	

$\tilde{w}$—exploiting the possibility to sell some assets short—an investor may in principle obtain a wealth process that is characterized by returns which have lower kurtosis (more generally, even-order moments) and higher or even positive skewness (more generally, odd-order moments) than what can be achieved under a restrictive pure-diffusion model. For instance, by shorting stocks with low skewness, high excess kurtosis, and average low coskewness with other stocks, an investor may obtain a portfolio return process that has favorable skewness and kurtosis properties. This seems to be possible for our parameter configuration in Table 8.3, since we see that $\tilde{w}$ implies a higher (or less negative) weight for all markets, than $\hat{w}$ does. As a result, the stocks may be more attractive under a jump-diffusion process than under a pure-jump one.

Table 8.4 also shows that while the differences between $\hat{w}$ and $\tilde{w}$ are moderate (but certainly not zero, with peaks of 5.1%) in total terms, they are massive when it comes to the composition of the risky (equity) portfolio only, with absolute differences in excess of 100% for Europe ex-United Kingdom and Japan, and close to 100% for the United Kingdom and EM Latin America. In terms of composition, taking jumps into account seems to slightly favor developed stock markets over emerging ones: while the weight of the developed stock markets increases by approximately 5%, emerging stock markets lose an identical percentage. In overall terms, under $\hat{w}$ the proportions of the equity portfolio to be invested in emerging market stocks is 59.6% and declines to 54.6% under $\tilde{w}$. This matches the common perception that once jumps and nonnormalities are taken into account in a portfolio choice framework, then the weight to be assigned to emerging markets is likely to decline. However, the size of such a drop in portfolio weight appears to be moderate. In particular, as one may expect in the light of the results in Tables 8.1 and 8.2, we observe that by going from $\hat{w}$ to $\tilde{w}$ the weight assigned to EM Latin America drastically declines (from 185% to 90%), and it is not completely compensated by the higher weights assigned to EM Asia (from −78% to −24%) and EM Europe and EM (from −47% to −11%).

8.5.2 Recursive Results: Should Jumps Have Mattered?

To validate the results reported in Section 8.5.1, we proceed to a recursive pseudo out-of-sample portfolio experiment (as in Guidolin and Na, 2008). For each of the 120 months between 1998:07 and 2008:06, we proceed to compute both pure- and jump-diffusion optimal weights at the end of each month and using recursively updated parameters estimates $\hat{\lambda}_t$, $\hat{\mu}_t$, and $\hat{v}_t$ for t = 1998:07, 1998:08,..., 2008:06. Instead of reporting and commenting on the recursive parameter estimates, Figures 8.1 and 8.2 plot the recursive means and volatilities for both developed and emerging markets (we omit covariances to save space) as well as sample and fitted skewness and excess kurtosis (here, we omit coskewness coefficients to save space). Notice that the benchmark pure-diffusion model is only based on recursive estimates of means, variances, and covariances, while the plots concerning higher-order moments show the sample skewness and excess kurtosis with continuous lines and the fitted moments with dotted lines, computed on the basis of the recursive estimates of the jump-diffusion model. While sample means for developed markets are remarkably stable and the oddity of the persistently negative Japanese mean return is confirmed, in the case of emerging markets recursive means reach a minimum between 2002 and 2003 and then rise up again to the levels of the late 1990s. The high mean returns given by the Latin American EMs also stand out. Recursively estimated volatilities are, instead, very stable and simply tend to slowly trend down over the pseudo out-of-sample period, although the change is moderate. Also, the plot concerning skewness shows that for developed markets, the moments have been generally stable over time, exhibiting negative skewness; some difficulties at matching the United Kingdom and continental European recursive sample skewness emerge. Similar remarks apply to the skewness of emerging markets, although in this case the jump-diffusion model does an excellent job by matching the statistics over time. Finally, some more time-variation is present in the dynamics of excess kurtosis of developed markets (with a trough in 2003–2004 and excess kurtosis recently on the rise), although in this case the jump-diffusion model provides a superlative fit over time.

Figure 8.3 plots and compares recursive optimal portfolio weights $\hat{\mathbf{w}}$ and $\tilde{\mathbf{w}}$ (expressed in terms of the total portfolio) both for each individual MSCI EM index/region and also for the aggregate of developed vs. emerging markets. We also report the plot of weight attributed to North American stocks as a representative of developed markets/regions. The weights are computed assuming γ = 3, although the results are similar for alternative values of γ. Starting from this last aspect, we notice that while the pure-diffusion weight to emerging markets should have remained roughly constant and close to a net overall weight of zero, when jumps are taken into account, the total net weight to emerging

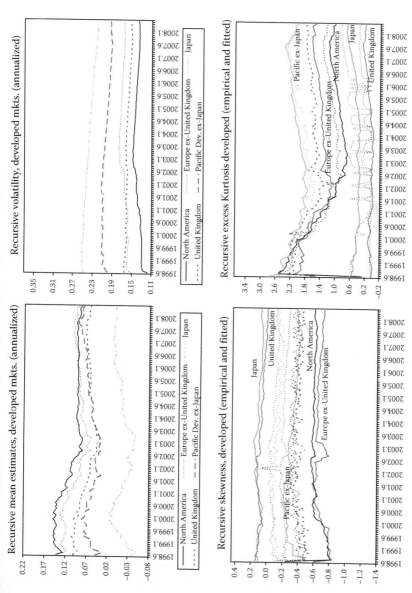

FIGURE 8.1   Recursive moment estimates vs. sample estimated—developed markets.

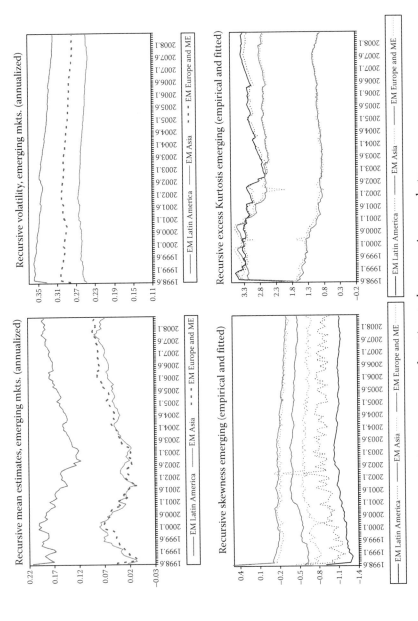

FIGURE 8.2   Recursive moment estimates vs. sample estimated—emerging markets.

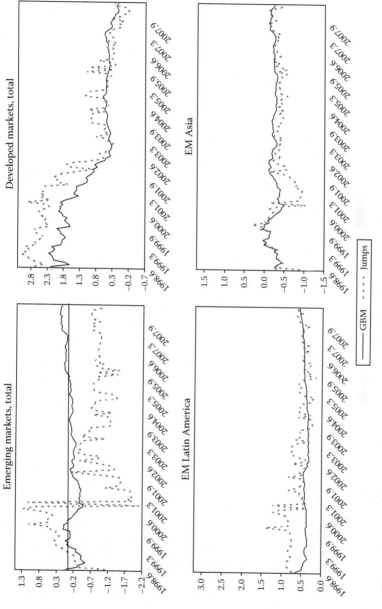

FIGURE 8.3   Recursive portfolio weights under pure-diffusion and jump-diffusion models.

(continued)

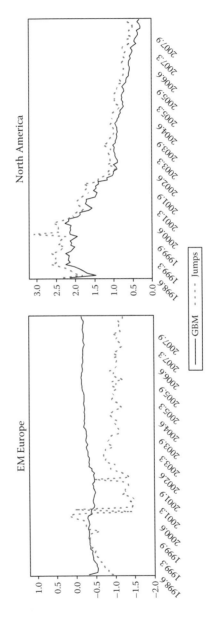

FIGURE 8.3 (continued)

markets increases at first between 1998 and early 2001 (from roughly −100% to 100%) and then rapidly drops (after 6 months of considerable variability) to a stable but negative level by the end of 2001. On the contrary, the differences between the subsets of $\hat{w}$ and $\tilde{w}$ concerning the developed markets are more modest and especially, they are important only between 1998 and 2001. In general, between 1998 and 2001, the jump-diffusion weights are above the pure-diffusion ones (with differences between 70% and 90%), while after 2002, the weights are very similar and no particular ranking may be established. These results confirm that taking jumps into account does change optimal equity portfolio choices and—at least after 2001–2002—that it tends to favor investments in equities issued in developed markets, over those issued in emerging markets. Even though this qualitative conclusion is similar to the comments expressed in Section 8.5.1, Figure 8.3 shows that the differences may in fact have taken massive proportions in real time, and that while taking jumps into account may have a modest (and difficult to sign) effect on the proportions invested in stocks issued in developed markets, the impact would have been strong (especially after 2004) for emerging markets.

The other panels of Figure 8.3 confirm these comments with reference to the individual MSCI indices, showing that the most important differences between $\hat{w}$ and $\tilde{w}$ concern Japan (here, $\hat{w} > \tilde{w}$ especially after 2006) and EM Europe and ME (again $\hat{w} > \tilde{w}$ especially after the end of 2001, with differences in excess of 50%). Table 8.5 concludes by reporting a few summary statistics concerning the time series of $\hat{w}$ and $\tilde{w}$ over the 120 recursive out-of-sample period. Interestingly, in this experiment the jump-diffusion model implies a higher percentage investment in the riskless asset, which fits the common (but not completely correct) perception that accounting for jumps must make stocks riskier than they otherwise are. This finding is different from the one reported in Section 5.1. The table clearly shows that the differences between $\hat{w}$ and $\tilde{w}$ in a recursive experiment are rather large, with absolute differences in excess of 20% for the North American, Japanese, U.K., and EM Europe and ME MSCI portfolios. As one may anticipate, jump-diffusion weights also tend to be more volatile than pure-diffusion ones. In Table 8.5, we have also computed empirical 90% confidence bands (as the sample mean ± twice the empirical standard deviation of the weights) and we find that the confidence bands are uniformly wider for jump-diffusion weights.

TABLE 8.5 Summary Statistics for Recursive Portfolio Weights

	Mean			Std. Dev.			90% CI—Lower		90% CI—Upper		Median		
	GBM	Jumps	Δ	GBM	Jumps	Δ	GBM	Jumps	GBM	Jumps	GBM	Jumps	Δ
North America	1.165	1.372	0.206	0.618	0.677	0.059	0.146	0.254	2.185	2.489	0.934	1.084	0.150
Europe ex-United Kingdom	0.978	0.964	−0.015	0.333	0.482	0.150	0.430	0.168	1.527	1.760	0.898	0.746	−0.153
Japan	−0.736	−1.010	−0.274	0.120	0.280	0.160	−0.935	−1.473	−0.538	−0.548	−0.704	−1.058	−0.354
United Kingdom	−0.309	−0.022	0.287	0.150	0.230	0.080	−0.556	−0.401	−0.061	0.358	−0.357	−0.087	0.270
Pacific Dev. ex-Japan	−0.203	−0.091	0.112	0.257	0.326	0.069	−0.627	−0.629	0.222	0.448	−0.084	−0.013	0.072
EM Latin America	0.416	0.581	0.165	0.063	0.225	0.162	0.312	0.209	0.520	0.952	0.414	0.543	0.129
EM Asia	−0.228	−0.357	−0.130	0.099	0.232	0.133	−0.391	−0.740	−0.064	0.025	−0.235	−0.344	−0.109
EM Europe and ME	−0.291	−0.893	−0.602	0.114	0.353	0.238	−0.479	−1.475	−0.103	−0.311	−0.269	−0.981	−0.712
Developed markets	0.886	1.212	0.326	0.664	1.007	0.343	−0.209	−0.449	1.982	2.874	0.505	0.825	0.320
Emerging markets	−0.103	−0.670	−0.566	0.181	0.700	0.519	−0.402	−1.825	0.196	0.486	−0.103	−0.801	−0.698
Total equities	0.783	0.543	−0.240	0.589	1.395	0.806	−0.188	−1.759	1.755	2.845	0.511	−0.102	−0.613
U.S. 1 month T-bills	0.217	0.457	0.240	0.589	1.395	0.806	−0.755	−1.845	1.188	2.759	0.489	1.102	0.613

8.6 CONCLUSION

In this chapter, we have shown that the emerging stock markets are indeed quite special, in that they are particularly vulnerable (in terms of both size and variability) to the occurrence of systemic events in the form of potentially large but infrequent price jumps. These jumps cause strong and persistent departures of emerging market stock returns from the classical benchmark of a multivariate IID Gaussian distribution in which only means, variances, and covariances characterize the joint distribution of asset returns. Importantly, the presence of jumps in the distribution of emerging market returns is not merely a statistical curiosity. We show that optimal international portfolio diversification computed under the assumption of the best fitting jump-diffusion model differs substantially from the portfolio structure implied by a suboptimal but simpler pure-diffusion model. In particular, the overall percentage of investment in emerging stock markets is lower and more variable over time when systemic jumps are taken into account. This is due to the fact that the emerging markets are more affected by jumps, in the sense that conditional on jumps occurrence, their size (in absolute value) or variability are both substantially larger than for developed markets. This is what we have found to be the special nature of the emerging stock markets.

ACKNOWLEDGMENT

Yu Man Tam provided excellent research assistance. All errors remain our own.

REFERENCES

Ball, C. A. and Torous, W. N. (1983) A simplified jump process for common stock returns. *Journal of Financial and Quantitative Analysis*, 18(1): 53–65.

Cvitanić, J., Polimenis, V., and Zapatero, F. (2008) Optimal portfolio allocation with higher moments. *Annals of Finance*, 4(1): 1–28.

Das, S. and Uppal, R. (2004) Systemic risk and international portfolio choice. *Journal of Finance*, 59(6): 2809–2834.

De Santis, G. and Gerard, B. (1997) International asset pricing and portfolio diversification with time-varying risk. *Journal of Finance*, 52(5): 1881–1912.

Guidolin, M. and Na, C. F. (2008) The economic and statistical value of forecast combinations under regime switching: An application to predictable US returns. In: M. Wohar and D. Rapach (Eds.), *Forecasting in the Presence of Structural Breaks and Model Uncertainty*. Emerald Publishing Ltd., Bingley, U.K.

Guidolin, M. and Nicodano, G. (2009) Small caps in international equity portfolios: The effects of variance risk. *Annuals of Finance*, 5(1): 15–48.

Guidolin, M. and Timmermann, A. (2008) International asset allocation under regime switching skew and kurtosis preferences. *Review of Financial Studies*, 21(2): 889–935.

Jondeau, E. and Rockinger, M. (2006) Optimal portfolio allocation under higher moments. *European Financial Management*, 12(1): 29–55.

Kallsen, J. (2000) Optimal portfolios for exponential Lévy processes. *Mathematical Methods of Operations Research*, 51(3): 357–374.

Ingersoll, J. E. (1987) *Theory of Financial Decision Making*. Rowman and Littlefield, Totowa, NJ.

Leippold, M. and Trojani, F. (2008) Asset pricing with matrix jump diffusions. Working Paper, University of St. Gallen, St. Gallen, Switerland.

Liu, J., Longstaff, F., and Pan, J. (2003) Dynamic asset allocation with event risk. *Journal of Finance*, 58(1): 231–259.

Merton, R. C. (1971) Optimum consumption and portfolio rules in a continuous-time model. *Journal of Economic Theory*, 3(4): 373–413.

Ossola, E. (2007) Asset Allocation under Non-Gaussian Returns, Jumps and Systemic Risk. Unpublished MPhil Dissertation, University of Insubria, Varese, Italy.

Pagan, A. (1996) The econometrics of financial markets. *Journal of Empirical Finance*, 3(1): 15–102

Singleton, K. (2006) *Empirical Dynamic Asset Pricing: Model Specification and Econometric Assessment*. Princeton University Press, Princeton, NJ.

CHAPTER 9

Overreaction Hypothesis in Emerging Balkan Stock Markets

Dimitris Kenourgios and Aristeidis Samitas

CONTENTS

9.1 INTRODUCTION

The efficient market hypothesis states that share prices fully reflect all available information, and no profit opportunities are left unexploited. The agents form their expectations rationally, and rapidly arbitrage away any deviations of the expected returns consistent with supernormal profits. However, a large body of literature challenges this hypothesis by showing that past prices can predict future movements in prices and that trading strategies based on past returns can generate risk-adjusted abnormal returns. These strategies are based on that markets exhibit a tendency to overreact and underreact at medium- and long-run horizons.

The stock market overreaction hypothesis states that investors overreact to unanticipated news, resulting in exaggerated movements in stock prices followed by corrections—prices revert to the mean (DeBondt and Thaler, 1985, 1987). In addition to this medium-term tendency toward reversals of trend, there is a shorter weak tendency toward momentum or contrarian stock prices' movements. Momentum strategies (buy winner stocks and sell loser stocks) stem from overreaction and the main cause of intermediate-term momentum but long-term overreaction is the heuristic-driven bias (Daniel et al., 1998). On the other hand, a contrarian portfolio strategy (buy undervalued stocks and sell overvalued ones) essentially exploits the time-varying conditional reverting behavior of the stock prices in the short run (Nam et al., 2001).

This chapter examines the overreaction hypothesis in four emerging Balkan stock markets (Bulgaria, Romania, Croatia, and Turkey), using average returns of the four developed markets (the United States, the United Kingdom, Germany, and Greece), during the period 2000–2007. Specifically, we test whether developed market returns exhibit the Balkan market returns, using Dimson's (1979) aggregated coefficients method upon the conventional market model. The acceptance of the aforementioned hypothesis implies overreaction to the movements of the Balkan markets due to developed ones.

However, many studies support that time-varying volatility in equity market returns exhibits an asymmetric effect (e.g., French et al., 1987; Nelson, 1991; Glosten et al., 1993). This means that a stock or a stock market index entails more risk following an unexpected negative return, than it does following an unexpected positive return of the same magnitude. Hence, asymmetric reverting patterns in return dynamics cannot be captured from the conventional autoregressive model restricted by the constant serial correlation coefficient. In this case, a nonlinear autoregressive model that allows serial correlation to the change in response to a prior positive and negative return shock is required. Therefore, we also examine the short-run interrelationships between the stock markets by testing the overreaction hypothesis using an asymmetric nonlinear smooth-transition generalized autoregressive conditional heteroskedasticity (ANST–GARCH) model and propose the most appropriate strategy to be followed by international asset managers, traders, and investors in the Balkan stock markets.

This chapter is inspired by a number of factors. First, there is no other study investigating the overreaction hypothesis in the Balkan stock markets to the best of our knowledge. Second, as Fama and French (1996) argue, analysis and tests on international data set contribute to the literature, establishing whether U.S. evidence is indicative of the general behavior or is a special case. Third, the importance of further investigation in the emerging markets is emphasized by the interest of fund managers in formulating portfolio strategies to exploit possible market reactions and earn potential large profits.

Our empirical analysis provides three main findings: (1) we find significant evidence supporting the overreaction hypothesis in the emerging Balkan markets, (2) we provide evidence on the existence of excess volatility with asymmetric mean reversion patterns in the Balkan markets, and (3) a momentum strategy from international investors and asset managers can be used in the short run, taking the advantage from information linkages and excess volatility in the emerging Balkan markets.

The structure of this chapter is organized as follows: Section 9.2 comprises a brief literature review; Section 9.3 provides an overview of the Balkan economic and financial environment; Section 9.4 analyzes methodological issues; Section 9.5 presents the data; the empirical results are reported in Section 9.6; and the final Section 9.7 contains the concluding remarks.

9.2 LITERATURE REVIEW

DeBondt and Thaler (1985) are among the first to find that stock prices behaved as if individual investors overreacted to the given information. They provide evidence of performance reversal observing that a portfolio of New York Stock Exchange (NYSE) stocks which perform worst (best) over an initial 3 year period (rank or portfolio formation period [PFP]) tend to perform best (worst) in the subsequent 3 year period (test period). This "winner–loser" effect is interpreted as investors' irrational behavior suggesting that they overreact, that excessive optimism or pessimism causes prices to be driven too high or too low from their fundamental values, and that overreaction is corrected in a subsequent period.

However, specific studies question the strong finding of DeBondt and Thaler about a stock market overreaction on grounds of varying beta, size differences between winner and loser stocks, and bid-ask spread bias (e.g., Chan, 1988; Zarowin, 1990; Conrad and Kaul, 1993). Moreover, in a review article, Fama (1998) claims that many of these studies' results are due to methodological problems, rejecting the suggestion that the evidence is consistent with a general tendency of markets to overreact and its interpretation as a manifestation of the investors' irrational behavior.*

Clare and Thomas (1995), Dissanaike (1997), and Campell and Limmack (1997) provide supportive results for the overreaction hypothesis in the U.K. stock markets. Da Costa (1994) and Bowman and Iverson (1998) support overreaction in Brazil and New Zealand, respectively. Other studies that use non-U.S. data investigate the overreaction hypothesis in a multicountry setting using national stock market indexes. Most of them provide evidence of overreaction (e.g., Richards, 1997; Baytas and Cakici, 1999), while others either report both the reactions (over/underreaction), depending on the size of a market's speculative bubble, the economic growth of that particular market, and the exchange rate volatility of the local currency (e.g., Schnusenberg and Madura, 2001), or reject the overreaction hypothesis (e.g., Lasfer et al., 2003).

Many studies support that momentum strategies work in international developed stock markets (e.g., Jegadeesh and Titman, 1993; Conrad and Kaul, 1998; Rouwenhorst, 1998; Schiereck et al., 1999). On the other hand, Nam et al. (2001) and Lasfer et al. (2003), among others, provide supportive results for contrarian profits in U.S. stock markets. Finally, several comprehensive studies document the momentum in the emerging stock markets (e.g., Rouwenhorst, 1999; Hameed and Kusnadi, 2002).

---

* However, the aforementioned critiques of overreaction have not gone unchallenged. D&T (1987), Chopra et al. (1992) and Loughran and Ritter (1996) reject the explanation that the "winner–loser effect" is explained by changes in beta, the size effect, and the bid-ask spread bias, respectively.

9.3 BALKAN ECONOMIES AND STOCK MARKETS

Since the 1990s, the Balkan economies are through a transitory phase of structural adjustment toward a market-oriented economic system (International Monetary Fund [IMF], 2000). Nevertheless, after 2000, the Balkan region displays robust growth rates (over 4%) expanding more rapidly than the EU average. Romania, Croatia, Bulgaria, and Turkey are among the top performers. According to the IMF (2000)

"... proximity to western Europe was associated with more favourable initial conditions, as the imprint of central planning was more limited, while rapid reorientation of trade to the more stable western European markets reduced these countries' exposure to external shocks. The resulting more favourable output performance was generally associated with more ambitious structural and institutional reforms, which in turn seems to have been partly a result of the external anchor provided by potential accession to the European Union. Macroeconomic stabilization was not a smooth process in all countries. Most economies succeeded in bringing down the high inflation rates that characterized the start of transition, reaching reasonable price stability by the mid-1990s. Despite that, several countries experienced a resurgence of inflation in the second half of the decade, since 2004, inflation rates converge to the EU average."

The simultaneous growth increase and inflation decrease appreciated capital inflows of foreign direct investments. While monetary policy is broadly satisfactory in the region as a whole, the picture is less clear with regard to the fiscal stance. General government finances vary considerably across the region, reflecting to some extent the different degree of structural transformation.

Globalization in trading systems and the new Internet-based tools enable international fund managers and investors to access the Balkan stock markets for portfolio diversification or speculative reasons. Stock prices in the Balkan markets increased on average by over 70% in terms of dollar during 2000–2006, compared to the 15% of the MSCI world market return. Despite the robust growth rates, the Balkan stock markets remain small in terms of capitalization, turnover, and liquidity compared to developed markets. Institutional investors are small in Balkan economies, with the exception of Istanbul Stock Exchange in which they own above 50% of the free float shares.

Overall, prospects for stock markets in the emerging Balkan economies seem to be prosperous. Due to the vast restructuring effort in public and private sectors, the encouragement for more investments, the increased value of institutional investors' assets, the improvement in investor protection, and the prospects of EU integration, the Balkan stock markets should continue to grow.

9.4 DATA

The data used in this chapter consist of the daily and monthly closing prices (in logs) in the four Balkan stock markets, the United States, the United Kingdom, Germany, and Greece. The stock market indexes of interest are SOFIX of Bulgaria, VANGUARD of Romania, CROBEX of Croatia, ISE NATIONAL 50 of Turkey, S&P500 of the United States, FTSE100 of the United Kingdom, Xetra DAX of Germany, and the Athens General Index of Greece.

All the national stock indexes are selected to guarantee representativeness of the domestic markets examined in this chapter. Furthermore, we use indexes expressed in national currencies. This restricts their changes to movements in the security prices and avoids any currency devaluation in the Balkan countries, which may have taken place during this period. Moreover, all Balkan daily closing prices are adjusted to inflation.

The high frequency data incorporated here include information on short-run market interactions that may be absent in lower frequency data. The data were obtained from national stock exchanges and Bloomberg database. The sample covers a period of 8 years, from January 2000 till December 2007, due to constraints on data availability. Because most markets are operating in the same time zone, the problem of nonoverlapping trading hours does not arise, except in the United States. Price changes in Europe are reflected the next day because trading in European Stock Exchanges is over by the time trading in the American market commences. Also, when a stock exchange is closed due to a national holiday, we use the previous day closing prices.

9.5 METHODOLOGICAL ISSUES

9.5.1 Dimson's Approach

The Dimson's (1979) aggregated coefficients method for the conventional market model is used to test the overreaction hypothesis in the emerging Balkan stock markets. We split the data into two equal parts (48 + 48 months) from January 2000 till December 2007. Also, we split our sample into two parts. The first part contains the average monthly return of all

Balkan stock markets ($R_{i,t}$). The second part includes the average monthly return of the mature markets ($R_{m,t}$) examined in this chapter. We follow the Dimson's (1979) approach employing the following regression equation:

$$R_{i,t} = a_i + \beta_{i,-1}(R_{m,t-1}) + \beta_{i,0}(R_{m,t}) + \beta_{i,+1}(R_{m,t+1}) + \varepsilon_{i,t} \tag{9.1}$$

Using α and β parameters generated by the market model, monthly expected returns are calculated as follows:

$$E(R_{i,t}) = a + \sum_{t=-1}^{t=+1} \beta_i(R_{m,t}) \tag{9.2}$$

where

$E(R_{i,t})$ is the expected return of the Balkan stock markets i in month t
$R_{m,t}$ is the return on the mature stock markets m in month t

These expected returns are then subtracted from the actual returns to generate monthly abnormal returns as follows:

$$AR_{i,t} = R_{i,t} - E(R_{i,t}) \tag{9.3}$$

where

$AR_{i,t}$ is the abnormal return
$R_{i,t}$ is the actual return
$E(R_{i,t})$ is the expected return

These abnormal returns are cumulated over the 8 year period, while cumulative abnormal returns are calculated as follows:

$$CAR_i = \sum_{t=1}^{48} AR_{i,t} \tag{9.4}$$

where
CAR_i is the cumulative abnormal return
$AR_{i,t}$ is the abnormal return

The paired two sample t-tests are employed to check if statistically significant differences exist in mean abnormal returns between the portfolio formation period (PFP) and the portfolio test period (PTP). If the abnormal returns experience a reversal in their portfolio returns, as the overreaction hypothesis predicts, then the differences in abnormal returns should be negative for periods of rises in the mature markets and positive for periods of declines in the mature markets.

In addition, we estimate the following regression equation using CARs:

$$PTPCAR_i = a + b(PFPCAR_i) + \varepsilon_i \qquad (9.5)$$

where
   $PTPCAR_i$ is the PTP CAR
   $PFPCAR_i$ is the PFP CAR

If there is a mean reverting pattern in abnormal returns, the coefficient b in Equation 9.5 should be negative and statistically significant.

9.5.2 ANST–GARCH Model

The autoregressive conditional heteroskedasticity (ARCH) and GARCH models, proposed by Engle (1982) and Bollerslev (1986), respectively, have allowed the magnitude of volatility to be predicted from past news and lagged conditional variance. However, many studies have documented that their first generation models are not well suited for capturing asymmetric volatility in stock returns series (e.g., French et al., 1987; Nelson, 1991; Glosten et al., 1993). Even the refined GARCH models (quadratic, exponential, threshold, and modified GARCHs) do not provide as consistent and flexible results as the ANST–GARCH model used in this chapter.

Following Nam et al. (2001), the modeling strategy is to capture asymmetry in both the conditional mean and variance process, asymmetric return reversals in the conditional mean equation, and the asymmetric volatility response in the conditional variance equation. For an excess return series R_t of a stock index, we specify the ANST–GARCH model.* The model used in this chapter, allowing asymmetry in conditional variance and return functions, has the following specifications†:

---

* There are several asymmetric GARCH models that allow an asymmetry in both the conditional mean and variance equations within the models. These are the modified model by Glosten et al. (1993) and Gonzalez-Rivera (1998), the sign- and volatility-switching ARCH (SVSARCH) model by Fornari and Mele (1997), the Markov switching volatility ARCH (MSVARCH) model by Hamilton and Susmel (1994), and the ANST–GARCH model by Nam et al. (2001). For an exposition of the ANST–GARCH models on financial and economic time series, see Anderson et al. (1999).

† Nam et al. (2001) define four nonlinear specifications to investigate the mean reverting pattern of monthly return indexes of the New York Stock Exchange (NYSE), American Stock Exchange (AMEX), and NASDAQ. Those allow asymmetry only in the conditional variance; the serial coefficient to vary with positive and negative returns shock; the investigation of the relationship between the predicted excess future volatility and its contemporaneous risk premium; and the investigation of the connection between the asymmetric mean reverting components in the expected short-horizon returns and mispricing behavior on the part of investors.

$$R_t = \mu + [\Phi_1 + \Phi_2 \cdot F(\varepsilon_{t-1})] \cdot R_{t-1} + \varepsilon_t \qquad (9.6a)$$

$$h_t = a_0 + a_1\varepsilon_{t-1}^2 + a_2h_{t-1} + [b_0 + b_1\varepsilon_{t-1}^2 + b_2h_{t-1}] \cdot F(\varepsilon_{t-1}) \qquad (9.6b)$$

where

$F(\varepsilon_{t-1}) = \{1 + \exp[-\gamma(\varepsilon_{t-1})]\}^{-1}$ is a smooth transition and continuous function of the value of ε_{t-1}

R_t is the excess return of a market index at time t

Φ_1 and Φ_2 are the time-varying serial correlations

h_t is the conditional variance, denoting a collective series of news at time t

ε_t is the white noise series of innovations, denoting a collective series of news at time t

The returns persistence in the model is governed by the sum of the coefficients $[\alpha_1 + b_1F(\varepsilon_{t-1})] + [\alpha_2 + b_2F(\varepsilon_{t-1})]$. The speed of transition between volatility regimes is governed by parameter γ. The greater the value of γ, the faster is the transition between volatility regimes. When $F(\varepsilon_{t-1}) = 0$ due to a large negative shock ($\varepsilon_{t-1} \ll 0$), returns persistence is measured by $(\alpha_1 + \alpha_2)$. When $F(\varepsilon_{t-1}) = 1$ due to a large positive shock ($\varepsilon_{t-1} \gg 0$), returns persistence is measured by $(\alpha_1 + b_1) + (\alpha_2 + b_2)$. The asymmetric effect implies that the value of $(\alpha_1 + \alpha_2)$ is greater than the value of $(\alpha_1 + b_1) + (\alpha_2 + b_2)$. Thus, $b_1 + b_2 < 0$ captures excess returns generated from asymmetric volatility response to positive or negative shocks. That is, $\varepsilon_{t-1} < 0$ yields a value of $F(\varepsilon_{t-1}) < 0.5$. By contrast, when $\varepsilon_{t-1} > 0$, then $F(\varepsilon_{t-1}) > 0.5$. If an asymmetric mean reverting pattern is a result of overreaction on the part of investors causing mispricing, then Φ_1 should be negative and Φ_2 positive.

Suppose that the dynamics of Balkan stock market return (R_t) evolve with the following nonlinear autoregressive process:

$$R_t = \mu + \varphi^+ R_{t-n} + \varepsilon_t, \quad \text{if } \varepsilon_{t-1} \geq 0.5 \qquad (9.7a)$$

$$R_t = \mu + \varphi^- R_{t-n} + \varepsilon_t, \quad \text{if } \varepsilon_{t-1} \leq -0.5 \qquad (9.7b)$$

where

$|\varphi^+| < 1$ and $|\varphi^-| < 1$ for stationarity condition of R_t

R_{t-n} is the excess return of developed markets, where n is the time horizon ($n = 1, 2, 3 \ldots$)

R_t is the Balkan market return (or reaction)

This specification allows a different autoregressive process for R_t under a prior positive and negative returns shock. Hence, $\mu + \phi^+ = X$ is a vector of explanatory variables available at time t when there is a positive shock, while $\mu + \phi^- = X$, when there is a negative shock. The returns used are daily average for all Balkan and developed markets. The performance of the U.S. market is used as the benchmark. A higher return from the benchmark implies that the Balkan portfolio provides excess returns. We also set r as the conditional asymmetric correlation which specifies the degree of persistence for the aforementioned models.

We search for abnormal price performance in the short-term window (up to 1 day after the developed markets excess return) following Balkan markets positive (negative) price shocks. It is important to mention here that if Balkan emerging markets do not react in the very first day after the excess return (return > 0.5 or < -0.5) exercised in the developed markets, we assume that Balkan market movements are not affected by the received information on developed markets. Thus, we focus our tests in capturing the very short-term reaction of Balkan stock markets.

9.6 EMPIRICAL EVIDENCE

Table 9.1 reports descriptive statistics in the Balkan stock market returns that are of prime interest to international portfolios. All stock price series show leptokurtosis, and there is evidence of negative skewness. Skewness is a particular feature of returns in the emerging Balkan markets. Significant kurtosis and negative skewness (long left tail) indicate rejection of normality in stock return distributions.

Table 9.2 reports the mean CARs for the PFP and the PTP, using the Dimson's method on the conventional market model.

The portfolio tests are all significant at conventional levels, and the coefficient b in Equation 9.5 is negative implying that there is mean reverting pattern in abnormal returns. The differences in abnormal returns show

TABLE 9.1   Descriptive Statistics

Countries	Min	Max	Std. Dev.	Skewness	Kurtosis
Turkey	−0.18	0.18	0.020	−0.2	8.04
Romania	−0.15	0.14	0.018	−0.14	6.93
Bulgaria	−0.17	0.21	0.022	−0.18	7.87
Croatia	−0.14	0.14	0.019	−0.09	7.16

TABLE 9.2   Balkan CARs for the PFP and the PTP

	Time Period					
	2000–2003			2004–2007		
Portfolio	PFP	PTP	Difference	PFP	PTP	Difference
	0.7459	−0.3284	−1.0743	0.7207	−0.1265	−0.8472
	(0.6230)	(−0.3817)	(−1.0047)*	(0.6812)	(−0.1170)	(−0.7982)*

Notes: The table reports mean (median) CARs at the end of PFP and PTP, together with the differences between PTP and PFP CARs for the two subperiods.
Average monthly return of all Balkan and developed stock markets is used.
* Statistical significance at the 5% level.

that there is a correlation between the size of the abnormal returns earned by the portfolio during the PFPs and those achieved in the test periods. In both subperiods, there are consistent positive abnormal returns (0.7459 for 2000–2003 and 0.7207 for 2004–2007) for the PFP. However, when we apply the PTP, the Balkan markets move back to their equilibrium levels, asymmetrically (−0.3284 for 2000–2003 and −0.1265 for 2004–2007). Hence, the Balkan market indexes initially deviate from the actual values implied by the new information received from the developed markets and then move back to their equilibrium levels, providing support to the overreaction hypothesis.

Table 9.3 reports the estimates of time-varying parameters for the Balkan daily average excess return, using the ANST–GARCH model (Equation 9.7a and b). As long as Φ_1 is negative and Φ_2 positive, the overreaction hypothesis is accepted for the Balkan stock markets. Our findings indicate that there are asymmetric mean reverting movements in the short horizon. This implies the existence of mispricing behavior from investors in the short run, which reverts to the mean (equilibrium level) in the long run.

While conditional asymmetric correlation r is positive with a strong statistical significance at 1% level, $\Phi_1 < 0$ confirms the positive effect the developed markets exercise on the Balkan markets. Hence, the hypothesis that Balkan stock markets indexes initially deviate from the actual values, implied by the new information received from the developed markets, and the indexes then move back to their equilibrium level, is not rejected giving rise to the overreaction hypothesis.

Consequently, short-run deviations from equilibrium can be expected to reverse (mean reversion pattern), thereby implying a degree of market predictability. A momentum strategy is consistent with the evidence from

TABLE 9.3   Estimation of Overreaction Using the ANST–GARCH
Model

Coefficients	Balkans vs. Developed
Φ_1	−0.0526 (−11.890)
Φ_2	0.0712 (12.042)
α_0	0.0174(0.748)
α_1	0.1268(5.829)
ρ-Value	0.0074
b_0	1.281 (4.273)
b_1	−1.4438 (−5.063)
b_2	−1.5006 (−5.947)
r	0.063
γ	157.89 (2.004)

Notes: 1. We specify the following ANST–GARCH model as the conditional mean and variance equations:

$$R_t = \mu + [\Phi_1 + \Phi_2 F(\varepsilon_{t-1})]\, R_{t-1} + \varepsilon_t$$
$$h_t = a_0 + a_1 \varepsilon_{t-1}^2 + a_2 h_{t-1} + [b_0 + b_1 \varepsilon_{t-1}^2 + b_2 h_{t-1}]F(\varepsilon_{t-1})$$

where

$F(\varepsilon_{t-1}) = \{1 + \exp[-\gamma(\varepsilon_{t-1})]\}^{-1}$ is a smooth transition and continuous
function of the value of ε_{t-1}

R_t is the excess return of a market index at time t

Φ_1 and Φ_2 are the time-varying serial correlations

h_t is the conditional variance, denoting a collective series of news at
time t

ε_t the white noise series of innovations, denoting a collective series
of news at time t

parameter γ governs the speed of adjustment between volatility
regimes

2. Daily average Balkan stock market return (R_t) evolves with the
following nonlinear autoregressive process:

$$R_t = \mu + \varphi^+ R_{t-n} + \varepsilon_t, \quad \text{if } \varepsilon_{t-1} \geq 0.5$$
$$R_t = \mu + \varphi^- R_{t-n} + \varepsilon_t, \quad \text{if } \varepsilon_{t-1} \leq -0.5$$

where

$|\varphi^+|<1$ and $|\varphi^-|<1$ for stationarity condition of R_t

R_{t-n} is the developed markets excess return, where n is the time horizon $(n = 1,2,3...)$

R_t is the Balkan market return

$\mu + \varphi^+ = X$ is a vector of explanatory variables available at time t
when there is a positive shock

$\mu + \varphi^- = X$ when there is a negative shock

r is the conditional asymmetric correlation which specifies the degree
of persistence for the models presented in note 2.

Values in parentheses are the Bollerslev–Wooldrige t-statistics.

developed markets supporting behavioral explanations. The most important behavioral bias is the overestimation of the information provided in the long-term.

This information causes prices to underreact to reliable information and overreact to unreliable information. Hence, the Balkan emerging markets outperform their developed counterparts in good times and underperform in bad times. Accordingly, the investor's confidence is moderated toward his prior expectation. Consequently, after taking into consideration the implemented time lag and risk adjustment, profitable trading can be developed to exploit this overreaction of reliability. Particularly, when investors are overly pessimistic (optimistic), the Balkan market indexes tend to move well below (above) that of their mature counterparts. When the excess pessimism (optimism) is gone, the indexes revert to get closer to their previous levels. Consequently, after taking into consideration the implemented time lag and risk adjustment, profitable trading can be developed to abuse this overreaction.

Indeed, in Table 9.3 we observe that α_0, α_1, and b_0 are positive, while b_1 is negative, providing support to a momentum portfolio strategy. Since $b_1 + b_2 < 0$, the model captures the asymmetric volatility response to positive or negative return shocks. Also, the high value of the parameter γ implies that the transition among volatility regimes occurs very quickly. At 1% critical level, $\Phi_1 < 0$ and $\Phi_2 > 0$ provide support to the asymmetric pattern of price reversals, where negative returns are more likely to revert to positive returns ($\Phi_1 + \Phi_2 > 0$), than positive returns to negative ones.

Table 9.4 reports the momentum strategy results applied in our sample. The findings support that the Balkan markets produce excess returns in a short-term window (days 0–3). We provide evidence of 1.2% excess return when allowing for 1-day implementation delay. In other words, Balkan markets receive the asymmetric information on day 1, while it was received in developed markets in day 0 (=currently). As a result, the emerging markets follow the movement of the developed markets from day 1 to day 2. On day 3, Balkan markets reverse to the mean.

Particularly, an international investor could hold a long position in the Balkan markets during the first day of the postshock in developed markets prices (=day 2). On the third day, the asset manager exercises his gains. Overall, a momentum strategy in the Balkan stock markets provides an efficient way for investors to generate excess returns. However, market failures that may prevent successful implementation of a momentum portfolio strategy in the Balkan stock markets are the lack of liquidity and low volume.

TABLE 9.4   Momentum Strategy

Day	Developed Markets Excess Return	Balkan Markets Excess Return	Strategy
0	≥0.5% or ≤0.5%	Position unchanged	Position unchanged
1	Reverse to the mean	1.2%	Long if return ≥0.5%, short if ≤0.5%
2	Reverse to the mean	0.2%	Hold position
3	Position unchanged	−0.6%	Close position

Notes: Daily average returns are used for all Balkan and developed markets. The performance of the U.S. stock market (S&P index) is used as the benchmark.

9.7 CONCLUSION

This chapter examines overreaction hypothesis in the four emerging Balkan stock markets (Bulgaria, Romania, Croatia, and Turkey), using average returns of four developed markets (the United States, the United Kingdom, Germany, and Greece), during the period 2000–2007.

The stock market overreaction hypothesis asserts that investors tend to overreact to new information which results in exaggerated movements in share prices; as a result, prices deviate from the actual values implied by the new information. Once investors considered the news in more detail, the overreaction wanes causing share prices to move back to their equilibrium levels. Hence, we test the overreaction hypothesis which provides information about the evolution of the market return paths over time. The hypothesis tested is that developed markets returns exhibit Balkan markets returns. The acceptance of the aforementioned hypothesis implies that overreaction to the Balkan markets movements is due to the developed ones.

Using the Dimson's (1979) aggregated coefficients method on the conventional market model, results show the existence of a strong mean reverting pattern in the Balkan markets portfolio based on past returns. Hence, the Balkan market indexes initially deviate from the actual values implied by the new information received from the developed markets, and then move back to their equilibrium levels, providing support to the overreaction hypothesis.

Using the ANST–GARCH model, the results support the overreaction hypothesis and the existence of asymmetric mean reverting patterns in the Balkan portfolio. As a result, the information which is currently available from the developed markets is not fully reflected in the Balkans future prices, providing the opportunity for investors to take advantage from the mean reverting movement.

The unexpected increase (decline) in the developed stock prices creates an unexpected raise (drop) in the Balkan stock prices in the short run. Following a momentum strategy, our findings show that an investor holding a Balkan markets portfolio could achieve excess return which is at least marginally higher than the benchmark (U.S. stock market return). This explains the time variation of stock returns in the Balkan region providing asymmetric information, to a great extent, predictable. The disparity and the asymmetry are largely attributable to mispricing behavior from investors, who consistently overreact to certain market news with optimism and pessimism unwarranted by time-varying market movements.

ACKNOWLEDGMENT

Dmitris Kenourgios acknowledges a research grant (Kapodistrias) from the University of Athens—Special Account for Research Grants.

REFERENCES

Anderson, H.M., Nam, K., and F. Vahid (1999) An asymmetric nonlinear smooth-transition GARCH model. In: P. Rothman (Ed.), *Nonlinear Time Series Analysis of Economic and Financial Data.* Kluwer Academic Publishers: Boston.

Baytas, A. and Cakici, N. (1999) Do markets overreact: International evidence. *Journal of Banking and Finance,* 23(7): 1121–1144.

Bollerslev, T. (1986) A generalized autoregressive conditional heteroscedasticity. *Journal of Econometrics,* 31(3): 307–327.

Bowman, R.G. and Iverson, S.D. (1998) Short-run overreaction in the New Zealand stock Market. *Pacific-Basin Finance Journal,* 6(5): 475–491.

Campell, K. and Limmack, R.J. (1997) Long term overreaction in the UK stock market and size adjustment. *Applied Financial Economics,* 7(5): 537–548.

Chan, K.C. (1988) On the contrarian investment strategy. *Journal of Business,* 61(2): 147–163.

Chopra, N., Lakonishok, J., and Ritter, J.R. (1992) Measuring abnormal performance: Do stock markets overreact? *Journal of Financial Economics,* 31(2): 235–268.

Clare, A. and Thomas, S. (1995) The overreaction hypothesis and the UK Stock market. *Journal of Business Finance and Accounting,* 22(7): 961–973.

Conrad, J. and Kaul, G. (1993) Long term market overreaction or bias in computed returns? *Journal of Finance,* 48(1): 39–63.

Conrad, J. and Kaul, G. (1998) An anatomy of trading strategies. *Review of Financial Studies,* 11(3): 489–519.

Da Costa, N.C.A. (1994) Overreaction in the Brazilian stock market. *Journal of Banking and Finance,* 18(4): 633–642.

Daniel, K., Hirshleifer, D., and Subrahmanyam, A. (1998) Investor psychology and security market under- and overreactions. *Journal of Finance*, 53(6): 1839–1885.

DeBondt, W.F.M. and Thaler, R. (1985) Does the stock market overreact? *Journal of Finance*, 40(3): 793–805.

DeBondt, W.F.M. and Thaler, R.H. (1987) Further evidence of investor overreaction and stock market seasonality. *Journal of Finance*, 42(3): 557–582.

Dimson, E. (1979) Risk measurement when shares are subject to infrequent trading. *Journal of Financial Economics*, 7(2): 197–226.

Dissanaike, G. (1997) Do stock market investors overreact? *Journal of Business Finance and Accounting*, 24(1): 27–49.

Engle, R. (1982) Autoregressive conditional heteroskedasticity with estimates of the variance of United Kingdom inflation. *Econometrica*, 50(4): 987–1008.

Fama, E. (1998) Market efficiency, long-term returns, and behavioral finance. *Journal of Business*, 73(2): 161–175.

Fama, E.F. and French, K.R. (1996) Multifactor explanation of asset pricing anomalies. *Journal of Finance*, 51(1): 55–84.

Fornari, F. and Mele, A. (1997) Sign- and volatility-switching ARCH models: Theory and applications to international stock markets. *Journal of Applied Econometrics*, 12(1): 49–65.

French, K., Schwert, G., and Stambaugh, R. (1987) Expected stock returns and volatility. *Journal of Financial Economics*, 19(1): 3–29.

Glosten, L., Jagannathan, R., and Runkle, D. (1993) On the relation between the expected value and the volatility of the nominal excess return on stocks. *Journal of Finance*, 48(5): 1779–1801.

Gonzalez-Rivera, G. (1998) Smooth-transition GARCH models. *Studies in Nonlinear Dynamics and Econometrics*, 3(1): 61–78.

Hameed, A. and Kusnadi, Y. (2002) Momentum strategies: Evidence from Pacific Basin stock markets. *Journal of Financial Research*, 25(3): 383–397.

Hamilton, J.D. and Susmel, R. (1994) Autoregressive conditional heteroskedasticity and changes in regime. *Journal of Econometrics*, 64(1–2): 307–333.

IMF (2000) Transition: Experience and policy issues. In: *World Economic Outlook, Focus on Transition Economies*, Chapter III, International Monetary Fund. Available at http://www.imf.org/External/Pubs/FT/weo/2000/02/pdf/chapter3.pdf

Jegadeesh, N. and Titman, S. (1993) Returns to buying winners and selling losers: Implications for stock market efficiency. *Journal of Finance*, 48(1): 65–91.

Lasfer, M.A., Melnik, A., and D.C. Thomas (2003) Short-term reaction of stock markets in stressful circumstances. *Journal of Banking and Finance*, 27(10): 1959–1977.

Loughran, T. and Ritter, J.R. (1996) Long-term market overreaction: The effect of low-priced stocks. *Journal of Finance*, 51(5): 1959–1970.

Nam, K., Pyun, C.S., and Avard, S.L. (2001) Asymmetric reverting behaviour of short-horizon stock returns: An evidence of stock market overreaction. *Journal of Banking & Finance*, 25(4): 807–824.

Nelson, D.B. (1991) Conditional heteroskedasticity in assets returns: A new approach. *Econometrica*, 59(2): 347–370.

Richards, A.J. (1997) Winners–loser reversal in national stock market indices: Can they be explained? *Journal of Finance*, 52(5): 2129–2144.

Rouwenhorst, K.G. (1998) International momentum strategies. *Journal of Finance*, 53(1): 267–284.

Rouwenhorst, K.G. (1999) Local return factors and turnover in emerging stock markets. *Journal of Finance*, 54(4): 1439–1464.

Schiereck, D., DeBondt, W., and Weber, M. (1999) Contrarian and momentum strategies in Germany. *Financial Analysts Journal*, 55(6): 104–116.

Schnusenberg, O. and Madura, J. (2001) *Global and Relative Over- and Underreaction in International Stock Market Indexes*. Mimeo, St. Joseph's University: Philadelphia.

Zarowin, P. (1990) Size, seasonality and stock market overreaction. *Journal of Financial and Quantitative Analysis*, 25(1): 113–125.

CHAPTER 10

Does Currency Risk Depress the Flow of Trade? Evidence from the European Union and Transition Countries

M. Nihat Solakoglu and Ebru Guven Solakoglu

CONTENTS

10.1 INTRODUCTION

Is currency risk important for traders in international markets? With the breakdown of Bretton Woods in early 1973, this question became important for researchers as well as traders. It was expected that the fluctuations in exchange rates would decline with the shift to floating exchange rates. However, the reality turned out to be different: currency risk, as measured by the volatility of exchange rates, was higher than before.

In price theory, it has been shown that price uncertainty leads to a decline in the output level for a perfectly competitive firm (e.g., Baron, 1970; Sandmo, 1971). On the other hand, the availability of hedging separates a firm's decision on optimum production from the uncertain market price. In other words, the decision on optimal production level does not depend on the utility function or the probability distribution function of the uncertain price. All production and export decisions are made on the basis of the forward price.*

This result in price theory provides the main theoretical justification for the impact of exchange rate risk on trade volumes. The works of Hooper and Kohlhagen (1978), Clark (1973), Ethier (1973), and Cushman (1986) provide some support for the theoretical studies by finding a negative relationship between currency risk and trade flows. Recently, however, some studies show necessary and sufficient conditions that lead to positive or ambiguous relationships between currency risk and international trade. Some examples are the works of De Grauwe (1988), Neumann (1995), Franke (1991), and Giovannini (1988).

This lack of consensus on theoretical framework has also led to a diverse, and sometimes unwieldy, empirical literature.† While some studies find a negative relationship between volatility and trade (e.g., Pozo, 1992; De Grauwe, 1988), others find a positive relationship or no relationship (e.g., Assery and Peel, 1991). Other examples of empirical works that investigate the relationship between exchange rate volatility and trade flows

---

* This concept is known as the "Separation Theorem." For more details, see Ethier (1973) or Kawai and Zilcha (1986) for the separation theorem in trade literature and Holthausen (1979) for the theory of the firm.

† By a lack of consensus, we mean that theoretical works do not clearly specify which variables should be included in the conditioning set, or the measure of volatility, or correct model specification. Solakoglu (2005) investigates the robustness of this so-called relationship by using extreme bound analysis of Leamer (Leamer and Leonard, 1983; Leamer, 1985) and finds that it is not robust. In a simulation, Gagnon (1993) also shows that this effect is too small to be detected.

include Pozo (1992), Bahmani-Oskooee (2002), Solakoglu et al. (2008), Baum et al. (2004), Cushman (1986), Hooper and Kohlhagen (1978), Gagnon (1993), Thursby and Thursby (1987), Kumar and Dhawan (1991), Grobar (1993), and Klein (1990).

This chapter also examines the relationship between currency risk and trade flows. However, our approach differs from earlier studies in two ways. First, we investigate this relationship by focusing on transition countries with their exports to European Union (EU) and North American Free Trade Agreement (NAFTA) member countries between 1995 and 2006. The EU discussed the accession of the Central and Eastern Europe (CEE) and the Baltic states into EU, conditional on their compliance with a set of economic criteria such as well-functioning market economy and leveled competitiveness along with political and administrative criteria in the early 1990s. Second, by taking into account the country categories within transition countries, we indirectly evaluate the effect of EU accession talks or EU membership on the relationship between currency risk and international trade.

The remainder of this chapter is organized as follows. Section 10.2 gives the model specification and the sources of data used in the analysis. In Section 10.3, we discuss our findings. Finally, Section 10.4 reports our main conclusions and suggestions for further research.

10.2 MODEL SPECIFICATION AND IMPLEMENTATION

We investigate the impact of currency risk on international trade by considering transition countries' exports to EU and member countries of the NAFTA. Our analysis is performed using annual data for the period between 1995 and 2006. The data for exports is obtained from the *Handbook of Statistics* published by United Nations Conference on Trade and Development (UNCTAD). The consumer price index (CPI) values at the country level are obtained from the *World Development Indicator* (WDI) database as provided by the World Bank.[*] The *Global Financial Data* Web site is utilized to get both U.S. and EU-25 CPI values.[†] This same source is also utilized to obtain monthly exchange rate information to estimate volatility of exchange rates to proxy currency risk. The gross domestic product (GDP) for the target regions, EU and the United States,[‡] are also taken from WDI database.

---

[*] http://devdata.worldbank.org/dataonline/
[†] www.globalfinancialdata.com
[‡] To represent the level of economic activity in NAFTA area, the GDP of the United States is used.

We consider 28 transition countries in our analysis. These countries can be grouped under the following categories: the new EU member countries (EU-N10), CEE countries,* and Commonwealth Independent States (CIS). EU-N10 countries include Cyprus, Czech Republic, Estonia, Hungary, Latvia, Lithuania, Malta, Poland, Slovakia, and Slovenia. The CEE countries include Albania, Bulgaria, Croatia, Macedonia, Romania, and Serbia and Montenegro. For the CIS countries, we focus on Armenia, Azerbaijan, Belarus, Georgia, Kazakhstan, Kyrgyz Republic, Moldova, Russia, Serbia and Montenegro, Tajikistan, Turkmenistan, Ukraine, and Uzbekistan. Because of data issues, we dropped Turkmenistan, Uzbekistan, and Serbia and Montenegro from the estimations.

The annual currency risk at the country level is estimated by the standard deviation of the monthly exchange rate returns within a year. For exchange rates, we utilized the exchange rate between local currency and U.S. dollar. The first reason for this choice is the high use of U.S. dollar in international transactions. The second reason is to test the impact of third-country effect on the volatility–trade relationship. That is, as argued by Cushman (1986), while a higher currency risk can lower the trade between two countries, it can also shift the trade to other countries with lower risk, thus, leaving total level of the trade unchanged. Figure 10.1 presents the

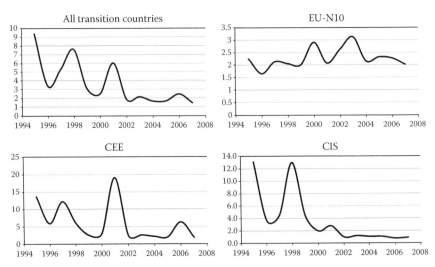

FIGURE 10.1   The behavior of exchange rate volatility over time.

---

* We included South Eastern countries in that category as well.

volatility estimates as an average of all transition countries in our sample, and as averages of three country groups over time.

For all countries in our sample, we observe a declining trend in exchange rate volatility over time. In particular, the level of risk is much smaller and stable after 2002 which corresponds to a stable environment for economies globally. The volatility estimates for CEE and CIS countries also follow similar pattern. However, there is a significant jump in volatility between 2000 and 2002, which is not surprising since this period corresponds to a crisis environment for most economies. On the other hand, the behavior of volatility series for EU-N10 countries is unexpected: there seems to be either no trend or a slight upward trend. At the same time, we observe a higher volatility, especially post-2002 period compared to CEE or CIS countries. Nevertheless, on the positive side, EU-N10 countries also observe a smaller shift in volatility during the 2000–2002 periods. Overall, it can be argued that EU membership or accession talks do not lead to lower volatility, but it can lower the effect of crisis on the level of volatility.

The analysis is performed using unbalanced panel data with the following specification:

$$\Delta \mathrm{Exp}_{it} = a + b_1 \Delta \mathrm{Relpr}_{it} + \beta_2 \Delta \mathrm{Gdp}_{it} + \beta_3 \Delta \mathrm{Exrate}_{it} + \beta_4 \mathrm{Vol}_{it-1}$$
$$+ \beta_4 \Delta \mathrm{Vol}_{it} + \varepsilon_{it} \tag{10.1}$$

where Exp_{it} is the change in the share of exports to the destination region for transition country i at time t. The right-hand side variables include a measure of economic activity in the importing country, a relative price measure expressed as the ratio of foreign to domestic prices, the bilateral exchange rate between U.S. dollar and the local currency, and measures of volatility. All the variables are in differences from the previous year, except the level of volatility in the previous year. Economic activity is measured by the GDP of either the United States or the European area depending on the destination region.[*] In calculating relative prices, we use CPI of exporting country and destination markets of again either the United States or EU. Exchange rate risk is estimated by the standard deviation of the monthly exchange rate returns in a particular year although various measures of volatility exist in the related empirical literature. We include both the level, with one lag, and the change in volatility in the equation. Hence, we try to capture the effect of a change as well as the level effect on trade flows.

---

[*] Constant 2000 USD GDP values are used for all countries.

10.3 RESULTS AND DISCUSSION

We estimated Equation 10.1 using ordinary least squares (OLS), one-way fixed and random effects model and presented the results in Table 10.1 for two dependent variables corresponding to two destination markets. The first dependent variable is the share of exports in total to the United States, Mexico, and Canada, hence to the members of the NAFTA. Given that the majority of export payments are denominated in U.S. dollars, we use the economic activity and CPI of United States to represent this region. Moreover, since we measure currency risk as the volatility of exchange rates between local economy and the United States, we expect volatility measures, both as a level and as the difference, to have a negative effect on the share of exports to NAFTA countries. Our second dependent variable is the share of exports to EU countries. If the U.S. dollar is still the choice of transaction currency in trade between the exporting country and destination countries, we still expect a negative relationship between currency risk and exports. However, if the transaction currency used is Euro or any other currency, we might expect to see the share of exports to EU members to increase because of the third-country effect as described by Cushman (1986). As a result, we do not form any expectations on the sign of the volatility coefficients when the destination region is EU countries.

In Table 10.1, we present both least squares (LS) and panel data estimation results. A high value of the Lagrange multiplier (LM) test, as reported in the table, indicates that generalized least squares (GLS) should be preferred over

TABLE 10.1   Estimation Results

Variable	Export to the United States			Export to EU		
	LS	Fixed	Random	LS	Fixed	Random
Relative price	0.71971	0.45677	0.63131	0.29758	0.23330	0.28467
	0.58557	0.69619	0.62673	0.23826	0.28858	0.25534
Economic activity	4.35616	3.78413	4.18718	0.02231	0.05461	0.02962
	4.28058	4.40252	4.34846	0.24868	0.26051	0.25601
Exchange rate	0.00224	0.00189	0.00208	0.00285**	0.00283**	0.00283**
	0.00296	0.00308	0.00303	0.00120	0.00127	0.00124
Volatility	0.00282	0.00223	0.00263	0.00141	0.00258	0.00176
	0.00657	0.00697	0.00677	0.00267	0.00288	0.00278
Δ (Volatility)	−0.00132	−0.00179	−0.00149	−0.00301	−0.00271	−0.00294
	0.01009	0.01077	0.01043	0.00409	0.00442	0.00426
LM test	0.55			2.80*		
Hausman χ^2	1.17			1.87		

Note: ***, **, * represent statistical significance at 1%, 5%, and 10%, respectively.

OLS.* In addition, the high values of Hausman χ^2 indicate that we should prefer the fixed effects model over the random effects model.† Based on the coefficient estimates in Table 10.1, we can argue that currency risk has no significant impact on share of exports, neither for exports to NAFTA members nor to EU member countries, for the entire estimation period. The existence of structural breaks in our sample period can be the source of this insignificant result.

Because of the several important events in global economies during the entire estimation period, it might be important to evaluate the hypothesis for subperiods. In particular, we can distinguish pre- and post-2002 periods from each other as world economies enjoyed a stable growth environment after 2002. Thus, we divide our sample into two subperiods corresponding to a volatile and stable environment. The volatile environment includes observations up to 2001, whereas stable environment includes post-2002 period. Table 10.2 presents estimation results. The results indicate that

TABLE 10.2 Coefficients Estimates on Volatility for Subperiods

		Export to the United States			Export to EU		
		LS	Fixed	Random	LS	Fixed	Random
Period 1	Volatility	-0.01180	-0.01166	-0.01225	0.00558	0.00664	0.00563
		0.01387	0.01563	0.01495	0.00692	0.00747	0.00696
	Δ (Volatility)	0.00235	0.00244	0.00258	-0.00413	-0.00326	-0.00408
		0.01329	0.01589	0.01469	0.00664	0.00761	0.00670
	LM test	2.43			0.00		
	Hausman χ^2	2.16			2.21		
Period 2	Volatility	0.00724	-0.00027	0.00689	-0.00068	0.00008	-0.00039
		0.00713	0.00840	0.00728	0.00148	0.00161	0.00147
	Δ (Volatility)	-0.01253	-0.02081	-0.01297	-0.00391	-0.00306	-0.00368
		0.01975	0.02238	0.02014	0.00405	0.00422	0.00397
	LM test	0.35			2.90*		
	Hausman χ^2	7.64			3.00		

Note: ***, **, * represent statistical significance at 1%, 5%, and 10%, respectively.

---

* The high values of the LM test suggest that exogenous factors that may be correlated with the dependent variable and possibly omitted from the model are not correlated with the right-hand side variables, which results in inefficient OLS estimates; whereas GLS gives efficient estimates.

† The null hypothesis states no correlation; thus, low values of Hausman's χ^2 test suggest statistical preference for a random effects model specification. It suggests that the differences between cross-section units are not simply parametric shifts of the regression function; hence, it is more appropriate to view country-specific constant terms as being randomly distributed across firms.

TABLE 10.3 Coefficients Estimates on Volatility for Subperiods

			Export to the United States					Export to EU				
			LS	Fixed	Random	LM Test	Hausman χ²	LS	Fixed	Random	LM Test	Hausman χ²
EU-N10	Period 1	Volatility	-0.10040*	0.10265	-0.08754	0.55	1.64	0.031212*	0.03475	0.03118	4.66**	0.71
			0.05382	0.13645	0.06511			0.01776	0.04592	0.02053		
		Δ(Volatility)	-0.29803**	-0.14224	-0.28348*			-0.10547**	-0.11230	-0.10596**		
			0.14423	0.17244	0.14792			0.05134	0.06791	0.05712		
	Period 2	Volatility	-0.00262	-0.00349	0.00070	0.32	2.50	0.00690	0.03053	0.00825	0.52	2.08
			0.09495	0.17472	0.11251			0.01518	0.02527	0.01622		
		Δ(Volatility)	0.00052	-0.00681	-0.00151			-0.00065	0.00154	-0.00047		
			0.01420	0.01776	0.01521			0.00249	0.00296	0.00259		
CEE	Period 1	Volatility	0.00174	-0.00240	0.00026	0.63	2.63	0.00308*	0.00302	0.00309*	1.23	1.5
			0.00685	0.00772	0.00724			0.00157	0.00187	0.00171		
		Δ(Volatility)	-0.01534	0.02672	-0.00172			-0.00296	0.00153	-0.00246		
			0.02172	0.03157	0.02532			0.00499	0.00725	0.00552		
	Period 2	Volatility	-0.00536	-0.01060	-0.00962	0.20	1.29	0.00065	0.00090	0.00084	0.03	0.41
			0.01313	0.01307	0.01296			0.00057	0.00058	0.00058		
		Δ(Volatility)	-0.96682	-1.18943*	-1.13434			0.0493*	0.06283*	0.05911**		
			0.61756	0.67665	0.65202			0.02655	0.03138	0.02910		

Period 1	Volatility	−0.01357	−0.01933	−0.01461	3.05*	0.01173	0.00534	0.01125	0.21	1.68
		0.03448	0.04076	0.03801		0.01727	0.02017	0.01787		
	Δ (Volatility)	−0.00141	0.00282	0.00011		−0.00689	−0.00338	−0.00596		
		0.02177	0.02716	0.02466		0.01119	0.01324	0.01186		
Period 2	Volatility	−0.09586	−0.09070	−0.09286	1.18	−0.01355	−0.00955	−0.01309	1.26	0.58
		0.13199	0.17396	0.14087		0.03280	0.03840	0.03503		
	Δ (Volatility)	−0.05975	−0.08946	−0.06164		−0.01868	−0.01288	−0.01575		
		0.05829	0.07266	0.06201		0.01452	0.01625	0.01513		

CIS

1.50   1.30

Note: ***, **, * represent statistical significance at 1%, 5%, and 10% respectively.

there is no significant relationship between currency risk and share of exports for both the destination regions and for both the subperiods.

Because we do not find any significant relationship between currency risk and trade, we segment our sample in detailed segments based on both the time periods and the country categories, EU-N10, CEE, or CIS. The results are presented in Table 10.3. We find that currency risk has a negative impact on share of exports to NAFTA member countries for the first period for the EU-N10 countries. Moreover, both the level and change in volatility have the depressing effect on trade. Consistent with our expectation, a higher currency risk, as expressed by the volatility of U.S. dollar and local currency, does not lead to a decline in the share of exports to EU member countries. Based on LM and Hausman tests, the preferred model is the one-way fixed effects model, and it shows that there is no significant relationship between currency risk and trade. However, given the widespread preference of LS specification in estimations, we can also argue that currency risk leads to an increase in the share of exports. This finding of larger share of exports might be caused by what Cushman (1986) calls as the third-country effect. That is, a higher risk between U.S. dollar and local currency can shift exports to other regions that have lower risk. For the second subperiod, however, there does not seem to be a significant relationship between risk and trade.

For the CEE countries, we do not find any effect of currency risk on share of exports to NAFTA countries. On the other hand, for exports to EU members, we again find some, though weak, effect of currency risk on share of exports. While the level of volatility impact the share of exports positively in the first period, the change in volatility, but not the level, has a positive impact in the second period. For the CIS countries, we do not find any relationship between currency risk and the share of exports. Overall, consistent with the existent literature, the role of currency risk on trade flows appears to be insignificant, or it is too small to be detected statistically as Gagnon (1993) argues. However, there is some weak evidence supporting the third-country effect. Moreover, it can be argued that currency risk and its effect on trade flows are not impacted by the EU membership as shown by our findings in Table 10.3.

10.4 CONCLUSION

We investigated the effect of currency risk, as measured by the volatility of exchange rates between U.S. dollar and local currency, on the share of exports to NAFTA and EU member countries by the transition economies.

To measure exchange rate volatility, we use monthly exchange rates between U.S. dollar and the local currency. The countries we include in the analysis can fall under three groups: EU-N10, CEE, and CIS. We argue that exchange rate risk will have a depressing effect on share of exports to NAFTA members, and it will have a positive or no effect on share of exports to EU member countries.

Consistent with earlier literature, we do not find a strong relationship between currency risk and share of exports. However, under country and time segments, we find that currency risk, for the EU-N10 country group, causes a decline in share of exports to NAFTA member countries for the first subperiod ending in 2001. On the other hand, there is a positive impact on share of exports to EU member countries for the same time period. We also find some weak evidence that the share of exports to EU member countries is positively impacted for CEE countries for both periods. There does not seem to be any relationship between risk and trade for the CIS countries. Overall, the role of currency risk on exports does not appear to be large enough to cause concerns on the traders' behavior. This might be either due to size of the effect or the existence of hedging tools.

REFERENCES

Assery, A. and Peel, D.A. (1991) The effects of exchange rate volatility on exports: Some new estimates. *Economics Letters*, 37(2): 173–177.

Bahmani-Oskooee, M. (2002) Does black market exchange rate volatility deter the trade flows? Iranian experience. *Applied Economics*, 34(18): 2249–2255.

Baron, D. (1970) Price uncertainty, utility, and industry equilibrium. *International Economic Review*, 11(3): 463–480.

Baum, C.F., Çağlayan, M., and Ozkan, N. (2004) Nonlinear effects of exchange rate volatility on the volume of bilateral exports. *Journal of Applied Econometrics*, 19(1): 1–23.

Clark, P.B. (1973) Uncertainty, exchange risk, and the level of international trade. *Western Economic Journal*, 11(3): 302–313.

Cushman, D.O. (1986) Has exchange risk depressed international trade? The impact of third country exchange risk. *Journal of International Money and Finance*, 5(3): 361–378.

De Grauwe, P. (1988) Exchange rate variability and the slowdown in growth of international trade. *International Monetary Fund Staff Papers*, 35(1): 63–84.

Ethier, W. (1973) International trade and the forward exchange market. *American Economic Review*, 63(3): 494–503.

Franke, G. (1991) Exchange rate volatility and international trading strategy. *Journal of International Money and Finance*, 10(2): 292–307.

Gagnon, J.E. (1993) Exchange rate variability and the level of international trade. *Journal of International Economics*, 34(3/4): 269–287.

Giovannini, A. (1988) Exchange rates and traded goods prices. *Journal of International Economics*, 24(3/4): 317–330.

Grobar, L.M. (1993) The effect of real exchange rate uncertainty on LDC manufactured exports. *Journal of Development Economics*, 41(2): 367–376.

Holthausen, D.M. (1979) Hedging and the competitive firm under price uncertainty. *American Economic Review*, 69(5): 989–995.

Hooper, P. and Kohlhagen, S.W. (1978) The effect of exchange rate uncertainty on the prices and volume of international trade. *Journal of International Economics*, 8(4): 483–511.

Kawai, M. and Zilcha, I. (1986) International trade with forward-futures markets under exchange rate and price uncertainty. *Journal of International Economics*, 20(1): 83–98.

Klein, M.W. (1990) Sectoral effects of exchange rate volatility on United States exports. *Journal of International Money and Finance*, 9(3): 299–308.

Kumar, R. and Dhawan, R. (1991) Exchange rate volatility and Pakistan's exports to the developed world, 1974–1985. *World Development*, 19(9): 1225–1240.

Leamer, E.E. (1985) Sensitivity analyses would help. *American Economic Review*, 75(3): 308–313.

Leamer, E.E. and Leonard, H. (1983) Reporting the fragility of regression estimates. *Review of Economics and Statistics*, 65(2): 306–317.

Neumann, M. (1995) Real effects of exchange rate volatility. *Journal of International Money and Finance*, 14(3): 417–426.

Pozo, S. (1992) Conditional exchange rate volatility and the volume of international trade: Evidence from the early 1900s. *The Review of Economics and Statistics*, 74(2): 325–329.

Sandmo, A. (1971) On the theory of the competitive firm under price uncertainty. *American Economic Review*, 61(1): 65–73.

Solakoglu, M.N. (2005) Exchange rate volatility and real exports: A sensitivity analysis. *Journal of Economic and Social Research*, 7(1): 1–30.

Solakoglu M.N., Solakoglu E.G., and Demirag, T. (2008) Exchange rate volatility and exports: A firm-level analysis. *Applied Economics*, 40(7): 921–929.

Thursby, J.G. and Thursby, M.C. (1987) Bilateral trade flows, the linder hypothesis, and exchange risk. *The Review of Economics and Statistics*, 69(3): 488–495.

CHAPTER 11

Stock Market Volatility of European Emerging Markets as Signals to Macroeconomic Activities

Bülent Köksal and Mehmet Orhan

CONTENTS

11.1 INTRODUCTION

The behavior of volatility in emerging markets deserves to be the subject matter of empirical studies since the asset allocation and investment decisions in these markets are closely related to their volatility. It is well known that the emerging market economies provide the investors with high levels of returns but at the expense of the risk associated therein. Table 11.1 reports the GDP per capita for the emerging markets in 2007 which reveals that these markets display huge differences in main macroeconomic indicators.

This chapter is devoted to the correlations between the volatilities of some selected European emerging market economies and developed countries as an attempt to improve our understanding of the simultaneous behavior of volatilities. Utilizing the results from the literature that documents the connection between real economies and the stock market volatilities, our purpose is to determine the strength of the connection between real economies by using the stock market volatilities. We begin by a literature survey of the relation between stock market volatility and macroeconomic variables.

11.1.1 Stock Market Volatility and Main Macroeconomic Variables

Main macroeconomic indicators have long been studied as determinants of the stock market volatility. It is plausible to consider that the stock market

TABLE 11.1   GDP Per Capita for the Emerging Markets in 2007

Asia		Latin America		Europe		Africa	
China	$2,483	Chile	$9,884	Poland	$11,072	Morocco	$2,422
Korea	$20,015	Mexico	$9,717	Hungary	$13,745	South Africa	$5,916
Taiwan	$16,697	Venezuela	$8,282	Turkey	$9,569	Jordon	$2,766
India	$942	Brazil	$6,938	Czech Republic	$16,956	Egypt	$1,739
Malaysia	$6,956	Colombia	$4,264	Israel	$23,579		
Indonesia	$1,925	Argentina	$6,609	Russia	$9,075		
Thailand	$3,732	Peru	$3,826				
Philippines	$1,626						
Pakistan	$909						
Sri Lanka	$1,623						

Source: IMF, World Economic Outlook Database, October 2008. With permission.

indices are shaped, at least partly, by economic activities of a country. There is a vast amount of literature on the relationship between stock markets and macroeconomic fundamentals. Beltratti and Morana (2006) attempts to find an answer to the question raised by Schwert (1989) addressing the factors leading to the stock market volatility. They analyze the macroeconomic causes of stock market volatility and conclude that the direction of causality is stronger from macroeconomic to stock market volatility while stock market volatility also affects macroeconomic volatility. The remarkable study by Schwert mainly examines the volatilities of the macroeconomic variables to figure out the preceding macro variables. Schwert finds that inflation and money growth volatility predict stock market volatility, but only for some subperiod of the analysis, and industrial output volatility predicts the stock market volatility weakly. On the other hand, Schwert reports that stock market volatility helps predict money growth and industrial output volatility and does not help predict inflation volatility. The findings of Schwert can be criticized due to the use of econometric techniques which are more than two decades old. These techniques could have suffered from bias and have to be revised. To this end, Beltratti and Morana (2006) find that the stochastic process volatility belonging to the U.S. stock market can be characterized by the structural change and long memory. With the help of the model including these features, they find that the causality from macroeconomic volatility to the stock market volatility is stronger than the otherwise direction. Furthermore, the fractional cointegration analysis they conduct indicates that the cointegrating vectors link output growth, money growth, stock market return, the Federal funds rate, and inflation volatility for the long run.

In econometrics, cointegration analysis is used to figure out the long run relations between variables. Binswanger (2004) makes use of the structural VAR model to claim that shocks to the real economy explain a substantial proportion of variability in real stock prices in the United States and Japan over 1960s and 1970s. The structural vector autoregression (VAR) approach is appropriate in analyzing the movements of the stock market prices after both fundamental and nonfundamental shocks. In an earlier study, Bittlingmayer (1998) focuses on political instability as a source of the stock market return volatility. He claims that the increase in stock market volatility is a consequence of output decline and uses the history of Germany over 1880–1940 to conclude that both output decline and stock market volatility are stemming from political instability. Lee (1992) makes use of a multivariate VAR model to study the causal relations and dynamic interactions among asset returns, real economic activity, and inflation in the postwar United States. Lee finds that stock returns Granger-cause real activity with the help of cross correlations. In a similar study, James et al. (1985) examine the causal links between stock returns, real activity, money supply, and inflation. They find evidence that stock returns lead to both changes in real activity and changes in money growth.

Boucher (2007) explores the adjustment of stock prices to their fundamental value in order to better predict the stock prices in a VAR framework. Boudoukh and Richardson (1993) report strong support for a positive relation between nominal stock returns and inflation at long horizons. In a more recent study, Chaudhuri and Smiles (2004) examine the relationship between stock market volatility and macro variables in Australia by using multivariate cointegration. They provide evidence on the long-run relationship between real GDP, private consumption, money, and price of oil.*

This chapter intends to examine the correlations between stock market volatilities of some EU countries and emerging European markets with a purpose of better understanding the connection between emerging and developed real economies of Europe as well as the United States.

11.1.2 GARCH Framework for Modeling Conditional Volatility

Generalized Autoregressive Conditional Heteroskedasticity (GARCH) models are used extensively in financial studies since these models are convenient in modeling the conditional volatility in the financial markets. Indeed, homoskedasticity—defined as the constant error variance—is a standard assumption of the classical model to be fulfilled, and in case it is violated, the Gauss–Markov theorem cannot guarantee the minimum variance of the least squares estimators. Since financial data usually proves to be heteroskedastic, use of GARCH models is almost the common practice in estimating the regression models of the financial markets. The volatility of the return series are clustered for many of the financial market series. Furthermore, the return distributions usually have fat tails, which can better be characterized by Student's *t*-distribution rather than the normal distribution.

---

* There is a vast literature about the relationship between macrovariables and the stock market prices which continue to grow. For other examples, see Fama (1981), French et al. (1983), Geske and Roll (1983), Chen et al. (1986), Diebold and Yilmaz (2008), and Engle et al. (2008).

In a typical univariate GARCH(p,q) model, the error term variances are expressed in terms of its past values as well as the past values of the error terms:

$$r_t = \mu + \varepsilon_t$$
$$\sigma_t^2 = \psi + \alpha(L)\,\varepsilon_t^2 + \beta(L)\,\sigma_t^2$$

where
r_t is the return series
μ is the mean of the return series
ε_t is the error term
σ_t^2 is the conditional variance of the error terms
ψ is the intercept parameter of the variance equation
$\alpha(L)$ and $\beta(L)$ are the lag operators, i.e., $\alpha(L) = \alpha_1 L + \alpha_2 L^2 + \alpha_3 L^3 + \cdots + \alpha_p L^p$ and $\beta(L) = \beta_1 L + \beta_2 L^2 + \beta_3 L^3 + \cdots + \beta_p L^p$

The choice of the lag period is very critical and one has to find the most appropriate time lag. Akaike's information criterion and Schwartz criterion are the most commonly used tools to select the optimal lag. The simplest GARCH model, GARCH(1,1) lags both squared error term and the variance of the error term for one period, i.e.,

$$\sigma_t^2 = \psi + \alpha_1\,\varepsilon_{t-1}^2 + \beta_1 \sigma_{t-1}^2$$

Several versions of the GARCH models are developed to handle specific features of financial markets. For instance, GARCH-in-mean (GARCH-M) model proposed by Engle et al. (1987) allows the return of a security to depend on its volatility by including the conditional variance term in the return equation. More recent extension of GARCH models attempt to attribute some different characteristics of financial time series and integrate them to the GARCH-M specifications. To this end, threshold-GARCH (TARCH), model by Zakoian (1994) help to identify the asymmetric effects of positive and negative shocks to the return series, whereas the component-GARCH (CGARCH), model decomposes the variation into short- and long-run components. Several other specifications include Quadratic-GARCH (QGARCH) by Sentana (1995), the GJR-GARCH model by Glosten et al. (1993), the nonlinear-GARCH (NGARCH), and the Integrated-GARCH (IGARCH).*

11.2 DATA AND EMPIRICAL MODEL

We obtain data for main stock market indices from Global insight databases and the Web sites of the country stock markets. Our sample period covers the years from 2002 to 2007. For each stock market index, we calculate the daily compounded index returns and model the conditional volatility of these returns for each year to examine the yearly changes between correlations of the real economies. Table 11.2 reports the indices and EU entry years of the countries we have used in this study. We estimate the correlations between conditional volatilities for each country pair, but we selectively report the results to save space as described in Section 11.3.

Many economic time series follow periods of unusual large volatility, followed by periods of relative stability. Stock market index is an example. Stock prices reflect the uncertainties and risks in the economy, and more uncertain periods are associated with more volatile stock markets, as the investors have difficulties to evaluate their estimates of current stock prices. Under these circumstances, the conditional variance of the stock market is a better indicator of the current economic conditions rather than the unconditional variance. In other words, an estimate of the stock market volatility over the last 30 days, for example, is a better indicator than the stock market volatility calculated by using the data from the last 10 years.

If two countries have a strong economic relationship, then their economies will be affected by similar factors. A recession that causes an increase in the unemployment in the first country will have a significant impact on the economy of the second country. Conversely, a completely closed economy is not affected by shocks to the economies of other countries. Engle (1982) shows how to model the mean and the variance of the economy. Bollerslev (1986) generalizes Engle's work by allowing the conditional variance to be an ARMA process. The literature continued to grow by extending these works to the case of vector processes. Early articles on multivariate extensions are Engle et al. (1986), Diebold and Nerlove (1989), and Bollerslev et al. (1988).

---

* See Bolleslev (2008) for a glossary of ARCH (GARCH) models.

TABLE 11.2   Countries, EU Entry Years, and Stock Market Indices

Country	EU Entry Year	Index
Belgium	1951	BEL-20
France	1951	CAC-40
Germany	1951	DAX
Italy	1951	MIB-30 (mibtel)
Luxembourg	1951	LuxX Index
The Netherlands	1951	AEX
Denmark	1973	(OMXC) (all share)
Ireland	1973	ISEQ Overall Index
United Kingdom	1973	FTSE-100
Greece	1981	Athex Composite
Portugal	1986	PSI-20
Spain	1986	IBEX-35
Austria	1995	ATX
Finland	1995	OMX Helsinki 25 (OMXH25)
Sweden	1995	Stockholm General (OMX)
Cyprus	2004	CySE GENERAL
Czech Republic	2005	PX-50
Estonia	2006	OMX Tallinn (OMXT)
Hungary	2007	Bux Composite Index
Latvia	2008	OMX Riga (OMXR)
Lithuania	2009	OMX Vilnius (OMXV)
Malta	2010	MSI
Poland	2011	WIG 20
Slovakia	2012	Sax
Slovenia	2013	SBI 20
Bulgaria	2007	Sofix
Romania	2007	Bet Index
Croatia	Candidate	Crobex
Macedonia	Candidate	MBI10
Turkey	Candidate	ISE-100 Index
Russia		RTS

We follow Bollerslev (1990) to determine the correlations between stock markets of two countries by using a multivariate GARCH model. Specifically, we estimate the following model for each pair of countries:

$$\mathbf{r}_t = \boldsymbol{\mu}_t + \boldsymbol{\varepsilon}_t \quad \text{(Mean model)}$$

$$\text{Var}(\boldsymbol{\varepsilon}_t \mid \mathbf{I}_{t-1}) = \Sigma_t \quad \text{(Volatility model)}$$

where

r_t is a 2×1 time-series vector of compounded stock market index returns

$\mu_t = E(r_t | I_{t-1})$ is the conditional expectation of r_t given the past information I_{t-1}

ε_t is the 2×1 vector of shocks at time t

Σ_t is the conditional covariance matrix of ε_t which is a 2×2 positive-definite matrix defined as $\Sigma_t = \text{Cov}(\varepsilon_t | I_{t-1})$

Let σ_{12t} denote the off-diagonal element in Σ_t. Then the correlation coefficient between ε_{1t} and ε_{2t} is given by $\rho_{12t} = \sigma_{12t} / \sqrt{\sigma_{11t}\sigma_{22t}}$ which is time-varying. Following Bollerslev (1990), we assume that $\mu_t = (\alpha_1, \alpha_2)$, the conditional variances follow a GARCH(1,1) structure and $\rho_{12t} = \rho_{12}$, i.e., the correlation is constant. With these assumptions, our model simplifies to

$$r_{it} = \mu_i + \varepsilon_{it}$$
$$\sigma_{iit} = \phi_i + \alpha_{i1}\varepsilon_{it-1}^2 + \beta_{i1}\sigma_{iit-1}$$
$$\sigma_{ijt} = \rho_{ij}(\sigma_{iit}\sigma_{jjt})^{1/2}, \tag{11.1}$$
$$i, j = \text{Country 1 and Country 2}$$

We estimate the model in Equation 11.1 by using the maximum likelihood method under the assumption of conditional normality for each pair of countries. The optimization method is the BFGS (Broyden, Fletcher, Goldfarb, and Shanno), which is described in Press et al. (1988). In theory, we could estimate a multivariate GARCH model for all return series in which the ρ's are time varying, or use a VECH model which allows complete interaction among the terms in the conditional variance, but this requires estimation of hundreds of parameters which is infeasible in practice. Besides it is not necessary for our analysis, because we would like to estimate the average correlation between the conditional volatilities of the country stock markets for each year. We are not interested in daily changes in these correlations.

11.3 RESULTS

Estimated correlations between volatilities are reported in Tables 11.3 through 11.9. From this point on, we will use the term "correlation" to mean the correlation between the *volatilities* of the stock markets. For parsimony, we only report the correlations between the United States,

TABLE 11.3   Correlations between Czech Republic and Other Countries

	2002	2003	2004	2005	2006	2007
Hungary	0.439***	0.323***	0.330***	0.547***	0.625***	0.587***
Poland	0.537***	0.294***	0.352***	0.484***	0.589***	0.596***
Russia	0.278***	0.167***	0.248***	0.461***	0.576***	0.642***
Turkey	0.179***	0.159***	0.194***	0.415***	0.472***	0.592***
France	0.394***	0.347***	0.365***	0.325***	0.507***	0.672***
Germany	0.369***	−0.002***	0.396***	0.273***	0.491***	0.645***
Italy	0.361***	0.286***	0.331***	0.271***	0.488***	0.622***
Spain	0.402***	0.320***	0.379***	0.327***	0.524***	0.645***
United Kingdom	0.403***	0.350***	0.297***	0.379***	0.549***	0.656***
United States	0.257***	0.164***	0.152***	0.080	0.224***	0.327***

TABLE 11.4   Correlations between Hungary and Other Countries

	2002	2003	2004	2005	2006	2007
Czech Rep.	0.439***	0.323***	0.330***	0.547***	0.625***	0.587***
Poland	0.487***	0.413***	0.341***	0.652***	0.656***	0.584***
Russia	0.346***	0.084***	0.120**	0.482***	0.591***	0.493***
Turkey	0.265***	0.159***	0.118*	0.372***	0.529***	0.522***
France	0.422***	0.249***	0.330***	0.291***	0.472***	0.565***
Germany	0.403***	0.276***	0.347***	0.230***	0.445***	0.561***
Italy	0.369***	0.200***	0.314***	0.264***	0.412***	0.528***
Spain	0.384***	0.279***	0.307***	0.293***	0.447***	0.522***
United Kingdom	0.364***	0.276***	0.287***	0.343***	0.463***	0.529***
United States	0.439***	0.323***	0.330***	0.547***	0.625***	0.587***

TABLE 11.5   Correlations between Poland and Other Countries

	2002	2003	2004	2005	2006	2007
Czech Rep.	0.537***	0.294***	0.352***	0.484***	0.589***	0.596***
Hungary	0.487***	0.413***	0.341***	0.652***	0.656***	0.584***
Russia	0.384***	0.230***	0.235***	0.495***	0.572***	0.571***
Turkey	0.179***	0.159***	0.194***	0.415***	0.472***	0.592***
France	0.352***	0.417***	0.356***	0.399***	0.455***	0.655***
Germany	0.360***	0.337***	0.355***	0.321***	0.433***	0.624***
Italy	0.362***	0.342***	0.362***	0.368***	0.423***	0.602***
Spain	0.386***	0.404***	0.411***	0.383***	0.428***	0.625***
United Kingdom	0.348***	0.331***	0.283***	0.464***	0.498***	0.649***
United States	0.537***	0.294***	0.352***	0.484***	0.589***	0.596***

TABLE 11.6 Correlations between Russia and Other Countries

	2002	2003	2004	2005	2006	2007
Czech Rep.	0.278***	0.167***	0.248***	0.461***	0.576***	0.642***
Hungary	0.346***	0.084***	0.120**	0.482***	0.591***	0.493***
Poland	0.384***	0.230***	0.235***	0.495***	0.572***	0.571***
Turkey	0.136**	0.104*	0.155***	0.382***	0.499***	0.671***
France	0.378***	0.246***	0.200***	0.272***	0.463***	0.602***
Germany	0.344***	0.203***	0.198***	0.224***	0.396***	0.596***
Italy	0.327***	0.207***	0.186***	0.261***	0.466***	0.566***
Spain	0.382***	0.232***	0.254***	0.272***	0.414***	0.569***
United Kingdom	0.393***	0.284***	0.194***	0.347***	0.499***	0.577***
United States	0.177***	0.181***	0.095	0.066	0.154**	0.261***

TABLE 11.7 Correlations between Turkey and Other Countries

	2002	2003	2004	2005	2006	2007
Czech Rep.	0.288***	−0.010	0.140**	0.405***	0.543***	0.684***
Hungary	0.265***	0.159***	0.118*	0.372***	0.529***	0.522***
Poland	0.179***	0.159***	0.194***	0.415***	0.472***	0.592***
Russia	0.136**	0.104*	0.155***	0.382***	0.499***	0.671***
France	0.204***	0.113*	0.167***	0.387***	0.402***	0.662***
Germany	0.267***	0.103**	0.180***	0.326***	0.370***	0.656***
Italy	0.207***	0.099*	0.144**	0.325***	0.321***	0.653***
Spain	0.240***	0.120**	0.237***	0.392***	0.402***	0.629***
United Kingdom	0.216***	0.117**	0.145**	0.351***	0.432***	0.668***
United States	0.179***	0.159***	0.194***	0.415***	0.472***	0.592***

TABLE 11.8 Correlations between France and Other Countries

	2002	2003	2004	2005	2006	2007
Czech Rep.	0.394***	0.347***	0.365***	0.325***	0.507***	0.672***
Hungary	0.422***	0.249***	0.330***	0.291***	0.472***	0.565***
Poland	0.352***	0.417***	0.356***	0.399***	0.455***	0.655***
Russia	0.378***	0.246***	0.200***	0.272***	0.463***	0.602***
Turkey	0.200	0.113*	0.167***	0.387***	0.402***	0.660***
Germany	0.860***	0.854***	0.922***	0.907***	0.947***	0.933***
Italy	0.895***	0.882***	0.867***	0.856***	0.894***	0.928***
Spain	0.862***	0.875***	0.856***	0.849***	0.903*	0.884***
United Kingdom	0.861***	0.826***	0.835***	0.838***	0.871***	0.920***
United States	0.524***	0.527***	0.433***	0.355***	0.544***	0.571***

TABLE 11.9   Correlations between Germany and Other Countries

	2002	2003	2004	2005	2006	2007
Czech Rep.	0.369***	−0.002	0.396***	0.273***	0.491***	0.640
Hungary	0.403***	0.276***	0.347***	0.230***	0.445***	0.561***
Poland	0.360***	0.337***	0.355***	0.321***	0.433***	0.624***
Russia	0.344***	0.203***	0.198***	0.224***	0.396***	0.596***
Turkey	0.267***	0.103***	0.180***	0.326***	0.370***	0.656***
France	0.860***	0.854***	0.922***	0.907***	0.947***	0.933***
Italy	0.836***	0.802***	0.876***	0.839***	0.894***	0.870***
Spain	0.765***	0.790***	0.844***	0.827***	0.888***	0.865***
United Kingdom	0.755***	0.690***	0.819***	0.760***	0.851***	0.884***
United States	0.643***	0.642***	0.463***	0.374***	0.587***	0.530***

five highest GDP EU countries, and five emerging market economies of Europe, i.e., Czech Republic, Hungary, Poland, Turkey, and Russia. We do not report the significance levels of the correlations but rather use the conventional notation of stars. As usual, (***), (**), and (*) indicate estimated correlation coefficients that are significant at 1%, 5%, and 10% levels, respectively.

We start with the correlations between the Czech Republic and other countries. Table 11.3 reports the results. It is interesting to see that almost all estimated correlation coefficients are positive except for the one with Germany in 2003 and significant at 1% level except for the one with the United States in 2005. This indicates that the conditional volatilities of the stock market indices of other countries and the stock market volatility of Czech Republic move in the same direction. The magnitudes of the correlations show the degree to which economies of these countries are affected by similar factors. As discussed in Section 11.3, the correlations between stock market volatilities are reduced form proxies for the connection between the real economies. Figure 11.1 displays graphs produced by using data from Table 11.3. In Figure 11.1, "Emerging average" and "EU average" refer to the yearly averages of the correlations for emerging and EU economies in Table 11.3 respectively.* The correlations between Czech Republic and the United States are generally less than the group of the EU countries. This is not surprising since establishing close economic ties with the United States is much more difficult because of the geographical

---

* Figures 11.3 through 11.9 in the following text are produced similarly.

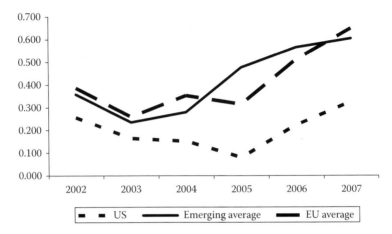

FIGURE 11.1   Correlations between Czech Republic and others.

location. The other point that deserves attention is the relatively high correlation with Hungary and Poland. Indeed, Czech Republic, Hungary, and Poland are following similar routes of liberalism in the aftermath of USSR's collapse. In addition, Czech Republic and Poland are neighbor countries, and Hungary is located closely to the Czech Republic. Another noteworthy observation from Figure 11.1 is that the connection between economies of the Czech Republic and other countries strengthen after the entry of the Czech Republic to the EU.

Table 11.4 reports the correlation coefficients between Hungary and other countries. Again almost all correlations are highly significant and positive implying that conditional variances of all stock market indices are moving in the same direction. The two exceptions are in 2004, with Russia at 5% and with Turkey at 10% significance levels. Figure 11.2 shows that Hungary developed close economic ties with the EU countries after its entry to the EU in 2004. Similar to the Czech Republic, Hungary has the highest average correlation with Poland and in contrast to the Czech Republic, it has developed deeper economic ties with the United States after its entry to the EU.

We report the correlation coefficients between Poland and other countries in Table 11.5. All estimated correlation coefficients are positive and significant at the 1% level. Note that the lowest levels of correlations in 2002, 2003, and 2004 are with Turkey. The year with the greatest average is 2007 (Figure 11.3). Similar to the Czech Republic and Hungary, joining the EU seems to have a positive effect on the Polish economy in terms of the EU integration.

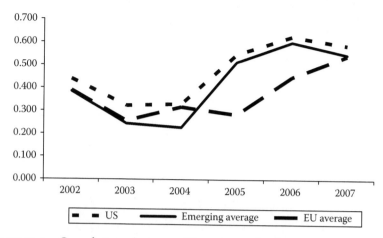

FIGURE 11.2   Correlations between Hungary and others.

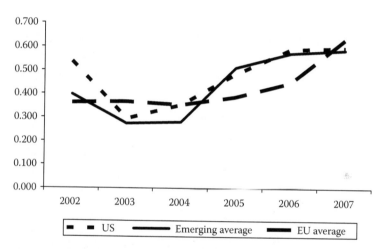

FIGURE 11.3   Correlations between Poland and others.

Recent global financial crisis that is triggered by the mortgage pay-back problem in the United States and which spread to the whole world instantly, have well proved that the Russian economy is not well organized and depends heavily on the energy exports. This fact makes the Russian economy very fragile to the surprises in the oil market and the exchange rate. Russian stock exchange had to be closed occasionally to prevent drastic falls beyond even the most pessimistic expectations. We report the correlations as well as the significance levels for Russia in Table 11.6. The first interesting point to note is the relatively lower correlation figures with the United States as well as the insignificant correlations in

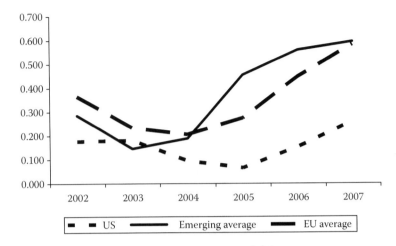

FIGURE 11.4 Correlations between Russia and others.

2004 and 2005. The lowest average correlations on the yearly basis are for 2003 and 2004. Figure 11.4 shows that although the correlation between Russia and the United States has increased since 2005, there still remains a large gap between EU average and the United States correlation.

The last emerging market economy we have in our agenda is Turkey. The Turkish economy had suffered so much from political instability and high and persistent inflation for more than three decades. The last economic crisis in 2001 had shrunk the Turkish GDP by about 25%. Besides, the majority of the Turkish land is situated in Asia the characteristics of which are somehow reflected in the Turkish culture and society. Nevertheless, Turkey started the formal negotiations with the EU on October 3, 2005, and closed its first chapter of negotiations in June 2006. Table 11.7 reveals that the correlations between Turkey and emerging markets are almost all positive and significant at 1% level. The unique exception is the insignificant correlation with the Czech Republic in 2003. The other point deserving attention is that the figures are somehow lower than the ones in the previous tables. Again the largest average belongs to 2007 (0.63) followed by 2006 (0.44) whereas the lowest average is in 2003 (0.11) and average correlations are very close for all countries (all ranging between 0.29 and 0.34). Figure 11.5 shows that in contrast to the new members of the EU that we have discussed earlier, we do not observe a significant difference between the United States and EU countries in terms of their correlations with Turkey. Finally, after the Justice and Development Part (AKP) assumed power in the 2002, we see stronger ties between Turkish and the

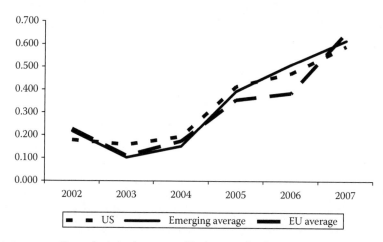

FIGURE 11.5   Correlations between Turkey and others.

EU economies as a result of a number of reforms AKP adopted which opened the way for increased economic and political stability.

Table 11.8 reports the results about the first EU giant economy, France. The first point we note is the set of high and significant correlations between France and other major economies of the EU. This figure is close to 90% throughout the whole period. We also note that the correlations with Turkey are relatively lower. As Figure 11.6 shows, the correlations between the United States are somewhat lower than that of the EU countries but still they are much larger compared to the emerging markets. This fact stems from the globalizations of the financial markets. Any factor

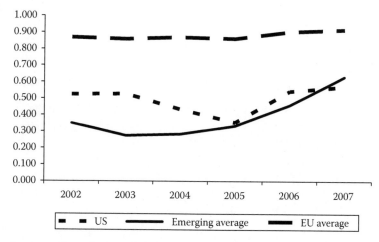

FIGURE 11.6   Correlations between France and others.

influencing the U.S. economy does influence the French economy as well and vice versa.

Table 11.9 reports the correlations between Germany, which has the highest GDP in Europe, and other countries. One can categorize the correlation coefficients as follows: The highest correlations are with the EU giants followed by the United States. The third category is the emerging market economies with the lowest correlations coming from the Turkish economy (see Figure 11.7).

We do not include the similar tables for Italy, Spain, and the United Kingdom since all display similar figures.

The correlations listed in Table 11.10 for the United States are somewhat different than those for the developed EU economies. First of all, all

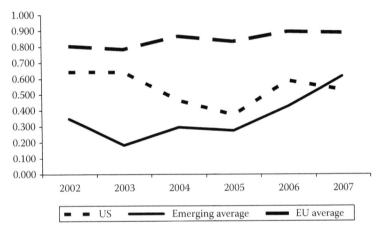

FIGURE 11.7 Correlations between Germany and others.

TABLE 11.10 Correlations between the United States and Other Countries

	2002	2003	2004	2005	2006	2007
Czech Rep.	0.257***	0.164***	0.152***	0.080	0.224***	0.327***
Hungary	0.439***	0.323***	0.330***	0.547***	0.625***	0.587***
Poland	0.537***	0.294***	0.352***	0.484***	0.589***	0.596***
Russia	0.278***	0.167***	0.248***	0.461***	0.576***	0.642***
Turkey	0.179***	0.159***	0.194***	0.415***	0.472***	0.592***
France	0.524***	0.527***	0.433***	0.355***	0.544***	0.571***
Germany	0.643***	0.642***	0.463***	0.374***	0.587***	0.530***
Italy	0.508***	0.526***	0.439***	0.394***	0.528***	0.546***
Spain	0.465***	0.515***	0.400***	0.365***	0.532***	0.478***
United Kingdom	0.481***	0.439***	0.414***	0.337***	0.504***	0.556***

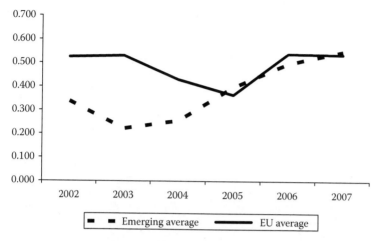

FIGURE 11.8 Correlations between the United States and others.

correlations are positive and significant at 1%. Secondly, the correlations are much lower. And finally, there is no clear-cut distinction between the correlations belonging to the developed EU countries and the emerging markets especially after 2004 (See Figure 11.8).

11.4 CONCLUSION

We can make the following remarks by using the empirical results that we discussed in Sections 11.1 through 11.3:

- The correlation coefficients indicate the coordinated behaviors of the countries. The behaviors of developed EU countries and the emerging market economies are having remarkably different correlations.

- Countries in the groups of developed EU countries and the emerging markets are characterized by similar behaviors whereas the United States is somewhat different.

- Almost all correlations are positive and significant which means that the stock indices of all countries included in this chapter exhibit similar volatility patterns.

- Figures 11.2 through 11.8 reveal that there is a pattern of convergence of volatility correlations between the economies. This can be seen as a consequence of the increased integration of the real economies.

- It would be interesting to extend the period of our study for a few years to see how the findings would be changing at times of global financial crises. Repeating our empirical work after a couple of years with the data covering the aftermath of the global crisis would fulfill a room on the topic.

REFERENCES

Beltratti, A. and Morana, C. (2006) Breaks and persistency: Macroeconomic causes of stock market volatility. *Journal of Econometrics*, 131(1–2): 151–177.

Binswanger, M. (2004) How important are fundamentals?—Evidence from a structural VAR model for the stock markets in the US, Japan and Europe. *Journal of International Financial Markets, Institutions and Money*, 14(2): 185–201.

Bittlingmayer, G. (1998) Output, stock volatility, and political uncertainty in a natural experiment: Germany, 1880–1940. *Journal of Finance*, 53(6): 2243–2257.

Bollerslev, T. (1986) Generalized autoregressive conditional heteroskedasticity. *Journal of Econometrics*, 31(3): 307–327.

Bollerslev, T. (1990) Modeling the coherence in short-run nominal exchange rates: A multivariate generalized ARCH model. *Review of Economics and Statistics*, 72(3): 498–505.

Bollerslev, T., Engle, R.F., and Wooldridge, J.M. (1988) A capital asset pricing model with time-varying covariances. *Journal of Political Economy*, 96(1): 116–131.

Bong-Soo Lee. (1992) Causal relations among stock returns, interest rates, real activity, and inflation. *Journal of Finance*, 47(4): 1591–1603.

Boucher, C. (2007) Asymmetric adjustment of stock prices to their fundamental value and the predictability of US stock returns. *Economics Letters*, 95(3): 335–347.

Boudoukh, J. and Richardson, M. (1993) Stock returns and inflation: A long-horizon perspective. *American Economic Review*, 83(5): 1346–1355.

Chaudhuri, K. and Smiles, S. (2004) Stock market and aggregate economic activity: Evidence from Australia. *Applied Financial Economics*, 14(2): 121–129.

Chen, N., Roll, R., and Ross, S.A. (1986) Economic forces and the stock market. *Journal of Business*, 59(3): 383–403.

Diebold, F.X. and Nerlove, M. (1989) Dynamic exchange rate volatility: A multivariate latent factor ARCH model. *Journal of Applied Econometrics*, 4(1): 1–21.

Diebold, F.X. and Yilmaz, K. (2008) Macroeconomic volatility and stock market volatility, worldwide, *National Bureau of Economic Research Working Paper Series No. 14269.*

Engle, R.F. (1982) Autoregressive conditional heteroskedasticity with estimates of the variance of United Kingdom inflation. *Econometrica*, 50(4): 987–1007.

Engle, R.F., Ghysels, E., and Sohn, B. (2008) On the economic sources of stock market volatility. Available at SSRN: http://ssrn.com/abstract=971310.

Engle, R.F., Granger, C.W.J., and Kraft, D. (1986) Combining competing forecasts of inflation using a bivariate ARCH model. *Journal of Economic Dynamics and Control*, 8(2): 151–165.

Engle, R.F., Lilien, D.M., and Robins, R.P. (1987) Estimating time varying risk Premia in the term structure: The ARCH-M model. *Econometrica*, 55(2): 391–407.

Fama, E.F. (1981) Stock returns, real activity, inflation and money. *American Economic Review*, 71(4): 545–565.

French, K.R., Ruback, R.S., and Schwert, G.W. (1983) Effects of nominal contracting on stock returns. *Journal of Political Economy*, 91(1): 70–96.

Geske, R. and Roll, R. (1983) The fiscal and monetary linkages between stock returns and inflation. *Journal of Finance*, 38(1): 1–31.

Glosten, L.R., Jagannathan, R., and Runkle, D.E. (1993) On the relation between the expected value and the volatility of the nominal excess return on stocks. *Journal of Finance*, 48(5): 1779–1801.

James, C., Koreisha, S., and Partch, M. (1985) A VARMA analysis of the causal relations among stock returns, real output, and nominal interest rates. *Journal of Finance*, 40(5): 1375–1384.

Press, W.H., Flannery, B.P., Teukolsky, S.A., and Vettering, W.T. (1988) *Numerical Recipes in C*. New York: Cambridge University Press.

Schwert, G.W. (1989) Why does stock market volatility change over time? *Journal of Finance*, 44(5): 1115–1153.

Sentana, E. (1995) Quadratic ARCH models. *Review of Economic Studies*, 62(4): 639–661.

Zakoian, J. (1994) Threshold heteroskedastic models. *Journal of Economic Dynamics Control*, 18(5): 931–944.

CHAPTER 12

Profitability of the Contrarian Strategy and the Overreaction Effect on the Istanbul Stock Exchange

Mehmet A. Civelek and M. Banu Durukan

CONTENTS

12.1 INTRODUCTION

The prediction of stock returns is one of the widely debated issues in the finance literature. The efficient markets hypothesis (EMH) states that the stock prices follow a random walk and fully reflect all the available information in the market. Thus, it is not possible to predict the prices based on any information. EMH rests the efficiency of the markets on two assumptions: the first one is that investors are rational, and the second is that the irrational investor behavior is eliminated by the rational investors who take advantage of the arbitrage opportunity created by such behavior (Shleifer, 2000).

Deviations from the EMH, named as anomalies such as the small firm effect and calendar effects, cannot be explained by the EMH and the expected return theory. Studies which present evidence that portfolio strategies formed based on these anomalies provide high returns can be found widely in the existing finance literature. One of these strategies is to short a portfolio made up of big and low book-to-market stocks and long a portfolio made up of small and high book-to-market stocks. Basu (1977) and Reinganum (1981) pioneered in the development of this strategy.

Another portfolio strategy suggests that portfolios should be formed based on the past price movements of stocks. There are two different approaches to this strategy. One approach argues that investors underreact to new information, thus the winner (loser) stocks continue to win (lose). It is also well documented in the literature that the strategy based on this approach formally known as the momentum strategy, which is to long past winner stocks and short past loser stocks, yields high returns in the short term (Chopra et al., 1992; Jegadeesh and Titman, 1995; Rouwenhorst, 1999; Jegadeesh and Titman, 2001).

On the other hand, the winner–loser strategy first originated from the work of De Bondt and Thaler (1985). They state that investors overreact to new information, so the prices greatly deviate from their fundamental values due to overoptimism or overpessimism. Thus, price reversals are experienced in the subsequent periods. Stated differently, prices of stocks that have gone down (up) in the past, that is, the prices of loser (winner) stocks, will reverse and consequently provide high (low) returns in the future. The overreaction hypothesis argues that past loser stock portfolios will outperform the past winner stock portfolios. Based on this argument, the overreaction hypothesis suggests that the contrarian strategy (to short past winners and to long past losers) results in high returns. Due to the ease of the applicability of this suggested strategy, the overreaction hypothesis continues to attract strong academic and practitioner research interest (Gaunt, 2000). Moreover, the debate continues as to whether the profits of contrarian portfolios are robust to different markets and are due to calendar, risk, and size effects or data snooping.

In light of the aforementioned explanations, this chapter aims to fill a gap in the literature by investigating the profitability of June-end contrarian portfolios in an emerging market, the Istanbul stock exchange (ISE), for the period from June 1988 to June 2008. This chapter will contribute to the literature in three ways: first, it expands on this topic by empirically investigating long-term overreaction within the context of an emerging market, which has a different development level and structure compared with the developed markets studied previously. Previous studies mainly focused on and provided evidence from the U.S. data. There were other studies which investigated price reversal and price continuation in developed markets: mainly the United States, the United Kingdom, Japan, and the European markets.

The evidence from emerging markets is scarce and crucial to study since, as stated by Swanson and Lin (2005), U.S. investors tend to employ these strategies in emerging markets for all horizons. As Jegadeesh and Titman (2001) argue, the study of the profitability of the contrarian/momentum strategies in different markets is essential to test their robustness. First, emerging markets are distinct from the developed markets in that they have different market structures and characteristics, such as thin trading, low liquidity, and possibly less-informed investors compared with developed markets. The behavior of stock prices may be influenced by these characteristics exhibiting different patterns.

Second, there is a limited number of studies on price reversal in the ISE, which attracts foreign investments widely and is one of the leading emerging markets. The findings of the study have practical implications for the existing and potential portfolio managers and investors who invest in the ISE with the aim of international diversification. It will endow them with an understanding of how stock returns behave at the ISE. To our knowledge, Bildik and Gulay (2007), Durukan (2004), and Yucel and Taskin (2007) are the only ones who have investigated the contrarian profits at the ISE.

Third, the study forms June-end portfolios to eliminate the January effect from the data, as suggested by Zarowin (1990) following Ball et al.

(1995) and Mun et al. (2001), who argue that June-end results are more reliable since the profits generated by the contrarian strategy can be attributed to the January effect. The January effect is also well documented at the ISE by Balaban (1995) and Bildik (2000).

The findings of the study suggest that the winner–loser portfolios formed based on prior returns provide excess subsequent returns; price reversal exists in the ISE even though it is asymmetric for loser and winner portfolios; the price reversal is the highest in the first year, and consequently the returns to the winner–loser portfolio are the highest; after the first year, even though the winner–loser portfolio return declines, the loser portfolio continues to provide higher returns compared with the winner portfolio; and the effect is a price-based effect rather than a size effect.

The rest of this chapter proceeds as follows. Section 12.2 begins with an explanation of the overreaction hypothesis and reviews the literature. Section 12.3 introduces the data and methodology based on the framework laid out in Section 12.2. Section 12.4 presents the empirical evidence. Finally, in Section 12.5 concluding comments are offered.

12.2 CONTRARIAN PORTFOLIOS AND THE OVERREACTION HYPOTHESIS

De Bondt and Thaler (1985) argue that in the long run gaining excess returns over the market return is possible by buying past losers and selling past winners, following a contrarian strategy. They have based the source of these profits on the price reversal in the long-term created by the overreaction of investors to information. They explain this situation by "what goes up must go down and what goes down must go up." Lehmann (1990), Lo and MacKinlay (1990), and Jegadeesh and Titman (1995) have cited that their evidence supports high returns due to the overreaction hypothesis and price reversals in the short run as well.

On the contrary, there are others who argue that the returns on the winner–loser portfolio strategies are not only due to overreaction but that they can be explained by the small firm and calendar effects or different risk levels (Lo and MacKinlay, 1990; Zarowin, 1990; Chopra et al., 1992; Fama, 1998). These studies argue that the loser stocks are the stocks of small firms with high risk levels and the profits are due to the January effect. To that end, the returns to the loser stocks can be explained by the high returns to small firm stocks.*

---

* See Power and Lonie (1993) for a detailed review of the studies on the overreaction hypothesis.

Jegadeesh and Titman (1995) respond to the aforementioned arguments by claiming that the methodology used by Lo and MacKinlay (1990) provides flawed results since it takes the delayed reaction into account twice. De Bondt and Thaler (1987) and Chopra et al. (1992) further present evidence that, even after controlling for risk and firm size, the winner–loser portfolio strategy yields high returns and these returns can be explained by overreaction. Moreover, Chopra et al. (1992) claim that, even though calendar effects are observed in the returns, the long-term overreaction can not be explained by this effect. They support their claim with their finding that 50% of the returns are gained in the non-January months. Dreman and Lufkin (2000) conclude that the price reversals occur due to psychological factors, namely, overreaction, since the fundamentals that prices are tied to show little change at times when returns experience significant changes.

Seyhun (1990) finds evidence that supports the overreaction effect in market pricing during and after the 1987 crash. Theobald and Yallup (2004) further argue that there is overreaction in the long run and stocks with higher capitalizations react sooner to information compared with the stocks with lower capitalizations. Thereby, big firms lead small firms. Nam et al. (2001) provide findings revealing that the price reversals are asymmetric and negative returns reverse to positive returns more quickly than positive returns reverse to negative returns. Consequently, they conclude that the asymmetry is due to the mispricing behavior of investors who overreact to information.

Conrad and Kaul (1993) challenge the overreaction hypothesis by arguing that the excess returns that are found by De Bondt and Thaler (1985, 1987) are due to their methodology. They argue that there is an upward bias in the returns calculated for the long-term contrarian strategies since the returns are calculated by cumulating monthly returns over long intervals instead of buy-and-hold returns. In their study, they use the holding period returns (buy–hold strategy), which are conceptually consistent and reduce statistical biases; they find no support for the overreaction hypothesis. Conversely, Loughran and Ritter (1996) contradict Conrad and Kaul (1993) and provide evidence that refutes their arguments. That is, they demonstrate that Conrad and Kaul's methodology needs correction and that their results are driven by survivorship bias and long-term reversion in the stock market rather than cross-sectional patterns on individual stocks. Fama (1998) supports Conrad and Kaul (1993), and states that "apparent anomalies are methodological illusions." He further argues that long-term returns should be calculated based on averages or sums of short-term abnormal returns rather than buy-and-hold abnormal returns due to theoretical and statistical considerations. Viewed as a whole, studies on the overreaction hypothesis accept and utilize both methodologies and the evidence produced confirms overreaction.*

Even though not all of them support the overreaction hypothesis, the number of studies that test it in different country markets is increasing. Some of these studies use weekly or daily return data to test short-term overreaction and some use monthly return data to test long-term overreaction. Along the same lines, these studies also test for the risk, price, small firm, and calendar effects as well as data snooping, bid–ask bias and survivor bias, which are pointed out as the sources of the high returns of the winner–loser portfolio strategy.

Studies that investigate long-term overreaction in developed markets make up the majority of the studies on this issue. Gunaratne and Yonesawa (1997), Iihara et al. (2004), Chiao and Hueng (2005), and Chou et al. (2007) provide evidence consistent with the overreaction hypothesis from the Tokyo stock exchange for the different time periods that they have investigated. They all agree that the high returns to the winner–loser portfolio strategy can not be explained by risk, small firm, or calendar effects.

Kryzanowski and Zhang (1992) contradict the supporting evidence of long-term contrarian profits by stating that their findings are robust to January and non-January and size-based portfolios, and for various performance measures in the Toronto stock exchange. They argue that the empirical evidence on the reversal of prices in the U.S. markets depends on the methodology used. Moreover, Assoe and Sy (2003) indicate that short-term contrarian investing is not economically profitable when transaction costs are accounted for.

Gaunt (2000), on the other hand, provides mixed results on the price reversal in the Australian market and presents evidence of the small firm effect. Lee et al. (2003) conduct a study on the short-term contrarian profits by using weekly data. Even though their findings exhibit that the winner–loser portfolio provides high returns, they also provide evidence that supports Gaunt (2000) that this return is attributable to the small firm effect and big firms lead small firms in the Australian market. Bowman and Iverson (1998) provide supporting evidence in New Zealand for short-term overreaction using weekly data.

Studies that investigate contrarian profits and provide supporting evidence for the contrarian profits in the United Kingdom include Mase (1999),

---

* See De Bondt (2000) for a detailed discussion of the studies on international markets.

Gregory et al. (2001), Antoniou et al. (2006), and Mazouz and Li (2007). Mun et al. (1999) provide evidence from the French and German markets for short-term overreaction, Forner and Marhuenda (2003) for the long-term overreaction from the Spanish market and Mengoli (2004) from the Italian market.

Kang et al. (2002), on the other hand, find positive returns for the Chinese stock exchange for the 1993–2000 period even though they claim that a momentum strategy for the intermediate term would be profitable. Chen and Wu (2007) present evidence from the Taiwanese market.

Besides the studies that focus on only one market, there are others that carry out comparable analyses of different markets. For instance, Poterba and Summers (1988) study 18 different markets and find that prices have a positive relationship in the short run and a negative one in the long run. Wong (1997) provides evidence inconsistent with short-run overreaction using daily data in nine countries (Hong Kong, Taiwan, Singapore, Thailand, Australia, the Philippines, Japan, the United States, and South Korea) in the Asia-Pacific region for the 1986–1995 period. Shen et al. (2005) present evidence that contradicts the overreaction hypothesis but confirms momentum profits.

Baytas and Cakici (1999) study seven developed markets (the United States, Canada, Japan, the United Kingdom, Germany, France, and Italy) in the 1983–1991 period and find evidence supporting long-term overreaction. Mun et al. (2000) confirm the results of Baytas and Cakici (1999) for the 1986–1996 period for the United States and Canada. In line with the overreaction hypothesis, they find evidence that the returns cannot be explained by calendar effects or risk. Balvers and Wu (2006) investigate the combination momentum–contrarian strategies in 18 developed markets and provide evidence that the combination strategies outperform pure momentum or pure contrarian strategies. Rouwenhorst (1999) further argues that emerging markets exhibit price continuation, small stocks outperform large stocks and value stocks outperform growth stocks, as in developed markets, based on the data from 31 emerging markets.

Even though the statistical results of the winner–loser portfolio strategy formed based on the overreaction hypothesis is now well established in the finance literature, the debate over the interpretation of the data still continues (De Bondt, 2000; Mun et al., 2001). Proponents of the EMH argue that this anomaly can be explained by the Fama–French (1996) three-factor model. Thereby, the anomalies in the market are random and as much as there is overreaction there is underreaction in the market.

The opponents of the EMH contradict the proponents, arguing that the high returns that arise from anomalies are due to the mistakes made by the investors in their decision-making process. Therefore, the rationality assumption does not hold and the investor behavior can be explained by the models that are based on psychology (Shleifer, 2000; Hirshleifer, 2001). In this respect, Fama (1998) challenges the proponents of the EMH to develop an alternative hypothesis that can replace EMH and that can explain the anomalies in the literature.* Along these lines, Daniel et al. (1998), Barberis et al. (1998), and Hong and Stein (1999) developed their alternative hypotheses. These hypotheses are based on the theories in the psychology literature. Since they are out of the scope of this chapter, they will not be reviewed.

In sum, the combined evidence leads to the ongoing debate over the sources of the contrarian profits. Thus, researchers confront the issue by accounting for transaction costs; data snooping and lead–lag, calendar, size, and risk effects; asymmetric lengths of formation and test periods; and different performance measures such as the cumulative abnormal returns and buy–hold returns. They examine whether the Fama and French (1996) three-factor model can explain this phenomenon.

12.3 DATA AND METHODOLOGY

The monthly returns of all stocks traded on the ISE for the period from June 1988 to June 2008 are retrieved from the ISE Web site. Following the studies, such as De Bondt and Thaler (1985, 1987) and Loughran and Ritter (1996), that test the long-run overreaction hypothesis, a 3 year portfolio formation period is used and the winner and loser portfolios are determined. The first portfolio formation period is taken as July 1988 to June 1991, which is the first possible formation period.

As Loughran and Ritter (1996) and Baytas and Cakici (1999) suggest, to increase the number of observations and the statistical significance of the analysis, overlapping portfolio formation periods are used. Moreover, the study forms June-end portfolios to eliminate the January effect from the data, as suggested by Zarowin (1990) following Ball et al. (1995) and Mun et al. (2001), who argue that June-end results are more reliable since the profits generated by the contrarian strategy can be attributed to the January effect. In other words, winner and loser portfolios are formed based on 15 three year

---

* See Fama (1998) for the properties and criteria that should be applied to this alternative hypothesis.

portfolio formation periods starting in July 1988, 1989,..., 2002. The testing periods are the 3 year periods that start in July 1991, 1992,..., 2005.

First, the abnormal monthly return for each stock ($AR_{i,t}$) over the market return ($R_{m,t}$) is calculated. Then, for each stock, the cumulative abnormal return (CAR_i) in the portfolio formation period is calculated by using the following formulas*:

$$AR_{i,t} = R_{i,t} - R_{m,t}$$

$$CAR_i = \sum_{t=-35}^{0} AR_{i,t}$$

The monthly return on the ISE 100 Index is used as the market return in the calculations. $AR_{i,t}$ denotes the abnormal return of stock i in month t; CAR_i denotes the cumulative abnormal return of stock i in the portfolio formation period.

For each year, at the beginning of July, the stocks are ranked based on their CAR_i in a descending order. To avoid survivor bias, if a stock is included in the contrarian portfolio based on its past performance and survives for less than the test periods in the future (because of trading suspension or being delisted), its test period return is calculated from time t up to its last trading month, following Chou et al. (2007).

Then, the winner and loser portfolios are formed from the stocks that fall in the highest and lowest deciles, respectively. It is assumed that each stock has equal weight in the portfolio. After the formation of the winner and loser portfolios, the cumulative abnormal return on each portfolio for each month in the test period is calculated:

$$CAR_{p,z,t} = \sum_{t} \left[(1/N) \sum_{i=1}^{N} AR_{i,t} \right]$$

where
  p is the loser (L) and winner (W) portfolios
  z is the portfolio formation period
  N is the number of stocks that make up the portfolio

---

* Fama (1998) provides a discussion of the long-term cumulative return calculation methods. The tests are also carried out using buy and hold returns and yield similar results.

The average cumulative returns (ACAR) of each loser and winner portfolio are calculated for each month in the test period:

$$\text{ACAR}_{p,t} = \frac{\sum_{z=1}^{Z} \text{CAR}_{p,z,t}}{Z}$$

To test the price reversals (H_1 and H_2), a t-test is used. The study also tests whether the winner–loser portfolio formed provides excess returns over the market return (H_3):

$$H_1 : \text{ACAR}_{L,t} = 0; \quad t = 1, 2, \ldots, 36$$

$$H_2 : \text{ACAR}_{W,t} = 0; \quad t = 1, 2, \ldots, 36$$

$$H_3 : \text{ACAR}_{L-W,t} = 0; \quad t = 1, 2, \ldots, 36$$

12.4 EMPIRICAL FINDINGS

12.4.1 Winner–Loser Portfolios

An overall evaluation of the cumulative abnormal returns of loser (3.01%; $t = 3.959$), winner (0.19%; $t = 0.356$), and loser–winner portfolios (2.82%; $t = 4.319$) implies that the 36 month average cumulative return of the loser portfolio is 3.01%, greater than the market return for the same period. On the other hand, the winner portfolio has an average cumulative abnormal return of 0.19% for the same period. As these findings suggest, the overreaction effect on loser and winner portfolios is asymmetric, that is, the overreaction effect is greater on the loser portfolios. The t-statistic for the average returns being higher for the loser portfolios also confirms this conclusion. The average cumulative abnormal return of the winner–loser portfolio ($\text{ACAR}_L - \text{ACAR}_W$) is found to be 2.82% for the 36 month period following the portfolio formation period. These findings can be interpreted as supporting evidence for the overreaction hypothesis and contrarian profits.

Figure 12.1 shows the monthly average cumulative abnormal returns for each portfolio. The winner portfolio, even though it does not have negative returns as in the other studies in the literature, has lower returns compared with the loser portfolio in the majority of the months. The evidence in Figure 12.1 is in line with the aforementioned.

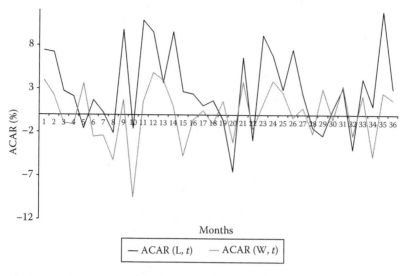

FIGURE 12.1 Monthly portfolio returns for the period following the portfolio formation period.

To analyze the overreaction effect further, Table 12.1 is prepared, which provides information on the breakdown of returns for each 6 month subperiod of the whole 36 month test period. Put differently, the ACARs of the loser and winner portfolios for the 6, 12, 18, 24, 30, and 36 months subsequent to the portfolio formation period are exhibited in Table 12.1.

As can also be seen from Table 12.1, the highest (lowest) returns are received after 12 months for the loser (winner) portfolios. To be concrete, an investor who expects to gain high returns from a contrarian strategy

TABLE 12.1　Average Cumulative Abnormal Returns in the Test Period

Time Passed in the Test Period	$ACAR_{L,t}$	$ACAR_{W,t}$	$ACAR_{L-W,t}$
6 months	3.25 (2.298)*	0.88 (0.782)	2.37 (1.548)
12 months	3.85*** (3.848)	−0.31 (0.405)	4.16*** (3.540)
18 months	3.73*** (3.742)	−0.26 (0.293)	3.99*** (4.764)
24 months	3.32*** (3.440)	0.04 (0.061)	3.28*** (4.177)
30 months	2.97*** (3.601)	0.14 (0.244)	2.83*** (3.924)
36 months	3.01*** (3.959)	0.19 (0.356)	2.82*** (4.319)
Portfolio formation period	−2.49*** (4.963)	8.66*** (8.614)	−11.16*** (18.333)

Note: The numbers in the parentheses are *t*-statistics.
*, Statistically significant at the 10% level; ***, statistically significant at the 1% level.

at the ISE should hold the portfolio for 12 months for a 4.16% return, for 18 months for a 3.99% return and for 36 months for a 2.82% return. Even though for all periods the strategy yields positive abnormal returns, the highest return is received for the 12 month period. That is, the winner–loser portfolio that is formed based on the overreaction hypothesis provides the highest return (4.16%) in the first year. Beyond 12 months, even though the loser portfolio continues to provide higher returns compared with the winner portfolio, the difference decreases. It should also be noted that the change in ACARs points to an asymmetric price reversal.

Table 12.2 presents the yearly returns of the winner and loser portfolios. The returns for each year are calculated separately in the table. The calculated returns confirm the conclusions drawn from the returns exhibited in Table 12.1. As can also be seen from Table 12.2, the highest (lowest) average cumulative abnormal returns are received in the first year for the loser (winner) portfolios. In the second and third years, the yearly returns decrease (increase) for the loser (winner) portfolios, leading to lower returns for the winner–loser portfolio compared with the first year. In sum, price reversal exists for the 3 year period after the portfolio formation period and it is, however, the highest in the first year.

The findings in this part confirm the following: (a) winner–loser portfolios formed based on prior returns provide excess subsequent returns, (b) price reversal exists in the ISE even though it is asymmetric for loser and winner portfolios, (c) the price reversal is the highest in the first year; consequently, the returns to the winner–loser portfolio are the highest, and (d) after the first year, even though the winner–loser portfolio return

TABLE 12.2   Average Cumulative Abnormal Returns in Each Year of the Test Period

	$ACAR_{L,t}$	$ACAR_{W,t}$	$ACAR_{L-W,t}$
Portfolio formation period	−2.49***	8.66***	−11.16***
	(4.963)	(8.614)	(18.333)
In the first year	3.85***	−0.31	4.16***
	(3.848)	(0.405)	(3.540)
In the second year	2.79**	0.40	2.39*
	(2.28)	(0.482)	(1.989)
In the third year	2.37	0.48	1.89
	(1.47)	(0.559)	(0.969)

Note: The numbers in the parentheses are t-statistics.
*, Statistically significant at the 10% level; **, statistically significant at the 5% level; ***, statistically significant at the 1% level.

declines, the loser portfolio continues to provide higher returns compared with the winner portfolio.

The findings support the findings of Bildik and Gulay (2007) and Yucel and Taskin (2007), who investigated contrarian profits using different returns and for different time periods. Thus, in the face of the combined evidence provided, it can be argued that long-term contrarian profits exist at the ISE and they are higher in the first 2 years.

12.4.2 Effect of Size and Price on Winner and Loser Stock Returns

Table 12.3 provides information on the size and price characteristics of the winner and loser portfolios at the portfolio formation date. As can be seen from the content of the table, the mean price of the loser portfolio is less than the winner portfolio and the t-statistic for equality of means is statistically significant at the 1% level. The size, measured by market capitalization (market value of equity), for each portfolio has similar characteristics. The loser portfolios are made up of smaller stocks compared with the winner portfolio stocks. The t-statistic for the equality of means of size is also found to be statistically significant at the 1% level (t-statistic: -3.612).

Based on the evidence in Table 12.3, a cross-sectional regression analysis is carried out to determine the relationship between the average cumulative abnormal return of each stock in the test period and the independent variables of (a) the price of stock on the portfolio formation date, (b) size calculated as the number of shares outstanding times the price of the stock on the portfolio formation date, and (c) the stock's portfolio formation (prior) period return.

$$ACAR_{3,it} = a_0 + a_1 \ln Price_{it} + a_3 \ln MV_{it} + a_4 Prior Return_{it}$$

TABLE 12.3 Price and Size Characteristics of Winner and Loser Portfolios

	Loser Portfolio	Winner Portfolio	t-Statistic
Price			
Mean	6,957	24,975	$-5.647*$
Standard deviation	10,788	32,315	
Size (market value of equity)			
Mean	13,700,000,000	185,000,000,000	$-3.612*$
Standard deviation	47,550,814,950	466,310,000,000	

* Statistically significant at the 1% level.

In this study, the results for the "Prior Return" instead of "ln(1 + Prior Return)" are presented.* This decision was made because many values were lost if the ln version of the variable was used due to negative returns greater than 100%. Moreover, since ln MV = ln Price + ln Outstanding Shares, the regressions are also run without the ln MV variable. The results of the regressions for the loser and winner stocks separately and as a whole are reported in Table 12.4.

TABLE 12.4   Regression Results

Equation Number	Sample	Constant	ln Price	ln MV	Prior Return	R^2	F-Statistic
				Independent Variables			
1	Losers	0.1350*** (3.770)	0.0011 (0.356)	−0.0068 (3.513)	−0.0201*** (4.924)	0.299	12.962***
2	Winners	0.0072* (1.935)	−0.0049** (1.865)	−0.0004 (0.353)	−0.0032 (1.100)	0.067	2.238*
3	All	0.0761*** (3.331)	−0.0047** (2.336)	−0.0009 (0.945)	−0.0023** (2.042)	0.131	9.513***
4	Losers	0.126*** (4.876)	−0.0013*** (4.450)		−0.0052 (1.163)	0.424	12.058***
5	Winners	0.0314 (1.284)	−0.0016 (0.636)		−0.0046** (2.089)	0.210	2.560*
6	All	0.0822*** (4.859)	−0.0074*** (3.813)		−0.0032*** (2.784)	0.174	23.542***
7	Losers	0.136*** (3.859)		−0.0065*** (3.989)	−0.0194*** (5.382)	0.298	19.565***
8	Winners	0.0431 (1.257)		−0.0049 (0.848)	−0.0011* (1.776)	0.032	1.577
9	All	0.0525** (2.533)		−0.0017* (1.881)	−0.0037*** (3.851)	0.106	11.277***
10	Losers	0.139*** (3.461)	−0.0058* (1.910)	−0.0033 (1.635)		0.113	5.842***
11	Winners	0.0556 (1.629)	−0.0059** (2.343)	0.0002 (0.251)		0.055	2.745*
12	All	0.0937*** (4.387)	−0.0069*** (4.024)	−0.0009 (0.896)		0.112	11.986***

Note: In all regressions, numbers in parentheses are *t*-statistics.
*, Statistically significant at the 10% level; **, statistically significant at the 5% level; ***, statistically significant at the 1% level.

---

* In either case, however, the regression results are consistent with each other. Therefore, the regression results for the ln(1 + Prior Return) variable are not provided in the study.

The price variable has a negative sign in all the regression results except in Equation 1. It is also statistically significant when it has a negative sign, which shows that, regardless of their prior performance, the returns to all stocks increase as the price declines. The size variable also has a negative sign in all the regression equations except Equation 11 but it is found to be significant when the price variable is excluded (Equations 7 and 9). Based on these results, it can be pointed out that the overreaction effect is a price-based effect.

The prior return variable has a negative sign in all the regression equations. It has statistically significant coefficients at the 5% or 1% levels in all the equations except Equations 2 and 4. Consequently, it can be argued that, as prior return decreases, subsequent return increases. This finding confirms the existence of the overreaction effect.

12.5 CONCLUSIONS

This chapter aimed to fill a gap in the literature by investigating the profitability of June-end contrarian portfolios in an emerging market, the ISE, for the period from 1988 to 2008. Its key contributions can be listed as (a) it expands on this topic by empirically investigating long-term overreaction within the context of an emerging market, which has a different development level and structure compared with the developed markets studied previously, (b) it investigates price reversal in the ISE, which attracts foreign investments widely for international diversification purposes and is one of the leading emerging markets, and (c) the chapter forms June-end portfolios to eliminate the January effect on the contrarian profits.

The chapter confirms that contrarian strategies work in the ISE and there is asymmetric price reversal in the ISE for winner and loser portfolios, which is in line with the findings of Bildik and Gulay (2007), and Yucel and Taskin (2007). The price reversal is the highest in the first year; even though the winner–loser portfolio returns decline, the loser portfolio continues to provide higher returns compared with the winner portfolio. Further analysis suggests that the effect is a price-based effect rather than a size effect. Hence, the evidence implies that the investment community can profit by pursuing contrarian strategies at the ISE.

This chapter focused on the long-term overreaction in the ISE. Since the highest return for the winner–loser portfolio is found to be in the first year by this chapter, the analysis of short-term overreaction certainly calls for investigation. Moreover, different and asymmetric formation and test periods should further be investigated.

It is instructive to note that the results of this chapter need to be confirmed by further analysis. That is, the main sources of these profits should be investigated in more detail. This chapter only focused on size and price as other characteristics that affect stock returns. However, the overreaction effect should also be tested by the three-factor model of Fama and French and for calendar effects. Put differently, this chapter discussed only two firm characteristics that are accepted to influence returns in the finance literature, and concluded that it is possible to attribute the abnormal returns to overreaction. Finally, it should be emphasized that overreaction plays a role in determining returns in the ISE but future research should focus (a) on the separation of the overreaction effect from the effect of other factors, such as the calendar, lead–lag, bid–ask, size, and risk effects and (b) on the short-term overreaction.

REFERENCES

Antoniou, A., Galariotis, E. C., and Spyrou, S. I. (2006) Short-term contrarian strategies in London stock exchange: Are they profitable? Which factors affect them? *Journal of Business Finance and Accounting*, 33(5–6): 839–867.

Assoe, K. and Sy, O. (2003) Profitability of the short run contrarian strategy in Canadian stock markets. *Canadian Journal of Administrative Sciences*, 20(2): 311–319.

Balaban, E. (1995). January effect, Yes! What about Mark Twain effect? Discussion Paper for the Central Bank of the Republic of Turkey.

Ball, R., Kothari, S. P., and Shanken, J. (1995) Problems in measuring portfolio performance: An application to contrarian investment strategies. *Journal of Financial Economics*, 38(1): 79–107.

Balvers, R. J. and Wu, Y. (2006) Momentum and mean reversion across national equity markets. *Journal of Empirical Finance*, 13(1): 24–48.

Barberis, N., Shleifer, A., and Vishny, R. (1998) A model of investor sentiment. *Journal of Financial Economics*, 49(3): 307–343.

Basu, S. (1977) Investment performance of common stocks in relation to their price-earnings ratios: A test of market efficiency. *Journal of Finance*, 32(3): 663–682.

Baytas, A. and Cakici, N. (1999) Do markets overreact: International evidence. *Journal of Banking and Finance*, 23(7): 1121–1144.

Bildik, R. (2000). *Hisse Senedi Piyasasında Dönemsellikler ve İMKB Üzerine Ampirik Bir Çalışma*. Istanbul, Turkey: İMKB Yayınları.

Bildik, R. and Gulay, G. (2007) Profitability of contrarian strategies: Evidence from the Istanbul stock exchange. *International Review of Finance*, 7(1–2): 61–87.

Bowman, R. G. and Iverson, D. (1998) Short-run overreaction in the New Zealand stock market. *Pacific-Basin Finance Journal*, 6(5): 475–491.

Chen, Y. and Wu, H. (2007) Investigation of the returns of the contrarian and momentum strategies in the Taiwanese equity market. *Journal of American Academy of Business*, 11(2): 143–150.

Chiao, C. and Hueng, C. J. (2005) Overreaction effects independent of risk and characteristics: Evidence from the Japanese stock market. *Japan and the World Economy*, 17(4): 431–455.

Chopra, N., Lakonishok, J., and Ritter, J. R. (1992) Measuring abnormal performance: Do stocks overreact? *Journal of Financial Economics*, 31(2): 235–268.

Chou, P., Wei, K. C. J., and Chung, H. (2007) Sources of contrarian profits in the Japanese stock market. *Journal of Empirical Finance*, 14(3): 261–286.

Conrad, J. and Kaul, G. (1993) Long-term market overreaction or biases in computed returns? *Journal of Finance*, 48(1): 39–63.

Daniel, K., Hirshleifer, D., and Subrahmanyam, A. (1998) Investor psychology and security under- and overreactions. *Journal of Finance*, 53(6): 1839–1885.

De Bondt, W. F. (2000) Psychology of underreaction and overreaction in World equity markets. In: D. Keim and Ziemba W. (Eds), *Security Market Imperfections in World Equity Markets*. Cambridge, United Kingdom: Cambridge University Press, pp. 65–89.

De Bondt, W. F. and Thaler, M. R. (1985) Does the stock market overreact? *Journal of Finance*, 40(3): 793–805.

De Bondt, W. F. and Thaler, M. R. (1987) Further evidence on investor overreaction and market seasonality. *Journal of Finance*, 42(3): 557–581.

Dreman, D. N. and Lufkin, E. A. (2000) Investor overreaction: Evidence that its basis is psychological. *Journal of Psychology and Financial Markets*, 1(1): 61–75.

Durukan, M. B. (2004) Aşırı Tepki Hipotezi: İstanbul Menkul Kıymetler Borsasından Kanıtlar. *Proceeding of the VIII National Finance Symposium*. Istanbul. pp. 137–142.

Fama, E. F. (1998) Market efficiency, long-term returns, and behavioral finance. *Journal of Financial Economics*, 49(3): 283–306.

Fama, E. F. and French, K. R. (1996) Multifactor explanations of asset pricing anomalies. *Journal of Finance*, 51(1): 55–84.

Forner, C. and Marhuenda, J. (2003) Contrarian and momentum strategies in the Spanish stock market. *European Financial Management*, 9(1): 67–88.

Gaunt, C. (2000) Overreaction in the Australian equity market: 1974–1997. *Pacific-Basin Finance Journal*, 8(3–4): 375–398.

Gregory, A., Harris, R. D. F., and Michou, M. (2001) An analysis of contrarian investment strategies in the UK. *Journal of Business Finance and Accounting*, 28(9–10): 1192–228.

Gunaratne, P. S. M. and Yonesawa, Y. (1997) Return reversals in the Tokyo stock exchange: A test of stock market overreaction. *Japan and the World Economy*, 9(3): 363–384.

Hirshleifer, D. (2001) Investor psychology and asset pricing. *Journal of Finance*, 56(4): 1533–1597.

Hong, H. and Stein, J. C. (1999) A unified theory of underreaction, momentum trading, and overreaction in asset markets. *Journal of Finance*, 54(6): 2143–2184.

Iihara, Y., Kato, H. K., and Tokunaga, T. (2004) The winner-loser effect in Japanese stock returns. *Japan and the World Economy*, 16(4): 471–485.

Jegadeesh, N. and Titman, S. (1995) Overreaction, delayed reaction, and contrarian profits. *Review of Financial Studies*, 8(4): 973–993.

Jegadeesh, N. and Titman, S. (2001) Profitability of momentum strategies: An evaluation of alternative explanations. *Journal of Finance*, 56(2): 973–993.

Kang, J., Liu, M. H., and Ni, S. X. (2002) Contrarian and momentum strategies in China stock market: 1993–2000. *Pacific-Basin Finance Journal*, 10(3): 243–265.

Kryzanowski, L. and Zhang, H. (1992) The contrarian investment strategy does not work in Canadian markets. *The Journal of Financial and Quantitative Analysis*, 27(3): 383–395.

Lee, D. D., Chan, H., Faff, R. W., and Kalev, P. S. (2003) Short-term contrarian investing—Is it profitable? ... Yes and No. *Journal of Multinational Financial Management*, 13(4–5): 385–404.

Lehmann, B. N. (1990) Fads, martingales, and market efficiency. *Quarterly Journal of Economics*, 105(1): 1–28.

Lo, A. W. and MacKinlay, A. C. (1990) When are contrarian profits due to stock market overreaction? *Review of Financial Studies*, 3(2): 175–205.

Loughran, T. and Ritter, J. R. (1996) Long-term market overreaction: The effect of low-priced stocks. *Journal of Finance*, 51(5): 1959–1970.

Mase, B. (1999) The predictability of short-horizon stock returns. *European Finance Review*, 3(2): 161–173.

Mazouz, K. and Li, X. (2007) The overreaction hypothesis in the UK market: Empirical analysis. *Applied Financial Economics*, 17(13): 1101–1111.

Mengoli, S. (2004) On the source of contrarian and momentum strategies in the Italian equity market. *International Review of Financial Analysis*, 13(3): 301–331.

Mun, J. C., Kish, R., and Vasconcellos, G. M. (2001) The contrarian investment strategy: Additional evidence. *Applied Financial Economics*, 11(6): 619–640.

Mun, J. C., Vasconcellos, G. M., and Kish, R. (1999) Tests of the contrarian investment strategy: Evidence from the French and German stock markets. *International Review of Financial Analysis*, 8(3): 215–234.

Mun, J. C., Vasconcellos, G. M., and Kish, R. (2000) The contrarian/overreaction hypothesis: An analysis of the US and Canadian stock markets. *Global Finance Journal*, 11(1): 53–72.

Nam, K., Pyun, C. S., and Avard, S. L. (2001) Asymmetric reverting behavior of short-horizon stock returns: An evidence of stock market overreaction. *Journal of Banking and Finance*, 25(4): 807–824.

Poterba, J. M. and Summers, L. H. (1988) Mean reversion in stock prices: Evidence and implications. *Journal of Financial Economics*, 22(1): 27–59.

Power, D. M. and Lonie, A. A. (1993) The overreaction effect: Anomaly of the 1980s? *British Accounting Review*, 25(4): 325–366.

Reinganum, M. R. (1981) Abnormal returns in small firm portfolios. *Financial Analyst Journal*, 37(2): 52–56.

Rouwenhorst, K. G. (1999) Local return factors and turnover in emerging stock markets. *Journal of Finance*, 54(4): 1439–1464.

Seyhun, H. N. (1990) Overreaction nor fundamentals: Some lessons from insiders' response to the market crash of 1987. *Journal of Finance*, 45(5): 1363–1388.

Shen, Q., Szakmary, A., and Sharma, S. C. (2005) Momentum and contrarian strategies in international stock markets: Further evidence. *Journal of Multinational Financial Management*, 15(3): 235–255.

Shleifer, A. (2000) *Inefficient Markets: An Introduction to Behavioral Finance.* Oxford, United Kingdom: Oxford University Press.

Swanson, P. E. and Lin, A. Y. (2005) Trading behavior of and investment performance of U.S. investors in global equity markets. *Journal of Multinational Financial Management*, 15(2): 99–115.

Theobald, M. and Yallup, P. (2004) Determining security speed of adjustment coefficients. *Journal of Financial Markets*, 7(1): 75–96.

Wong, M. C. S. (1997) Abnormal stock returns following large one-day advances and declines: Evidence from Asia-Pacific markets. *Financial Engineering and the Japanese Markets*, 4(2): 171–177.

Yucel, A. T. and Taskin, D. (2007) Overreaction hypothesis and evidence from Istanbul stock exchange. *İktisat İşletme ve Finans*, 22(260): 26–37.

Zarowin, P. (1990) Size, seasonality, and stock market overreaction. *Journal of Financial and Quantitative Analysis*, 25(1): 113–125.

CHAPTER 13

What Determines Going Public in Latin America?

Samuel Mongrut, Aaron Garay,
and Alonso Valenzuela

CONTENTS

13.1 INTRODUCTION

There are advantages and disadvantages attached to the firm's decision to go public. The advantages for the issuer are that stocks placement are a source of financial capital, provide greater negotiation power with respect to credit entities, and have a wider diffusion among public investors and financial institutions. Likewise, a public company becomes more prestigious, influential, and easier to form alliances, negotiate contracts with suppliers, clients, as well as credit entities. In addition, the firm obtains publicity with its clients and the stock price can be perceived as an indicator of performance while establishing a mechanism to monitor value creation within the firm (Holmstrong and Tirole, 1993).

However, international evidence shows that placing an initial public offering (IPO) is not easy and it requires a well-coordinated effort with external advisory parties from commercial, legal, and accounting perspectives. Among the disadvantages of an IPO is the high cost of going public. For example, in Argentina, the average cost of going public is approximately $300,000 with a 7%–10% commission fee that is not included for the writer, whereas a private offer cost in the order of $100,000 (Gattás, 2001). Moreover, the company that goes public is obligated to disclose information that could lead to a loss of confidentiality, for instance, disclosure about technology developments and marketing plans as pointed out by Campbell (1979), Yosha (1995), and Maksimovic and Pichler (2001). A public firm also faces strong pressure toward its performance in the short term, whereas a private firm may display only long-term objectives and exhibit controlled growth. In this sense, the public firm is hard-pressed by stockholders to grow earnings and display positive results reflected in a stock price increase with higher dividends distributions.

A common limitation in IPO studies is the data shortage available for private firms. Despite this drawback, there are numerous studies done on countries such as Sweden (Rydqvist and Hoghölm, 1995), Italy (Pagano et al., 1998), Germany (Fischer, 2000), the United Kingdom (Gill de Albornoz and Pope, 2004), and Taiwan (Shen and Wei, 2007), which conclude that market conditions are the most important factors when going public, followed by the life cycle stage of the firm (firm size). However, the question remains whether these factors are also the same for firms that operate in Latin America. Common sense indicates that the same critical factors are not necessarily similar due to the lack of market transparency and the presence of country risk.

The main objective of this chapter is to determine what factors influence the "going public" decision of firms in Latin America. This chapter is organized as follows: in Section 13.2, we summarize the advantages and disadvantages of going public whereas in Section 13.3 we describe the behavior of firms going public in Latin America. We then focus on seven Latin American emerging markets: Argentina, Brazil, Chile, Colombia, Mexico, Peru, and Venezuela. In Section 13.4, we discuss prior empirical evidence and in Section 13.5 we present our methodology, the variables, and models with the results following in Section 13.6 and the last Section 13.7, provides conclusions.

13.2 ADVANTAGES AND DISADVANTAGES OF GOING PUBLIC

Table 13.1 summarizes the different advantages and disadvantages of going public that have been addressed in the literature.

TABLE 13.1   Advantages and Disadvantages of the Going Public Decision

Advantages	Disadvantages
• Better financial alternatives	• Adverse selection
✓ Main advantage: minimizes the cost of capital and maximizes the value of the company.	✓ Information asymmetry may cause underpricing.
✓ Reduces the cost of financing through avoiding the intervention of financial intermediaries.	✓ Younger and smaller firms are more affected by adverse selection costs.
✓ Pecking order: Internal financing, debt, external equity.	✓ There could be a window dressing.
• Diversification	• Loss of confidentiality
	✓ May discourage companies, which base their advantages in R&D, to go public.
• Stocks liquidity and company selling	• Meaningful initial costs of going public
• Monitoring managerial decisions and getting publicity	• Loss of control over company's decisions
• Window opportunity	
✓ IPO waves, reason: stock price does not reflecting (temporarily) the true value of the companies, there is an overvaluation. Firms take advantage of this window of opportunity.	

In terms of advantages, the need for financing growth has been viewed as one of the most critical factors of going public. The advantage of going public is to obtain financing minimize the cost of capital and maximize firm value (Modigliani and Miller, 1963). Likewise, when going public, companies reduce the cost of financing because it avoids the interference by financial intermediaries. Choe et al. (1993) and Nanda (2002) sustain that companies raise public equity when they reach an extreme point at the business' growth cycle whereby the need for external capital continues to grow.

Although, there is some evidence that U.S. firms go public to finance their expansion (Mikkelson et al., 1997), this is not the case in Italy (Pagano et al., 1998) and in Germany (Fischer, 2000) suggesting that financial needs are not a critical factor in the decision of going public.

Many authors (Pagano, 1993; Zingales, 1995; Stoughton Zechnev, 1998; Chemmanur and Fulghieri, 1999) affirm that firm's stockholders aim to diversify their wealth by rebalancing their portfolios. For example, Rydqvist and Hoghölm (1995) argue that the main reason for a Swedish firm going public is its wish for diversification. However, portfolio diversification can be achieved in a direct way by disinvesting in the firm that becomes public and investing in other assets; hence the motivation could come from either side.

The liquidity of stock reduces the high transaction costs that a stockholder must face when the company is not quoted in the capital market. Given that the stock liquidity of a firm is an increasing function of its volume, the benefit obtained of having more liquidity will only benefit firms that are large enough to gain from this advantage and go public. A firm's recognition and reputation is increased if it is quoted on a stock exchange (Maksimovic and Pichler, 2001). As shown by Merton (1987) in his landmark paper that the greater the number of investors who are aware of a publicly traded firm, the higher its stock price will be.

One documented anomaly in the literature is the "hot issue markets" where the average first month performance of new IPOs is unusually high. One of the reasons of this pattern may be that the price of the firm does not reflect (temporarily) its fundamental value, thus resulting in mispricing (an overvaluation) creating an incentive for firms to become public in order to make the most of this "window of opportunity." Ritter (1991) maintains that there are periods in which investors are too optimistic about young firms' potential growth and firms take advantage of the "window of opportunity," which explains the IPO's waves. Nevertheless, the empirical evidence is not conclusive with respect to the window of opportunity. For instance, Shen and Wei (2007) find no window of opportunity for Taiwanese firms while and Brau and Fawcett (2006) observe that financial managers define the window of opportunity in terms of the general market conditions and industry as a group, instead of the IPO's market conditions.

Asymmetric information clearly exists among the potential investors and issuers resulting in a disadvantage. There is a justified assumption that issuers are able to get more information about the firm's fundamental value than investors which can negatively affect the sale price of firms (Leland and Pyle, 1977) and reduce the price of the IPO. This firm's undervaluation seems to be necessary in order to sell stocks at the break even point (Rock, 1986).

It is apparent that the costs derived from the adverse selection problem are greater for small and young firms than for corporate firms because of their reduced performance and recognition profile. Therefore, small and young firms would be less prone to go public. However, as sustained by Diamond (1991), a factor that could help companies to overcome the problem of adverse selection is profitability. High profits might be seen as a sign of quality of the firm, thus one would expect a positive relation with the likelihood of going public. In addition, it is necessary to be careful since the positive relation could be due to other reasons. According to Ritter (1991), firms can take advantage of increases in their profitability and go public to take advantage of the current investors' perceptions of high profits in the distant future that will overestimate their stocks.

Companies are typically reluctant to disclose information about their technological developments or marketing plans to their competitors. Therefore, it is expected that firms of industries intensive in research and development (R&D) are less likely to go public than others in more conventional businesses. For example, Brau and Fawcett (2006) find that confidentiality loss is an important obstacle in going public.

Apart from underpricing, firms that go public face other explicit initial costs when issuing an IPO. For instance, underwriters' fees and the registrations costs imposed by the supervisor (not related to firm size), are more significant for small companies making them less inclined to go public. Empirical evidence by Pagano et al. (1998) finds that firm size is positive and significantly related to the likelihood of going public for Italian firms. In subsequent studies Gill de Albornoz and Pope (2004) and Shen and Wei (2007) reach similar results for United Kingdom and Taiwanese firms respectively.

TABLE 13.2   Initial Public Offerings in Latin America during 1990–2007

Country	Situation
Argentina	Most of IPO's are concentrated during 1990–1995 period with 1992 displaying a maximum of 35 new public offerings. In addition, after 1995, the number of IPOs decreased, reaching an average of two IPOs per year.
Brazil	There is not outstanding period of IPOs; however, the annual average number of IPOs is the highest in the region with 19.
Chile	There are 7 IPOs issued per year on average, without any outstanding increase in any specific period. There were 91 IPOs in and a maximum number of 159 in 1986.
Colombia	Most of IPO's are concentrated during 1990–1995 period, with the maximum number of IPOs occurring in 1990 with 20. In 1994 there were 15 and subsequent to that the number of annual IPO's number decreased, to 2–3 per year.
Mexico	There is no "IPO's wave," but the number of IPO's amounts to 11 per year on average during the 1990–2007.
Peru	The maximum number of IPOs was 49 in 1990 with an annual average of 9.
Venezuela	The annual average number of IPOs during the 1990–2007 period was 2 and in 1992 the number of IPOs reached a maximum at 18.

13.3 PATTERN OF INITIAL PUBLIC OFFERINGS IN LATIN AMERICA[*]

From 1990 to 2007, there were 63 IPOs per year (on average), with a high concentration of IPO's during the first 5 years of the 1990s with a maximum of 161 occurring in 1992. However, the following years there was a decline and a downward trend during the 2000–2005 period with little recovery in recent years. A description of the IPOs trend during analysis period is shown in Table 13.2. If one examines the behavior of the IPOs across seven Latin American markets, one can conclude that the number of IPOs was highly volatile from the mid-1980s until mid-1990s. Thereafter, the number of IPOs decreased substantially until 2004 and from that year on the number of IPOs started to increase considerably.

13.4 PREVIOUS EMPIRICAL EVIDENCE

Many U.S. studies on IPOs in the aftermarket performance have been published in the last 5–10 years (Jain and Kini, 1994; Lowry and Schwert, 2002). Even though it is complicated to evaluate several theories of why

---

[*] In this chapter, we only take into account private firms that meet the requirements of issuing stock in the Latin American markets.

firms go public, this issue was addressed in the pioneering study of Pagano et al. (1998) which set the standard. Their study (p. 27) highlights that firms go public "... not to finance future investments and growth, but rather to rebalance their accounts after a period of high investment and growth."

In later years, numerous studies examined IPO determinants in Sweden (Rydqvist and Hoghölm, 1995), the United Kingdom (Gill de Albornoz and Pope, 2004), Germany (Fischer, 2000), and Taiwan (Shen and Wei, 2007). Table 13.3 shows that firm size has a positive impact in going public, especially for Italy. In addition, firm size in the United Kingdom has a positive effect on going public, as well as, for surviving, acquired and failed firms. Moreover, in Taiwan there is evidence to support that large companies are more prone to go public. For Italian companies, greater investment opportunities and a high growth ratio in prior years before going public, had a positive effect in going public. For subsidiaries, such variables do not have any significant effect due to their lower investment levels than the independent firms. In addition, we observe that after 2 years of issuing the IPO, investment levels suffer a decrease that remains in later years. For U.K. firms, growth opportunities do not have a significant effect on the likelihood of going public, however, for Taiwanese firms investment opportunities have a little significant effect of going public.

As for leverage in the Italian case, larger underwriting negatively influences the likelihood of issuing an IPO. Firms that are more leveraged than others are more likely to raise public equity if they are independent, and less likely to do it if they are subsidiaries. After an IPO, there is a significant decrease in leveraged levels of independent firms. Contrary to the Italian case, U.K. firms do not go public because they want to reduce underwriting levels while in Taiwan the firm's leverage level negatively influences the likelihood of going public.

The more profitable is a firm, the more is the likelihood of going public in Italy. However, there is also a decrease on the profitability after going public (showing up as a greater effect for subsidiaries). In the United Kingdom, firm's profitability has a negative and significant impact on the likelihood of going public.

A better valuation by the industry is one of the main factors of the going public in Italy (the effect is stronger for subsidiaries), which validates the "window of opportunity" hypothesis. In the United Kingdom, the firms' motivation to go public can also be the result of a bullish market in that particular industry the firm operates in. Furthermore, the market-to-book

TABLE 13.3 Relevant Empirical Evidence

Country	Italy (1982–1992)	United Kingdom	Taiwan (1989–2000)
Author	Pagano et al. (1998)	Gill de Albornoz and Pope (2004)	Shen and Wei (2007)
Size	Positive impact (important for independents, nonsignificant for subsidiaries).	Positive impact (in the four formulated models, and for surviving firms, acquired and failed firms)	Positive impact (firms size in terms of sales)
Investment opportunities	Positive impact (not significant especially for subsidiaries). Two years later of issuing IPO, investment level of the independent ones suffers a sharp fall, but for subsidiaries there is an increase recorded.	Positive impact	
Growth		Nonsignificant effect (for the surviving ones this effect is greater than for the rest of the samples)	No significant effect
Underwriting	Negative impact (low significance). Greater effect in independents than in subsidiaries. After IPO issuance there is a significant decrease in underwriting of independent firms.	Negative impact	Negative impact
Profitability	Positive impact (the impact of this variable is 50% higher on subsidiaries than on independents). It is a permanent decrease in the profitability after IPO issuance (greater effect on subsidiaries)	Negative impact	Positive impact

Industry market to book ratio	Positive impact (one of the main factors, stronger in the subsidiaries case)	Positive impact (for all types of regressions)	Negative Impact
Firm risk		Positive impact. The positive signal of CAPEX for the group of surviving firms, suggests that the sample of the period includes many risky firms that face financing constraints and are not able to raise enough internal funds to go public only because they need to raise funds to finance their investments.	
Financing needs	Nonsignificant effect	Positive impact (even this is not one of the main factors)	No significant effect
Industry type (retail)		Positive impact (firms with larger client databases are the ones belonging to the retail industry and are more prone to go public)	

Source: Pagano, M., Panetta, F., and Zingales, L., *J. Finance*, 53(1), 27, 1998; Gil de Albornoz, B. and Pope, P., Determinants of the going public decision: Evidence from the U.K. Working Paper, Universitat Jaume I, Castellon Spain, 2004; Shen, Y. and Wei, P., *J. Econ. Finance*, 31(3), 359, 2007. With permission.

ratio presents a positive and significant effect on the dependent variable, for all regression models, which is consistent with the found evidence for Italy. Nonetheless, for the case of Taiwan, in contrast with European studies, the market to book ratio has a negative effect on the likelihood of going public. In the United Kingdom, riskier firms are more prone to go public, suggesting that many risky firms facing financial constraints are not able to raise enough internal funds and therefore go public with the goal to access funds. Moreover, financing needs are not relevant to going public for U.K. firms. In accordance with theory, the benefits of raising public equity are greater in firms with large client database. For Taiwanese firms R&D expenses do not have a significant effect on going public due to the confidentiality loss.

13.5 METHODOLOGY, VARIABLES, AND MODEL

In order to obtain the final sample of IPOs, we compiled information from the Bolsa de Comercio de Buenos Aires–Argentina (BCBA), la Sociedad Operadora del Mercado de Activos (SOMA) and la Bolsa de Valores de Sao Paulo (Bovespa) in Brazil, la Bolsa de Comercio de Santiago (BCS) in Chile, la Bolsa Mexicana de Valores (BMV) in Mexico, and la Bolsa de Valores de Lima (BVL) in Peru. The analysis was conducted for the period: March 1986–December 2007. Table 13.4 shows the number of firms in the final sample per country.

The main data source used for this chapter is Economatica and Bloomberg. Quarterly information was extracted from both databases in order to obtain the necessary variables to run our regression models. In order to complete and double-check our data we consulted with ISI Emerging Markets and the *Emerging Markets Data Base*, as well as information from the stock exchanges and national regulatory entities. The final sample of firms was obtained after applying certain filters: firms that

TABLE 13.4   Number of Firms in the Final Sample

Country	Initial Quarter	Final Quarter	Number of Firms	Private Firms (%)	Public Firms (%)
Argentina	3Q 91	4Q 07	18	28	72
Brazil	2Q 86	4Q 07	95	30	70
Chile	1Q 91	4Q 07	42	48	52
Mexico	1Q 89	4Q 07	70	21	79
Peru	1Q 95	4Q 07	19	63	37

operate in the financial sector were discarded as well as firms that did not have at least eight quarters of reported financial information before going public.

The database used in this chapter combines time series and cross-sectional data for modeling. We then set up a panel data model. The following model (Equation 13.1) is used to evaluate the probability that firms have of going public. The selected variables for the model represent a clear and intuitive relation with the formulated hypothesis and also represent data availability. The Panel–Logit model to estimate for each country is specified as follows:

$$\Pr\text{ob}(IPO_{it}) = F(\alpha_0 + \alpha_1 SIZE_{it} + \alpha_2 CAPEX_{it} + \alpha_3 GROWTH_{it} + \alpha_4 LEVERAGE_{it} + \alpha_5 PROFIT_{it} + \alpha_6 MTB_{it}) \quad (13.1)$$

In addition, there will be pool country estimations, for which the variable SS is a surrogate of country risk:

$$\Pr\text{ob}(IPO_{it}) = F(\alpha_0 + \alpha_1 SIZE_{it} + \alpha_2 CAPEX_{it} + \alpha_3 GROWTH_{it} + \alpha_4 LEVERAGE_{it} + \alpha_5 PROFIT_{it} + \alpha_6 MTB_{it} + \alpha_7 SS) \quad (13.2)$$

where
   IPO is a binary variable that equals zero, if the firm i opts to remain private in period t and equals 1 if the firm goes public
   $F(\cdot)$ is the function of cumulative probability of a logistic variable
   SIZE is the lagged value of the total assets logarithm
   CAPEX is the lagged value of fixed assets growth
   GROWTH is the lagged value of operative revenue growth
   LEVERAGE is the lagged value of net financial debt over total assets
   PROFIT is the lagged value of EBIT over total assets average
   MBT is the lagged value of market to book ratio's median of the industry's public firms that are quoted in the capital market, as per the classification provided by the Economatica software

The country risk is incorporated through the lagged value of Stripped Spread del EMBI+ (SS). In each t period, the sample consists of all firms that would go public, which are removed from the sample after they go public. Table 13.5 describes the hypotheses to be tested by the proposed models, the variables and expected signs.

TABLE 13.5 Description of Variables and Hypotheses

Variable	Description	Hypothesis
SIZE (+)	Logarithm of total assets Ln(TA)	Pecking order of financing implies larger companies are more prone to going public
		Only large companies may take full advantage of the liquidity gains of going public
CAPEX (+)	Capital expenditures over total fixed assets	Proxy for investment opportunities
GROWTH (+)	Operative income growth	Measures the firm growth
LEVERAGE (?)	Net financial debt/total assets	Highly levered firms that invest intensively may find it attractive to finance using public equity
		IPOs: solution to borrowing constraints in high levered firms. Investors on the other hand may perceive a high levered firm as more risky and ask for higher returns on their investment, aggravating the problem of underpricing
PROFITABILITY (?)	EBIT/total assets	More profitable companies may be overvalued, which makes them to go public. On the other hand, the firm may have less financing needs due to their own internal financing
MEDIAN MTB (+)	Median market to book ratio for public companies in the same industry	Opportunist behavior of companies is to take advantage of temporary overvaluation. Also indicates investment and growth opportunities
COUNTRY RISK (−)	Stripped spread EMBI+	Riskier countries, higher return leads to underpricing

13.6 RESULTS

Table 13.6 shows the estimated coefficients of maximum likelihood from the Panel–Logit model regressions for Brazil, Chile, and Mexico and for Latin America as a group (first the three countries are considered and then Argentina and Peru are added).

The first relevant result is that firm size is a significant determinant in the going public for Latin America. In the specific analysis for Brazil and Mexico, a change in the standard deviation of the assets logarithm generates an impact of 9% and 11% points on the likelihood of raising

TABLE 13.6   Determinants of the Going Public Decision in Latin America

Variable	Brazil	Chile	Mexico	Latin America[a]	Latin America[b]
Constant	−6.7459***	−5.5838***	−2.7288***	—	—
	(1,6886)	(1.6109)	(1.3112)		
SIZE	0.6057***	0.0464	0.3027*	2.6345***	2.9508***
	(0.2325)	(0.2451)	(0.201)	(1.1196)	(0.9757)
	[0.04250]	[0.0018]	[0.0477]		
CAPEX	0.5919**	−10.6902**	0.1393	−0.2953	−1.0569
	(0.3383)	(5.6698)	(0.3894)	(0.6624)	(0.7941)
	[0.0415]	[−0.4332]	[0.0219]		
GROWTH	0.8757**	−0.3178	−1.5065	4.3482***	3.3899***
	(0.4497)	(6.9447)	(1.2533)	(1.7170)	(1.6067)
	[0.0614]	[−0.0128]	[−0.2377]		
LEVERAGE	1.1159	9.3805***	−4.8170***	5.2541***	3.7857
	(2.0012)	(4.7524)	(2.2937)	(2.5947)	(2.5385)
	[0.0782]	[0.3801]	[−0.7601]		
PROFIT	0.8759	3.8131***	−0.7003*	9.2505***	8.2370***
	(0.6554)	(1.3387)	(0.4675)	(2.4936)	(2.3572)
	[0.0614]	[0.1545]	[−0.1105]		
MTB	0.2485***	0.0033	0.5345***	0.8210***	0.9817***
	(0.1047)	(0.0329)	(0.2376)	(0.37191)	(0.3669)
	[0.0174]	[0.0001]	[0.0843]		
Stripped spread EMBI+	—	—	—	−32.3989***	−26.0916***
				(14.7925)	(12.0057)
Number of obs.	548	164	463	481	557
Log likelihood	−123.23	−34.17	−133.16	−60.60	−69.89

Note: Standard errors () and marginal effects [].
[a] Includes Brazil, Chile, and Mexico.
[b] Includes Argentina, Brazil, Chile, México, and Peru.
***, **, and * indicate significance at 5%, 10%, and 15%, respectively.

public equity, respectively.* Likewise, this effect is statistically significant for Brazil (5%) and Mexico (15%). For the group of models, the impact is again positive and significant.[†]

This result confirms the theories discussed in Section 13.4 with respect to firm size, where larger firms face less costs derived from adverse selection. Therefore, the greater their size, the more is the share in the initial portfolio generating incentives to issue an IPO for diversification purposes. Moreover, when comparing the studies of other countries outside Latin America the results present a remarkable similarity confirming a positive and significant relation of this variable with the probability of going public.

Investment opportunities represented by means of the CAPEX variable do not show a significant impact for Mexican companies neither for the region as a Group (negative impact in both samples) and a similar result was found for Italy and Taiwan. In contrast, for the Brazilian case a change in the CAPEX standard deviation generates an increase of 3% points on the likelihood of going public as well for the U.K. case. Furthermore, as sales increase the proxy of a firm's financing needs displays a positive and significant impact as much for the group models as for Brazil. In the Mexican and Chilean cases, the variable has a negative impact, but is not significant.

The proxy variable of leverage has a positive impact for the group model (even if the significance level disappears when including Argentina and Peru). As for Brazil there is a positive impact but it is not significant. These results contrast with the ones obtained in earlier studies, but are similar to the specific case of Mexico for which a standard deviation change on leverage generates a decrease of 10% points on the likelihood of going public (effect statistically significant at 5%). These results highlight the fact that the Latin American region presents much heterogeneity within the firms' characteristics that choose to issue stocks for first time. Profitability presents a positive impact for the models as a group and for Brazil at the individual level, even though in this case the variable is not significant. For Mexico, the result is the inverse, given that the impact is negative, even significant at the 15% level. The comparison with the studies of other countries shows that the positive and significant effect is similar to the results

---

* These impact effects are evaluated for the average value of the considered variables.
† For models estimated with fixed effects, it is not possible to calculate marginal effects (Greene, 2003, p. 699)

found in Italy and Taiwan. On the other hand, Mexico presents similar results to the United Kingdom. An inverse relation between the probability of going public and the profitability could suggest that Mexican firms that go public do not raise enough funds to finance large investments.

For Latin America as a group, the positive sign suggests that firms make the most of their large profits by going public since they can get a higher price for their stocks and the CAPEX sign is negative, therefore, the main reason to go public is not to finance investments for Latin America as a group.

The medium variable MTB presents for all cases a positive and a significant effect. For Brazil and Mexico an increase in the standard deviation results in an increase from 7% to 10% points on the likelihood of going public for a firm in the same industry. It is important to mention that for the group regressions, this variable is the second most important (after firm size) on the likelihood of going public. In order to run group regressions, an extra variable was added, the Embi+ Stripped Spread which allows one to capture a differenced effect per country. Results for both models were negative and significant. This shows that Latin American firms are conscious that the greater the country risk, the higher return is required for the investor, which worsens the mispricing problem of new issues.

13.7 CONCLUSIONS

Firms in Latin America go public to take advantage of a window of opportunity, to raise capital at the right time when there are factors mitigating the underpricing problem, i.e., when they have higher profits and when there is low country risk which can possibly explain the recent patterns of IPOs in the Latin American region. From the mid-1990s up until the beginning of the twenty-first century, country risk in the region was increasing due to numerous Latin American financial crises in Mexico, Brazil, and Argentina. In recent years the country risk has been diminishing consistently and the number of IPOs has risen substantially. Hence, the so-called IPO wave could be related not only to firm's fundamentals, but also to the business environment in the country and of the region.

Another result is due to heterogeneity across Latin American countries of going public because for almost all variables one finds a country in which a particular effect is different from the others or even from the region. This is expected because the Latin American emerging markets are partially integrated signifying that investors may take advantage of different risk and return patterns across countries. Given that corporate

Latin American firms seem to use the window opportunity argument to go public, they wait until the country environment is particularly positive, as compared to the region, to go public to attract investors.

The last remark concerns the case of institutional investors where the diversification is an important issue. The issuance of new IPOs at the "right" moment can be seen as an opportunity to further diversify institutional portfolios given the foreign investment limit large players such as pension funds encounter.

This chapter reveals that not only corporate firms in emerging markets are more prone to go public when the likelihood of underpricing decreases, but also small firms in these markets face substantial barriers. By stimulating this group of smaller firms to enter the market not only could help them gain accessibility to financial resources at a lower cost, but also could help institutional investors find a different "menu" of financial assets to invest in. Hence, further research must concentrate in explaining the factors related to small firms as data become available.

REFERENCES

Brau, J. and Fawcett, S. (2006) Initial public offerings: An analysis of theory and practice. *The Journal of Finance*, 61(1): 399–436.

Campbell, T. (1979) Optimal investment financing decisions and the value of confidentiality. *Journal of Financial and Quantitative Analysis*, 14(5): 913–924.

Chemmanur, T. and Fulghieri, P. (1999) A theory of the going public decision. *The Review of Financial Studies*, 12(2): 249–279.

Choe, H., Masulis, R., and Nanda, V. (1993) Common stock offerings across the business cycle: Theory and evidence. *Journal of Empirical Finance*, 1(1): 3–31.

Diamond, D. (1991) Monitoring and reputation: The choice between bank loans and directly placed debt. *Journal of Political Economy*, 99(4): 689–721.

Fischer, C. (2000) Why do companies go public? Empirical evidence from Germany's Neuer Markt. Working Paper, Munich University, Munich, Germany.

Gattás, A. (2001) Oferta Pública Inicial un Instrumento de Financiación para Empresas Entrepreneur, Working Paper, Universidad del Cema, Buenos Aires, Argentina.

Gill de Albornoz, B. and Pope, P. (2004) Determinants of the going public decision: Evidence from the UK. Working Paper, Universitat Jaume I, Castellon, Spain.

Greene, W. (2000) *Econometric Analysis*. Prentice Hall: New York.

Holmstrong, B. and Tirole, J. (1993) Market liquidity and performance monitoring. *Journal of Political Economy*, 101(4): 678–709

Jain, B. and Kini, O. (1994) The post-issue operating performance of IPO firms. *The Journal of Finance*, 49(5): 1699–1726.

Leland, H. and Pyle, D. (1977) Informational asymmetries, financial structure, and financial intermediation. *The Journal of Finance*, 32(2): 371–387.

Lowry, M. and Schwert, W. (2002) IPO market cycles: Bubbles or sequential learning? *The Journal of Finance*, 57(3): 1171–1200.

Maksimovic, V. and Pichler, P. (2001) Technological innovation and initial public offerings. *Review of Financial Studies*, 14(2): 459–494.

Merton, R. (1987) A simple model of capital market equilibrium with incomplete information. *The Journal of Finance*, 42(3): 483–510.

Mikkelson, W., Partch, M., and Shah, K. (1997) Ownership and operating performance of firms that go public. *Journal of Financial Economics*, 44(3): 281–307.

Modigliani, F. and Miller, M. (1963) Corporate income taxes and the cost of capital: A correction. *The American Economic Review*, 53(3): 433–443.

Nanda, V. (2002) Internal capital markets and corporate refocusing. *Journal of Financial Intermediation*, 11(2): 176–211.

Pagano, M. (1993) The flotation of companies on the stock market: A coordination failure model. *European Economic Review*, 37(5): 1101–1125.

Pagano, M., Panetta, F., and Zingales, L. (1998) Why do companies go public? An empirical analysis. *The Journal of Finance*, 53(1): 27–64.

Ritter, J. (1991) The long-run performance of initial public offerings. *The Journal of Finance*, 46(1): 3–27.

Rock, K. (1986) Why new issues are underpriced. *Journal of Financial Economics*, 15(1–2): 187–212.

Rydqvist, K. and Hoghölm, K. (1995) Going public in the 1980s: Evidence from Sweden. *European Financial Management*, 1(3): 287–315.

Shen, Y. and Wei, P. (2007) Why do companies choose to go IPOs? New results using data from Taiwan. *Journal of Economics and Finance*, 31(3): 359–367.

Stoughton, N. and Zechner, J. (1998) IPO-mechanisms, monitoring, and ownership structure. *Journal of Financial Economics*, 49(1): 45–77.

Yosha, O. (1995) Information disclosure costs and the choice of financing source. *Journal of Financial Intermediation*, 4(1): 3–20.

Zingales, L. (1995) Insider ownership and the decision to go public. *The Review of Economic Studies*, 62(3): 425–448.

CHAPTER 14

Preholiday Effect and Stock Returns in Mexican Financial Markets

Dermott Tennyson and Begoña Torre Olmo

CONTENTS

14.1 INTRODUCTION

There has always been a discussion surrounding whether or not the stock market's behavior is implicitly subject to a set of rules. In other words there are investors who act according to an irrational way and this pattern is important enough for it not to be eliminated by the price system so that those who are aware of these behavioral patterns can take advantage of them. Investors possessing the knowledge of these behavioral patterns can take advantage of them. A growing concern has appeared in relation to this, not just amongst the brokers who takepart in the markets, but also in the academic world because of the attempt to model the price behavior of stocks. This has led to various studies, of which the efficient markets hypothesis of Fama (1970) has achieved the greatest importance. The concept of efficiency is characterized by the lack of any ex-post regularity, behavior patterns that repeat themselves after the occurrence of a certain event a key factor when attempting to study the market's "anomalies." These are phenomena that are difficult to reconcile with what is expected from an efficient market. In this regard, Lakonishok and Smidt (1988) state that the numerous anomalies documented in the stock market's daily performance question the theory of Fama because they enable expected performance to be predicted. In other words, it is neither random nor does it depend on the arrival of new information.

One of the most evident and persistent anomalies in the international stock markets is the preholiday effect. It is defined as the existence of extraordinary performance from the time the market closes on the day before the preholiday to the market close on the preholiday day. This chapter analyzes the preholiday effect at Mexican stock exchange (BMV), in the period between January 2, 1980 and December 31, 2004. The preholiday's magnitude and level of significance in addition to its persistence throughout time are analyzed. Research is also undertaken as to whether the reward associated with the preholiday effect is associated with greater risk or volatility. Finally, the existence or not of a dependency with the same effect in the U.S. capital markets is also studied.

This chapter is structured as follows: In Section 14.2, we review the research published with regard to the preholiday effect. Special emphasis is made on its possible existence over time as well as its interrelationship on the international stock markets. In Section 14.3, there is an empirical analysis, description of the data and methodology used as well as outlining the results. Finally, in Section 14.4 the main conclusions are outlined.

14.2 PREHOLIDAY EFFECT IN STOCK MARKETS

With the initial study on the Dow Jones Industrial Average (DJIA) index published by Fields (1934), the preholiday effect has been widely documented on world stock markets. To measure its size, Fosback (1976) calculated the S&P 500's daily performance between 1928 and 1975 and found an accumulated performance on preholiday days of 102.6% and on preholiday days of 333.3%. The accumulation for both days was 778%, quite higher to the 414% attained throughout the study's total period, excluding dividend payments. The author states that if two investors had started the period under study with $10,000 each and the first had bought on preholiday days to then sell on postholiday days and the other had bought on postholiday days to sell on preholiday days, the former investor would have finished the study period with $87,787 whereas the latter investor would have finished with $5,855. Obviously, these results assume that the transaction cost is zero.

The magnitude of the preholiday effect is measured by the average performance ratio on preholiday days and average performance on the rest of the days. The level of significance is measured with statistical models that compare the average performance of various samples. The results of different studies regarding the extent of the preholiday effect on international markets are outlined in Tables 14.1 through 14.4.

Most international stock markets show a preholiday effect, and it is more significant in the case of the U.S. and Asian markets. The effect exists in European markets, especially in England, Italy, and Spain. The magnitude of the preholiday effect in the U.S. and Spanish markets is especially noteworthy. The average performance ratio for the former varies between 9 and 30 and for the latter, between 14 and 16. The classic asset evaluation models such as CAPM predict greater expected performance in exchange for greater risk. Nevertheless, Ariel (1990) and Meneu and Pardo (2001) found that the volatility on preholiday days was less than the rest of the days for the U.S. and Spanish markets, respectively. This indicates that superior performance on preholiday days is not accompanied by greater risk, as would be expected.

TABLE 14.1   Preholiday Effect in the U.S. Markets

Author	Sample	Country	Period	Coc
Lakonishok and Smidt (1988)	DJIA	United States	1897–1986	23
Ariel (1990)	DJIA	United States	1963–1982	30
Ariel (1990)	CRSP	United States	1963–1982	14
Ariel (1990)	CRSP	United States	1963–1982	8.9
Cadsby and Ratner (1992)	CRSP	United States	1962–1987	7.5
Cadsby and Ratner (1992)	CRSP	United States	1962–1987	10
Pettengill (1989)	CRSP	United States	1962–1986	7
Liano and Marchland (1992)	NASDAQ	United States	1973–1989	10.5
Liano and Marchland (1992)	NASDAQ	United States	1973–1989	6.5
Kim and Park (1994)	NASDAQ	United States	1973–1986	11
Liano and White (1994)	NASDAQ	United States	1972–1991	14
Vergin and McGinnis (1999)	NASDAQ	United States	1987–1996	3.4
Vergin and McGinnis (1999)	AMEX	United States	1987–1996	9
Kim and Park (1994)	AMEX	United States	1963–1986	27
Vergin and McGinnis (1999)	NYSE	United States	1987–1996	1.7
Kim and Park (1994)	NYSE	United States	1963–1986	9
Liano and White (1994)	S&P 500	United States	1962–1991	11
Kim and Park (1994)	S&P 500	United States	1972–1987	15.4
Pettengill (1989)	S&P 500	United States	1962–1986	13.5
Vergin and McGinnis (1999)	S&P 500	United States	1987–1996	0.9
Keef and Roush (2005)	S&P 500	United States	1930–1987	15
Cadsby and Ratner (1992)	TSE	Canada	1975–1987	2.6
Cervera and Keim (1999)	IPyC	Mexico	1980–1987	4.5

Source: Author's compilation.

TABLE 14.2   Preholiday Effects in British Markets

Author	Sample	Country	Period	Coc
Arzad and Coutts (1997)	FT 30	United Kingdom	1935–1994	14
Kim and Park (1994)	FT 30	United Kingdom	1972–1987	5.6
Mills and Coutts (1995)	FTSE 100	United Kingdom	1986–1992	6.8
Cadsby and Ratner (1992)	FTSE500	United Kingdom	1983–1988	(Neg)

Source: Author's compilation.

Agrawal and Tandon (1994) studied the preholiday effect in several international indexes. In the case of Mexico, the authors document an average accumulated performance of 1% for the Christmas and New Year's holiday celebrations between 1977 and 1988. Cervera and Keim (1999)

TABLE 14.3   Preholiday Effect in European Markets

Author	Sample	Country	Period	Coc
Van der Sar (2003)	CBSTRI	Amsterdam	1981–1998	1.6
Cadsby and Ratner (1992)	SBCII	Switzerland	1980–1989	3.5
Cadsby and Ratner (1992)	CI	Germany	1980–1989	0.6
Cadsby and Ratner (1992)	CACGI	France	1980–1989	0.7
Barone (1990)	MIB	Italy	1975–1989	28
Cadsby and Ratner (1992)	BCI	Italy	1980–1989	3.4
Meneu and Pardo (2001)	IBEX-35	Spain	1990–2000	14
Meneu and Pardo (2001)	IBEX-C	Spain	1990–2000	47
Lauterbach and Ungar (1992)	TASE	Israel	1977–1991	1.2
Mills et al. (2000)	GIASE	Greece	1986–1997	3.9

Source: Author's compilation.

TABLE 14.4   Preholiday Effect in Asian Markets

Author	Sample	Country	Period	Coc
Cadsby and Ratner (1992)	NIKKEI	Japan	1979–1988	4.5
Kim and Park (1994)	NIKKEI	Japan	1972–1987	4.4
Ziemba (1991)	NIKKEI	Japan	1949–1988	5
McGuinness (2005)	HIS	Hong Kong	1995–2005	28
Cadsby and Ratner (1992)	HIS	Hong Kong	1980–1989	13
Cervera and Keim (1999)	HIS	Hong Kong	—	6
Wong and Yuanto (1999)	JCI	Indonesia	1983–1997	6.7
Easton (1990)	Sydney	Australia	1958–1980	11
Easton (1990)	Melbourne	Australia	1963–1980	3.5
Cadsby and Ratner (1992)	AOI	Australia	1980–1989	6

Source: Author's compilation.

documented an average performance ratio of 4.5 for the IPyC (*Mexican price index*) between 1980 and 1987 (cf. Table 14.1).

14.2.1 Existence of the Preholiday Effect over Time

Fama (1998) observes that most long-term anomalies disappear when there are reasonable changes in the methodology used to study them. He states that once they are discovered, most anomalies tend to disappear over time. Investors presumably add this information to the decision-making process. Nevertheless, the persistence of the extraordinary performance associated with certain anomalies throughout different test periods and in different capital markets make us wonder about the validity of the Efficient Markets Theory. As an example, Haugen and Jorion (1996) state the well-known case of the January effect and its persistence over time, many years after the date on which the effect was outlined in scientific publications.

By applying this to the case of the preholiday effect, Lakonishok and Smidt (1988) studied the DJIA daily performance between 1897 and 1986. The authors found that the average performance of the preholiday days becomes insignificant in the last subsample of the period under study between 1976 and 1986. Vergin and McGinnis (1999) compared the magnitude and level of significance of the holiday effect in the studies by Pettengill (1989) and Kim and Park (1994) performed on samples prior to 1987 with the effect's size and level of significance in the same indexes between 1987 and 1996. The average performance ratio fell from 13.5, documented by Pettengill and 15.4 documented by Kim and Park, to a mere 0.9 in the case of the study's second period (cf. Table 14.1). Researchers discovered that the preholiday effect was only persistent and significant at the 1% level on the AMEX, an index of small companies. Keef and Roush (2005) analyzed the daily performance on the S&P 500 index between 1930 and 1999. The average performance ratio was 15 until 1987. However, after this date, it was reduced to 2, and the effect was no longer statistically significant. Finally, Vergin and McGinnis (1999) documented a Labor Day effect for the Mondays before the holiday. The average performance of the Mondays before Labor Day was 0.579% compared to the average performance of 0.216% for the rest of the preholiday days that fell on a Monday. Nevertheless, the effect disappears after 1987.

14.2.2 International Preholiday Effect

There is empirical evidence which supports the relationship between the behavior of the different international stock markets. Finnerty et al. (1990) found that the U.S. market's behavior is transferred within one day to the Japanese market. The S&P index's performance on day $t-1$ explained between 7% and 25% of the Nikkei index's performance of day t. This observation is confirmed by Ko and Lee (1991) for the Singapore, Hong Kong, and Japanese markets. The authors documented correlation coefficients of +0.425, +0.233, and +0.460, respectively, between the daily performance of these markets and those of the U.S. markets. Mc Guinness (2005) found that the DJIA performance on day $t-1$ explained 20% of the Hong Kong HSI index on day t between 1975 and 1990. However, the influence disappears in the second subperiod.

Arellano (1993) studied the influence that the U.S. stock market has on the Mexican stock market. The author found that between 1980 and 1990, the IPyC movements reflected 20% of the DJIA index. During the subperiod of 1986–1990, the IPyC movements reflected 45% of the movements.

Moving on to the analysis of the international holiday effect, Cadsby and Ratner (1992) studied several anomalies in the daily performance of 11 international stock markets and Brockman and Michayluck (1998) focused their attention in the persistence analysis. Regarding the Hong Kong (Hang Seng index) index, the authors found a significant additional preholiday effect between 1980 and 1989 explained by the same effect on the U.S. markets. A similar result was also found for the case of the Canadian TSX index between 1975 and 1987. Kim and Park (1994) compared the preholiday effect in three international indexes between 1972 and 1987: S&P 500, FT30, and the Nikkei. However, they did not find a significant influence of the U.S. preholiday effect on the other two markets. Meneu and Pardo (2001) analyzed the preholiday effect in five Spanish indexes between 1990 and 2000. They discovered that the local preholiday effect in all indexes, except for one, is explained both by the holidays exclusive to Spain and the holidays exclusive to North America. Upon adjusting for the preholiday effect on the U.S. markets, the local preholiday effect remains significant. Similarly, Lucey (2005) documented the effect of a preholiday day in several Irish market indexes between 1979 and 1998. He showed that it is independent of the same effect on U.K. markets.

This evidence indicates that the local preholiday effect is partly explained by the preholiday effect on the U.S. markets in the case of the Hong Kong and Canadian markets, at least until 1985. This observation indicates a greater dependence of the Hong Kong and Canadian markets' on the U.S. markets until this year. Should a preholiday effect be identified in the case of the Mexican market, it would not be unusual to find a dependence on the U.S. market, especially if we consider the Mexican market's high degree of dependence on this neighboring market.

14.3 EMPIRICAL ANALYSIS
14.3.1 Hypothesis, Data, and Methodology
In accordance with the theoretical arguments that have previously been outlined, we explain the hypothesis in the following text which we intend to examine in this chapter:

H1: There is a preholiday effect in the Mexican stock market (BMV) in the period from 1980 to 2004.

H2: The preholiday effect in the Mexican stock market tends to disappear over time.

H3: The preholiday effect in the Mexican stock market depends on the same effect in North American markets.

For this study, data have been taken from the closing of the Share Price Index from the Mexican Stock Exchange, referred to as IPyC obtained from the *Economática* database. IPyC is an index weighted by market value so that assets with greater capitalized value have more weight in the calculation of the index. In the 1980's the companies with the greatest capitalized value were in the mining sector while after the 1990s, they came from the telecommunications and construction industries. This change is due to Mexico's stronger commercial relations together with the privatization of the Mexican banking sector and Telmex. In the 1980s and 1990s, the stock exchange operations of the BMV were conducted by bidding on the auctions floor. On January 11, 1999, the SENTRA system was gradually introduced into the BMV, a mechanism for automating stock exchange operations.

We have sought to frame the periods under examination in the context of certain important developments in the Mexican economy. As such, the results may reflect the relationship between investor behavior and the economic conditions affecting their individual decisions. The 1980's in Mexico were characterized by an economy that was closed to the outside world, a nationalized banking system and a strict control of foreign currency exchange. The first half of the 1990's was characterized by commercially opening up to the outside world, the privatization of the banking sector and the freeing up of capital. Nonetheless, this opening up and relaxation of controls led to a financial crisis at the end of 1994, forcing the Mexican government to seek the help of the United States Federal Reserve by requesting an emergency loan. The subsequent years were characterized by the rationalization of public expenditures, a hard monetary policy and the restructuring of public debt. This is why the period under examination is divided into three subperiods: 1980–1989, 1990–1994, and 1995–2004. This decision also helps to establish a contrast with the hypothesis concerning the persistence of the preholiday effect over time.

Following the methodology of other studies, the daily performance of the IPyC was calculated using a logarithm based on the index for two consecutive days. There are a total of 6231 records for the entire period under examination. Figure 14.1 shows the distribution of the frequency of daily performance and the descriptive statistics. The nonparametric Kolmogorov–Smirnov test was applied to the entire sample and the hypothesis of normality was accepted for the distribution of the daily performance at a 5% level of significance.

Two samples of daily returns were created: the first is composed by the prices during preholiday days, and the second by the performance for other business days. In the selection of the preholiday days, we have replicated the methodology of other research studies, not distinguishing between holidays that fall during the week and those that fall on the weekend, so that the days in question are the following:

- January 1st, New Year

- February 5th, Anniversary of the Constitution

- March 21st, Birthday of Benito Juárez

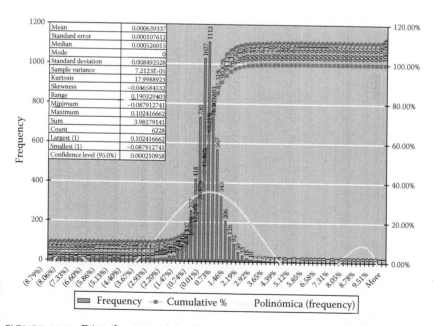

FIGURE 14.1 Distribution of the frequency of daily performance and the descriptive statistics.

- Holy Thursday and Good Friday

- May 1st, Labor Day

- May 5th, Anniversary of the Battle of Puebla (a nonworking day until 2002)

- July 31st, last day of summer (a nonworking day until 1980)

- September 1st, Presidential Address (variable holiday)

- September 16th, Anniversary of Independence

- October 12th, day of the Race (a nonworking day until 1993)

- November 2nd, Day of the Dead

- November 20th, Anniversary of the Revolution

- December 1st, day the president takes the oath of office (celebrated every 6 years)

- December 12th, Festival of the Virgin of Guadalupe

- December 25th, Christmas

Excluded from the first sample are the prices for the days prior to six unscheduled market closings in the period under examination. Similarly, September 11, 2001 was eliminated because it was more notable for the events of terrorism than because it was a day-before the holiday in Mexico. In fact, the stock market remained closed from Wednesday September 12th until Friday September 14th, and in a normal scenario the day before the holiday would have fallen on Friday the 14th. Similarly, August 31, 1982 was included in the sample of days before a holiday, the day before the presidential address when José López Portillo announced the nationalization of the Mexican banking sector. The capital market stayed closed from September 1st until the 20th, and in fact, the news given on the actual day of September 1st was a complete surprise to the markets.

The sample of preholiday returns has a total of 306 observations and the sample of performance for the other days has a total of 5925 observations. To contrast the various hypotheses put forth we have used variance equality tests and mean tests as well as a regression analysis using a dummy variable for the preholiday days. The use of nonparametric tests eliminates the reservation associated with parametric testing that does not adjust for the well-known autocorrelation of daily performance in the capital markets.

14.3.2 Results

When analyzing the performance for all days and for preholiday days, we find that the preholiday days explain 15% of the total performance for the period under examination and 22.6% of the total performance for the 1980–1989 subperiod (Table 14.5). This second result is comparable to the findings of Lakonishok and Smidt (1988) who studied the preholiday effect during a period of 90 years in the DJIA index.

Upon analyzing the average performance for the preholidays for certain particular holidays, we find average of 0.37% and 0.57% for the preholiday days of Christmas and New Years, respectively, between 1980 and 1989. This result translates into an average cumulative of 0.95% for the two preholidays and it is comparable to the results of Agrawal and Tandon (1994) referred to previously.

The results in Table 14.6 indicate that the volatility on preholidays is less than the remaining days, for all of the period under examination and two of the three subperiods. This observation coincides with the results of other researchers and we can confirm that, if there is a preholiday effect in these periods, it is not accompanied by a greater risk. The results indicate that in the last subperiod between 1995 and 2004, the trend is reversed and the volatility on preholidays is greater than the volatility on the remaining days. This change probably has an explanation in the possible causes of the preholiday effect.

The variance ratio is within the permissible range for the statistics in the entire period under examination and in the 1990–1994 and 1995–2005 subperiods. The hypothesis is only accepted in those cases and is rejected for the 1980–1989 subperiod. A t-test can be applied to contrast the hypothesis of the equality of average performance for the entire period under examination and in the last two subperiods.

Table 14.7 presents the results of the t-test to contrast the equality of the averages of the two samples. The daily performance for the preholidays in the entire period under examination is 0.197%, that is, 3.45 times the average performance of the other on the remaining days of the year. The value p in the first row of Table 14.7 indicates that the difference in average performance is significant and confirms the existence of a preholiday effect in the whole period under examination. The average daily performance for the preholiday days in the 1980–1989 period was 0.424% and 5.2 times the daily performance of the other days. This result is comparable to the results of Cervera and Keim (1999) for the IPyC between 1980 and 1987 (Table 14.1).

TABLE 14.5 Comparative Performance for All Days and for Preholidays in the IPyC, 1980–2004

Period	Total Days	Total Performance (%)	Total Performance (%)	Preholidays	Daily Performance Preholidays (%)	Total Performance Preholidays (%)	Ratio Preholidays/ Total (%)
1980–2004	6231	0.064	398.18	306	0.196	59.95	15.06
1980–1989	2472	0.101	249.27	135	0.417	56.25	22.57
1990–1994	1248	0.060	75.36	60	0.056	3.36	4.46
1995–2004	2511	0.029	73.54	111	0.053	5.85	7.96

TABLE 14.6   Results of Nonparametric Testing to Compare Variances in Daily Performance; Preholidays and Other Days for the IPyC 1980–2004

Period	Preholidays	Other Days	Preholiday Var	Other Var	Var Quotient	Holiday Statistics above	Holiday Statistics below
1980–2004	303	5927	0.634	0.724	1.141	1.185	0.854
1980–1989	132	2339	0.680	1.053	1.549	1.302	0.790
1990–1994	60	1188	0.319	0.431	1.350	1.498	0.711
1995–2004	111	2400	0.650	0.547	0.842	1.334	0.775

TABLE 14.7   Result of t-Test to Compare Average Performance of Preholidays and the Other Days for the IPyC, 1980–2004

Period	Preholiday Performance	Other Performance	Performance Quotient	t	p
1980–2004	0.197	0.057	3.447	2.969	0.003
1980–1989	0.424	0.083	5.128	4.559	0.000
1990–1994	0.056	0.061	0.925	−0.061	1.048
1995–2004	0.003	0.031	0.099	−0.352	1.275

Although the value for p indicates the existence of a significant preholiday effect in the subperiod 1980–1989, for the reasons noted in the previous paragraph, it is necessary to validate the hypothesis through nonparametric testing.

Table 14.8 shows the result of comparing the ratio for days with positive progress in the two samples composed of preholidays and the remaining days. For the 1980–1989 subperiod were 76.5% and 55.2%, respectively.

TABLE 14.8   Results of Nonparametric Testing to Compare Ratios for Positive Progress in the Daily Performance for Preholidays and the Remaining Days for the IPyC, 1980–2004

Period	Preholidays	Other Days	Preholiday Quotient	Other Quotient	t	p
1980–2004	303	5927	0.630	0.533	3.402	0.001
1980–1989	132	2339	0.765	0.552	5.566	0.000
1990–1994	60	1188	0.567	0.530	0.554	0.579
1995–2004	111	2400	0.505	0.517	−0.259	1.204

The values for p in the first two rows of Table 14.8 indicate the existence of a significant preholiday effect for the entire period under examination and in the 1980–1989 subperiod, but not in the other subperiods.

Finally, in Table 14.9 we present the results of the contrast between the average daily performance through a linear regression model with a dummy variable for the preholidays estimated as following:

$$R_{i,t} = c_i + \alpha_{i\,\text{PRE}}\, D_{\text{PRE}} + \mu_i$$

where

$R_{i,t}$ is the return of IPC

D_{PRE} is the dummy variable which equals 1 if the return occurs on a day before holiday and 0 otherwise

μ_i is the random disturbance term

The levels of significance associated with the preholiday variable in the last column confirm the existence of a preholiday effect for the entire period under examination and in the 1980–1989 subperiod, but not in the last two subperiods. We conclude that there is a significant preholiday effect in the entire period under examination and in the 1980–1989 subperiod.

Upon analyzing the results in Table 14.7, we find that the average preholiday performance in the 1990–1994 subperiod has practically the same level as the performance for the remaining days and the difference is not significant. The average preholiday performance in the 1995–2004 subperiod is only 10% of the average performance for the remaining days, but it also does not involve a statistically significant difference. We cannot

TABLE 14.9 Results of Linear Regression Model to Contrast Daily Performance with a Dummy Variable for Preholidays in IPyC, 1980–2004

Period	Constant (Signif.)	Preholiday (Signif.)
1980–2004	0.057 (0.000)	0.136 (0.006)
1980–1989	0.083 (0.000)	0.333 (0.000)
1990–1994	0.0606 (0.001)	−0.0046 (0.958)
1995–2004	0.031 (0.044)	−0.0270 (0.703)

Note: Levels of significance in parentheses (significant analysis at 5% level).

affirm that the holiday effect in the stock exchange is persistent over the course of time; furthermore, it tends to disappear after 1989. This result is consistent with the findings of Vergin and McGinnis (1999), and Keef and Roush (2005) for the North American markets, and can be explained by the fact that the investors who operate within the stock market have already assimilated this information in their decision making after 1989. This date coincides with an increase in Mexico's commercial relations with the outside world, as we noted previously.

Finally, Table 14.10 shows the results of the contrast of a linear regression model of daily performance that assign dummy variables to preholidays that are exclusively Mexican ($D_{PRE-MEX}$), exclusively of the United States ($D_{PRE-USA}$), and common to both markets ($D_{PRE-MEX-USA}$).

$$R_{i,t} = c_i + \alpha_{i\,PRE}\, D_{PRE-MEX} + \beta_{i\,PRE}\, D_{PRE-USA} + \lambda_{i\,PRE}\, D_{PRE-MEX-USA} + \mu_i$$

Despite the magnitude of the average performance for the preholidays held in common by Mexico and the United States (Christmas and New Year), a statistically significant average performance is only recorded for the preholiday days that are exclusively Mexican. The foregoing can be discerned by the levels of significance in the second column of the Table 14.10, for the entire period under examination and the 1980–1989 subperiod.

TABLE 14.10   Results of Linear Regression Model to Vontrast Daily Performance with Dummy Variables for Preholidays That Are Exclusively Mexican, Exclusively the United States, and Common to Both Markets, 1980–2004

Period	Constant (Signif.)	Mexican (Signif.)	United States (Signif.)	Shared (Signif.)
1980–2004	0.056	0.131	0.0700	0.1660
	(0.000)	(0.016)	(0.370)	0.1450
1980–1989	0.081	0.332	0.121	0.3540
	(0.000)	(0.001)	(0.423)	0.0900
1990–1994	0.063	−0.022	−0.1030	0.0740
	(0.001)	0.812	(0.463)	0.7230
1995–2004	0.028	−0.032	0.100	0.001
	(0.065)	(0.693)	(0.330)	(0.996)

This conclusion coincides with the results of Kim and Park (1994) for British and Japanese markets, and with those of Meneu and Pardo (2004), in the case of Spanish markets.

If there is some degree of dependency among the Mexican and U.S. financial markets in general, as stated by Arellano (1993), it is not possible to affirm that there is a relation of causality between the preholiday effect in the United States and Mexico. We have already noted that the 1980s in Mexico were characterized by protectionist economic policies, currency exchange controls, and very limited commercial relations of Mexico with the United States. However, the preholiday effect has been identified in Mexico precisely during this decade. If the Mexican preholiday effect is not evident in the 1990s, in a scenario of a greater interrelationship between the two financial markets, then we could affirm with more certainty that the Mexican preholiday effect is not a manifestation of the same effect in the U.S. markets. We do affirm, then, that the preholiday effect in the Mexican markets exists independently of the same effect in the North American markets.

14.4 CONCLUSION

Within the context of a financial economy, one of the most thoroughly researched aspects has been the search for patterns in the behavior of prices with the purpose of endeavoring to determine the degree of market efficiency. This is indeed a classic subject in literature, and has increased in importance from the end of the 1990's to the beginning of the present decade. This has particularly been a result of a number of financial scandals in the most developed financial markets.

Our research focuses on the preholiday effect and offers a review of the more recent literature along with an empirical analysis of the Mexican market, placing our focus mainly on the influence that the United States market may exercise upon it.

The results obtained show, upon studying the daily performance of the IPyC between 1980 and 2004, that a significant preholiday effect was found in the entire period under examination and in the 1980–1989 subperiod. This effect is comparable to the same effect documented by Arellano (1993) in the case of the IPyC between 1980 and 1987. The premium associated with the day-before-the holiday effect is accompanied by a greater risk, as one would expect. The magnitude of the preholiday effect declines and becomes insignificant over time, which is also comparable with what occurred in the North American markets. The foregoing observations point to a certain degree of influence of the North American financial markets on the Mexican markets, an aspect that has been extensively discussed in financial literature. Nevertheless, our results indicate that the preholiday effect in Mexican markets does not depend on the same effect in North American markets.

Research into the possible causes of the preholiday effect in the Mexican market is a field of investigation for future studies. Such a study is necessary if one wishes to explain certain characteristics of the aforementioned anomaly such as the lower volatility with respect to the rest of the days in the sample and the fact that it does not persist over time.

REFERENCES

Agrawal, A. and Tandom, D. (1994) Anomalies or illusions? Evidence from stock markets in eighteen countries. *Journal of International Money and Finance,* 13(1): 83–106.

Arellano, R. (1993) Relación de Largo Plazo del Mercado Bursátil Mexicano con el Estadounidense: Un Análisis de Cointegración. *El Trimestre Económico,* 237(LX): 91–112.

Ariel, R. (1990) High stock returns before holidays: Existence and evidence on possible causes. *The Journal of Finance,* 7(5): 1611–1626.

Arzad, Z. and Coutts, J.A. (1997) Security price anomalies in the London international stock exchange: A 60-year perspective. *Applied Financial Economics,* 7(5): 455–464.

Barone, E. (1990) The Italian stock market: Efficiency and calendar anomalies. *Journal of Banking and Finance,* 14(2): 483–510.

Brockman, P. and Michayluck, D. (1998) The persistence holiday effect: Additional evidence. *Applied Economics Letters,* 5(2): 205–209.

Cadsby, C.B. and Ratner, M. (1992) Turn-of-month and pre-holiday effects on stock markets. *Journal of Banking and Finance,* 16(3): 497–509.

Cervera, A. and Keim, D.B. (1999) High stock returns before holidays: International evidence and additional tests. In: D.B. Keim and W.T. Ziemba (Eds.), *Security Market Imperfections in World Wide Equity Markets.* Cambridge University Press: Cambridge, United Kingdom.

Easton, S. (1990) Returns to equity before and after holidays: Australian evidence and tests of plausible hypotheses. *Australian Journal of Management,* 15(2): 281–297.

Fama, E. (1970) Efficient capital markets; A review of theory and empirical work. *The Journal of Finance,* 25(2): 383–417

Fama, E. (1998) Market efficiency, long-term returns and behavioural finance. *The Journal of Financial Economics,* 49(3): 283–306.

Fields, M. (1934) Security prices and stock exchange holidays in relation to short-selling. *Journal of Business,* 7(3): 334.

Finnerty, J., Becker, K., and Gupta, M. (1990) The international relation between the US and Japanese stock markets. *The Journal of Finance,* 45(4): 1297–1306.

Fosback, N. (1976) Stock market logic. Working Paper, Dearnborn Financial Publishing Inc./Institute for Econometric Research: Fort Lauderdale, FL.

Haugen, R. and Jorion, P. (1996) The January effect; Still there after all these years. *Financial Analysts Journal,* 52(1): 27–31.

Keef, S. and Roush, M. (2005) Day-of-the-week effects in the pre-holiday returns of the Standard & Poors 500 stock index. *Applied Financial Economics* 15(2): 107–119.

Kim, C.W. and Park, K. (1994) Holiday effects and stock returns: Further evidence. *Journal of Financial and Quantitative analysis,* 29(1): 145–157.

Ko, K. and Lee, S. (1991) A comparative analysis of the daily behaviour of stock returns: Japan, the US and the Asian NICs. *Journal of Business Finance & Accounting,* 18(2): 219–234.

Lakonishok, J. and Smidt, S. (1988) Are seasonal anomalies real? A ninety-year perspective. *Review of Financial Studies,* 1(4): 403–425.

Lauterbach, B. and Ungar M. (1992) Calendar anomalies: Some perspectives from the behaviour of the Israeli stock market. *Applied Financial Economics,* 2(1): 57–60.

Liano, K., Marchand, P.H., and Huang, G. (1992) The holiday effect in stock returns: Evidence from the OTC market. *Review of Financial Economics,* 2(1): 45–54.

Liano, K. and White, L. (1994) Business cycles and the pre-holiday effect in stock returns. *Applied Financial Economics,* 4(3): 171–175.

Lucey, B. (2005) Are local or international influences responsible for the pre-holiday behaviour of Irish equities? *Applied Financial Economics,* 15(6): 381–389.

McGuinness, P. (2005) A re-examination of the holiday effect in stock returns: The case of Hong Kong. *Applied Financial Economics,* 15(16): 1107–1123.

Meneu, V. and Pardo, A. (2001) El Efecto Día Festivo en la Bolsa Española. *Moneda y Crédito,* 213: 97–127.

Meneu, V. and Pardo, A. (2004) Pre-holiday effect, large trades and small investor behaviour. *Journal of Empirical Finance,* 11(2): 231–246.

Mills, T.C. and Coutts, J.A. (1995) Calendar effects in the London stock exchange FTSE indices. *European Journal of Finance,* 1(4): 79–93.

Mills, T.C., Siriopoulos, C., Markellos, R., and Harizanis, D. (2000) Seasonality in the Athens stock exchange. *Applied Financial Economics,* 2(1): 137–142.

Pettengill, G. (1989) Holiday closings and security returns. *Journal of Financial Research,* 12(1): 57–67.

Van der Sar, N. (2003) Calendar effects on the Amsterdam stock exchange. *De Economist,* 151(3): 271–292.

Vergin, R.C. and McGinnis, J. (1999) Revisting the holiday effect; Is it on holiday? *Applied Financial Economics,* 9(5): 477–482.

Wong, K. and Yuanto, K. (1999) Short-term seasonalities on the Jakarta stock exchange. *Review of Pacific Basin Financial Markets & Policies,* 2(3): 375–398.

Ziemba, W. (1991) Japanese security market regularities: Monthly, turn-of-the-month and year, holiday and golden week effects. *Japan and the World Economy,* 3(2): 119–146.

CHAPTER 15

Business-Cycle and Exchange-Rate Fluctuations in Emerging Market Economies in Asia, Latin America, and Central and Eastern Europe

Marcelo Sánchez

CONTENTS

15.1 INTRODUCTION

The macroeconomic literature on emerging market economies (EMEs) has substantially expanded on both theoretical and empirical fronts. This concerns many key research areas such as business cycles, exchange-rate determination, exchange-rate pass-through, current account movements, and the impact of oil prices. The latest thrust of related theoretical studies is illustrated by small open-economy microfounded dynamic stochastic general equilibrium models, which trace macroeconomic developments to a number of economically interpretable shocks. The empirical literature on EMEs has remained broad, being to varying degrees linked to the latest theoretical research. While some authors attempt to identify underlying forces affecting specific markets and spreading through the economy, a variety of observable features are often tackled by researchers somewhat in isolation from the wider context of macroeconomic interactions.

Especially relevant to this title are empirical studies that separate out the influence of domestic and external factors on a country's economy. Many studies have found evidence that external factors are very important. Genberg (2003) finds that they are responsible for over 75% of business cycles in Hong Kong, and Canova (2005) estimates the corresponding average share for Latin America at almost 90% with 50% being the United States-driven. Canova's study attributes most of the foreign impact to a financial transmission channel, with a large contribution of U.S. monetary shocks. Even for larger open economies, results have tended to attach a large share to external factors. Cushman and Zha (1997)'s study on Canada estimates the United States to contribute by over 70% to business-cycle dynamics. Results for small industrial economies tend to be consistent with that for

Canada (see Dungey and Pagan, 2000, for Australia, and Buckle et al., 2003, for New Zealand). Using sign-restricted VAR models for individual countries, Rüffer et al. (2008) find that extraregional developments tend to play a large role in driving business-cycle movements in emerging East Asia. In contrast, Hoffmaister and Roldós (1997) find that external factors account for a limited fraction of macroeconomic fluctuations in EMEs (20% and 30% at the very maximum, respectively). Similarly, Kose et al.'s (2003) dynamic factor analysis indicates that macroeconomic fluctuations in these two regions are largely explained by domestic factors, while extraregional and especially intraregional developments play a very modest role.

I investigate the determinants of EME's business cycles and real exchange-rate developments. VAR models are estimated for 15 EME countries and sign restrictions are applied in line with a large number of macroeconomic models. The use of sign restrictions draws on work by Faust (1998), Canova and De Nicolò (2002), and Uhlig (2005) for advanced economies.*

The remainder of this chapter is organized as follows: Section 15.2 presents my econometric methodology; Section 15.3 briefly describes the data; Section 15.4 discusses the results of this chapter, including the reaction of macroeconomic variables to a number of structural shocks as well as variance decomposition analysis; and Section 15.5 provides some concluding remarks.

15.2 METHODOLOGY

This section consists of two parts. The first part outlines my identification strategy. The second part describes the key features of my VAR setup.

15.2.1 Choice of Variables and Sign Restrictions

I model each EME using four macroeconomic variables: real output (Y), consumer prices (P), real exchange rates (Q), and real imports (Y^{m}). I characterize the dynamics of the economy in terms of responses to global shocks as well as four domestic structural disturbances: a technology shock, a preference shock, a monetary policy shock, and a risk premium shock.

---

* Related approaches also include Canova and De Nicolò (2003), Peersman (2005), and Peersman and Straub (2009). Canova (2005) uses an approach similar to the one employed here to identify U.S. structural shocks by means of sign-restricted VARs, then estimates the impact of these shocks on Latin American economies.

I postulate the following sign restrictions:

	Y	P	Y^m	Q
Technology shock	+	−	+	+
Preference shock	+	+	+	−
Monetary shock	−	−	−	−
Risk premium shock	?	+	−	+

A technology disturbance is seen to drive real output and real imports upward, while it pushes inflation down and triggers a real exchange-rate depreciation. A preference shock yields a rise in inflation, real output, and real imports, as well as real exchange-rate appreciation. A monetary shock induces all four variables to fall (meaning a real appreciation in the case of the exchange rate). The risk premium shock generates an increase in inflation and the real exchange rate generates a fall in real imports and an indeterminate impact on real output.

The signs reported earlier are broadly in line with findings in the literature. For example, Ambler et al. (2003) obtained comparable signs on impact for impulse responses of all six variables considered here to a wide variety of disturbances, including technology and monetary shocks.* McCallum and Nelson (2000) studied the impact of monetary and risk premium shocks, obtaining exactly the same sign for contemporaneous responses of all four baseline variables analyzed here. Specifically, McCallum and Nelson (1999) reported responses of variables including real output and inflation to monetary and risk premium shocks. In only one out of the four results involved, the contemporaneous response is not strictly the same as the one reported here, namely, the response of inflation to risk premium shocks. McCallum and Nelson (1999) reported a contemporaneous lack of response of inflation to the risk premium shock, in light of their assumption that prices are predetermined. In practice, this difference plays no role in this chapter given that the probability that responses be exactly zero is negligible. Finally, Galí and Monacelli (2005) examined the impact of a technology shock on several macroeconomic variables under four different setups. The results are consistent with the signs postulated here, except for consumer prices in one of the four scenarios studied by the authors, namely, that of a pegged exchange rate. In the latter case, the

---

* Ambler et al. (2003) also report responses to a government spending shock that are comparable to those associated with a preference disturbance here. Moreover, they find reactions to a foreign interest rate shock that are in line with the consequences of a risk premium disturbance in this chapter.

authors predict consumer prices to fall after a favorable technology shock. While many countries in my sample adopted bilateral exchange-rate pegs for some time, Galí and Monacelli (2005) studied a case corresponding to a peg in effective terms which is rare in practice. Even in Singapore—which officially targets the effective exchange rate—the target has not been fixed over time, the (undisclosed) weights are deemed to be time varying and a band is considered around the target. I thus decided to use a sign restriction with a negative response of consumer prices to the technology shock, which is also in line with other studies.

The risk premium shock deserves special discussion. The decline in real imports induced by the shock can be interpreted as resulting from a substitution effect that is not fully offset by a possible favorable income effect (or even compounded by the fall in real output under some parameter values). This is a standard prediction in the related literature. The ambiguous sign for the real output response mirrors the debate in the literature concerning the expansionary or contractionary effect of a depreciation. The empirical literature for EMEs suggests that a weakening in the exchange rate as arising from higher risk premia is contractionary.* However, this "contractionary depreciation" result relies on the role of the domestic economy's net borrower position and should not be taken for granted. Indeed, depreciations also yield an increase in real exports that may more than offset the adverse forces set in motion. This favorable effect appears to be strong enough for the calibration used by Céspedes et al. (2003, 2004), even if these authors pay attention to the balance sheet effects arising from liability dollarization. In my empirical investigation, I will leave the sign of the real output response to the risk premium shock unrestricted, thereby allowing the data to determine such sign for each economy.

15.2.2 Vector Autoregressive Model Setup

Estimation proceeds in three steps. First, I set up a VAR model as outlined in this subsection. Second, I used the sign restrictions discussed earlier in order to identify structural shocks. Third, I computed impulse response and variance decomposition analyses. Domestic macroeconomic series are used as endogenous variables. I controlled "global" exogenous variables, which are assumed to follow AR(1) processes with error terms denoted by x_t. The reduced form model can be written as

---

* See, e.g., Ahmed (2003) regarding the related empirical literature. Eichengreen (2005) and Sánchez (2007 and 2008) analyze how differently an economy displaying contractionary depreciations responds to financial and real shocks.

$$A(L)y_t = G(L)x_t + \varepsilon_t \quad \text{with } \varepsilon_t \overset{D}{\sim} \text{WN}(0,\Sigma) \tag{15.1}$$

where

y_t is an $n \times 1$ vector of domestic variables
x_t is a $k \times 1$ vector of global shocks
ε_t is a vector of white noise errors
$A(L)$ and $G(L)$ are the polynomials of orders p and q, respectively

In my setup, $n = 4$. Model (18) can be estimated using OLS equation by equation.

The VAR model in Equation 15.1 can be rewritten in the Wold form:

$$y_t = H(L)x_t + B(L)\varepsilon_t$$

where

$H(L) = A(L)^{-1} G(L)$
$B(L) = A(L)^{-1}$

The structural form of the system expresses endogenous variables in terms of exogenous variables and economically interpretable disturbances. The latter can be represented by a vector ω_t of structural shocks that satisfies

$$\omega_t \overset{D}{\sim} \text{WN}(0,I_n) \quad \text{and} \quad \varepsilon_t = C\omega_t \tag{15.2}$$

This implies that $CC' = \Sigma$. The Wold representation for the structural form becomes

$$y_t = H(L)x_t + B(L)C\omega_t \tag{15.3}$$

This chapter employs impulse responses for identification purposes. The orthogonalized impulse response of the ith variable to one unit deviation of the jth shock after s periods is

$$\frac{\partial y_{t+s|t}}{\partial \omega'_{jt}} = B_s c_j \tag{15.4}$$

where

$B_s = \partial y_{t+s|t}/\partial \varepsilon'_t$ can be obtained from $B(L)$
c_j is the jth column of C

Variance decompositions split the mean square error (MSE) of endogenous variables' forecasts due to domestic shocks from that determined by external variables. From Equation 15.3, jth domestic shock ω_j contributes to the MSE of the s-period-ahead forecast of y_{it} by

$$D_{ij} = B_s^i c_j c_j' B_s^{i'}$$

where B_s^i is the ith row of B_s. The expression for exogenous variables (each indexed by l) is

$$F_i = \sum_{l=1}^{k} F_{il} = H_s^i H_s^{i'}$$

where

H_s^i is the ith row of H_s

$H_s^{i'} = \partial y_{t+s|t}/\partial x_t'$ can be obtained from $H(L)$

15.3 DATA DESCRIPTION

The database consists of monthly series for 15 EME countries over 1990: 1–2005:5.* Exogenous variables used capture global effects outside the EMEs, including world economic activity, consumer prices, and interest rates, as well as oil and nonoil commodity prices. For global activity and interest rates, I constructed G7 measures of industrial production, consumer price indices (CPI), and short-term interest rates. As with Canova and De Nicolò (2002), I linearly detrended and seasonally adjusted all series, as well as checked visually whether the transformed data showed signs of nonstationarity. Overall, I did not detect any evidence of stochastic nonstationarity.[†]

Real output is measured by using industrial production data from IMF's *International Financial Statistics* (IFS) except in China, Hong Kong, and Taiwan (national statistics). CPI is from IFS, except in China, Hong Kong, and Taiwan (national statistics). Import data are from IFS, except in Poland (national statistics). World economic output is given by G7 industrial production (country data from IFS, weighted using national accounts data in US dollars from OECD). The same weights allowed me to construct

---

[*] A slightly shorter sample size is available for China (start in 1991:12) and the Czech Republic (start in 1991:1). My Chinese VAR model is in year-on-year growth terms as industrial production and CPI data are provided on this basis.

[†] The usefulness of formal tests for stationarity is constrained by the relatively short number of years in the present samples.

(a) a G7 CPI index from individual countries' indices (data from IFS) and (b) a measure of G7 interest rates from short-term interest rates (from IFS). Brent oil prices in U.S. dollars are from IFS. Nonoil commodity prices in U.S. dollars (OECD country weights) are from the Hamburg Institute of International Economics.

15.4 EMPIRICAL RESULTS

This section presents impulse responses (Section 15.4.1) and variance decompositions (Section 15.4.2). I used Akaike information criteria to jointly select the lags of endogenous variables and exogenous disturbances (p and q, respectively) and the set of dummies entering the VAR model. I constrained the maximum values of p and q to 24. Lag selection tests normally suggest optimal values of p no larger than 12 and q equal to 0. Thus, I used lags of the endogenous variables not going beyond 1 year back in time, while only the contemporaneous level of the exogenous shocks enters the model significantly. For each Asian EME, I tried consecutive monthly impulse dummies from 1997:7 through 1998:12. I limited dummies to a maximum of 2, choosing the ones, if any, that are most significant. In practice, allowing extra dummies does not yield substantial gains in goodness of fit. In the cases of Argentina and Brazil, estimation starts in 1990:4 to avoid the first quarter of that year, in which both countries experienced extreme nominal volatility, with inflation rates above all other realizations among the samples used here.

For each country, the final choices made for the reduced and structural forms of the model are reported in Tables 15.1 and 15.2, respectively. Table 15.2 presents technical details underlying my identification search, including the quarters over which sign restrictions hold and ways in which decompositions C of Σ in Equation 15.2 are achieved (number of Monte Carlo draws, angle grid, and number of identifying rotations). Based on the relevant decomposition matrices, I calculated the statistics of interest (means and medians) reported for impulse responses and variance decompositions.

15.4.1 Impulse Responses

In Tables 15.3 through 15.6, I report the results obtained for impulse responses to unit shocks. Responses are shown for the first quarter as well as for the end of the first and second years. Identification of all four shocks is achieved in all countries. Among other results, signs of impulse

TABLE 15.1   Reduced Form Specifications for Model with Exchange Rates

Countries	Lags of Endogenous Variables	Asian Crisis Dummies	
Asia			
China	4	1997:7	1997:11
Hong Kong	8	1997:11	1998:10
Korea	8	1997:8	1998:9
Malaysia	12	1997:10	1998:6
Singapore	10	1997:11	1998:4
Taiwan	8	1997:11	1998:10
Thailand	10	1997:9	1997:11
Latin America			
Argentina	6		
Brazil	7		
Chile	8		
Mexico	8		
NMS and Turkey			
Czech Republic	4		
Hungary	8		
Poland	11		
Turkey	7		

responses are normally found not to deviate over time from the restrictions set around the end of the first quarter, while the reaction of endogenous variables to unit shocks appear to be rather small. Moreover, impulse responses tend to die out by the end of the second year.

The quantitatively bigger responses concern consumer prices and—depending on the shock—real imports. Consumer prices react more to technology and risk premium shocks, which induce persistent and increasing inflation, respectively. Such responses are especially strong in two Latin American countries (namely, Argentina and Brazil) and less so in Poland. Turkey also ranks high among the countries whose consumer prices are influenced by unexpected developments, with the exception of preference shocks. Real imports continue to react most strongly to technology and monetary shocks, with Chinese real imports being particularly affected. Otherwise, the effect on real imports appears to be rather spread out, with real imports from Argentina and Turkey reacting relatively more strongly to risk premium shocks.

TABLE 15.2   Structural Form Specifications for Model with Exchange Rates

Countries	Angle Grid	Monte Carlo Draws	Sign Restrictions on Quarters	Number of Identifying Rotations
Asia				
China	7	1100	2 through 4	1267
Hong Kong	3	1000	2 through 3	1011
Korea	3	1200	1 through 5	1114
Malaysia	3	2000	2 through 4	1796
Singapore	8	1000	2 through 3	1009
Taiwan	3	1000	3 through 4	2452
Thailand	4	1000	1 through 4	1606
Latin America				
Argentina	4	1000	1 through 3	1189
Brazil	3	2000	2 through 3	1248
Chile	3	2000	1 through 3	1987
Mexico	3	1000	1 through 3	1040
NMS and Turkey				
Czech Republic	4	1000	3	2581
Hungary	5	1000	2 through 4	1182
Poland	4	1000	3 through 4	1088
Turkey	3	1000	1 through 3	2004

Among the reactions of a relatively smaller magnitude, the reaction of real output to the risk premium shock continues to be positive in a number of emerging Asian countries (being negative only at the end of the first quarter in Singapore). The response of real output to the risk premium shock for the Czech Republic is also positive on impact, while the corresponding ones for Brazil and Turkey are negative instead. With regard to the real exchange rate, the largest impact is registered for China (under monetary and risk premium shocks), Argentina (under preference and risk premium shocks), Brazil (under preference shocks), and Turkey (under risk premium shocks).

Finally, I report the degree of exchange-rate pass-through to consumer prices (Table 15.7). The degree of pass-through (in response to domestic shocks) can be computed as $\hat{P}_t/\hat{S}_t = \hat{P}_t/(\hat{Q}_t + \hat{P}_t)$, where hats denote deviations from the no-shock path. One caveat to these pass-through estimates is that while they can be interpreted economically in terms of each structural

TABLE 15.3 Impulse Responses of Real Output to Unit Shocks

	EMEs[a]	Asia								Latin America				EU NMS			
		China	Hong Kong	Korea	Malaysia	Singapore	Taiwan	Thailand	Argentina	Brazil	Chile	Mexico	Czech Rep.	Hungary	Poland	Turkey
(A) Responses to a technology shock																
1 quarter	0.01	0.01	0.01	0.00	0.01	0.00 [0.01]	0.01	0.02	0.02	0.01	0.01	0.01	0.02	0.01	0.01	0.00
4 quarters	0.01	0.02	0.01	0.00	0.01	0.00	0.00	0.01	0.01	0.01	0.00	0.00 [0.01]	0.01	0.01	0.01	0.00
8 quarters	0.00	0.02 [0.01]	0.00	0.00	0.00	0.00	0.00	0.00	0.00	0.00	0.00	0.00	0.01	0.01	0.01	0.00
(B) Responses to a monetary shock																
1 quarter	−0.01	−0.01	−0.01	−0.01 [0.00]	−0.01 [−0.02]	0.00	−0.01	−0.01	0.00	−0.01	−0.01	−0.01 [0.00]	0.00	0.00 [0.01]	0.00	0.00
4 quarters	0.00	−0.01 [−0.02]	−0.01	0.00	−0.01	0.00	0.00 [0.01]	−0.01	−0.01 [0.00]	0.00	0.00	0.00	0.00	0.00 [0.01]	0.00	0.00
8 quarters	0.00	−0.02 [−0.01]	−0.01	−0.01	−0.01	0.00	0.00	−0.01	0.00	0.00	0.00	0.00	0.00	0.00 [0.01]	0.00	0.00

(continued)

TABLE 15.3 (continued)   Impulse Responses of Real Output to Unit Shocks

	EMEs[a]	Asia							Latin America				EU NMS			
		China	Hong Kong	Korea	Malaysia	Singapore	Taiwan	Thailand	Argentina	Brazil	Chile	Mexico	Czech Rep.	Hungary	Poland	Turkey
(C) Responses to a preference shock																
1 quarter	0.01	0.01	0.01	0.01 [0.02]	0.01	0.00	0.01	0.00	0.01	0.00	0.01	0.00	0.00	0.00	0.00	0.02
4 quarters	0.00	0.02	0.00	0.01	0.01	0.00	0.01	0.00	-0.01	0.00	0.00	0.00	0.00	0.00	0.00	0.00
8 quarters	0.00	0.02 [0.01]	0.00	0.01	0.00	0.00	0.00	0.00	-0.01	0.00	0.00	0.00	0.00	0.00	0.00	0.00
(D) Responses to a risk premium shock																
1 quarter	0.00	0.01	0.01	0.00	0.03 [0.02]	-0.01	0.00	0.01	0.00	-0.01	0.00	0.00	0.01	0.00 [-0.01]	0.00	-0.01
4 quarters	0.00	0.01	0.00	0.00	0.01	0.00	0.00	0.00	0.00 [-0.01]	0.00	0.00	-0.01	0.01	0.00 [-0.01]	0.00	0.00
8 quarters	0.00 [0.00]	0.01	0.00	0.00	0.01	0.00 [0.01]	0.00 [0.01]	0.00	0.00 [-0.01]	0.00	0.00	-0.01	0.00	0.00 [-0.01]	0.00	0.00

Notes: This table reports estimated accumulated responses at the end of the corresponding period (in percent). Medians are reported in between square brackets when different from the respective means.

[a] Values for this grouping are arithmetic averages over all individual countries included.

TABLE 15.4 Impulse Responses of Consumer Prices to Unit Shocks

	Asia								Latin America				EU NMS			
	EMEs[a]	China	Hong Kong	Korea	Malaysia	Singapore	Taiwan	Thailand	Argentina	Brazil	Chile	Mexico	Czech Rep.	Hungary	Poland	Turkey
(A) Responses to a technology shock																
1 quarter	-0.02	0.00	0.00	0.00	0.00	0.00	0.00	0.00	-0.09	-0.12	-0.01	-0.02	-0.01	-0.01	-0.05	-0.03
4 quarters	-0.03	0.01	0.00	0.00	0.00	0.00	0.00	0.00	-0.08	-0.22	-0.02	-0.03	-0.01	-0.01	-0.03	-0.03
8 quarters	-0.03	0.02	0.00	0.00	0.00	0.00	0.00	0.00	-0.07	-0.31 [-0.35]	-0.02	-0.03 [-0.04]	-0.01	-0.01	-0.02	-0.04
(B) Responses to a monetary shock																
1 quarter	-0.01	-0.01 [-0.02]	-0.01	-0.01	0.00	0.00	0.00	0.00	-0.09 [-0.11]	-0.04 [-0.03]	0.00	0.00	0.00	0.00	-0.02 [-0.03]	-0.01
4 quarters	-0.02	-0.02 [-0.03]	-0.01	-0.01	0.00	0.00	0.00	0.00	-0.14 [-0.19]	-0.06	0.00	-0.01	0.00	0.00	-0.02 [-0.04]	-0.02
8 quarters	-0.02	-0.02 [-0.03]	-0.01	-0.01	0.00	0.00	0.00	0.00	-0.17 [-0.23]	-0.06	0.00	-0.01	0.00	0.00	-0.02 [-0.04]	-0.04

(continued)

TABLE 15.4 (continued)   Impulse Responses of Consumer Prices to Unit Shocks

	EMEs[a]	Asia							Latin America				EU NMS			
		China	Hong Kong	Korea	Malaysia	Singapore	Taiwan	Thailand	Argentina	Brazil	Chile	Mexico	Czech Rep.	Hungary	Poland	Turkey
(C) Responses to a preference shock																
1 quarter	0.01	0.01	0.00	0.00	0.00	0.00	0.00	0.00	0.02 [0.01]	0.07	0.01	0.02	0.00	0.00	0.03 [0.04]	0.00
4 quarters	0.02	0.01	0.00	0.00	0.00	0.00	0.00	0.00	0.03 [0.00]	0.11 [0.13]	0.01	0.03	0.00	0.00	0.03	0.01
8 quarters	0.01	0.02	0.00	0.00	0.00	0.00	0.00	0.00	0.04 [0.00]	0.08 [0.12]	0.01	0.03	0.00	0.00	0.02 [0.01]	0.00
(D) Responses to a risk premium shock																
1 quarter	0.01	0.01	0.00	0.00	0.00	0.00	0.00	0.00	0.08 [0.07]	0.04 [0.03]	0.00	0.02	0.00	0.00	0.01 [0.00]	0.01
4 quarters	0.03	0.01	0.00	0.00	0.00	0.00	0.00	0.00	0.21 [0.19]	0.08 [0.07]	0.01	0.05	0.00	0.00	0.01	0.02
8 quarters	0.04 [0.00]	0.01	0.00	0.00	0.00	0.00	0.00	0.00	0.32 [0.30]	0.15 [0.11]	0.01	0.07 [0.06]	−0.01	0.00	0.01	0.04

Notes: This table reports estimated accumulated responses at the end of the corresponding period (in percent). Medians are reported in between square brackets when different from the respective means.

[a]   Values for this grouping are arithmetic averages over all individual countries included.

TABLE 15.5 Impulse Responses of Real Exchange Rates to Unit Shocks

		Asia							Latin America				EU NMS			
	EMEs[a]	China	Hong Kong	Korea	Malaysia	Singapore	Taiwan	Thailand	Argentina	Brazil	Chile	Mexico	Czech Rep.	Hungary	Poland	Turkey
(A) Responses to a technology shock																
1 quarter	0.01	0.01	0.01	0.00	0.00	0.00	0.00	0.01	0.00 [0.03]	0.00	0.02	0.01 [0.00]	0.01	0.00	0.00	0.01
4 quarters	0.00	0.00	0.01	0.00	0.00	0.00	0.00	0.00	0.01 [0.04]	0.01	0.01	0.01	0.01	0.00	−0.01	0.01
8 quarters	0.00	0.00	0.00	0.00	0.00	0.00	0.00	0.00	0.00 [0.01]	0.00	0.02	0.01	0.00	0.00	0.01	0.00
(B) Responses to a monetary shock																
1 quarter	−0.01	−0.06 [−0.07]	−0.01	0.00	−0.01	0.00	−0.01	−0.01	−0.02	−0.01	−0.01	−0.01 [−0.02]	0.00	0.00 [−0.01]	0.00	−0.02
4 quarters	0.00	−0.02	0.00	0.00	0.00 [−0.01]	0.00	0.00	0.00	0.00	0.00 [−0.01]	0.00	−0.01	−0.01	0.00 [−0.01]	−0.01	−0.01
8 quarters	0.00	−0.02 [−0.01]	0.00	0.00	0.00	0.00	0.00 [−0.01]	0.00	0.00	0.00	0.00	−0.01	0.00	0.00 [−0.01]	0.00	−0.01

(continued)

TABLE 15.5 (continued) Impulse Responses of Real Exchange Rates to Unit Shocks

	Asia								Latin America				EU NMS			
	EMEs[a]	China	Hong Kong	Korea	Malaysia	Singapore	Taiwan	Thailand	Argentina	Brazil	Chile	Mexico	Czech Rep.	Hungary	Poland	Turkey
(C) Responses to a preference shock																
1 quarter	-0.02	-0.01 [-0.02]	-0.01	0.00	-0.01	-0.01	-0.01	-0.02	-0.04 [-0.05]	-0.03 [-0.04]	-0.01	-0.02 [-0.03]	-0.02	-0.01	-0.01	-0.03
4 quarters	-0.01	-0.01	-0.01	0.00	-0.01	0.00	-0.01	0.00	-0.01	-0.02	-0.01	-0.02	0.00	-0.01	0.00	-0.01
8 quarters	0.00	-0.01	-0.01	0.00	0.00	0.00	0.00 [-0.01]	0.00	0.00	0.00	-0.01 [0.00]	-0.01	0.00	-0.01	0.01	-0.01
(D) Responses to a risk premium shock																
1 quarter	0.02	0.06 [0.05]	0.00	0.03	0.01	0.00	0.00	0.01	0.03 [0.02]	0.02	0.00	0.01	0.02	0.00	0.01	0.03
4 quarters	0.01	0.01	0.00	0.02	0.01	0.00	0.00	0.01	0.02	0.01	0.01	0.02	0.02	0.00	0.00	0.02
8 quarters	0.01	0.00	0.00	0.01	0.01	0.00	0.00	0.01 [0.00]	0.01	0.00	0.01	0.01	0.01	0.00	0.00	0.01

Notes: This table reports estimated accumulated responses at the end of the corresponding period (in percent). Medians are reported in between square brackets when different from the respective means.

[a] Values for this grouping are arithmetic averages over all individual countries included.

TABLE 15.6 Impulse Responses of Real Imports to Unit Shocks

	Asia								Latin America				EU NMS			
	EMEs[a]	China	Hong Kong	Korea	Malaysia	Singapore	Taiwan	Thailand	Argentina	Brazil	Chile	Mexico	Czech Rep.	Hungary	Poland	Turkey
(A) Responses to a technology shock																
1 quarter	0.02	0.02	0.01	0.02	0.01	0.00 [0.01]	0.01 [0.00]	0.02	0.03	0.01 [0.02]	0.03	0.01	0.02	0.02 [0.03]	0.01 [0.02]	0.01
4 quarters	0.01	0.03 [0.04]	0.01	0.02	0.01	0.00 [0.01]	0.01	0.03	0.04	0.01 [0.02]	0.03	0.01	0.00	0.00	0.02	0.00
8 quarters	0.01	0.02 [0.01]	0.01	0.01	0.01	0.00	0.01 [0.00]	0.03	0.01	0.01	0.01	−0.01	0.00	0.00	0.00 [0.01]	0.00
(B) Responses to a monetary shock																
1 quarter	−0.02	−0.03 [−0.04]	−0.01 [0.00]	−0.01	−0.02	−0.02 [−0.01]	−0.02 [−0.03]	−0.03	−0.01	−0.01	−0.02	−0.02	−0.01	0.00	0.00	−0.03
4 quarters	−0.02	−0.03 [−0.04]	−0.01 [0.00]	−0.01	−0.02	−0.02	−0.02 [−0.03]	−0.04	−0.02 [0.00]	−0.01	−0.02	−0.01 [−0.02]	−0.01	0.00 [−0.01]	0.00 [−0.01]	−0.01
8 quarters	−0.01	−0.02	−0.02	−0.01 [−0.03]	−0.02 [−0.03]	−0.01 [−0.02]	−0.01 [−0.02]	−0.03	−0.01 [0.00]	−0.01	−0.01	0.00	0.00	0.00 [−0.01]	0.00 [−0.01]	0.00

(continued)

TABLE 15.6 (continued) Impulse Responses of Real Imports to Unit Shocks

	EMEs[a]	Asia							Latin America				EU NMS			
		China	Hong Kong	Korea	Malaysia	Singapore	Taiwan	Thailand	Argentina	Brazil	Chile	Mexico	Czech Rep.	Hungary	Poland	Turkey
(C) Responses to a preference shock																
1 quarter	0.01	0.01	0.00	0.02	0.01	0.02	0.02	0.01	0.01	0.02	0.02	0.01 [0.00]	0.02	0.01	0.01	0.02
4 quarters	0.01	0.01 [0.02]	0.00	0.02	0.01	0.01	0.01	0.01	−0.02	0.02 [0.01]	0.00	0.01	0.00	0.01	0.00	0.00
8 quarters	0.00	0.01	−0.01	0.01	0.00	0.00	0.01	−0.01	−0.02 [−0.01]	0.01	−0.01	0.01 [0.00]	0.00	0.01	0.01	−0.01
(D) Responses to a risk premium shock																
1 quarter	−0.02	−0.06 [−0.05]	−0.01	−0.01	−0.01	−0.01	−0.02	−0.01	−0.05	−0.02 [−0.01]	−0.02	−0.02 [−0.03]	−0.03	−0.01	0.00 [−0.01]	−0.04
4 quarters	−0.02	−0.01	0.00	−0.01	−0.01	0.00	−0.01	−0.01 [0.00]	−0.06 [−0.07]	−0.02 [−0.01]	−0.03	−0.03 [−0.04]	−0.03	0.00	0.00 [−0.01]	−0.02
8 quarters	−0.01	−0.01	0.00	0.00	−0.01	0.01	0.00	0.00	−0.04 [−0.05]	−0.01 [0.00]	−0.02	−0.02	−0.02	0.00	0.00 [−0.01]	−0.01

Notes: This table reports estimated accumulated responses at the end of the corresponding period (in percent). Medians are reported in between square brackets when different from the respective means.

[a] Values for this grouping are arithmetic averages over all individual countries included.

shock, both, the magnitudes and signs, of the implied impulse responses can adopt various features. With this in mind, I constrained my analysis to a comparison with the recent SVAR study by Ca' Zorzi et al. (2007). The latter reports pass-through estimates for an identification scheme in which the exchange rate is the most exogenous variable, being allowed to react to the rest of the system only with a lag. Under such identification strategy, the unexpected exchange-rate component does not depend on the state of the macroeconomy in the very short run, which could be connected with the exchange rate being driven by exogenous factors relating to "noise trading" or imperfect information considerations. In this regard, it makes sense to compare Ca' Zorzi et al. (2007)'s estimates with those reported in Table 15.7 under the column of risk premium shock. One point in common between the two sets of results is that, among the group of EMEs, Latin American countries (except for Chile) and, to some extent, Turkey exhibit high pass-through in the range of 50%–100% at business-cycle frequencies.

15.4.2 Variance Decomposition Results

Variance decompositions are reported in Tables 15.8 through 15.11. Business-cycle fluctuations as well as movements in real exchange rates and real imports are mostly dominated by domestic shocks. Foreign shocks account on average for no more than 10% of the variation in the endogenous variables (real output, consumer prices, real exchange rates, and real imports).

Each domestic disturbance explains a considerable fraction of the variability in the endogenous variables under study. Focusing on above-average contributions, the following is worth emphasizing. For real output, the technology disturbance exceeds the contributions of the other shocks by some margin, in light of results for Latin America (Brazil, Chile, and Mexico) and, less so, emerging Asia (Thailand and especially Korea). Instead, new member states' (NMS) real output display an above-average contribution from monetary shocks (driven by Poland), whereas this place is occupied by preference shocks in the case of Turkey. With regard to consumer prices, the most noticeable relatively larger contribution is that of technology shocks in NMS (due to Czech Republic and Hungary) and Turkey. In the latter country, an above-average fraction of consumer price variability can also be attributed to preference disturbances. In the case of real imports, I do not detect shocks that have a widespread role across regional groupings.

TABLE 15.7 Degree of Exchange-Rate Pass-Through to Consumer Prices

	EMEs[a]	Asia							Latin America				EU NMS			
		China	Hong Kong	Korea	Malaysia	Singapore	Taiwan	Thailand	Argentina	Brazil	Chile	Mexico	Czech Rep.	Hungary	Poland	Turkey
(A) Under technology shocks																
1 quarter	0.54	−0.34	−0.75	0.31	0.42	0.17	0.69	0.27	1.00	0.98	0.43	1.40	0.50	1.21	0.97	0.77
		[−1.53]		[−0.68]	[0.37]	[0.32]	[0.59]	[0.25]		[0.97]		[0.82]		[1.00]	[1.08]	
4 quarters	0.15	0.95	−8.44	−0.85	0.75	0.06	0.56	0.70	0.93	1.03	0.53	1.31	0.53	2.52	0.85	0.83
		[0.86]	[−8.39]	[−2.57]	[0.67]	[0.03]	[0.57]	[0.69]				[1.17]		[1.70]	[0.83]	
8 quarters	0.88	0.84	1.26	2.04	−0.56	0.33	0.36	−0.05	0.94	1.00	0.58	1.31	0.69	1.82	1.73	0.95
				[0.15]	[−0.59]	[0.25]	[0.53]	[−0.08]				[1.27]		[1.40]	[1.47]	
(B) Under monetary shocks																
1 quarter	0.38	0.17	0.49	0.62	0.04	0.21	−0.06	0.14	0.83	0.80	0.26	0.25	0.63	−0.07	1.10	0.36
				[0.64]	[0.02]	[0.17]	[0.09]	[0.13]	[0.84]	[0.82]		[0.19]		[0.03]	[0.78]	
4 quarters	0.43	0.49	0.76	0.83	0.19	−3.11	0.59	4.43	1.00	1.09	0.57	0.53	−2.29	−0.18	0.80	0.71
		[0.57]		[0.81]	[0.14]	[−1.23]	[0.93]	[4.12]	[0.99]	[1.10]		[0.41]		[−0.20]	[0.86]	
8 quarters	0.69	0.61	0.94	0.70	1.68	−0.79	−0.76	2.54	1.01	1.08	1.20	0.69	−0.34	−0.03	0.98	0.85
		[0.79]			[0.79]	[−0.98]	[−1.00]	[2.68]	[1.00]			[0.67]		[−0.30]		

(C) Under preference shocks

1 quarter	-0.36	-1.10 [-0.38]	-0.06	0.77 [0.49]	-0.11 [-0.12]	-0.29 [-0.27]	-0.45 [-0.48]	-0.03	-1.05 [-0.27]	2.05 [1.88]	-1.23	-4.28 [-2.11]	-0.24	-0.14 [-0.11]	0.84 [0.74]	-0.11
4 quarters	1.39	2.46 [3.80]	-0.12	1.25 [1.02]	-0.03	-0.45	-0.66 [-0.75]	-0.05 [-0.10]	1.64 [-0.38]	1.30 [1.23]	12.93 [12.18]	2.59 [2.21]	0.14	-0.01 [-0.03]	0.93 [0.99]	-1.10
8 quarters	0.60	2.47 [1.80]	-0.08	0.97 [1.09]	-0.56 [-0.62]	-0.17 [-0.13]	-0.76 [-0.83]	1.07 [1.08]	0.96 [0.64]	1.08 [1.04]	1.86 [1.85]	0.67 [1.57]	0.35	0.28 [0.26]	1.49 [22.79]	-0.65

(D) Under risk premium shocks

1 quarter	0.64	0.18 [0.20]	-0.91	0.07 [0.05]	0.05 [0.06]	0.56 [0.38]	0.53 [0.51]	0.03	0.72 [0.73]	0.68 [0.64]	0.41	0.64 [0.65]	0.00	2.56 [1.12]	3.76 [0.28]	0.27
4 quarters	1.15	2.44 [-90.83]	0.37	0.08 [0.07]	0.05 [0.04]	0.17	-0.60 [-0.62]	0.47 [0.48]	0.91	0.89 [0.88]	0.48	0.76 [0.74]	-0.35	10.20 [-9.57]	0.81 [0.86]	0.53
8 quarters	1.00	5.09 [-0.02]	0.37	-0.04 [-0.09]	-0.16 [-0.18]	0.65 [0.64]	1.79 [1.55]	-0.14 [0.82]	0.96	0.99 [0.98]	0.54	0.87 [0.84]	-0.91	3.13 [2.91]	0.97 [1.13]	0.82

Notes: This table reports estimated accumulated responses at the end of the corresponding period (in percent). Medians are reported in between square brackets when different from the respective means.

a Values for this grouping are arithmetic averages over all individual countries included.

TABLE 15.8   Variance Decomposition of Real Output at the 3-Year Horizon

	Technology	Preference	Monetary	Risk Premium	Foreign	Total
Emerging markets[a,b]	31.5	18.5	22.5	21.0	6.5	100.0
Asia[b]	24.1	22.4	20.8	24.4	8.3	100.0
China	25.2	24.9	21.2	21.1	7.6	100.0
	[20.8]	[18.1]	[13.5]	[14.3]	[6.9]	
Hong Kong	1.7	60.8	3.3	13.0	21.2	100.0
		[61.2]	[3.0]	[12.8]		
Korea	64.6	3.3	22.0	3.1	7.0	100.0
	[87.3]	[1.7]	[1.4]	[2.7]	[6.6]	
Malaysia	20.3	8.2	34.8	33.0	3.7	100.0
	[20.0]	[7.7]	[15.1]	[50.3]		
Singapore	12.0	16.6	27.3	33.2	10.9	100.0
	[6.8]	[6.9]	[12.6]	[25.1]	[9.4]	
Taiwan	12.0	16.8	18.2	48.1	4.9	100.0
	[3.9]	[17.1]	[10.4]	[53.8]		
Thailand	33.1	26.2	18.9	19.0	2.8	100.0
	[33.5]	[2.1]	[17.2]	[16.4]	[2.7]	
Latin America[b]	49.0	12.2	17.6	16.7	4.7	100.0
Argentina	0.4	14.7	40.8	42.6	1.5	100.0
	[0.1]		[43.3]	[43.5]		
Brazil	61.3	10.4	17.1	8.6	2.6	100.0
	[60.3]	[9.3]	[16.1]	[6.7]		
Chile	72.9	2.5	6.9	9.8	7.9	100.0
	[73.5]	[1.1]	[6.7]	[9.5]	[4.3]	
Mexico	61.3	21.0	5.6	5.6	6.7	100.0
	[61.5]	[22.3]	[3.6]	[5.0]	[6.8]	
EU NMS[b]	16.7	21.8	35.1	20.5	6.0	100.0
Czech Republic	33.0	18.6	17.7	18.6	12.1	100.0
	[39.1]	[16.0]	[12.1]	[14.0]	[11.8]	
Hungary	5.3	28.9	32.2	32.1	1.5	100.0
	[2.4]	[27.7]	[29.6]	[27.8]	[1.0]	
Poland	11.7	18.0	55.3	10.7	4.3	100.0
	[1.3]	[14.2]	[76.7]	[4.3]	[3.9]	
Turkey	58.4	6.2	15.9	16.5	2.9	99.9
	[57.9]	[5.7]	[5.7]	[7.7]	[3.0]	

Notes: The values reported in this table are averages over all plausible identifications by type of shock and are in percentage terms. Median values are reported in brackets (only for individual countries) when different from the respective means.

[a]   The values for this grouping are the unweighted average of countries in Asia, Latin America, and EU NMS, to which Turkey is added.

[b]   The values for these regions are computed as the simple average of the countries listed under each of them.

TABLE 15.9 Variance Decomposition of Consumer Prices at the 3-Year Horizon

	Technology	Preference	Monetary	Risk Premium	Foreign	Total
Emerging markets[a,b]	30.9	23.5	17.0	20.5	8.1	100.0
Asia[b]	28.2	26.9	18.8	19.2	6.9	100.0
China	22.4	25.5	22.3	22.3	7.5	100.0
	[17.5]	[19.7]	[15.7]	[15.0]	[6.9]	
Hong Kong	1.2	78.5	3.0	8.8	8.5	100.0
				[8.7]		
Korea	61.0	4.2	19.8	8.1	6.9	100.0
	[80.4]	[3.2]	[3.4]	[3.0]	[6.7]	
Malaysia	47.8	3.3	19.6	24.4	4.9	100.0
	[47.2]	[2.6]	[24.5]	[17.3]	[4.8]	
Singapore	23.7	29.5	15.0	20.1	11.7	100.0
	[16.8]	[22.0]	[9.6]	[12.3]	[10.2]	
Taiwan	24.1	22.8	25.7	24.1	3.3	100.0
	[21.6]	[21.1]	[25.2]	[20.5]	[2.9]	
Thailand	17.3	24.5	26.3	26.3	5.6	100.0
	[17.1]	[1.8]	[25.8]	[25.1]	[4.7]	
Latin America[b]	27.8	24.9	19.3	24.3	3.7	100.0
Argentina	0.9	15.6	40.2	41.9	1.4	100.0
	[0.7]	[1.2]	[47.6]	[47.3]		
Brazil	58.2	0.4	9.3	25.7	6.3	100.0
	[75.1]	[0.1]	[10.0]	[2.6]		
Chile	36.2	48.1	11.0	2.0	2.7	100.0
		[48.2]				
Mexico	15.7	35.6	16.6	27.6	4.5	100.0
	[15.6]	[37.9]	[14.3]	[28.1]		
EU NMS[b]	40.0	6.2	12.3	24.0	17.5	100.0
Czech Republic	56.4	4.4	13.8	19.4	6.0	100.0
	[71.8]	[1.7]	[5.7]	[7.9]	[5.0]	
Hungary	60.2	9.2	11.2	14.0	5.4	100.0
	[64.8]	[3.2]	[4.7]	[4.5]	[4.8]	
Poland	3.5	4.9	11.9	38.5	41.2	100.0
	[1.7]	[2.8]	[3.8]	[46.6]	[42.3]	
Turkey	35.5	46.6	9.0	3.8	5.1	100.0
	[35.3]	[46.4]	[8.3]	[2.8]		

Notes: The values reported in this table are averages over all plausible identifications by type of shock and are in percentage terms. Median values are reported in brackets (only for individual countries) when different from the respective means.

[a] The values for this grouping are the unweighted average of countries in Asia, Latin America, and EU NMS, to which Turkey is added.

[b] The values for these regions are computed as the simple average of the countries listed under each of them.

TABLE 15.10   Variance Decomposition of Real Exchange Rates at the 3-Year Horizon

	Technology	Preference	Monetary	Risk Premium	Foreign	Total
Emerging markets[a,b]	16.6	21.6	30.8	20.8	10.2	100.0
Asia[b]	21.5	18.4	33.1	17.1	9.9	100.0
China	29.8	22.9	19.8	19.0	8.5	100.0
	[26.5]	[14.5]	[11.7]	[10.4]	[8.0]	
Hong Kong	7.6	13.8	57.7	8.6	12.3	100.0
Korea	56.6	3.3	20.6	3.7	15.8	100.0
	[89.2]	[1.7]	[1.4]	[1.8]	[6.5]	
Malaysia	7.0	11.1	47.2	31.0	3.7	100.0
	[5.8]	[9.0]	[34.3]	[46.0]		
Singapore	21.7	26.5	25.2	15.1	11.5	100.0
	[15.9]	[20.4]	[20.6]	[7.8]	[10.2]	
Taiwan	8.5	38.1	35.1	12.5	5.8	100.0
	[3.1]	[33.3]	[31.6]	[5.5]	[4.1]	
Thailand	19.3	13.2	26.4	29.5	11.6	100.0
	[18.7]	[12.0]	[27.8]	[33.8]	[11.1]	
Latin America[b]	10.5	25.1	36.2	22.0	6.2	100.0
Argentina	0.4	14.8	40.8	42.5	1.5	100.0
	[0.3]	[0.1]	[43.0]	[42.9]		
Brazil	1.1	9.3	60.4	19.1	10.1	100.0
	[0.5]	[9.1]	[63.6]	[14.9]		
Chile	30.9	44.1	14.8	2.0	8.2	100.0
			[14.1]	[1.6]		
Mexico	9.6	32.1	28.7	24.4	5.2	100.0
	[9.4]	[34.3]	[27.3]	[24.5]		
EU NMS[b]	17.3	19.1	22.5	23.6	17.5	100.0
Czech Republic	26.1	20.5	21.2	23.3	8.9	100.0
	[25.7]	[17.3]	[17.1]	[16.1]	[7.1]	
Hungary	4.3	25.6	28.7	33.9	7.5	100.0
	[1.3]	[18.9]	[21.3]	[27.5]	[7.6]	
Poland	21.4	11.3	17.6	13.6	36.1	100.0
		[3.6]	[15.8]	[9.1]	[39.2]	
Turkey	4.9	38.0	17.5	33.2	6.4	100.0
	[4.4]	[38.1]	[15.1]	[31.3]		

Notes: The values reported in this table are averages over all plausible identifications by type of shock and are in percentage terms. Median values are reported in brackets (only for individual countries) when different from the respective means.

a   The values for this grouping are the unweighted average of countries in Asia, Latin America, and EU NMS, to which Turkey is added.

b   The values for these regions are computed as the simple average of the countries listed under each of them.

TABLE 15.11   Variance Decomposition of Real Imports at the 3-Year Horizon

	Technology	Preference	Monetary	Risk Premium	Foreign	Total
Emerging markets[a,b]	20.1	17.5	24.7	28.9	8.7	100.0
Asia[b]	24.0	24.9	21.4	20.4	9.3	100.0
China	27.7	22.7	21.6	20.0	8.0	100.0
	[23.2]	[15.1]	[14.4]	[12.0]	[7.6]	
Hong Kong	2.2	62.4	1.0	13.0	21.4	100.0
	[2.1]		[0.8]	[12.9]		
Korea	58.4	4.2	26.3	5.3	5.8	100.0
	[80.4]	[3.2]	[13.5]	[4.6]	[5.3]	
Malaysia	11.3	4.3	29.3	49.3	5.8	100.0
		[1.3]	[41.8]	[37.3]		
Singapore	33.7	20.1	17.6	21.5	7.1	100.0
	[20.0]	[7.9]	[5.5]	[14.4]	[5.8]	
Taiwan	6.8	39.4	35.2	14.5	4.1	100.0
	[6.2]	[16.2]	[5.7]	[3.0]	[3.7]	
Thailand	27.9	21.1	18.8	19.3	12.9	100.0
	[30.3]	[6.6]	[10.9]	[13.3]		
Latin America[b]	9.9	10.2	27.1	48.4	4.4	100.0
Argentina	0.6	14.8	40.6	42.5	1.5	100.0
	[0.5]	[0.1]	[43.4]	[43.5]	[1.5]	
Brazil	26.7	24.2	23.9	21.1	4.1	100.0
	[20.5]	[24.8]	[21.9]	[22.0]	[4.0]	
Chile	1.3	1.1	16.8	74.3	6.5	100.0
	[0.8]	[0.8]	[16.9]	[74.5]	[6.3]	
Mexico	11.0	0.6	27.2	55.5	5.7	100.0
	[10.4]		[28.9]	[54.5]		
EU NMS[b]	19.4	11.9	29.4	24.8	14.5	100.0
Czech Republic	20.4	8.7	28.6	26.6	15.7	100.0
	[8.7]	[6.6]	[19.7]	[17.1]	[15.9]	
Hungary	10.3	20.6	27.5	35.8	5.8	100.0
	[7.0]	[8.4]	[18.0]	[29.8]		
Poland	27.5	6.4	32.1	12.0	22.0	100.0
	[25.6]	[2.3]	[34.2]	[7.7]	[23.0]	
Turkey	35.8	12.6	24.6	22.7	4.3	100.0
	[35.3]	[12.7]	[19.9]	[15.6]		

Notes: The values reported in this table are averages over all plausible identifications by type of shock and are in percentage terms. Median values are reported in brackets (only for individual countries) when different from the respective means.

[a] The values for this grouping are the unweighted average of countries in Asia, Latin America, and EU NMS, to which Turkey is added.

[b] The values for these regions are computed as the simple average of the countries listed under each of them.

Above-average single contributions can be detected for risk premium shocks in Latin America, monetary shocks in NMS, and technology shocks in Turkey. Concerning the real exchange rate, monetary shocks exhibit the biggest single contribution in the cases of emerging Asia (owing to Hong Kong, Malaysia, and Taiwan) and Latin America (owing to Argentina and Brazil). Instead, in Turkey, technology disturbances play a larger-than-fair role in driving the real exchange rate.

15.5 CONCLUSIONS

This chapter focuses on the sources of business-cycle and exchange-rate fluctuations in emerging markets. Macroeconomic developments tend to vary across countries, time horizons, and shocks considered, with some patterns being, however, identified. Signs of impulse responses do not normally deviate over time from those imposed on impact, with responses tending to die out after the second year. Impulse responses are often rather muted. Among other findings, it is worth emphasizing the following. Consumer prices and, depending on the shock, real imports are the most reactive endogenous variables. Consumer prices are mostly driven by technology and risk premium shocks. Latin America (owing to Brazil and Argentina) and Poland show above-average consumer price responses in the baseline approach. Monetary shocks elicit muted reactions of consumer prices and real output. The above-average responses of real exchange rates to these shocks may suggest undesirable side effects of unanticipated monetary policy.

Turning to variance decomposition analysis, foreign shocks driving developments in advanced economies and commodity prices affect emerging markets little, on average explaining no more than 10% of the variation in endogenous variables. This is on the low side of the spectrum of estimates found in the literature, which does not by itself imply that external forces have a small influence on emerging economies. As long as a sufficiently large part of world developments is predictable, my estimates are in line with the conventional wisdom that small open economies are quite reactive to global factors. Taking this into account, my results are broadly in line with other studies pointing to a modest contribution of external determinants in emerging economies' fluctuations (Hoffmaister and Roldós, 1997; Kose et al., 2003). Among domestic disturbances, technology shocks play an above-average role in explaining consumer prices, driven by results for EU NMS and Turkey. Real imports fail to display a clear cross-regional pattern, with a different shock playing the key role in each region considered.

ACKNOWLEDGMENT

I gratefully acknowledge discussions with Balázs Vonnák and Felix Hammermann, as well as comments made by seminar participants at the European Central Bank (ECB). The analysis presented here may not reflect the views of the ECB. All errors are the author's responsibility.

REFERENCES

Ahmed, S. (2003) Sources of economic fluctuations in Latin America and implications for the choice of exchange rate regimes. *Journal of Development Economics*, 72(1): 181–202.

Ambler, S., Dib, A., and Rebei, N. (2003) Nominal rigidities and exchange rate pass-through in a structural model of a small open economy. Bank of Canada Working Paper No. 29.

Buckle, R., Kim, K., and McLellan, N. (2003) The impact of monetary policy on New Zealand business cycles and inflation variability. New Zealand Treasury Working Paper No. 9.

Canova, F. (2005) The transmission of US shocks to Latin America. *Journal of Applied Econometrics*, 20(2): 229–251.

Canova, F. and De Nicolò, G. (2002) Monetary disturbances matter for business cycles fluctuations in the G-7. *Journal of Monetary Economics*, 49(6): 1131–1159.

Canova, F. and De Nicolò, G. (2003) On the sources of business cycles in the G-7. *Journal of International Economics*, 59(1): 77–100.

Ca' Zorzi, M., Hahn, E., and Sánchez, M. (2007) Exchange rate pass-through in emerging markets. *ICFAI Journal of Monetary Economics*, 5(4): 84–102.

Céspedes, L., Chang, R., and Velasco, A. (2003) IS-LM-BP in the Pampas. *IMF Staff Papers*, 50(special issue): 143–156.

Céspedes, L., Chang, R., and Velasco, A. (2004) Balance sheets and exchange rates. *American Economic Review*, 94(4): 1183–1193.

Cushman, D. and Zha, T. (1997) Identifying monetary policy in a small open economy under flexible exchange rates. *Journal of Monetary Economics*, 39(3): 433–448.

Dungey, M. and Pagan, A. (2000) A structural VAR model of the Australian economy. *The Economic Record*, 76(235): 321–342.

Eichengreen, B. (2005) Can emerging markets float? Should they inflation target? In: R. Driver, P. Sinclair, and C. Thoenissen (Eds.), *Exchange Rates, Capital Movements and Policy*. Routledge: London.

Faust, J. (1998) On the robustness of the identified VAR conclusions about money. *Carnegie-Rochester Conference Series on Public Policy*, 49(1): 207–244.

Galí, J. and Monacelli, T. (2005) Monetary policy and exchange rate volatility in a small open economy. *Review of Economic Studies*, 72(3): 707–736.

Genberg, H. (2003) Foreign versus domestic factors as sources of macroeconomic fluctuations in Hong Kong. HEI Working Paper No. 5.

Hoffmaister, A. and Roldós, J. (1997) Are business cycles different in Asia and Latin America? IMF Working Paper No. 9.

Kose, M., Otrok, C., and Whiteman, C. (2003) International business cycles: World, region, and country-specific factors. *American Economic Review*, 93(4): 1216–1239.

McCallum, B. and Nelson, E. (1999) Nominal income targeting in an open economy optimizing model. *Journal of Monetary Economics*, 43(3): 553–578.

McCallum, B. and Nelson, E. (2000) Monetary policy for an open economy: An alternative framework with optimizing agents and sticky prices. *Oxford Review of Economic Policy*, 16(Winter): 74–91.

Peersman, G. (2005) What caused the early millennium slowdown? Evidence based on vector autoregressions. *Journal of Applied Econometrics*, 20(2): 185–207.

Peersman, G. and Straub, R. (2009) Technology shocks and robust sign restrictions in a Euro area SVAR. *International Economic Review* (forthcoming).

Rüffer, R., Sánchez, M., and Shen, J.-G. (2008) Emerging Asia's growth and integration: How autonomous are business cycles? *ICFAI Journal of Monetary Economics*, 6(2): 50–78.

Sánchez, M. (2007) How does information affect the comovement between interest rates and exchange rates? *Rivista Internazionale di Scienze Sociali*, 4/2007: 547–562.

Sánchez, M. (2008) The link between interest rates and exchange rates: Do contractionary depreciations make a difference? *International Economic Journal*, 22(1): 43–61.

Uhlig, H. (2005) What are the effects of monetary policy on output? Results from an agnostic identification procedure. *Journal of Monetary Economics*, 52(2): 381–419.

CHAPTER 16

Institutional Factors behind Capital Structure: Evidence from Chilean Firms

Viviana Fernandez

CONTENTS

16.1 INTRODUCTION

Several regularities in capital structure have been observed throughout the world (see Megginson, 1997, Chapter 7). First, capital structures vary across countries. For instance, American, German, and Canadian firms have lower book debt ratios than their counterparts in other industrialized nations such as Japan, France, and Italy (e.g., Rajan and Zingales, 1996). In addition, there are differences in the correlation between long-term leverage ratios and firms' profitability, size, growth, and riskiness across countries due to differences in tax policies and agency costs (e.g., Wald, 1999).

Second, capital structures display industry patterns that are similar around the world. Utilities, transportation companies, and capital-intensive manufacturing firms have high debt-to-equity ratios as opposed to service firms, mining companies, and technology-based manufacturing firms which employ very little long-term debt, if at all they employ some. Third, within industries, leverage is inversely associated with profitability. This evidence contradicts the tax-based capital structures theories which predict that more profitable firms should borrow more intensively to reduce their tax load. One interpretation of this pattern is that capital structure may not necessarily arise from a deliberate policy choice but may be rather an artifact of historic profitability and dividend policy.

A fourth stylized fact of capital structure is that taxes are important but not crucial to determine debt usage. Evidence from the United States shows that capital structures for American firms have remained fairly constant over the period 1929–1980 despite major changes in tax rates and regulatory structures that took place over that time period. Fifth, leverage ratios seem to be negatively correlated with perceived costs of bankruptcy and financial distress. For instance, firms rich in collateralizeable assets (e.g., commercial real state and transportation) are able to tolerate higher debt ratios than firms whose principal assets are human capital, brand image, or intangible assets. Sixth, several empirical studies have shown that when a firm announces a leverage-increasing event (e.g., debt-for-equity exchange offers, debt-financed share repurchases), its stock price rises. Conversely, leverage-decreasing events (e.g., new stock offerings) are most of the time associated with a decline in stock prices.

Moreover, the change in the cost of issuing new debt and equity securities has had little effect on capital structure despite its declining trend over time worldwide. On the other hand, capital structure appears to be influenced by ownership structure. For instance, managers who place a high value on the personal benefits associated with controlling a firm will favor debt over equity in order to minimize dilution of ownership stake. Finally, when a firm deviates from its preferred capital structure, it tends to return to it over time. In general, firms operate with target leverage zones, and they issue new equity when debt ratios get too high and issue debt if they get too low.

There are three major theoretical models to explain the choice of capital structure: the trade-off/agency cost model, the pecking-order theory, and the free-cash flow theory (see Myers, 2001). The trade-off/agency cost model has evolved from modifications of the Modigliani and Miller capital structure irrelevance hypothesis. It states that capital structure is the result of an individual firm's trading-off the benefits of increased leverage (e.g., a tax shield) against the potential financial distress caused by heavy indebtedness. Financial distress includes the costs of bankruptcy or reorganization and the agency costs that arise when the firm's solvency is called into question. Accordingly, the trade-off theory predicts moderate debt ratios.

However, as Jensen and Meckling's (1976) pioneering work showed, firms will seek target debt ratios even in the absence of taxes or bankruptcy costs. The reason is that a firm's expected cash flows are not independent of the ownership structure. In particular, if a fraction α is sold to outside investors, corporate managers are responsible for only a fraction $1 - \alpha$ of their actions (i.e., the agency cost of the outside equity). Therefore, they have an incentive to consume perquisites. External debt can overcome this agency cost because the cost of excessive perk consumption will make corporate managers lose control of the firm in the event of default.

Agency costs may be also associated with the issuance of new debt. Given that equity is a residual claim, managers might be tempted to shift to riskier operating strategies to transfer wealth from debt to stock holders. Given that debt investors are aware of this conflict of interest, debt covenants will restrict excessive borrowing. And, therefore, firms will operate at a conservative debt ratio.

The empirical support for the trade-off theory is mixed. Bradley et al. (1984) have developed a model where optimal leverage is inversely related to expected costs of financial distress and nondebt tax shields. For a sample of 20 year average leverage ratios of over 800 firms, they have found that the volatility of firm earnings and the intensity of R&D and advertising expenditures are inversely related to leverage. This is consistent with the trade-off theory. But, they have surprisingly found a strong and positive relation between firm leverage and the amount of nondebt tax shields.

Further evidences on the trade-off theory are in MacKie-Mason (1990), who has found that companies with low marginal rates are more likely to issue equity, and Graham (1996), who has concluded that changes in the long-term debt are positively related to the firm's effective marginal tax rate. More recently, Graham and Harvey (2001) surveyed over 300 chief financial officers and found that 44% of them reported that their firms had target capital structures, as the trade-off theory would predict. Tax deductibility of interest payments, cash flows volatility, and flexibility were mentioned as relevant factors to set target debt ratios.

However, Graham (2000) has found that firms' leverage is persistently conservative. This holds, in particular, for large, profitable, and liquid firms in stable industries that face low *ex ante* costs of distress. Nevertheless, those firms also have growth options and relatively few tangible assets. Debt conservatism is also positively related to excess cash holdings. Graham (2003) has pointed out that more research is called for to understand the underlevered paradox. In particular, nondebt tax shields, such as employee stock options deductions and accumulated foreign tax credits, might be an explanation to such underleverage.

Myers and Majluf (1984)'s pecking-order theory—which is further discussed in Myers (1984)*—offers an alternative framework for understanding the driving forces of corporate leverage. The pecking-order theory is based on the assumptions that managers are better informed about the

---

* The pecking-order theory falls into the category of signaling hypotheses, which assume that market prices do not reflect all information, in particular that which is not publicly available. Changes in capital structure are then a signaling device to convey information to the market. The first signaling model based on asymmetric information problems between well-informed managers and poorly informed investors was developed by Ross (1977). In order to differentiate itself from competitors, a highly valuable company will use a costly and credible signal; a high levered capital structure. Less valuable firms are unwilling to use so much debt because they are more likely to go bankrupt. Ross shows that there is a separating equilibrium where high-value firms are highly levered, and low-value firms rely more heavily on equity financing.

firm's investment opportunities than outsiders, and corporate managers act in the best interest of existing shareholders. Myers and Majluf have showed that, under these assumptions, firms will sometimes forego positive net present value projects if accepting them requires issuing new equity at a price that does not reflect the true value of the firm's investment opportunities. This helps explain why firms value financial slack (e.g., cash and marketable securities) and unused debt capacity.

The pecking-order hypothesis has received attention because it is able to explain some regularities observed empirically, which we referred to earlier: (1) debt ratios and profitability are inversely related; (2) markets react negatively to new equity issues, and managers resort to such issues only when they do not have any other choice or they think that equity is overvalued; and (3) managers sometimes choose to hold more cash and issue less debt than the trade-off theory would predict. While the trade-off theory is good at explaining observed corporate debt levels (i.e., static viewpoint), the pecking-order hypothesis is more suitable for explaining observed changes in capital structure (i.e., dynamic viewpoint).

Shyam-Sunder and Myers (1999) have compared the pecking-order theory with the trade-off theory. The former predicts that the change in debt each year depends on the funds flow deficit of that year; if the deficit is positive, the firm issues debt whereas if the deficit is negative, the firm retires debt. The latter, by contrast, predicts that changes in debt will revert toward the firm's target debt ratio. The authors have found that the speed of adjustment toward the target debt ratio is too slow to support the trade-off theory whereas the evidence strongly favors the pecking-order theory. Shyam-Sunder and Myers' conclusions were later challenged by Chirinko and Singha (2000). In turn, Fama and French (2002) have found support for both the theories while analyzing dividend and debt policies.

More recently, Frank and Goyal (2003) tested the pecking-order theory for a sample of publicly traded U.S. firms for 1971–1998 and found little support for it. First, net equity issues track the financing deficit more closely than net issues. In addition, when estimating leverage regressions—in the trade-off theory's spirit—they found that the financing deficit has some explanatory power but it does not annihilate the effect of conventional variables such as tangibility, size, and profitability.

Very recent contributions in the area of capital structure are by Leary and Roberts (2005), who have analyzed whether firms engage in dynamic rebalancing of their capital structures while allowing for costly adjustment, Molina (2005), who has studied the effect of firms' leverage on default probabilities and the consequent impact of leverage on the *ex ante* costs of financial distress, and Hennessy and Whited (2005), who have developed a dynamic trade-off model with endogenous choice of leverage.

The contribution of this chapter is twofold. First, the literature on capital structure has focused primarily on developed economies. Some exceptions are the international comparisons that include emerging economies. For instance, Booth et al. (2001) have analyzed the determinants of capital structures of 10 developing countries, including two Latin American countries, Brazil and Mexico. Their database, however, lacks information on sources and uses of funds statements. This is essential to contrast the trade-off and the pecking-order theories.

Fan et al. (2003), in turn, have carried out a more ambitious study where they have analyzed a sample of 35 countries that also included emerging countries (e.g., Chile, Indonesia, and Peru). Fan et al.'s database included only 16 Chilean firms for a 10 year period. By contrast, our database has complete information for 64 firms. Furthermore, their database did not have sources and uses-of-funds statement information either.

Second, we resort to panel data models for uncensored data and devise specification tests for nonnested random-effect models. Most of the literature on capital structure focuses on the cross-section variation of the data by averaging observations over time. Or, when using panel data models, the bias of fixed-effect estimates, under a dynamic specification, is usually neglected.

This chapter is organized as follows: Section 16.2 discusses our econometric specification; Section 16.3 presents descriptive statistics of the data and our estimation results; Section 16.4 discusses the potential importance of tax and monetary policies to determine firm capital structure; and Section 16.5 provides conclusions to this chapter.

16.2 ECONOMETRIC MODEL

16.2.1 Random-Effect Model*

Our econometric specification is based on Kim and Maddala's (1992) model who have studied the determinants of dividend policy for firms in the U.S. manufacturing sector. Given that firms do not necessarily pay dividends in all periods, Kim and Maddala utilize a censored panel data model. Specifically, they propose a random-effect model of the form

$$y_{it} = \beta' \mathbf{x}_{it} + \varepsilon_{it} \qquad (16.1)$$

---

* This section draws from Fernandez (2006).

where

$$\varepsilon_{it} = \upsilon_{it} + \omega_{it}$$

with

υ_{it} and ω_{it} being independent normal
$$\text{var}(\upsilon_{it}) = \sigma_i^2$$
$$\text{var}(\omega_{it}) = \theta_t^2$$

That is, errors are heteroskedastic, with firm- and time-specific components,* but uncorrelated:

$$E(\varepsilon_{it}\varepsilon_{js}) = \begin{cases} \sigma_i^2 + \theta_t^2 & i = j, t = s; i, j = 1, \dots, N; t, s = 1, \dots, T \\ 0 & \text{otherwise} \end{cases} \qquad (16.2)$$

Kim and Maddala have chosen this specification because it circumvents the problem of having to use numerical integration to maximize the log-likelihood function of the data in the presence of censored data.

Under the usual specification of the random-effect model, errors are homoskedastic and equicorrelated. That is, $\varepsilon_{it} = \upsilon_i + \omega_{it}$, $E(\varepsilon_{it}\varepsilon_{js}) = \sigma_\upsilon^2$ for $i = j$ and $t = s$, $E(\varepsilon_{it}\varepsilon_{js}) = \sigma_\upsilon^2$ for $i = j$ and $t \neq s$, and $E(\varepsilon_{it}\varepsilon_{js}) = 0$ otherwise.

Kim and Maddala have focused on the case where y_{it} is censored at zero. We have extended Kim and Maddala's model for the case in which y_{it} is uncensored. In this case, the log-likelihood function boils down to

$$\ln L \propto -\frac{1}{2}\sum_{i=1}^{N}\sum_{t=1}^{T}\ln(\sigma_i^2 + \theta_t^2) - \frac{1}{2}\sum_{i=1}^{N}\sum_{t=1}^{T}\frac{(y_{it} - \beta'\mathbf{x}_{it})^2}{\sigma_i^2 + \theta_t^2} \qquad (16.3)$$

The first-order conditions are given by

$$\frac{\partial \ln L}{\partial \beta} = \sum_{i=1}^{N}\sum_{t=1}^{T}\frac{(y_{it} - \beta'\mathbf{x}_{it})\mathbf{x}_{it}}{\sigma_i^2 + \theta_t^2} = 0 \qquad (16.4a)$$

$$\frac{\partial \ln L}{\partial \sigma_i^2} = -\frac{1}{2}\sum_{i=1}^{T}\frac{1}{\sigma_i^2 + \theta_t^2} + \frac{1}{2}\sum_{i=1}^{T}\frac{(y_{it} - \beta'\mathbf{x}_{it})^2}{(\sigma_i^2 + \theta_t^2)^2} = 0 \quad i = 1, 2, \dots, N \qquad (16.4b)$$

---

* Kim and Maddala also consider a multiplicative heteroskedastic specification of Anderson (1986), in which $E(\varepsilon_{it}\varepsilon_{js}) = \sigma_i^2\theta_t^2$ for $i = j$, $t = s$, and $i, j = 1, \dots, N$; and $s, t = 1, \dots, T$; 0 otherwise.

$$\frac{\partial \ln L}{\partial \theta_t^2} = -\frac{1}{2} \sum_{i=1}^{N} \frac{1}{\sigma_i^2 + \theta_t^2} + \frac{1}{2} \sum_{i=1}^{N} \frac{(y_{it} - \boldsymbol{\beta}' \mathbf{x}_{it})^2}{(\sigma_i^2 + \theta_t^2)^2} = 0 \quad t = 1, 2, \dots, T \qquad (16.4c)$$

The number of parameters to be estimated is $k + N + T$, where k is the dimension of $\boldsymbol{\beta}$. In order to reduce the dimension of the parameter space, we have followed a line of reasoning similar to Kim and Maddala's and first obtained estimates of σ_i^2 and θ_t^2 as*

$$\hat{\sigma}_i^2 = \frac{1}{T} \sum_{t=1}^{T} (y_{it} - \boldsymbol{\beta}' \mathbf{x}_{it})^2 \quad i = 1, 2, \dots, N$$

$$\hat{\theta}_t^2 = \frac{1}{N} \sum_{i=1}^{N} (y_{it} - \boldsymbol{\beta}' \mathbf{x}_{it})^2 \quad t = 1, 2, \dots, T$$

$$(16.5)$$

These estimates are substituted in Equation 16.3, and we have maximized the concentrated log-likelihood function with respect to $\boldsymbol{\beta}$. The number of parameters to be estimated is reduced to k. After obtaining a new estimate of $\boldsymbol{\beta}$, we have recomputed the estimates of σ_i^2 and θ_t^2 and maximized the concentrated log-likelihood with respect to $\boldsymbol{\beta}$. This iterative procedure is repeated until convergence is reached. In order to start up the iterations, we have used the pooled ordinary least-squares estimate of $\boldsymbol{\beta}$.

The parameter estimates and their variance–covariance matrix can be obtained as

$$\hat{\boldsymbol{\beta}} = \left(\sum_{i=1}^{N} \mathbf{X}_i' \hat{\boldsymbol{\Sigma}}_{ii}^{-1} \mathbf{X}_i \right)^{-1} \left(\sum_{i=1}^{N} \mathbf{X}_i' \hat{\boldsymbol{\Sigma}}_{ii}^{-1} \mathbf{Y}_i \right), \quad \text{Var}(\hat{\boldsymbol{\beta}}) = \left(\sum_{i=1}^{N} \mathbf{X}_i' \hat{\boldsymbol{\Sigma}}_{ii}^{-1} \mathbf{X}_i \right)^{-1} \qquad (16.6)$$

where $\mathbf{X}_i = (\mathbf{x}_{i1} \, \mathbf{x}_{i2} \dots \mathbf{x}_{iT})'$, $\mathbf{Y}_i = (y_{i1} \, y_{i2} \dots y_{iT})'$, $\hat{\boldsymbol{\Sigma}}_{ii} = \begin{pmatrix} \hat{\sigma}_i^2 + \hat{\theta}_1^2 & 0 & \cdots & 0 \\ 0 & \hat{\sigma}_i^2 + \hat{\theta}_2^2 & \cdots & 0 \\ \cdots & \cdots & \ddots & \cdots \\ 0 & 0 & \cdots & \hat{\sigma}_i^2 + \hat{\theta}_T^2 \end{pmatrix}$,

$i = 1, \dots, N$, and $\hat{\sigma}_i^2$ and $\hat{\theta}_t^2$ are obtained from Equation 16.5.

---

* In the absence of a time-specific component, σ_i^2 can be directly obtained as $\hat{\sigma}_i^2 = \frac{1}{T} \sum_{t=1}^{T} (y_{it} - \boldsymbol{\beta}' \mathbf{x}_{it})^2$ from Equation 16.4b. In turn, in the absence of a specific firm-component, θ_t^2 can be obtained as $\hat{\theta}_t^2 = \frac{1}{N} \sum_{i=1}^{N} (y_{it} - \boldsymbol{\beta}' \mathbf{x}_{it})^2$ from Equation 16.4c.

16.2.2 Specification Tests

Besides the Kim–Maddala estimator, we have also computed the conventional random- and fixed-effect models from the specification

$$y_{it} = \beta' x_{it} + \alpha_i + v_{it} \tag{16.7}$$

where

$\alpha_i = z_i' \alpha$ for the fixed-effect model

$\alpha_i = \alpha + \mu_i$ for the random-effect model

An asymptotically equivalent way of carrying out Hausman's specification test of random versus fixed effects is by using the following augmented regression (see Baltagi, 2001, Chapter 4):

$$y^* = X^* \beta + \tilde{X} \gamma + \omega \tag{16.8}$$

where

$$y^* = \begin{pmatrix} y_1^* \\ y_2^* \\ \cdots \\ y_N^* \end{pmatrix}, \quad y_i^* = \begin{pmatrix} y_{i1} - \phi \bar{y}_{i.} \\ y_{i2} - \phi \bar{y}_{i.} \\ \cdots \\ y_{iT} - \phi \bar{y}_{i.} \end{pmatrix}, \quad X^* = \begin{pmatrix} x_1^{*(1)} & x_1^{*(2)} & \cdots & x_1^{*(k)} \\ x_2^{*(1)} & x_2^{*(2)} & \cdots & x_2^{*(k)} \\ \cdots & \cdots & \cdots & \cdots \\ x_N^{*(1)} & x_N^{*(2)} & \cdots & x_N^{*(k)} \end{pmatrix},$$

$$x_i^{*(j)} = \begin{pmatrix} x_{i1}^{(j)} - \phi \bar{x}_{i.}^{(j)} \\ x_{i2}^{(j)} - \phi \bar{x}_{i.}^{(j)} \\ \cdots \\ x_{iT}^{(j)} - \phi \bar{x}_{i.}^{(j)} \end{pmatrix},$$

$$\phi = 1 - \frac{\sigma_v}{\sqrt{\sigma_v^2 + T \sigma_\mu^2}}, \quad \sigma_v^2 = E(v_{it}^2), \quad \sigma_\mu^2 = E(\mu_i^2),$$

$$\tilde{X} = \begin{pmatrix} \tilde{x}_1^{(1)} & \tilde{x}_1^{(2)} & \cdots & \tilde{x}_1^{(k)} \\ \tilde{x}_2^{(1)} & \tilde{x}_2^{(2)} & \cdots & \tilde{x}_2^{(k)} \\ \cdots & \cdots & \cdots & \cdots \\ \tilde{x}_N^{(1)} & \tilde{x}_N^{(2)} & \cdots & \tilde{x}_N^{(k)} \end{pmatrix}, \quad \text{and} \quad \tilde{x}_i^{(j)} = \begin{pmatrix} x_{i1}^{(j)} - \bar{x}_{i.}^{(j)} \\ x_{i2}^{(j)} - \bar{x}_{i.}^{(j)} \\ \cdots \\ x_{iT}^{(j)} - \bar{x}_{i.}^{(j)} \end{pmatrix} \quad i = 1, \ldots, N; j = 1, \ldots, k,$$

The notation $x_i^{*(j)}$ indicates regressor j, where $j = 1, \ldots, k$ and for unit i, $i = 1, \ldots, N$. There are T observations for each regressor within each unit. Similarly, for $\tilde{\mathbf{x}}_i^{(j)}$.

Under the null hypothesis of random effects, $\boldsymbol{\gamma} = \mathbf{0}$. The advantage of this formulation is that one circumvents the problem that $\hat{\boldsymbol{\psi}} \equiv \hat{\mathrm{Var}}(\hat{\boldsymbol{\beta}}_{FE}) - \hat{\mathrm{Var}}(\hat{\boldsymbol{\beta}}_{RE})$ has usually a rank less than k in the Wald criterion, $(\hat{\boldsymbol{\beta}}_{FE} - \hat{\boldsymbol{\beta}}_{RE})\,\hat{\boldsymbol{\psi}}^{-1}\,(\hat{\boldsymbol{\beta}}_{FE} - \hat{\boldsymbol{\beta}}_{RE}) \xrightarrow{d} \chi^2(k)$, where k is the number of slopes, FE stands for fixed effects, and RE for random effects.

The conventional random effects and the Kim–Maddala models are nonnested. Therefore, in order to compare them, we have used both Davidson and Mackinnon's (1981, 1982) J-test and Cox's (1962) test. Let us first consider the J-test when the null hypothesis is Kim–Maddala's specification:

$$H_0 : \mathbf{y} = \mathbf{X}\boldsymbol{\beta} + \boldsymbol{\varepsilon}$$

where

$$E(\boldsymbol{\varepsilon}\boldsymbol{\varepsilon}') = \begin{pmatrix} \Sigma_{11} & \mathbf{0} & \cdots & \mathbf{0} \\ \mathbf{0} & \Sigma_{22} & \cdots & \mathbf{0} \\ & & \vdots & \\ \mathbf{0} & \mathbf{0} & \cdots & \Sigma_{NN} \end{pmatrix}, \quad \boldsymbol{\Sigma}_{ii} = \begin{pmatrix} \sigma_i^2 + \theta_1^2 & 0 & \cdots & 0 \\ 0 & \sigma_i^2 + \theta_2^2 & \cdots & 0 \\ \cdots & \cdots & \ddots & \cdots \\ 0 & 0 & \cdots & \sigma_i^2 + \theta_T^2 \end{pmatrix},$$

$$i = 1, \ldots, N.$$

$$H_1 : \mathbf{y} = \mathbf{X}\boldsymbol{\beta} + \boldsymbol{\eta}$$

where $\eta_{it} = \mu_i + v_{it}$, $E(v_{it}^2) = \sigma_v^2, E(\mu_i^2) = \sigma_\mu^2$, $i = 1, \ldots, N;\ t = 1, \ldots, T.$

$$E(\boldsymbol{\eta}\boldsymbol{\eta}') = \begin{pmatrix} \Sigma & \mathbf{0} & \cdots & \mathbf{0} \\ \mathbf{0} & \Sigma & \cdots & \mathbf{0} \\ & & \vdots & \\ \mathbf{0} & \mathbf{0} & \cdots & \Sigma \end{pmatrix} = \mathbf{I}_N \otimes \boldsymbol{\Sigma}, \quad \boldsymbol{\Sigma} = \begin{pmatrix} \sigma_v^2 + \sigma_\mu^2 & \sigma_\mu^2 & \cdots & \sigma_\mu^2 \\ \sigma_\mu^2 & \sigma_v^2 + \sigma_\mu^2 & \cdots & \sigma_\mu^2 \\ \cdots & \cdots & \ddots & \cdots \\ \sigma_\mu^2 & \sigma_\mu^2 & \cdots & \sigma_v^2 + \sigma_\mu^2 \end{pmatrix}.$$

We test whether $\lambda = 0$ in the following compound model:

$$\tilde{\mathbf{y}} = (1 - \lambda)\tilde{\mathbf{X}}\boldsymbol{\beta} + \lambda\hat{\mathbf{y}}^* + \xi \tag{16.9}$$

where

$$
\tilde{\mathbf{y}} = \begin{pmatrix} \tilde{\mathbf{y}}_1 \\ \tilde{\mathbf{y}}_2 \\ \cdots \\ \tilde{\mathbf{y}}_N \end{pmatrix}, \tilde{\mathbf{y}}_i = \begin{pmatrix} \dfrac{y_{i1}}{\sqrt{\sigma_i^2 + \theta_1^2}} \\[2ex] \dfrac{y_{i2}}{\sqrt{\sigma_i^2 + \theta_2^2}} \\[2ex] \cdots \\[2ex] \dfrac{y_{iT}}{\sqrt{\sigma_i^2 + \theta_T^2}} \end{pmatrix}, \tilde{\mathbf{X}} = \begin{pmatrix} \tilde{\mathbf{x}}_1^{(1)} & \tilde{\mathbf{x}}_1^{(2)} & \cdots & \tilde{\mathbf{x}}_1^{(k)} \\ \tilde{\mathbf{x}}_2^{(1)} & \tilde{\mathbf{x}}_2^{(2)} & \cdots & \tilde{\mathbf{x}}_2^{(k)} \\ \cdots & \cdots & \cdots & \cdots \\ \tilde{\mathbf{x}}_N^{(1)} & \tilde{\mathbf{x}}_N^{(2)} & \cdots & \tilde{\mathbf{x}}_N^{(k)} \end{pmatrix}, \tilde{\mathbf{x}}_i^{(j)} = \begin{pmatrix} \dfrac{x_{i1}^{(j)}}{\sqrt{\sigma_i^2 + \theta_1^2}} \\[2ex] \dfrac{x_{i2}^{(j)}}{\sqrt{\sigma_i^2 + \theta_2^2}} \\[2ex] \cdots \\[2ex] \dfrac{x_{iT}^{(j)}}{\sqrt{\sigma_i^2 + \theta_T^2}} \end{pmatrix}
$$

$$
\hat{\mathbf{y}}^* = \begin{pmatrix} \hat{\mathbf{y}}_1^* \\ \hat{\mathbf{y}}_2^* \\ \cdots \\ \hat{\mathbf{y}}_N^* \end{pmatrix}, \ \hat{\mathbf{y}}_i^* = \begin{pmatrix} y_{i1} - \hat{\phi}\bar{y}_{i.} \\ y_{i2} - \hat{\phi}\bar{y}_{i.} \\ \cdots \\ y_{iT} - \hat{\phi}\bar{y}_{i.} \end{pmatrix}, \text{ and } \hat{\phi} = 1 - \frac{\hat{\sigma}_v}{\sqrt{\hat{\sigma}_v^2 + T\hat{\sigma}_\mu^2}}, \quad i = 1, \ldots, N.
$$

Given that σ_i^2 and θ_i^2 are unknown, we plugged in their maximum-likelihood estimates. In order to test the random-effect model against Kim–Maddala's specification, we just reversed the roles of H_0 and H_1.

In order to obtain the functional form of the Cox test for this particular case, we have followed Pesaran's (1974, pp. 156–158) line of reasoning.[*] Under the null hypothesis that the Kim–Maddala model is true,

$$
\frac{c_0}{\sqrt{\hat{V}(c_0)}} \xrightarrow{d} N(0,1) \tag{16.10}
$$

where

$$
c_0 = \frac{NT}{2} \ln\left(\frac{\hat{\sigma}_{x^*}^2}{\hat{\sigma}_{\tilde{x}}^2 + \dfrac{1}{NT}\hat{\boldsymbol{\beta}}_0' \tilde{\mathbf{X}}' \mathbf{M}_{x^*} \tilde{\mathbf{X}} \hat{\boldsymbol{\beta}}_0} \right) = \frac{NT}{2} \ln\left(\frac{\hat{\sigma}_{x^*}^2}{\hat{\sigma}_{x^*\tilde{x}}^2} \right)
$$

---

[*] The functional form of the Cox test for a linear regression model is reproduced in Greene (2003, Chapter 8).

$$\hat{V}(c_0) = \frac{\hat{\sigma}_{x^*}^2}{\hat{\sigma}_{x^*\tilde{x}}^4} \, \hat{\beta}_0' \tilde{X}' M_{X^*} M_{\tilde{x}} M_{X^*} \tilde{X} \hat{\beta}_0$$

$$M_{X^*} = I - X^* (X^{*'}X^*)^{-1} X^{*'} \quad M_{\tilde{x}} = I - \tilde{X}(\tilde{X}'\tilde{X})^{-1} \tilde{X}'$$

$$\hat{\beta}_0 = (\tilde{X}'\tilde{X})^{-1} \tilde{X}' \tilde{y}$$

$$\hat{\sigma}_{x^*}^2 = \frac{e_{x^*}{}' e_{x^*}}{NT} \text{Mean-squared residual in the regression of } \tilde{y} \text{ on } X^*$$

$$\hat{\sigma}_{\tilde{x}}^2 = \frac{e_{\tilde{x}}{}' e_{\tilde{x}}}{NT} \text{Mean-squared residual in the regression of } \tilde{y} \text{ on } \tilde{X}$$

$$\hat{\sigma}_{x^*\tilde{x}}^2 = \hat{\sigma}_{\tilde{x}}^2 + \frac{\hat{\beta}_0' \tilde{X}' M_{X^*} \tilde{X} \hat{\beta}_0}{NT}.$$

$\tilde{y}$, X^*, and $\tilde{X}$ are as previously defined in Equations 16.8 and 16.9.

Similar to the J-test, for testing the random-effect model against Kim–Maddala's specification, we reversed the roles of H_0 and H_1.

An additional diagnostic test that we have used to discriminate between models is Pesaran's (2004) test of cross-section dependence. Pesaran has pointed out that Breusch and Pagan's (1980) Lagrange multiplier (LM) statistic for testing cross-equation error correlation is likely to present considerable size distortions for large N and small T—which is usually the case in panels. Therefore, he has proposed the following alternative LM statistic:

$$\sqrt{\frac{2T}{N(N-1)}} \left(\sum_{i=1}^{N-1} \sum_{j=i+1}^{N} \hat{\rho}_{ij} \right) \xrightarrow{d} N(0,1) \tag{16.11}$$

where $\hat{\rho}_{ij} = \hat{\rho}_{ji} = \dfrac{\sum_{t=1}^{T} e_{it} e_{jt}}{\left(\sum_{t=1}^{T} e_{it}^2\right)^{1/2} \left(\sum_{t=1}^{T} e_{jt}^2\right)^{1/2}}$ is the sample pair-wise correlation of residuals.

16.3 ESTIMATION RESULTS

As discussed in the introduction, the implications of the pecking-order theory are that firms prefer internal financing in the first place. They adapt their target dividend payout ratios to their investment opportunities so as to avoid sudden changes in dividends. In case the uses exceed the sources of funds, firms issue the safest security first (i.e., debt), then bonds, and then use equity issues as the last resort. Conversely, if the sources exceed the uses of funds, firms pay off debt, invest on marketable securities, or repurchase equity.

Frank and Goyal (2003) have used the following accounting cash-flow identity for the financing deficit:

$$DEF_t = DIV_t + I_t + \Delta W_t - C_t = \Delta D_t + \Delta E_t \qquad (16.12)$$

where

DEF_t is the financing deficit in year t
DIV_t is the cash dividends in year t
I_t is the net investment in year t
ΔW_t is the change in working capital in year t
C_t is the cash flow after interest and taxes in year t

The gap between the uses and sources of funds is filled by net debt issues (ΔD_t) and/or net equity issues (ΔE_t).

Table 16.1 shows average figures for each year of the period 1990–1996 on the items in identity (Equation 16.12). Our sample was taken from quarterly balance-sheet data gathered by the Chile Superintendency of Securities and Insurance (Superintendencia de Valores y Seguros, SVS) in the Uniformly Coded Statistical Record (Ficha Estadistica Codificada Uniforme, FECU).* All figures are scaled by net assets (total assets minus current liabilities). For the sample period, the financial deficit averaged 3.7% of total assets and was covered primarily by net equity issues (3.1% of total assets). This information is depicted in Figure 16.1. This evidence already questions the validity of the pecking-order theory, so we next turn to more testing.

If the pecking-order theory holds, then when running a regression of the net debt issued on the financing deficit, the slope of such a regression should be statistically equal to 1. Likewise, if the dependent variable is, in turn, the gross debt issued or the change in the long-maturity debt ratio.

---

* The information on firm cash flows was discontinued between 1997 and 2001.

TABLE 16.1   Average Cash Flows and Financing: 1990–1996

	1990	1991	1992	1993	1994	1995	1996
Cash dividends[1]	0.077	0.080	0.083	0.076	0.073	0.069	0.064
Net Investment[2]	0.121	0.086	0.117	0.121	0.116	0.098	0.108
Δ Working capital[3]	0.000	0.016	−0.026	0.027	0.017	−0.018	0.004
Internal cash flow[4]	0.177	0.174	0.144	0.160	0.147	0.129	0.119
Financing deficit[1]+[2]+[3]−[4]	0.022	0.008	0.030	0.063	0.058	0.020	0.058
Net debt issues	0.003	−0.005	−0.004	0.024	0.009	0.005	0.010
Net equity issues	0.018	0.013	0.034	0.039	0.049	0.015	0.048
Net external financing	0.022	0.008	0.030	0.063	0.058	0.020	0.058

Notes: (i) All variables are scaled by net assets (total assets minus current liabilities). (ii) Figures are averages of December of each year for the whole sample of 64 Chilean firms. The data was taken from quarterly balance-sheet data gathered by the Chile Superintendency of Securities and Insurance (SVS) in the Uniformly Coded Statistical Record (FECU).

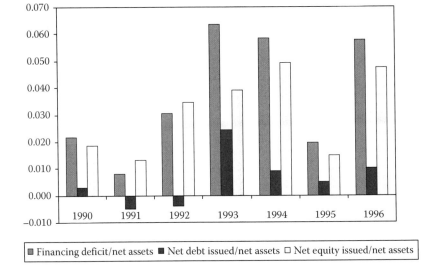

FIGURE 16.1   Financing deficit and external financing of Chilean firms: 1990–1996. (Author's own elaboration based on funds-flow statements gathered by the Superintendency of Securities and Insurance. Figures are averages of December of each year for the whole sample of 64 firms.)

Table 16.2 addresses this point. Table 16.2 (Panel a) shows that slopes are substantially lower than 1, in particular for the change in the long-maturity ratio. (For the debt net issued, neither the J-test nor the Cox test can discriminate between the models at the 1% significance level. In turn, the

TABLE 16.2 Testing the Pecking-Order Hypothesis

Regressor	Fixed Effects (FE)			Random Effects (RE)			Kim–Maddala (KM)		
	Coeff.	t-Test	p-Value	Coeff.	t-Test	p-Value	Coeff.	t-Test	p-Value
(a) Test of the pecking-order									
Net debt issued									
Financing deficit	0.473	5.48	0.000	0.464	16.19	0.000	0.519	21.18	0.000
Adjusted R^2	0.350			0.367			0.368		
Observations	448			448			448		
Cross-correlation test	8.463	p-Value	0.000	8.537	p-Value	0.000	8.111	p-Value	0.000
H_0: RE H_1: FE		p-Value		1.306	p-Value	0.253			
J-test									
H_0: KM H_1: RE							0.060	p-Value	0.952
H_0: RE H_1: KM							0.820	p-Value	0.412
Cox test									
H_0: KM H_1: RE							2.536	p-Value	0.011
H_0: RE H_1: KM							-3.031	p-Value	0.024
Gross debt issued									
Financing deficit	0.353	4.75	0.000	0.412	10.63	0.000	0.422	14.96	0.000
Adjusted R^2	0.329			0.237			0.237		
Observations	448			448			448		
Cross-correlation test	-0.704	p-Value	0.481	0.308	p-Value	0.758	-0.232	p-Value	0.817
H_0: RE H_1: FE		p-Value		9.081	p-Value	0.003			

(continued)

TABLE 16.2 (continued) Testing the Pecking-Order Hypothesis

Regressor	Fixed Effects (FE)			Random Effects (RE)			Kim–Maddala (KM)		
	Coeff.	t-Test	p-Value	Coeff.	t-Test	p-Value	Coeff.	t-Test	p-Value
J-test									
H_0: KM H_1: RE							3.483	p-Value	0.001
H_0: RE H_1: KM							3.321	p-Value	0.001
Cox test									
H_0: KM H_1: RE							−2.542	p-Value	0.011
H_0: RE H_1: KM							0.842	p-Value	0.400
Δ Long-maturity debt ratio									
Financing deficit	0.174	2.989	0.003	0.152	5.457	0.000	0.120	4.691	0.000
Adjusted R^2	0.078			0.072			0.069		
Observations	384			384			384		
Cross-correlation test	0.897	p-Value	0.369	0.988	p-Value	0.323	1.154	p-Value	0.249
H_0: RE H_1: FE				3.250	p-Value	0.071			
J-test									
H_0: KM H_1: RE							−0.438	p-Value	0.661
H_0: RE H_1: KM							−1.067	p-Value	0.285
Cox test									
H_0: KM H_1: RE							−3.096	p-Value	0.020
H_0: RE H_1: KM							3.272	p-Value	0.001

(b) Components of the flow fund deficit

Net debt issued

Cash dividends	0.471	3.63	0.000	0.493	7.89	0.000	0.604	15.30	0.000
Net investment	0.437	5.84	0.000	0.429	10.65	0.000	0.494	16.98	0.000
Δ Working capital	0.497	5.40	0.000	0.490	14.63	0.000	0.539	18.62	0.000
Internal cash flow	-0.447	-3.90	0.000	-0.457	-11.93	0.000	-0.568	-17.33	0.000
Adjusted R^2	0.349			0.367			0.365		
Observations	448			448			448		
Cross-correlation test	7.711	p-Value	0.000	7.412	p-Value	0.000	3.892	p-Value	0.000
H_0: RE H_1: FE		p-Value		1.518	p-Value	0.823			
J-test									
H_0: KM H_1: RE							-0.349	p-Value	0.727
H_0: RE H_1: KM							-0.390	p-Value	0.696
Cox test									
H_0: KM H_1: RE							3.186	p-Value	0.001
H_0: RE H_1: KM							-3.796	p-Value	0.000

random-effect model is preferable to the fixed-effect model. For the gross debt and the Δ long-maturity ratio equations, the fixed-effect model gets more support, at least at the 10% level in the latter case).

Table 16.2 (Panel b) disaggregates the financial deficit into its components. The dependent variable in this case is the net debt issued. If the pecking-order hypothesis holds, then the coefficients of the change in the working capital and net investment would be close to unity. Intuitively, after controlling for internal cash flows, investments in fixed assets and working capital should be entirely financed by net debt issues. However, no model specification supports this hypothesis for our data set. (The Hausman test gives support to the random-effect model but the J- and Cox tests are inconclusive about whether we should prefer this to Kim–Maddala's. Nonetheless, based on the cross-correlation test, one would be more inclined to pick Kim–Maddala's model).

We further investigate the validity of the pecking-order theory. Table 16.3 (Panel a) shows a leverage regression in first differences in which the financing deficit is an additional explanatory variable. The dependent variable in this case is the change in the leverage ratio, defined as total interest-bearing liabilities to net assets. First differences are used given the dynamic content of the pecking-order theory. If the latter was true, the financing deficit would wipe out all the explanatory powers of the other variables used in conventional leverage regressions. But this is not the case. In fact, the (lagged) financing deficit has explanatory power in the fixed-effect and random-effect regression models but not under Kim–Maddala's specification—which, according to the cross-correlation test, would get more support than the random-effect model.*

Table 16.3 (Panel b) reports a leverage regression, in first differences, where the lagged change in leverage is an additional regressor. The fixed- and random-effect models are shown just for illustrative purposes given that they yield biased estimates. The lagged financing deficit has no explanatory power in Kim–Maddala's model whereas the lagged difference in leverage does. The first differences in nondebt tax shields, the equity ratio, and size are all highly significant.

In short, our estimation results give little support to the pecking-order hypothesis. According to our figures, firms favored equity over debt issues to cover their financing deficit. We think the explanation might be found in Chile's tax and monetary policies. We address this point in the next section.

---

* The lagged financing deficit gives a better fit than its current value.

TABLE 16.3 Change in the Leverage Ratio and Financing Deficit

Regressor	Fixed Effects			Random Effects			Kim–Maddala		
	Coeff.	t-Test	p-Value	Coeff.	t-Test	p-Value	Coeff.	t-Test	p-Value
(a) Without lagged Δ leverage ratio as a regressor									
Δ Leverage ratio									
Δ Tangibility	0.913	14.41	0.000	0.906	22.73	0.000	0.816	24.71	0.000
Δ Nondebt tax shields	−0.250	−1.62	0.106	−0.243	−2.48	0.013	−0.155	−2.24	0.026
Δ Equity ratio	−0.292	−2.68	0.008	−0.275	−3.83	0.000	−0.290	−5.10	0.000
Δ Market-to-book	0.002	1.08	0.280	0.002	0.82	0.412	0.001	1.00	0.318
Δ Profitability	−0.337	−1.67	0.097	−0.328	−2.41	0.016	−0.205	−2.04	0.042
Δ Size	0.239	2.93	0.004	0.203	5.00	0.000	0.123	4.12	0.000
Lagged financing deficit	0.188	2.59	0.010	0.126	2.58	0.010	0.056	1.44	0.150
Adjusted R^2	0.633			0.664			0.724		
Observations	320			320			320		
Cross-correlation test	2.566	p-Value	0.010	2.060	p-Value	0.039	0.413	p-Value	0.680
H_0: RE H_1: FE				9.825	p-Value	0.199			
J-test									
H_0: KM H_1: RE							4.145	p-Value	0.000
H_0: RE H_1: KM							−2.319	p-Value	0.020
Cox test									
H_0: KM H_1: RE							−6.882	p-Value	0.000
H_0: RE H_1: KM							9.566	p-Value	0.000

(continued)

TABLE 16.3 (continued) Change in the Leverage Ratio and Financing Deficit

Regressor	Fixed Effects			Random Effects			Kim–Maddala		
	Coeff.	t-Test	p-Value	Coeff.	t-Test	p-Value	Coeff.	t-Test	p-Value
(b) With lagged Δ leverage ratio as a regressor									
Δ Leverage ratio									
Δ Tangibility	0.795	10.63	0.000	0.797	18.23	0.000	0.801	21.51	0.000
Δ Nondebt tax shields	−0.314	−1.88	0.061	−0.291	−3.19	0.001	−0.194	−2.92	0.004
Δ Equity ratio	−0.250	−2.70	0.008	−0.269	−4.13	0.000	−0.281	−5.01	0.000
Δ Market-to-book	0.005	0.99	0.322	0.009	1.46	0.145	0.011	2.18	0.030
Δ Profitability	−0.189	−0.94	0.350	−0.255	−1.73	0.083	−0.181	−1.45	0.148
Δ Size	0.169	3.30	0.001	0.179	4.79	0.000	0.134	4.76	0.000
Lagged financing deficit	−0.050	−0.67	0.503	−0.027	−0.57	0.567	−0.007	−0.19	0.849
Lagged Δ leverage ratio	−0.258	−3.37	0.001	−0.240	−7.46	0.000	−0.189	−5.60	0.000
Adjusted R^2	0.726			0.780			0.780		
Observations	256			256			256		
Cross-correlation test	0.026	p-Value	0.979	1.998	p-Value	0.046	2.129	p-Value	0.033

16.4 TAX AND MONETARY POLICY ISSUES

16.4.1 Corporate and Personal Taxes in Chile

Historically, the corporate tax rate has been much lower than the highest marginal personal tax rate in Chile. In 2001, an amendment to the Income Tax Law reduced the gap between corporate and personal tax rates in order to prevent tax avoidance. Still, the gap between corporate and personal tax rates is large when compared with other countries in the world (Table 16.4). For example, as of 2002, Argentina has the same (consolidated) tax rate for corporations and individuals whereas in Mexico, the gap between the two rates is only 5% points. Relative to OECD countries, Chile has a tax policy similar to Ireland's.

If we recall Miller's (1977) model on corporate and personal taxes, the value of the levered firm (V_L) is given by

$$V_L = V_u + \left(1 - \frac{(1-T_E)(1-T_c)}{1-T_D}\right)D \tag{16.13}$$

where

V_u is the value of the unlevered firm
T_c is the corporate tax rate
T_E is the effective personal tax on equity income
T_D is the personal tax rate on interest income
D is the amount of debt held by the firm

The relative tax advantage of debt over equity is

$$\frac{1-T_D}{(1-T_E)(1-T_c)} \tag{16.14}$$

The tax rate applicable to equity income in Chile will depend on whether we are dealing with capital gains or dividends. Capital gains are taxed at the corporate tax rate (17%). However, by the 2001's amendment of the Law of Capital Markets, sales of highly liquid stocks are exempted from the capital-gain tax. But they are subject to the personal tax rate (40% for the upper-income bracket), the same as dividends and interest income (i.e., bank deposits and corporate and government bonds). Nevertheless, in the case of stockholders, the amount paid in corporate taxes can be used as a credit in the annual personal tax statement, according to the percent of firm ownership.

TABLE 16.4   Consolidated Tax Rates in Some Selected
Countries: 2002

Country	Corporate	Personal
G7		
Germany	26.38	51.17
Canada	33.9	46.4
United States	[15–47]	49.5
France	36.33	52.75
Italy	40.3	47.3
Japan	[24.2–39.2]	37
United Kingdom	30	40
Other OECD		
Australia	30	47
Spain	35	48
Finland	29	56
Ireland	16	42
New Zealand	33	39
The Netherlands	[29–34.5]	52
Sweden	28	56
Latin America		
Argentina	35	35
Brazil	[15–25]	27.5
Chile	17	40
Mexico	35	40
Asia-Pacific		
Korea	[16.5–9.7]	39.6
Indonesia	[10–30]	35
Malaysia	28	28
Singapore	24.5	26

Source: Tax load and tax rates: Chile vis-à-vis other countries, Research Department at the Chile Internal
Revenue Service, July 2003. The document is available at www.sii.cl

Consequently, it will be generally the case that $T_E < T_D$. For instance, if $T_E \le 27\%$ and $T_D = 40\%$, then $(1 - T_D) < (1 - T_E)(1 - T_c)$. Then equity will have a relative tax advantage over debt for the investor facing the maximum marginal personal tax rate. So the pattern depicted in Figure 16.1 might be, to a certain degree, explained by tax policy. Nevertheless, part of the explanation might lie in the stance of monetary policy. In particular, periods of loose monetary policy usually make bond issues more attractive. We discuss this next.

16.4.2 Monetary Policy and Firm Financing

In the past few years, Chile has enjoyed a one-digit annual inflation rates. Accordingly, the Central Bank of Chile has kept the stance of monetary policy rate at very low levels. Figure 16.2 depicts the evolution of monetary

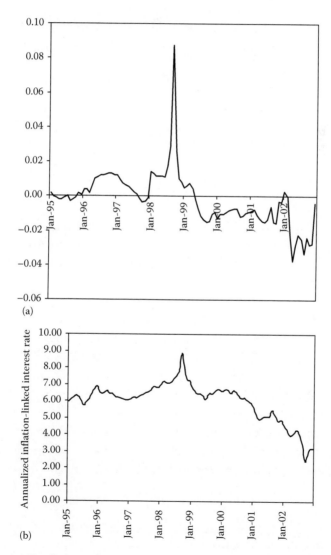

FIGURE 16.2   Evolution of monetary policy in Chile: 1995–2002. (a) Spread of 90 day and 8 year interest rates. (b) Evolution of the 8 year interest rate. (Data from Central Bank of Chile.)

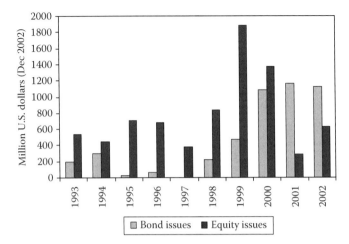

FIGURE 16.3 Announcement of bond and equity issues: 1993–2002. *Note:* The figures of announced issues by year are the sum of all future issues registered at the Superintendency of Securities and Insurance in that particular year.

policy tightness over 1995–2002. Following the Asian crisis outbreak, monetary policy became extremely tight and a slow down in economic activity followed for about 2 years. In order to reverse this process, the Central Bank started reducing interest rates from 1999 onward. As we see, approximately from mid-1999, the spread between the short (90 days) and the long (8 years) interest rates became systematically negative. On the other hand, the long interest rate went down from about 9% per year in September 1998 to 3.2% per year in December 2002.

The decrease in interest rates had an impact on the announcements of bonds and equity issues (Figure 16.3). Indeed, from 1999 onward, the number of future bond issues registered at the Superintendency of Securities and Insurance increased noticeably relative to future equity issues. Furthermore, in 2001 and 2002, the former was larger in monetary value than the latter, a pattern that differs from the pattern observed in the early 1990s.

16.5 CONCLUSIONS

This chapter analyzes the driving forces of capital structure in Chile for the period 1990–1996. Our findings are more congruent with the trade-off theory than the pecking-order hypothesis. In particular, in recent years, equity issues have followed firms' financing deficits more closely than net debt issues. Frank and Goyal (2003) also conclude that financing deficit is less important in explaining net debt issues over time. Nevertheless, they do not attempt to explain why this might be the case. For Chile, we conjecture that tax and monetary policies might have been the driving forces.

The contribution of this chapter is twofold. First, the literature on capital structure has focused primarily on developed economies. Some exceptions are international comparisons that include emerging economies. But their databases usually cover short time spans and lack information on sources and uses of funds statements, which is an essential piece of information to contrast the trade-off and pecking-order theories.

Second, we resort to an extension of Anderson (1986) and Kim and Maddala's (1992) work to panel data models for uncensored data and devise specification tests for nonnested random-effect models. Most of the literature on capital structure focuses on the cross-section variation of the data by averaging observations over time. Or, when using panel data models, the bias of fixed-effect estimates, under a dynamic specification, is usually neglected.

ACKNOWLEDGMENT

Financial support from FONDECYT Grant No. 1070762, and from an institutional grant of the Hewlett Foundation to CEA is greatly acknowledged. An extended version of the article was published as Determinants of firm leverage in Chile: Evidence from panel data. *Estudios de Administracion* 12(1), 41–85.

REFERENCES

Anderson, G. (1986) An application of the Tobit model to panel data: Modeling dividend behavior in Canada. Working Paper No. 85–22, McMaster University, Ontario, Canada.

Baltagi, B. (2001) *Econometric Analysis of Panel Data*, 2nd edn. John Wiley & Sons: Chichester, United Kingdom.

Booth, L., V. Aivazian, A. Demirgue-Kunt, and V. Maksimovic (2001) Capital structure in developing countries. *Journal Finance*, 56(1): 97–129.

Bradley, M., G. Jarrell, and E. Kim. (1984) On the existence of an optimal capital structure: Theory and evidence. *Journal of Finance*, 39(3): 857–878.

Breusch, T. and A. Pagan (1980) The LM test and its applications to model specification in econometrics. *Review of Economic Studies*, 47: 239–254.

Chirinko, R. and A. Singha (2000) Testing static tradeoff against pecking order models of capital structure: A critical comment. *Journal of Financial Economics*, 58: 417–425.

Cox, D. (1962) Further results on tests of separate families of hypothesis. *Journal of the Royal Statistical Society, Series B,* 24: 406–424.

Davidson, R. and J. Mackinnon (1981) Several tests for model specification in the presence of alternative hypotheses. *Econometrica,* 49(3): 781–793.

Davidson, R. and J. Mackinnon (1982) Some non-nested hypothesis test and the relations among them. *Review of Economic Studies,* 49(4): 551–565.

Fama, E. and K. French (2002) Testing tradeoff and pecking order predictions about dividends and debt. *Review of Financial Studies,* 15: 1–33.

Fan, J., Titman, S., and G. Twite (2003), An international comparison of capital structure and debt maturity choices. Manuscript Presented at the 2003 Financial Management Association (FMA) Meeting, Denver, CO.

Fernandez, V. (2006) Specification tests for a parsimonious random-effects model. *Applied Economics Letters,* 13(15): 1009–1012.

Frank, M. and V. Goyal (2003) Testing the pecking order theory of capital structure. *Journal of Financial Economics,* 67(2): 217–248.

Graham, J. (1996) Debt and the marginal tax rate. *Journal of Financial Economics,* 41(1): 41–73.

Graham, J. (2000) How big are the tax benefits of debt? *Journal of Finance,* 55(5): 1901–1941.

Graham, J. (2003) Taxes and corporate finance. *The Review of Financial Studies,* 16(4): 1075–1129.

Graham, J. and C. Harvey (2001) The theory and practice of corporate finance: Evidence from the field. *Journal of Financial Economic,* 60: 187–243.

Greene, W. (2003) *Econometric Analysis,* 5th edn. Prentice Hall: Upper Saddle River, NJ.

Jensen, M. and W. Meckling (1976) Theory of the firm: Managerial behavior, agency costs and ownership structure. *Journal of Financial Economics,* 3(4): 305–360.

Hennessy, C. and T. Whited (2005) Debt dynamics. *The Journal of Finance,* 60(3): 1129–1165.

Kim, B. and G. Maddala (1992) Estimation and specification analysis of models of dividend behavior based on censored panel data. *Empirical Economics,* 17: 111–124.

Leary, M. and M. Roberts (2005) Do firms rebalance their capital structures? *The Journal of Finance,* 60(6): 2575–2619.

Mackie-Mason, J. (1990) Do taxes affect corporate financing decisions? *Journal Finance,* 45(5): 1471–1493.

Megginson, W. (1997) *Corporate Finance Theory.* Addison-Wesley Educational Publishers Inc: Reading, MA.

Miller, M. (1977) Debt and taxes. *Journal of Finance,* 32(2): 261–275.

Molina, C. (2005) Are firms underleveraged? An examination of the effect of leverage on default probabilities. *The Journal of Finance,* 60(3): 1427–1459.

Myers, S. (1984) The capital structure puzzle. *Journal of Finance,* 39: 575–592.

Myers, S. (2001) Capital structure. *Journal of Economic Perspectives,* 15(2): 81–102.

Myers, S. and N. Majluf (1984) Corporate financing and investment decisions when firms have information investors do not have. *Journal of Financial Economics,* 13(2): 187–221.

Nilsen, J. (2002) Trade credit and the bank lending channel. *Journal of Money, Credit, and Banking*, 34(1): 226–253.

Pesaran, H. (1974) On the general problem of model selection. *The Review of Economic Studies*, 41(2): 153–171.

Pesaran, H. (2004), General diagnostic tests for cross section dependence in panels. IZA Discussion Paper No. 1240.

Rajan, R. and L. Zingales (1996) What do we know about capital structure? Some evidence from international data. *Journal of Finance*, 50: 1421–1460.

Ross, S. (1977) The determination of financial structure: The incentive-signaling approach. *Bell Journal of Economics*, 8: 23–40.

Shyam-Sunder, L. and S. Myers (1999) Testing static tradeoff against pecking order models of capital structure. *Journal of Financial Economics*, 51: 219–244.

Wald, J. (1999) How firm characteristics affect capital structure: An international comparison. *Journal of Financial Research*, 22(2): 161–187.

CHAPTER 17

Private Equity in the MENA Region: An Exploratory Analysis

Thomas Lagoarde-Segot and Laurence Le Poder

CONTENTS

17.1 INTRODUCTION

Fifteen years after the launching of the Barcelona process, the emerging transition economies of the MENA* region are still contending with dysfunctional financial systems. At an internal level, financial structures are heavily bank oriented, with bank assets accounting for 85% of total financial assets vs. 48% in emerging Asian countries, 41% in emerging Europe, and 33% globally (Abed and Soueid, 2005). At an external level, these countries remain net capital exporters, oil-related surpluses being channeled abroad through OECD financial intermediaries rather than invested domestically (OECD, 2006). This dynamic takes place in a context of massive investment needs as 22 million new jobs must be created before 2020 in order to stabilize the region's unemployment levels at their current rate of 15% (FEMISE, 2006).

Taking this into account, the development of a local private equity industry may be viewed as a necessary component of financial sector modernization. Four theoretical mechanisms indeed unite private equity with economic development. First, the private equity industry brings lenders and borrowers together where asymmetric information and uncertainty costs exist, thereby allowing the riskiest projects to obtain financing (Bonini and Alkan, 2006). Second, stage financing usually implies a tight control on a firm's operations. This helps to ensure productive efficiency in small business entities often characterized by a lack of management expertise (Gorman and Sahlman, 1989). Third, private equity industries participate in the creation of a knowledge-based economy. Private equity investors play an important social role in the innovation process through their involvement in four embedded networks: financial markets, entrepreneurs, services to business, and labor market professionals (Hellmann and Puri, 2000). Fourth, a large number of family owned companies operate in emerging markets. Although often profitable and employing large numbers of workers, these companies usually do not comply with international norms of transparency, corporate governance, and investor protection and often operate at the borderline of informal economy. In this context, private

---

* This chapter focuses on 10 MENA countries: Morocco, Egypt, Tunisia, Turkey, Israel, Algeria, Jordan, Lebanon, Libya, and Syria. These countries are referred to as the MEDA group by the European Commission.

equity markets may act as a bridge from traditional proprietary companies to modern listed companies (OECD, 2006). Overall, private equity market development in the MENA region could (1) channel greater investment flows into domestic economies, (2) diversify financing sources for local firms, and (3) increase productivity through managerial externalities.

A growing awareness of these issues has led most MENA countries to undertake significant financial reforms over the last decade. Foreign investment is liberalized, and the region's stock markets are active and developing (Lagoarde-Segot and Lucey, 2008). However, virtually nothing is known about private equity in the region. As shown in Table 17.1, the only available aggregated data highlights that the region is still lagging behind Emerging Asia and Eastern Europe ($5,027 billion invested in 2007 vs. $28,668 billion and $14,629 billion, respectively). In addition, Israel accounts for more than half of the region's private equity investment flows (ANIMA, 2008).

Taking this into account, the objective of this chapter is to explore the MENA private equity market development dynamic by juxtaposing local investors' perceptions and the observed institutional reform process. We thus first discuss investors' perception of the region through a modified questionnaire. We then compare MENA countries to other emerging markets in terms of institutional reforms using cross-country data. We finally consider the intersection of these results and discuss the gap between investor's perceptions and institutional developments, which allows us to raise a set of conjectures for the conduct of policy making.

The remainder of this chapter is structured as follows: Section 17.2 reviews the determinants of private equity market development; Section 17.3 describes the questionnaire and discusses investor's perceptions of the MENA region; Section 17.4 develops a battery of attractiveness indices and

TABLE 17.1 Emerging Private Equity Fundraising Total, 2003–2007 (U.S.$ Billions)

	Emerging Asia	CEE/Russia	Latin America	Sub-Saharan Africa	MENA
2003	2,200	406	417	NA	NA
2004	2,800	1,777	714	NA	NA
2005	15,446	2,711	1,272	791	1,915
2006	19,386	3,272	2,656	2,353	2,946
2007	28,668	14,629	4,419	2,340	5,027

Source: Emerging Private Equity Association 2007.

proceeds to a cluster analysis for a panel of emerging markets; and Section 17.5 brings together our conclusions.

17.2 EMERGING MARKETS CHARACTERISTICS AND PRIVATE EQUITY DEVELOPMENT

The size of the domestic economy is clearly a major determinant for the development of private equity investment whose volumes are usually significantly correlated with GDP growth (Romain and Van Pottelsberghe, 2004). Gompers and Lerner (1998) indeed pointed out that there are more attractive opportunities for entrepreneurs in large and dynamic economies. Although hard to quantify, political risk is another key variable for emerging market investment as it determines the risk premium associated with local projects (Chuhan, 1992). Local financial development is also essential. Black and Gilson (1998) suggested a positive relationship between financial development and private equity investment levels. Gompers and Lerner (1998) also emphasized that risk capital flourishes in countries with deep and liquid stock markets while the maturity of the private equity market itself may also attract foreign investors.

In addition, the overall business environment may also play a significant role in determining private equity investment levels. For instance, Jeng and Wells (2000) found that labor market rigidities, the level of IPOs, entrepreneurship climate, and bankruptcy procedures explained a large part of cross-country variations in private equity activity. Focusing on fiscal factors, Poterba (1989) argued that lower tax rates prompt employees to become entrepreneurs, leading to more demand for private equity funds. This was confirmed by Gompers and Lerner (1998) who found that lower capital gains tax rates have strong effect on the amount of venture capital (VC) investments supplied. Similarly, legal development is an important factor. Cumming et al. (2006) indeed suggested that the quality of a country's legal system has a stronger impact on private equity activity than the size of its stock market while Johnson et al. (1999) emphasized the importance of the protection of property rights for private equity markets.

High levels of human capital are also necessary for the development of private equity markets. Schertler (2003) emphasized that the number of employees in the R&D field and the number of patents have a positive impact on the development of private equity activity. Along the same lines, Farag et al. (2004) highlighted that the quality of management ranks as a primary reason for private equity investment failure in Central Europe. Finally, social environment may also have a role to play.

For instance, Baughn and Neupert (2003) argued that national attitudes toward entrepreneurial activity determine the development of a local risk-capital culture and affect the set of investment opportunities for international investors.

Based on this literature review, we classify institutional underpinnings of private equity market development into eight criteria: (1) economic activity, (2) business opportunities, (3) favorable taxation environment, (4) political stability, (5) capital market development, (6) human capital, (7) legal investor protection, and (8) social environment.

17.3 LOCAL INVESTOR'S PERCEPTIONS

17.3.1 Data

In an effort to measure investor's perceptions of the MENA private equity markets, the following questions were asked to a panel of private equity investors:

1. How do you regard the attractiveness of the following emerging markets for private equity investors (7 = excellent, 1 = poor)?

2. How important is each of the attractiveness criteria in your decision to invest in a given country in general (7 = very important, 1 = not important at all)?

3. How attractive do you consider the MENA region according to the same criteria (7 = very attractive, 1 = not attractive at all)?

4. How attractive are the MENA countries for you as an investor (7 = very attractive, 1 = not attractive at all)?

This short questionnaire was first sent by e-mail to 1500 private equity investors worldwide, using e-mail contacts from European Venture Capital Association, Gulf Venture Capital Association, and the African Venture Capital Association Web sites. This approach yielded 13 responses. MENA markets private equity investors were directly interviewed by phone, yielding 22 responses.* Finally, the questionnaire was distributed to investors participating in the second EUROMED Capital Forum held in Tunis, April 24–25, 2008, yielding 25 additional responses. In total, we obtained

---

* Thanks to Raphaël Botiveau from the ANIMA Investment Network (http://www.animaweb. org) for gracefully providing data.

60 responses, a reasonable sample for our exploratory purposes. By comparison, Groh et al. (2008) considered a sample of 75 responses in a study focusing on Eastern Europe.

17.3.2 Results

As shown in Table 17.2, respondents are quite optimistic about the region's ability to attract further private equity investment. Taken as a whole, the MENA region (5.09) is indeed ranked first among emerging countries, ahead of Asia (4.64), Central Europe (4.63), sub-Saharan Africa (4.45),

TABLE 17.2   Investors' Perceptions

	No. Obs.	Mean	SD	Min	Max
Question 1: How do you regard the attractiveness of the following emerging markets for private equity investors?					
MEDA	55.00	5.09	1.54	1.00	7.00
Asia	53.00	4.64	1.82	1.00	7.00
Central Europe	51.00	4.63	1.60	1.00	7.00
Sub-Saharan Africa	53.00	4.45	1.38	2.00	7.00
Latin America	48.00	4.13	1.47	1.00	7.00
Question 2: How attractive are the following MEDA countries for you?					
Morocco	57.00	5.18	1.68	1.00	7.00
Tunisia	57.00	5.05	1.62	1.00	7.00
Turkey	53.00	4.98	1.69	2.00	7.00
Egypt	55.00	4.89	1.51	2.00	7.00
Jordan	55.00	4.51	1.14	1.00	7.00
Algeria	57.00	4.25	1.89	1.00	7.00
Israel	49.00	4.00	2.00	1.00	7.00
Libya	55.00	3.85	1.94	1.00	7.00
Lebanon	55.00	3.82	1.59	1.00	7.00
Syria	54.00	3.65	1.75	1.00	7.00
Question 3: How important are the following criteria in your decision to invest in a given country in general?					
Business opportunities	56.00	5.39	2.02	1.00	7.00
Investor protection	56.00	5.18	2.22	1.00	7.00
Political risk	53.00	5.17	2.04	1.00	7.00
Human capital	55.00	5.13	1.72	1.00	7.00
Economic activity	57.00	5.02	1.88	1.00	7.00
Capital market development	55.00	4.87	1.49	2.00	7.00
Taxation	56.00	4.79	1.82	1.00	7.00
Social environment	54.00	4.63	1.42	2.00	7.00

(continued)

TABLE 17.2 (continued)   Investors' Perceptions

Question 4: How attractive do you consider the MEDA region according to the same criteria?				
Business opportunities	54.00	5.33	1.78	1.00	7.00
Economic activity	54.00	5.30	1.72	1.00	7.00
Political risk	49.00	5.08	1.59	1.00	7.00
Taxation	54.00	4.80	1.50	1.00	7.00
Investor protection	54.00	4.70	1.56	1.00	7.00
Human capital	51.00	4.69	1.44	1.00	7.00
Capital market development	54.00	4.61	1.42	1.00	7.00
Social environment	53.00	4.45	1.50	1.00	7.00

and Latin America (4.13). Investors' perceptions on the region as a whole appear relatively homogeneous as standard deviation (1.54) is third lowest, behind Sub-Saharan Africa (1.38) and Latin America (1.47).

Turning to an intraregional assessment of private equity attractiveness, investors ranked Morocco first (5.18), followed by Tunisia (5.05), Turkey (1.69), Egypt (4.89), and Jordan (4.51). Finally, Algeria (4.25), Israel (4.00), Libya (3.85), Lebanon (3.82), and Syria (3.65) constitute a third group of countries. The low ranking of Israel is somewhat surprising given that this country has one of the world's most developed private equity markets. However, our respondents were all based in the MENA region, whereas the Israeli private equity market is relying on national and global investors (especially U.S. investors). The low ranking of Israel might, thus, reflect a low intraregional economic integration. It may also reflect negative local perceptions due to the persistence of the Middle East conflict.

Turning to attractiveness criteria, investors seem to adopt a holistic approach to country assessment as all criteria obtain average scores higher than 4 from business opportunities (5.33) to social environment (4.63). Investor protection (5.18) and political stability (5.17) obtain very close scores, suggesting that these are deeply connected in this region. These are followed by human capital (5.13), economic activity (5.02), capital market development (4.87), and other important factors for private equity development. Taxation (4.79) and social environment (4.63) rank at the bottom. Overall, this suggests that labor costs are not as important as economic opportunities and legal guarantees in the allocation of international private equity investment flows.

Interestingly, the MENA countries' attractiveness for private equity stems mostly from business opportunities (5.33), economic activity (5.30), and political stability (5.08). This may result from a strong policy commitment to economic reforms, which has resulted in a significant privatization program and a relatively high rate of capital accumulation; most MENA countries are experiencing economic growth rates in excess of 4%. However, areas of improvements can be identified in human capital (4.69), capital market development (4.61), and social environment (4.45).

17.4 ATTRACTIVENESS INDICES

17.4.1 Data

Our dataset covers the 53 countries classified as either "emerging" or "frontier" markets by the Standard & Poor's rating agency. We gather data from the CEPII's 2006 *Institutional Profile* (IP) database and the World Bank's 2006 *World Development Indicators* (WDI) database. The IP database is developed by means of a questionnaire addressed by French embassies in 86 countries and offers a very comprehensive analysis of international institutional arrangements. The WDI database offers key economic variables as well as a set of institutional ratings developed by benchmark agencies. We consider the cross-section of these databases and identify a set of variables reflecting the eight chosen components of private equity attractiveness: (1) economic activity, (2) business opportunities, (3) political stability, (4) capital market development, (5) investor protection, (6) social environment, (7) tax environment, and (8) human capital. Merging these two databases leaves a total of 42 countries in the sample. In many cases, scale, direction, and magnitude of each variable differ. We thus rescaled and normalized raw indices so that variables range from 0 to 1, a higher score indicating higher attractiveness. Selected index components and sources are described in Appendix 17.A.1.

17.4.2 Methodology

We generate a set of synthetic indexes reflecting the criteria described in Table 17.2. These indexes can be described in Table 17.3.

For each index, weights are comprised between 0 and 1 and derived based on a nonparametric bootstrap technique. The process is described next. We first generate 10,000 random combinations of uniformly distributed weights adding up to unity in the interval [0,1]. The corresponding indexes are calculated for each of these combinations, and the selected index value corresponds to the 50th percentile of the associated cumulative distribution. This methodology allows us to derive a significance

TABLE 17.3   Synthetic Indexes Generated

$ECOINDEX_i = \alpha_i ECO1 + \beta_i E\,CO2 + \chi_i ECO3 + \delta_i ECO4 + \phi_i ECO5 + \gamma_i ECO6$

$BUSINDEX_i = \alpha_i BUS1 + \beta_i BUS2 + \chi_i BUS3 + \delta_i BUS4 + \phi_i BUS5$

$POLINDEX_i = \alpha_i POL1 + \beta_i POL2 + \chi_i POL3 + \delta_i POL4 + \phi_i POL5 + \gamma_i POL6 + \mu_i POL7$
$\qquad + \sigma_i POL8 + \rho_i POL9$

$CAPINDEX_i = \alpha_i CAP1 + \beta_i CAP2 + \chi_i CAP3 + \delta_i CAP4 + \phi_i CAP5 + \gamma_i CAP6 + \mu_i CAP7$ $\qquad\qquad$ (17.1)

$TAXINDEX_i = \alpha_i TAX1 + \beta_i TAX2 + \chi_i TAX3$

$INVINDEX_i = \alpha_i INV1 + \beta_i INV2 + \chi_i INV3 + \delta_i INV4 + \phi_i INV5 + \gamma_i INV6 + \mu_i INV7$

$SOCINDEX_i = \alpha_i SOC1 + \beta_i SOC2 + \chi_i SOC3 + \delta_i SOC4 + \phi_i SOC5 + \gamma_i SOC6$

$HUMINDEX_i = \alpha_i HUM1 + \beta_i HUM2 + \chi_i HUM3 + \delta_i HUM4 + \phi_i HUM5 + \gamma_i HUM6$
$\qquad + \mu_i HUM7 + \sigma_i HUM8$

---

level for the index without relying on strong distributional assumptions of investor's preferences.[*]

To refine our understanding of institutional development in the MENA region, we then analyzed those indexes with a hierarchical clusters based on Ward's (1963) linkage. Within this framework, the squared Euclidean distance is used as a measure of dissimilarity. For each cluster, the means for all the variables are computed. Then, for each object, the squared Euclidean distance to the cluster means is calculated. These distances are summed for all the objects. At each stage, the two clusters with the smallest increase in the overall sum of squares within cluster distances are combined. The recurrence formula is the following:

$$d_{k(i,j)} = \frac{\eta_i + \eta_k}{\eta_i + \eta_j + \eta_k} d_{ki} + \frac{\eta_j + \eta_k}{\eta_i + \eta_j + \eta_k} d_{kj} - \frac{\eta_k}{\eta_i + \eta_j + \eta_k} d_{ij} \qquad (17.2)$$

where
   η_i, η_j, and η_k are the numbers of observations contained in groups i, j, and k, respectively
   d_{ij} is the distance between clusters i and j
   $d_{k(i,j)}$ is the distance between cluster k and the new cluster formed by joining clusters i and j

---

[*] The distribution of normalized indexes is not shown for space-saving considerations but is available upon request.

The optimal number of clusters is identified based on the pseudo F index (Calinsky and Harabasz, 1974) which is defined as $F = \text{Trace}[B/(k-1)]/\text{Trace}[W/(n-k)]$ where n is the number of observations in a sample, K is the number of clusters, B is the distance between cluster sum of squares and cross product matrix, and W is the pooled within cluster sum of squares and cross products matrix. Using this method, the optimal number of clusters is determined by plotting the F index against the number of clusters. An inspection of the repartition of clusters across the scatterplot matrix provides insight into their respective characteristics.

17.4.3 Results

Country positions are shown in Figures 17.1 through 17.8 that highlight that the MENA region is very heterogeneous, countries being scattered evenly across the emerging markets universe. As shown in Figure 17.1, levels of economic activity are extremely variable in the MENA region. Israel (3.38), the first MENA country, comes third in the entire sample, after China (3.40) and South Korea (3.52). It is followed by Turkey (3.24) which may be compared to Chile (3.25). Next are Algeria (3.16), which can be compared to India (3.15), and Tunisia (3.04), Egypt (3.08), and Lebanon (3.02), which are close to the Philippines (3.03). Finally, Jordan (2.90) and Morocco (3.00) are lagging behind and can be compared to Sri Lanka (2.92).

Business opportunity indices are charted in Figure 17.2. Jordan (0.82) and Israel (0.82) come first in the entire sample, hence confirming investor's claim that the MENA region is the most attractive among emerging markets. Morocco (0.65) comes third and can be compared to Brazil (0.66) and Poland (0.65). Then come Turkey (0.59) and Egypt (0.57) which are ranked ahead of Bulgaria (0.56) and Chile (0.56). Tunisia (0.45) and Algeria (0.41) are significantly lower in our ranking and can be compared to Mexico (0.45) and Malaysia (0.42). Finally, Lebanon (0.32) and Syria (0.27) seem to offer the least attractive business opportunities in the MENA region and can be compared to Argentina (0.31) and Botswana (0.27).

Inspection of Figure 17.3 suggests that the MENA region does not compare favorably with other emerging market areas in terms of political stability. The MENA countries are indeed located in the lower segment of the figure. Israel (0.67), nevertheless, comes first in the MENA region and can be compared to Ukraine (0.68). It is followed by Jordan (0.58), Tunisia (0.55), and Turkey (0.54) which can be compared to India (0.58), Botswana (0.57), and Peru (0.53), respectively. Morocco (0.51) and Algeria

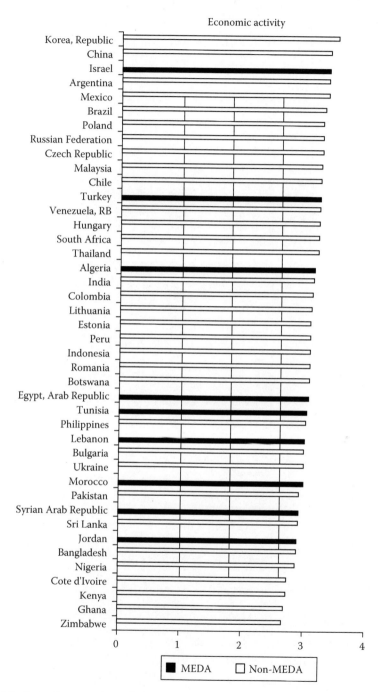

FIGURE 17.1 Economic activity.

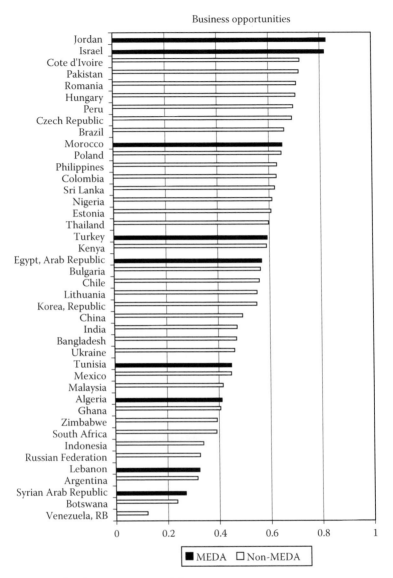

FIGURE 17.2   Business opportunities.

(0.46) can be compared to Bangladesh (0.48) and Indonesia (0.47), respectively. Finally, Lebanon (0.39), Egypt (0.35), and Syria (0.34) are lagging behind and can be compared to Venezuela (0.38) and Zimbabwe (0.32), respectively.

Turning to capital market development, Figure 17.4 shows that Israel (0.82) comes first in the entire sample. Within the MENA region, it is

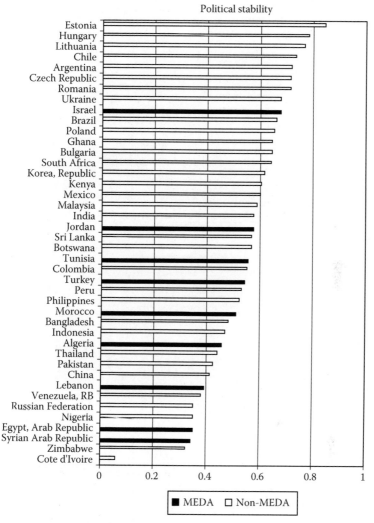

FIGURE 17.3   Political stability.

followed by Turkey (0.58). Then come Lebanon (0.56) and Tunisia (0.55) which can be compared to Indonesia (0.54). Egypt (0.49) and Morocco (0.47) follow and can be compared to Thailand (0.48) and Peru (0.47), respectively. Jordan (0.37), Algeria (0.35), and Syria (0.20) are lagging at the lower end of the ranking.

Taxation environment is described in Figure 17.5, which highlights that Israel (0.77) is the most fiscally competitive country in the entire sample, well ahead of Czech Republic (0.67). Morocco (0.46) and Jordan (0.44)

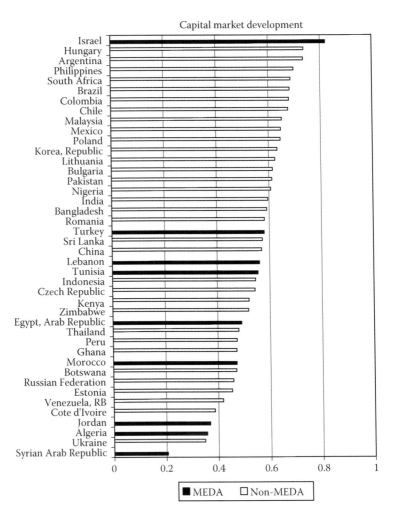

FIGURE 17.4   Capital market development.

seem to constitute an intermediate group that is comparable to Argentina (0.45). By contrast, Algeria (0.38), Tunisia (0.34), Syria (0.33), Lebanon (0.26), Egypt (0.21), and Turkey (0.14) are at the lower end of the ranking and compare unfavorably with other emerging markets.

Social environment constitutes an interesting case. As shown in Figure 17.6, Israel (0.72) clearly outperforms most emerging markets as it is ranked just behind the Czech Republic (0.73). Algeria (0.56), Morocco (0.56), Lebanon (0.53), and Tunisia (0.53), which are all civil law countries, constitute a very homogeneous group and are ranked just above Indonesia (0.50) and Thailand (0.50). Syria (0.48), Jordan (0.48), and Turkey (0.44) follow, while Egypt (0.34) is lagging behind.

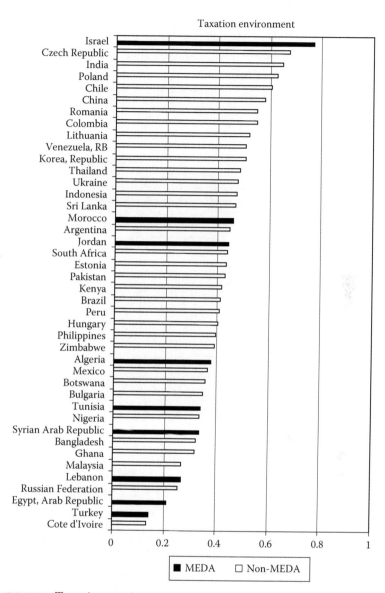

Taxation environment

FIGURE 17.5 Taxation environment.

Inspection of Figure 17.7 highlights that the highest level of investor protection can be observed in Israel (0.87), which is ranked just behind Chile (0.88). Jordan (0.65) comes second in the MENA region and can be compared to Korea (0.65). Tunisia (0.61), Morocco (0.56), Turkey (0.54), and Algeria (0.54) constitute a relatively homogeneous group while Lebanon (0.50), Egypt (0.40), and Syria (0.38) appear to be the region's least investor-friendly countries.

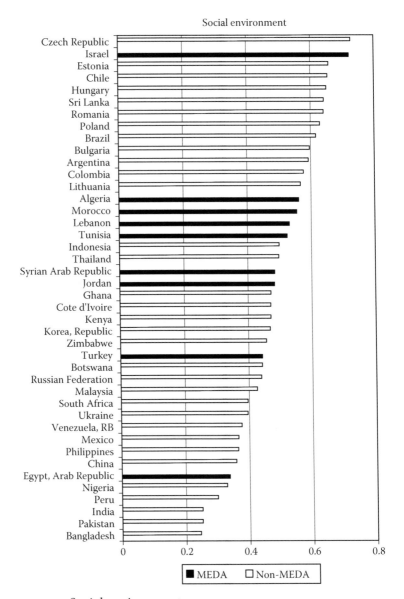

FIGURE 17.6 Social environment.

Finally, human capital levels are also very heterogeneous. As shown in Figure 17.8, Israel (0.85) offers the most educated workforce in the entire sample, ahead of Estonia (0.81). It is followed by Lebanon (0.58) and Jordan (0.57) which have attained levels similar to Brazil (0.58) and Sri Lanka (0.55), respectively. Turkey (0.53) and Tunisia (0.51) are close to

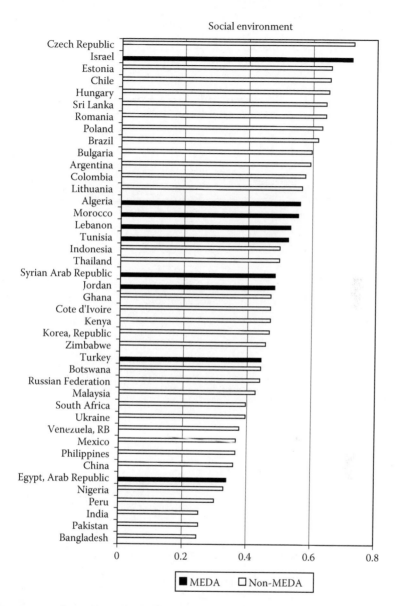

FIGURE 17.7   Investor protection.

one another in ranking while Algeria (0.42), Syria (0.39), Morocco (0.28), and Egypt (0.28) are ranked at the bottom level. Overall, this analysis suggests that Israel is one of the most attractive emerging private equity markets, an observation in line with high private equity activity in this country. It also highlights the existence of a real success story in the MENA region.

Human capital

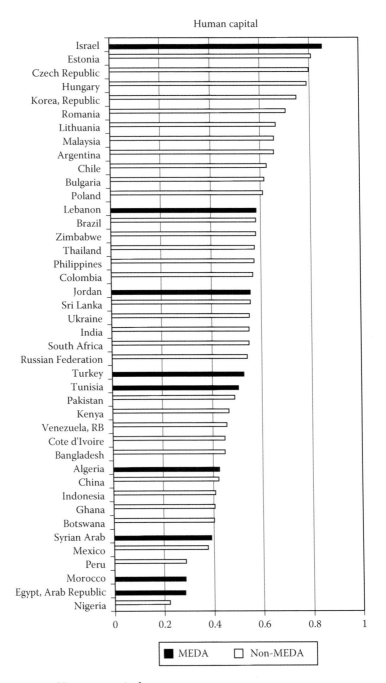

FIGURE 17.8 Human capital.

As shown in Table 17.4, most indices are positively and significantly correlated. This suggests that institutional development reforms in the MENA region should be coordinated. Interestingly, although significantly correlated with investor protection, the business opportunities index does not appear directly related to other components of private equity market attractiveness, suggesting that improving institutional climate requires a specific reform program.

Comparing objective attractiveness levels with investor perception constitutes an interesting question. To this end, we calculate a composite attractiveness index. The latter is defined as a weighted average of our eight attractiveness indices. Such weights are determined by average score of question 3 "How important are the following criteria in your decision to invest in a given country in general?" and hence directly reflect investor's preferences. Results are shown in Table 17.5.

Looking first at regional averages for each index, the comparative advantages of the MENA region seem to be business opportunities and social environment as the region is ranked second after Central Europe in both criteria. Economic activity, political stability, and taxation environment constitute areas of improvements as the MENA region is ranked behind Asia, Central Europe, and Latin America in each criterion, respectively. Finally, the region's weakest points are human capital, investor protection, and capital market development. The MENA region is indeed ranked last in each of these criteria.

Interestingly, there appears to be a significant gap between local investors' perception and the region's attractiveness; investors optimistically perceive the MENA region as the most attractive of emerging market areas. However, within our composite index, this region is ranked behind Central Europe, Latin America, and Asia. This suggests that MENA private equity markets benefit from a home bias, which could be attributed to geographical and cultural proximity from the Euro-Mediterranean area and the Gulf countries.

Turning to country level ranking, investors converge with the composite indices in the case of Turkey (ranked third in both), Algeria (ranked sixth in both), Syria (ranked last in both), and Lebanon (ranked eighth by investors and ninth in our composite index). Investors may be overly optimistic in the case of Morocco (ranked first by investors and fifth in our index), Tunisia (ranked second by investors and fourth in the composite index), and Egypt (ranked fourth by investors and eighth in the composite index). By contrast, investors may be overly pessimistic in the case of

TABLE 17.4 Index Correlation Matrix

	ECOINDEX	SOCINDEX	BUSINDEX	TAXINDEX	POLINDEX	CAPINDEX	HUMINDEX	INVINDEX
ECOINDEX	1.0000							
SOCINDEX	0.2236	1.0000						
	(0.1546)							
BUSINDEX	−0.0398	0.2785	1.0000					
	(0.8023)	(0.0741)						
TAXINDEX	0.3947**	0.3666	0.2487	1.0000				
	(0.0097)	(0.0170)	(0.1122)					
POLINDEX	0.3766**	0.5053**	0.2247	0.5020**	1.0000			
	(0.0140)	(0.0006)	(0.1525)	(0.0007)				
CAPINDEX	0.4539**	0.2063	0.2825	0.3389*	0.4547**	1.0000		
	(0.0025)	(0.1900)	(0.0699)	(0.0281)	(0.0025)			
HUMINDEX	0.3976**	0.6323**	0.2708	0.4505**	0.6074**	0.4473**	1.0000	
	(0.0091)	(0.0000)	(0.0829)	(0.0028)	(0.0000)	(0.0030)		
INVINDEX	0.0849	0.5267**	0.4681**	0.4174**	0.6213**	0.5492**	0.5008**	1.0000
	(0.5928)	(0.0003)	(0.0018)	(0.0060)	(0.0000)	(0.0002)	(0.0007)	

Note: p-Values are in brackets.
* and ** indicate significance at the 5% and 1% level, respectively.

TABLE 17.5 Indices and Questionnaires

	Perception	Composite	ECO	SOC	BUS	TAX	POL	CAP	HUM	INV
MENA	5.09 (1)	1.517 (4)	3.08	0.52	0.55	0.37	0.49	0.49	0.49	0.56
Asia	4.64 (2)	1.533 (3)	3.14	0.40	0.53	0.46	0.51	0.59	0.54	0.61
Central Europe	4.63 (3)	1.568 (1)	3.15	0.59	0.58	0.48	0.68	0.56	0.68	0.63
Sub-Saharan Africa	4.45 (4)	1.491 (5)	2.81	0.45	0.50	0.34	0.45	0.52	0.49	0.59
Latin America	4.13 (5)	1.550 (2)	3.25	0.50	0.49	0.47	0.60	0.62	0.51	0.64
Morocco	5.18 (1)	1.514 (5)	3.00	0.56	0.65	0.46	0.51	0.47	0.29	0.56
Tunisia	5.05 (2)	1.520 (4)	3.05	0.52	0.45	0.34	0.55	0.55	0.51	0.61
Turkey	4.98 (3)	1.522 (3)	3.24	0.44	0.59	0.14	0.54	0.58	0.53	0.54
Egypt	4.89 (4)	1.459 (8)	3.08	0.34	0.57	0.21	0.35	0.49	0.28	0.40
Jordan	4.51 (5)	1.536 (2)	2.90	0.48	0.82	0.44	0.58	0.37	0.56	0.65
Algeria	4.25 (6)	1.499 (6)	3.16	0.56	0.41	0.38	0.46	0.36	0.43	0.54
Israel	4 (7)	1.651 (1)	3.39	0.72	0.82	0.77	0.68	0.82	0.85	0.87
Lebanon	3.82 (8)	1.490 (7)	3.02	0.53	0.32	0.26	0.39	0.56	0.58	0.49
Syria	3.65 (9)	1.427 (9)	2.93	0.48	0.27	0.33	0.34	0.20	0.39	0.37

Note: This table shows regional averages and country scores for investor perception, the composite index (using a logarithmic scale), and each of the bootstrapped indices. Numbers in brackets denote region and country ranks.

Jordan (ranked fifth by investors and second in the composite index) and Israel (ranked seventh by investors and first in the composite index). This suggests that private equity investment decisions do not reflect institutional development levels in the region and may be affected by psychological factors.

The cluster analysis allows us to further analyze the attractiveness of the MENA private equity markets. As shown in Table 17.6, the MENA countries are scattered in four different clusters. Egypt, Lebanon, Syria, and Algeria belong to cluster A. In spite of relatively favorable taxation levels (taxation index is the second highest), this cluster seems to gather the least attractive emerging private equity markets. This cluster is the second lowest in terms of human capital, social environment, and economic activity. It also gathers countries with the lowest investor protection, political stability, and capital market development, indicating areas of improvement for these four countries.

Morocco and Jordan belong to cluster B that displays relatively good performance in business opportunities and investor protection and echoes previous results. However, these countries perform relatively poorly in terms of political stability, capital market development, and taxation and also have the lowest human capital, economic activity, and social environment scores. Considering that business opportunities and investor

TABLE 17.6   Cluster Analysis

Cluster A	Cluster B	Cluster C	Cluster D
India	Kenya	Romania	Argentina
China	Bangladesh	Tunisia	South Africa
Syrian Arab Republic	Jordan	Hungary	Korea, Republic
Indonesia	Cote d'Ivoire	Israel	Turkey
Russia	Nigeria	Czech Republic	Malaysia
Egypt	Ghana	Chile	Mexico
Lebanon	Peru	Estonia	
Botswana	Zimbabwe	Lithuania	
Ukraine	Pakistan	Bulgaria	
Algeria	Philippines	Thailand	
Venezuela	Morocco	Sri Lanka	
		Brazil	
		Poland	
		Colombia	

protection are necessary but not sufficient conditions for the development of a private equity market, this suggests areas of improvements.

Tunisia and Israel belong to cluster C that gathers mostly Central European markets and is ranked first in terms of political stability, business opportunities, taxation environment, investor protection, and social environment. It comes second in terms of economic activity and capital market development. These two countries thus seem the most attractive private equity markets of the MENA region.

Finally, Turkey belongs to cluster D, gathering advanced emerging markets with the highest economic activity and capital market development (Table 17.6). This cluster comes second in terms of human capital, social environment, and political stability (Table 17.7). Areas of improvements include business opportunities, investor protection, and taxation environment. For Turkey, one priority should be to improve the overall business climate.

17.5 CONCLUSIONS

The objective of this chapter was to conduct an exploratory analysis of private equity market development in the MENA region. We focused on a set of eight criteria, namely, (1) economic activity, (2) business opportunities (3) political stability, (4) capital market development, (5) investor protection, (6) tax environment, (7) social environment, and (8) human capital, and proceeded to a set of international comparisons. In doing so, we analyzed answers obtained from a questionnaire and developed a set of comprehensive attractiveness indices. Considering the intersection of these analyses permits a set of conjectures.

First, as shown in the questionnaire, local investors rank the MENA region ahead of other emerging market areas. In addition, we observe a gap between investor's perceptions and a set of quantitative attractiveness indices. Our indices indeed suggest that taken as a whole, the MENA region is, in fact, less attractive than Central Europe, Asia, or Latin America. These positive investors' perceptions could be interpreted as evidence of a Mediterranean "home bias." In addition, investors appear overly optimistic in the case of Morocco, Tunisia, and Egypt and overly pessimistic in the case of Jordan and Israel. This suggests a possible interference of psychological factors in the allocation of private equity investment to this region.

A cluster analysis also revealed that the MENA countries may be divided into three main groups. Israel, Tunisia, and Turkey seem to be converging toward the most attractive emerging private equity markets. Morocco and

TABLE 17.7 Descriptive Statistics

Cluster	ECOINDEX	POLINDEX	BUSINDEX	CAPINDEX	TAXINDEX	INVINDEX	SOCINDEX	HUMINDEX
A	11	11	11	11	11	11	11	11
	3.13	0.45	0.37	0.46	0.41	0.46	0.43	0.46
	0.14	0.11	0.13	0.12	0.14	0.10	0.09	0.09
B	11	11	11	11	11	11	11	11
	2.87	0.46	0.61	0.52	0.37	0.61	0.40	0.43
	0.15	0.17	0.14	0.10	0.09	0.06	0.11	0.12
C	14	14	14	14	14	14	14	14
	3.17	0.66	0.63	0.62	0.51	0.72	0.62	0.66
	0.13	0.11	0.09	0.10	0.12	0.11	0.07	0.11
D	6	6	6	6	6	6	6	6
	3.33	0.62	0.45	0.66	0.36	0.61	0.45	0.58
	0.11	0.06	0.10	0.05	0.14	0.06	0.08	0.13

Note: This table shows descriptive statistics for the four clusters described in Table 17.4. In each cell, the first row displays the number of observations, the second row shows the cluster average, and the third row displays standard deviation.

Jordan display strong business and investor protection but have low human capital, economic activity, and social environment scores. Finally, Egypt, Lebanon, Syria, and Algeria belong to the least attractive segment of emerging private equity markets. From a policy point of view, this highlights potential improvement areas for each country. Interestingly, our attractiveness indices are significantly correlated, suggesting that policy reforms should be coordinated if the region is to attract higher investment levels.

It should be noted, however, that the developmental effects of private equity flows depend upon their sectoral allocation. For a positive impact to be reached, these flows must be channeled toward the riskiest and most innovative segment of investment projects rather than traditional sectors (such as real estate, textile, and manufacturing). In the latter case, the high returns demanded by private equity investors could drain productive capital unnecessarily, especially if profits are repatriated abroad or used as collaterals to borrow from local bank and invest internationally. Thus, more research is needed in order to disentangle the nature and impact of private equity investment on economic growth in this region. In this context, recent institutional developments regarding the *Union pour la Méditerranée* could provide new avenues for research.

APPENDIX 17.A.1 Institutional Data

INDEX 1: Economic activity	Source
ECO1: 2006 log GDP (constant 2000 U.S.$)	WDI database
ECO2: 2006 log GDP per capita (constant 2000 U.S.$)	WDI database
ECO3: 2006 log GNI per capita, PPP (current international $)	WDI database
ECO4: 2006 Gross fixed capital formation (% of GDP)	WDI database
ECO5: 2006 Gross domestic savings (% of GDP)	WDI database
ECO6: 2006 GDP growth	WDI database
INDEX 2: Business opportunities	
BUS1: Price liberalization	CEPII database
BUS2: Reforms and privatization of nonfinancial institutions between 2001 and 2006	CEPII database
BUS3: Implementation of privatization program	CEPII database
BUS4: Openness of privatization program	CEPII database
BUS5: Weight of institutional shareholders	CEPII database
INDEX 3: Political stability	
POL1: Political rights and functioning of political institutions	CEPII database
POL2: Change in political rights over the last 3 years	CEPII database
POL3: Public freedom and civil society development	CEPII database
POL4: Change in public freedoms over the last 3 years	CEPII database

(continued)

APPENDIX 17.A.1 (continued)   Institutional Data

POL5: Internal public security	CEPII database
POL6: External public security	CEPII database
POL7: Change in security levels over the last 3 years	CEPII database
POL8: Corruption	CEPII database
POL9: Performance of judicial system	CEPII database
INDEX 4: Capital market development	
CAP1: Weights of small shareholders	CEPII database
CAP2: Venture capital and innovation	CEPII database
CAP3: Insurance, pension funds	CEPII database
CAP4: Traditional credit systems	CEPII database
CAP5: Disclosure requirement	CEPII database
CAP6: Financial system regulation reforms over the last 3 years	CEPII database
CAP7: Openness to foreign equity and loans	CEPII database
INDEX 5: Taxation environment	**CEPII database**
TAX1: Centralization vs. Fiscal autonomy	CEPII database
TAX2: Fiscal efficiency	CEPII database
TAX3: Fiscal reforms over the last 3 years	WDI database
INDEX 6: Investor protection	
INV1: Enforcement of traditional property rights	CEPII database
INV2: Formal property rights	CEPII database
INV3: Nature of private contracts	CEPII database
INV4: Enforcement of private contracts	CEPII database
INV5: Enforcement of governmental contracts	CEPII database
INV6: Financial information	CEPII database
INV7: Respect of intellectual property	CEPII database
INDEX 7: Social environment	
SOC1: Labor market rigidity	CEPII database
SOC2: Labor market reforms over the last 3 years	CEPII database
SOC3: Informal labor market	CEPII database
SOC4: Protection of workers	CEPII database
SOC5: Labor contract protection	CEPII database
SOC6: Social dialog	CEPII database
INDEX 8: Human capital	
HUM1: Education and health—basic public goods	CEPII database
HUM2: Attitude toward change and innovation	CEPII database
HUM3: Investment for future generations	CEPII database
HUM4: Equity in access to public goods	CEPII database
HUM5: Training of elite	CEPII database
HUM6: Diffusion of innovation	CEPII database
HUM7: Adult professional training	CEPII database
HUM8: Social mobility	CEPII database

REFERENCES

Abed, G.T. and Soueid, M.M. (2005) Capital markets in the Middle East and North Africa. Working Paper, European Investment Bank—FEMIP Experts Committee, Luxembourg.

ANIMA (2008) Med funds: Panorama du Capital-Investissement dans la Région MEDA. ANIMA Investment Network, Notes & Documents 26, Marseilles.

Baughn, C.C. and Neupert, K.E. (2003) Culture and national conditions facilitating entrepreneurial start-ups. *Journal of International Entrepreneurship*, 1(3): 313–330.

Black, B.S. and Gilson, R.J. (1998) Venture capital and the structure of capital markets: Banks versus stock markets. *Journal of Financial Economics*, 47(2): 243–277.

Bonini, S. and Alkan, S. (2006) The macro and political determinants of venture capital investments around the world. Working Paper, University of Bocconi, Bocconi. Electronic copy available at: http://ssrn.com/abstract=945312.

Calinsky, T. and Harabasz, J. (1974) A dendrite method for cluster analysis. *Communications in Statistics*, 3(1): 1–27.

Chuhan, P. (1992) Sources of portfolio investment in emerging markets. Working Paper, World Bank, International Economics Department, Washington, D.C.

Cumming, D.J. (2006) The determinants of venture capital portfolio size: Empirical evidence. *Journal of Business*, 79(3): 1083–1126.

Farag, H., Hommel, U., Witt, P., and Wright, M. (2004) Contracting, monitoring, and exiting venture investments in transitioning economies: A comparative analysis of Eastern European and German markets. *Venture Capital*, 6(4): 257–282.

FEMISE (2006) Annual Report on the Euro-Mediterranean partnership. European Commission, Marseilles, France.

Gompers, P. and Lerner, J. (1998) What drives venture fundraising? *Brookings Papers on Economic Activity: Microeconomics*, July: 149–192.

Gorman, M. and Sahlman, W.A. (1989) What do venture capitalists do? *Journal of Business Venturing*, 4(4): 231–248.

Groh, A., Liechtenstein, H., and Canela, M.A. (2008) Limited partners' perceptions of the Central Eastern European venture capital and private equity market Working Paper IESE Business School No. 727, Barcelona, Spain.

Hellmann, T. and Puri, M. (2000) The interaction between product market and financing strategy: The role of venture capital. *Review of Financial Studies*, 13(4): 959–984.

Jeng, L.A. and Wells, P.H.C. (2000) The determinants of venture capital funding: Evidence across countries. *Journal of Corporate Finance*, 6(3): 241–289.

Johnson, S.H., McMillan, J., and Woodruff, C.M. (1999) Property rights, finance and entrepreneurship. CESifo Working Paper Series No. 212, Munich, Germany. Available at SSRN: http://ssrn.com/abstract=198409.

Lagoarde-Segot, T. and Lucey, B. (2008) The emerging MENA equity markets. Situation and characteristics. *Emerging Markets Finance and Trade* 44(5): 68–81.

OECD (2006) Challenges for reform in financial markets in MENA countries. Working Group 4 Discussion Paper, Paris, France.

Poterba, J.M. (1989) Venture capital and capital gains taxation. Working Paper No. 2832, NBER, Cambridge, MA.

Romain, A. and Van Pottelsberghe, B. (2004) The economic impact of venture capital. Working Papers CEB 04-014.RS, Université Libre de Bruxelles, Solvay Business School, Centre Emile Bernheim (CEB), Bruxelles, Belgium.

Schertler, A. (2003) Driving forces of venture capital investments in Europe: A dynamic panel data analysis. European Integration, Financial Systems and Corporate Performance (EIFC). United Nations University, Working Paper 03-27, Tokyo, Japan.

Ward, J.H. (1963) Hierarchical grouping to optimize an objective function. *Journal of the American Statistical Association*, 58(301): 236–244.

CHAPTER 18

Examining the Implications of Linear and Nonlinear Dependencies on Efficiency and Conditional Volatility of MENA Markets: The Case of Egypt and Tunisia

Imed Chkir, Lamia Chourou, and Samir Saadi

CONTENTS

18.1 INTRODUCTION

Random walk hypothesis (RWH) has been widely used in empirical finance literature to test the weak-form efficiency market hypothesis (EMH). Fama (1970) argues that if a market is weak-form efficient, the historical information of past prices cannot be used to generate regular abnormal positive returns. The behavior of stock prices should therefore follow a random walk. Campbell et al. (1997) define three versions of the random walk model: Random Walk 1, Random Walk 2, and Random Walk 3. The most restrictive one is the Random Walk 1, which states that asset returns should be independent and identically distributed (*i.i.d*) increments. If the assumption of *i.i.d* increments is relaxed and independent but nonidentical distributed increments are assumed instead, we obtain the Random Walk version 2. The least restrictive version, Random Walk 3, is achieved by dropping the two assumptions of the Random Walk 1, where the only restriction on asset price changes is that they must be serially uncorrelated. It is clear that version 1, although it does not allow linear correlation in returns series, overlooks the possibility of nonlinear dependence that could be inherent in financial time series.

Most of the empirical studies that have examined the weak-form EMH have investigated whether stock price changes follow a Random Walk 3. Given the assumption underlying this version, they have employed statistical techniques aimed at detecting linear structure in time series, such as the autocorrelation test of Box and Pierce (1970). However, as emphasized by Granger and Anderson (1978), absence of linear dependence does not necessarily imply independency, but merely a lack of linear autocorrelation. Furthermore, several empirical studies have applied tests, which were originally designed for fields such as physics, capable of detecting linear as well as nonlinear patterns in data, and report evidence of nonlinear structure in economics and financial data (see, e.g., Hsieh, 1989; Scheinkman and LeBaron, 1989).

Financial literature proposes several reasons to explain the nonlinear dynamics in asset price changes. For instance, Campbell et al. (1997), among others, argue that market imperfections and some features of market microstructure may lead to a delay of response to new information, implying nonlinearity in share price changes. Given thin trading and the imperfections in emerging stock markets, nonlinear patterns seem to be even more evident in these markets compared to developed ones. Therefore, to investigate the RWH for emerging markets, we need to employ tests that are able to detect both linear and nonlinear dependencies in financial time series.

This chapter examines two of the most active stock markets in the MENA region: the Egyptian and Tunisian markets. Testing the RWH for these equities markets has important implications for asset pricing modeling, especially for traders and practitioners who are searching for patterns in prices and betting on the emerging markets using these patterns. For more than a decade, these two countries have engaged in extensive economic reforms in the form of structural adjustment programs aimed to increase their degree of financial liberalization and integration. In order to attract foreign capital flows, these reforms aimed at enhancing means of trading and ensuring sound information disclosure mechanism.

After the Asian crisis in the late 1990s, international investors seeking portfolio diversification and higher returns began to express interest in nontraditional capital markets located in the MENA region. Their steady economic growth and their political stability have allowed the Egyptian and Tunisian markets to gain special attention from international economic and financial institutions during the last few years. For instance, in 2004, the World Economic Forum ranked Tunisia 2nd for competitiveness in the African continent and 34th worldwide. Given their increasing weight in the world market, this chapter aims to examine the characteristics and dynamics of the Egyptian and Tunisian stock returns. Specifically, we examine the RWH for these markets by testing for linear and nonlinear dependence in daily returns of their major market indices: Cairo and Alexandria Stock Exchanges (CASE 30) and Bourse des Valeurs Mobilières de Tunis (BVMT). Clearly, investigating the Egyptian and Tunisian stock markets will enhance the financial literature of linear and nonlinear dynamics in stock markets, especially emerging ones where there is still a tremendous lack of research at both the empirical and theoretical levels.

The remaining of this chapter is organized as follows: Section 18.2 describes the motivations for examining nonlinear dependency in financial time series, particularly within emerging stock markets; Section 18.3 describes the properties of the dataset; Section 18.4 describes the issues to be examined and methodology for analyzing the data generating process, using various linear and nonlinear tests; Section 18.5 presents the empirical results; and Section 18.6 concludes this chapter.

18.2 IMPLICATION OF NONLINEARITY DYNAMICS FOR THE FINANCIAL FIELD

Examining nonlinearity in financial time series and in particular in the Egyptian and Tunisian stock markets raises several questions. Why examine nonlinear dependency at the first place? What are the implications through the field of finance for finding evidence of nonlinearity? Why do we examine emerging stock markets?

The importance of examining nonlinear dynamics in financial time series is better appreciated through its implications for the field of finance at the theoretical and empirical levels. Evidence of nonlinear dependence has very important implications for academicians and practitioners. For academicians, the existence of nonlinearity in financial series casts serious doubts on the statistical adequacy of statistical models of asset pricing that implicitly take a linear form as well as empirical tests of the weak-form market efficiency, tests of causality, tests of stationarity, and tests of cointegration. For practitioners, evidence of nonlinear dependency directly affects the widely debated issue of predictability of financial time series, which has been examined mainly through a linear approach. Moreover, nonlinear models have important implications for portfolio management techniques, hedging and pricing of derivatives (such as volatility index), and allow superior out-of-sample forecasts of financial series. Furthermore, since the test for nonlinearity is also a test for adequacy of employing linear models, evidence of nonlinearity show that it is inappropriate to use linear methods. For instance, Bera et al. (1993) question the ability of the ordinary least square model in estimating the optimal hedge ratio using future contracts and find that, compared to autoregressive conditional heteroskedasticity (ARCH) hedge ratios, the conventional model leads to too many or too few short-selling of future contracts. Another aspect that this chapter deals with is to gauge the level of complexity in the structure of nonlinearity in emerging markets compared to developed ones. For example, several studies show that generalized autoregressive conditional heteroskedasticity [GARCH(1,1)] works very well in developed stock markets, but how well does a GARCH(1,1) perform in emerging markets?

Evidence of nonlinearity in developing markets would boost our understanding of the linkages between assets markets in these markets. Using sophisticated analyses that are able to uncover and model the complexity of true data generating process within developing markets would enhance our knowledge regarding the level of interaction or independence between those markets which enhance asset allocation, portfolio diversification, and predictability. Furthermore, evidence of nonlinearity allow investors and policy makers to anticipate better and measure the effect of spillover and volatility transmission between developed and developing economies, especially during global financial crisis.

18.3 DATA DESCRIPTION AND SUMMARY STATISTICS

Choosing the appropriate time interval and data frequency generates great debate in applied financial economics literature; however, it remains an unresolved issue. In fact, empirical study needs a sample size large enough to allow sound empirical results. Taking a large time interval or using ultra-high frequency data can overcome this. However, long time-interval data tend to be nonstationary, especially in emerging markets, where structural changes are quite frequent. Working with high frequency data is not safe either, since ultra-high frequency data capture some artificial dependencies (Hsieh, 1991) which are even more persistent in emerging stock markets due to thin trading. To cope with this issue, this chapter considers daily closing prices of the main indices of the Tunisian Stock Market, the BVMT, and CASE 30 from January 1998 to October 2004. Data are provided from capital market authorities. Note that while BVMT includes all the listed companies on Tunisia Stock Exchange (TSE), CASE 30 includes only the top 30 companies in terms of liquidity and activity in CASE.

Figure 18.1 illustrates the daily market index prices as well as daily returns over the sample period. It is clear that both index returns exhibit volatility clustering, with some structural breaks in the price series. Table 18.1 provides the basic properties of the returns series in question. Daily returns are calculated using the natural logarithm of the index price. While the two indices have almost the same average returns over the time interval, there is a significant difference in terms of departure from normal distribution. Distribution of BVMT returns series seems to be closer to Gaussian distribution than CASE 30 returns.

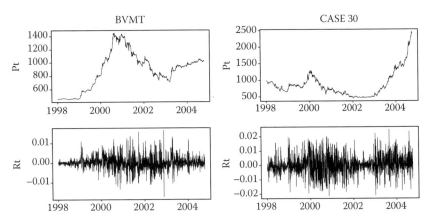

FIGURE 18.1   BVMT and CASE 30 prices and returns for January 1998 to October 2004.

TABLE 18.1   Univariate Summary Statistics for Daily Returns Market Indices

Statistics	CASE 30	BVMT
Mean	0.00020	0.00021
Standard deviation	0.00652	0.00356
Minimum	−0.02135	−0.01742
Maximum	0.02478	0.01643
Skewness	0.15216	0.19921
Kurtosis	0.87929	2.28498
Jarque–Bera	323.4*	47.32*
Number of observation	1691	1695

Note: Daily returns are defined as $r_t = \mathrm{Ln}\, P_t - \mathrm{Ln}\, P_{t-1}$, where P_t is the index price at the end of period t.
* Significant at 5% level.

18.4 METHODOLOGY

18.4.1 Test for Random Walk Hypothesis

The EMH asserts that share prices should always fully reflect all available and relevant information. Successive share price changes are therefore independent and identically distributed. As a result, future shares prices are unpredictable and fluctuate only in response to the random flow of news. Fama (1970) defines three versions of EMH depending on what information set the asset price reflects: weak form, semistrong form, and strong form. In particular, the weak form, the most tested version of EMH, states that stock price reflects all information found from market trading data or historical trends. Therefore, we cannot use technical analysis to predict and beat a market.

To test the weak-form EMH, most empirical studies assume that stock prices follow a random walk. There are numerous ways in which we can state the random walk model, but the most common form is

$$p_t = p_{t-1} + \varepsilon_t, \tag{18.1}$$

where
  p_t is the natural logarithm of a share price P_t at time t
  p_{t-1} is the natural logarithm of a share price P_{t-1} at time $t-1$
  ε_t is an *i.i.d* random variable or strict white noise

As such, successive one-period returns, $r_t \equiv \Delta p_t = \varepsilon_t$, are strict white noise. As expected price changes can be nonzero, to test for RWH, we use the random walk with drift version, defined as follows:

$$p_t = \mu + p_{t-1} + \varepsilon_t. \tag{18.2}$$

Or equivalently

$$r_t = \mu + \varepsilon_t, \tag{18.3}$$

where μ is the expected price change or drift.

For nearly three decades, testing for weak-form EMH has been traditionally associated with the application of standard statistical tests, such as the serial correlation test and run test. However, standard tests may not be appropriate, especially when examining the efficiency of emerging stock markets. First, evidence of linear dependence can be the result of spurious autocorrelation. Spurious autocorrelation may exist due to institutional factors such as nonsynchronous trading, which may generate price-adjustment delays in the trading process. Lo and Mackinlay (1990) argue that evidence of spurious positive autocorrelation can find its way into market index when individual stock prices trade at different frequencies. Several empirical studies (such of Amihud and Mendelson, 1987; McInish and Wood, 1991) report serial autocorrelation in equity markets, which are discernible by high liquidity, reliable information, and sophisticated investors. Therefore, as thin stock markets, the Tunisian and Egyptian markets are expected to exhibit significant autocorrelation coefficients in their index returns.

As evidence of linear relationship in returns series does not necessarily imply market inefficiency, absence of serial correlation need not be proof of market efficiency either. Harvey (1993) states that a nonlinear and identically distributed model can exhibit the properties of a white noise process while being dependent. Brooks (1996) also adds that if we employ standard linear and spectral tests to financial returns time series, the latter may seem to follow a random walk process; however, it is possible to uncover a nonlinear structure if a more sophisticated procedure (such as BDS test) is used. In other words, absence of linear dependence does not necessarily mean independency, but merely a lack of linear autocorrelation (Granger and Anderson, 1978). The bottom line is that the test for EMH should be undertaken using powerful statistical techniques that are capable of detecting both linear and nonlinear dependencies.

18.4.2 Tests for Nonlinear Serial Dependence

The study of nonlinear dynamics and chaos theory has successfully helped describe important natural phenomena in several fields such as physics and chemistry. Low-dimension deterministic nonlinear processes have the ability to mimic random walk behavior and to allow for unpredictable fluctuations of big magnitudes such as those seen in major equity market crashes. As a result, several authors have applied nonlinear analysis to economics and financial data (see, among others, Scheinkman and LeBaron, 1989; Serletis and Dormaar, 1996).

As Campbell et al. (1997, p. 467) explain, "... many aspects of economic behavior may not be linear. Experimental evidence and casual introspections suggest that investors' attitudes toward risk and expected return are nonlinear. The terms of many financial contracts such as options and other derivative securities are nonlinear. In addition, the strategic interactions among market participants, the process by which information incorporates into security prices and the dynamics of economy-wide fluctuation are all inherently nonlinear. Therefore, a natural frontier for financial econometrics is the model of nonlinear phenomena." While current empirical studies show mixed evidence for chaos behavior in stock markets, there is increasing literature reporting nonlinearity dynamics in asset returns series. For instance, while Vaidyanathan and Krehbiel (1992) and Mayfield and Mizrach (1992) find evidence of chaos behavior in the S&P 500 index, Abhyankar et al. (1997) and Serletis and Shintani (2003) reject the null hypothesis of low-dimensional chaos and report evidence of nonlinear dependency in S&P 500 and Dow Jones Industrial Average.

Most of the studies investigating nonlinear dynamics in stock markets focus on developed ones, such as those of the United States, the United Kingdom, Japan, and Germany (see, among others, Brock et al., 1992; Abhyankar et al., 1997; Serletis and Shintani, 2003). Only few studies have examined nonlinearity in emerging markets. Sewell et al. (1993) find evidence of nonlinear structure in weekly indices of Korea and Taiwan. Yadav et al. (1996) also report nonlinearity in daily stock index returns in Hong Kong, Singapore, and Japan markets. Poshakwale (2002) finds evidence of nonlinear behavior in daily returns of Indian stock market. Finally, using both linear and quadratic logistic smooth transition autoregressive (QLSTAR) nonlinear models, McMillan (2005) finds evidence of nonlinear dynamics in a series of four developed markets and two developing ones, namely Malaysia and Singapore, and show that it is due to the presence of noise traders.

By examining linear as well as nonlinear dependencies in the major index of the Egyptian and Tunisian stock markets, this chapter will significantly contribute to the literature of nonlinear dynamics in stock markets and help narrow the gap in literature between emerging and developing markets. To test for nonlinearity in Egyptian and Tunisian stock markets we will use three techniques: the McLeod–Li test, the Engle test, and the BDS test. It is important to note that the BDS test statistic is sensitive to the choice of the embedding dimension m and the bound ε. As mentioned by Scheinkman and LeBaron (1989), if we attribute a value that is too small for ε, the null hypothesis of a random $i.i.d$ process will be accepted too often irrespective of it being true or false. Also, it is not safe to choose very large values for ε. To deal with this problem, Brock et al. (1992) suggest that, for a large sample size ($T > 500$), ε should be 0.5, 1.0, 1.5, and 2 times the standard deviations of the series in question. As for the choice of the relevant embedding dimension m, Hsieh (1989) suggests considering a broad range of values from 2 to 10 for this parameter. Following recent studies of Barnett et al. (1995), we implement the BDS test for the range of m-values from 2 to an upper bond of 8.

A rejection of the null hypothesis is consistent with some type of dependence in the returns that could result from a linear stochastic process, nonstationarity, a nonlinear stochastic process, or a nonlinear deterministic system.* Linear dependence can be ruled out by prior fitting of an Akaike

---

* The simulation studies by Brock et al. (1992) show that the BDS test has power against a variety of linear and nonlinear processes, including, for example, GARCH and exponential GARCH (EGARCH) processes.

information criterion (AIC)-minimizing autoregressive moving average (ARMA) model. In addition, since we are using daily observations over a relatively short time period, nonstationarity is not likely to be the cause of nonlinearity, an assumption that will be tested using unit root tests.[*]

18.4.3 Testing for Stationarity

To see whether the series are stationary, we employ the augmented Dickey–Fuller (ADF) and Philips–Perron (PP) unit root tests on the price levels and the first differences of the CASE 30 and BVMT daily data. We should note, however, that despite their popularity, the ADF and PP tests have been criticized in the unit root literature. Perron (1989) shows that ADF is subject to misspecification bias and size distortion when the series involved have undergone structural shifts leading to spurious acceptance of the unit root hypothesis. As for the PP test, it has also been criticized because it suggests determining the breakpoints exogenously. Zivot and Andrews (1992) demonstrate that endogenously determining the time of structural breaks may reverse the results of the unit root hypothesis. To overcome the limitations of ADF and PP tests, we also utilize the unit root test proposed by Zivot and Andrews (1992) where one endogenously estimated structural change is allowed.

The null hypothesis in Zivot–Andrews test is that the variable under investigation contains a unit root with a drift that excludes any structural break, while the alternative hypothesis is that the series is a trend stationary process with a one-time break in the trend variable occurring at an unknown point in time.

Let T_B be a potential breaking point in $\{p_t\}$, the Zivot–Andrews test starts by estimating the following three equations:

$$\text{Model (A): } \Delta p_t = \mu^A + \gamma^A t + \beta^A \, \mathrm{DU}_t + \alpha^A p_{t-1} + \sum_{j=1}^{k} \phi_j^A \Delta p_{t-j} + \varepsilon_t, \quad (18.4)$$

$$\text{Model (B): } \Delta p_t = \mu^B + \gamma^B t + \lambda^B \, \mathrm{DT}_t + \alpha^B p_{t-1} + \sum_{j=1}^{k} \phi_j^B \Delta p_{t-j} + \varepsilon_t, \quad (18.5)$$

$$\text{Model (C): } \Delta p_t = \mu^C + \gamma^C t + \lambda^C DT_t + \beta^C DU_t + \alpha^C p_{t-1} + \sum_{j=1}^{k} \phi_j^C \Delta p_{t-j} + \varepsilon_t, \quad (18.6)$$

---

[*] Nonstationarity is assumed to be the result of structural change, such as policy changes, technological and financial innovation, etc.

where DU_t is a sustained dummy variable capturing a shift in the intercept, and DT_t representing a shift in the trend occurring at time T_B. In other words, DU_t and DT_t are defined as follows:

$$DU_t = \begin{cases} 1 & \text{if } t > T_B \\ 0 & \text{otherwise} \end{cases} \quad \text{and} \quad DT_t = \begin{cases} 1 & \text{if } t = T_B + 1 \\ 0 & \text{otherwise} \end{cases}$$

We estimate Models (A), (B), and (C) using an ordinary least squares regression. As can be seen, Model (A) allows a one-time shift in the intercept; Model (B) is used to test for stationarity of the series around a broken trend; while Model (C) accommodates the possibility of a change in the intercept as well as a broken trend. For each model, we determine the number of extra repressors, k, following a sequential downward t-test on all lags (for more details, see Campbell and Perron, 1991).

A rejection of the *i.i.d* assumption using stationary filtered data may result from a nonlinear stochastic process or a low-dimension deterministic nonlinear system. Since we are dealing with thin emerging stock markets, it is unlikely that the rejection of the *i.i.d* assumption would be the result of chaos, and instead we should examine the nonlinear stochastic process. Given the high volatility in emerging markets, we assume that the nonlinearity enters through the variance. To test this assumption, we fit a GARCH type model and then employ the $Q_{ML}(k)$ test, Engle test, and BDS test to its standardized residuals. Values of test statistics less than the significance level will support the view that the evidence of nonlinearity dependencies is due to heteroscedasticity, which the chosen model captures well.

18.5 RESULTS

Table 18.2 reports the Ljung–Box test for returns series of CASE 30 and BVMT index up to lag 40. We can see that although the daily returns resemble white noise (see Figure 18.1), the Ljung–Box test rejects the null hypothesis of no autocorrelation in all the lags for the Egyptian as well as Tunisian stock market. Figure 18.2 displays the autocorrelation coefficients and the QQ-plot. It confirms the presence of linear dependencies and non-normal distribution of the returns series in question. Note, however, that linear dependence in the returns series should not be considered irrefutable evidence against market efficiency. As expressed earlier, spurious auto-correlation could be the cause of rejection of the null hypothesis of no autocorrelation. Therefore, before testing for the RWH, we should remove

TABLE 18.2   Ljung–Box Statistics for Daily Returns
Market Indices

$Q_{LB}(k)$	CASE 30	BVMT
$Q_{LB}(5)$	106.45*	255.83*
$Q_{LB}(10)$	122.42*	280.18*
$Q_{LB}(20)$	137.17*	312.49*
$Q_{LB}(30)$	155.89*	335.37*
$Q_{LB}(40)$	165.51*	354.43*

Note: $Q_{LB}(k)$ is the Ljung–Box statistic at lag k.
* Significant at 5% level.

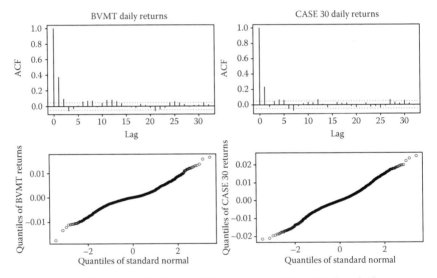

FIGURE 18.2   Autocorrelation coefficients and QQ-plot for daily returns market.

the existent linear relationship (since it could be spurious) by prior fitting of AIC-minimizing ARMA model. ARMA models provide a flexible approach to control linear dependencies in time series. Table 18.3 shows that an AR(4) and AR(5) are successful in removing the serial correlation in the CASE 30 and BVMT daily returns series, respectively. In fact, Ljung–Box test statistics shown in Table 18.3 cannot reject the null hypothesis of no autocorrelation up to lag 40 for AR(4) and AR(5) models.

Removing the linear dependencies from CASE 30 and BVMT returns series should not imply that the residuals of the AR(k) models follow a random walk process. Lack of linear dependence does not necessarily mean a lack of dependence. The residuals can behave as white noise, however,

TABLE 18.3   Ljung–Box Statistics for AR(k)
Models

$Q_{LB}(k)$	CASE 30	BVMT
k	4	5
AIC	0.00	0.00
$Q_{LB}(5)$	0.01	0.02
$Q_{LB}(10)$	0.13	0.20
$Q_{LB}(20)$	5.52	4.49
$Q_{LB}(30)$	21.78	17.44
$Q_{LB}(40)$	34.23	34.23

Note: k is the order of autoregression, AIC is the Akaike information criteria, and $Q_{LB}(k)$ is the Ljung–Box statistic at lag k for the AR(k) residuals.

they are nonlinearly dependent. Table 18.4 displays the McLeod–Li test statistics up to lag 40 and Engle test statistics up to lag 12 applied to the residuals of the AR(4) and AR(5) models. The test statistics are significant at the 1% level, which indicates evidence of nonlinear dependencies in the residuals of the AR(k) models. Thus, although the residuals of the

TABLE 18.4   McLeod–Li and LM Statistics for AR(k) Models

	CASE 30	BVMT
k	AR(4)	AR(5)
McLeod–Li Test		
ML(5)	333.62**	417.38**
ML(10)	451.28**	468.80**
ML(20)	613.44**	499.28**
ML(30)	713.67**	528.44**
ML(40)	783.42**	603.88**
Engle Test		
LM(1)	50.42**	297.59**
LM(2)	141.54**	297.40**
LM(3)	169.07**	304.76**
LM(4)	177.65**	304.63**
LM(12)	205.78**	312.13**

Note: ML(k) is the McLeod–Li test at lag k and LM(k) is Engle test for lag k.
** Significant at 1% level.

AR(4) and AR(5) are white noise, their squared values exhibit some kind of dependencies possibly with nonlinear structures.

To confirm the presence of nonlinearity in the CASE 30 and BVMT daily returns series, we use the BDS test, which has more power than both the McLeod–Li and Engle tests. Tables 18.5 and 18.6 provide the BDS statistics for embedding dimensions 2–8 and for epsilon values starting from 0.5 to 2 times the standard deviation of the residuals of the AR(4) and AR(5) models, respectively. BDS test statistics reject the *i.i.d* assumption at 5% and 1% level of significance. According to Hsieh (1991), the non-*i.i.d* behavior could be due to linear dependence, nonstationary, nonlinear stochastic processes or chaos. Since the AR(4) and AR(5) capture all the

TABLE 18.5   BDS Statistics for the Standardized Residuals of the AR(4) Model

m	ε/σ		ε/σ		ε/σ		ε/σ	
1	0.5	20.179**	1	19.764**	1.5	18.781**	2	17.199**
2	0.5	27.567**	1	26.306**	1.5	23.162**	2	20.013**
3	0.5	34.570**	1	31.405**	1.5	25.472**	2	21.294**
4	0.5	42.653**	1	36.765**	1.5	27.400**	2	21.970**
5	0.5	53.758**	1	43.584**	1.5	29.621**	2	22.700**
6	0.5	69.172**	1	52.085**	1.5	32.043**	2	23.371**
7	0.5	91.616**	1	63.440**	1.5	34.897**	2	24.191**
8	0.5	123.739**	1	78.293**	1.5	37.979**	2	24.906**

Note: m is embedding dimension and ε is the bound.
** Significant at 1% level. The critical values for BDS test are 1.96 for 5% and 2.58 for 1%.

TABLE 18.6   BDS Statistics for the Residuals of the AR(5) Model

m	ε/σ		ε/σ		ε/σ		ε/σ	
1	0.5	9.610**	1	9.282**	1.5	8.247**	2	6.912**
2	0.5	16.450**	1	14.900**	1.5	13.100**	2	11.184**
3	0.5	24.246**	1	19.525**	1.5	15.984**	2	13.393**
4	0.5	33.901**	1	24.004**	1.5	18.150**	2	14.780**
5	0.5	48.454**	1	29.145**	1.5	19.974**	2	15.735**
6	0.5	69.831**	1	35.193**	1.5	21.754**	2	16.475**
7	0.5	108.119**	1	43.361**	1.5	23.760**	2	17.262**
8	0.5	180.839**	1	54.158**	1.5	25.966**	2	17.994**

Note: m is embedding dimension and ε is the bound.
** Significant at 1% level. The critical values for BDS test are 1.96 for 5% and 2.58 for 1%.

correlation in the indices returns series, we can rule out the linear dependence. Therefore, we concentrate on the remaining causes.

Given that we are dealing with emerging markets, the frequent structural changes (results of economic and political reforms) may cause nonstationarity in the index daily return series, thus causing a rejection of the *i.i.d* assumption. To see whether the series are stationary, we employ the ADF and PP on the price levels and the first differences of the CASE 30 and BVMT daily data. We also use Zivot–Andrews unit root test to overcome the shortcomings of ADF and PP tests discussed earlier. Table 18.7 displays the outcomes from conventional ADF and PP unit root tests along with Zivot–Andrews sequential procedure. The results from unit root tests suggest that the returns series are nonstationary in levels and stationary in first difference at 5% degree of significance. Therefore, we can rule out the nonstationarity cause and investigate whether the nonlinear stochastic processes or the chaos is causing the rejection of the *i.i.d* assumption. Several empirical studies attempt to investigate whether the evidence of nonlinearity structure reported in developed financial markets could be due to low-dimension deterministic nonlinear process. But, to date, the results remain inconclusive. Therefore, it is unlikely that a low-dimension deterministic nonlinear process could generate market index return series of thin and immature stock markets, such as that of Egypt or Tunisia. Thus, we focus on the stochastic modeling of the nonlinearity dependencies exhibited by both time series.

In order to model the nonlinearity dependence in CASE 30 and BVMT daily returns series, it is crucial to look for the source of nonlinearity. Nonlinearity can enter through the mean of a return-generating process (additive dependence), as in the case of threshold autoregressive model, or through the variance (multiplicative dependence), as in the case of ARCH model proposed by Engle (1982). Nonlinearity can be both additive and multiplicative as in the case of Arch-in-Mean (ARCH-M) model. Although very powerful in detecting several nonlinear structures, BDS test cannot discriminate between additive and multiplicative stochastic dependence. However, since the McLeod–Li and Engle tests use squared residuals of the AR(k) models, we can say that volatility clustering in the data may be the cause of nonlinearity structure in the daily returns, and therefore we should choose models that are able to capture multiplicative dependence in time series.

To select the models that best fit the revealed nonlinear returns structures, we have examined several GARCH type models. The identification of model is based on three criteria: the AIC, the Bayes information criterion (BIC), and the maximum likelihood. Table 18.8 provides the outcomes of several experimentations, showing model types along with corresponding

TABLE 18.7 Unit Root Tests

PP and ADF Tests	Ln (Pt)		Rt			Zivot–Andrews Test		
	Trend	No Trend	Trend	No Trend		Model (A)	Model (B)	Model (C)
Panel A: BVMT								
ADF(1)	−1.97	2.29	−27.85*	−27.79*	α	−0.348*	−0.478	−0.245*
ADF(2)	−1.75	1.52	−24.25*	−24.17*				
ADF(3)	−1.75	1.59	−22.76*	−22.67*	β	−0.347*		−0.214*
ADF(4)	−1.80	1.75	−19.31*	−19.21*				(−2.96)
ADF(5)	−1.79	1.69	−17.12*	−17.02*	λ		−0.004*	−0.011*
PP	−1.77	1.7	−27.42*	−27.33*				
Panel B: CASE 30								
ADF(1)	1.43	1.39	−32.49*	−32.48*	α	−0.465*	−0.544	−0.196*
ADF(2)	0.82	1.12	−27.71*	−27.68*				
ADF(3)	1.07	1.23	−21.39*	−21.36*	β	−0.147		−0.266*
ADF(4)	0.90	1.16	−18.27*	−18.24*				
ADF(5)	0.74	1.09	−15.99*	−15.96*	λ		−0.013*	−0.021*
PP	0.93	1.15	−32.49*	−32.48*				

Note: PP is the Phillips–Perron test and ADF is the augmented Dickey–Fuller test.
Ln (pt): the natural logarithm of the daily index price.
Rt: daily returns.
* Significant at 5% level.

TABLE 18.8 Modeling Nonlinear Return Structures

	GARCH (1,1)	EGARCH (1,1)	PGARCH (1,1)	TGARCH (1,1)	GARCH-2-Comp.	FIEGARCH (1,d,1)	FIGARCH (1,d,1)
CASE 30							
AIC	2921	2920	2924	2921	2916	2917	2918
BIC	3015	3017	3022	3019	3019	3015	3022
Likelihood	−1443	−1440	−1444	−1443	−1439	−1440	−1440
BVMT							
AIC	355.2	297.6	328.8	357.3	231.6	253.2	309
BIC	436.7	384.6	415.8	444.2	324	345.6	396
Likelihood	−162.6	−132.8	−148.4	−162.6	−98.8	−109.6	−138.5

Note: FIEGARCH = Fractional integrated exponential GARCH.

model selection criteria. It is clear from Table 18.8 that model selection criteria suggest that we should model the CASE 30 daily returns series as an AR(5)-FIEGARCH(1,1) and the BVMT daily return series as AR(4)-GARCH-two-components.* FIEGARCH and GARCH-two-components models are usually utilized to capture high persistence of conditional volatility in time series. Table 18.9 shows the results of estimating AR(5)-FIEGARCH(1,1) and AR(4)-GARCH-two-components models. The lagged terms in the AR(k) models are not insignificantly different from zero, which is evidence against RWH for both the Egyptian and Tunisian stock markets. The GARCH effects and high persistent conditional volatility are undeniably evident since the coefficients λ_1, θ_1, d, λ_2, and θ_2 are statistically significant. The result from estimating the FIEGARCH model also shows that there is a leverage effect in the BVMT daily return series, captured by the coefficient δ.

Table 18.9 reports some diagnostic test statistics for the standardized residuals. Jarque–Bera (JB) test for normality shows some interesting results. The p-values of the JB test statistics suggest that standardized residuals are now normally distributed. This is consistent with others

---

* AR(k)-FIEGARCH (p,d,q) of Bollersev and Mikkelsen (1996) can be expressed as follows:

$$r_t = \beta_0 + \sum_{i=1}^{k} \beta_i r_{t-i} + \omega_t$$

$$[1-\theta(L)]^d \phi(L) \ln(\sigma_t^2) = \eta + \sum_{j=1}^{q} \left(\theta_j |\xi_{t-j}| + \rho_j \xi_{t-j} \right),$$

where based on information set up to time $t-1$, ω_t is an *i.i.d* random variable with mean 0 and variance σ_t^2.

ξ_{t-j} is the standardized residuals, $\xi_{t-j} = \omega_t/\sigma_t$, L is the lag operator, $\phi(L) = 1 - \lambda(L) - \theta(L)$, $\lambda(L) = \sum_{i=1}^{p} \lambda_i L^i$, $\theta(L) = \sum_{i=1}^{q} \theta_i L^i$, λ, θ, d, ρ, are the ARCH, GARCH, integration, leverage parameters.

As for the GARCH-two-components, it decomposes conditional variance into two components: $\sigma_t^2 = q_t + s_t$ where q_t is a highly persistent long run component and s_t is a transitory short run component. The general form of the GARCH-two-components model is provided by Ding and Granger (1996):

$$\sigma_t^d = q_t^d + s_t^d, \text{ where } q_t^d = \lambda_1 |\varepsilon_{t-1}|^d + \theta_1 q_{t-1}^d \text{ and } s_t^d = \lambda_1 |\varepsilon_{t-1}|^d + \theta_1 s_{t-1}^d.$$

Note that the long-run components q_t follow a highly persistent PGARCH(1,1) model, and the transitory component s_t follows another PGARCH(1,1) model.

TABLE 18.9   Modeling Nonlinear Return Structures

Coefficient	CASE 30 AR(5)-FIEGARCH(1,1)	p-Value	BVMT AR(4)-GARCH-2-Comp.	p-Value
β_0	0.0098	0.2175	0.0012	0.3774
β_1	0.2704	0.0000	0.2779	0.0000
β_2	−0.0881	0.0007	0.0360	0.1167
β_3	0.0663	0.0087	−0.0521	0.0228
β_4	0.0181	0.2599	−0.0184	0.2521
β_5	0.0542	0.0217	—	—
η	−0.2322	0.0000	0.0000	0.0000
λ_1	0.2933	0.0000	0.0018	0.0018
θ_1	0.4876	0.0000	0.9913	0.0000
δ	0.0444	0.0040	—	—
d	0.5349	0.0000	—	—
λ_2	—	—	0.4833	0.0000
θ_2	—	—	0.5177	0.0000
JB	166.8	0.000	226.4	0.000
$Q_{LB}(10)$	4.893	0.898	15.771	0.1064
$Q_{LB}(20)$	9.993	0.968	25.664	0.177
$Q_{LB}(30)$	23.006	0.815	36.456	0.193
$Q_{LB}(40)$	37.967	0.562	45.831	0.243

Note: λ, θ, d, ρ, δ are the ARCH, GARCH, integration, leverage, and risk premium parameters, respectively. JB is the Jarque–Bera test for normality, and $Q_{LB}(k)$ is the Ljung–Box statistic at lag k for the standardized residuals series.

empirical studies in the literature, which show that when we account for ARCH effects, the evidence of nonnormality of returns diminish. If FIEGARCH(1,1) and GARCH-two-components models are specified correctly, then the residuals standardized by the conditional standard deviation, should be white noise. From Table 18.9, we find that Ljung–Box statistics, $Q_{LB}(k)$, up to lag 40 fail to reject the null hypothesis of absence of autocorrelation in the standardized residuals of both models. To see whether the selected models have accounted for all the revealed nonlinear structures, we have first applied the McLeod–Li and Engle tests to the squared residuals. Results from Table 18.10 show that the test statistics are insignificant at all lags, which suggests that AR(5)-FIEGARCH(1,1) and

TABLE 18.10   McLeod–Li and LM Statistics for AR(k) Models

	CASE 30	BVMT
	AR(5)-FIEGARCH(1,1)	AR(4)-GARCH-2-Comp.
McLeod–Li test		
ML(5)	1.3932	8.2722
ML(10)	7.7542	11.5723
ML(20)	21.5015	20.0092
ML(30)	29.1411	32.9233
ML(40)	37.1005	41.0913
Engle test		
LM(1)	0.6995	0.0822
LM(2)	0.6930	2.0446
LM(3)	0.7330	3.5966
LM(4)	1.3430	7.9162
LM(12)	8.5612	13.3591

Note: ML(k) is the McLeod–Li test at lag k and LM(k) is Engle test for lag k.

AR(4)-GARCH-two-components models have captured all the linear as well as nonlinear dependencies in the daily return series.

Tables 18.11 and 18.12 provide the BDS test statistics for standardized residuals of AR(5)-FIEGARCH(1,1) and AR(4)-GARCH-two-components models, respectively. The results from the McLeod–Li and Engle tests are corroborated by the BDS test for CASE 30 Index; however, for BVMT the *i.i.d* assumption is rejected for bound, ε/σ, of 0.5 with embedding

TABLE 18.11   BDS Statistics for the Standardized Residuals of the AR(5)-FIEGARCH Model

m	ε/σ		ε/σ		ε/σ		ε/σ	
1	0.5	−0.2374	1	−0.3817	1.5	−0.5982	2	−0.9690
2	0.5	−0.6397	1	−0.7714	1.5	−1.0697	2	−1.5773
3	0.5	−0.5772	1	−0.9119	1.5	−1.3286	2	−1.9711
4	0.5	−0.8799	1	−1.4015	1.5	−1.8352	2	−2.5603
5	0.5	−0.7386	1	−1.4520	1.5	−2.1086	2	−2.9672
6	0.5	−0.4172	1	−1.1779	1.5	−2.0231	2	−2.8938
7	0.5	1.0382	1	−0.8644	1.5	−1.8469	2	−2.7158
8	0.5	4.4427**	1	−0.4905	1.5	−1.6690	2	−2.5902

Note: m is embedding dimension and ε is the bound.
** Significant at 1% level. The critical values for BDS test are 1.96 for 5% and 2.58 for 1%.

TABLE 18.12   BDS Statistics for the Standardized Residuals of the AR(4)-GARCH-2-Comp. Model

m	ε/σ		ε/σ		ε/σ		ε/σ	
1	0.5	0.7935	1	0.7560	1.5	0.8651	2	1.0429
2	0.5	1.6403	1	1.3743	1.5	1.1056	2	0.8477
3	0.5	2.1296**	1	1.9101	1.5	1.4242	2	0.8605
4	0.5	2.4401**	1	2.1464**	1.5	1.4802	2	0.6845
5	0.5	2.7220**	1	2.4234**	1.5	1.5901	2	0.5699
6	0.5	2.7259**	1	2.6125**	1.5	1.5838	2	0.3732
7	0.5	2.5080**	1	3.0279**	1.5	1.7097	2	0.3077
8	0.5	3.1818**	1	3.5108**	1.5	1.7901	2	0.1737

Note: m is embedding dimension and ε is the bound.
** Significant at 1% level. The critical values for BDS test are 1.96 for 5% and 2.58 for 1%.

dimension, m, ranging from 3 to 8 and for bound of 0.5 with m ranging from 4 to 8. Considering these empirical findings, we state that AR(5)-FIEGARCH(1,1) has successfully captured nonlinear structure in CASE 30 daily return series, whereas inherent nonlinearity dependence in BVMT daily return series seems too complex to be *fully* captured by a GARCH-two-components model and therefore, a more flexible model should be used. It is interesting to mention how the McLeod–Li and Engle tests are incapable of detecting all kinds of nonlinear structures in time series and, consequently, should be limited to diagnostic use only. These results clearly exhibit how crucial it is to rely on powerful techniques to examine the RWH and avoid those designed to detect only linear structures.

18.6 CONCLUSION

This chapter examines the Egyptian and Tunisian markets by testing for linear and nonlinear dependencies in daily returns of their major market indices: CASE 30 and BVMT. Taken together, the findings of this chapter reject the RWH for the Egyptian and Tunisian indices. The rejection is due to inherent nonlinear dependencies in the returns series, explained by high persistent conditional heteroscedasticity, a feature that seems to characterize thin emerging stock markets.* The results of this chapter have

---

* It is noteworthy that rejection of the RWH should not *necessarily* imply market inefficiency, because nonlinearity does not necessarily mean predictability. As noted by Abhyankar et al. (1997), future price changes can be predictable, but only in a small window of time, not enough to allow for excess profits. In addition, the relative high transaction costs in emerging markets are likely to discourage investors from engaging in speculative trading.

■ Emerging Markets: Performance, Analysis and Innovation

importantimplicationsforresearchersandpractitionerswhoareincreasingly interested in nontraditional emerging markets for several reasons, ranging from modeling high-persistent volatility to portfolio diversification. This empirical study also helps expand the literature of nonlinear dynamics in financial markets, especially in emerging markets, where the number of studies is very limited compared to developed markets.

REFERENCES

Abhyankar, A.H., Copeland, L.S., and Wong, W. (1997) Uncovering nonlinear structure in real-time stock-market indices: The S&P 500, the DAX, the Nikkei 225, and the FTSE-100. *Journal of Business and Economic Statistics,* 15(1): 1–14.

Amihud, Y. and Mendelson, H. (1987) Trading mechanisms and stock returns: An empirical investigation. *Journal of Finance,* 42(3): 533–556.

Barnett, W.A., Gallant, A.R., Hinish, M.J., Jungeilges, J., Kaplan, D., and Jensen, M.J. (1995) Robustness of non-linearity and chaos tests to measurement error, inference method, and sample size. *Journal of Economic Behaviour and Organization,* 27(1): 301–320.

Bera, A., Bubnys, E., and Park, H. (1993) ARCH effects and efficient estimating of hedge ratios for stock index futures. *Advances in Futures and Options Research,* 6(3): 313–328.

Box, G.E.P. and Pierce, D.A. (1970) Distribution of residual autocorrelations in autoregressive-integrated moving average time series models. *Journal of American Statistical Association,* 65(3): 1509–1526.

Brock, W.A., Hsieh, D.A., and LeBaron, B. (1992) *Nonlinear Dynamics, Chaos and Instability: Statistical Theory and Economic Evidence.* MIT Press: Boston, MA.

Brooks, C. (1996) Testing for non-linearity in daily sterling exchange rates. *Applied Economics Letters,* 5(4): 719–722.

Campbell, J.Y., Lo, A.W., and MacKinlay, A.C. (1997) *The Econometrics of Financial Markets.* Princeton University Press: Princeton, NJ.

Campbell, J.Y. and Perron, P. (1991) Pitfalls and opportunities: What macroeconomists should know about unit roots. NBER Technical Working Papers 0100. National Bureau of Economic Research: Cambridge, MA.

Ding, Z. and Granger, C.W.J. (1996) Modeling volatility persistence of speculative returns: A new approach. *Journal of Econometrics,* 73(1): 185–215.

Engle, R.F. (1982) Autoregressive conditional heteroskedasticity with estimates of the variance of United Kingdom inflation. *Econometrica,* 50(2): 987–1007.

Fama, E. (1970) Efficient capital markets: A review of theory and empirical work. *Journal of Finance,* 25(2): 383–417.

Granger, C. and Anderson, A.P. (1978) *An Introduction to Bilinear Time Series Models.* Vandenhoeck and Ruprecht: Gottingen.

Harvey, A.C. (1993) *Time Series Models.* Harvest Wheatsheaf: New York, London.

Hsieh, D.A. (1989) Testing for nonlinearity dependence in daily foreign ex-change rate. *Journal of Business,* 62(3): 339–368.

Hsieh, D.A. (1991) Chaos and nonlinear dynamics: Application to financial markets. *Journal of Finance*, 46(5): 1837–1877.

Lo, A.W. and Mackinlay, A.C. (1990) An econometric analysis of non-synchronous trading. *Journal of Econometrics*, 45(2): 181–211.

Mayfiedl, E.S. and Mizrach, B. (1992) On determining the dimension of real-time stock-price data. *Journal of Business and Economic Statistics*, 10(3): 367–374.

McInish, T.H. and Wood, R.A. (1991) Autocorrelation of daily index returns: Intraday-to-intraday versus close-to-close intervals. *Journal of Banking and Finance*, 15(1): 193–207.

McMillan, D.G. (2005) Non-linear dynamics in international stock market returns. *Review of Financial Economics*, 14(1): 81–91.

Perron, P. (1989) The great crash, the oil price shock, and the unit root hypothesis. *Econometrica*, 57(6): 1361–1401.

Poshakwale, S. (2002) The random walk hypothesis in emerging Indian stock market. *Journal of Business Finance and Accounting*, 29(9–10): 1275–1299.

Scheinkman, J.A. and LeBaron, B. (1989) Non-linear dynamics and stock returns. *Journal of Business*, 62(3): 311–337.

Serletis, A. and Dormaar, P. (1996) Chaos and nonlinear dynamics in futures markets. In: W.A. Barnett, A.P. Kirman, and M. Salmon (Eds.), *Nonlinear Dynamics and Economics*, pp. 113–132. Cambridge University Press: Cambridge, United Kingdom.

Serletis, A. and Shintani, M. (2003) No evidence of chaos but some evidence of dependence in the U.S. stock market. *Chaos Solitons and Fractals*, 17(2): 449–454.

Sewell, P.S., Stansell, S.R., Lee, I., and Pan, M.S. (1993) Nonlinearities in emerging foreign capital markets. *Journal of Business Finance and Accounting*, 20(2): 237–248.

Vaidyanathan, R. and Krehbiel, T. (1992) Does the S&P500 futures mispricing series exhibit non-linear dependence across time? *Journal of Futures Markets*, 112(6): 659–677.

Yadav, P.K., Paudyal, K., and Pope, P.F. (1996) Non-linear dependence in daily stock index returns: Evidence from Pacific basin markets. *Advances in Pacific Basin Financial Markets*, 2(1): 349–377.

Zivot, E. and Andrews, D.W.K. (1992) Further evidence on the great crash, the oil-price shock, and the unit-root hypothesis. *Journal of Business and Economic Statistics*, 10(3): 251–270.

CHAPTER 19

Study of Market Integration, Share Price Responses, and Global Portfolio Investments in the MENA Region

Mohamed El Hedi Arouri
and Duc Khuong Nguyen

CONTENTS

19.1 INTRODUCTION

This chapter examines the regional and global market integration of six major stock markets in the Middle East and North Africa (MENA) region by investigating how they are linked to each other and in turn to the world stock market. More precisely, we employ, as measure of the market integration degree, the bilateral conditional correlations from empirically implementing a multivariate dynamic conditional correlation GARCH (DCC-GARCH) model which allows for CAPM effects. This methodological choice permits the discussions of empirical patterns that were observed on the dynamics of market integration both in regional and global dimensions as well as the share price volatility in studied markets, conditionally on their integration degree. Further, we examine the extent to which financial linkages among markets are subjected to structural changes based on an intuition that ongoing policy reforms, enhanced structural adjustments, and tremendous increase in stock market activity in MENA markets over the recent period are likely to have significantly impacted return behavior and their comovements.

Using stock market data at daily frequency, our results display two intriguing facts. First, market integration among MENA markets becomes felt, but its level still remains low. Second, the comovements of these markets with world stock market, albeit being significant only in two cases, are subject to structural breaks with respect to local, regional, and global market events.

The rest of this chapter is organized as follows: Section 19.2 gives a brief overview of MENA stock markets and their main characteristics; Section 19.3 describes the empirical method to test whether MENA markets are integrated with each other and into world financial system, as well as a structural break test in linear framework that helps exploring the time-varying characteristics of financial linkages among markets is also presented; Section 19.4 provides a description of the data used and their properties; Section 19.5 reports and discusses the results; and concluding remarks are given in Section 19.6.

19.2 FINANCIAL AND ECONOMIC CHARACTERISTICS OF MENA STOCK MARKETS

Stock markets in the MENA region have been widely ignored by global investors and academic researchers due to imposed restrictions on foreign stock ownership, lack of market transparency and appropriate accounting standards, and economic and political instability. They also share several common patterns such as small market capitalization relative to the economy, few listed companies, and low trading volume; and access to these markets for direct investment in equities until recently was given only to domestic investors. We observed, however, dramatic changes in recent years. Indeed, capital requirements to fund budget deficits and economic development have driven the MENA country governments to embark on capital market liberalization, privatization, and broad-ranging structural reforms, allowing foreign investors greater access to the stock markets, which may have increased stock market linkages within the MENA region and between the MENA markets and the developed markets.

Compared to developed countries, MENA countries present other common patterns. With the exception of Oman, inflation-adjusted economic growth rates exceed those of most developed economies, and the average GDP growth rate which exceeds 5% in the region is among the highest in the world. Nevertheless, a glance at MENA region shows that economic conditions across MENA economies contain several dissimilarities (Omran and Gunduz, 2001; Girard and Ferreira, 2004; Bley, 2007). For example, on a purchasing power parity basis, the GDP per capita in the MENA region ranges from $3,710 (Egypt) to $40,800 (Qatar). MENA stock markets differ tremendously in terms of market capitalization and liquidity. While Lebanon and Tunisia have market capitalizations of $2.2 billion and $2.4 billion, respectively, the market value of equity listed at Saudi Arabia's stock exchange reached almost $238 billion in 2004. If we take a close look at the capital market growth rates, they are significantly much higher for Gulf Cooperation Council (GCC) countries than for non-GCC stock markets of the MENA region.* For instance, from September

---

* The GCC was established in 1981, and it includes six countries, namely, Saudi Arabia, Kuwait, Oman, United Arab Emirates, Qatar, and Bahrain.

2002 to September 2004, total GCC market capitalization increased by 161% against less than 15% for the non-GCC markets.

These singularities of MENA markets in terms of investment opportunities, growth, organization, trading mechanisms, international openness, and liquidity may imply different linkages and transmission patterns than those in the developed and major emerging stock markets. Early studies on MENA stock markets could not identify significant cross-linkages within the MENA region and with developed markets (Abraham et al., 2001; Omran and Gunduz, 2001; Girard et al., 2003; Maghyereh, 2006). This effectively raises the question of measuring and quantifying the cross-border linkages within MENA markets and between them and the world stock market. Giving appropriate answers to this question is of great interest for both global portfolio managers and policymakers, because based on the actual degree of market integration, the latter could adjust their liberalization policies to avoid harmful attacks of free capital flows while the former would be able to rebalance their worldwide diversified holdings. This chapter is a contribution to fill the gap.

19.3 DYNAMIC MODEL OF INTERNATIONAL STOCK MARKET LINKAGES

Assessing the degree of market integration is always a difficult task given its dependences on many factors such as home bias, exchange rate, and market differences in terms of microstructure, trading system, and investor's risk preferences. We can cite, for instance, empirical measures from international asset pricing models (see, Errunza et al., 1992; Bekaert and Harvey, 1995), closed-end country fund premiums in excess of its net asset value, and financial asset correlations. If the first two approaches permit directly to test the market integration hypothesis (i.e., the equality of market risk premium across financial markets), they are often subject to critics which mainly concern the lack of explanatory power of asset pricing models in an international context. The correlation approach that provides a measure of real stock market linkages (or also referred to as comovements) constitutes an alternative way to infer the trends toward market integration. According to the meanings of correlation coefficients, one should interpret higher comovements as increased integration among international stock markets. This approach is employed in this chapter, but we focus on conditional correlations to essentially count for the impact of past and current market conditions on the observed comovements, which is not possible with simple correlation analysis.

Indeed, we implement a multivariate DCC-GARCH model as in Engle (2002) to investigate the dynamic linkages among sample markets. In addition to capturing the time-varying feature of market comovements, this model is advantageous in that it helps control the conditional heteroscedasticity in variances of stock returns data and reduce the departure from normal distribution, a condition required for using various financial models. Moreover, compared with vector autoregressive model, it enables to quantify the amount of market interdependencies. Finally, we can examine the reactions of studied markets to diverse market and economic events over the study period through performing tests of structural breaks in the time-paths of estimated conditional correlations.

Specifically, we consider the following model:

$$R_t = \mu_t + \varepsilon_t, \quad \varepsilon_t | I_t \rightarrow N(0, H_t) \tag{19.1}$$

where

- $R_t = [R_1, R_2, \ldots, R_6, R_w]'$ is a $(k \times 1)$ vector of realized returns on market indices at daily frequency conditional on available information set at time t.

- $\mu_{i,t} = \delta_{0i} + \delta_{1i} R_{i,t-1} + \delta_{2i} R_{w,t} + \delta_{3i} R_{w,t-1}$ for individual market i; $\mu_{w,t} = \delta_{0w} + \delta_{1w} R_{w,t-1}$ for the world stock market with R_w denoting its return.

- ε_t is a $(k \times 1)$ vector of zero mean return innovations.

- H_t refers to the conditional variance–covariance matrix of return innovations and is defined by $H_t \equiv D_t \tilde{R}_t D_t$, where $\tilde{R}_t$ is the $(k \times k)$ conditional correlation matrix, and D_t is a $(k \times k)$ diagonal matrix with elements on its main diagonal being the conditional standard deviations of the returns on each individual market. Therefore, $[[H_t]_{ij} = h_{ij}, [D_t]_{ij} = \mathrm{diag}\left(\sqrt{h_{ij}}\right)$ with $i = j$, and $\tilde{R}_t = (\mathrm{diag}\, Q_t)^{-1/2}\, Q_t (\mathrm{diag}\, Q_t)^{-1/2}$, where $Q_t = (1 - \alpha - \beta)\overline{Q} + \alpha u_{t-1} u'_{t-1} + \beta Q_{t-1}$ referring to a $(k \times k)$ symmetric positive definite matrix with $u_{it} = \varepsilon_{it} / \sqrt{h_{iit}}$. $\overline{Q}$ is the $(k \times k)$ unconditional variance matrix of u_t, and α and β are nonnegative scalar parameters satisfying $\alpha + \beta < 1$. It should be noted that h_{iit} are assumed to follow a univariate GARCH (1,1) process.

Globally speaking, our model assumes that returns on each individual market depend on five components: a common-trend factor (δ_0), a one-period-lagged country factor (δ_1), a global market factor (δ_2), a one-period-lagged global market factor (δ_3), and a country-specific disturbance (ε). Compared to other studies on stock market linkages, we particularly introduce the current value of the global market return in the conditional mean equation in order to control the linear asset pricing relationship within considered country. These unknown parameters as well as those of the multivariate GARCH process are estimated using the method of quasi-maximum likelihood (QML) estimation proposed by Bollerslev and Wooldrige (1992). The log-likelihood function of the observations on ε_t to be maximized is given by

$$L = -\frac{1}{2}\sum_{t=1}^{T}\left(n\log(2\pi)+\log\left|D_t\tilde{R}_t D_t\right|+\varepsilon_t' D_t^{-1}\tilde{R}_t^{-1}D_t^{-1}\varepsilon_t\right)$$

Once conditional correlations between sample markets become available, the investigation of possibly structural breaks in time-varying cross-market comovements is straightforward based on Bai and Perron's (2003) procedure. The test aims at determining the number and location of breaks in a linear regression model. Let us suppose that there are m breaks $(n_1,...,n_m)$ in the time-path of the dependant variable, the problem of dating structural breaks turns to find the breakpoints $(\tilde{n}_1,...,\tilde{n}_m)$ that minimize the objective function $(\tilde{n}_1,...,\tilde{n}_m) = \arg\min(n_1,...,n_m) \text{RSS}(n_1,...,n_m)$, with RSS_n is the resulting residual sum of squares based on the m linear regressions of the form

$$y_t = \beta x_t^{\mathrm{T}} + \varepsilon_t (t = 1, ..., n) \tag{19.2}$$

In Equation 19.2, y_t is the conditional correlation series that we obtained from estimating our multivariate DCC-GARCH model, $x_t = (1, y_{t-1})^{\mathrm{T}}$ is the (2×1) vector of observations of the independent variables with the first component equal to unity, β is the (2×1) vector of regression coefficients, and ε_t is assumed to follow an independent and identical distribution with zero mean and variance of σ^2. Then, the null hypothesis of "no structural break" is tested against the alternative that the regression coefficients change over time. We use the Bayesian information criteria (BIC) to select breakpoints and the optimal number of breaks which precisely corresponds to the one with the lowest BIC score.

19.4 DATA AND STATISTICAL PROPERTIES

To conduct the study, we use daily total U.S. dollar natural log returns computed from Morgan Stanley Capital International total return indices. In addition to the world market index, the sample markets include six major stock markets in the MENA region: Bahrain, Egypt, Jordan, Morocco, Oman, and United Arab Emirates. The data cover the period from June 1, 2005 to September 22, 2008, and are date-matched across the seven markets. Most equity markets in MENA region are closed on Thursdays and Fridays, whereas the other markets are closed for trading on Saturdays and Sundays. Therefore, matching of price-data from the three markets results in the loss of some data points. Descriptive statistics and stochastic properties of daily returns are presented in Table 19.1.

Panel A of Table 19.1 reveals a number of interesting facts. Compared to the world market, MENA markets have higher volatilities, but not necessarily high returns. Morocco has the highest daily return followed by Egypt. Two markets, Bahrain and UAE, have negative daily returns. UAE has the highest risk followed by Egypt. The Engle (1982) test for conditional heteroscedasticity rejects the null hypothesis of no ARCH effects for all the markets we study, except for Bahrain. Skewness is generally negative, and kurtosis is above three. The Jarque–Bera (JB) test statistic strongly rejects the hypothesis of normality. These facts support our decision to use the QML approach of Bollerslev and Wooldridge (1992) to estimate and test the model. The null hypothesis of no autocorrelation of order 12 is rejected for all the markets, except for Jordan.

The autocorrelations of market index returns from the first to the sixth order are reported in Panel B of Table 19.1. As they are relatively high and significant for many countries, we then think that the inclusion of an autoregressive correction in the mean equations is necessary in explaining the time variations of expected returns.

Panel C of Table 19.1 reports the unconditional correlations among markets. As we can see, cross-market correlations are low and range from 2.83% (between Bahrain and Morocco) to 20.67% (between Oman and UAE). Moreover, correlations between MENA and world stock markets are not large and negative for two countries (Bahrain and Jordan). This is indicative of the facts that the MENA markets in our study are generally disconnected from the world market trends and that global investors can still benefit from adding financial assets of this region in their diversified portfolios.

TABLE 19.1 Descriptive Statistics

	Bahrain	Egypt	Jordan	Morocco	Oman	UAE	World
Panel A: Summary statistics							
Mean (% per day)	-0.016	0.069	0.001	0.127	0.011	-0.052	0.012
Std. deviation (% per day)	1.010	1.693	1.392	1.148	1.224	1.971	0.812
Skewness	1.706*	-0.793*	-0.522*	-0.253*	-0.750*	-0.603*	-0.019
Kurtosis	24.923*	4.621*	5.361*	2.002*	9.971*	6.200*	4.081*
JB	22781.902†	859.725†	1074.081†	153.580†	3660.511†	1436.617†	599.728†
Q(12)	26.517†	27.443†	12.674	66.322†	20.078††	36.664††	19.836††
ARCH(6)	4.431	86.984†	77.978†	65.093†	44.603†	102.817†	192.564†
Panel B: Autocorrelations							
Lag	Bahrain	Egypt	Jordan	Morocco	Oman	UAE	World
1	0.117	0.080	0.039	0.266	0.077	0.035	0.102
2	0.024	0.019	-0.058	0.020	-0.017	-0.015	-0.020
3	0.045	0.112	0.029	-0.015	0.057	0.129	-0.044
4	0.055	0.014	-0.002	-0.007	-0.031	0.011	0.005
5	0.033	-0.024	-0.001	-0.004	0.022	0.128	-0.022
6	0.038	0.020	0.054	-0.016	0.074	-0.020	-0.063

Panel C: Unconditional correlations of stock market returns (in percentage)

	Bahrain	Egypt	Jordan	Morocco	Oman	UAE	World
Bahrain	100						
Egypt	3.197	100					
Jordan	4.178	18.210	100				
Morocco	2.832	18.257	9.293	100			
Oman	13.608	17.715	14.441	3.801	100		
UAE	14.134	28.790	17.558	3.818	29.667	100	
World	-2.445	9.410	-1.878	18.347	1.216	0.891	100

Note: The test for kurtosis coefficient has been normalized to zero. JB is the Jarque–Bera test for normality based on excess skewness and kurtosis. Q(12) is the Ljung–Box test for autocorrelation of order 12. ARCH is the Engle's (1982) test for conditional heteroscedasticity.

★, ★★, and ★★★ indicate significance of coefficients at the 1%, 5%, and 10%, respectively.

†, ††, and ††† indicate rejection of the null hypotheses of no autocorrelation, normality, and homoscedasticity at the 1%, 5%, and 10% levels of significance, respectively, for statistical tests.

19.5 EMPIRICAL RESULTS

19.5.1 Estimation Results of the DCC-GARCH Model

Table 19.2 contains parameter estimates and a number of diagnostic tests for DCC-GARCH model. The coefficients relating the return series to the one-lag local and world market returns (Panel A of Table 19.2) are

TABLE 19.2 Estimation Results

	Bahrain	Egypt	Jordan	Morocco	Oman	UAE	World
Panel A: Parameter estimates: mean equations							
δ_0	0.002	0.001***	0.002	0.001*	0.000	0.000	0.001**
	(0.004)	(0.001)	(0.004)	(0.000)	(0.000)	(0.001)	(0.001)
δ_1	0.175*	0.013	0.050	0.174*	0.126*	0.039	0.126*
	(0.063)	(0.033)	(0.040)	(0.033)	(0.039)	(0.037)	(0.037)
δ_2	−0.042	0.024	0.005	−0.053	0.068	0.067	—
	(0.046)	(0.054)	(0.055)	(0.044)	(0.051)	(0.976)	
δ_3	0.090**	0.485*	0.078***	0.147*	0.129*	0.354*	—
	(0.039)	(0.061)	(0.046)	(0.045)	(0.043)	(0.070)	
Panel B: Parameter estimates: GARCH process							
$\varpi_0 \, (10^{-3})$	0.331*	0.002*	0.002**	0.111*	0.068*	0.061*	0.001*
	(0.063)	(0.000)	(0.001)	(0.021)	(0.013)	(0.011)	(0.000)
ϖ_1	0.125*	0.040*	0.063*	0.116*	0.109*	0.094*	0.095*
	(0.027)	(0.007)	(0.007)	(0.023)	(0.015)	(0.014)	(0.018)
ϖ_2	0.565*	0.954*	0.933*	0.797*	0.855*	0.896*	0.893*
	(0.070)	(0.008)	(0.008)	(0.034)	(0.019)	(0.012)	(0.020)
α	0.001**						
	(0.001)						
β	0.753*						
	(0.208)						
Panel C: Robust tests for model standardized residuals							
Skewness	2.186*	−0.638*	−0.285*	−0.028	−0.809*	−0.917*	−0.364*
Kurtosis	35.399*	4.465*	3.505*	1.232*	7.225*	5.451*	1.154*
JB	45746.621†	775.702†	453.529†	54.727	1971.493†	1189.883†	67.031†
Q(12)	13.874	6.846	12.554	7.114	8.807	30.544†	8.691
ARCH(6)	0.367	12.252††	18.186†	1.790	4.263	13.563††	5.236

Note: Bollerslev and Wooldridge's (1992) robust standard errors are given in parentheses. ϖ_0, ϖ_1, and ϖ_2 refer to the parameters of a GARCH(1,1) process. The test for kurtosis coefficient has been normalized to zero. JB is the Jarque–Bera test for normality based on excess skewness and kurtosis. Q(12) is the Ljung–Box test for autocorrelation of order 12. ARCH is the Engle (1982) test for conditional heteroscedasticity.

*, **, and *** indicate significance of coefficients at the 1%, 5%, and 10%, respectively.

†, ††, and ††† indicate rejection of the null hypotheses of no autocorrelation, normality, and homoscedasticity at the 1%, 5%, and 10% levels of significance, respectively, for statistical tests.

significant in most cases. The coefficients relating the return series to the current market returns are, however, insignificant. There is then no world CAPM effect in MENA markets.

The ARCH coefficients and GARCH coefficients reported in Panel B of Table 19.2 are significant for all the countries. In addition, the ARCH coefficients are relatively small in size, which indicates that conditional volatility does not change very rapidly. However, the GARCH coefficients are large, indicating gradual fluctuations over time. On the other hand, the estimates of α and β satisfy the stationary conditions for all the variance and covariance processes.

Diagnostics of standardized residuals are provided in Panel C of Table 19.2. One can remark that the indices of kurtosis are often lower than those for the returns. However, the JB test statistics for normality indicates that the unconditional distribution of the conditionally normal GARCH process is not sufficiently fat-tailed to accommodate the excess kurtosis in the data. This result justifies, once again, the use of the QML procedure.

We also compute the Ljung–Box statistics to test the null hypothesis of absence of autocorrelation and the Engle's (1982) test of absence of ARCH effects. The results show that the specification we use is flexible enough to capture a significant part of the dynamics of the conditional first and second moments.

19.5.2 Conditional Volatilities and Time-Varying Patterns of Comovements

The daily volatility of selected MENA countries and the world stock market index, as measured by the conditional variance, is depicted in Figure 19.1. As expected, stock market in UAE displays the highest level of conditional volatility followed by Egypt. The volatility picks in UAE and most other studied markets are observed in July 2005, May 2006, and February and September 2008 reflecting respectively the crisis of 2005–2006, the rising oil prices, and the recent market instability related to global credit crisis. These findings testify that the MENA stock markets have reacted to local, regional, and global events.

Dynamic conditional correlations within some MENA markets and with the world market as well as their 95% confidence intervals are plotted together in Figure 19.2. If zero line is located between the lower and upper bounds of the confidence interval, the statistical insignificance of the conditional correlation coefficients cannot be rejected. An in-depth

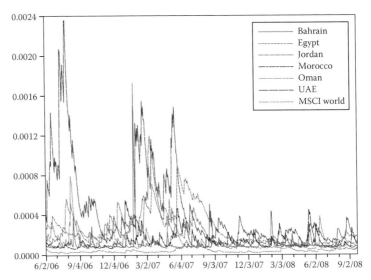

FIGURE 19.1 Time-varying conditional variance of selected countries.

analysis of the results shows that the conditional correlations within MENA markets are relatively low, 14.34% on average, and they vary considerably over time and from a couple of countries to another. It is also observed that the evolution of these correlations witnesses some periods of negative values. Further, most of them have slightly increased at the end of sample period indicating a slight increase in market comovements in the MENA region. Finally, we find insignificant linkages within MENA markets for the following couples of markets over almost the entire study period: Bahrain–Egypt, Bahrain–Jordan, Bahrain–Morocco, Morocco–Oman, and Morocco–UAE.

On the other hand, the average of conditional correlations between the MENA markets that we study and the world market is low, 2.83%. More importantly, except for some very short periods, the comovements of four MENA markets (Bahrain, Jordan, Oman, and UAE) with the world are not statistically different from zero. This empirical finding generally indicates a very low independence of MENA markets on world stock market fluctuations and should mean substantial diversification opportunities from investing in MENA region stock markets. As a prime example, they had a little exposure to the global economic downturns trigged by the subprime crisis. Note, however, that the remaining markets including Egypt and Morocco tend to comove closely with the world stock market.

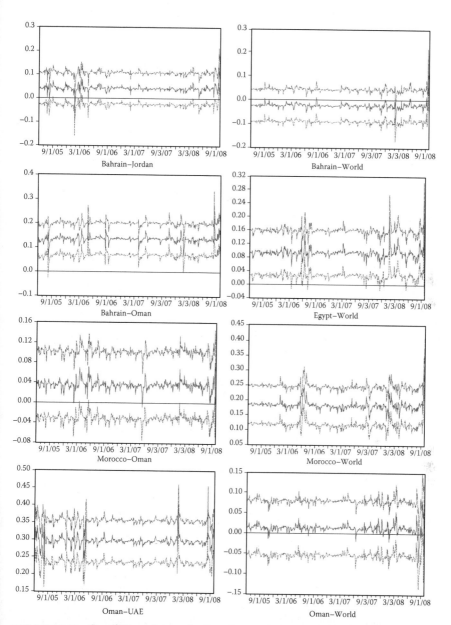

FIGURE 19.2 Conditional correlation dynamics of selected markets.

19.5.3 Perspective of Structural Changes

We now turn to interpret and discuss the empirical results issued from Bai and Perron's (2003) structural test that was applied to the conditional correlation series between the MENA markets and the world market over the sample period. The obtained optimal breakpoints for each market and their 95% confidence intervals are reported in Table 19.3. In Figure 19.3, we present the BIC and residual sum of squares issued from the test of structural breaks in the comovements between MENA and world stock markets.

Accordingly, the null hypothesis of stability is rejected for all the studied markets since the Bai and Perron's (2003) test detects breakpoints in the time-path of the comovements between each of the six MENA stock markets under consideration and the world market. Two significant breakpoints are obtained for Bahrain, Egypt, Jordan, and Morocco, and only one breakpoint is obtained for Oman and UAE. Breakpoints are also obtained for the correlations within the MENA markets. These breaks essentially happened in 2006 as a response to financial liberalization and reforms occurred in the MENA region and in 2007 and 2008 as a result of the subprime crisis.

TABLE 19.3 Optimal Number of Structural Breaks in the Conditional Correlations with the World Market

Market	Optimal Number of Breakpoints	Estimated Break Dates	95% Confidence Intervals for Break Dates
Bahrain	2	August 15, 2007	[March 06, 2007–September 10, 2007]
		March 25, 2008	[January 11, 2006–April 15, 2008]
Egypt	2	July 10, 2007	[May 24, 2007–August 9, 2007]
		January 16, 2008	[October 11, 2007–January 25, 2008]
Jordan	2	May 17, 2006	[March 6, 2006–June 5, 2006]
		November 27, 2006	[September 7, 2006–February 26, 2007]
Morocco	2	September 7, 2006	[August 8, 2006–October 24, 2006]
		January 21, 2008	[September 12, 2007–February 01, 2008]
Oman	1	March 18, 2008	[October 31, 2006–April 8, 2008]
United Arab Emirates	1	December 12, 2007	[May 23, 2007–January 08, 2008]

Note: The breakpoint selection procedure is based on the BIC (see Bai and Perron, 2003). First, we arbitrarily set the maximum number of breaks to be 5. If the effective number of breaks is equal to 5, a higher number of breaks will be chosen so that the testing procedure captures all possible breakpoints. In principle, a model's optimal number of breakpoints is the one associated with the minimum BIC. For the countries considered in this present study, none of the volatility series has more than five breakpoints.

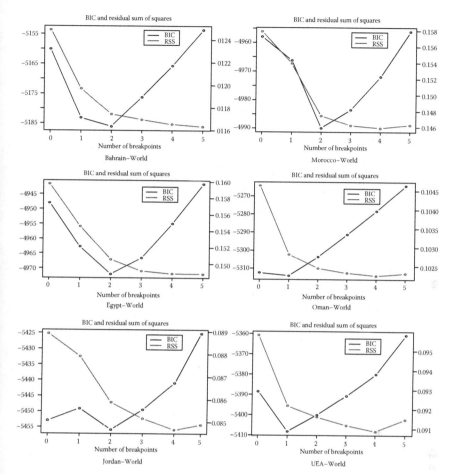

FIGURE 19.3 BIC and residual sum of squares (RSS) for models with *m* breakpoints.

It is, however, important to mention that the impact of these events in the conditional correlations is low in magnitude and that there is no clear upward trend in market comovements over the sample period. Thus, recent market reforms and liberalization have not increased the financial integration within the MENA region and between MENA markets and the world market. Using different methodologies, other studies reached the same conclusions (Darrat et al., 2000; Abraham et al., 2001; Achy, 2005). Clearly, if the risk sharing and diversification opportunities are desirable attributes, policymakers of MENA countries should pursue more liberalization of their capital markets to attract more foreign direct and portfolio investments.

19.6 CONCLUSION

The objective of this chapter was to explore the time-varying characteristics of stock market linkages in the MENA region. From the practical point of view, the obtained results pointed out important implications in asset choices and portfolio allocations in the MENA region. First, we observe that the overall level of market integration within the MENA region is increasing but is still relatively low. Diversification within the MENA region is still beneficial, because market return behavior is, as indicated by our empirical results, far from homogeneous. Second, the linkages between MENA markets and the world market are generally weak and significant only in two cases. There are then substantial opportunities for global investors to improve their portfolio risk-return performance. This may, however, change temporarily as financial markets currently embrace more foreign investments.

To sum up, the results of this study indicate that the changing dynamics and contemporaneous stock market interactions within the MENA region yield substantial intraregional diversification benefits and suggest the inclusion of regional equity in a global portfolio.

REFERENCES

Abraham, A., Seyyed, F.J., and Al-Elg, A. (2001) Analysis of diversification benefits of investing in the emerging Gulf equity markets. *Managerial Finance*, 27(10–11): 47–57.

Achy, L. (2005) Financial liberalization, savings, investment, and growth in MENA countries. *Research in Middle East Economics*, 6: 67–94.

Bai, J. and Perron, P. (2003) Computation and analysis of multiple structural change models. *Journal of Applied Econometrics*, 18(1): 1–22.

Bekaert, G. and Harvey, C. (1995) Time-varying world market integration. *Journal of Finance*, 50(2): 403–444.

Bley, J. (2007) How homogeneous are the stock markets of the Middle East and North Africa? *Quarterly Journal of Finance and Business*, 46(3): 3–24.

Bollerslev, T. and Wooldridge, J.M. (1992) Quasi maximum likelihood estimation and inference in dynamic models with time-varying covariances. *Econometric Review*, 11(2): 143–172.

Darrat, A., Elkhal, K., and Hakim, S. (2000) On the integration of emerging stock markets in the Middle East. *Journal of Economic Development*, 25(2): 61–78.

Engle, R.F. (1982) Autoregressive conditional heteroscedasticity with estimates of the variance of UK inflation. *Econometrica*, 50(4): 987–1008.

Engle, R.F. (2002) Dynamic conditional correlation: a new simple class of multivariate GARCH models. *Journal of Business and Economic Statistics*, 20(3): 339–350.

Errunza, V., Losq, E., and Padmanabhan, P. (1992) Tests of integration, mild segmentation and segmentation hypotheses. *Journal of Banking and Finance*, 16(5): 949–972.

Girard, E. and Ferreira, E.J. (2004) On the evolution of inter- and intra-regional linkages to Middle East and North African capital markets. *Quarterly Journal of Business and Economics*, 43(1–2): 21–43.

Girard, E., Omran, M., and Zaher, T. (2003) On risk and return in MENA capital markets. *International Journal of Business*, 8(3): 285–314.

Maghyereh, A. (2006) Regional integration of stock markets in MENA countries. *Journal of Emerging Market Finance*, 5(1): 59–94.

Omran, M. and Gunduz, L. (2001) Stochastic trends and stock prices in emerging markets: The case of Middle East and North Africa region. *Istanbul Stock Exchange Review*, 5: 3–16.

CHAPTER 20

Empirical Analysis of Herding Behavior in Asian Stock Markets

Thomas C. Chiang and Lin Tan

CONTENTS

20.1 INTRODUCTION

Herding is often used to describe the correlation in trade resulting from interactions between investors. In the emerging financial markets, herding activity among investors is perceived to be more apparent. The reason is straightforward, since investors in the emerging markets lack the financial knowledge to assess economic/financial data that can be used to formulate a timely, rational investment strategy. Thus, mimicking financial gurus or seeking advice from successful investors appears to be an optimal strategy. Doing so minimizes the transaction costs. The consequence of this herding behavior is "a group of investors trading in the same direction over a period of time (Nofsinger and Sias, 1999)." Empirically, this may lead to some observable time series patterns that are correlated across individuals, bringing about systematic, erroneous decision making by the entire population (Bikhchandani et al., 1992). As a result, herding behavior can cause stock prices to depart from their underlying economic fundamentals, frustrating the efficient functioning of the markets.

Empirical investigations of herding behavior in financial markets can be divided into two approaches.* The first approach is to employ dynamic time series models to explore whether stock index returns display comovements. For instance, using correlation analysis, Boyer et al. (2006) estimated and compared the degree to which accessible and inaccessible stock index returns comove with the index returns of the country in crisis. Their study suggests that in emerging stock markets, there is greater comovement during high-volatility periods, indicating that crises that spread through the asset holdings of international investors are mainly due to contagion rather than to a change in fundamentals. Chiang et al. (2007) applied a dynamic conditional-correlation model (Engle, 2002) to examine daily stock-return data from six Asian markets from 1990 to 2003. Their empirical evidence suggests that the contagion effect takes place during the early stage of the Asian financial crisis and that herding behavior dominates the later stage of the crisis.

---

* A number of research papers on herding behavior of economic agents, from mutual fund managers to institutional analysts, have been studied. For instance, Grinblatt et al. (1995) find evidence of herding activity in mutual fund markets because fund managers tend to buy securities that can make a profit. Welch (2000) finds that the most recent revisions of investment recommendations have a positive influence on the next analyst's revision. Lakonishok et al. (1992) report that pension managers are buying (selling) the same stocks that other managers buy (sell) and follow a positive-feedback trading strategy. Wermers (1999) finds more evidence of herding in trades of small stocks and in trades by growth-oriented funds.

The second approach for examining herding behavior focuses on the cross-sectional correlation dispersion in stock returns in response to excessive movements in market conditions. By extending Christie and Huang's analysis (1995), Chang et al. (2000) studied international herding behavior and found significant evidence of herding in South Korea and Taiwan, and partial evidence of herding in Japan. However, there is no evidence of herding on the part of market participants in the United States and Hong Kong. Following the same approach, Demirer and Kutan (2006) test whether investors in Chinese markets, in making their investment decisions, are following market consensus rather than private information during periods of market stress. Their testing results find no evidence of herd formation, suggesting that market participants in the Chinese stock markets make investment choices rationally. However, Tan et al. (2008) studied the Chinese stock markets and reported that herding occurs under both rising and falling market conditions. By dividing the data between A share and B share markets, they find that herding behavior is more apparent in A share investors. Thus, the evidence emerging from the Chinese market is mixed.

Although the aforementioned studies have provided some evidence of herding behavior, the coverage of the markets is rather limited. More crucially, the evidence is very diverse. This motivates us to form a testable model to examine the herding behavior for a group of Asian markets, including China, Indonesia, Thailand, South Korea, Hong Kong, and Japan. There are good reasons for choosing these Asian markets. First, these markets consist of both emerging and advanced markets. It provides a forum for comparing whether participants in both types of markets display a similar behavior while responding to extreme market conditions. Second, earlier studies of herding behavior find mixed empirical evidence in these markets. Employing cross-country data in the same region allows us to derive consistent investment behavior in relation to the existing literature.

The remainder of this chapter is organized as follows: Section 20.2 presents the estimation procedure for examining herding behavior; Section 20.3 describes the data; Section 20.4 reports the empirical evidence of herding behavior; Section 20.5 tests the information of return dispersions in the mean-variance framework; and Section 20.6 contains conclusions.

20.2 LITERATURE REVIEW

Christie and Huang (1995) noted that the investment decision-making process used by market participants depends on overall market conditions. In particular, during normal periods, rational asset-pricing models predict that the dispersion in cross-sectional returns will move with the absolute value of the market returns, since individual investors are trading based on their own private information, which is expected to be diverse. However, during periods of extreme market movements, individuals tend to suppress their own information, and their investment decisions are more likely to mimic collective actions in the market. Individual stock returns under these situations tend to cluster around the overall market return. Thus, herding will be more apparent during periods of market stress. To measure the return dispersion, Christie and Huang (1995) propose the cross-sectional standard deviation (CSSD) method, which is expressed as

$$\text{CSSD}_t = \sqrt{\frac{\sum_{i=1}^{N}(R_{i,t} - R_{m,t})^2}{(N-1)}} \qquad (20.1)$$

where
    N is the number of firms in the portfolio
    $R_{i,t}$ is the observed stock return of firm or industry i at time t
    $R_{m,t}$ is the cross-sectional average stock of N returns in the portfolio at time t

Alternatively, Chang et al. (2000) suggest the cross-sectional absolute deviation (CSAD) as the return dispersion (RD_t):

$$RD_t \equiv \text{CSAD}_t = \frac{1}{N}\sum_{i=1}^{N}\left|R_{i,t} - R_{m,t}\right| \qquad (20.2)$$

To conduct an empirical test for herding activity, we specify a regression model in the spirit of Chang et al. (2000) as

$$RD_t = \beta_0 + \beta_1 R_{m,t} + \beta_2 \left|R_{m,t}\right| + \beta_3 R_{m,t}^2 + \varepsilon_t \qquad (20.3)$$

where CSAD_t will be used as a measure of return dispersion and $R_{m,t}$ is the equally weighted average stock return over all industries. The rationale of Chang et al.'s (2000) model is based on a simple formulation that there is a linear relationship between the dispersion in individual asset returns and the return on a market portfolio. As the absolute value of the market return increases, so should the dispersion in individual asset returns. During periods of relatively large movements in market prices, investors may react in a more uniform manner, exhibiting herding activity. This behavior is likely to increase the correlation among asset returns, and the corresponding dispersion among returns will decrease or, at least, increase at a less-than-proportional rate with the market return. For this reason, a nonlinear market return is included in the test equation. If a testing result comes up with a significantly negative coefficient of the market return squared, β_3, the evidence would be consistent with the occurrence of herding behavior.

20.3 DATA

Daily data employed in this chapter consist of industry and market price indices. The samples cover China (CN), Hong Kong (HK), Indonesia (ID), Japan (JP), South Korea (KR), and Thailand (TH). The data sample varies from market to market, depending on the availability of daily data. The starting time is January 3, 1996, for China; November 1, 1990, for Hong Kong; November 1, 1991, for Indonesia; May 3, 1988, for Japan and South Korea; and November 6, 1991, for Thailand. The ending period for the data is April 24, 2008. The exception is China, which ends on April 28, 2007. The stock return is calculated as $R_t = 100 \times [\log(P_t) - \log(P_{t-1})]$, where P_t denotes either the market index or the industrial stock index. All of the data are taken from the *Datastream International*.[*]

20.4 EMPIRICAL EVIDENCE

20.4.1 Preliminary Result

Equation 20.3 is estimated by using the Newey–West consistent estimator and the results are reported in Table 20.1.[†] The estimated statistics include markets in China (CN), Hong Kong (HK), Indonesia (ID), South Korea (SK), Thailand (TH), and Japan (JP).

The evidence indicates that with the exception of the Chinese market in β_1, the coefficients of both market return and absolute return are positive and statistically significant. As stated earlier, a negative value of the coefficient of $R_{m,t}^2$ is consistent with the existence of herding behavior. The evidence in Table 20.1 suggests that β_3 bears a negative sign and is statistically significant. This finding holds true for all markets under investigation. Our

---

[*] Anastasios Pisimisis and DaZhi Zheng provided assistance in collecting the data from *DataStream International*.

[†] Both CSAD and CSSD measures were used to generate the results. Since using a squared formula tends to be more sensitive to outliers (Schwert and Sequin, 1990), only CSAD results are reported to save space.

TABLE 20.1   Estimates of Cross-Sectional Return Dispersions and Market Returns

Market	β_0	β_1	β_2	β_3	$\bar{R}^2$
CN	2.1532	−0.0199	0.3388	−0.0221	0.09
	(42.88)***	(−1.49)	(7.62)***	(2.91)***	
HK	0.9048	0.0543	0.4933	−0.0093	0.39
	(44.29)***	(4.60)***	(18.40)***	(2.00)**	
ID	0.8276	0.0348	0.8326	−0.0289	0.51
	(40.00)***	(2.77)***	(26.33)***	(4.83)***	
KR	0.9239	0.0109	0.4227	−0.0220	0.29
	(41.95)***	(1.53)	(18.42)***	(5.36)***	
TL	0.9156	0.0251	0.5367	−0.0189	0.38
	(46.26)***	(2.24)**	(20.45)***	(3.29)***	
JP	0.6223	0.0250	0.3245	−0.0219	0.28
	(50.36)***	(5.50)***	(22.37)***	(6.18)***	

Notes: The estimations of this table are based on daily industry indices. The starting dates of the data vary for different markets and are subject to limitation or availability. The starting dates are 5/5/1998, 8/22/1990, 5/03/1988, 5/03/1988, 5/05/1988, and 1/1/1996 for Hong Kong (HK), Indonesia (ID), Japan (JP), South Korea (SK), Thailand (TL), and China (CN), respectively; the ending date is 4/24/2008, except for China, which ends on 4/30/2007. The estimated equation is expressed as follows:

$$RD_t = \beta_0 + \beta_1 R_{m,t} + \beta_2 |R_{m,t}| + \beta_3 R_{m,t}^2 + \varepsilon_t$$

where $RD_t = \frac{1}{N}\sum_{i=1}^{N}|R_{i,t} - R_{m,t}|$, $R_{m,t}$ is the equally weighted value of realized returns of all industry indexes on date t. The numbers in the parentheses are the absolute values of t-statistics. The statistics are obtained from the Newey–West consistent estimator.

***,**,* denote that the coefficient is significant at the 1%, 5%, and 10% levels, respectively.

evidence is in contrast to the results reported by Chang et al. (2000), who find no evidence of herding in the HK market and only partial evidence of herding in the JP market. The evidence also conflicts with the findings of Demirer and Kutan (2006), which indicate no evidence of herding behavior and conclude that market participants in the Chinese stock markets are motivated by rational behavior in making investment decisions. The result, however, is consistent with the evidence presented by Tan et al. (2008), who find evidence of herding in the Chinese dual listing firm data.

20.4.2 Empirical Refinement

Although Equation 20.3 is able to capture the spirit of herding behavior proposed by Chang et al. (2000), some empirical refinements are necessary. First, the return dispersion variable, RD_t, is likely to present some time series patterns. It is more convenient to specify the RD_t series in an autoregressive process. Second, outliers are often shown in the data. If we do not take care of the outliers, the estimated results can be biased. To incorporate these elements into the model, the test equation can be rewritten as

$$(1 - \phi_1 B - \phi_2 B^2 - \cdots - \phi_s B^p) RD_t = \beta_0 + \beta_1 R_{m,t} + \beta_2 |R_{m,t}| + \beta_3 R_{m,t}^2 + \lambda_i D_i + \varepsilon_t \tag{20.4}$$

where

B denotes a backward-shift operator, such that $B^n X_t \equiv X_{t-n}$

ϕ_s, β_j, and λ_i are constant parameters

The term $\lambda_i D_i$ is an indicator variable for capturing an outlier on date i. We set these dates to be unity if an outlier is present, and zero otherwise.* This specification in Equation 20.4 is considered to be a more general model, since by imposing $\phi_s = \lambda_i = 0$, Equation 20.4 will reduce to Equation 20.3. By construction, Equation 20.3 is nested in Equation 20.4. Alternatively, we can view lagged RD_t and D_i as the control variables that ensure the robustness of the test equation.

Empirical estimates of Equation 20.4 are presented in Table 20.2.† Our statistical results are consistent with previous findings. As noted earlier, a negative value of the coefficient of $R_{m,t}^2$ is consistent with the existence of herding behavior. The evidence of β_3 in Table 20.2 confirms that herding is present in all the markets, while controlling the lagged variables of RD_t and outliers, D_i. As may be seen in the last column, the values of the adjusted R-squared for different markets have increased substantially compared with those reported in Table 20.1. Two points are worth noting. First, relative to Equation 20.3, all of the incremental variables are statistically significant, suggesting that the original specification proposed by Chang et al. (2000) may suffer from a missing variable problem and lead to biased estimates. Specifically, Equation 20.4 can be rewritten as

$$RD_t = \frac{\beta_0}{(1 - \varphi(B))} + \frac{\beta_1}{(1 - \varphi(B))} R_{m,t} + \frac{\beta_2}{(1 - \varphi(B))} |R_{m,t}| + \frac{1}{(1 - \varphi(B))} \sum_{i=1}^{k} \lambda_i D_i + v_t \tag{20.5}$$

---

* The procedures for identifying and estimating outliers can be found in Tsay (1988) and Peña (2001, pp. 147–151).
† We also estimate the dummy variables on the intercept and slope of market stock returns squared for testing the impact of the Asian crisis for the period from July 2, 1997, through December 24, 1998. The evidence (not reported) indicates a shift in the intercept dummy. However, we cannot find statistical significance on the slope of $R_{m,t}^2$.

TABLE 20.2 Estimates of Cross-Sectional Return Dispersions and Market Returns with AR(p) and Outliers

Market	β_0	β_1	β_2	β_3	ϕ_1	ϕ_2	ϕ_3	λ_1	λ_2	λ_3	$\bar{R}^2$
CN	2.4202	−0.0543	0.1188	−0.0181	0.4566	0.2547	0.1358	2.7523	3.0008	3.2363	0.58
	(33.08)***	(6.46)***	(4.33)***	(2.96)***	(22.61)***	(11.13)***	(7.16)***	(12.19)***	(10.88)***	(15.52)***	
HK	1.0042	0.0514	0.3776	−0.0071	0.3325	0.1890		9.6634	7.6735	4.5963	0.55
	(46.43)***	(6.08)***	(17.11)***	(2.05)**	(17.16)***	(9.48)***		(62.60)***	(53.19)***	(48.57)***	
ID	1.0062	0.0251	0.6437	−0.0242	0.3272	0.1537	0.1668	6.7049	6.3311	5.6942	0.63
	(27.54)***	(2.55)**	(20.83)***	(4.89)***	(13.70)***	(6.56)***	(9.00)***	(44.97)***	(55.48)***	(54.98)***	
KR	1.0638	0.0067	0.2944	−0.0196	0.3416	0.2889		5.9732	4.4987	3.4944	0.48
	(40.87)***	(1.07)	(13.90)***	(4.62)***	(18.27)***	(15.69)***		(28.10)***	(14.14)***	(17.48)***	
TL	1.0529	0.0187	0.3955	−0.0151	0.3017	0.1616	0.2277	2.5314	3.1330	2.5939	0.56
	(31.99)***	(2.03)**	(14.79)***	(2.46)**	(12.06)***	(6.42)***	(10.72)***	(2.59)***	(5.49)***	(5.41)***	
JP	0.6789	0.0187	0.2454	−0.0196	0.3945	0.2825		1.0587	1.6569	1.5958	0.52
	(51.24)***	(5.43)***	(20.49)***	(6.52)***	(24.31)***	(15.17)***		(6.40)***	(21.40)***	(8.52)***	

Notes: The estimated equation is as follows: $(1 - \phi_1 B - \phi_2 B^2 - \cdots - \phi_p B^p) RD_t = \beta_0 + \beta_1 R_{m,t} + \beta_2 |R_{m,t}| + \beta_3 R_{m,t}^2 + \lambda_i D_i + \varepsilon_t$, where $RD_t = \dfrac{1}{N} \sum_{i=1}^{N} |R_{i,t} - R_{m,t}|$, $R_{m,t}$ is the equally weighted value of realized returns of all industry indexes on date t, D_i is an indicator variable, and i denotes the top three outliers. The numbers in the parentheses are absolute values of t-statistics. The statistics are obtained from the Newey–West consistent estimator.

***, **, * denote significance at the 1%, 5%, and 10% levels, respectively.

where $(1 - \phi(B)) = (1 - \phi_1 B - \phi_2 B^2 - \cdots - \phi_s B^p)$. In terms of Equation 20.5, the corresponding values of β_3 are $-0.0193, -0.0124, -0.073, -0.0280, -0.0258,$ and -0.0266 for China (CN), Hong Kong (HK), Indonesia (ID), South Korea (KR), Thailand (TH), and Japan (JP), respectively.[*] It is evident that the difference in the estimated coefficient, β_3, is due to the significance of the lagged variables of RD_t.

Second, the coefficients of outliers are highly significant. As noted by Peña (2001), if the outliers are not removed, both estimated parameters and residuals will be affected. Fortunately, the large sample used in our estimation will decrease the threat of the outlier effect.

Third, the cross-sectional return dispersion has substantial information as reflected in the explanatory power of the estimated equation. As the model stands, stock return dispersion contains information influenced by market return, market volatility, outliers, and lagged shocks. The RD_t variable should serve as a proxy for risk and should be priced in the asset valuation.

20.5 RETURN DISPERSION AND RISK

As stated in the textbooks on finance, the risk-return apparatus plays a central role in advising portfolio decisions. Merton's (1980) pioneer research on the intertemporal capital asset pricing model (CAPM) (ICAPM) postulates a positive relation between expected excess returns and risk. Following Merton's theoretical prediction, French et al. (1987), Baillie and DeGennaro (1990), Scruggs (1998), and Lundblad (2007) tested the null hypothesis by relating expected stock return to conditional variance and found evidence for a positive relation between stock return and risk. However, Campbell (1987), Breen et al. (1989), Nelson (1991), and Glosten et al. (1993) tested the same hypothesis and documented a negative relation. Thus, the empirical evidence on the risk-return trade-off is somewhat inconclusive.

In a recent research paper, Bekaert and Harvey (1997) investigated cross-country variations in market-level stock volatility and documented that a higher return dispersion is associated with higher market volatility for more developed markets. They suggest that dispersion may reflect the magnitude of industry-level information flow for these markets. Motivated by these empirical studies, we examine the effect of the RD in relation to stock returns.

---

[*] For instance, the coefficient of β_3 for Japan is $[-0.0196/(1-0.0187-0.2454)] = -0.0266$. The same calculation applies to other markets.

Following the conventional approach, we specify both the mean and the variance equations as

$$R_{m,t} = \delta_0 + \delta_1 \sigma_t + \delta_2 \, RD_t + \varepsilon_t \tag{20.6}$$

$$\sigma_t^2 = \omega_0 + \omega_1 \varepsilon_{t-1}^2 + \omega_2 \sigma_{t-1}^2 + \omega_3 I[\varepsilon_{t-1} < 0]\varepsilon_{t-1}^2 \quad \text{with } I = 1 \text{ if } \varepsilon_{t-1} < 0,$$
$$\text{and 0 otherwise} \tag{20.7}$$

where Equation 20.6 is the mean equation; $R_{m,t}$ is the aggregate stock returns for (CN), Hong Kong (HK), Indonesia (ID), South Korea (KR), Thailand (TH), and Japan (JP), respectively; the conditional standard deviation, σ_t, is generated from a variance equation represented by Equation 20.7; RD_t is the return dispersion; δ and ω are constant parameters; and ε_t is a random-error term. In Equation 20.7, good news is associated with $\varepsilon_{t-1} > 0$, and bad news with $\varepsilon_{t-1} < 0$. The different signs have differential effects on the conditional variance; good news has an impact captured by ω_1, while bad news has an impact of $\omega_1 + \omega_3$. If $\omega_3 > 0$, bad news increases stock-return volatility. This asymmetric GARCH-in-mean model was popularized by Glosten et al. (1993), Koopman and Uspensky (2002), and Cappiello et al. (2006), among others.

In sum, taking Equations 20.6 and 20.7 together allows us to test a number of hypotheses by examining the following restrictions. First, the asymmetric effect can be tested by examining the restriction of $\omega_3 = 0$. Second, the risk-return trade-off hypothesis can be tested by examining whether $\delta_1 > 0$ and statistically significant in Equation 20.6. Third, an incremental efficiency of RD_t can be tested by testing $\delta_2 = 0$. In order to account for the risk factor, we require the coefficient to be positive and statistically significant.

Equations 20.6 and 20.7 are estimated twice, one with a restriction of $\delta_2 = 0$, for which the estimates are reported in Table 20.3, and one without that restriction, for which the results are reported in Table 20.4. Since both the tables yield similar results and Table 20.4 contains more information, our explanation will focus on Table 20.4. The findings are consistent with most AGARCH(1,1)-in-mean estimates; the coefficients of the variance equation, in all of the cases, are highly significant, indicating that stock-return volatilities are characterized by a heteroscedastic process. With the exception of the Chinese market, the estimated value of ω_3 is

TABLE 20.3   AGARCH Regression Results for Asian Markets

Market	δ_0	δ_1	ω_0	ω_1	ω_2	$\omega_3[\varepsilon_{t-1} < 0]$	$-$Log LL
CN	−0.0932	0.0451	0.0956	0.1101	0.8499	0.0167	4998
	(2.12)**	(2.60)***	5.81)***	(8.10)***	(94.19)***	(0.92)	
HK	0.0581	−0.0044	0.0829	0.0567	0.8390	0.1162	7131
	(2.76)***	(0.33)	(7.07)***	(3.74)***	(55.31)***	(6.00)***	
ID	−0.0001	0.0000	0.0853	0.0890	0.8376	0.0950	7224
	(0.00)	(0.00)	(6.24)***	(6.16)***	(53.17)***	(4.41)***	
KR	−0.0009	0.0064	0.0454	0.0619	0.8887	0.0756	9201
	(0.04)	(0.65)	(5.32)***	(5.95)***	(84.25)***	(5.00)***	
TL	−0.0001	0.0000	0.0853	0.0890	0.8376	0.0950	7224
	(0.00)	(0.00)	(6.24)***	(6.16)***	(53.17)***	(4.41)***	
JP	0.0022	0.0100	0.0274	0.0347	0.8778	0.1215	7051
	(0.12)	(0.52)	(7.35)***	(4.89)***	(98.21)***	(10.98)***	

Notes: The estimated equations are $R_{m,t} = \delta_0 + \delta_1\sigma_t + \varepsilon_t$ and $\sigma_t^2 = \omega_0 + \omega_1\varepsilon_{t-1}^2 + \omega_2\sigma_{t-1}^2 + \omega_3 I + [\varepsilon_{t-1} < 0]\varepsilon_{t-1}^2$ with $I = 1$ if $\varepsilon_{t-1} < 0$, and 0 otherwise. The first equation is the mean equation and the second equation is the variance equation; $R_{m,t}$ is the market stock return; σ_t is the conditional standard deviation generated from a variance equation. The numbers in parentheses are absolute values of z-statistics. $-$Log LL = $-$log likelihood statistic.

AGARCH = asymmetric generalized autoregressive conditional heteroscedasticity.

***, **, * denote that the coefficient is significant at the 1%, 5%, and 10% levels, respectively.

positive and statistically significant, suggesting that stock returns react more profoundly to bad news than to good news, which is consistent with the asymmetric hypothesis. However, the coefficient of the conditional standard deviation in the mean equation for each market does not produce a significant sign. The only exception is the Chinese market. This means that the trade-off hypothesis holds true for the Chinese market but not for other Asian markets. Apparently, the statistics in Table 20.4 cannot find supporting evidence for the risk-return trade-off. This finding is not surprising, since the evidence is consistent with the findings reported by Breen et al. (1989), Nelson (1991), and Koopman and Uspensky (1999).

However, inspecting the statistics for the return dispersion, we find that, with the exception of China, the coefficient of market returns and RD_t is positive and highly significant. This finding holds true for the equation of five Asian markets, supporting the risk-return trade-off hypothesis. However, the risk factor here consistent with Connolly and Stivers (2006) is associated with the cross-sectional return dispersion rather than the conditional standard deviation derived from the generalized autoregressive conditional heteroscedasticity (GARCH) process. This result is

TABLE 20.4 AGARCH Regression Results for Asian Markets

Market	δ_0	δ_1	δ_2	ω_0	ω_1	ω_2	$\omega_3[\varepsilon_{t-1} < 0]$	$-$Log LL
CN	0.0328	0.0327	−0.0181	0.3200	0.3062	0.7991	−0.0192	5174
	(0.94)	(5.79)***	(1.35)	(3.11)***	(3.85)***	(27.06)***	(0.31)	
HK	−0.0033	−0.0056	0.1451	0.0914	0.0532	0.8267	0.1194	7081
	(0.11)	(−0.43)	(7.01)***	(7.19)***	(2.85)***	(51.43)***	(5.49)***	
ID	0.0188	−0.0211	0.0636	0.0207	0.0667	0.9026	0.0537	7782
	(0.64)	(1.74)*	(5.91)***	(9.78)***	(13.83)***	(249.39)***	(7.87)***	
KR	−0.0503	0.0020	0.1256	0.0455	0.0634	0.8856	0.0704	9179
	(1.40)	(0.18)	(4.32)***	(5.43)***	(5.68)***	(83.32)***	(4.78)***	
TL	0.01163	−0.0136	0.0865	0.0766	0.0881	0.8451	0.0817	7210
	(0.35)	(1.13)	(3.45)***	(5.95)***	(5.75)***	(55.38)***	(4.04)***	
JP	−0.0635	−0.0126	0.1381	0.0266	0.0303	0.8796	0.1215	7040
	(2.12)**	(0.60)	(4.01)***	(7.49)***	(4.29)***	(101.85)***	(11.40)***	

Notes: The estimated equations are $R_{m,t} = \delta_0 + \delta_1 \sigma_t + \delta_2\,RD_t + \varepsilon_t$, and $\sigma_t^2 = \omega_0 + \omega_1 \varepsilon_{t-1}^2 + \omega_2 \sigma_{t-1}^2 + \omega_3 I + [\varepsilon_{t-1} < 0]\varepsilon_{t-1}^2$ with $I = 1$ if $\varepsilon_{t-1} < 0$, and 0 otherwise. The first equation is the mean equation and the second equation is the variance equation; $R_{m,t}$ is the market stock return; σ_t is the conditional standard deviation generated from a variance equation; RD_t is stock-return dispersions. The numbers in parentheses are absolute values of z-statistics. $-$Log LL = $-$log likelihood statistic.

***, **, * denote that the coefficient is significant at the 1%, 5%, and 10% levels, respectively.

logical and convincing based on our earlier analysis, since the RD_t has substantial information about market variability. As shown in the previous section, the RD_t variable contains risk due to extreme market movements or market variability resulting from sector rotations or portfolio adjustments across different industries. The level of conditional return volatility derived from the traditional GARCH-type model fails to carry the information arising from cross-sectional variability.

20.6 CONCLUSIONS

This chapter examines the herding activity of the investors, for a group of Asian stock markets, including China, Indonesia, Thailand, South Korea, Hong Kong, and Japan. We examine national and industrial stock indices from 1988 through 2008 and test whether these markets exhibit herding behavior. The evidence from our study strongly suggests that herding is present in each Asian market under study. The result stands in contrast to the earlier literature, which finds no evidence of herding on the part of participants in the Hong Kong market (Chang et al., 2000). Our evidence also conflicts with the findings of Demirer and Kutan (2006), which indicates no evidence of herding behavior for investors in the Chinese stock markets. Our testing results also uncover two drawbacks for the conventional herding equation: a missing lagged, dependent-variable problem and the exclusion of outliers. These specification errors lead to biased estimates.

This chapter also contributes to the study of asset pricing by presenting evidence to justify the role of cross-sectional return dispersion. Because of the informational of RD_t for modeling risk, we show that, with the exception of the Chinese market, RD_t is capable of outperforming σ_t, the conditional standard deviation derived from an asymmetric GARCH-in-mean model.

REFERENCES

Baillie, R. and R. DeGennaro (1990) Stock returns and volatility. *Journal of Financial and Quantitative Analysis*, 25(2): 203–214.

Bekaert, G. and C. Harvey (1997) Emerging equity market volatility. *Journal of Financial Economics*, 43(1): 29–77.

Bikhchandani, S., D. Hirshleifer, and I. Welch (1992) A theory of fads, fashion, custom, and cultural change as informational cascades. *Journal of Political Economy*, 100(5): 992–1026.

Boyer, B., T. Kumagai, and K. Yuan (2006) How do crises spread? Evidence from accessible and inaccessible stock indices. *Journal of Finance*, 61(2): 957–1003.

Breen, W., L. Glosten, and R. Jagannathan (1989) Economic significance of predictable variations in stock index returns. *Journal of Finance*, 44(5): 1177–1189.

Campbell, J. (1987) Intertemporal asset pricing without consumption data. *American Economic Review*, 83(3): 487–512.

Cappiello, L., R. Engle, and K. Sheppard (2006) Asymmetric dynamics in the correlations of global equity and bond returns. *Journal of Financial Econometrics*, 4(4): 537–572.

Chang, E.C., J.W. Cheng, and A. Khorana (2000) An examination of herd behavior in equity markets: An international perspective. *Journal of Banking and Finance*, 24(10): 1651–1679.

Chiang, T.C., B.N. Jeon, and H. Li (2007) Dynamic correlation analysis of financial contagion: Evidence from Asian markets. *Journal of International Money and Finance*, 26(7): 1206–1228.

Christie, W.G. and R.D. Huang, (1995) Following the pied piper: Do individual returns herd around the market? *Financial Analysts Journal*, 51(4): 31–37.

Connolly, R. and C. Stivers (2006) Information content and other characteristics of daily cross-sectional dispersion in stock returns. *Journal of Empirical Finance*, 13(1): 79–112.

Demirer, R. and A.M. Kutan (2006) Does herding behavior exist in Chinese stock markets? *Journal of International Financial Markets, Institutions and Money*, 16(2): 123–142.

Engle, R.E. (2002) Dynamic conditional correlation: A simple class of multivariate generalized autoregressive conditional heteroskedasticity models. *Journal of Business and Economic Statistics*, 20(3): 339–350.

French, K., W. Schwert, and R. Stambaugh (1987) Expected stock returns and volatility. *Journal of Financial Economics*, 19(1): 3–29.

Glosten, L.R., R. Jagannathan, and D. Runkle (1993) On the relation between the expected value and the volatility of the normal excess return on stocks. *Journal of Finance*, 48(5): 1779–1801.

Grinblatt, M., S. Titman, and R. Wermers (1995) Momentum investment strategies, portfolio performance, and herding: A study of mutual fund behavior. *American Economic Review*, 85(5): 1088–1105.

Koopman, S. and E. Uspensky (2002) The stochastic volatility in mean model: Empirical evidence from international stock markets. *Journal of Applied Econometrics*, 17(6): 667–689.

Lakonishok, J., A. Shleifer, and R.W. Vishny (1992) The impact of institutional trading on stock prices. *Journal of Financial Economics*, 32(1): 23–44.

Lundblad, C. (2007) The risk return tradeoff in the long run: 1836–2003. *Journal of Financial Economics*, 85(1): 123–150.

Merton, R. (1980) On estimating the expected return on the market: An exploratory investigation. *Journal of Financial Economics*, 8(4): 323–361.

Nelson, D. (1991) Conditional heteroskedasticity in asset returns: A new approach. *Econometrica*, 59(2): 347–370.

Newey, W. and K. West (1987) A simple, positive semi-definite, heteroskedasticity and autocorrelation covariance matrix. *Econometrica*, 55(3): 703–708.

Nofsinger, J. and R. Sias (1999) Herding and feedback trading by institutional and individual investors. *Journal of Finance*, 54(6): 2263–2295.

Peña, D. (2001) Outliers, influential observations, and missing data. D. Peña, G.C. Tiao, and R.S. Tsay (Eds.), *A Course in Time Series Analysis*, John Wiley: New York, pp. 136–170.

Schwert, G.W. and P.J. Sequin (1990) Heteroskedasticity in stock returns. *Journal of Finance*, 45(4): 1129–1155.

Scruggs, J. (1998) Resolving the puzzling intertemporal relation between the market risk premium and conditional market variance: A two-factor approach. *Journal of Finance*, 52(2): 575–603.

Tan, L., T.C. Chiang, J. Mason, and E. Nelling (2008) Herding behavior in Chinese stock markets: An examination of A and B shares. *Pacific-Basin Finance Journal*, 16(1 and 2): 61–77.

Tsay, R. (1988) Outliers, level shifts, and variance changes in time series. *Journal of Forecasting*, 7(1): 1–20.

Welch, I. (2000) Herding among security analysts. *Journal of Financial Economics*, 58(3): 369–396.

Wermers, R. (1999) Mutual fund herding and the impact on stock prices. *Journal of Finance*, 54(2): 581–622.

CHAPTER 21

Institutions and Investment Activities in the Venture Capital Industry: Evidence from China, Hong Kong, and India

Anson L.K. Wong and Michael C.S. Wong

CONTENTS

21.1 INTRODUCTION

Over the past three decades, there has been much debate on the role of venture capital (VC) in promoting technological innovation and economic growth. The supply of funds and value-added services are viewed as the most significant functions of venture capitalists. The financial system, government policies, the existence of small- and medium-sized enterprises, the stage of development of an economy, and availability of a strong legal system are the factors regarded as important determinants of the success of the VC industry. India and China have the combined key elements needed for venture investment opportunities: a large mass manufacturing base, engineering and entrepreneurial talent, a large domestic market, capitals supplied, and innovations (Forer and Yonge, 2004).

Although there have been many papers on the sources of funds and the investment activities of the VC industry (Gomper and Lerner, 1998; Gilson and Schizer, 2002; Cornelli and Yosha, 2003), their focus is placed on developed countries. There are few insights into how their results or theories can be applied to developing countries. Also, there are theoretical discussions on the relationship between capital structure, the financial systems, and investment activities (see, for instance, Black and Gilson, 1998; Hellman and Puri, 2002). However, they do not have strong supports from empirical evidence. This chapter attempts to fill the following gaps in the current literature. First, we discuss and provide empirical evidence on the development of the VC industry in three Asian economies, namely, China, India, and Hong Kong. Second, we compare how institutional forces impact and shape venture capitalists' investment activities, such as deal selection and preferred investment stages, and distributions of types of VC firms in the three economies. We find that China and India have VC deals concentrated at early stage, while Hong Kong has VC deals concentrated on turnaround stage. Our regression analysis suggests that government support is the only significant factor that shapes the choice of VC on financing stages. Other factors, such as investee industry and investor types, do not have significant impact on the choice.

The rest of this chapter is structured as follows: Section 21.2 introduces our theoretical framework in this chapter. Section 21.3 describes the backgrounds of VC development in the three economies. Section 21.4 discusses the data, the hypotheses, and the methodology involved. Section 21.5 analyzes the empirical results. Section 21.6 concludes this chapter.

21.2 THEORETICAL FRAMEWORK

There is increasing interest in applying institutional theory in managerial economics and investment activities because of its explanatory power to address issues relevant in international markets (see, for instance, Scott, 1995; Hoskisson et al., 2000). Scott (1995) differentiates the institutional forces into three categories: normative, regulatory, and cognitive. The classification has been used widely in the general international entrepreneurial literature (Busenitz et al., 2000). The status and impact of the various institutions in each of the three economies is summarized in Table 21.1.

TABLE 21.1 The Institution of Venture Capital in Hong Kong, China, and India

	Normative	Regulatory	Cognitive
Hong Kong	Industry coped with strong normative values from United States due to training interconnections in the industry (Wright et al., 2002)	Highly efficient and transparent; mature stock markets; effectively enforced laws provide high shareholder protection (La Porta et al., 2000)	Status of entrepreneurs is moderate (Reynolds et al., 2002); emphasis on social networks but ability to produce economic value is more critical
China and India	The same as above	Immaturely developed; insufficient law enforcement; bank-centered financial markets (Bruton et al., 2003)	Status of entrepreneurs is moderate (Reynolds et al., 2002); reliance on social networks stronger than in Hong Kong (Tsang and Walls, 1998)

21.3 VC DEVELOPMENT BACKGROUND AND GOVERNMENT ROLES

Prior VC research in the United States relies heavily on the agency theory to explain how VC firms operate and attributes VC activities to economically driven investment decisions. However, other countries are different from the United States in many aspects (Wright et al., 2002). Therefore, a richer theoretical model, beyond the agency theory, should be employed to examine VC internationally.

To maintain economic growth, developing countries attempt to develop plans to ensure constant inflow of capitals into various industries. In the 1980s, both Chinese and Indian governments recognized the importance of the VC industry as a means of supporting the development of competitive information technology (IT) ventures and promoting economic progress through technological advancements. Consequently, the two governments initiated VC as a significant mechanism for encouraging scientific and technological capabilities (see, for instance, White et al., 2004). In 1985, as a government initiative, China's State Science and Technology Commission (SSTC) and Ministry of Finance (MoF) jointly set up and funded the first local VC company in China, the China New Technology Venture Capital Investment Corporation.

Apart from injecting capital in the VC industry and developing science and technology industrial parks, both governments provided tax incentives and implemented policies and regulations for the VC industry in the 1990s. Under such favorable environments, the first foreign VC fund established in China was The Pacific Technology Venture Investment Fund, founded by the International Data Group (IDG) of the United States in 1992. There are a number of successful firms incubated by the VCs. In China, for example, AsiaInfo, a new tech-based venture enterprise, received $6.5 million from several foreign VC firms in 1997.

Hong Kong is the second largest VC center in Asia. Its laissez-faire policy, efficient governance, a highly transparent and well-established stock market and a simple tax regime are favorable factors for the development of the VC industry. The tax exemption on capital gains from disposal of portfolio investments and the absence of any exchange control are two key incentives for venture capitalists to invest in Hong Kong.

Unlike Hong Kong, both China and India have a history of state-directed institutional development. Their governments have tight control on their economies. They have large numbers of trained and skilled engineers, scientists, and technicians, and have successfully boosted their home-grown software industries since the 1980s. They recognize that vibrant stock markets and a low rate of capital gains tax are factors favorable to the development of the VC industry.

Table 21.2 shows a comparison of VC systems of the United States, China, Hong Kong, and India in terms of fundamentals, history, and government policies. Like the United States, China and India implemented policies to facilitate development of VC, even though they were state-planned economies. In contrast, there was no government-led industrial policy in Hong Kong. Similar to NASDAQ in the United States, Hong Kong developed a Growth Enterprise Market (GEM) for young ventures to raise capital by initial public offers (IPOs).

21.4 HYPOTHESES AND METHODOLOGY

The data of this chapter come from the *Asian VC Journal*. The study period ranges from 1998 to 2004. The dataset includes 316 VC investment deals in China, 573 deals in India, and 420 deals in Hong Kong. For individual VC deals, information is collected on the types of VC firms, the stage of financing, the industrial sector of the ventures, and years of the investment.

21.4.1 Hypotheses

It is commonly known that VC funding in early stages of ventures has been a key driver of economic development (Gompers and Lerner, 2001; Reynolds et al., 2002; Allen and Song, 2003). Black and Gilson (1998) find that there is more early stage investing in stock market–centered markets as VC firms have easy mechanisms for exit from their investment in these markets. In contrast, some find that bank-owned VC firms tend to invest more in later-stage ventures, as compared to individual- and corporate-backed VC firms (Mayer et al., 2004; Hellman et al., 2004). We have the first two hypotheses set as follows:

Hypothesis 1: Early-stage ventures are more likely to be selected for VC funding in societies that have well-established stock markets.

Hypothesis 2: Finance-associated VC firms are less likely to invest at early stage of financing.

Sorenson and Stuart (2001) and Hurry et al. (1992) found that venture capitalists usually make use of social networks to generate prospective investment possibilities as there is a strong reliance on social networks. Also, Fried and Hisrich (1994) find that referrals are important to get the deal beyond the VC firms' generic screening in the United States, but it is not a major factor in deal evaluation. When VC managers assess the return and

TABLE 21.2 Comparison of VC Systems in the United States, Japan, India, China, and Hong Kong

	United States	Japan	India	China	Hong Kong
Fundamentals					
Financial system	Market-based/separation of commercial and investment banking	Bank-based (with *Keiretsu* ties)/ separation of commercial and investment banking	Bank-based, with state ownership	Bank-based, with state ownership	Market-based/ separation of commercial and investment banking
Primary focus on industrial innovation	Research universities, companies	Research universities, companies	Research universities, companies	Research universities, companies	Research universities, companies
History and Government Support					
First public effort to foster enterprise creation	1958: Small Business Investment Act	1963: Small and Medium Enterprise Law	1988: Government announced guidelines for an institutional structure for VC	1999: Venture Capital Law and Regulation on start-up funds under deliberation	Not available
First venture capital organization involving nonprofit institutions	1946: American Research and Development Corporation	1975: Center for promotion of R&D intensive business (an industrial group coordinated by MITI)	1975: Risk Capital Foundation—RCF (semipublic venture capital firm)	1985: China New Technology Start-up Investment Company	Not available

First private venture capital firm	1958	1973	1995	1992	Not available
Date of creation of first public equity market dedicated to high-growth companies	1971: NASDAQ	1999: JASDAQ	Not available	Not available	2002: Growth Enterprise Market
Government policies favorable to the development of venture capital industry	Tax policy to reduce capital gains taxes and set up strict regulations on stock market	Tax incentive policy, direct funding support, revision of law and regulations favorable for small business creation, e.g., SME Creation Law	Tax incentive policy, allow mutual funds to invest in venture capital companies, establishment of high-tech zones	Tax incentive policy, direct funding support, high-tech zones	Not available

Sources: Adapted from Kuemmerle, W., Comparing catalysts of change: Evolution and institutional difference in the venture capital industries in the US, Japan, and Germany, in Burgleman, R.E. and Chesbrough, H. (Eds.), *Comparative Studies of Technological Evolution*. Jai Press, New York, 2001, 227–261; Rausch, L., Venture Capital Investment Trends in the United States and Europe, Division of Science Resources Studies, Directorate for Social, Behavioral and Economic Science, NSF, 1998, 99–303. Available at http://www.nsf.gov/statistics/issuebrf/sib99303.htm; White, S., Gao, J., and Zhang, W., in Mani, S. and Bartzokas, A. (Eds.), *The Case of China in Financial Systems, Corporate Investment in Innovation and Venture Capital*. Edward Elgar Publishing, Cheltenham, U.K., 2004, 159–198.

MITI, The Ministry of International Trade and Industry.

risk of the proposed ventures, they not only consider ventures' business plans but also assess whether the legal system is good enough to protect their interests against agency problems with the entrepreneurs (Sapienza and Gupta, 1994; Bruton et al., 2000). In developing countries, social networks could be more important than legal systems in investor protection. Perkins (2000) proposes a substitution effect of social networking for the legal system in Asian countries. A substitute for an effective legal system is a social network that is a personal relationship based on family ties. Through the network (guanxi), the business transactions can be carried out according to a given set of rules. The relationships become alternatives to enforce contracts under the legal system. Guanxi is particularly important to the economic development in Chinese societies as it provides a basis for its economic development like a social capital (Pye, 2000).

Nevertheless, government support can be important in shaping VC activities. In the United States, Japan, Israel, the United Kingdom, and other Organisation for Economic Co-operation and Development (OECD) countries, governments implement financial programs to support the development of VC industries and IT firms. Hence, the third and fourth hypotheses are proposed:

> *Hypothesis 3: Early stage ventures are more likely to be selected for VC funding in societies with high reliance on social networks.*

> *Hypothesis 4: High-tech ventures are more likely to be selected for VC funding in societies with government financial incentives for establishment of high-tech ventures or VC firms.*

In the study, we apply the chi-square test and multinomial logit régression to identify factors affecting VCs' preference on the financing stages. The dependent variable is financing stages. The independent variables include all investee industries, all types of VC firms, and government support (a dummy). The specification model is written as follows:

$$\text{Financing stage} = f[\beta_0 + \sum \beta_{1,j}(\text{Investee\_ind})_j + \sum \beta_{2,h}(\text{Investor\_g})_h + \beta_3(\text{Government\_p})]$$

where
Investee_ind is the industries of ventures
Investor_g is the types of VC firms

Government_p is the dummy variable of government program support for VC firms or young ventures (1 indicates availability of government program and otherwise is 0)

21.5 EMPIRICAL RESULTS

Tables 21.3 through 21.5 show the distribution of VC deals in the three economies. There is an obvious difference in the stages of financing and types of VC firms in three markets. China and India concentrate their VC deals in the segments of the early stage and high-tech ventures as their governments support high-tech industry. In Hong Kong, most VC funding was arranged at later stages of financing. It has VC deals strongly associated with corporate-affiliated and independent limited partnership VC firms. Gompers and Lerner (1998) also have similar findings in other developed economies.

21.5.1 Regions: China, India, and Hong Kong

Table 21.6 applies two-way contingency table analysis on regions and financing stages. The Pearson statistics is significant at 1%, suggesting that the two factors have impacts on the VC deals. Interestingly, China has more VC deals (261) at early stage than they are expected (192), while Hong Kong has less VC deals (140) at early stage than they are expected (255). In contrast, VC deals in Hong Kong are more concentrated on later stages of investment. It has a well-established stock market in Asia but has less VC deals at the early stage. The result does not support Hypothesis 1 that early-stage ventures are more likely to be selected for VC funding in societies that have well-established stock markets.

21.5.2 Types of VC Firms

Table 21.7 applies two-way contingency table analysis on types of VC firms and financing stages. The Pearson statistics suggesting both the two factors have statistically significant impacts on VC deals. Finance-associated VC firms have more than 58% of their deals concentrated on the early-stage financing. This fails to hold the Hypothesis 2 that finance-associated VC firms are less likely to invest at an early stage of financing. Obviously all types of VC firms are keen on early-stage financing.

21.5.3 Reliance on Social Networks

Table 21.8 applies two-way contingency table analysis on reliance on social network and financing stages. It shows that China and India are classified as having "High" reliance on social network, while Hong Kong is classified

TABLE 21.3 VC Investments by Financing Stage, Industry, and Types of VC Firms (Hong Kong)

Investee Region	Financing Stage	Industry Sector of Ventures	Corporate-Affiliated VC Firms	Finance-Associated VC Firms	Independent Limited Partnership VC Firms	State-Owned VC Firms	Total
Hong Kong	Early stage	Electronics	3	1	4		8
		IT	52	14	30		96
		Life science			1		1
		Manufacturing heavy	2	2			4
		Manufacturing light	2	1		1	4
		Service	7	9	11		27
	Total		66	27	46	1	140
	Expansion	Electronics	1	2	1		4
		IT	13	3	10		26
		Manufacturing heavy	1				1
		Manufacturing light	1	2			3
		Service	4	4	4		12
	Total		20	11	15		46

Mezzanine	IT	2	1	3		6
	Manufacturing light	1				1
	Service		1	1		2
Total		3	2	4		9
Turnaround	Electronics	10	4	3		17
	IT	69	34	47	2	152
	Life science			1		1
	Manufacturing heavy	2		1		3
	Manufacturing light	4	3	3		10
	Service	20	16	6		42
Total		105	57	61	2	225
Hong Kong total		194	97	126	3	420

TABLE 21.4 VC Investments by Financing Stage, Industry, and Types of VC Firms (China)

Investee Region	Financing Stage	Industry Sector of Ventures	Corporate-Affiliated VC Firms	Finance-Associated VC Firms	Independent Limited Partnership VC Firms	State-Owned VC Firms	Total
China	Early stage	Electronics	6	1	4	2	13
		IT	40	18	78	13	149
		Life science		4	5	6	15
		Manufacturing heavy	7	1	7	2	17
		Manufacturing light	1	2	11		14
		Primary		1	1	1	3
		Service	8	5	29	8	50
	Total		62	32	135	32	261
	Expansion	IT	12	3	17	5	37
		Life science	1		1		2
		Manufacturing heavy			2		2
		Service			2		2
	Total		13	3	22	5	43
	Mezzanine	IT	4	1	7		12
	Total		4	1	7		12
China total			79	36	164	37	316

TABLE 21.5 VC Investments by Financing Stage, Industry, and Types of VC Firms (India)

Investee Region	Financing Stage	Industry Sector of Ventures	Corporate-Affiliated VC Firms	Finance-Associated VC Firms	Independent Limited Partnership VC Firms	State-Owned VC Firms	Total
India	Early stage	Electronics	4	2	5		11
		IT	44	49	109		202
		Life science		1	13		14
		Manufacturing heavy	2	5	12	1	20
		Manufacturing light	2	2	9		13
		Service	51	44	98	1	194
	Total		103	103	246	2	454
	Expansion	Electronics		2			2
		IT	32	15	45		92
		Life science		2	3		5
		Manufacturing heavy	2	1	1		4

(continued)

TABLE 21.5 (continued) VC Investments by Financing Stage, Industry, and Types of VC Firms (India)

Investee Region	Financing Stage	Industry Sector of Ventures	Corporate-Affiliated VC Firms	Finance-Associated VC Firms	Independent Limited Partnership VC Firms	State-Owned VC Firms	Total
		Manufacturing light	1	1	3		5
		Service	18	8	21	1	48
	Total		53	29	73	1	156
	Mezzanine	IT	3	5	15		23
		Life science			2		2
		Manufacturing heavy	1				1
		Manufacturing light	1	3	2		6
		Service	5	3	12		20
	Total		10	11	31		52
	Turnaround	IT	3	4	1		8
		Life science			1		1
		Service		1	1		2
	Total		3	5	3		11
India total			169	148	353	3	673

TABLE 21.6   Financing Stages in Individual Regions: Hong Kong, China, and India

			Regions			
			China	**Hong Kong**	**India**	**Total**
Financing stage	Early stage	Count	261	140	454	855
		Expected count	192	255	408	855
		% within financing stage	30.53	16.37	53.10	100.00
		% within regions	82.59	33.33	67.46	60.68
		% of total	18.52	9.94	32.22	60.68
	Expansion	Count	43	46	156	245
		Expected count	55	73	117	245
		% within financing stage	17.55	18.78	63.67	100.00
		% within regions	13.61	10.95	23.18	17.39
		% of total	3.05	3.26	11.07	17.39
	Mezzanine	Count	12	9	52	73
		Expected count	16	22	35	73
		% within financing stage	16.44	12.33	71.23	100.00
		% within regions	3.80	2.14	7.73	5.18
		% of total	0.85	0.64	3.69	5.18
	Turnaround	Count	0	225	11	236
		Expected count	53	70	113	236
		% within financing stage	0.00	95.34	4.66	100.00
		% within regions	0.00	53.57	1.63	16.75
		% of total	0.00	15.97	0.78	16.75
Total		Count	316	420	673	1409
		Expected count	316	420	673	1409
		% within financing stage	22.43	29.81	47.76	100.00
		% within regions	100.00	100.00	100.00	100.00
		% of total	22.43	29.81	47.76	100.00

Note: Pearson chi-square test two-sided p-value: 0.

as having "Low" reliance on social network. This classification is strongly associated with the development of legal systems. Both China and India have less developed legal system, while Hong Kong has a well-developed legal system. The Pearson statistics is significant at 1% level, suggesting that two factors have an impact on the VC deals. At the early stage of financing, the "High" (715) category has more count than the expected number (600), while the "Low"

TABLE 21.7   Relationship between Types of VC Firms and Financing Stages

			Types of VC Firms (Investor_gc)					
			Corporate-Affiliated VC Firms	Finance-Associated VC Firms	Independent Limited Partnership VC Firms	Other VC Firms	State-Owned VC Firms	Total
Financing stage (fin_s)	Early stage	Count	231	162	427	32	3	855
		Expected Count	268	171	390	22	4	855
		% within fin_s	27.02	18.95	49.94	3.74	0.35	100.00
		% within Investor_gc	52.26	57.65	66.41	86.49	50.00	60.68
		% of total	16.39	11.50	30.31	2.27	0.21	60.68
	Expansion	Count	86	43	110	5	1	245
		Expected count	77	49	112	6	1	245
		% within fin_s	35.10	17.55	44.90	2.04	0.41	100.00
		% within Investor_gc	19.46	15.30	17.11	13.51	16.67	17.39
		% of total	6.10	3.05	7.81	0.35	0.07	17.39
	Mezzanine	Count	17	14	42	0	0	73
		Expected count	23	15	33	2	0	73

	% within fin_s	23.29	19.18	57.53	0.00	0.00	100.00
	% within Investor_gc	3.85	4.98	6.53	0.00	0.00	5.18
	% of total	1.21	0.99	2.98	0.00	0.00	5.18
Turnaround	Count	108	62	64	0	2	236
	Expected count	74	47	108	6	1	236
	% within fin_s	45.76	26.27	27.12	0.00	0.85	100.00
	% within Investor_gc	24.43	22.06	9.95	0.00	33.33	16.75
	% of total	7.67	4.40	4.54	0.00	0.14	16.75
Total	Count	442	281	643	37	6	1409
	Expected count	442	281	643	37	6	1409
	% within fin_s	31.37	19.94	45.64	2.63	0.43	100.00
	% within Investor_gc	100.00	100.00	100.00	100.00	100.00	100.00
	% of total	31.37	19.94	45.64	2.63	0.43	100.00

Note: Pearson chi-square test two-sided p-value: 0.

TABLE 21.8   Relationship between VC Reliance on Social Networks
and Financing Stages

| | | | Reliance on Social Networks (social_n) | | |
			High	Low	Total
Financing stage (fin_s)	Early stage	Count	715	140	855
		Expected count	600	255	855
		% within fin_s	83.63	16.37	100.00
		% within social_n	72.30	33.33	60.68
		% of total	50.75	9.94	60.68
	Expansion	Count	199	46	245
		Expected count	172	73	245
		% within fin_s	81.22	18.78	100.00
		% within social_n	20.12	10.95	17.39
		% of total	14.12	3.26	17.39
	Mezzanine	Count	64	9	73
		Expected count	51	22	73
		% within fin_s	87.67	12.33	100.00
		% within social_n	6.47	2.14	5.18
		% of total	4.54	0.64	5.18
	Turnaround	Count	11	225	236
		Expected count	166	70	236
		% within fin_s	4.66	95.34	100.00
		% within social_n	1.11	53.57	16.75
		% of total	0.78	15.97	16.75
Total		Count	989	420	1409
		Expected count	989	420	1409
		% within fin_s	70.19	29.81	100.00
		% within social_n	100.00	100.00	100.00
		% of total	70.19	29.81	100.00

Note: Pearson chi-square test two-sided p-value: 0.

(140) category has less count than the expected number (255). The "High" category has around 72% VC investments at the early stage, while "Low" category has only 33%. This supports the Hypothesis 3 that early-stage ventures are more likely to be selected for VC funding in regions with high reliance on social networks. As mentioned earlier, "Low" is correlated with "well-developed legal system." Many attribute the success of Hong Kong an international banking center and well-established common law system. However, it seems that this legal system does not result in more early stage financing.

21.5.4 Government Financial Incentives and Industries of Ventures

Table 21.9 applies a two-way contingency table analysis to government financial incentives and industries of ventures. Hong Kong is classified as "No" (i.e., without government financial incentives), while both China and India are classified as "Yes" (i.e., with government financial incentives"). The Pearson statistics is significant at 1% level and supports Hypothesis 4 (i.e., "Yes"). VC deals are concentrated on IT (around 53%) and services (around 32%). Without government supports (i.e., "No"), VC deals are also concentrated on IT (around 67%) and services (around 19%). It seems that "No" has higher concentration than "Yes" on VC deals related to high-tech ventures. Life science belongs to another type of high-tech ventures. In this category, "Yes" has 4% VC deals, while has "No" has only 0.5% VC deals. As a whole, the aforementioned results fail support the Hypothesis 4 that high-tech ventures are more likely to be selected for VC funding in societies with government financial incentives for establishment of high-tech ventures or VC firms.

21.5.5 Multinomial Logit Regression Results

Table 21.10 shows the results of multinomial logit regression mentioned in Section 21.3. The results show that the presence government programs to support VC significantly affects the financing stages.

21.6 CONCLUSIONS

This chapter has reached the following interesting conclusions. First, the presence of well-established stock market does not lead to concentration of early stage financing. Hong Kong has a well-developed stock market but VC funding is mostly provided at later stages of development of ventures. Both China and India have less-developed stock market but their VC funding is focused on early stage ventures. Second, contrary to previous findings, financial-associated VC firms are keen on early stage financing in the Asian economies. This may be due to government programs or participation of state-owned banks.

Third, this chapter finds that regions with high reliance on social networks, such as China and India, are associated with early stage financing. In such an environment, ventures have to rely on social networks to get funds. Without strong protection from the legal system, VC firms tend to feel safer to invest in ventures via social networks. This echoes the substitution effect of social networks on legal system.

Fourth, this chapter concludes that no strong evidence the government supports will result in higher chance for IT ventures to get VC funds.

TABLE 21.9 Relationship between Types of Industries of Ventures and Government Financial Incentives for Establishment of High-Tech Ventures or VC Firms

		Government Financial Incentive for Establishment of High-Tech Ventures or VC Firms (g_finance)		
		Yes	No	Total
Types of industries of ventures (investee_ind)	Electronics			
	Count	26	29	55
	Expected count	39	16	55
	% within investee_ind	47.27	52.73	100.00
	% within g_finance	2.63	6.90	3.90
	% of total	1.85	2.06	3.90
	IT			
	Count	523	280	803
	Expected count	564	239	803
	% within investee_ind	65.13	34.87	100.00
	% within g_finance	52.88	66.67	56.99
	% of total	37.12	19.87	56.99
	Life Science			
	Count	39	2	41
	Expected count	29	12	41
	% within investee_ind	95.12	4.88	100.00
	% within g_finance	3.94	0.48	2.91
	% of total	2.77	0.14	2.91
	Manufacturing heavy			
	Count	44	8	52
	Expected count	36	16	52

	% within investee_ind	84.62	15.38	100.00
	% within g_finance	4.45	1.90	3.69
	% of total	3.12	0.57	3.69
	Expected count	39	17	56
	% within investee_ind	67.86	32.14	100.00
	% within g_finance	3.84	4.29	3.97
	% of total	2.70	1.28	3.97
Primary	Count	3	0	3
	Expected count	2	1	3
	% within investee_ind	100.00	0.00	100.00
	% within g_finance	0.30	0.00	0.21
	% of total	0.21	0.00	0.21
Service	Count	316	83	399
	Expected count	280	119	399
	% within investee_ind	79.20	20.80	100.00
	% within g_finance	31.95	19.76	28.32
	% of total	22.43	5.89	28.32
Total	Count	989	420	1409
	Expected count	989	420	1409
	% within investee_ind	70.19	29.81	100.00
	% within g_finance	100.00	100.00	100.00
	% of total	70.19	29.81	100.00

Note: Pearson chi-square test two-sided *p*-value: 0.

TABLE 21.10   Likelihood Ratio Tests on the Multinomial Logit Regression

Effect	Model Fitting Criteria	Likelihood Ratio Tests		
	−2 Log Likelihood of Reduced Model	Chi-Square	df	Significance (p-Level)
Intercept	269.139(a)	.000	0	
Investor_g	288.192	19.053	12	.087
Investee_ind	294.547	25.409	18	.114
Government_p	761.400	492.261	3	.000

Hong Kong does not have many government supports on IT industry but VC deals are concentrated on IT firms (around 66%). The result indicates that, with government supports, VC firms are active in early stage financing regardless of investee industries and types of VC firms.

This chapter has analyzed how institutional forces shape the investment activities of VC firms in Asia. Future research can gain an insight on the relationship between industrial development and role of the financial institutions. This chapter has discussed India, China, and Hong Kong. Other Asian countries, like Taiwan and South Korea, may have remarkable differences that need to be explored. Also, the institutions are changing very fast in Asia, especially in China and India, which are developing countries. It is necessary to keep an update of differences in their institutions and their impact on the VC industry.

REFERENCES

Allen, F. and Song, W.-L. (2003) *Venture Capital and Corporate Governance*. Wharton Financial Institutions Center Working Paper 03–05.

Black, B.S. and Gilson, R.J. (1998) Venture capital and the structure of capital markets: Banks versus stock markets. *Journal of Financial Economics*, 47(3): 243–277.

Bruton, G.D., Ahlstrom, D., and Wan, J.C.C. (2003) Turnaround in Southeast Asian firms: Evidence from ethnic Chinese communities. *Strategic Management Journal*, 24(6): 519–540.

Bruton, G.D., Fried, V.H., and Hisrich, R.D. (2000) CEO dismissal in venture capital backed firms: Further evidence from an agency perspective. *Entrepreneurship Theory and Practice*, 24(4): 69–77.

Busenitz, L.W., Gomez, C., and Spencer, J.W. (2000) Country institutional profiles: Interlocking entrepreneurial phenomena. *Academy of Management Journal*, 43(5): 994–1003.

Cornelli, F. and Yosha, O. (2003) Stage financing and convertible debt. *Review of Economic Studies*, 70(242): 1–32.

Forer, G. and Yonge, J.D. (2004) Focus China why China? Why now? Annual Venture Capital Insights Report, Ernst & Young, Hong Kong.

Fried, V.H. and Hisrich, R.D. (1994) Toward a model of venture capital investment decision making. *Financial Management*, 23(3): 28–37.

Gilson, R. and Schizer, D. (2002) Understanding venture capital structure: A tax explanation for convertible preferred stock. *Harvard Law Review*, 116(3): 878–916.

Gompers, P.A. and Lerner, J. (1998) What drives venture capital fundraising? Brookings Papers on Economic Activity. *Microeconomics*, 149–204.

Gompers, P.A. and Lerner, J. (2001) *The Money of Invention: How Venture Capital Creates New Wealth*. Boston, MA: Harvard Business School Press.

Hellman, T., Lindsey, L., and Puri, M. (2004) Building relationships early: Banks in venture capital. NBER Working paper No: 10535.

Hellman, T. and Puri, M. (2002) The interaction between product market and financial strategy: The role of venture capital. *Review of Financial Studies*, 13(4): 959–984.

Hoskisson, R.E., Eden, L., Lau, C.M., and Wright, M. (2000) Strategy in emerging economies. *Academy of Management Journal*, 43(3): 249–267.

Hurry, D., Miller, A.T., and Bowman, E.H. (1992) Calls on high-technology: Japanese exploration of venture capital investments in the United States. *Strategic Management Journal*, 13(2): 85–102.

Kuemmerle, W. (2001) Comparing catalysts of change: Evolution and institutional difference in the venture capital industries in the US, Japan and Germany. In: R.A. Burgleman and H. Chesbrough (Eds.), *Comparative Studies of Technological Evolution* (pp. 227–261). Greenwich, Conn.: JAI Press.

La Porta, R., Lopez-de Silanes, F., Shleifer, A., and Vishny, R.W. (2000). Investor protection and corporate governance *Journal of Financial Economics*, 58(1): 3–27.

Mayer, C., Schoors, K., and Yafeh, Y. (2004) Sources of funds and investment activities of venture capital funds: Evidence from Germany, Israel, Japan and the UK. *Journal of Corporate Finance*, 11(3): 586–608.

Perkins, D.H. (2000) Law, family ties and the East Asian way of business. In: L. Harrison and S. Huntington (Eds.), *Culture Matters: How Values Shape Human Progress* (pp. 232–243). New York: Basic Books.

Pye, L.C. (2000) Asian values: From dynamos to dominoes. In: L. Harrison and S. Huntington (Eds.), *Culture Matters: How Values Shape Human Progress* (pp. 244–255). New York: Basic Books.

Rausch, L. (1998) *Venture Capital Investment Trends in the United States and Europe*, Division of Science Resources Studies, Directorate for Social, Behavioral and Economic Science, NSF, pp. 99–303. Available at http://www.nsf.gov/statistics/issuebrf/sib99303.htm

Reynolds, P.D., Bygrave, W.D., Autio, E., Cox, L.W., and Hay, M. (2002) Global entrepreneurship monitor: 2002 executive report. London: Babson College, Ewing Marion Kauffman Foundation and London Business School.

Sapienza, H.J. and Gupta, A. K. (1994) The impact of agency risks and task uncertainty in venture capitalist–CEO interaction. *Academy of Management Journal*, 37(6): 1618–1632.

Scott, W.R. (1995) *Institutions and Organizations*. Thousand Oaks, CA: Sage Publications.

Sorenson, O. and Stuart, T.E. (2001) Syndication networks and the spatial distribution of venture capital investments. *American Journal of Sociology*, 106(6): 1546–1588.

Tsang, W.K. (1998) Can Guanxi be a source of competitive advantage for doing business in China. *Academy of Management Executive*, 12(2): 64–73.

Tsang, E. and Walls, W. (1998) Can *Guanxi* be a source of competitive advantage for doing business in China. *Academy of Management Executive*, 2(2): 64–73.

White, S., Gao, J., and Zhang, W. (2004) The Case of China, In: S. Mani, and A. Bartzokas (Eds.), *Financial Systems, Corporate Investment in Innovation, and Venture Capital*. Edward Elgar Publishing, 159–198.

Wright, M., Lockett, A., and Pruthi, S. (2002) Internationalization of western venture capitalists into emerging markets: Risk assessment and information in India. *Small Business Economics*, 19(1): 13–29.

CHAPTER 22

Rating Skewness Spillovers in Equity and Currency Markets: Evidence from the Pacific Rim

Sirimon Treepongkaruna and Eliza Wu

CONTENTS

22.1 INTRODUCTION

Credit rating agencies are specialist information providers in international financial markets and are expected to facilitate market efficiency. Yet, the informational value of ratings and the role of rating agencies in the international financial system remains widely debated. Do sovereign ratings have significant and timely impacts on the higher moments of asset returns? Are their impacts equal across different financial markets?

This chapter aims to examine the effect of sovereign credit rating events on the realized third moment of stock and currency market returns for five advanced markets in the Asia-Pacific region—Australia, Hong Kong, Japan, Korea, and Singapore. As credit rating agencies have often been criticized for their inability to forewarn market participants and also for their delayed reactions to international financial crises (see Mora, 2006), it is important to assess the wider impacts of rating agencies' guidance on the stability of stock and foreign exchange (FX) markets, as measured by higher moments of their realized return distributions. In particular, we focus on the rating impacts on realized skewness measures over the period from 1997 to 2001, covering major episodes of financial crises arising from East Asia (1997), Russia (1998), and other parts of the world. This chapter is motivated by market participants' and policy makers' concern for downside risks and its contagious effects in international financial markets.

Sovereign credit ratings provide publicly available information on a national government's ability and willingness to service its debts in full and in a timely manner and are primarily determined by a country's economic fundamentals (see Cantor and Packer, 1996; Afonso, 2003). To date, the full extent of the impacts of agency ratings in the financial system are not well understood. This chapter complements existing studies and adds a significantly new dimension to the academic literature on rating impacts in international financial markets. While the significant impacts of sovereign credit ratings on stock and debt market returns are established in the ratings literature (see inter alia Cantor and Packer, 1996; Kaminsky and Schmukler, 1999, 2002; Reisen and Von Maltzan, 1999;

Brooks et al., 2004; Gande and Parsely, 2005; Ferreira and Gama, 2007; Pukthuanthong-Le et al., 2007), the effects on the skewness of asset returns and currency markets have never been explicitly examined. News on sovereign debt ratings may affect both stock and currency markets as ratings information provide signals on future economic conditions within a rated country and a rating change may cause the national government to implement policies that affect companies' future cash flows, thereby affecting stock returns as well as affecting general investor confidence and buying and selling pressures on the countries' currency. Furthermore, as the asymmetric and spillover effects of ratings are established in the extant literature (Reisen and Von Maltzan, 1999; Brooks et al., 2004; Gande and Parsely, 2005; Ferreira and Gama, 2007), it is only natural to examine whether there are also asymmetries and spillovers in the rating impacts on higher moments of stock and currency market returns.

The existing studies on rating impacts predominantly use the event study methodology to examine the cumulative abnormal returns of stock markets in a time window of several days after a rating announcement to determine the impact of rating changes (see, e.g., Brooks et al., 2004; Ferreira and Gama, 2007). Instead, we first use high-frequency currency and stock market data to compute realized skewness and then examine their financial linkages and the impact of ratings events within a vector auto regression (VAR) and panel data regression framework. The differential impacts on currencies and stock markets in the Asia-Pacific region during the 1997–1998 Asian financial crisis (AFC) present a good natural experiment for ascertaining the impact of sovereign ratings events on realized skewness measures.

Overall, we find that currency and stock market skewness react heterogeneously to ratings announcements with stock market skewness being more responsive to rating news than currency markets in both VAR and panel regression analyses. The regional AFC only marginally affected currency market skewness. Rating effects are asymmetric as rating upgrades (downgrades) increase (decrease) realized skewness. Moreover, outlooks impact on both stock and currency markets whereas actual rating changes are anticipated by stock market participants and hence not significantly reflected in realized stock market skewness. The impulse response functions (IRF) indicate that rating shocks of all sample countries immediately affect both stock and currency market skewness and the effects last for several days. Rating shocks generally stimulate stock market skewness to a greater degree. However, we find only weak ratings spillover effects from Korea to other countries' realized skewness measures with particularly weaker results in currency markets.

The contributions of our chapter are as follows. First, this is one of the few studies to provide high-frequency evidence on the financial market impact of sovereign credit ratings. The advantages of using daily measures computed from intraday market data over day-to-day closing prices is that they provide a better representation and more robust estimate of actual price behavior. Daily close-to-close measures are unable to capture the intraday price fluctuations, which can be substantial particularly during times of financial distress. Second, we empirically investigate the impacts of sovereign credit ratings on stock and currency market skewness for the first time. In doing so, we shed new light on the impacts of sovereign ratings on the higher moments of asset returns.

This research has serious implications in light of the increased role of sovereign credit ratings under the new Basel II banking regulatory framework. As financial assets are marred by downside risks, a clearer understanding of rating impacts on stock and currency market skewness will not only be beneficial for risk management by corporate treasurers, portfolio investors, and financial institutions managers but also system stability management by policy makers.

The organization of this chapter is as follows. In Section 22.2, we provide the data description followed by the empirical modeling in Section 22.3. In Section 22.4, we discuss our findings before concluding with Section 22.5.

22.2 DATA DESCRIPTION

The data set used in this chapter consists of the Bid–Ask quote prices for both currencies traded and stock market indexes in five advanced countries in the Asia-Pacific region, namely, Australia, Hong Kong, Japan, Korea, and Singapore. Our sample period starts on January 2, 1997 and ends on August 31, 2001.

The currency market data used in this chapter consists of the tick-by-tick exchange rates from Olsen and Associates for the following currencies: Australian dollar (AUD), Hong Kong dollar (HKD), Japanese yen (JPY), Korean won (KRW), and Singaporean dollar (SGD). All currencies are quoted against the U.S. dollar (USD). The most liquid currency traded in our sample is the JPY with the average number of quotes being 6923 quotes a day while KRW is the least liquid rate with the average number of quotes being 369 quotes a day. The stock market index data are captured from

Reuters' terminals and is provided by Securities Industry Research Centre of Asia-Pacific (SIRCA) in their TACTIQ database for capital market microstructure data. These indices include the Australian S&P/ASX100* (ATO1), Hong Kong Hang Seng Index (HSI), Japan Nikkei index (Nikkei), Korean KOSPI 200 Index (KS200), and Singaporean Strait Times Index (SSI). All indices are denominated in local currencies. The KS200 is the most liquid with the average number of quotes being 1308 quotes a day while SS1 is the least liquid with the average number of quotes being 22 quotes a day.

Although the FX market is a nonstop trading market, the stock market is not a nonstop trading market. Hence, we only consider part of the day where stock markets in the five sample countries are open. We therefore define our trading hours for all currency and stock markets considered as 23:00 GMT to 09:00 GMT, excluding weekends. For example, the period of Sunday 23:00 GMT to Monday 09:00 GMT is considered as our Monday sample (i.e., Monday for Australasia).

In addition, we use the history of foreign currency sovereign credit ratings and credit outlooks and watches from Standard and Poors (S&P). We focus only on foreign currency sovereign ratings assessments provided by S&P as previous studies have found these exert the greatest impact on market returns and are less anticipated (Reisen and von Maltzan, 1999; Brooks et al., 2004). S&P ratings announcements are generally made local AM time but the exact timing is not consistent within announcement dates. As the timing of rating announcements is not consistent, we focus on daily (rather than intraday) impacts of rating announcements. Following the approaches of Gande and Parsley (2005) and Ferreira and Gama (2007), we linearly transform the actual ratings and outlook and credit watch guidances (on imminent rating changes) into comprehensive credit rating series to provide a sovereign debt uncertainty measure (DUM). Both forms of ratings guidance are intended to be forward-looking measures of the perceived ability and willingness of sovereign debt issuers to service their financial obligation. However, actual rating changes reflect perceived permanent changes in credit quality in the long term whereas credit outlooks and watches indicate imminent changes in ratings over the short term. We define a "rating event" as a nonzero change in the DUM series. There are a total of 18 rating events in our overall sample, with Korea and Hong Kong being the most actively re-rated countries, contributing 11 and 5 of those events, respectively.

---

* As an alternative benchmark stock market index for Australia, we also analyzed the All Ordinaries index and our conclusions remain qualitatively unchanged.

Based on the works of Andersen and Bollerslev (1998), Barndorff-Nielsen and Shephard (2001), and Andersen et al. (2003), we argue that daily realized measures calculated based on intraday returns provide more consistent and efficient measures than those computed from close to close prices. In this chapter, the intraday return is calculated as the log difference of the midpoint at time t and midpoint at time $t-1$. We use the midpoint quote between the Bid and Ask price to minimize the effect of Bid–Ask bounce, as suggested by Roll (1984). To minimize microstructural bias and sampling error, we use the daily realized skewness measures computed from 30 min intervals for our empirical estimations.[*]

Following Hutson et al. (2008), we compute the daily "down-to-up-volatility" skewness measure defined as follows:

$$DU_t = \log\left(\frac{(D_u - 1)\sum_{d=1}^{D_d} R^2_{\mathrm{down}d,t}}{(D_d - 1)\sum_{d=1}^{D_u} R^2_{\mathrm{up}d,t}} \right) \tag{22.1}$$

where
$R_{\mathrm{down}d,t}$ denotes a dth 30 min return during day t that is less than the average return for this particular day
$R_{\mathrm{up}d,t}$ denotes a dth 30 min return during day t that is greater than the average return for this particular day
D_d and D_u are the daily totals of the corresponding returns

It should be noted that $D = D_d + D_u$. This is a log ratio of the standard deviations of returns below and above the mean return. A higher value of this measure corresponds with more left (negatively) skewed return distributions.

22.3 EMPIRICAL MODELING

22.3.1 Vector Auto Regression Analyses

We first employ Granger causality tests and impulse response analyses within a VAR framework to examine the interrelationships between rating changes and equity and currency market skewness. A VAR structure is in the following equation:

---

[*] As a robustness check, we also ran regressions with measures computed from other sampling intervals in the day. Our results in both VAR and panel regression analyses remain qualitatively the same. We also performed preliminary volatility and skewness signature plots to support our selection of 30 min intervals.

$$y_t = A_1 y_{t-1} + \cdots + A_p y_{t-p} + Bx_t + \varepsilon_t \qquad (22.2)$$

where

y_t is a k vector of endogenous variables (i.e., realized skewness for all countries' stock and currency returns and their sovereign ratings)

x_t is a d vector of exogenous variables (i.e., constants in our case)

$A_1, \ldots, A_p$ and B are the matrices of coefficients to be estimated

ε_t is a vector of innovations that may be contemporaneously correlated but are uncorrelated with their own lagged values and uncorrelated with all of the right-hand side variables

Using this framework, we ascertain the direction of potential causal relationships between skewness and sovereign ratings and analyze IRF from rating shocks.

22.3.2 Panel Regression Analyses

To further investigate the impacts of ratings announcements on the realized skewness of currency and stock market returns, we utilize a framework similar to that adopted by Christiansen and Ranaldo (2007) for studying intraday news effects in the U.S. stock and bond markets. However, instead of using straightforward dummy variables for capturing announcement effects during the trading day, we adopt the credit rating event variables similar to those used in Gande and Parsley (2005) and Ferreira and Gama (2007) for studying rating spillover effects from other countries in international debt and stock markets, respectively. In this methodological fusion, we introduce a more flexible framework for investigating the skewness impacts of international financial crises and different types of ratings information on the day of release using more efficient and consistent daily realized skewness measures.

Using pooled (panel) regression analysis, we estimate the following general model with fixed country and time effects to account for financial crises:

$$Y_{i,t} = \alpha_i + \beta_1 \text{Event}_{i,t} + \beta_2 \text{Event}_{i,t-1} + \beta_3 \text{DUM}_{i,t} + \beta_4 \text{CRISIS}_t + \varepsilon_{i,t} \qquad (22.3)$$

where

$Y_{i,t}$ is the realized skewness for stock indices and currency returns for country i on day t

$\text{DUM}_{i,t}$ is the country's ratings level

$\text{Event}_{i,t}$ is the rating event

CRISIS is a set of dummy variables included one at a time to capture various periods of international financial crises [AFC, Russian debt crisis (RFC) Global financial crises (GFC)] which is the sum of the AFC, RFC, as well as the Brazilian and Turkish financial crises (BFC and TFC) occurring during our sample period].*

The financial crises dummy variables are defined as one on days during international financial crises and zero otherwise based on dates in Kaminsky and Schmukler (2002) and Kaminsky et al. (2003). The main variable of interest is Event and the DUM variable controls for nonlinearities in market reaction relative to the position of each country on the rating scale.

This empirical framework is sufficiently flexible to allow the base model specification to be extended for additional tests on the market impacts of different types of ratings information—specifically, downgrades and upgrades; outlook and rating changes; and rating spillovers.

First, to separately compare the impact of downgrade and upgrade phases in sovereign ratings, the following model was estimated:

$$Y_{i,t} = \alpha_i + \beta_1 \text{Event}_{i,t} + \beta_2 \text{Event}_{i,t-1} + \beta_3 \text{DUM}_{i,t} + \beta_4 \text{CRISIS}_t + \beta_5 I_t + \varepsilon_{i,t}$$
(22.4)

where I_t is an indicator variable for downgrades [DG] (upgrades [UG]) and takes a value of one in the period from a negative (positive) to positive (negative) Event and zero otherwise. The bulk of existing rating studies find that rating downgrades have more significant impact on market returns than upgrades (see, e.g., Brooks et al., 2004).

Second, to identify the potential differential market reactions to short-term outlook and long-term rating changes, the model was augmented to

$$Y_{i,t} = \alpha_i + \beta_1 \text{DUM}_{i,t} + \beta_2 \text{CRISIS}_t + \beta_3 \text{Outch}_t \times \text{Event}_t$$
$$+ \beta_4 \text{Ratch}_t \times \text{Event}_t + \varepsilon_{i,t}$$
(22.5)

where

Outch$_t$ is a dummy variable defined as one when there is a change in sovereign outlook or credit watch and zero otherwise

Ratch$_t$ is similarly defined for actual ratings changes

---

* Dynamic panel data estimations with ΔY and instrumented Y_{t-1} were not appropriate. Preliminary augmented Dickey Fuller (ADF) tests rejected the existence of a unit root in the time series of daily realized skewness for both stock and currency markets. Hence, we analyze realized skewness in levels.

Both of these variables are then interacted with the ratings Event variable to compare the separate impacts of outlook versus actual rating events

Third, in the spirit of Gande and Parsley (2005) and Ferreira and Gama's (2007) ratings spillover studies, we also replace the ratings Event variable for country i with all other countries excluding i to determine the rating spillover effects to other sample countries' stock and currency market skewness in the Asia-Pacific region. Hence, the following model specification in Equation 22.6 was also estimated:

$$Y_{j,t} = \alpha_i + \beta_1 \text{Event}_{i,t} + \beta_2 \text{Event}_{i,t-1} + \beta_3 \text{DUM}_{i,t} + \beta_4 \text{DUM}_{j,t}$$
$$+ \beta_5 \text{CRISIS}_t + \varepsilon_{j,t}, \forall j \notin i \tag{22.6}$$

22.4 EMPIRICAL FINDINGS*

We discuss the results with respect to first the VAR then panel data regression analyses. Finally, we examine the rating spillover effects into other markets within the Asia-Pacific region.

22.4.1 Vector Auto Regression Results

To investigate the impact of sovereign ratings on our realized skewness measures, we fit a multivariate vector auto regressive (VAR(1)) system for sovereign credit ratings and the realized skewness of stock indices and currency returns. The lag length tests (sequential modified Log-Ratio test statistic, Final Prediction Error, and Akaike Information Criterion) all indicate that a one day lag is appropriate for our VAR system. Overall, we find that the spillover effects from ratings to realized skewness is particularly weak in currency markets compared with stock markets. We find lagged ratings in Australia positively affect the realized skewness of the Korean stock indices. We also find lagged ratings for Hong Kong has a weak positive impact on the realized skewness of the Hong Kong index and Korean stock index. Furthermore, we also find that lagged ratings for HK positively affect the realized skewness of the Singapore stock market and AUD. The lagged value of ratings in Korea negatively affects the realized skewness of the KRW. Finally, the one day lag of Singapore's and Japan's ratings does not have significant affects on skewness measures. This suggests that rating changes for these two highly

---

* To conserve space, we do not report the VAR results and Granger causality tests but these results are available upon request.

creditworthy countries do not impart any significant effects on either downside or upside risks within the Asia-Pacific region.

When we specifically conduct Granger causality tests for our system of sovereign ratings and realized skewness series we find there are weak causal relationships across countries but significant ones within countries, particularly for stock markets.

Figure 22.1 plots the IRF for our VAR system of ratings and stock and currency market realized skewness along with asymptotic standard errors. In general, we find a ratings shock for each country in the VAR system not only has immediate effects on its own stock and currency market skewness but also that of other countries and these may last for up to four or five trading days afterward. A ratings shock will mostly increase the degree of realized skewness in stock markets but may reduce skewness in currency markets within the Asia-Pacific region (with the exception of AAA rated Singapore). The impacts on stock markets are generally of a larger magnitude than in currency markets. Again, we see evidence of a heterogeneous response to ratings information in stock and currency markets.

22.4.2 Panel Regression Results

Table 22.1 reports the estimates of the panel regression models represented in Equations 22.3 through 22.5 for realized stock and currency market skewness as measured by the Hutson et al. (2008) "down-to-up" (DU) skewness measure. Consistent with their interpretation, a higher value of this measure corresponds with more left (negatively)-skewed return distributions.

We find evidence that rating events have significant impacts on the third moments of both stock and currency returns. However, there is a different relationship in the two asset markets as rating events are negatively related to stock market skewness but positively related to currency market skewness and the effect is clearly more persistent in the former. We find evidence of heterogeneous responses in these two different asset markets. Interestingly, the skewness of neither asset markets is affected by financial crises with the exception of the currency market being significantly affected by the AFC at the 10% significance level. This suggests that region-specific financial crises play a greater role than general international financial crises. In terms of asymmetries, stock market skewness responds significantly to upgrade phases but currency skewness responds asymmetrically. The signs are consistent across asset markets in that

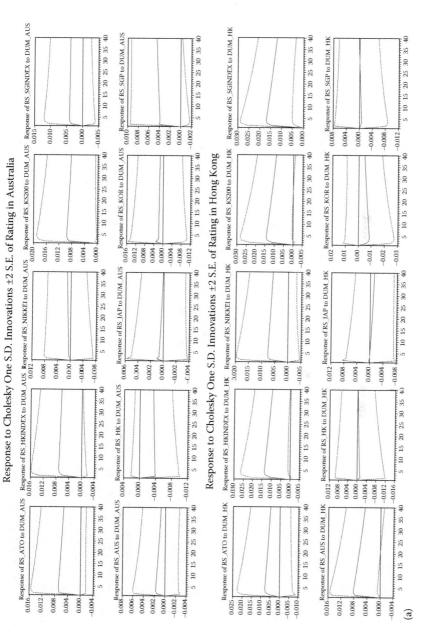

FIGURE 22.1 IRFs for rating shocks on stock and currency market realized skewness.

(continued)

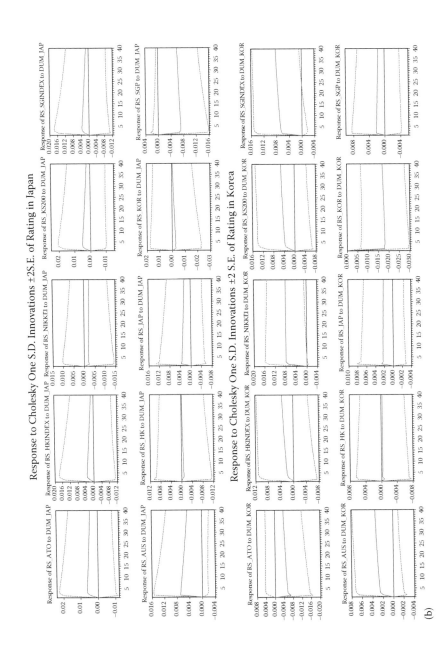

FIGURE 22.1 (continued)

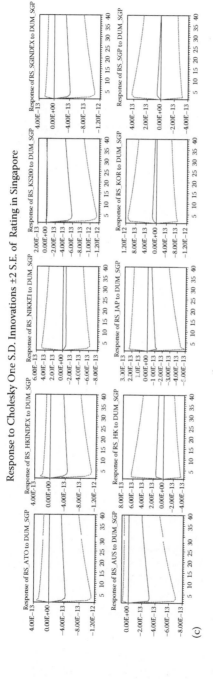

FIGURE 22.1 (continued)

TABLE 22.1   Impact of Sovereign Ratings on the Realized Skewness of Stock and Currency Markets

	Stock Market Skewness			Currency Market Skewness		
	(1)	**(2)**	**(3)**	**(1)**	**(2)**	**(3)**
Const	−1.744***	−1.8534***	−1.7927***	1.9640***	2.3767***	1.9858***
	{0.0019}	{0.0011}	{0.0017}	{0.0004}	{0.0001}	{0.0001}
Event	−0.7976***	−0.8150***		0.5567***	0.5313***	
	{0.0047}	{0.0048}		{0.0033}	{0.0089}	
Lag Event	−1.1129***	−1.1304***		−0.7520	−0.7775	
	{0.0006}	{0.0006}		{0.3962}	{0.3852}	
DUM	0.0994***	0.1024***	0.1021***	−0.1133***	−0.1352***	−0.1146***
	{0.0021}	{0.0016}	{0.0019}	{0.0004}	{0.0001}	{0.0001}
AFC	−0.0744	−0.0565	−0.0596	−0.1082*	−0.1077*	−0.0997
	{0.5693}	{0.6636}	{0.6499}	{0.0980}	{0.0985}	{0.1467}
GFC	0.1148			−0.0587		
	{0.1877}			{0.2299}		
RFC	0.0979			−0.0619		
	{0.6259}			{0.4927}		
DG		0.1333			0.2873***	
		{0.2555}			{0.0001}	
UG		−0.1913**			−0.1383**	
		{0.0213}			{0.0348}	
Outch × Event			−1.0220***			0.3494*
			{0.0010}			{0.0530}
Ratch × Event			−0.2357			0.8453**
			{0.4616}			{0.0112}
Adj. R-square	0.0048	0.0057	0.0032	0.0099	0.1361	0.0082
No. of observations	6065	6065	6070	6065	6065	6065

Notes: This table presents the panel estimation results for stock and currency market realized skewness over the sample from 7/1/1997 to 30/8/2001. Model specifications (1)–(3) are based on Equations 22.3 through 22.5. The crisis periods are from 1/7/1997 to 30/1/1998 (AFC); 1/8/1998 to 30/10/1998 (RFC), and the GFC includes the sum of the Asian, Russian, Brazilian (1/2/1999 to 28/2/1999), and Turkish (1/2/2001 to 28/2/2001) financial crises. The GFC, RFC, and UG coefficients are estimated from a separate regressions to avoid collinearity issues. *, **, and *** denote significance at the 10%, 5%, and 1% levels.

upgrades reduce skewness while downgrades increase skewness toward the left. The results suggest that most rating downgrades may already be anticipated by stock but not currency market participants. We find that outlook changes are also significant on market skewness measures, albeit more so for stock market skewness. Interestingly, currency market skewness is more significantly affected by actual ratings changes (5% level) than outlook changes (10% level). Again, this presents evidence of heterogeneous market responses to agency ratings guidance.

22.4.3 Rating Spillover Effects on Realized Skewness

Table 22.2 presents the panel estimation results for Equation 22.6. We find that within our sample, the other markets in the Asia-Pacific region were marginally affected by Korea's rating events but there were no spillovers from the other four markets' rating events into stock and currency markets.[*] This is not surprising given that of the more developed Asian financial markets studied, Korea was the worst affected during the AFC. In spite of positive Granger causality results from the VAR analyses, we find Hong Kong's rating events did not have consistent spillover effects across all other sample countries. We only find evidence that realized skewness in

TABLE 22.2   Rating Spillover Effects from Korea to Other Countries' Stock and Currency Market Skewness

	RS Stocks	RS Currency
Constant	−4.0567**	0.0077
	{0.0106}	{0.9919}
Event_Korea	−0.5731***	0.0591
	{0.0005}	{0.4585}
Lag Event_Korea	−0.0809	−0.0246
	{0.6246}	{0.7581}
Dum_Korea	0.0125	0.0086
	{0.3238}	{0.1569}
Dum_others	0.2097**	−0.0076
	{0.0147}	{0.8549}
AFC	0.0064	−0.0131
	{0.9479}	{0.7804}
Adj. R-square	0.0031	0.0015
No. obs.	4852	4852

Notes: This table presents the rating spillover effects from Korea to other sample countries in the Asia-Pacific region. The model specification is based on Equation 22.6. The AFC crisis period is from 1/7/1997 to 30/1/1998.
*, **, and *** denote significance at the 10%, 5%, and 1% levels.

---

[*] Only rating spillover results from Korea are presented for brevity.

stock markets were particularly responsive to rating spillover effects from Korea. However, the market impact of ratings spillovers are economically and statistically less significant than own country rating effects discussed earlier. Interestingly, Korean rating events had no spillover effects on other advanced Asian currency markets. These findings suggests that while the ratings events of advanced markets in the Asia-Pacific are generally interpreted by market participants as country-specific news, there were common rating information spillovers from Korea into the other developed Asian stock markets. As Korea's sovereign rating performance declined, the perception of riskiness in other Asian stock markets also increased.

22.5 CONCLUSION

We have examined the impact of different types of sovereign rating announcements on the realized skewness of stock and currency returns in the Asia-Pacific region over 1997–2001 using high-frequency data. We study the cross-country and same country rating impacts on market return skewness using VAR and panel regression analyses.

We find evidence of heterogeneous market responses to agency ratings guidance in currency and stock markets with the latter being more responsive and experiencing more persistent effects. Changes on sovereign credit outlooks have more significant impact on the realized skewness of stock markets but actual rating changes are more important in currency markets. We also find clear evidence that rating events have significant and asymmetric impacts on higher moments of both asset market returns. That is, realized skewness increases with downgrades and declines with upgrades. Further, we find mute effects of global financial crises on the realized skewness of stock and currency returns in the Asia-Pacific and only marginal effects of the 1997–1998 regional AFC on the realized skewness of currency returns. Finally, there were marginal rating spillover effects from Korea on other markets' realized measures. More developed and stable financial markets are less inclined to impart rating spillover effects into other asset markets in the region.

In summary, we find new evidence that national sovereign rating events have significant impacts on the higher moments of stock and currency returns. Future research into the impacts of credit ratings on international financial markets need to recognize and account for this to fully capture the true extent of rating influence on asset returns.

REFERENCES

Afonso, A. (2003) Understanding the determinants of sovereign debt ratings: Evidence for the two leading agencies. *Journal of Economics and Finance*, 27(1): 56–74.

Andersen, T. G. and Bollerslev, T. (1998) Answering the skeptics: Yes, standard volatility models do provide accurate forecasts. *International Economic Review*, 39(4): 885–905.

Andersen, T., Bollerslev, T., Diebold, F., and Labys, P. (2003) Modeling and forecasting realized volatility. *Econometrica*, 71(2): 579–625.

Barndorff-Niesen, O. E. and Shephard, N. (2001) Non-Gaussian Ornstein–Uhlenbeck-based models and some of their uses in financial economics. *Journal of the Royal Statistical Society: Series B (Statistical Methodology)*, 63(2): 167–241.

Brooks, R., Faff, R., Hillier, D., and Hillier, J. (2004) The national market impact of sovereign rating changes. *Journal of Banking and Finance*, 28(1): 233–250.

Cantor, R. and Packer, F. (1996) Determinants and impact of sovereign credit ratings. *Federal Reserve Bank of New York Economic Policy Review*, 2(2): 37–53.

Christiansen, C. and Ranaldo, A. (2007) Realized bond-stock correlation: Macroeconomic announcement effects. *Journal of Futures Markets*, 27(5): 439–469.

Ferreira, M. A. and Gama, P. M. (2007) Does sovereign debt ratings news spill over to international stock markets? *Journal of Banking and Finance*, 31(10): 3162–3182.

Gande, A. and Parsley, D. C. (2005) News spillovers in the sovereign debt market. *Journal of Financial Economics*, 75(3): 691–734.

Hutson, E., Kearney, C., and Lynch, M. (2008) Volume and skewness in international equity markets. *Journal of Banking and Finance*, 32(7): 1255–1268.

Kaminsky, G., Reinhart, G., and Vegh, C. (2003) The unholy trinity of financial contagion. *Journal of Economic Perspectives*, 17(4): 51–74.

Kaminsky, G. and Schmukler, S. L. (1999) What triggers market jitters? A chronicle of the Asian crisis. *Journal of International Money and Finance*, 18(4): 537–560.

Kaminsky, G. and Schmukler, S. L. (2002) Emerging market instability: Do sovereign ratings affect country risk and stock returns? *World Bank Economic Review*, 16(2): 171–195.

Mora, N. (2006) Sovereign credit ratings: Guilty beyond reasonable doubt? *Journal of Banking and Finance*, 30(7): 2041–2062.

Pukthuanthong-Le, K., Elayan, F. A., and Rose, L. C. (2007) Equity and debt market responses to sovereign credit ratings announcement. *Global Finance Journal*, 18(1): 47–83.

Reisen, H. and Von Maltzan, J. (1999) Boom and bust and sovereign ratings. *International Finance*, 2(2): 273–293.

Roll, R. (1984) A simple implicit measure of the effective bid–bsk spread in an efficient market. *Journal of Finance*, 39(4): 1127–1139.

CHAPTER 23

Dealing with East Asian Equity Market Contagion: Some Policy Implications

Thomas J. Flavin and Ekaterini Panopoulou

CONTENTS

23.1 INTRODUCTION

East Asian equity markets have suffered many episodes of high volatility over the past two decades. These episodes are often costly and have repercussions for both the investment community and the domestic economy. The effects of market turbulence are magnified by the presence of contagious effects between countries and can lead to regionwide crises. These crises lead to large monetary losses for investors who are subsequently more reluctant to commit money to risky assets. The risk is even greater for foreign investors who also bear exchange rate risk. Therefore equity market crises reduce the ability of entrepreneurs to access capital markets, with ensuing harmful ramifications for economic growth.

There is now considerable evidence that contagion has been a feature of East Asian equity markets over the recent past. Caporale et al. (2005), Bekaert et al. (2005), Chiang et al. (2007), and Flavin and Panopoulou (2008), using a variety of techniques, all find evidence of contagion between many pairs of Asian markets.* Therefore, a key issue facing policy makers is how to curb the spread of contagion between regional markets. However, formulating appropriate policy responses is only possible with a full understanding of the transmission channels through which contagion operates. There are many definitions of contagion in the literature (for an overview, see Pericoli and Sbracia, 2003) so we need to be explicit about our use of the contagion terms. Flavin and Panopoulou (2008) develop a test for two distinct types of contagion—shift and pure contagion—within a unified framework. Shift contagion occurs when the interdependencies between pairs of markets change significantly during a crisis and the transmission of common shocks undergoes a significant change. The presence of shift contagion between markets implies that this "normal"

---

* Other studies, e.g., Forbes and Rigobon (2002), fail to reject the hypothesis of "no contagion." For a more complete review of the literature, the reader is referred to Dungey et al. (2006) and references therein.

relationship becomes unstable during episodes of high volatility. On the other hand, pure contagion reflects excess contagion suffered during a crisis that cannot be accounted for by market fundamentals or common shocks. Such contagion is due to idiosyncratic shocks being transmitted from the "ground zero" market to other (often neighboring) countries through channels that do not exist before the event. It is important to correctly identify the type of contagion that is present in markets before prescribing policy to deal with it. For example, if markets decline due to the effects of pure contagion, then policies such as capital controls aimed at breaking market linkages are unlikely to be successful and may even aggravate the effects of the crisis. A better strategy would be to introduce policies aimed at reducing country-specific risks.

This chapter builds on the analysis of Flavin and Panopoulou (2008) in analyzing the occurrences of both shift and pure contagion between pairs of East Asian equity markets. In particular, we focus on the successes and failures of different market pairs in limiting the spread of contagious effects. Based on these observations, we propose a number of policy measures for the countries under consideration and the wider international financial community aimed at counteracting the extent of future financial crises. Many studies of Asian equity markets focus exclusively on the crisis of 1997–1998 but we take a broader view of contagion and analyze a sample of over two decades. An advantage of this approach is that our analysis is free from the problems associated with having very small crisis samples, often leading to low power in empirical tests (Dungey et al., 2007). Furthermore, Ito and Hashimoto (2005) document that turbulent episodes in Asian equity markets are not confined to the 1997–1998 period.

The remainder of this chapter is organized as follows. Section 23.2 briefly reviews the methodology employed in the original study and discusses the data. Section 23.3 reexamines the main results on a country-by-country basis, while Section 23.4 investigates their implications for formulating policy. Finally, Section 23.5 presents our concluding remarks.

23.2 ECONOMETRIC METHODOLOGY AND DATA

23.2.1 Model

Flavin and Panopoulou (2008) extend the methodology of Gravelle et al. (2006) to test for both shift and pure contagion within a unified framework. The model is bivariate in nature with regime-switching volatilities and belongs to the family of factor models widely used in financial economics. Returns (r_{it}) are decomposed into expected, μ_i, and unexpected, u_{it}, components, with the latter further broken down into common and idiosyncratic shocks, denoted by z_{ct} and z_{it}, $i = 1,2$, respectively. Both the common and the idiosyncratic shocks are allowed to switch between two states—high and low volatility. Their impacts on asset returns are measured by σ_{cit} and σ_{it}, $i = 1,2$. The regime paths are Markov switching and consequently are endogenously determined. The expected component varies with the state of the common shock only. The model can be summarized as follows:

$$r_{it} = \mu_i(1 - S_{ct}) + \mu_i^* S_{ct} + \sigma_{cit} z_{ct} + \sigma_{it} z_{it}, \quad i = 1,2 \tag{23.1}$$

and

$$\sigma_{it} = \sigma_i(1 - S_{it}) + \sigma_i^* S_{it}, \quad i = 1, 2$$
$$\sigma_{cit} = \sigma_{ci}(1 - S_{ct}) + \sigma_{ci}^* S_{ct}, \quad i = 1, 2 \tag{23.2}$$

where $S_{it} = (0, 1)$, $i = 1,2,c$ are state variables that take the value of zero in normal and unity in turbulent times. Variables belonging to the high-volatility regime are labeled with an asterisk.

Furthermore, the idiosyncratic shock of the "ground-zero" country is allowed to potentially influence the return of the second country over and above that captured by the common shock during periods of market turbulence. This describes the fact that pure contagion occurs when a country-specific shock becomes a global factor during a crisis. It is captured by augmenting the return equation of country 2 with the idiosyncratic shock of country 1 during the crisis period. This results in the following set of equations for the regime in which the idiosyncratic shock of the ground-zero country experiences high volatility:

$$r_{1t} = \mu_1^* + \sigma_{c1}^* z_{ct} + \sigma_1^* z_{1t}$$
$$r_{2t} = \mu_2^* + \sigma_{c2}^* z_{ct} + \sigma_2^* z_{2t} + \delta \sigma_1^* z_{1t} \tag{23.3}$$

A final assumption of normality of the structural shocks enables us to estimate the full model via maximum likelihood along the lines of the methodology for Markov switching models (see Hamilton, 1989).*

23.2.2 Testing for Shift Contagion

Shift contagion occurs when the transmission of common shocks changes between regimes. To empirically test the null hypothesis of "no shift

---

* For a more detailed description of the econometric model, the reader is referred to Flavin and Panopoulou (2008) and references therein.

contagion," we conduct a likelihood ratio test specifying the null and alternative as follows:

$$H_0: \frac{\sigma_{c1}^*}{\sigma_{c2}^*} = \frac{\sigma_{c1}}{\sigma_{c2}} \quad \text{versus} \quad H_1: \frac{\sigma_{c1}^*}{\sigma_{c2}^*} \neq \frac{\sigma_{c1}}{\sigma_{c2}} \quad (23.4)$$

The test statistic has a χ^2 distribution with one degree of freedom corresponding to the restriction of equality of the ratio of coefficients between the two regimes.

23.2.3 Testing for Pure Contagion

From Equation 23.3, the final term in the return generating process of country 2 measures the impact of the "ground-zero" market shock on its return and hence measures the effect of pure contagion. This term only becomes active when the idiosyncratic shock of the latter market is in the high-volatility regime. Now, our test for pure contagion is a simple t-test on the coefficient δ, where under the null $\delta = 0$ and there is no pure contagion.

23.3 DATA

We employ weekly closing stock index prices from nine East Asian markets: Japan, Korea, Indonesia, Malaysia, the Philippines, Singapore, Taiwan, Thailand, and Hong Kong. We choose to work with weekly rather than daily data in order to dilute the effects of nonsynchronous trading between market pairs.* All indices are computed by Datastream International. They are value-weighted and denominated in U.S. dollars. Our sample extends for over 17 years from April 4, 1990 to September 13, 2007, yielding a total of 910 observations. Returns are computed as the log change between two consecutive trading periods. Conducting the analysis with U.S. dollar denominated returns allows us to take the perspective of a global investor or institution that is concerned with possible contagion effects within the region.

Table 23.1 presents descriptive statistics for our weekly returns. There is considerable variation in mean returns across countries, ranging

---

* Forbes and Rigobon (2002) employ a 2 day moving-average return but this introduces serial correlation into the return generating process. Since we focus on episodes of high volatility over a longer time period and are consequently less restricted by sample size, we work with weekly returns.

TABLE 23.1 Summary Descriptive Statistics

	Japan	Korea	Indonesia	Malaysia	Philippines	Singapore	Taiwan	Thailand	Hong Kong
Mean	0.063	0.248	0.257	0.185	0.169	0.165	0.094	0.189	0.292
Median	0.000	0.176	0.071	0.275	0.213	0.161	0.145	0.099	0.441
Maximum	12.50	30.73	70.92	36.24	17.34	16.96	29.42	26.47	15.12
Minimum	−12.14	−44.13	−41.52	−32.28	−25.46	−20.34	−21.98	−24.11	−18.25
Standard deviation	3.139	5.129	5.244	4.057	3.965	2.887	4.710	4.999	3.337
Skewness	0.375	−0.053	2.410	0.344	−0.218	−0.285	0.507	0.298	−0.247
Kurtosis	4.526	13.957	44.614	22.657	7.316	8.553	8.011	6.684	5.922
Jarque Bera	109.5	4547.8	66469.7	14652.6	712.8	1180.4	990.1	527.4	332.5
	(0.000)	(0.000)	(0.000)	(0.000)	(0.000)	(0.000)	(0.000)	(0.000)	(0.000)

from 0.063% in Japan to 0.292% in Hong Kong. Korea and Indonesia were the most volatile over this period while the Singaporean market enjoyed relative stability. Normality of returns is rejected for all markets, which is consistent with the presence of both skewness and excess kurtosis. Specifically, return distributions are negatively skewed for many countries with Singapore being the most skewed. In contrast, the most positively skewed return is Indonesia. All returns, but particularly Indonesia, Korea, and Malaysia exhibit considerable leptokurtosis. This is consistent with shocks of a large magnitude being a characteristic of the distribution of equity returns. Combined with the rejection of normality, it suggests that returns may be best modeled as a mixture of distributions, which is consistent with the existence of a number of volatility regimes.

23.4 EMPIRICAL RESULTS

To formulate empirical tests for pure as well as shift contagion, it is necessary to select a "ground-zero" market from which to test if its idiosyncratic risk is transmitted to other markets during periods of high volatility. We choose Hong Kong, which is often put forward as the shock source for studies focusing on the 1997–1998 crisis (see Forbes and Rigobon, 2002; Chiang et al., 2007 among others). It is also potentially interesting as it experienced a great deal of country-specific volatility in both financial markets and politically over the sample period. Furthermore, as it is a well-developed financial center in the region, it may transmit shocks to more peripheral markets in the manner described by Kaminsky and Reinhart (2007).

We test for the presence of both shift and pure contagion using the model of Section 23.2 and the results are summarized in Table 23.2.* The most striking feature of our results is that both types of contagion appear to have been prevalent in the East Asian equity markets—often simultaneously. For five markets—Japan, Singapore, Korea, Thailand, and the Philippines—we find strong statistical evidence that both shift and pure contagion occurred over the sample period. Therefore, market linkages with Hong Kong for all these countries were highly unstable. The presence of shift contagion implies that when common

---

* For a full presentation of results, complete with diagnostic tests, see Flavin and Panopoulou (2008).

TABLE 23.2   Results of Tests for Shift
and Pure Contagion

Country	Shift	Pure
Japan	Yes	Yes
Korea	Yes	Yes
Indonesia	No	No
Malaysia	No	Yes
Philippines	Yes	Yes
Singapore	Yes	Yes
Taiwan	No	No
Thailand	Yes	Yes

shocks experience high volatility, then the degree of market interdependence undergoes a structural change. On the other hand, the existence of pure contagion indicates that when the country-specific shock of Hong Kong entered a state of turbulence, it spilled over to its partner market, thus becoming another common factor. Interestingly, Malaysia was only exposed to pure contagion, while both Taiwan and Indonesia were immune to contagious effects from Hong Kong over the period. Contrasting the fortunes of the sampled countries in combating contagion allows us to deduce appropriate policy responses for the future. Therefore we proceed to look at the experiences of markets with similar contagion patterns.

23.4.1 Taiwan and Indonesia

Let us initially focus on those equity markets that remained immune from contagious effects. Neither Taiwan nor Indonesia suffered from shift or pure contagion. However, the reasons for this immunity and the consequences of the crisis appear to be very different. Compared to its neighbors, Taiwan escaped relatively unscathed from the financial crisis of 1997–1998. Chow (2000) states that based on a misery index of stock market and currency declines, Taiwan fared best of all the East Asian markets. From Table 23.3, we see that the Taiwan's idiosyncratic shock rarely moved into the high-volatility state—it spent less than 10% of time in this regime. Furthermore, even if the common shock with Hong Kong experienced many episodes of turbulence, it did not alter the interdependence between the two markets. Many commentators attribute this resistance to contagion to the healthy position of the economy at the

TABLE 23.3 Proportion of Time Each Shock Spends
in High-Volatility Regime

	Common Shock (%)	Hong Kong Idiosyncratic (%)	Own Idiosyncratic (%)
Japan	50	48	33
Korea	51	41	3
Indonesia	40	38	12
Malaysia	34	50	11
Philippines	30	17	21
Singapore	58	26	56
Taiwan	48	68	9
Thailand	54	12	45

outset of the crisis. Wang (2000) reports that Taiwan had higher levels of economic growth, relatively low foreign debt, and high levels of foreign exchange reserves compared to its regional counterparts. Furthermore, its banking sector was less influenced by industrial powers and together with competition from foreign banks, this led to greater efficiency in loan issuance. Another feature of the Taiwanese market was that market liberalization had been introduced gradually leading to a relatively low ratio of foreign investment in equities to foreign direct investment. Many of Taiwan's industries were small- to medium-sized enterprises and open to direct investment. This meant the domestic equity market was less vulnerable to external shocks as these capital flows were not easily reversible.

On the other hand, while Indonesia remained unaffected by the crisis in Hong Kong, it certainly was hit hard by domestic events. The common shock is in the high-volatility regime approximately 40% of the time (Table 23.3) but there is no evidence that the transmission process changes between regimes. The idiosyncratic shock of Hong Kong rarely enters the turbulent regime and hence never spills over to its neighbor but clearly Indonesia was still not a stable market. Its own idiosyncratic risk factor is very often in the turbulent regime. In this case, we have simultaneous independent shocks causing volatility in both equity markets but no contagion. Indonesian troubles stemmed from a number of sources but are mostly due to an extremely weak banking sector, a culture of crony capitalism and political instability. Radelet (2000) argues that the crisis was exacerbated by bad management of the situation by both the International

Monetary Fund (IMF) and President Suharto. In particular, the imposition of inappropriate policy by the IMF and the international market reaction to the Indonesian response to these measures caused a severe lack of confidence among international investors. Indonesia suffered further country-specific problems due to low world oil prices and a drought, which badly affected its agricultural produce. Both events served to adversely affect the value of Indonesian exports. As the economy and financial markets of Indonesia went into freefall, the country was thrown into further turmoil with mass rioting and violence, which eventually led to the end of President Suharto's 30 year rule. The associated instability and ensuing political vacuum added to financial crisis. Therefore, Indonesian woes appear to have been largely resulting from domestic problems and not from contagion from Hong Kong. Interestingly, the two countries at the extremes—Taiwan and Indonesia—did not suffer from contagion but their experiences during the 1997–1998 period are markedly different.

23.4.2 Malaysia

At the outset of the crisis, Malaysia was a popular market among international investors and its market capitalization was four times its gross domestic product (GDP). Up to mid-2007, pre-crisis levels have not been achieved since the crisis. Many domestic investors were borrowing to invest in equity and these leveraged positions led to an inflated market. However, its economic situation was vulnerable with a current account deficit of about 5% and deteriorating fundamentals. Our results show that Malaysia was hit by pure contagion from Hong Kong but the diffusion of common shocks between the two markets was unaffected by high volatility in the common shock, i.e., no shift contagion. From Table 23.3, we observe that the idiosyncratic shock of Hong Kong is very often in the "crisis" regime, while Malaysia only enters into this state around the 1997–1998 period. The high-volatility shock of Hong Kong does, however, spillover and affects Malaysian equity during periods of turbulence. Interestingly, Malaysia pursued a different strategy to many of its neighbors in attempting to curb the spread of market volatility in the late 1990s. While many markets, such as Thailand, Korea, Indonesia, and the Philippines, sought external aid packages from the IMF, Malaysia unilaterally decided to regulate both inward and outward capital flows. It would appear that this approach offered protection against shift contagion but its proximity to Hong Kong still resulted in pure contagion effects. There is great debate as to whether or not these controls were successful but Johnson and Mitton

(2003) show that the imposition of capital controls led to a recovery in stock market values, albeit mainly for politically favored firms. However, it could also be argued that the capital controls introduced by Malaysia may have come a little too late as restrictions were only adopted on September 1, 1998. However, the success in preventing shift contagion was also offset by the loss of reputation and subsequent downgrading of government debt by international rating agencies.

23.4.3 Japan, Singapore, Korea, and the Philippines

All of these markets endured the damaging effects of both types of contagion. While Japan and Singapore suffered contagion, these relatively developed markets recovered quite well. On the other hand, Korea was badly hit by the financial crisis. Korea found itself in a vulnerable position as the crisis loomed. Financial liberalization in Korea resulted in the development and growth of large industrial conglomerates—the chaebols—which enjoyed political favor. The lifting of restrictions on capital flows allowed these firms to take on large levels of debt. Yanagita (2000) reports that debt–equity ratios in these firms reached approximately 400% and the incentives were such that most of this was in short-term instruments. This was allowed to happen by a weak banking sector and poor corporate governance. Furthermore, Yanagita (2000) claims that IMF policy exacerbated the problems of those debt-ridden corporations through the imposition of high interest rates. Therefore, Korea was highly susceptible to the shocks of the late 1990s and its largest firms were not sufficiently flexible to react in the required fashion. The Philippines suffered many similar problems and in common with Korea was poorly served by its weak banking sector and underdeveloped stock market.

It is clear from Table 23.3 that all markets, with the exception of Korea, experienced many episodes of high market volatility. A more in-depth analysis of our results shows that the shift contagion suffered was due entirely to changes in the response of Hong Kong to moving to the high-volatility state of the common shock. This led to instability in the levels of interdependence between market pairs during the crisis, which is particularly bad news for investors hoping to exploit the benefits of international portfolio diversification. Increased comovement during periods of high volatility is likely to reduce such benefits just when they are most needed. Therefore, for these countries, it appears to be the domestic circumstances of the "ground-zero" market that caused the change in interdependence. Therefore the turbulence suffered in Hong Kong not only led to pure contagion, i.e., its idiosyncratic shock spilling over to other markets, but also increased its own sensitivity to common shocks, resulting in shift contagion.

23.4.4 Thailand

Again, Thailand was exposed to both shift and pure contagion from Hong Kong. The first signs of crisis in Asian financial markets manifested itself with a speculative attack on the Thai currency. The Thai baht had followed a fixed exchange regime with the U.S. dollar. Thai firms had built up large foreign-denominated short-term debt obligations but had failed to adequately hedge foreign exchange risk. With the onset of currency attacks, the government responded by attempting to defend the currency but this proved costly and ultimately futile. The IMF intervention coupled with a peaceful change of government helped to restore confidence to financial markets but not before Thailand was exposed to significant levels of contagion.

Our evidence shows frequent periods of turbulence in all shocks for this pair. However, unlike other countries, the shift contagion experienced between Thailand and Hong Kong resulted from changed responses of both markets to a volatility switch in the common shock. Both markets exhibited increased sensitivity. This may be due to the fact that Thailand has also been identified as the "ground-zero" market for the 1997–1998 financial crisis by some authors, e.g., Baur and Schulze (2005), and was itself in a state of turbulence. Therefore it may be that in this pairing, i.e., Hong Kong–Thailand, there is contagion operating in both directions.*

23.5 FORMULATING POLICY

It is obvious that the pattern of contagion experienced by our sampled markets has differed considerably. The strategies employed to respond to the problems also varied across countries. While all countries would like to replicate the response of Taiwan to the crisis, this may not be possible given its unique position at the outset of the crisis. However, it does stress the importance of sound macroeconomic fundamentals, which we will return to in the following discussion.

Many markets suffered the effects of pure contagion so therefore it is imperative that the ground-zero market needs to be helped to curb excessive levels of idiosyncratic equity return risk. A number of policies aimed

---

* Flavin and Panopoulou (2008) also employ Thailand as the "ground-zero" market and results are largely consistent and would not lead to any different policy prescription.

at these domestic markets should help their stability. Firstly, the health of the stock market is closely linked to the overall health of the economy, so it is crucial that countries pursue sound macroeconomic policy. Schwert (1989) was the first to document that equity market volatility increases during recessions. As demonstrated by Taiwan, a strong economy is the best way to fend off the threat of contagion. Investor recognition of sound fundamentals led to less outward flows of capital. Key policies would appear to be the choice of exchange rate regime and the structure of debt. A credible exchange rate regime must be in place to assure foreign investors that capital gains in equity markets will not be eroded by foreign exchange losses. The choice of appropriate exchange rate needs to be addressed on a country-by-country basis but in general, there is a danger to completely fixing the rate against a strong currency. Singapore, for example, fared much better with a managed rate that had some flexibility to react to events.

Chang and Majnoni (2001) warn against the dangers of having too much short-term debt, which must be rolled over during "crisis" periods in which liquidity may be absent. Following market liberalization, many of the East Asian economies became overly burdened with short-term debt, which made them highly vulnerable to changes to liquidity and interest rates. Much of this debt was denominated in foreign currency with inherent exchange rate risk, which many countries had failed to hedge, thus increasing their exposure to external events. It is important to balance the term structure of debt and where possible to hedge against adverse currency fluctuations.

Secondly, a key policy for East Asian markets is to strengthen their financial sector. Fragility among the region's biggest banks was a common factor in many of the East Asian countries. Banks need to be decoupled from government and large domestic industries must reduce the culture of "crony capitalism" and promote confidence among international market participants. Regulation of the banking sector should seek to encourage the entrance of foreign banks to increase competition. The development and increased diversification of the region's stock markets would also be beneficial. It is imperative that stock market firms become more open to investors. An aggravating factor in the recent crisis has been the prevalence of closely held shares through firms having widespread "cross-holdings," which tend to multiply the effects of any downturn. Claessens et al. (2000) report "extensive family control" in a majority of East Asian firms. This is a feature of many markets but particularly in Indonesia, the Philippines and Thailand. There is also much anecdotal evidence that such dominance results from privileges conferred by the government. Hence, it is important that stock markets insist on listing firms that are willing to have their shares widely held.

Thirdly, the corporate governance regime also needs to be improved. Johnson et al. (2000) reveal that country-specific measures of corporate governance are more useful in explaining the depreciation of emerging stock markets during the crisis than often employed macroeconomic variables. In this regard, the adoption of, and adherence to, international disclosure and accounting standards would increase the transparency of the overall financial system. A regional version of the U.S. Securities and Exchange Commission (SEC) could serve to provide greater regulation and supervision of all markets and firms, providing more confidence in the financial system. Mitton (2002) shows that firms who already adhered to better protection of shareholders by issuing American depository receipts (ADRs) or using internationally renowned auditors outperformed firms who did not during the Asian crisis. In particular, firms with high disclosure requirements, higher levels of external ownership, and firms that concentrated on core operations (rather than large conglomerates) were more successful in preserving firm value during the period of financial turbulence. Rajan and Zingales (1998) claim that it was the lack of investor protection that fueled capital flow reversals in the region as the crisis gathered pace. The Bank of International Settlements reported a sharp reversal in capital flows between 1996 and 1997 for Indonesia, Thailand, Korea, Malaysia, and the Philippines. Net capital inflows of $95 billion turned to a net capital outflow of $12 billion. Therefore, the adoption of internationally accepted accounting standards and a monitoring body with power to discipline transgressions is vital for the health of firms and domestic stock markets.

Fourthly, we have also seen that regulation of capital flows can potentially be successful in stemming shift contagion. However, as pointed out by Abdelal and Alfaro (2003), this can be difficult for many countries to implement unilaterally due to the high reputation costs associated by opting for market de-liberalization. In the case of Malaysia, the rating agencies downgraded their sovereign risk and the subsequent increase in the required risk premium persisted long after the crisis. This policy is only advisable in extreme circumstances when all other alternatives have been exhausted. It is noteworthy that many Latin American countries who adopted such capital restrictions during the 1980s have not reemployed them during more recent crises that have hit the region.

Our final proposal is aimed more at the international financial community and centers on the need for the provision of liquidity to markets that may be in crisis. The provision of liquidity could potentially reduce the spread of the crisis as investors respond to portfolio devaluations by selling off "good" assets in other countries to maintain sufficient levels of liquidity in their business. A timely injection of liquidity may be enough to assuage the fears of investors and restore stability to affected markets. However, this could also suffer from problems of "moral hazard" and any institution charged with deciding if and when to provide liquidity would have to closely monitor each market and attach conditions to any aid packages given.

We believe that the adoption and enforcement of financial regulations can be more efficiently achieved by the creation of an "Asian Financial Regulatory" body. While macroeconomic policy will continue to be the responsibility of individual governments, the regulation and discipline of financial markets with disclosure and accounting standards on par with international norms should be handed over to an independent and objective regional institution. This would enhance the reputation of participating markets and reassure investors that they are sufficiently protected. Furthermore, this institution could have the power to create a liquidity reserve that may be used to stabilize markets in crisis periods.

23.6 CONCLUSIONS

We investigate the propagation of shocks that give rise to contagious effects in East Asian equity markets. We test for both shift and pure contagion within a unified framework and analyze the factors that caused the contagion patterns observed between different country pairs. These observations give us insight into formulating appropriate policy in the fight to curb the spread of financial crises.

Many country pairs are exposed to both forms of contagion. However, it is also clear that high-volatility country-specific shocks are most important in transmitting contagion between markets. These are mostly associated with pure contagion but the increased sensitivity of the "ground-zero" country during these crises often causes an increased response to common shocks thereby also triggering shift contagion between markets. Therefore, policy aimed at eliminating contagion must always focus on reducing volatility in the "ground-zero" country.

Based on the experiences of the sampled countries, we stress the importance of a number of policies for the East Asian financial markets. Firstly, the ability of stock markets to withstand the spread of contagion is inextricably linked to the health of the domestic economy. Therefore, it is imperative that governments pursue sound macroeconomic policy.

Secondly, a key policy must be to strengthen all aspects of the financial system. Stock market development, greater transparency in the banking sector, and improved corporate governance will all serve to reassure investors of the health of the financial sector, thereby bringing stability to these markets. Thirdly, we take the view that capital controls can be successful in temporarily stopping the transmission of contagion but should only be used in extreme circumstances due to their high costs in terms of reputation loss and persistent high premia required on sovereign debt. Finally, a regional institution should be created to regulate financial markets. Such a multilateral institution would be independent of any political lobbying and thereby restore investor confidence. Furthermore, it should have the power to create a liquidity reserve which would be administered in a strict and objective manner. This fund could be used to inject liquidity into markets during a crisis to stop its spread as investors seek to raise cash to meet their liabilities.

ACKNOWLEDGMENTS

We would like to thank Mardi Dungey and other members of the Centre for Financial Analysis and Policy, University of Cambridge for helpful comments and suggestions on an earlier version of this chapter. E.P. thanks the Irish Higher Education Authority for providing research support under the North South Programme for Collaborative Research.

REFERENCES

Abdelal, R. and Alfaro, L. (2003) Capital and control: Lessons from Malaysia. *Challenge*, 46(4): 36–53.

Baur, D. and Schulze, N. (2005) Co-exceedances in financial markets: A quantile regression analysis of contagion. *Emerging Markets Review*, 6(1): 21–43.

Bekaert, G., Harvey, C., and Ng, A. (2005) Market integration and contagion. *Journal of Business*, 78(1): 39–69.

Caporale, G., Cipollini, A., and Spagnolo, N. (2005) Testing for contagion: A conditional correlation analysis. *Journal of Empirical Finance*, 12(3): 476–489.

Chang, R. and Majnoni, G. (2001) International contagion: Implications for policy. In: S. Claessens and K. Forbes (Eds.), *International Financial Contagion*. Kluwer Academic, Boston, MA.

Chiang, T.C., Jeon, B.N., and Li, H. (2007) Dynamic correlation analysis of financial contagion: Evidence from Asian markets. *Journal of International Money and Finance*, 26(7): 1206–1228.

Chow, P.Y. (2000) What we have learned from the Asian financial crisis. In: P.Y. Chow and B. Gill (Eds.), *Weathering the Storm: Taiwan, Its Neighbors and the Asian Financial Crisis*. Brookings Institutions Press, Washington DC.

Claessens, S., Djankov, S., and Lang, L. (2000) The separation of ownership and control in East Asian corporations. *Journal of Financial Economics*, 58(1–2): 81–112.

Dungey, M., Fry, R., Gonzalez-Hermosillo, B., and Martin, V. (2007) Sampling properties of contagion tests. Unpublished manuscript, University of Cambridge, Cambridge, United Kingdom.

Dungey, M., Fry, R., and Martin, V. (2006) Correlation, contagion and Asian evidence. *Asian Economic Papers*, 5(2): 32–72.

Flavin, T. and Panopoulou, E. (2008) Detecting shift and pure contagion in East Asian equity markets: A unified approach. Working Paper, NUI Maynooth, Maynooth, Ireland.

Forbes, K.J. and Rigobon, R.J. (2002) No contagion, only interdependence: Measuring stock market comovements. *Journal of Finance*, 57(5): 2223–2261.

Gravelle, T., Kichian, M., and Morley, J. (2006) Detecting shift-contagion in currency and bond markets. *Journal of International Economics*, 68(2): 409–423.

Hamilton, J.D. (1989) A new approach to the economic analysis of nonstationary time series and the business cycle. *Econometrica*, 57(2): 357–384.

Ito, T. and Hashimoto, Y. (2005) High-frequency contagion between exchange rates and stock prices during the Asian currency crisis. In: M. Dungey and D.M. Tambakis (Eds.), *Identifying International Financial Contagion: Progress and Challenges*. Oxford University Press, New York.

Johnson, S., Boone, P., Breach, A., and Friedman, E. (2000) Corporate governance in the Asian financial crisis, 1997–98. *Journal of Financial Economics*, 58(1–2): 141–186.

Johnson, S. and Mitton, T. (2003) Cronyism and capital controls: Evidence from Malaysia. *Journal of Financial Economics*, 67(2): 351–382.

Kaminsky, G. and Reinhart, C.M. (2007) The center and the periphery: The globalization of financial turmoil. In: C. Reinhart, C. Vegh, and A. Velasco (Eds.), *Capital Flows, Crisis, and Stabilization, Essays in Honor of Gulliermo Calvo*. MIT Press, Cambridge, MA.

Mitton, T. (2002) A cross-firm analysis of the impact of corporate governance on the East Asian financial crisis. *Journal of Financial Economics*, 64(2): 215–241.

Pericoli, M. and Sbracia, M. (2003) A primer on financial contagion. *Journal of Economic Surveys*, 17(4): 571–608.

Radelet, S. (2000). Indonesia: Long road to recovery. In: P.Y. Chow and B. Gill (Eds.), *Weathering the Storm: Taiwan, Its Neighbors and the Asian Financial Crisis*. Brookings Institutions Press, Washington, DC.

Rajan, R. and Zingales, L. (1998) Which capitalism? Lessons from the East Asian crisis. *Journal of Applied Corporate Finance*, 11(3): 40–48.

Schwert, G.W. (1989) Why does stock market volatility change over time? *Journal of Finance*, 44(5): 1115–1153.

Wang, J.C. (2000) Taiwan and the Asian financial crisis: Impact and response. In: P.Y. Chow and B. Gill (Eds.), *Weathering the Storm: Taiwan, Its Neighbors and the Asian Financial Crisis*. Brookings Institutions Press, Washington DC.

Yanagita, T. (2000) International Monetary Fund conditionality and the Korean economy in the late 1990s. In: P.Y. Chow and B. Gill (Eds.), *Weathering the Storm: Taiwan, Its Neighbors and the Asian Financial Crisis*. Brookings Institutions Press, Washington, DC.

CHAPTER 24

Response of Indian Equities to U.S. Stock Market Movements of the Prior Trading Day

Silvio John Camilleri

CONTENTS

24.1 INTRODUCTION

The aim of this chapter is to glean empirical evidence about the interconnections between emerging and developed markets. This issue has attracted much academic and practitioner interest given that as emerging markets become more integrated with established ones, their traditional diversification potential may change. Specifically, this analysis considers the connections between Indian and U.S. markets in terms of how the former market reacts to the prior U.S. trading day price changes. Analyzing the Indian markets offers the advantage of access to a large base of shares, which are considerably more liquid as compared to those of other emerging markets.

This chapter first tackles the price (returns) and volatility (squared returns) responses between the above two markets. The analysis also delves into the nature of the Indian responses to U.S. market movements by considering asymmetries and whether the response timing may be deemed consistent with an efficient market. The final investigation relates to whether the Indian response to U.S. market fluctuations is affected by liquidity factors such as trading activity and transaction size.

24.2 INDIAN SECURITIES MARKETS AND RELATED DATA

Indian capital markets went through a regulatory reform in the early 1990s, moving away from a policy where share issues were controlled by the government. Subsequent improvements included efforts to enhance transparency and settlement systems and curbing market manipulation.

This analysis uses data from the National Stock Exchange (NSE) of India. NSE is one of the major Indian exchanges, together with the Bombay Stock

Exchange (BSE). NSE trading activity commenced in 1994 and around 1630 equity issues were trading in 2008. Most major stocks are quoted on both NSE and BSE and therefore these exchanges compete both for listings and for order flow. On average, around 5.5 million transactions were processed on each trading day at NSE in January 2008. Brokers interact through an automated limit order book and there are no designated market makers.

The NSE Nifty Index (N) comprises the 50 most liquid stocks whereas the Nifty Junior Index (NJ) includes the next 50 liquid stocks, jointly accounting for a substantial part of market capitalization. The data comprise N and NJ daily observations from January 1998 to May 2008. The data were filtered by deleting those Indian trading days when the market opened in the absence of a prior U.S. trading day, which was yet unaccounted for on the Indian markets. For instance if July 4 is a U.S. trading holiday, the Indian observations on July 5 were deleted, since one would expect no information spillovers from the United States in such instances. The final data thus consisted of 2514 daily observations. The Standard and Poor's (S&P) 500 Index was used as a proxy for daily U.S. market movements. Subsidiary data included volume statistics for NSE. Intraday data for the Indian indices were available for the period March 1999 until March 2000, comprising 263 trading days. One should note that due to time zone differences, Indian and U.S. trading hours do not overlap. Figure 24.1 shows a plot of the daily closing values (levels) of the indices.

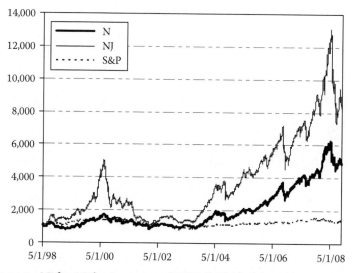

FIGURE 24.1   Nifty, Nifty Junior, and S&P daily index values.

24.3 INTERCONNECTIONS BETWEEN INDIAN AND U.S. STOCK MARKETS

This section investigates the price and volatility connections between the Indian and U.S. markets. We first study whether the Indian markets have become more integrated with U.S. markets over the years. We then estimate vector autoregression (VAR) models to test for spillovers across markets using the daily observations.

24.3.1 Level of Market Integration over the Years

Market integration may be thought of as the tendency for emerging markets to become similar to developed ones and moving more in line with the latter as different assets across markets command the same expected risk-adjusted return. One factor that makes markets more interconnected is the cross-listing of stocks on overseas exchanges. Purfield et al. (2006) presented statistics that show that the number of Indian cross-listings increased in absolute value since the year 2000, yet it has decreased in relative terms when considered as a proportion of local market capitalization. When considering simple correlation results shown in Table 24.1 Panel A, we get an idea that correlation between U.S. and Indian markets increased in the post-2000 period.

We further investigate this issue to inquire whether the level of integration between Indian and U.S. markets increased after the year 2000. We estimated the model shown in Equation 24.1 where $r_{i,t}$ is the log return of the respective Indian Index on day t, r_{St} is the S&P log return, and $\varepsilon_{i,t}$ is a

TABLE 24.1   Integration of Indian Markets with U.S. Markets

Panel A: Correlation between Indian and U.S. daily returns

	1998–2000	2001–2008
Correlation: N–S&P	0.187	0.255
Correlation: NJ–S&P	0.117	0.191

Panel B: Regression estimations with integration variable

	α	β	γ [DSP]	R^2
N Model	0.0006	0.2736	0.0900	0.0536
	(1.70)	(5.90)	(1.55)	
NJ Model	0.0007	0.2221	0.0946	0.0266
	(1.80)	(3.89)	(1.32)	

Notes: DSP is the product of the S&P daily log return and D_1, a dummy variable with a value of 1 after the year 2000 and 0 otherwise.

t-ratios are shown in brackets underneath the respective coefficients.

random error term. We created a dummy variable D_I which takes a value of 1 after the year 2000, and zero otherwise. In order to gauge whether integration between the U.S. and Indian markets increased, we use a variable DSP_t, being the product of D_I and r_{St}:

$$r_{i,t} = \alpha_i + \beta_i r_{S,t-1} + \gamma_i DSP_{t-1} + \varepsilon_{i,t} \tag{24.1}$$

Results shown in Table 24.1 Panel B indicate that DSP is positive yet insignificant.*

The resulting price connections due to market integration may be of a long-term nature and therefore should not necessarily be tested for through the comovement of daily returns. For instance, cointegration tests may be more appropriate. Different tests were specified to infer whether the U.S. and Indian indices are cointegrated. Such tests uniformly failed to reject the null hypothesis of no cointegration at the 95% confidence level and therefore are not being reported. Overall, these results suggest that while the interconnections between the Indian and U.S. markets may be increasing, India is still fairly autonomous. This may possibly be attributed to the relatively low participation rate on part of institutional investors whose actions tend to materialize in higher integration. For instance, Purfield et al. (2006) argue that Indian assets held by institutional investors are still relatively low and this may be due to restrictions over the types of assets that such investors may purchase.

24.3.2 Daily Volatility Spillovers (Squared Returns VAR)

We now test the daily connections between the U.S. and Indian markets through the estimation of VAR models using seemingly unrelated regression equations (SURE) methodology. Various studies such as Koutmos and Booth (1995) report a more significant relationship across markets when investigating volatility responses as compared to price responses. This might be due to the possibility that larger markets affect overseas ones at times in the same direction and occasionally in the opposite direction depending on the nature of particular events. For instance, adverse news from the United States may have

---

* Given that research suggests that the level of integration may change when markets go through a crisis period (Yang et al., 2003; Gębka and Serwa, 2006), the observations coinciding with the Asian Financial Crisis of 1998 were eliminated and the models were reestimated using the data for 1999–2008. Results (unreported) showed that while DSP significance increased, it was still below the 95% confidence level.

negative impacts on the Indian markets due to a possible reduction in demand for Indian exports. Conversely, adverse U.S. news may lead investors to sell U.S. stocks and seek shelter in alterative markets, causing a price change in the opposite direction. Such intricacies do not apply when analyzing squared returns since in such cases the direction of the responses is irrelevant.

A preliminary VAR(5) model was estimated as suggested by the Schwarz Bayesian criterion. More compact models were then estimated, eliminating some insignificant lags. Results are reported in Table 24.2. The S&P volatility of the prior U.S. trading session is highly significant in the NJ equation, yet it is insignificant in the N equation. Similarly the NJ model

TABLE 24.2   SURE Estimations on Squared Returns

N–S&P Estimation			NJ–S&P Estimation		
N Equation			**NJ Equation**		
Intercept	0.0002	(9.73)	Intercept	0.0002	(8.24)
N $(t-1)$	0.2351	(11.87)	NJ $(t-1)$	0.2602	(13.18)
N $(t-2)$	0.0323	(1.60)	NJ $(t-2)$	0.0837	(4.12)
N $(t-3)$	0.1005	(5.11)	NJ $(t-3)$	0.0975	(4.95)
S&P $(t-1)$	0.0593	(1.24)	S&P $(t-1)$	0.2415	(3.62)
Explanatory Statistics:			Explanatory Statistics:		
R-Bar-Squared		0.0835	R-Bar-Squared		0.1275
F-Statistic (4, 2503)		58.12	F-Statistic (4, 2503)		92.59
S&P Equation			**S&P Equation**		
Intercept	0.0001	(9.16)	Intercept	0.0001	(9.19)
S&P $(t-1)$	0.0957	(4.81)	S&P $(t-1)$	0.0946	(4.74)
S&P $(t-2)$	0.0958	(4.86)	S&P $(t-2)$	0.0936	(4.72)
S&P $(t-3)$	0.0983	(4.99)	S&P $(t-3)$	0.0981	(4.95)
S&P $(t-4)$	0.0629	(3.19)	S&P $(t-4)$	0.0630	(3.17)
S&P $(t-5)$	0.1180	(6.00)	S&P $(t-5)$	0.1207	(6.10)
N (t)[a]	0.0074	(0.96)	NJ (t)[a]	0.0046	(0.85)
Explanatory Statistics:			Explanatory Statistics:		
R-Bar-Squared		0.0711	R-Bar-Squared		0.0710
F-Statistic (6, 2501)		32.98	F-Statistic (6, 2501)		32.94

Notes: *t*-ratios are shown in brackets next to the respective coefficients. For all the models, the *F*-Statistics reject the null hypothesis that all the regressors (except the intercept) are zero at the 99% confidence level.

[a] The N and NJ variables in the S&P equations are labeled as contemporaneous since they occur on the same trading day as the S&P observations. Yet, since the Indian trading session typically terminates by the time U.S. markets open, this is really a lagged relationship.

has a better explanatory power than the N model. One would usually expect the most liquid companies to be more affected by overseas factors as compared to less liquid ones, since for instance, overseas investors might confine their holdings to the more liquid stocks. This is not the case with the former results; it might be that stocks comprising the NJ are more prone to international movements due to the nature of their business. The models also show that Indian Index volatility is insignificant in the S&P equations. These notions were confirmed through four Wald-tests where the null hypothesis that the S&P volatility had no impact on the N model could not be rejected and the null hypotheses that the N and NJ volatility had no impact on the S&P models could not be rejected. Yet, the null that the S&P volatility had no impact on the NJ model was rejected at the 99% level of confidence. Overall, the explanatory power of the models is meagre and largely emanates from the lagged observations of the dependent variable rather than from interconnections between the U.S. and Indian markets.

24.3.3 Daily Price Connections (Returns VAR)

A similar approach was taken for the estimation of the returns VAR. We started with a VAR(5) model as suggested by the Aikaike information criterion, however a more compact model was reestimated, eliminating insignificant lags. We ended up with the models shown in Table 24.3. Looking at the models for the Indian indices, we should start by cautioning about the negative explanatory power statistic. The first S&P lag is negative and significant, which indicates that the Indian markets tend to move in the opposite direction of the U.S. ones. Yet, the third S&P lag is significantly positive, suggesting that the Indian markets might initially overreact to the U.S. movement and this is subsequently corrected.

Considering the S&P models, the first lag of the respective Indian index is significant and has a positive sign, which suggests that U.S. markets are not indifferent to the Indian market. This is somewhat in line with the findings of Cuardo Sáez et al. (2007) who documented that developed markets are sensitive to emerging market fluctuations, yet the authors specified that U.S. markets are more sensitive to Latin American markets as compared to Asian ones. It should also be noted that the same Indian Index lag was insignificant in the squared returns model (Table 24.2). This indicates that while the U.S. markets react to Indian (or emerging) market movements, there is no substantial volatility spillover from the Indian to the U.S. market. Indeed, a look at Figure 24.1 confirms that the Indian markets are more volatile than U.S. ones.

TABLE 24.3   SURE Estimations on Returns

N–S&P Estimation			NJ–S&P Estimation		
N Equation			**NJ Equation**		
Intercept	0.0007	(1.80)	Intercept	0.0009	(1.70)
N (t–1)	0.0785	(3.17)	NJ (t–1)	0.1675	(6.56)
N (t–2)	−0.0706	(−3.55)	NJ (t–2)	−0.0507	(−2.54)
S&P (t–1)	−0.7071	(−25.3)	S&P (t–1)	−1.1100	(−32.47)
S&P (t–2)	0.0181	(0.51)	S&P (t–2)	−0.0041	(−0.09)
S&P (t–3)	0.1062	(3.01)	S&P (t–3)	0.1114	(2.52)
Explanatory Statistics:			Explanatory Statistics:		
R-Bar-Squared	−0.4522		R-Bar-Squared	−0.5721	
F-Statistic (5, 2502)	Nil		F-Statistic (5, 2502)	Nil	
S&P Equation			**S&P Equation**		
Intercept	0.0002	(0.68)	Intercept	0.0002	(0.71)
S&P (t–1)	−0.0378	(−1.85)	S&P (t–1)	−0.0299	(−1.48)
S&P (t–2)	−0.0396	(−1.99)	S&P (t–2)	−0.0389	(−1.95)
S&P (t–3)	−0.0379	(−2.36)	S&P (t–3)	−0.0403	(−2.59)
S&P (t–4)	−0.0108	(−0.68)	S&P (t–4)	−0.0136	(−0.88)
S&P (t–5)	−0.0517	(−3.24)	S&P (t–5)	−0.0482	(−3.11)
N (t)[a]	0.0516	(3.63)	NJ (t)[a]	0.0318	(2.74)
Explanatory Statistics:			Explanatory Statistics:		
R-Bar-Squared	0.0079		R-Bar-Squared	0.0055	
F-Statistic (6, 2501)	4.31		F-Statistic (6, 2501)	3.33	

Notes: t-ratios are shown in brackets next to the respective coefficients. In case of the S&P equations, the F-Statistics reject the null hypothesis that all the regressors (except the intercept) are zero at the 99% confidence level.

[a] The N and NJ variables in the S&P equations are labeled as contemporaneous since they occur on the same trading day as the S&P observations. Yet, since the Indian trading session typically terminates by the time U.S. markets open, this is really a lagged relationship.

A series of Wald tests was conducted on the variables, which relate to the interconnections between the Indian and U.S. markets. In case of the N and NJ equations, the null hypotheses that the first lag of the S&P has no impact on the model were rejected at the 99% confidence level. Similarly, the Wald tests rejected the null hypothesis that the respective Indian Index had no impact on the S&P models at the 99% level of confidence. Overall, while the squared returns VAR has a higher explanatory power as compared to the returns VAR, the latter model highlights the interconnections between markets more clearly when considering the statistical significance of U.S. (Indian) lags in the Indian (U.S.) model.

24.4 NATURE OF INDIAN RESPONSES TO U.S. STOCK PRICE MOVEMENTS

This section delves into select characteristics of the Indian response to U.S. price movements in terms of asymmetric properties and promptness.

24.4.1 Asymmetric Responses

We now investigate whether the response to negative U.S. returns is larger than the response to U.S. positive returns. We use a dummy variable D_A, which takes a value of 1 when the prior day U.S. return is negative and a value of 0 otherwise. The model shown in Equation 24.2 was estimated using N and NJ data:

$$r_{i,t}^2 = \alpha_i + \beta_i r_{S,t-1}^2 + \gamma_i D_A + \varepsilon_{i,t} \tag{24.2}$$

Results are presented in Table 24.4. The dummy is positive in both estimations, indicating that the response to negative U.S. returns is larger than the response to positive ones. The dummies are significant at the 95% and 90% confidence levels, respectively. The pronounced volatility following negative overseas returns is in line with other research papers including Koutmos and Booth (1995) in the context of other markets.

24.4.2 Promptness of Indian Responses to U.S. Market Movements

Following market efficiency arguments, Indian stocks should react to the previous U.S. trading session early during the day and subsequent returns should be unrelated to U.S. movements. This notion was tested by investigating the relationship between the Indian return during the first six trading minutes and the prior U.S. trading session return. We then test for the relationship between the Indian return from the sixth to the twentieth minute and the prior U.S. trading session return; these returns should be

TABLE 24.4   Asymmetric Properties of the Indian Responses

	α	β	Y [D_A]	R^2
N Model	0.0002	0.3966	0.0001	0.0269
	(9.31)	(8.02)	(2.04)	
NJ Model	0.0003	0.5365	0.0001	0.0242
	(10.05)	(7.62)	(1.84)	

Notes: D_A is a dummy variable taking a value of one when the prior day U.S. return is negative and zero otherwise. t-ratios are shown in brackets underneath the respective coefficients.

unrelated in an efficient market. Similarly the Indian return from the sixth minute to the closing should be unrelated to the prior U.S. return.

We thus used the data period for which intraday observations were available (March 1999 to March 2000) and estimated the model:

$$r_{i,t} = \alpha_i + \beta_i r_{S,t-1} + \varepsilon_{i,t} \tag{24.3}$$

where Indian returns $r_{i,t}$ were defined as the first six minute return in the first estimation, the return between the sixth to the twentieth minute in the second estimation, and the return from the sixth minute to the closing in the third estimation.

Results are reported in Table 24.5. Column A shows a significant relationship between the Indian return during the first six minutes and the prior U.S. trading day return. Column B shows that the Indian return between the sixth and twentieth trading minute is unrelated to the returns of the prior U.S. trading session. Column C yields similar indications when the Indian return between the sixth and the closing is considered. Despite this, the S&P return becomes significant at the 90% confidence level in case of the NJ estimation. One should note that while the initial Indian returns are in the same direction as the U.S. prior day return, the subsequent Indian returns are in the opposite direction of the U.S. market. This may suggest that the initial return may constitute an overreaction to the U.S. market movement, which is subsequently reversed during the rest of the trading day. Indeed, a look at the first S&P lag in the N and NJ models shown in Table 24.3 (estimated on the whole sample of 10 years of

TABLE 24.5 Connection between Indian Intraday Returns and U.S. Returns

	Column A: First 6 min		Column B: 6th to 20th min		Column C: 6th min to Closing	
	α	β	α	β	α	β
N Models	0.0040	0.1530	−0.0009	−0.0068	−0.0036	−0.0628
	(6.06)	(2.80)	(−2.45)	(−0.21)	(−3.47)	(−0.73)
	$R^2 = 0.0291$		$R^2 = 0.0002$		$R^2 = 0.0020$	
NJ Models	0.0042	0.1141	−0.0009	−0.0125	−0.0030	−0.1894
	(5.71)	(1.89)	(−2.17)	(−0.35)	(−2.30)	(−1.76)
	$R^2 = 0.0135$		$R^2 = 0.0005$		$R^2 = 0.0118$	

Notes: t-ratios are shown in brackets underneath the respective coefficients.

data) points that the Indian markets would have significantly fluctuated in the opposite direction of the U.S. markets by the end of the day.

The nature of the intraday Indian volatility was investigated further through generalized autoregressive conditional heteroscedasticity (GARCH) estimations on high-frequency data. The data set consisted of 13 continuous trading days (June 9–25, 1999) sampled at two minute intervals, yielding 2149 observations. The modeling of the Indian response is particularly tricky, since this coincides with the typically highly volatile period at the beginning of the day. Higher opening volatility is well documented in market microstructure literature (e.g., Wood et al., 1985) and one should endeavor to separate the opening volatility from the response to the U.S. return. Two dummy variables were thus created: an opening dummy (D_O) and a response dummy (D_R). A visual inspection of the data set indicated that the pronounced opening volatility takes place during the first six minutes of trading. In this way, D_O took a value of 1 for the first three observations of each trading day and a value of 0 for the rest of the day. The former investigation suggested that the Indian response to U.S. volatility took place *within* six trading minutes, and therefore D_R took a value of 1 for the first two observations of each trading day and 0 (zero) for the rest of the day.[*] In estimating GARCH models, we did not include D_R directly in the equation, but the variable was multiplied by the U.S. prior trading day return. In this way, the new variable ($D_R r_S$) captured the dummy effect and it was also sensitive to the magnitude of the U.S. return. A note is warranted regarding the possibility that the inclusion of D_O and $D_R r_S$ induces multicollinearity. One symptom of the latter is that the correlated variables would be individually insignificant in the model, yet jointly significant. This is *not* the case with the results we obtained.

Tests for asymmetric volatility following Engle and Ng (1993) indicated that it was not necessary to account for this feature.[†] In this way, we estimated GARCH (1,1) models where returns are modeled as an AR(1) process, while heteroskedasticity is modeled as shown in the following equation:

$$h_{i,t} = \omega_i + \alpha_i \varepsilon_{i,t-1}^2 + \beta_i h_{i,t-1} + \gamma_i D_O + \rho_i D_R r_{S,t-1} \qquad (24.4)$$

---

[*] Specifying DR, which takes a value of 1 during the first three observations, did not lead to materially different results.

[†] Tests for ARCH effects following the Engle (1982) methodology indicated that the null hypothesis of no ARCH effects cannot be rejected; nonetheless the estimation of GARCH models was proceeded with.

TABLE 24.6   GARCH (1,1) Estimations

N Model			NJ Model		
Log Return AR(1) Process			**Log Return AR(1) Process**		
Intercept	**Lag**	**R^2**	**Intercept**	**Lag**	**R^2**
0.000005	0.2667	0.2861	−0.000011	0.2284	0.0528
(0.31)	(11.67)		(−0.87)	(9.82)	
Conditional Variance Equation			**Conditional Variance Equation**		
ω	α	β	ω	α	β
0.0000	0.1819	0.5267	0.0000	0.1207	0.7246
(0.11)	(6.83)	(23.32)	(0.03)	(5.88)	(41.09)
$\gamma[D_O]$	$\rho[D_R r_S]$		$\gamma[D_O]$	$\rho[D_R r_S]$	
0.000014	−0.0003		0.000004	−0.0001	
(7.50)	(−2.96)		(1.89)	(−38.46)	
R-Bar-Squared = 0.0263			R-Bar-Squared = 0.0506		
F-Statistic (5,2141) = 12.61			F-Statistic (5,2141) = 23.88		

Notes: *t*-ratios are shown in brackets underneath the respective coefficients.
The F-statistics reject the null hypothesis that all the regressors (except the intercept)
are equal to 0 at the 99% confidence level.

where

$h_{i,t}$ is the conditional variance of the respective index, which depends
on past information

$\varepsilon_{i,t}$ is the unexpected return observed during period t

GARCH estimations are shown in Table 24.6. The variable $D_R r_S$ is highly
significant in both models, yet the negative coefficient is in the unexpected
direction since this indicates that the response to the U.S. returns induces
a lower conditional variance. Perhaps this might be interpreted as a sign
that the information from U.S. markets is a yardstick that reduces uncertainty at the opening, although this explanation would conflict with the
former observation that the initial response to U.S. movements might
constitute an overreaction as it tends to reverse subsequently.

24.5 LIQUIDITY-RELATED FACTORS

This section considers whether the Indian response to U.S. fluctuations is
affected by liquidity factors. In particular, when a stock does not trade or
if there are only a few transactions in a particular stock, the latter might
fail to reflect the impact of recent news or overseas developments. In this

way, one would expect a more significant Indian response when the number of companies traded and the number of transactions on the exchange are higher. Similarly, we investigate whether the Indian response is more intense when the trading day is characterized by larger transactions. This might be due to the possibility that the typically larger transactions of professional fund managers may be more likely to reflect overseas developments if the former diversify their portfolios internationally.

24.5.1 Response to U.S. Market Movements and Trading Activity

The effect of the number of companies traded (c_t) on the Indian response was investigated by estimating the following ordinary least squares (OLS) regression on N and NJ daily data:

$$r_{i,t}^2 = \alpha_i + \beta_i r_{S,t-1}^2 + \gamma_i c_t + \varepsilon_{i,t} \tag{24.5}$$

The variable c_t was highly significant and positive as shown in Table 24.7 Panel A, implying that the number of companies traded positively impacts on the transmission of volatility.

TABLE 24.7   Effects of Liquidity Factors on Indian Responses

Panel A: Number of companies traded (c_t)

	α	β	$\gamma[c_t]$	R^2
N Model	−0.0001	0.3942	0.0000003	0.0313
	(−1.16)	(7.99)	(3.94)	
NJ Model	−0.0003	0.5299	0.000001	0.0348
	(−2.59)	(7.57)	(5.57)	

Panel B: Number of transactions per company traded (t_t)

	α	β	$\gamma[t_t]$	R^2
N Model	0.0002	0.4060	0.0000	0.0257
	(7.83)	(8.13)	(1.03)	
NJ Model	0.0003	0.5397	−0.0000	0.0229
	(9.15)	(7.60)	(−0.02)	

Panel C: Average transaction size (z_t)

	α	β	$\gamma[z_t]$	R^2
N Model	0.0002	0.3879	0.0176	0.0290
	(5.25)	(7.84)	(3.08)	
NJ Model	0.0001	0.5088	0.0484	0.0366
	(3.34)	(7.26)	(5.98)	

Note: t-ratios are shown in brackets underneath the respective coefficients.

A similar approach was adopted when investigating the effect of the number of transactions in each company. The average number of transactions per company was computed by dividing the total number of transactions on the exchange for the particular day by the number of companies traded during the same day. We then estimated the model:

$$r^2_{i,t} = \alpha_i + \beta_i r^2_{S,t-1} + \gamma_i t_t + \varepsilon_{i,t} \qquad (24.6)$$

where t_t is the average number of transactions per company traded on day t. As shown in Table 24.6 Panel B, the coefficients of t_t are insignificant in both estimations. This suggests that the Indian response to U.S. volatility is independent of the number of transactions per company and somewhat contradicts the well-established notion that volume may constitute an additional response to information apart from price changes (Verrechia, 1981). The reason behind such result might be that most of the Indian response to U.S. volatility takes place during the first six minutes of trading (Section 24.4.2) and therefore it materializes even after a relatively few transactions have been executed.

24.5.2 Response to U.S. Market Movements and Transaction Size

We now consider whether the Indian response is more intense when the average transaction size is larger. The total traded value was divided by the number of transactions for each trading day in order to obtain the average transaction size z_t. We then estimated the model shown in the following equation:

$$r^2_{i,t} = \alpha_i + \beta_i r^2_{S,t-1} + \gamma_i z_t + \varepsilon_{i,t} \qquad (24.7)$$

Results in Table 24.7 Panel C show that transaction size is highly significant, suggesting that larger transactions may be associated with higher volatility responses. Overall, the results point that volatility spillovers from the United States to India become more intense as the number of traded companies and average transaction size on the latter market increase.

24.6 CONCLUSION

This analysis investigated the price and volatility connections between the Indian and U.S. markets. Indian markets are sensitive to U.S. price changes particularly when considering volatility spillovers (rather than the price effect). This might be due to the possibility that U.S. fluctuations affect

Indian markets at times in the same direction, while at times the markets move in the opposite direction depending on the nature of the event to which they are responding. The Indian markets react fairly early in the trading day to U.S. price changes; the reaction is asymmetric and its intensity is also affected by liquidity factors. We also found evidence that U.S. markets are not indifferent to Indian markets, although this may possibly constitute a reaction to mutual factors across emerging markets or the possibility that both markets respond to common news. The low explanatory power of most models suggests that the majority of price fluctuations on the Indian markets may not be explained by U.S. factors. When coupled with the cointegration tests where the null hypothesis of no cointegration was not rejected, the results point that the Indian markets should have offered considerable diversification potential for U.S. investors over the sample period.

A note about the limitations of this study is warranted. Due to time zone differences, the markets' reaction to a given event might at times occur on the same trading date or at times on different trading days. Therefore, an international event occurring during Indian trading hours would firstly be accounted for by the Indian markets and subsequently by the U.S. ones giving the impression that the former markets influence the latter. Conversely, an event occurring during the U.S. trading hours is priced by the Indian markets on the subsequent day, which might be mistaken for direct U.S. influence on the Indian markets. Another limitation emanates from the fact that the analysis involves stock market data spanning over a long period of time. This implies that the conditions which underlie the pricing process are likely to change due to possible modifications in trading protocols and other factors which are unaccounted for.

As for future research topics, the efficiency of the Indian response to U.S. movements may be analyzed in further detail. While the above results suggest that Indian markets respond early in the day to U.S. fluctuations, the initial response may also constitute an overreaction since it seems to reverse at a subsequent stage.

REFERENCES

Cuardo Sáez, L., Fratzscher, M., and Thinmann, C. (2007) The transmission of emerging market shocks to global equity markets. Working Paper 724/2007, European Central Bank, Frankfurt, Germany.

Engle, R.F. (1982) Autoregressive conditional heteroscedasticity with estimates of the variance of U.K. inflation. *Econometrica*, 50(4): 987–1007.

Engle, R.F. and Ng, V.K. (1993) Measuring and testing the impact of news on volatility. *The Journal of Finance*, 48(5): 1749–1778.

Gębka, B. and Serwa, D. (2006) Are financial spillovers stable across regimes? Evidence from the 1997 Asian crisis. *International Financial Markets, Institutions and Money*, 16(4): 301–317.

Koutmos, G. and Booth, G.G. (1995) Asymmetric volatility transmission in international stock markets. *Journal of International Money and Finance*, 14(6): 747–762.

Purfield, C., Hiroko, O., Kramer, C., and Jobst, A. (2006). Asian equity markets: Growth, opportunities, and challenges. Working Paper, International Monetary Fund.

Verrechia, R.E. (1981) On the relationship between volume reaction and consensus of investors: Implications for interpreting tests of information content. *Journal of Accounting Research*, 19(1): 271–283.

Wood, R.A., McInish, T.H., and Ord, J.K. (1985) An investigation of transactions data for NYSE stocks. *The Journal of Finance*, 40(3): 723–739.

Yang, J., Kolari, J.W., and Min, I. (2003) Stock market integration and financial crises: The case of Asia. *Applied Financial Economics*, 13(7): 477–486.

CHAPTER 25

Asset Pricing with Higher-Order Co-Moments and Alternative Factor Models: The Case of an Emerging Market

Javed Iqbal, Robert D. Brooks, and Don U.A. Galagedera

CONTENTS

25.1 INTRODUCTION AND REVIEW OF THE LITERATURE

The failure of the conventional capital asset pricing model (CAPM) to explain cross-sectional variation of risky asset return adequately has spurred alternative explanations of asset pricing. The arbitrage pricing theory (APT) of Ross (1976) is one such alternative. APT stipulates that under no arbitrage the expected returns of risky assets are expressed as a linear function of certain common factors though the theory does not specify the factors themselves. Many studies follow this lead. For example, based primarily on statistical considerations, Fama and French (1992) advocate inclusion of two factors mimicking the size and book-to-market value of the assets besides the systematic beta risk. Kraus and Litzenberger (1976) and Dittmar (2002) emphasize inclusion of higher co-moments namely, co-skewness and co-kurtosis as explanatory variables of expected returns.

Barone-Adesi (1985) and Barone-Adesi et al. (2004) (henceforth referred to as BA and BAGU, respectively) recast the covariance–co-skewness CAPM as the APT restriction on the system of quadratic market model for which the multivariate methodology of Gibbons (1982) is readily applicable. Their testing approach avoids the errors-in-variables and multicollinearity problems of utility-based asset pricing and makes better use of available information by employing the contemporaneous covariance among the asset returns in a multivariate setting. This approach of APT testing is also efficient. BA and BAGU approach uses the information on the return on the stocks and the market portfolio only thereby being less dependent on external macroeconomic data unlike the prespecified macroeconomic approach of APT testing such as Chen et al. (1986). Relevance of co-moments of order greater than 2 and their likely impact on expected returns are known to be different in emerging and developed markets. For example, Aggarwal et al. (1999) observe that generally the skewness in the return distribution is positive for emerging market indices and negative for developed markets. Non-normality of returns is an important consideration when modeling emerging market returns as their microstructure and relatively turbulent political and economic environment make the normality assumption difficult to justify.

Stylized facts of emerging markets returns suggest that co-kurtosis may be an appropriate measure of systematic risk in such markets. In a sample of 17 emerging markets including Pakistan, Hwang and Satchell (1999) show that co-kurtosis of portfolio returns has at least as much explanatory power as co-skewness. Da Silva (2006) provides similar evidence in the Brazilian market. Intuitively, the widespread evidence of outliers in emerging market returns suggests that the extreme outcomes have a high probability of occurrence in emerging markets. BAGU point out the possibility of a missing systematic factor in their pricing model. They did not consider co-kurtosis as a potential explanatory variable of asset returns as their specification tests do not support a cubic market factor.

The APT does not prescribe the factors that need to be included in the factor space. The BA and BAGU provide a heuristic approach of linking the quadratic market model with APT. Therefore, the linear and quadratic market returns are deemed as APT factors. Fama and French (1993) suggest size and book-to-market portfolio returns as potential common factors. For developed capital markets, several studies have compared the Fama–French factors and higher-order co-moments in explaining asset returns. For example, in the U.S. market using Center for Research in Security Prices (CRSP) portfolios, Chung et al. (2006) find that Fama–French factors ceased to be effective in explaining asset returns when the first 10 co-moments are included in the return-generating process. Consequently, they conclude that Fama–French factors may proxy for higher-order co-moments. Using Fama–French size portfolios, BAGU reports that the size factor anomaly is resolved by incorporating co-skewness in the pricing model. On the other hand, for a sample of U.K. data, Hung et al. (2004) provide limited evidence in favor of the higher-order market factors associated with co-skewness and co-kurtosis compared to the Fama–French factors. For emerging markets, studies comparing alternative factor models involving higher-order co-moments are extremely rare. An exception is Da Silva (2006) for the Brazilian stock market.

The purpose of this chapter is twofold. First we generalize the multivariate methodology of BAGU to incorporate co-kurtosis in the asset pricing model that can be more relevant for emerging market data. We provide empirical evidence from an emerging market. Second we intend to provide evidence of comparing the explanatory power of systematic co-moments and fundamental factors for emerging markets and investigate whether recent U.S. evidence in favor of systematic co-moments continue to hold in emerging markets, particularly Pakistan. Unlike in the Brazilian study by Da Silva (2006), we provide both time series and cross-section evidence in this regard.

In the empirical investigations, we consider the Karachi stock market, which is the largest stock exchange in Pakistan.* In 2002, it was declared as the best performing stock market globally in terms of the percentage increase in the local market index. We investigate whether an asset pricing model with higher co-moments is able to explain risk–return relation in this emerging market.

Iqbal and Brooks (2007a,b) report statistical evidence using daily, weekly, and monthly data from Pakistan's market that certain risk variables including skewness explain individual stock returns. In Iqbal and Brooks (2007b), the risk–return relationship appears to be nonlinear. A possible implication of this nonlinearity is the relevance of systematic co-moments of order greater than 2. This chapter focuses on multifactor asset pricing models that incorporate fundamental variables and higher-order co-moments.

The rest of this chapter is organized as follows. Section 25.2 describes estimation and inference for a higher-order co-moment model. Section 25.3 discusses the data employed in this chapter. Section 25.4 investigates the empirical evidence on the asset pricing model with co-moments. The multifactor model based on co-moments and fundamental factors

---

* Karachi Stock Exchange is the largest of the three stock markets in Pakistan. In June 2007, the market capitalization was $66.4 billion, which is 46% of Pakistan's GDP for the fiscal year 2006–2007 (Pakistan Economic Survey, 2006–2007).

are compared in Section 25.5 and Section 25.6 provides some concluding remarks.

25.2 FRAMEWORK FOR ESTIMATION AND INFERENCE ON HIGHER-ORDER CO-MOMENTS MODEL

We consider a specification of the return-generating process with quadratic and cubic market return factors. The framework that we outline
here is an extension of the BAGU approach on a return-generating process
with a quadratic term. Let R_t denote an $N \times 1$ vector of N asset returns at
time t and R_{mt} and R_{ft} represent the return of the market portfolio and the
risk-free rate, respectively. The cubic market model can be expressed as

$$r_t = \alpha + \beta r_{mt} + \gamma q_{mt} + \delta c_{mt} + \varepsilon_t \qquad (25.1)$$

where

$r_t = R_t - R_{ft}$ is the vector of excess returns
$r_{mt} = R_{mt} - R_{ft}$
$q_{mt} = R_{mt}^2 - R_{ft}$
$c_{mt} = R_{mt}^3 - R_{ft}$

The N intercepts are collected in vector α and each of β, γ, and δ are $N \times 1$
vector of sensitivities. The ε_t is the vector of error term, which is assumed
to satisfy

$$E(\varepsilon_t | I_t) = 0 \quad \text{and} \quad E(\varepsilon_t \varepsilon_t' | I_t) = \Sigma \qquad (25.2)$$

The information set I_t includes all current and past lagged values of R_m
and R_f. Although γ does not exactly correspond to the usual definition of
co-skewness, BAGU argue that it is a good proxy for co-skewness. It can be
similarly argued that δ provides a proxy for co-kurtosis. The cubic market
model is a statistical description of the return-generating process consistent
with the four-moment CAPM. See for example, Hwang and Satchell (1999).

Following BA, the expected asset returns under APT is given by the
following linear specification:

$$E(r_t) = \beta \lambda_1 + \gamma \lambda_2 + \delta \lambda_3 \qquad (25.3)$$

where λ_1, λ_2, and λ_3 are expected excess return on portfolios whose return
are perfectly correlated with r_m, q_m, and c_m, respectively. A similar four-
moment expected return model is derived by Hwang and Satchell (1999,
Equation 11) who show that the expected sign of λ_1 is positive according

to usual risk–return trade-off. The sign of λ_2 is opposite of the market skewness, i.e., if skewness of market portfolio is negative as in our case, then λ_2 is expected to be positive. The sign of λ_3 is positive. The APT approach of BA involves minimal assumptions about the investor's utility function.

Now applying expectations to Equation 25.1 and equating with Equation 25.3 results in the following APT-imposed restriction on the coefficients of the cubic market model:

$$\alpha = \gamma v_1 + \delta v_2 \qquad (25.4)$$

where
$$v_1 = [\lambda_2 - E(q_m)]$$
$$v_2 = [\lambda_3 - E(c_m)]$$

Therefore the arbitrage equilibrium consistent with co-skewness and co-kurtosis results in the following restricted model:

$$E(r_t) = \beta r_{mt} + \gamma q_{mt} + \delta c_{mt} + v_1 \gamma + v_2 \delta \qquad (25.5)$$

The parameters in Equation 25.5 can be estimated by a Quasi Maximum Likelihood (QML) approach. In the present context, the essential idea of the QML approach is that consistent and asymptotically normally distributed estimators of the parameters are obtained by correctly specifying the first two moments of the error distribution given in Equation 25.2. The normal log likelihood function for the restricted model can then be constructed to estimate the parameters and perform the inference. The consistency and asymptotic normality of estimators is guaranteed even if the likelihood is misspecified. Thus this approach does not rely upon the assumption of normality of the errors. The widespread evidence of nonnormality of the returns and the compelling reasons to include the higher moments dictate the importance of this normality-robust feature in estimation and inference in an emerging market context.

Let $\hat{B} = [\hat{\alpha}\,\hat{\beta}\,\hat{\gamma}\,\hat{\delta}]$ be the $N \times 4$ matrix of the estimates of the parameters. Then QML implies that under assumption (Equation 25.2)

$$\sqrt{T}(\hat{B} - B) \xrightarrow{d} N(0, \Sigma \otimes E(F_t F_t')) \qquad (25.6)$$

where $F_t = [1\ r_m\ q_m\ c_m]$. The constrained model (Equation 25.5) involves cross-equation restrictions. The restricted parameters are

$$[\hat{\beta}' \, \hat{\gamma}' \, \hat{\delta}']' = \left[\sum_{t=1}^{T} r_t \hat{H}_t'\right] \left[\sum_{t=1}^{T} \hat{H}_t \hat{H}_t'\right] \tag{25.7}$$

$$\hat{H}_t = [r_{mt} \, q_{mt} + \hat{v}_1 \, c_{mt} + \hat{v}_2]', \ \ \hat{Z} = [\hat{\gamma} \, \hat{\delta}] \tag{25.8}$$

$$[\hat{v}_1 \ \hat{v}_2]' = (\hat{Z}' \hat{\Sigma}^{-1} \hat{Z})^{-1} \hat{Z}' \hat{\Sigma}^{-1} (\overline{r}_t - \hat{\beta} \overline{r}_{mt} - \hat{\gamma} \overline{q}_{mt} - \hat{\delta} \overline{c}_{mt}) \tag{25.9}$$

These parameter formulas are generalizations of those given in BAGU and can be estimated by nonlinear feasible generalized least square with starting values provided by their unrestricted counterparts.

The restriction (Equation 25.4) is tested using an asymptotic least square statistic:

$$W_2 = (T - N/2 - 3/2) \frac{(\hat{\alpha} - \tilde{v}_1 \hat{\gamma} - \tilde{v}_2 \hat{\delta})' \hat{\Sigma}^{-1} (\hat{\alpha} - \tilde{v}_1 \hat{\gamma} - \tilde{v}_2 \hat{\delta})}{1 + \tilde{\lambda}' \hat{\Sigma}_f^{-1} \tilde{\lambda}} \xrightarrow{d} \chi^2(N-2) \tag{25.10}$$

where

$$\tilde{\lambda} = \hat{\mu} + [0 \ \tilde{v}_1 \ \tilde{v}_2]'$$
$$\hat{\mu} = [\overline{r}_{mt} \ \overline{q}_m \ \overline{c}_{mt}]'$$
$$[\tilde{v}_1 \ \tilde{v}_2]' = (\hat{Z}' \hat{\Sigma}^{-1} \hat{Z})^{-1} \hat{Z}' \hat{\Sigma}^{-1} \hat{\alpha}$$

The unrestricted system has $4N$ parameters and the restricted system has $3N+2$ parameters. The APT therefore imposes $N-2$ restrictions, which are employed as the degrees of freedom. In the QML approach, the moments must be correctly specified. This essentially translates into specifying the return-generating process correctly. To this end, we consider two other alternative specifications of the return-generating process: one that considers only the co-skewness and another only the co-kurtosis in addition to systematic beta risk. To select the most appropriate return-generating specification a joint Wald test on the parameters of unrestricted system (Equation 25.1) is performed.

25.3 DATA

25.3.1 Description of the Data

The tests discussed in Section 25.2 are applied to portfolios formed from a sample of stocks listed on the Karachi Stock Exchange (KSE). The sample period spans 13.5 years from October 1992 to March 2006 and includes 162 monthly observations. The data consisting of monthly closing prices of 101 stocks and the Karachi Stock Exchange 100 index (KSE-100) are collected from the DataStream database. Stocks selection was based on the availability of time series data on active stocks for which the prices have been adjusted for dividend, stock split, merger, and other corporate actions. The KSE-100 is a market capitalization weighted index. It comprises top companies from each sector of KSE in terms of their market capitalization. The rest of the companies are picked on the basis of market capitalization without considering their sector. We consider the KSE-100 as a proxy for the market portfolio. The 101 stocks in the sample account for approximately 80% of the market in terms of capitalization. Market capitalization data is not routinely available for all firms in the database. However the financial daily, the *Business Recorder* (www.businessrecorder.com.pk), reports some information over the recent past. We selected the market capitalization of all selected stocks at the beginning of July 1999, which corresponds roughly to the middle of the sample period considered in this chapter. We use monthly data and compute raw returns assuming continuous compounding. The 30-day repurchase option rate is used as a proxy for the risk-free rate.

25.3.2 Formation of Portfolios

To investigate robustness in the empirical results, we consider several sets of portfolios. These are based on sorting stocks on size, beta, industry, co-skewness, and co-kurtosis. We construct 17 equally weighted size portfolios. First, the stocks are ranked on market capitalization in the ascending order. The first portfolio consists of the first five stocks while the rest consist of six stocks each. The portfolio return is calculated as the equally weighted average return of the stocks in the portfolio. We similarly construct 17 beta, co-skewness, and co-kurtosis portfolios, which are based on ranking of the stocks on these sample characteristics. The beta for sorting stocks into portfolios is estimated through the market model. The co-skewness and co-kurtosis are estimated as in Harvey and Siddique (2000) i.e.,

$$\text{Co-skewness} = \frac{E(\varepsilon_{it}\varepsilon_{mt}^2)}{\sqrt{E(\varepsilon_{it}^2)}E(\varepsilon_{mt}^2)}, \quad \text{Co-kurtosis} = \frac{E(\varepsilon_{it}\varepsilon_{mt}^3)}{\sqrt{E(\varepsilon_{it}^2)}E(\varepsilon_{mt}^3)} \quad (25.11)$$

where
  ε_{it} are the residual from regressing the excess stock return on excess market return
  ε_{mt} is the residual from regressing excess market return on a constant

For the industry portfolios, the stocks are classified into 16 major industrial sectors. The sector sizes range from two stocks in the transport sector and 13 stocks in the communication sector. The industry sectors employed are auto and allied, chemicals, commercial banks, food products, industrial engineering, insurance, oil and gas, investment banks and other financial companies, paper and board, pharmacy, power and utility, synthetic and rayon, textile, textile spinning and weaving, transport and communication and other/miscellaneous firms that include tobacco, metal, and building material companies.

Table 25.1 presents some descriptive statistics for excess returns on the size portfolios and the market portfolio for monthly portfolio data from October 1992 to March 2006. The last two columns report the Jarque–Bera normality test statistic and the associated *p*-value. The skewness of the market return is negative. The returns are quite volatile as observed by their standard deviations. It is generally observed that the source of nonnormality is the excess kurtosis.

TABLE 25.1   Descriptive Statistics of Returns of 17 Size-Sorted Portfolios

Portfolio	Mean	SD	Skewness	Kurtosis	Jarque–Bera	*p*-Value (JB)
1 (smallest)	−0.701	8.066	0.559	3.919	14.16	0.001
2	−0.490	13.29	0.183	13.09	689.5	0.000
3	−0.564	7.941	0.707	5.341	50.50	0.000
4	−0.688	7.632	−1.037	8.969	269.5	0.000
5	0.029	9.361	0.202	2.790	1.400	0.495
6	−0.376	8.231	0.104	3.557	2.390	0.301
7	0.233	8.380	0.938	7.094	137.0	0.000
8	0.490	6.950	0.379	2.821	4.100	0.129
9	0.199	9.561	0.363	3.700	6.870	0.032
10	0.053	7.040	−0.169	3.429	2.020	0.364
11	0.074	9.725	0.222	3.009	1.330	0.517
12	−0.279	7.897	0.565	3.754	12.47	0.002
13	0.407	7.090	0.388	3.267	4.550	0.102
14	0.076	10.34	0.263	3.146	2.030	0.363
15	0.383	9.433	0.398	3.319	4.980	0.083
16	0.057	11.52	−0.178	3.548	2.880	0.236
17 (largest)	−0.239	11.72	0.130	3.709	3.850	0.145
Market portfolio	0.535	9.823	−0.437	4.787	26.740	0.000

25.3.3 Construction of Fama–French Factors

Construction of the Fama–French factors requires firm level data on shareholder equity, number of outstanding stocks, and market capitalization. The State Bank of Pakistan's document "Balance sheet analysis of joint stock companies" publishes annual data on balance sheet items for nonfinancial firms. For financial firms, the data are obtained from other unpublished sources in the State Bank.* The data related to the market capitalization and the number of outstanding stocks is collected through the financial daily, the *Business Recorder*. As the accounting and capitalization data are not available for the full sample period, the data employed correspond roughly to the middle of the sample period. The book value is obtained as the net assets of the firms excluding any preferred stocks. The mimicking portfolios of the size and book-to-market are constructed according to the Fama and French (1993) methodology. The stocks are allocated into two size portfolios (small and large) depending on whether their market equity is above or below the median. A separate sorting of the stocks classifies them into three portfolios formed using the break points of the lowest 30%, middle 40%, and the highest 30% based on their book-to-market value. From these independent sorting, we construct six portfolios from the intersection of two size and three book-to-market portfolios (S/L, S/M, S/H, B/L, B/M, B/H). Equally weighted portfolios are constructed for the full sample range. The SMB factor is the return difference between the average returns on the three small-firms portfolios, (S/L + S/M + S/H)/3 and the average of the returns on three big-firms portfolios, (B/L + B/M + B/H)/3. In a similar way, the HML factor is the return difference in each time period between the return of the two high book-to-market portfolios, (S/H + B/H)/2 and the average of the returns on two low book-to-market portfolios, (S/L + B/L)/2. The construction in this way ensures that the two constructed factors represent independent dimensions in relation to the stock returns.

25.4 EMPIRICAL ANALYSIS OF PRICING MODELS WITH HIGHER-ORDER CO-MOMENTS

Initially, we consider three return-generating process associated with higher-order systematic co-moments. They are

---

* We are thankful to Mazhar Khan and Kamran Najam for their helpful cooperation in the balance sheet data access.

Model 1 (covariance – co-skewness): $r_t = \alpha + \beta r_{mt} + \gamma q_{mt} + \varepsilon_t$ (25.12)

Model 2 (covariance – co-kurtosis): $r_t = \alpha + \beta r_{mt} + \delta c_{mt} + \varepsilon_t$ (25.13)

Model 3 (covariance – co-skewness – co-kurtosis): $r_t = \alpha + \beta r_{mt} + \gamma q_{mt}$
$$+ \delta c_{mt} + \varepsilon_t$$
(25.14)

The aim here is to select the most appropriate model that fits the data for further analysis. Table 25.2 presents some goodness-of-fit measures of the three alternative systems of unrestricted seemingly unrelated regression equations for the size, industry, beta, co-skewness, and co-kurtosis portfolios.

The results show that the model with co-kurtosis (Model 2) has a higher overall average adjusted r-square compared to the model with co-skewness only (Model 1). The model with both co-skewness and co-kurtosis (Model 3) has a slightly higher explanatory power than Models 1 and 2 according to Glahn's (1969) squared composite correlation coefficient. These observations

TABLE 25.2   Goodness-of-Fit Measures of the Alternative System of Higher-Order Co-Moments Models

	Model 1	Model 2	Model 3
Panel A: Size portfolios			
Average $\bar{R}^2$	0.3203	0.3483	0.3486
System R^2	0.3535	0.3789	0.3831
Panel B: Industry portfolios			
Average $\bar{R}^2$	0.3138	0.3363	0.3362
System R^2	0.3725	0.3922	0.3957
Panel C: Beta portfolios			
Average $\bar{R}^2$	0.2923	0.3146	0.3155
System R^2	0.3625	0.3858	0.3896
Panel D: Co-skewness portfolios			
Average $\bar{R}^2$	0.3331	0.3561	0.3632
System R^2	0.3324	0.3532	0.3639
Panel E: Co-kurtosis portfolios			
Average $\bar{R}^2$	0.3171	0.3541	0.3541
System R^2	0.3317	0.3717	0.3753

are made in all five types of portfolios. Therefore, in terms of the goodness-of-fit, Model 3 is the preferred model.*

Now we discuss the results of the QML-based test of the restriction imposed by arbitrage equilibrium on Model 3. Table 25.3 reports the QML statistic subject to APT restrictions on the cubic market model. The test statistic is asymptotically distributed as Chi-square with $N - 2$ degrees of freedom. The results reported in Table 25.3 reveal that arbitrage restrictions are not rejected in all five types of portfolios, suggesting the appropriateness of higher-order co-moments (co-skewness and co-kurtosis) for the emerging market under investigation. This evidence is stronger in co-skewness and co-kurtosis portfolios. A similar analysis reveals that arbitrage equilibrium is supported in Model 2 as well. These results are not reported to save space. A detailed analysis of Model 3 parameter estimates at portfolio level is discussed next.

In Table 25.4, we report the estimates of the parameters of the cubic market model subject to the restriction of the three-factor arbitrage equilibrium for the size portfolios. The estimate of the coefficient of linear market return (β) is generally less than 1 except those in the two largest size portfolios. The coefficient of quadratic market return (γ) is significantly different from zero in only two of the 17 size portfolios. The evidence on the significance of δ (coefficient of cubic term) is very strong. In all portfolios with the exception of the two smallest in size, δ is significant at the 10% level. This observation clearly highlights the importance of co-kurtosis in modeling time series of asset returns in the emerging

TABLE 25.3   QML Test Statistic of the APT
Restrictions on Cubic Market Model

Portfolio	QML Test Statistic	p-Value
Size	11.221	0.736
Industry	12.991	0.527
Beta	12.292	0.656
Co-skewness	9.245	0.864
Co-kurtosis	10.311	0.799

---

* As a diagnostic check we investigated β, γ, and δ in Model 3 estimated using SUR. The results reveal that γ is not significant in any of the 17 portfolios. However, δ is significantly different from zero in 12 portfolios (i.e., in 70% portfolios) at the 5% level. Moreover, multivariate tests of joint zero restrictions across the system rejects co-skewness and do not reject co-kurtosis. The general conclusion obtained with the size portfolios is valid with other sets of portfolios such as beta and industry sorted portfolios. The results are not reported here to save space.

TABLE 25.4   Parameter Estimates for the Restricted Cubic Market Model
for the Size Portfolios

Portfolio	Market (β)	Co-Skewness ($\gamma \times 100$)	Co-Kurtosis ($\delta \times 1000$)
1	0.493*	−0.621*	−0.463*
(smallest)	(5.40)	(−1.803)	(−2.841)
2	0.321*	0.296	−0.165
	(1.998)	(0.50)	(−0.576)
3	0.370*	−0.443	−0.181
	(4.054)	(−1.311)	(−1.116)
4	0.286*	−0.526	−0.277*
	(3.124)	(−1.541)	(−1.696)
5	0.740*	−0.119	−0.466*
	(7.954)	(−0.340)	(−2.809)
6	0.656*	−0.327	−0.547*
	(7.747)	(−1.016)	(−3.611)
7	0.546*	−0.494	−0.487*
	(5.881)	(−1.417)	(−2.939)
8	0.453*	0.059	−0.245*
	(6.100)	(0.213)	(−1.851)
9	0.873*	−0.293	−0.834*
	(9.359)	(−0.795)	(−4.964)
10	0.623*	−0.294	−0.374*
	(9.651)	(−1.208)	(−3.243)
11	0.799*	−0.527	−0.654*
	(8.113)	(−1.401)	(−3.713)
12	0.675*	−0.544*	−0.543*
	(8.633)	(−1.818)	(−3.880)
13	0.568*	−0.042	−0.306*
	(8.281)	(−0.163)	(−2.499)
14	0.992*	−0.450	−0.662*
	(10.979)	(−1.298)	(−4.093)
15	0.940*	0.177	−0.381*
	(14.183)	(0.686)	(−3.201)
16	1.176*	−0.352	−0.391*
	(16.670)	(−1.324)	(−3.102)
17	1.135*	−0.071	−0.300*
(largest)	(14.859)	(−0.250)	(−2.201)

Note: The *t*-statistics of the parameter estimates are reported in the parenthesis.
* Indicates significance at 10% level.

market under investigation here. Further, both quadratic and cubic market return coefficients are negative. Brooks and Faff (1998) and Holmes and Faff (2004) invoke the literature from market timing ability of managed funds to provide an interpretation of the sign of the coefficients in the higher-order market model. Consider that the fund's (in this case size portfolios) time-varying beta is related to the market return and the squared market return. The coefficient γ measures the exposure of the portfolio or the managed fund to the market movements. The exposure is higher when the market returns are high and is lower when the market returns are low. The assets with such positive market timing ability are therefore attractive. Similarly the δ (coefficient of cubic term) measures the volatility timing ability of the asset. A negative delta implies that investors do not experience any return compensation during high-volatility periods and asset managers should seek to avoid market exposure during these times. In most cases in Table 25.4, the gamma coefficients are negative and delta coefficients are negative and significant too. Interestingly, a similar pattern was observed for a majority of funds in Holmes and Faff (2004) and for a majority of countries in the international asset pricing study in Brooks and Faff (1998).

25.5 COMPARISON WITH THREE FACTOR FAMA–FRENCH MODEL

In the literature, the following three-factor Fama and French (1992) model is advocated as an alternative to the CAPM where the size (SMB) and the book-to-market (HML) factors are stipulated as a set of APT factors:

$$r_{it} = \alpha_i + \beta_i r_{mt} + s_i SMB_t + h_i HML_t + \varepsilon_{it} \qquad (25.15)$$

The source of the underlying risk in the size and book-to-market factors is still unresolved. Some of the alternative explanations for the unknown risk factors are firm distress, momentum, leverage effect, and bias in earning growth. Chung et al. (2006) present another explanation based on higher-order co-moments. They demonstrate that the size and book-to-market factors are proxies for the missing higher-order systematic co-moments that are ignored in the return-generating process specified in Equation 25.15. It will be of interest to investigate this explanation in markets with characteristics that are different from the U.S. market. We investigate this issue in an emerging market, namely, Pakistan, which, according to Khawaja and Mian (2005), has typical characteristics of an emerging market.

TABLE 25.5   Goodness-of-Fit Measures of the Alternative System
of the Higher-Order Co-moment and Fama–French Models

	Co-moment Model	**Fama–French Model**
Panel A: Size portfolio		
Average $\bar{R}^2$	0.3486	0.4072
System R^2	0.3831	0.4498
Panel B: Industry portfolio		
Average $\bar{R}^2$	0.3362	0.3750
System R^2	0.3957	0.4321
Panel C: Beta portfolio		
Average $\bar{R}^2$	0.3155	0.3750
System R^2	0.3896	0.4407
Panel D: Co-skewness portfolios		
Average $\bar{R}^2$	0.3632	0.3878
System R^2	0.3639	0.3958
Panel E: Co-kurtosis portfolios		
Average $\bar{R}^2$	0.3541	0.3886
System R^2	0.3753	0.4135
Panel F: Book-to-market portfolios		
Average $\bar{R}^2$	0.2682	0.3363
System R^2	0.3864	0.4755

Table 25.5 presents goodness-of-fit measures of the two competing factor models. To allow maximum variation in portfolio returns with respect to their book-to-market ratio, we construct an additional set of portfolios by sorting stocks on the book-to-market ratio. The results reveal that the Fama–French alternative performs slightly better in explaining the variation in portfolio returns in all six types of portfolios. The average coefficient of determination and Glahn's composite correlation coefficient are generally higher by about 5% for the Fama–French model compared to the higher co-moment alternative. In the book-to-market portfolios, the Fama–French model records the largest increment in explanatory power relative to the systematic higher-order co-moment model.

25.5.1 Risk Exposure of Higher-Order Co-Moments and Fama–French Factors

In this section, we investigate the effect of including systematic co-moments on the size and book-to-market factors. This is done by comparing the results of the Fama–French model (Equation 25.15) with the augmented Fama–French model given as

$$r_{it} = \alpha_i + \beta_i r_{mt} + s_i SMB_t + h_i HML_t + \gamma_i q_{mt} + \delta_i c_{mt} + \varepsilon_{it} \qquad (25.16)$$

Table 25.6 presents the time series regressions for 16 portfolios sorted on size and book-to-market. The first half of each panel reports the results obtained in the three-factor Fama–French model (Equation 25.15) and the second half gives the results obtained in the augmented Fama–French model (Equation 25.16). It appears that the addition of higher-order systematic co-moments has no effect on the statistical significance of the size and book-to-market factors. The γ (exposure to co-skewness) is significant only in 3 out of the 16 portfolios whereas θ (exposure to co-kurtosis) is significant in 14 (87%) portfolios.

The absolute numerical value of book-to-market factor only marginally decreases as the systemic co-moments are introduced. In many cases, the coefficient on size factor increases in magnitude marginally. In two portfolios ("third size-second book-to-market" and "third size–third book-to-market") the size factor becomes even more statistically significant when the two co-moments are introduced. In only one case (the largest size-largest book-to-market portfolio) does the statistical significance of the book-to-market factor vanish. In most of the portfolios, the explanatory power of the Fama–French factors only marginally increases when the higher systematic co-moments are introduced. The results with portfolios formed by simultaneous sorting on co-skewness and co-kurtosis are largely similar to the size and book-to-market case. The same conclusion is reached when higher-order systematic co-moments up to order 10 are included in Equation 25.16. The details are not reported for the sake of brevity.

25.5.2 Risk Premia of Higher-Order Co-Moments and Fama–French Factors

The analysis so far has investigated only the risk exposure of the two types of factors to time series variation of portfolio returns. In this section, we investigate whether or not the factors considered in the models are priced. First we resort to visual inspection. Figure 25.1 displays the scatter plot of average excess portfolio return against the beta in 16 portfolios sorted according to co-skewness and co-kurtosis. In Figure 25.1, the thick lines join the co-skewness portfolios in the largest co-kurtosis category and the

TABLE 25.6 Coefficient of Factor Model with both Fama–French and Higher-Order Systematic Co-Moments for Two-Way Portfolios Formed from Size and Book-to-Market

Portfolio	Constant	β	s	h	γ(×100)	δ(×1000)	Adj R^2
Book-to-market 1							
Size 1	−0.536	0.640*	1.513*	−1.190*			0.628
Size 2	−0.253	0.598*	0.797*	−0.787*			0.241
Size 3	−0.062	0.559*	1.043	−0.464*			0.377
Size 4	−0.574	0.440*	0.677*	0.203*			0.324
Size 1	−0.369	0.556*	1.497*	−1.205*	−0.117	0.173	0.627
Size 2	−0.101	0.670*	0.807*	−0.776*	−0.202	−0.160	0.233
Size 3	−0.060	0.723*	1.070	−0.438*	−0.107	−0.350*	0.389
Size 4	−0.341	0.638*	0.706*	0.232*	−0.364	−0.420*	0.384
Book-to-market 2							
Size 1	0.221	0.448*	0.352*	−0.707*			0.486
Size 2	0.340	0.463*	0.347*	−0.297			0.449
Size 3	−0.046	0.380*	0.287	−0.080			0.119
Size 4	−0.041	0.472*	0.447*	0.246*			0.153
Size 1	0.201	0.665*	0.388*	−0.672*	−0.120	−0.460*	0.520
Size 2	0.425	0.685*	0.382*	−0.261	−0.230	−0.480*	0.507
Size 3	0.411	0.676*	0.330**	−0.035	−0.660	−0.650*	0.170
Size 4	−0.028	0.789*	0.498*	0.297*	−0.218	−0.680*	0.233

(continued)

TABLE 25.6 (continued)   Coefficient of Factor Model with both Fama–French and Higher-Order Systematic Co-Moments for Two-Way Portfolios Formed from Size and Book-to-Market

Portfolio	Constant	β	s	h	γ(×100)	δ(×1000)	Adj R^2
Book-to-market 3							
Size 1	-0.480	0.376*	-0.255	-0.913*			0.509
Size 2	-0.041	0.491*	-0.367*	-0.254*			0.477
Size 3	0.268	0.360*	0.111	-0.175**			0.274
Size 4	0.150	0.419*	-0.002	0.134**			0.366
Size 1	-0.148	0.774*	-0.194	-0.851*	-0.597**	-0.860*	0.598
Size 2	-0.199	0.717*	-0.328*	-0.216*	0.016	-0.480*	0.515
Size 3	0.801	0.479*	0.125*	-0.160**	-0.624*	-0.280*	0.289
Size 4	0.137	0.527*	0.016	0.152**	-0.056	-0.230*	0.378
Book-to-market 4							
Size 1	-0.104	0.444*	-0.555*	-0.782*			0.664
Size 2	-0.063	0.539*	-0.430*	-0.398*			0.695
Size 3	-0.450	0.731*	-0.230*	-0.141*			0.809
Size 4	0.106	0.878*	-0.541*	0.132*			0.666
Size 1	0.063	0.641*	-0.525*	-0.752*	-0.298	-0.430*	0.682
Size 2	-0.411	0.640*	-0.410*	-0.379*	0.292	-0.200**	0.709
Size 3	-0.231	0.891*	-0.206*	-0.117*	-0.327*	-0.350*	0.825
Size 4	0.389	0.701*	-0.572*	-0.101	-0.177	0.366*	0.681

*,** indicate significance at 5% and 10% levels respectively.

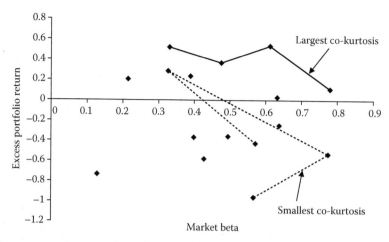

FIGURE 25.1 Scatter plot of average portfolio returns against beta. The connected lines indicate co-skewness portfolios under each co-kurtosis category.

dotted lines join the co-skewness portfolios in the smallest co-kurtosis category. Figure 25.1 also reveals that the range of the beta in the portfolios joined by the thick and dotted lines are almost the same. This is an indication that when controlled for the beta, co-kurtosis, and average excess returns may have a positive relationship.

Now to assess whether visually apparent return differences are associated with statistically significant risk premia, we estimate the following cross-sectional regression models:

$$\bar{r}_i = \gamma_0 + \gamma_1 \beta_i + \gamma_2 s_i + \gamma_3 h_i + v_i \tag{25.17}$$

$$\bar{r}_i = \gamma_0 + \gamma_1 \beta_i + \gamma_2 s_i + \gamma_3 h_i + \gamma_4 \gamma_i + \gamma_5 \delta_i + v_i \tag{25.18}$$

where

β_i, s_i, h_i, γ_i, and δ_i are the factor loadings estimated in the time series regression model given in Equations 25.15 and 25.16

$\bar{r}_i$ is the average excess return in portfolio i

Table 25.7 presents the coefficients estimated in the cross-sectional regressions of average excess portfolios reestimated on the factor loadings for six different types of portfolios: size, book-to-market, co-skewness, co-kurtosis, and two sets of portfolios obtained by simultaneous sorting

TABLE 25.7   Estimates of Risk Premia from Cross-Section Regression
of Average Portfolio Returns on the Factor Loadings of Higher Co-Moments
and Fama–French Factors

γ_0	γ_1	γ_2	γ_3	γ_4	γ_5	Adj R^2
Size portfolios						
0.106	−0.099	−0.386*	0.123			0.153
0.140	−0.717	−0.262	0.383	1.505	0.112	0.373
Book-to-market portfolios						
0.892*	0.264	−0.688*	0.792*			0.460
1.068**	0.034	−0.707	0.795*	0.657	−0.070	0.383
Two-way size and book-to-market portfolios						
0.193	0.217	−0.468*	0.749*			0.715
0.811	−0.546	−0.523*	0.750*	0.758	−0.106	0.718
Co-skewness portfolios						
−0.908*	2.309*	−0.591*	0.527**			0.510
−0.844	2.149**	−0.696*	0.473	0.780	−0.215**	0.421
Co-kurtosis portfolios						
0.248	0.152	−0.440**	0.920*			0.374
0.549	−0.206	−0.453**	0.983*	1.488	−0.122	0.371
Two way co-skewness and co-kurtosis portfolios						
−0.220	1.236	−0.091	1.459*			0.345
−0.142	0.392	0.168	1.487*	0.377	0.055	0.432

**,* indicate significance at 5% and 10% levels respectively.

by co-skewness and co-kurtosis and size and book-to-market. Only the size and book-to-market factors offer significant risk premia in most of the cases. In only one case namely, co-skewness portfolios, does the co-kurtosis appear to be a priced factor.

The market beta is priced in only co-skewness sorted portfolios. In general, none of the three co-moments appear to have significant risk premia. Even though the risk premium for co-skewness is not significant under all types of portfolios the premium is positive. This is an expected result given that the distribution of the market portfolio returns is negative. On the other hand, the sign of the premium for co-kurtosis in four of the six cases is contrary to the expectation. Hence it appears that co-kurtosis is not associated with any significant risk premia in the cross-section. In contrast, there is evidence that risk premia of both Fama–French factors are statistically significant and have the expected sign. For the Brazilian stock market (an emerging market), Da Silva (2006), using time series regressions, reveals that the co-kurtosis is a more important co-moment and that the factors associated with co-moments are ineffective relative to the fundamental factors. Overall, it appears that although co-kurtosis as a measure of systematic risk may not be ignored in emerging markets, higher-order co-moments do not have sufficient explanatory power to render Fama–French factors redundant.

25.6 CONCLUSION

This chapter extends the multivariate test of Barone-Adesi et al. (2004) for arbitrage pricing with co-skewness to incorporate co-kurtosis in the asset pricing model and provides empirical evidence from an emerging market. The empirical results support the arbitrage pricing argument where market beta, co-skewness, and the co-kurtosis are considered as factors. A comparison of the risk exposure of co-skewness and co-kurtosis through a system of the cubic market model under the arbitrage pricing restrictions reveals that co-kurtosis may be an important common factor while co-skewness is not.

In the literature, the Fama–French factors are strongly supported as an empirically useful set of common factors of asset returns. This chapter compares the relative merit of the two types of factor models in the Pakistan's stock market. The empirical analysis favors the model with Fama–French factors to the systematic higher-order co-moment model. However, the explanatory power of the latter model is only slightly less than the former model. This conclusion differs from a recent study of Chung et al. (2006) on the U.S. market where the fundamental factors are no longer significant once the first 10 co-moments are employed in the cross-sectional analysis. Our results are consistent with a recent study in the Brazilian markets by Da Silva (2006). Co-kurtosis risk is not associated with any significant risk premia in the cross-section regression. Thus, it appears that the appropriateness of higher-order systematic co-moments as factors of pricing may be different in emerging and developed markets. There may be many reasons for this finding. A feature that differs notably in the two types of markets is volatility in the returns.

ACKNOWLEDGMENT

We thank participants of the 2007 Australasian Meeting of the Econometric Society, Brisbane Australia, and the 12th Doctoral Research Conference Faculty of Business and Economics, Monash University for their helpful comments on earlier versions of this chapter.

REFERENCES

Aggarwal, R., Inclean, C., and Leal, R. (1999) Volatility in emerging stock markets. *The Journal of Financial and Quantitative Analysis*, 34(1): 33–55.

Barone Adesi, G. (1985) Arbitrage equilibrium with skewed asset returns. *Journal of Financial and Quantitative Analysis*, 20(3): 299–313.

Barone Adesi, G., Gagliardini, P., and Urga, G. (2004) Testing asset pricing models with coskewness. *Journal of Business and Economic Statistics*, 22(4): 474–485.

Brooks, R. D. and Faff, R. (1998) A test of two-factor APT based on the quadratic market model: International evidence. *Journal of Studies in Economics and Econometrics*, 22(2): 65–76.

Chen, N. F., Roll, R., and Ross, S. (1986) Economic forces and the stock market. *Journal of Business*, 59(3): 383–403.

Chung, Y. P., Johnson, H., and Schill, M. (2006) Asset pricing when returns are non-normal: Fama–French factors versus higher-order systematic comoments. *The Journal of Business*, 79(2): 923–940.

Da Silva, A. C. (2006) Modelling and estimating a higher systematic co-moment asset pricing model in the Brazilian stock market. *Latin American Business Review*, 6(1): 85–101.

Dittmar, R. (2002) Nonlinear pricing kernels, kurtosis preference, and evidence from cross section of equity returns. *Journal of Finance*, 57(1): 369–343.

Fama, E. and French, K. R. (1992) The cross-section of expected stock returns. *Journal of Finance*, 48(1): 26–32.

Fama, E. and French, K. R. (1993) Common risk factors in the returns on stocks and bonds. *Journal of Financial Economics*, 33(1): 3–56.

Gibbons, M. R. (1982) Multivariate tests of financial models: A new approach. *Journal of Financial Economics*, 10(1): 3–56.

Glahn, H. (1969) Some relationships derived from canonical correlation theory. *Econometrica*, 37(2): 252–256.

Harvey, C. R. and Siddique, A. (2000) Conditional skewness in asset pricing tests. *Journal of Finance*, 55(2): 1263–1295.

Holmes, K. and Faff, R. (2004) Stability, asymmetry and seasonality of fund performance: An analysis of Australian multi-sector managed funds. *Journal of Business Finance and Accounting*, 31(3–4): 539–578.

Hung, D. C., Shackleton, M., and Xu, X. (2004) CAPM, higher co-moment and factor models of UK stock returns. *Journal of Business Finance and Accounting*, 31(1–2): 87–112.

Hwang, S. and Satchell, S. E. (1999) Modeling emerging market risk premia using higher moments. *International Journal of Finance and Economics*, 4(4): 271–296.

Iqbal, J. and Brooks, R. D. (2007a) Alternative beta risk estimators and asset pricing tests in emerging markets: The case of Pakistan. *Journal of Multinational Financial Management*, 17(1): 75–93.

Iqbal, J. and Brooks, R. D. (2007b) A test of CAPM on Karachi stock exchange. *International Journal of Business*, 12(4): 429–444.

Khawaja, A. I. and Mian, A. (2005) Unchecked intermediaries: Price manipulation in an emerging stock market. *Journal of Financial Economics*, 78(1): 203–241.

Kraus, A. and Litzenberger, R. (1976) Skewness preference and the valuation of risk assets. *Journal of Finance*, 31(3): 1085–1100.

Ross, S. A. (1976) Arbitrage theory of capital asset pricing. *Journal of Economic Theory*, 13: 341–360.

CHAPTER 26

Market Risk Management for Emerging Markets: Evidence from the Russian Stock Market

Dean Fantazzini

CONTENTS

26.1 INTRODUCTION

Over the past years, Russian financial markets have attracted a large amount of domestic and international investors because of its rapid growth and economic incentives offered by the Russian policy makers. Besides, Russian market is important for international investors who are interested in diversifying their portfolios geographically. Therefore, Russian equity market is worth investigating. In this chapter, we examine and compare different multivariate parametric models with the purpose of estimating the value at risk (VaR) for a high-dimensional portfolio composed of Russian financial assets. First, we propose a unified framework for multivariate GARCH models by means of Copula functions, with constant and dynamic conditional correlation models (CCC and DCC models, respectively). We consider DCC-GARCH models, as they allow us to account for a dynamic structure in the assets correlations, reducing the computational effort in case of very large number of variables. Second, we compare different multivariate distributions, which allow to account for the excess kurtosis and dynamic dependence.

The competing models are then evaluated by comparing their VaR out-of-sample forecasts with different tests and statistical techniques. We implement the superior predictive ability (SPA) test by Hansen (2005) and, for completeness, Kupiec's unconditional coverage test and also Christoffersen's conditional coverage test (see Kupiec, 1995; Christoffersen, 1998). The SPA test allows to compare $m > 2$ competing forecasting models. In Hansen's framework, m alternative forecasts are compared to a benchmark forecast, where the predictive abilities are defined by the expected loss. The choice of the loss function to evaluate the predictive ability of the various volatility models is a fundamental step in our analysis. As our objective of this chapter is the conditional-quantile of the asset portfolio's distribution, we use the asymmetric linear loss function discussed in Gonzalez-Rivera et al. (2006) and in Giacomini and Komunjer (2005).

Another point that plays a crucial role is the forecasting scheme. Following Gonzalez-Rivera et al. (2006), Hansen (2005), Giacomini and White (2006), and Amisano and Giacomini (2008), we use a rolling-window estimation scheme since it may be more robust to a possible parameter variation and therefore ensures us to avoid a violation of Assumption 1 in Hansen (2005).

Our extensive out-of-sample analysis with Russian stocks suggests that, if one is interested in forecasting the extreme quantiles, particularly at the 1% and 99% levels (which is the usual case for regulatory purposes), then a Student's t GARCH model with *any* copula delivers quite precise VaR estimates. The evidence that the type of copula plays a minor role should not be considered a surprise, given previous empirical evidence with American and European stocks (see Ané and Kharoubi, 2003; Junker and May, 2005; Fantazzini, 2008a) and particularly given the simulation evidence in Fantazzini (2009b). Besides, our empirical evidence finds out that if normal marginals are used, then models with dynamic dependence deliver statistically significant (and more precise) VaR estimates than models with constant dependence. However, if Student's t marginals are used, the differences are much smaller and no more significant. This confirms again that marginal misspecification may result in significant misspecified dependence structures.

The rest of this chapter is organized as follows. In Section 26.2, we provide an outline of multivariate modeling and propose a unified approach by means of copula theory. In Section 26.3, we describe the models used for the analysis and present the main empirical findings. We conclude with a brief discussion in Section 26.4.

26.2 MULTIVARIATE MODELING

While univariate VaR estimation has been widely investigated, the multivariate case has been dealt with only in a limited and recent literature about the forecasting of correlations between assets. Empirical works that deal with this issue are those by Engle and Sheppard (2001), Giot and Laurent (2003), Bauwens and Laurent (2005), Rosenberg and Schuermann (2006), and Fantazzini (2009). When we use parametric methods, VaR estimation for a portfolio of assets can become very difficult due to the complexity of joint multivariate modeling. Moreover, computational problems arise when the number of assets increases.* As a consequence of this complexity, two models seem to have gained the greatest attention by practitioners and researchers so far:

---

* See the review of multivariate GARCH models by Bauwens et al. (2006) for a discussion of these issues.

- The constant conditional correlation (CCC) model by Bollerslev (1990)

- The dynamic conditional correlation (DCC) model by Engle (2002)

In the next two subsections, we will show that these models can be presented as special cases within a more general copula approach.

26.2.1 Copula Modeling

The study of copulas originated with the seminal papers by Höffding (1940) and Sklar (1959) and has seen various applications in the statistics literature. For more details, we refer the interested reader to the recent methodological overviews by Joe (1997) and Nelsen (1999), while Cherubini et al. (2004) provide a detailed discussion of copula techniques for financial applications.

Copula theory provides an easy way to deal with the (otherwise) complex multivariate modeling. The "…essential idea of the copula approach is that a joint distribution can be factored into the marginals and a dependence function called a copula. The term 'copula' means 'link': the copula couples the marginal distributions together in order to form a joint distribution. The dependence relationship is entirely determined by the copula, while scaling and shape (mean, standard deviation, skewness, and kurtosis) are entirely determined by the marginals…" (Rosenberg and Schuermann, 2004, p. 12). Copulas can therefore be used to obtain more realistic multivariate densities than the traditional joint Normal one, which is simply the product of a Normal copula and Normal marginals: for example, the Normal dependence relation can be preserved using a Normal copula, but marginals can be entirely general, e.g., Student's t marginals.

The "… class of elliptical distributions provides useful examples of multivariate distributions because they share many of the tractable properties of the multivariate Normal distribution. Furthermore, they allow to model multivariate extreme events and forms of non-normal dependencies. Elliptical copulas are simply the copulas of elliptical distributions (see Fang et al. (1990) for a detailed treatment of elliptical distributions)…" (Romano, 2001, p. 5).

We present two copulas belonging to the elliptical family that will be later used in empirical applications: the Gaussian and Student's t-copula. By applying Sklar's theorem and using the relationship between the distribution and the density function, we can derive their density functions.

1. The copula of the multivariate Normal distribution is the *Normal-copula* and its probability density function is

$$
c^{\text{Normal}}(\Phi_1(x_1), \ldots, \Phi_n(x_n); \theta_0) = \frac{f^{\text{Normal}}(x_1, \ldots, x_n)}{\prod\limits_{i=1}^{n} f_i^{\text{Normal}}(x_i)}
$$

$$
= \frac{\dfrac{1}{(2\pi)^{n/2} |\mathbf{R}|^{1/2}} \exp\left(-\dfrac{1}{2} x' \mathbf{R}^{-1} x\right)}{\prod\limits_{i=1}^{n} \dfrac{1}{\sqrt{2\pi}} \exp\left(-\dfrac{1}{2} x_i^2\right)}
$$

$$
= \frac{1}{|\Sigma|^{1/2}} \exp\left(-\frac{1}{2} \zeta'(\mathbf{R}^{-1} - I)\zeta\right) \qquad (26.1)
$$

where $\zeta = (\Phi^{-1}(u_1), \ldots, \Phi^{-1}(u_n))'$ is the vector of univariate Gaussian inverse distribution functions, $u_i = \Phi(x_i)$, while $\mathbf{R}$ is the correlation matrix.

2. On the other hand, the copula of the multivariate Student's t-distribution is the *Student's t-copula*, and its density function is

$$
c^{\text{Normal}}(t_{v_1}(x_1), \ldots, t_{v_n}(x_n)) = \frac{f^{\text{Student}}(x_1, \ldots, x_n)}{\prod\limits_{i=1}^{n} f_i^{\text{Student}}(x_i)}
$$

$$
= \frac{\Gamma\big((\upsilon+n)/2\big)\big[\Gamma(\upsilon/2)\big]^n \big(1 + \zeta^T \mathbf{R}^{-1}\zeta\big)^{-(\upsilon+n)/2}}{|\mathbf{R}|^{1/2}\, \Gamma(\upsilon/2)\big[\Gamma\big((\upsilon+1)/2\big)\big]^n \prod\limits_{i=1}^{n}\big(1 + \zeta_i^2/\upsilon\big)^{-(\upsilon+1)/2}}
$$

$$
(26.2)
$$

where $\zeta = (t_{\upsilon 1}^{-1}(u_1), \ldots, t_{\upsilon n}^{-1}(u_n))'$ is the vector of univariate Student's t inverse distribution functions, υ_i are the degrees of freedom for each marginal i, $u_i = t_\upsilon(x_i)$, while $\mathbf{R}$ is the correlation matrix.

Both these copulas belong to the class of elliptical copulas. An interesting extension that we will consider in our empirical analysis is the grouped-t copula introduced by Daul et al. (2003). The grouped-t copula can be considered as a copula imposed by a kind of multivariate-t distribution where m distinct groups of assets have m different degrees of freedom. Like the previous two copulas, it can be easily applied to high-dimensional portfolios and it is a model, which is no more difficult to calibrate than the

t-copula but allows for subgroups with different dependence structures. Let $\mathbf{Z}|F_{t-1} \sim \mathbf{N_n}(0, \mathbf{R}_t)$, $t = 1, \ldots, T$, given the conditioning set F_{t-1}, where $\mathbf{R}_t$ is the $n \times n$ conditional linear correlation matrix with a dynamic structure and $\bar{\mathbf{R}}$ is the unconditional correlation matrix. Furthermore let $U \sim$ Uniform $(0, 1)$ be independent of $\mathbf{Z}$. Let G_υ denote the distribution function of $\sqrt{\upsilon/\chi_\upsilon}$, where χ_υ is a chi-square distribution with υ degrees of freedom, and partition $1, \ldots, n$ into m subsets of sizes $s_1, \ldots, s_m$. Set $W_k = G_{\upsilon_k}^{-1}(U)$ for $k = 1, \ldots, m$ and then $\mathbf{Y}|F_{t-1} = (W_1 Z_1, \ldots, W_1 Z_{s1}, W_2 Z_{s1+1}, \ldots, W_2 Z_{s1+s2}, \ldots, W_m Z_n)$, so that $\mathbf{Y}$ has a so-called grouped t distribution. Finally, define

$$\mathbf{U}|F_{t-1} = (t_{\upsilon 1}(Y_1), \ldots, t_{\upsilon 1}(Y_{s1}), t_{\upsilon 2}(Y_{s1+1}), \ldots, t_{\upsilon 2}(Y_{s1+s2}), \ldots, t_{\upsilon m}(Y_n)) \quad (26.3)$$

$\mathbf{U}$ has a distribution on $[0, 1]^n$ with components uniformly distributed on $[0, 1]$. We call its distribution function the dynamic grouped t-copula. Note that $(Y_1, \ldots, Y_s)$ has a t distribution with υ_1 degrees of freedom and in general for $k = 1, \ldots, m - 1$, $(Y_{s1+\ldots+sk+1}, \ldots, Y_{s1+\ldots+sk+1})$ has a t distribution with υ_{k+1} degrees of freedom. Similarly, subvectors of $\mathbf{U}$ have a t-copula with υ_{k+1} degrees of freedom, for $k = 0, \ldots, m - 1$. In this case, no elementary density has been given. See Daul et al. (2003) and Demarta and McNeil (2005) for the case where the correlation matrix $\mathbf{R}_t = \mathbf{R}$ is constant, whereas Fantazzini (2009) for the case of dynamic dependence.

26.2.2 Unified Approach

Given the previous background, it is possible to show that the CCC and DCC models can be easily represented as special cases within a more general copula framework function. If we consider a general model for the conditional means and variances, the two models can be restated as follows:

$$Y_t = E[Y_t \,|\, \mathfrak{I}_{t-1}] + D_t \eta_t$$
$$\eta_t \sim H(\eta_1, \ldots, \eta_n) \equiv C^{\text{Normal}}(F_1^{\text{Normal}}(\eta_1), \ldots, F_n^{\text{Normal}}(\eta_n); \mathbf{R}_t)$$

where

Y_t is a vector stochastic process of dimension $N \times 1$, $D_t = \text{diag}(\sigma_{11,t}, \ldots, \sigma_{nn,t})$
$\sigma_{ii,t}$ is defined as a univariate GARCH model and the Sklar's Theorem was used

Furthermore, the two-step DCC estimation procedure described in Engle and Sheppard (2001) corresponds to the inference for margins (IFM) method first proposed by Joe and Xu (1996) for copula estimation. According to the IFM method, the parameters of the marginal distributions are estimated in a first stage, while the parameters of the copula are estimated separately in a second stage. Like the one-step ML estimator, it verifies the properties of asymptotic normality, but the covariance matrix must be modified (see Joe and Xu, 1996; Joe, 1997):

$$\sqrt{T}(\hat{\theta}_{IFM} - \theta_0) \rightarrow N(0, V(\theta_0))$$

where

θ_0 is a vector of marginals and copula parameters
$V(\theta_0) = D^{-1}M(D^{-1})'$ is the so-called "Godambe" information matrix, where
$D = E[\partial g(\theta)'/\partial\theta]$, $M = E[g(\theta)'g(\theta)]$, and $g(\theta)$ is the score function

This asymptotic result corresponds to the one reported in Engle and Sheppard (2001) for the two-step DCC estimation.

Therefore, if we consider the CCC model, this implies estimating n univariate GARCH models of any type with a Normal distribution at a first stage. The Normal cumulative distribution functions of the standardized residuals $u_{i,t} = \Phi(\eta_{i,t})$ are then used as arguments within the Normal copula density (Equation 26.1) with constant correlation matrix $R_t = R$. However, since $\zeta_t = (\Phi^{-1}(u_{1,t}), \ldots, \Phi^{-1}(u_{n,t}))'$ in Equation 26.1 is a vector of univariate Normal inverse distribution functions, the estimated constant correlation matrix corresponds to the estimated correlation matrix of the standardized residuals in the CCC model. In a similar way, if we consider a DCC model instead, the Normal cumulative degrees of freedom and inverse functions cancel out each other and the log-likelihood of the copula density is maximized assuming the following dynamic structure for the correlation matrix R_t:

$$R_t = (\text{diag}\, Q_t)^{-1/2} Q_t (\text{diag}\, Q_t)^{-1/2} \tag{26.4}$$

$$Q_t = \left(1 - \sum_{l=1}^{L}\alpha_l - \sum_{l=1}^{S}\beta_s\right)\bar{Q} + \sum_{l=1}^{L}\alpha_l \eta_{t-l}\eta_{t-l}' + \sum_{l=1}^{S}\beta_s Q_{t-s} \tag{26.5}$$

where

$\bar{Q}$ is the $n \times n$ unconditional correlation matrix of η_t $\alpha_l(\geq 0)$
$\beta_s(\geq 0)$ are the scalar parameters satisfying $\sum_{l=1}^{L}\alpha_l + \sum_{s=1}^{S}\beta_s < 1$

These conditions are needed to have $\mathbf{Q}_t > 0$ and $\mathbf{R}_t > 0$. $\mathbf{Q}_t$ is the covariance matrix of η_t, since $q_{ii,t}$ is not equal to 1 by construction. Then, $\mathbf{Q}_t$ is transformed into a correlation matrix by Equation 26.5. If $\theta_1 = \theta_2 = 0$ and $\bar{q}_{ii} = 1$, the CCC model is obtained. See Engle (2002) for more details about DCC modeling.

26.2.3 Some Extensions: Student's t Marginals and Dynamic Copulas

A well-known deviation from normality is leptokurtosis and a classical marginal distribution, which is used to allow for excess kurtosis is the Student's t. Therefore, a multivariate model that allows for marginal kurtosis and Normal dependence can be expressed as follows:

$$Y_t = E[Y_t \mid \mathfrak{I}_{t-1}] + D_t \eta_t$$
$$\eta_t \sim H(\eta_1, \ldots, \eta_n) \equiv C^{\text{Normal}}(F_1^{\text{Student's } t}(\eta_1), \ldots, F_n^{\text{Student's } t}(\eta_n); \mathbf{R}_t)$$

where
$F^{\text{Student's } t}$ is the cumulative distribution function of the marginal Student's t
$\mathbf{R}_t$ can be made constant or time-varying, as in the standard CCC and DCC models, respectively

If the financial assets present symmetric tail dependence, we can use a Student's t copula, instead:

$$Y_t = E[Y_t \mid \mathfrak{I}_{t-1}] + D_t \eta_t$$
$$\eta_t \sim H(\eta_1, \ldots, \eta_n) \equiv C^{\text{Student's } t}(F_1^{\text{Student's } t}(\eta_1), \ldots, F_n^{\text{Student's } t}(\eta_n); \mathbf{R}_t; \upsilon)$$

where υ are the Student's t copula degrees of freedom, while we can use a grouped t copula if the financial assets may be separated in m distinct groups:

$$Y_t = E[Y_t \mid \mathfrak{I}_{t-1}] + D_t \eta_t$$
$$\eta_t \sim H(\eta_1, \ldots, \eta_n) \equiv C^{\text{Grouped } t}(F_1^{\text{Student's } t}(\eta_1), \ldots, F_n^{\text{Student's } t}(\eta_n); \mathbf{R}_t; \upsilon_1, \ldots, \upsilon_m)$$

Similar approaches are proposed by Patton (2004), Jondeau and Rockinger (2006), and Granger et al. (2006). However, they focus on bivariate applications only, and no VaR measurement is made.

26.3 EMPIRICAL ANALYSIS

26.3.1 Model Specifications and Case Studies

The aim of this Chapter is to examine and compare different multivariate parametric models for estimating the VaR for a high-dimensional portfolio composed of 30 Russian stocks. We chose the 30 most liquid Russian assets with at least 2000 historical daily data ranging between 5/01/2000 and 23/05/2008, quoted at the RTS and MICEX Russian markets.

For sake of simplicity, we suppose to invest an amount $M_i = 1\$, i = 1, ..., n$ in every asset. We do this choice for sake of interest, since it represents the most common case in the financial literature (see, e.g., Junker and May, 2005) and because DeMiguel et al. (2009) have recently shown that a wide range of models is no consistently better than the simple equally weighted portfolio (the $1/n$ rule) in terms of Sharpe ratio, certainty-equivalent return or turnover, indicating that, out of sample, the gain from optimal diversification is more than offset by estimation error.

Based on our previous analysis, four elements are considered:

1. The *marginal distribution*: We compare the standard Normal and the standardized Student's t.

2. The specification of the *conditional moments of the marginals*. We chose to use a AR(1)–GARCH(1,1) model for the continuously compounded returns $y_t = 100[\log(P_t) - \log(P_{t-1})]$:

$$\begin{cases} y_t = \mu + \phi_1 y_{t-1} + \varepsilon_t \\ \varepsilon_t = \sigma_t \eta_t, \quad \eta_t \sim i.i.d. (0,1) \\ \sigma_t^2 = \omega + \alpha \varepsilon_{t-1}^2 + \beta \sigma_{t-1}^2 \end{cases}$$

Other GARCH models (like FIGARCH, FIEGARCH, APARCH, etc.) as well as other marginal distributions (Skewed t, Laplace, etc.) were not considered due to poor numerical convergence properties. Russian stocks are noisier and less liquid than European or American stocks and the range of models which can be used is limited.

3. The *type of copula*: We compare the Normal copula, the t-copula, and the Grouped-t copula. In order to compute the latter copula, different criteria can be used to classify the variable at hand in m different groups:

a. Geographical location as in Daul et al. (2003)

b. Credit rating as in Fantazzini (2009)

c. If none of the previous criteria is available or if there is only partial information, like for our data set, one may resort to hierarchical cluster analysis based on L^2 dissimilarity measure and "dendrograms": "...dendrograms graphically present the information concerning which observations are grouped together at various levels of (dis)similarity. The height of the vertical lines and the range of the (dis)similarity axis give visual clues about the strength of the clustering..." (http://www.stata.com). See, for example, Kaufman and Rousseeuw (1990) for more details about cluster analysis.

Given the evidence in Figure 26.1, we decided to consider $m = 3$ groups.

4. The specification of the *conditional copula parameters*: We consider a Normal copula both with a constant correlation matrix **R** and a dynamic $\mathbf{R}_t$, where in the latter case we use a DCC(1,1) model. We consider also a t-copula with constant correlation matrix **R** and degrees of freedom υ, as well as with a dynamic $\mathbf{R}_t$ and constant υ.*

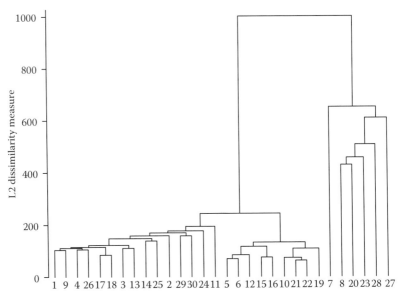

FIGURE 26.1   Dendrograms for the 30-asset portfolio composed of Russian stocks.

---

* We discarded a dynamic specification for υ since the numerical maximization of the log-likelihood failed to converge most of the time or the dynamic coefficients were not significant.

As for the grouped t copula, we consider both a specification with constant correlation matrix and with dynamic dependence as in Fantazzini (2009).

As a result of these four elements, we consider 12 different models as reported in Table 26.1.

26.3.2 VaR Estimation

Banks and financial institutions face the everyday problem of measuring the market risk exposure of their assets. If we use a probabilistic framework and assume to be at time t we want to assess the risk of a financial position for the next l periods. The VaR is the most widely used measure of risk and it has become the benchmark risk measure thanks to the Basle II agreements (see Basel Committee on Banking Supervision, 2005).

TABLE 26.1   Multivariate Distribution Specifications

	Marginal Distribution	Moment Specification	Copula	Copula Parameters Specification
Model 1	Normal	AR(1) T-GARCH(1,1)	Normal	Constant correlation
Model 2	Normal	AR(1) T-GARCH(1,1)	Normal	DCC(1,1)
Model 3	Normal	AR(1) T-GARCH(1,1)	t-Copula	Constant correlation Constant D.o.F.
Model 4	Normal	AR(1) T-GARCH(1,1)	t-Copula	DCC(1,1) Constant D.o.F.
Model 5	Normal	AR(1) T-GARCH(1,1)	Grouped t	Constant correlation Const. D.o.F.s
Model 6	Normal	AR(1) T-GARCH(1,1)	Grouped t	DCC(1,1) Constant D.o.F.s
Model 7	Student's t	AR(1) T-GARCH(1,1) Constant D.o.F.	Normal	Constant correlation
Model 8	Student's t	AR(1) T-GARCH(1,1) Constant D.o.F.	Normal	DCC(1,1)
Model 9	Student's t	AR(1) T-GARCH(1,1) Constant D.o.F.	t-COPULA	Constant correlation Constant D.o.F.
Model 10	Student's t	AR(1) T-GARCH(1,1) Constant D.o.F.	t-COPULA	DCC(1,1) Constant D.o.F.
Model 11	Student's t	AR(1) T-GARCH(1,1) Constant D.o.F.	Grouped t	Constant correlation Constant D.o.F.s
Model 12	Student's t	AR(1) T-GARCH(1,1) Constant D.o.F.	Grouped t	DCC(1,1) Constant D.o.F.s

Note: D.o.F., degrees of freedom.

VaR is simply defined as the worst expected loss of a financial position over a target horizon with a given confidence level.*

As indicated in Christoffersen and Diebold (2000) and Giot and Laurent (2003), volatility forecastability decays quickly with the time horizon of the forecasts. An immediate consequence is that volatility forecastability is relevant for short-time horizons (such as daily trading) but not for long-time horizons. Therefore, we focused on daily returns and VaR performances for daily trading portfolios, only. As our goal is to compare the forecast ability of the models, we generated portfolio VaR forecasts at the 0.25%, 0.5%, 1%, and 5% probability levels, that is VaR levels for long positions, and at the 95%, 99%, 99.5%, and 99.75% probability levels, that is for short positions too. The predicted one-step-ahead VaR forecasts were then compared with the observed portfolio losses and both results were recorded for later assessment.

A general algorithm for estimating the 0.25%, 0.5%, 1%, 5%, 95%, 99%, 99.5%, and 99.75% VaR over a 1 day holding period for a portfolio P of n assets with invested positions equal to M_i, $i = 1, \ldots, n$ is the following:

1. Simulate $j = 100,000$ scenarios for each asset log-returns, $\{y_{1,t}, \ldots, y_{n,t}\}$, over the time horizon $[t - 1, t]$, using a general multivariate distribution as in Table 26.1, by using the following procedure:

    a. First, generate an n random variate $(u_{1,t}, \ldots, u_{n,t})$ from the copula $\hat{C}_t$ forecast at time t, which can be Normal, Student's t, or Grouped-t.

    b. Second, get a vector $n \times 1$ $\boldsymbol{T}_t$ of standardized asset log-returns $\eta_{i,t}$ by using the inverse functions of the forecast marginals at time t, which can be normal, or Student's t:

    $$\boldsymbol{T}_t = (\eta_{1,t}, \ldots, \eta_{n,t}) = \left(F_1^{-1}(u_{1,t}; \hat{\alpha}_1), \ldots, F_n^{-1}(u_{n,t}; \hat{\alpha}_n) \right)$$

    c. Third, rescale the standardized assets log-returns by using the forecast means and variances, estimated with AR-GARCH models:

    $$\{y_{1,t}, \ldots, y_{n,t}\} = \left(\hat{\mu}_{1,t} + \eta_{1,t} \cdot \sqrt{\hat{h}_{1,t}}, \ldots, \hat{\mu}_{n,t} + \eta_{n,t} \cdot \sqrt{\hat{h}_{n,t}} \right)$$

    d. Finally, repeat this procedure for $j = 100,000$ times.

---

* See Jorion (2000) for an introduction to Value at Risk as well as a detailed discussion about its estimation.

2. By using these 100,000 scenarios, the portfolio P is reevaluated at time t, that is:

$$P_t^j = M_{1,t-1} \cdot \exp(y_{1,t}) + \cdots + M_{n,t-1} \cdot \exp(y_{n,t}), \quad j = 1 \ldots 100,000$$

3. Portfolio losses in each scenario j are then computed*:

$$\text{Loss}_j = P_{t-1} - P_t^j, \quad j = 1 \ldots 100,000$$

4. The calculus of the 0.25%, 0.5%, 1%, 5%, 95%, 99%, 99.5%, and 99.75% VaR is now straightforward:

   a. Order the 100,000 Loss_j in increasing order;

   b. The pth VaR is the $(1-p)$ 100,000th ordered scenario, where $p = \{0.25\%, 0.5\%, 1\%, 5\%, 95\%, 99\%, 99.5\%, 99.75\%\}$. For example, the 0.25% VaR is the 99,750th ordered scenario.

26.3.3 VaR Evaluation

Given the widespread use of VaR by banks and regulators, it is of interest to evaluate the accuracy of the different models used to estimate VaR. We perform an out-of-sample backtesting analysis by dividing the sample size T into a training part of size m and an out-of-sample part of size n, so that $T = m + n$. The in-sample part is used to estimate the models while the evaluation is performed forecasting the VaR over the remaining out-of sample part. There are two different methods by which forecasts can be produced: a fixed or a rolling window forecasting scheme. The first scheme of forecasting involves estimating the parameters only once on the first m observations using these estimates to produce all of the forecasts for the out of sample period $t = m + 1, \ldots, T$. The second scheme (rolling window), on the contrary, implies reestimating the parameters at each out-of-sample point t using an estimation sample containing the m most recent observations, that is, the observations from $t - m + 1$ to t. Following Giacomini and Komunjer (2005) and Gonzalez-Rivera et al. (2006), we use a rolling forecasting scheme because it may be more robust to a possible parameter variation; in our case, we have 2000 observations so we split the sample in this way: 1000 observations for the estimation window and 1000 for the out-of-sample evaluation.

---

* Possible profits are considered as negative losses.

In order to evaluate the forecasting performance of the models we implement the SPA test by Hansen (2005). This test compares the performances of two or more forecasting models by evaluating the forecasts with a prespecified loss function, e.g., the function described in Equation 26.6. The best forecast model is the model that produces the smallest expected loss. The evaluation of the predictive ability of the models under study is linked to a suitable loss function. Different types of loss functions may be found in the literature, each of these corresponding to different optimal forecast. For example, if our aim is the optimal forecast of the conditional mean a quadratic loss function is used. If, on the other hand, the aim is the forecast of the conditional median of the distribution, the appropriate loss function is the absolute value. Since our objective is to analyze the VaR of different portfolios (the conditional quantile of the distribution), the optimal corresponding loss function is the asymmetric linear loss function as discussed in Gonzalez-Rivera et al. (2006) and Giacomini and Komunjer (2005). This is defined as follows:

$$T_\alpha(e_{t+1}) \equiv (\alpha - \mathbf{1}(e_{t+1} < 0)e_{t+1}) \qquad (26.6)$$

where $e_{t+1} = L_{t+1} - \widehat{VaR}_{t+1|t}$, L_{t+1} is the realized loss, while $\widehat{VaR}_{t+1|t}$ is the VaR forecast at time $t + 1$ on information available at time t. This type of loss function penalizes more heavily, with weight $(1 - \alpha)$, the observations for which $y_{t+1} < \widehat{VaR}^\alpha_{t+1|t}$. For completeness, we also employ the Kupiec's unconditional coverage test (1995) and the Christoffersen's conditional coverage test (1998), given their importance in the empirical literature. However, we remark that their power can be very low. As for the Kupiec's test, following binomial theory, the probability of observing N failures out of T observations is $(1 - p)^{T-N} p^N$, so that the test of the null hypothesis $H_0: p = p^*$ is given by a LR test statistic:

$$\text{LR}_{UC} = 2 \cdot \ln[(1 - p^*)^{T-N} p^{*N}] + 2 \cdot \ln[(1 - N/T)^{T-N} (N/T)^N]$$

As for the Christoffersen's test, its main advantage over the previous statistic is that it takes into account of any conditionality in our forecast: for example, if volatilities are low in some period and high in others, the VaR forecast should respond to this clustering event:

$$\text{LR}_{CC} = -2 \ln[(1 - p)^{T-N} p^N] + 2 \ln[(1 - \pi_{01})^{n00} \cdot \pi_{01}^{n01} \cdot (1 - \pi_{11})^{n10} \cdot \pi_{11}^{n11}]$$

where

n_{ij} is the number of observations with value i followed by j for $i, j = 0, 1$

$\pi_{ij} = n_{ij} / \sum_j n_{ij}$

26.3.4 VaR Out-of-Sample Results

We analyzed a high-dimensional portfolio composed of 30 Russian assets, with daily data taking into consideration the very volatile period between 5/01/2000 and 23/05/2008. We chose the 30 most liquid assets with at least 2000 historical daily data quoted at the RTS and MICEX Russian markets.

Tables 26.2 and 26.3 report the actual VaR exceedances N/T, the p-values p_{UC} of Kupiec's unconditional coverage test and the p-values p_{CC} of Christoffersen's conditional coverage test for the VaR forecasts at all probability levels. Table 26.4 reports the p-values of the SPA test for all the quantiles and both for long and short positions, while Table 26.5 the asymmetric loss Equation 26.6.

The previous tables show that if one is interested in forecasting the extreme quantiles, particularly at the 1% and 99% levels (which is the usual case for regulatory purposes), then using a Student's t GARCH model with *any* copula does a good job. The fact that the type of copula plays a minor role is not a surprise, given previous empirical evidence with American and European stocks (see, e.g., Ané and Kharoubi, 2003; Junker and May, 2005; Fantazzini, 2008a). Furthermore, simulation evidence in Fantazzini (2009b) highlights that copula misspecification is overshadowed by marginal misspecification when dealing with small-to-medium-sized samples. Besides, copula misspecification is large only in case of negative dependence, while much smaller with positive dependence. In the latter case, different models may deliver quite close VaR estimates (given the same marginals are used).

It is interesting to note that if normal marginals are used, then models with dynamic dependence deliver statistically significant (and more precise) VaR estimates than models with constant dependence. If Student's t marginals are used, the differences are much smaller and not significant. This confirms that marginal misspecification may result in significant misspecified dependence structure, as shown by Bouye et al. (2001).

Moreover, it is possible to note that the models based on Normal marginals tend to have better results than competing models at the 5% level with long positions, while they result to be too conservative at the 5% level with short positions. This may be due to lack of a parameter modeling excess kurtosis, as explained by Junker and May (2005).

TABLE 26.2 Actual VaR Exceedances N/T, Kupiec's and Christoffersen's Test p-Values (Long Positions)

	Long Position											
	0.25%			0.50%			1%			5%		
Models	N/T%	p_{UC}	p_{CC}	N/T%	p_{UC}	p_{CC}	N/T%	p_{UC}	p_{CC}	N/T%	p_{UC}	p_{CC}
Model 1	0.40	0.38	0.67	0.70	0.40	0.67	1.40	0.23	0.29	6.50	**0.04**	**0.02**
Model 2	0.40	0.38	0.67	0.70	0.40	0.67	1.50	0.14	0.16	6.50	**0.04**	**0.02**
Model 3	0.40	0.38	0.67	0.80	0.22	0.44	1.80	**0.02**	**0.05**	7.70	**0.00**	**0.00**
Model 4	0.40	0.38	0.67	0.70	0.40	0.67	1.60	0.08	0.11	6.70	**0.02**	**0.02**
Model 5	0.50	0.16	0.37	1.00	**0.05**	0.13	1.90	**0.01**	**0.03**	7.70	**0.00**	**0.00**
Model 6	0.50	0.16	0.37	0.60	0.66	0.88	1.70	**0.04**	0.07	6.70	**0.02**	**0.00**
Model 7	0.10	0.28	0.56	0.30	0.33	0.62	1.20	0.54	0.71	7.90	**0.00**	**0.00**
Model 8	0.10	0.28	0.56	0.30	0.33	0.62	1.10	0.75	0.84	7.80	**0.00**	**0.00**
Model 9	0.10	0.28	0.56	0.30	0.33	0.62	1.10	0.75	0.84	8.00	**0.00**	**0.00**
Model 10	0.20	0.74	0.94	0.30	0.33	0.62	1.00	1.00	0.90	7.80	**0.00**	**0.00**
Model 11	0.10	0.28	0.56	0.30	0.33	0.62	1.20	0.54	0.71	7.90	**0.00**	**0.00**
Model 12	0.20	0.74	0.94	0.30	0.33	0.62	1.10	0.75	0.84	7.90	**0.00**	**0.00**

Note: p-Values smaller than .05 are reported in bold font.

TABLE 26.3 Actual VaR Exceedances N/T, Kupiec's and Christoffersen's test p-Values (Short Positions)

	Short Position											
	0.25%			0.50%			1%			5%		
Models	N/T%	p_{UC}	p_{CC}	N/T%	p_{UC}	p_{CC}	N/T%	p_{UC}	p_{CC}	N/T%	p_{UC}	p_{CC}
Model 1	0.70	**0.02**	0.06	1.00	**0.05**	0.07	1.30	0.36	0.24	3.40	**0.01**	**0.00**
Model 2	0.80	**0.01**	**0.02**	1.00	**0.05**	0.07	1.30	0.36	0.56	3.50	**0.02**	**0.01**
Model 3	0.80	**0.01**	**0.02**	1.00	**0.05**	0.08	1.40	0.23	0.20	4.00	0.13	**0.02**
Model 4	0.70	**0.02**	0.06	1.00	**0.05**	0.07	1.30	0.36	0.56	3.50	**0.02**	**0.01**
Model 5	0.80	**0.01**	**0.02**	1.10	**0.02**	0.08	1.50	0.14	0.16	4.10	0.18	**0.01**
Model 6	0.70	**0.02**	0.06	1.00	**0.05**	0.07	1.30	0.36	0.56	3.50	**0.02**	**0.01**
Model 7	0.30	0.76	0.95	0.50	1.00	0.98	1.00	1.00	0.90	4.70	0.66	**0.02**
Model 8	0.30	0.76	0.95	0.50	1.00	0.98	0.90	0.75	0.87	4.70	0.66	**0.02**
Model 9	0.20	0.74	0.94	0.50	1.00	0.98	0.90	0.75	0.87	4.80	0.77	**0.03**
Model 10	0.30	0.76	0.95	0.50	1.00	0.98	1.00	1.00	0.90	4.80	0.77	**0.03**
Model 11	0.30	0.76	0.95	0.50	1.00	0.98	1.10	0.75	0.84	4.80	0.77	**0.03**
Model 12	0.30	0.76	0.95	0.50	1.00	0.98	1.00	1.00	0.90	4.90	0.88	**0.03**

Note: p-Values smaller than .05 are reported in bold font.

TABLE 26.4   Hansen's SPA Test for the Portfolio Consisting of 30 Dow Jones Stocks

Benchmark	Long Position				Short Position			
Models	**0.25%**	**0.50%**	**1%**	**5%**	**0.25%**	**0.50%**	**1%**	**5%**
Model 1	0.138	0.172	0.730	0.981	0.461	0.318	0.061	**0.011**
Model 2	0.864	1.000	0.902	0.186	0.364	0.337	0.064	**0.003**
Model 3	0.990	0.537	0.120	**0.025**	0.076	0.146	0.060	**0.002**
Model 4	0.898	0.429	0.957	0.065	0.400	0.288	**0.048**	**0.005**
Model 5	0.238	0.213	0.060	**0.023**	0.065	0.093	0.067	**0.006**
Model 6	0.196	0.259	0.188	0.167	0.341	0.325	**0.040**	**0.005**
Model 7	0.268	0.274	0.680	0.580	0.911	0.808	0.892	0.427
Model 8	0.304	0.298	0.937	0.628	0.506	0.234	0.192	0.909
Model 9	**0.000**	0.155	0.589	0.233	0.390	0.600	0.477	0.327
Model 10	**0.000**	**0.000**	0.723	0.594	0.797	0.858	0.709	0.725
Model 11	**0.000**	0.296	0.867	0.057	0.906	0.945	0.446	0.227
Model 12	0.077	0.235	0.816	0.180	0.180	0.390	0.560	0.373

Note: p-Values smaller than .05 are reported in bold font.

TABLE 26.5   Asymmetric Loss Functions (Equation 26.6)

	Long Position				Short Position			
Models	**0.25%**	**0.50%**	**1%**	**5%**	**0.25%**	**0.50%**	**1%**	**5%**
Model 1	2.360	4.275	7.830	**29.852**	10.408	13.276	18.107	45.527
Model 2	2.332	**4.239**	7.811	29.918	10.430	13.269	18.047	45.451
Model 3	**2.329**	4.283	7.956	30.412	10.500	13.373	18.170	45.343
Model 4	2.334	4.257	**7.807**	29.957	10.421	13.288	18.046	45.453
Model 5	2.376	4.347	8.089	30.399	10.702	13.492	18.245	45.349
Model 6	2.346	4.267	7.844	29.939	10.384	13.260	18.008	45.405
Model 7	2.480	4.428	7.858	30.135	9.648	12.471	**17.178**	44.059
Model 8	2.480	4.414	7.837	30.142	9.681	12.531	17.273	**44.001**
Model 9	2.546	4.448	7.870	30.212	9.695	12.512	17.238	44.075
Model 10	2.551	4.491	7.853	30.134	9.614	**12.448**	17.198	44.012
Model 11	2.498	4.415	7.841	30.268	**9.611**	12.473	17.241	44.111
Model 12	2.505	4.432	7.852	30.226	9.686	12.530	17.230	44.067

Note: The smallest value is reported in bold font.

26.4   CONCLUSIONS

This chapter aimed at examining and comparing different multivariate parametric models with the purpose of estimating the VaR for a high-dimensional portfolio composed of Russian financial assets. To achieve this goal, we introduced a general multivariate framework by means of copulas to unify past approaches and propose new extensions. We then analyzed a high-dimensional portfolio composed of 30 Russian stocks over the time interval 2000–2008. By using a rolling window estimation scheme, we compared different multivariate models by looking at their VaR forecasts with different tests and statistical techniques.

We found out that if one is interested in forecasting the extreme quantiles, particularly at the 1% and 99% levels (which is the usual case for regulatory purposes), then a Student's t GARCH model with *any* copula delivers quite precise VaR estimates. The evidence that the type of copula plays a minor role should not be considered a surprise, given previous empirical evidence with American and European stocks (see Ané and Kharoubi, 2003; Junker and May, 2005; Fantazzini, 2008a) and particularly given the simulation evidence in Fantazzini (2009b). Besides, our empirical evidence finds out that if normal marginals are used, then models with dynamic dependence deliver statistically significant (and more precise) VaR estimates than models with constant dependence. However, if Student's t marginals are used, the differences are much smaller and no more significant. This confirms again that marginal misspecification may result in significant misspecified dependence structures.

An avenue for future research is in more sophisticated methods to separate the assets into homogenous groups when using the grouped-t copula. Finally, an alternative to DCC modeling for high-dimensional portfolios could be the semiparametric and nonlinear techniques proposed in Hafner et al. (2005) and Pelletier (2006).

REFERENCES

Amisano, G. and Giacomini, R. (2007) Comparing density forecasts via weighted likelihood ratio tests. *Journal of Economics Business and Statistics*, 25(2): 177–190.

Ané, T. and Kharoubi, C. (2003) Dependence structure and risk measure. *The Journal of Business*, 76(3): 411–438.

Basel Committee on Banking Supervision (1996) Supervisory Framework for the Use of Backtesting in Conjunction with the Internal Models Approach to Market Risk Capital Requirements. Bank for International Settlements, Basel, Switzerland.

Basel Committee on Banking Supervision (2005) Amendment to the Capital Accord to Incorporate Market Risks. Bank for International Settlements, Basel, Switzerland.

Bauwens, L. and Laurent, S. (2005) A new class of multivariate skew densities, with application to GARCH models. *Journal of Business and Economic Statistics*, 23(3): 346–354.

Bollerslev, T. (1990) Modelling the coherence in short-run nominal exchange rates: A multivariate generalized ARCH model. *The Review of Economics and Statistics*, 72(3): 498–505.

Bouye, E., Durrleman, V., Nikeghbali, A., Riboulet, G., and T. Roncalli (2001) *Copulas for Finance: A Reading Guide and Some Applications*. Groupe de Recherche Operationnelle, Credit Lyonnais.

Chen, X., Fan, Y., and Patton, A. (2004) Simple tests for models of dependence between multiple financial time series with applications to U.S. equity returns and exchange rates. FMG Technical Report n.483, LSE.

Cherubini, U., Luciano, E., and Vecchiato, W. (2004) *Copula Methods in Finance*. Wiley, Chichester, United Kingdom.

Christoffersen, P. (1998) Evaluating interval forecasts. *International Economic Review*, 39(4): 841–862.

Christoffersen, P. and Diebold, F.X. (2000) How relevant is volatility forecasting for financial risk management? *Review of Economics and Statistics*, 82(1): 12–22.

Daul, S., De Giorgi, E., Lindskog, F., and McNeil, A. (2003) The grouped *t*-copula with an application to credit risk. *Risk*, 2(3): 73–76.

Demarta, S. and McNeil, A. (2005) The *t* copula and related copulas. *International Statistical Review*, 73(1): 111–129.

DeMiguel, V., Garlappi L., and Uppal, R. (2009) Optimal versus naive diversification: How inefficient is the 1/N portfolio strategy? *Review of Financial Studies*, in press.

Engle, R.F. (2002) Dynamic conditional correlation—A simple class of multivariate GARCH models. *Journal of Business and Economic Statistics*, 20(3): 339–350.

Engle, R.F. and Sheppard, K. (2001) Theoretical and empirical properties of dynamic conditional correlation multivariate GARCH. NBER Working Papers, no. 8554, Cambridge, MA.

Fang K.T., Kotz S., and Ng K.W. (1990) *Symmetric Multivariate and Related Distributions*. Chapman and Hall, London.

Fantazzini, D. (2008a) Dynamic copula modelling for value at risk, *Frontiers in Finance and Economics*, 5(2): 1–36.

Fantazzini, D. (2009b) The effects of misspecified marginals and copulas on computing the value at risk: A Monte Carlo study. *Computational Statistics and Data Analysis*, in press.

Fantazzini, D. (2009) A dynamic grouped-*t* copula approach for market risk management, In: G. Gregoriou (Ed.), *A VaR Implementation Handbook*. McGraw-Hill, New York.

Giacomini, R. and Komunjer, I. (2005) Evaluation and combination of conditional quantile forecasts. *Journal of Business and Economic Statistics*, 23(4): 416–431.

Giacomini, R. and White, H. (2006) Tests of conditional predictive ability. *Econometrica*, 74(6): 1545–1578.

Giot, P. and Laurent, S. (2003) VaR for long and short positions. *Journal of Applied Econometrics*, 18(6): 641–664.

Gonzalez-Rivera, G., Lee, T., and Santosh, M. (2006) Forecasting volatility: a reality check based on option pricing, utility function, value-at-risk, and predictive likelihood. *International Journal of Forecasting*, 20(4): 629–645.

Granger, C., Patton, A., and Terasvirta, T. (2006) Common factors in conditional distributions for bivariate time series. *Journal of Econometrics*, 132(1): 43–57.

Hafner, C., Van Dijk, D., and Franses, P. (2005) Semi-parametric modelling of correlation dynamics. Econometric Institute Research Report 2005-26, Erasmus University, Rotterdam, the Netherlands.

Hansen, P. (2005) A test for superior predictive ability. *Journal of Business and Economic Statistics*, 23(4): 365–380.

Höffding, D. (1940) Masstabinvariante Korrelationstheorie. *Schriften des Mathematischen Seminars und des Instituts für Angewandte Mathematik der Universität*, 5(1): 181–233.

Kaufman, L. and Rousseeuw, P.J. (1990) *Finding Groups in Data: An Introduction to Cluster Analysis.* Wiley, New York.

Kupiec, P. (1995) Techniques for verifying the accuracy of risk measurement models. *Journal of Derivatives*, 2: 173–184.

Joe, H. (1997) *Multivariate Models and Dependence Concepts.* Chapman and Hall, London.

Joe, H. and Xu, J. (1996) The estimation method of inference functions for margins for multivariate models. Department of Statistics, University of British Columbia, Technical Report n. 166, Vancouver, BC.

Jondeau, E. and Rockinger, M. (2006) The copula-GARCH model of conditional dependencies: An international stock-market application. *Journal of International Money and Finance*, 25(5): 827–853.

Jorion, P. (2000) *Value at Risk: The New Benchmark for Managing Financial Risk,* 2nd ed. McGraw Hill, New York.

Junker, M. and May, A. (2005) Measurement of aggregate risk with copulas. *Econometrics Journal*, 8(3): 428–454.

McNeil, A. and Demarta, S. (2005) The t-copula and related copulas. *International Statistical Review*, 73(1): 111–129.

Nelsen, R.B. (1999) *An Introduction to Copulas. Lecture Notes in Statistics 139,* Springer, New York.

Patton, A. (2004). On the out-of-sample importance of skewness and asymmetric dependence for asset allocation. *Journal of Financial Econometrics*, 2(1): 130–168.

Patton, A. (2006a) Estimation of copula models for time series of possibly different lengths. *Journal of Applied Econometrics*, 21(2): 147–173.

Patton, A. (2006b) Modelling asymmetric exchange rate dependence. *International Economic Review*, 47(2): 527–556.

Pelletier, D. (2006) Regime switching for dynamic correlations. *Journal of Econometrics*, 127(1–2): 445–473.

Romano, C. (2001) Applying copula function to risk management. Working Paper, Banca di Roma, Rome, Available at www.gloriamundi.org/picsresources/cr04.pdf.

Rosenberg, J.V. and Schuermann, T. (2006) A general approach to integrated risk management with skewed, fat-tailed risks. *Journal of Financial Economics*, 79(3): 569–614.

Sklar A. (1959) Fonctions de Repartition a N Dimensions Et Leurs Marges. *Publications of the Institute of Statistics*, 8(1): 229–231.

CHAPTER 27

Microstructure of the Bid–Ask Spreads of Russian Sovereign Bonds (1996–2000): Spreads as Indicators of Liquidity

Peter B. Lerner and Chunchi Wu

CONTENTS

27.1 INTRODUCTION

Liquidity is one of the most elusive concepts in finance. While it is present in most contexts of financial analysis, it escapes a single quantitative definition. The influence of liquidity on more familiar indicators such as returns, yields, and volatility remains obscure. In this chapter, we turn to a subject of study for which the problem of liquidity is inseparable from asset pricing, namely, to emerging market bonds.

The classic treatment of liquidity (Amihud and Mendelson, 1991; Green and Elton, 1998) deals with reaction of prices to the trading volume. However, the information on volume may not be available for most fixed-income securities, which are closely bought and sold, or for some sovereign emerging market bonds, which further conflate open market transactions with the transfers of bonds between the Treasury and the Central Bank. Therefore, we study bid–ask spreads, which may be more numerically reliable despite all the theoretical shortcomings of using them as a proxy for liquidity.

Stoll (2003) separates bid–ask spread into three components: trade execution cost, inventory maintenance cost, and information asymmetry or adverse selection cost. The origin of trade execution cost is that traders cannot organize the market for themselves for regulatory reasons or because of externalities and so they turn to a special agent, who acts as a near-monopolist with respect to individual traders and can extract nonzero profits for his or her services (He and Wu, 2005). Inventory maintenance costs originate in the fact that market makers, to assure smooth execution of trades by market participants, hold their own portfolios of securities, which bear financial risk. In the absence of a physical market maker, an electronic trading platform bears inventory risk as long as the exchange provides statutory guarantee of trade execution.

Finally, there is an information asymmetry or adverse selection component. Even if there is a complete code of silence around nonpublic information, insiders inadvertently leak this information when they trade upon it (Glosten and Milgrom, 1985; Kyle, 1985). These leakages impose a cost on the market maker, who must compensate for them through an increased spread.

We selected Russian Federation bonds for our study for the following reasons. First, the bid–ask spreads were numerically large, reaching 16%–17% of the price at times close to the Russian default. Second, these bonds were traded relatively infrequently, sometimes not every day, and to observe effects of similar magnitude in developed markets, one must turn to high-frequency data, which is not available for the most fixed-income instruments. Third, the bonds were traded in large blocks between sophisticated traders such as large investment banks, hedge funds, and international monetary organizations, where the effect of adverse selection must be most pronounced (Dittmar and Yuan, 2005).

27.2 DESCRIPTION OF THE DATASET

We use a data set that contains eight long-term bond issues by the Russian Ministry of Finance beginning from November 27, 1996—the date on which the Russian Federation began to issue Eurobonds—and ending on October 25, 2000. During this period, Russian financial markets were impacted by the East Asian financial crisis (1997–1998), selective default and abandonment of the currency exchange band (August 1998), failure of the hedge fund Long-Term Capital Management (LTCM) and, finally the round of parliamentary (December 1999) and presidential (March 2000) elections. The bonds included in our data set are listed in Table 27.1.

We exclude two short series (Bonds 5 and 6) and thus are left with six series of bonds that all encompass some predefault history and also were traded throughout the entire postdefault period in our database. Bond issues in our data set constitute assets with the face value of $34.47 billion, which are all long-term publicly issued dollar-denominated debt of the Russian Federation at the time.* However, our analysis includes bonds

---

* The Russian Federation inherited $105 billion of the debt of the former USSR (Gaidar, 2003), which is regarded as institutionally different by Russian authorities, a perception which, according to econometric analysis by Duffie et al. (2003), is shared by the world's financial markets. The borderline between the issues of Soviet and post-Soviet debt is not sharp because the main economic reason for the Russian Federation's borrowings was to cover interest payments on Russian debt inherited from the Soviet Union.

TABLE 27.1 Russian Sovereign Bond Issues

No.	Our Abbreviation	Coupon (%)	ID Number	Records (Predefault)	Records (Postdefault)	Observations Begin	Bond Launched On	Due	Amount Issued ($ bn.)
1	Minfin1175	11.75	USX74344CZ29	45	550	6/11/1998	6/10/1998	6/10/2003	1.25
2	Minfin11	11	XS0089375249	14	550	7/27/1998	7/24/1998	7/24/2018	3.47
3	Minfin1275	12.75	XS0088543193	35	550	6/25/1998	6/24/1998	6/24/2028	2.65
4	Minfin875	8.75	XS0089372063	14	550	7/27/1998	7/24/1998	7/24/2005	2.97
5	Ru825	8.25	XS0114295560	0	45	8/25/2000	3/31/2000	3/31/2010	2.53
6	Rustepup	var. 2.25–7.5	XS0114288789	0	45	8/25/2000	3/31/2000	3/31/2030	18.20
7	Minfin10	10	XS0077745163	283	550	6/27/1997	6/26/1997	6/26/2007	2.40
8	Minfin95	9.25	—	428	550	11/29/1996	11/27/1996	11/27/2001	1.00

Notes: These debt issues were preceded by several 3% OGVZZ (Bazarkin, 1998) bonds, which were nicknamed Minfin bonds in international trading practice and were issued against the debt of the former USSR. The data set provided in Duffie et al. (2003) overlaps with our data set by the two last issues in our table. They abbreviate them as Eurobond-01 and Eurobond-07, respectively.

with the face value of only \$13.74 billion because we exclude the data for the largest bond issue, Rustepup, because of its variable coupon.

27.3 PREDICTABLE COMPONENT OF THE SPREADS

Significant empirical evidence indicates that bid–offer spreads for different classes of securities obey some modification of the AR(1) process (He and Wu, 2005). One can prove that, if the bid–ask spread is formed entirely due to insider trading, it approximately obeys AR(1) dynamics, but in a local time (Karatzas and Shreve, 1997; Borodin and Salminen, 2005), both in the framework of the Kyle (1985) and Glosten–Milgrom (1985) type models (Lerner (2007)). In this framework, the unpredictability of the spreads results from a microstructure noise (Ait-Sahalia et al., 2005) and the stochastic nature of the local time.

To eliminate the predictable component of the bid–ask spreads, we perform VAR(1) regression according to the following equation:

$$S_{t+1} = \alpha + \hat{A} S_t + U_{t+1} \qquad (27.1)$$

where $S_t = \{s_1, s_2, \ldots, s_6\}'$ is the bid–offer vector of spreads of the bonds from Table 27.1, $\hat{A}$ is a six-vector of constants, $\hat{A}$ is a 6×6 matrix of regression coefficients, and U_t is the six-vector of residuals. The regression of Equation 27.1 is performed for 565 days of the common existence of six bonds in our database. The data on the coefficient matrix $\hat{A}$, spread volatility matrix $\hat{\Sigma} = E[U_t U_t']$, and statistical significance of their elements are provided in Tables 27.2 and 27.3.

TABLE 27.2   Predictable Component of the Bond Spreads

	A_{i1}	A_{i2}	A_{i3}	A_{i4}	A_{i7}	A_{i8}
A_{1j}	0.193	0.104	0.092	0.110	0.044	0.298
A_{2j}	−0.026	0.214	0.086	0.141	0.060	0.107
A_{3j}	0.051	0.098	0.241	0.107	0.058	0.102
A_{4j}	0.126	0.158	0.152	0.165	0.053	0.037
A_{7j}	0.031	0.092	0.140	0.131	0.107	0.139
A_{8j}	0.106	0.073	0.071	0.032	0.085	0.463

Notes: We present VAR(1) matrix for the bid–offer spreads of the six Russian Federation bonds (1–4 and 7–8) from Table 27.1. The elements of VAR matrix significant at 5% are boldfaced. All diagonal elements are significant.

TABLE 27.3   Covariance Matrix for the Innovations in VAR(1) Regression of Equation 27.1

	Σi1	Σi2	Σi3	Σi4	Σi7	Σi8
Σ1j	0.232	0.053	0.054	0.038	0.057	0.086
Σ2j	0.053	0.180	0.052	0.052	0.027	0.030
Σ3j	0.054	0.052	0.188	0.044	0.054	0.040
Σ4j	0.038	0.052	0.044	0.181	0.041	0.033
Σ7j	0.057	0.027	0.054	0.041	0.167	0.040
Σ8j	0.086	0.030	0.040	0.033	0.040	0.219

Note: Cross-correlation terms are generally small.

Each spatial component of the vector U_t represents the part of the spread that cannot be explained by VAR(1) regression. These components may depend hypothetically on the trading fundamentals of a particular bond or may be entirely random. If the former is true then, according to commonly accepted views (O'Hara, 1995; Stoll, 2003), the residuals of the Equation 27.1 must include daily surprises in trade execution cost, inventory maintenance cost, and information asymmetry cost. Alternatively, these residuals could be a pure noise. To investigate this hypothesis, we subject the residuals of Equation 27.1 to a univariate ordinary least squares (OLS) regression. The choice of proxies for execution cost, informed trading, and inventory maintenance is far from unique and we describe it in the next section.

27.4 SEPARATION OF TRADE EXECUTION, INFORMATION ASYMMETRY, AND INVENTORY MAINTENANCE COMPONENTS OF THE BID–ASK SPREADS

Each of the three components of the bid–ask spread identified by Stoll (2003) exhibits a different dynamic. We expect that the trade execution component evolves in nearly constant positive or negative steps occurring at random times, reflecting the cost of maintenance of the trading platform. The informed trading component should behave randomly because it is fed by quasirandom leakages of economic news before their public announcement and dissemination of this information through the market order updates, which we might call "feed-forward" as opposed to feedback.* On the other hand, the inventory control process should be intermittent, being characterized by relatively long periods of gradual

---

* Liquidity traders who react to price signals from informed traders reflect in their behavior an event that will become public only in the future.

adjustment and relatively short periods of quick transition when market perceptions change quickly and large order imbalances may arise.

To analyze the components of the bid–ask spread, we round off the price quotes and place them into 10 ranks according to a rule which we describe in the following text. The logic behind using rounded-off prices is as follows. There is no reason to expect that when a bond price changes from $100 to $101, the trade execution cost would change in any proportion to 1%. However, there is a plausible argument that when a bond price changes from $100/par to $20/par, the trade execution cost will change because a nominal $1000 investment will result in the purchase of 50 rather than 10 bonds. The aforementioned argument suggests that the spreads follow only "significant" rather than all fluctuations in price.

Our choice of significant price corrections, those that change the price deciles, is based on our analysis of the autocorrelation of spreads. Empirically, spreads' autocorrelation falls off roughly by half at 45–50 days (and almost completely at 100 days).* Consequently, we tested several obvious choices for the price round-offs (none, 50¢, $1, $2, $5, $10, and $20), so that the average duration of price staying in a rank would be comparable or slightly lower than 45–50 days. The round-off to the nearest $10 complies with this choice for the longest price series in our sample (Bond 8). Thus we define "significant" price changes as jumps (up or down) across the $10 ranks.

Bonds priced between par and $90 are assigned the first rank, between $90 and $80 the second rank, and so on. Rank-ordered bond prices form the series of dummy variables, which we denote as $D_{p,t}$. We use the previous day's rank-order price variables as indicators of execution cost. Hence, the execution cost is treated as some constant plus a fraction of a past rounded-off bond price.

Our choice for the price variable is not unique. For instance, one may use the exact dollar price of the bond but this choice ignores the fact that the bid–ask spread typically changes in discrete amounts. Our definition takes into account the discrete structure of the market bids and offers in the real markets. For the Russian bonds between 1996 and 2000, bid and offer quotes were updated in increments of at least $0.25, rarely in $0.1 increments. An alternative choice would be to use the price/spread ratio, but the impact of a given spread (for instance, $1) would vary widely because in

---

* Here we omit this analysis, but it can be found in our working papers on www.ssrn.com

1996–2000, the bonds were traded in the range of 20¢ to $1 per dollar of face value. The continuous percentage change is exact as to the change in the price. However, this choice would not only obscure the discrete structure of the updates but also would amplify the spread shocks. The periods with the lowest bond prices, i.e., the highest credit risk, were typically accompanied by the highest spreads, i.e., the lowest liquidity. Regression results are not influenced much by the particular choice of measure and we select the exact dollar amount of the spread because of the transparency of interpretation of its components in dollars and cents.

The most recent change in price variable $\Delta D_{p,t}$ is taken as an indicator of informed trading. The fact, whether the change in bond price in the next trading session will be significant, is obscured to liquidity traders, but the insider might possess this information. Another parameter of importance is the average duration of price stability. According to the option-based theory for the bid–ask spreads (Bollen et al., 2004), spreads must be approximately proportional to the square root of the duration.

We use two different proxies for duration: one that reflects the information set of informed traders and another that reflects the information set of liquidity traders. For liquidity traders and the market maker, the best proxy for duration is the time $(T_{c,t})$, during which the price already stayed within its rank. This is the only information on duration they possess in an efficient market. For informed traders, the proxy for duration is the overall duration $(Time_t)$ between price corrections because they allegedly can foresee large future movements of prices. With respect to sovereign bonds, the insiders might know the future date on which the national government, finance ministry, or central bank will release the next batch of important news.

Two additional variables that describe the inventory maintenance component are, first, the sign of the previous large correction (up or down) $SIGN_t$ and, second, the product of the sign of change and the square root of duration after the last correction:

$$SIGNT_t = SIGN_t \times (T_{c,t})^{1/2}$$

The latter variable was added to account for possible asymmetry between the periods following positive or negative corrections.

All the variables except changes in price range $\Delta D_{p,t}$ and $Time_t$ can be directly observed by outsiders (liquidity traders and the market maker) before the start of the trading session. Only the factors known in advance can influence their decision. However, insiders can sometimes predict whether a large price correction will happen in an upcoming session and its direction, as well as the duration before the next public release of significant news. Our regression for each bond series is as follows:

$$u_{it} = a_{i0} + P_i \cdot D_{p,t-1} + Q_i \cdot \Delta D_{p,t} + \Lambda_i \cdot \sqrt{\text{Time}_t}$$
$$+ N_i \cdot \sqrt{T_{c,t}} + \theta_i \cdot SIGN_t + Z_i \cdot SIGNT_t + e_{it} \qquad (27.2)$$

where u_{it} are the residuals of the VAR(1) process in Equation 27.1 for each bond, $i = 1\text{–}6$. The first two terms in the regression of Equation 27.2 describe the order execution costs. This is a constant term plus some fixed portion of past day's rounded-off price. The next two terms are the contribution of information asymmetry or insider trading. The spread anticipates the time *before* the next significant correction of the bond price (factor Λ) and the expected magnitude of the correction (factor Q).

The last three factors measure the inventory maintenance component. For frequent jumps between ranks, the market maker might maintain a larger inventory to cope with potential imbalances in the face of receiving large "buy" or "sell" orders. Or, on the contrary, they might decide to hold a small inventory, believing that the next price correction may cancel out the imbalance. The cost of inventory maintenance can be asymmetric with respect to the sign of the price change, which is reflected in the last two regression factors.

The results of the OLS model of Equation 27.3 are provided in Table 27.4. Only for Bond 1, the regression explains a significant part of the variation. Moreover, this bond has the highest contribution of informed trading and the most significant informed trading factors. We might consider that this bond was a preferred instrument for execution by informed traders.

The signs of statistically significant coefficients in Equation 27.2 are intuitive. For instance, the trade execution cost rises in some proportion to the average past price ($P > 0$). Informed traders foresee tightening liquidity for higher yields and increasing liquidity for lower yields ($Q < 0$). If the past correction was positive, a market maker increases their inventory in anticipation of subsequent "buy" orders ($\theta > 0$), but over time, after a large correction, their inventory declines ($Z < 0$).

On the basis of Equation 27.2, we can deduce the relative contribution of the effects of trade execution costs, informed trading, and inventory maintenance into the formation of the spread. Because of the contribution of

TABLE 27.4 Univariate Analysis of the Residuals of the VAR(1) Regression of the Bid–Ask Spreads of the Six Russian Federation Bonds According to Equation 27.3

Instrument	Coefficient in Equation 27.2	Bond 1	Bond 2	Bond 3	Bond 4	Bond 5	Bond 6
Intercept	a_0	**-0.882**	0.072	0.072	0.036	0.119	-0.084
		(-11.88)	(0.85)	(0.90)	(0.46)	(1.15)	(-0.98)
D_{t-1}	P	**0.473**	0.001	0.014	0.013	-0.011	**0.033**
		(24.68)	(0.12)	(1.64)	(1.55)	(-0.69)	**(3.31)**
ΔD_{t-1}	Q	**-0.183**	-0.074	-0.069	-0.035	0.005	-0.154**
		(-16.55)	(-1.27)	(-1.13)	(-0.61)	(0.09)	(-2.26)
$(Time_t)^{1/2}$	Λ	-0.015	0.010	-0.016	-0.015	-0.010	-0.009
		(-1.51)	(0.96)	(-1.12)	(-1.14)	(-0.89)	(-0.95)
$(T_{c,t})^{1/2}$	N	-0.020	-0.028**	-0.019	-0.010	-0.003	0.001
		(-0.18)	(-2.36)	(-1.11)	(-0.68)	(-0.20)	(0.05)
$SIGN_t$	θ	0.086*	**0.142**	0.142*	0.034	-0.007	-0.070
		(2.76)	**(3.54)**	(3.07)	(0.81)	(-0.17)	(-1.15)
$SIGNT_t$	Z	-0.025*	-0.016	-0.030**	-0.091	0.008	0.008
		(-2.73)	(-1.76)	(-2.20)	(-0.88)	(0.67)	(0.32)
R^2_{adj} (%)		53.7	3.8	4.1	1.2	0.0	6.0

Notes: Student's *t*-statistic is given in parentheses. For all bonds, except Bond 1, regressions lack predictive power. Coincidentally, this is the only bond for which the insider trading term is statistically significant, and which has a nonzero-biased distribution of the VAR(1) residual ($a_0 \neq 0$). We can view P and Q as price impact coefficients, Λ and N as duration impact coefficients, and θ and Z as inventory leverage impact coefficients. The coefficients significant at $P < 0.1\%$ are boldfaced.

* marks the coefficients significant at 1% and ** mark coefficients significant at 5%.

de-trended processing costs, informed trading and inventory maintenance components can become negative; we use the average sums of the
squares as a measure of relative contribution. The results are listed in
Table 27.5.

From Table 27.5, we observe that the trade execution cost is the largest component of the spread. This agrees with the results of various
authors (referenced in the table) for different classes of securities,
primarily stocks, and futures. The readers are cautioned, however,
that these authors used diverse methodologies to separate the spread
components.

TABLE 27.5 Quadratic Mean $s_k = \sqrt{1/T \sum_{i=1}^{T} s_{ik}^2}$ Contribution of (1) Trade Execution
Cost, (2) Informed Trading, and (3) Inventory Maintenance Costs to the VAR(1)
De-Trended Spread of Six Russian Bonds in the Period 1996–2000

Security/Market	Trade Execution (%)	Informed Trading (%)	Inventory (%)	Source
Bond 1	52.0	46.0	2.0	This work
Bond 2	24.1	21.7	54.2	This work
Bond 3	32.9	20	47.1	This work
Bond 4	47.8	35.6	16.6	This work
Bond 7	46.6	34.4	19	This work
Bond 8	38.6	43.5	17.9	This work
Russian bonds average	**40.3**	**33.5**	**26.1**	This work
NYSE	65.0	35.0	0.0	Glosten and Harris (1988)
NASDAQ NMS	47.0	43.0	10.0	Stoll (1989)
NASDAQ NMS	87.0–92.0	8.0–13.0	10.0	George et al. (1991)
NASDAQ	76 to 102	−2 to 24	0.0	Porter and Weaver (1996)
NYSE	62.7	9.6	28.7	Huang and Stoll (1997)
SEHK (Hong Kong)	44.8	32.6	0.0	Brockman and Chung (1999)
LSE	30.1–61.9	20.9–47.0	0.0–22.8	Menyah and Paudyal (2000)

Note: Criteria used by the cited authors may vary. In compiling this table, we have used
Strother et al. (2002).

27.5 INFORMATION CONTENT OF THE INFORMED TRADING SERIES

To corroborate our identification of the informed trading series, these series must display independent, identically distributed (i.i.d.) properties. Indeed, the fundamental nonpublic events that drive informed trading and are leaked to the market through compound orders of informed and liquidity traders must arrive unexpectedly to be substantial news.

To confirm or disprove the i.i.d. hypothesis for the arrival of unexpected information, we use the Broock–Dechert–Scheinkman (BDS) test (1996), which was first proposed in 1983 by Grassberger and Procaccia (1983). This test is widely used in the statistical literature (see Kanzler, 1999) and the null hypothesis is usually interpreted as the absence of nonlinear dynamics effects responsible for the trading patterns. It is based on the fact that sequences of consecutive random numbers must have the same distribution independently of their location within the time series. On the other hand, if there is a hidden pattern, distribution must somehow depend on the grouping of events within the series.

We apply the BDS test to the "change-of-insider sentiment" variable defined as follows:

$$\Delta s_{it} = (\hat{\alpha}_{i2} \cdot \Theta'_{i2})_t - (\hat{\alpha}_{i2} \cdot \Theta'_{i2})_{t-1} \tag{27.3}$$

where $i = 1-6$ is the number of the bond, $\alpha_2 = \{Q, \Lambda\}$, $\Theta_2 = \{\Delta D_{p,t-1}, \sqrt{\text{Time}_t}\}$, are two-dimensional vectors and the hat above the variable signifies its OLS estimator. The first term in the right-hand side of Equation 27.3 is an insider contribution into the dollar bid–ask spread on day t. The second term is the same contribution of insider trading on the previous day. The choice of using a differenced series was informed by our desire to avoid a unit root problem (Greene, 2000).

The results of the BDS tests are given in Table 27.6. The null hypothesis H_0 is that "there is no underlying nonlinear pattern behind the informed trading events." We observe that the null hypothesis for the informed component of the spreads—the pattern is absent—cannot be rejected for Bonds 1, 2, 3, and 8, but we can reject it for Bonds 4 and 7 at $P < 1\%$ in most tests. This indicates that insiders may have traded only certain bond issues. In particular, we cannot rule out the i.i.d. innovations of changes in insider sentiment (Equation 27.3) for Bond 1, which we singled out in Section 27.4 as the most probable target for informed trading.

TABLE 27.6   BDS Statistic for the Informed Part of Spreads

	BDS Statistic
$m = 2$	$-0.79, -0.055, -0.098, -1.80, -1.05, -0.071$
$m = 3$	$-1.25, -1.78, \ -0.25, \ \mathbf{-3.71}, \mathbf{-2.48}, -0.33$
$m = 4$	$-0.53, \mathbf{-3.17}, \ -0.38, \ \mathbf{-6.14}, \mathbf{-4.47}, -0.54$
$m = 5$	$-0.74, \mathbf{-5.26}, \ -0.50, \ \mathbf{-9.27}, \mathbf{-7.24}, -0.74$

Broock, Dechert, and Schenikman proved, in 1996, that the quantity $q = T^{1/2}(C(m,\varepsilon,T)-C(1,\varepsilon,T)^m)/\sigma(m,\varepsilon,T)$ (see note) for the i.i.d. time series is distributed like a normal random variable. We provide results of the tests for informed trading events. The results of the tests for all six bonds are listed in the order of the bonds in the Table 27.1. Deviations from normal distribution of q are usually interpreted as the presence of nonlinear effects. All tests were performed for $\varepsilon = 2\sigma$ (σ is a standard deviation of the series). The duration of all series was chosen as 574 days for comparability, independent on the actual date of bond issue. The results significant at 1% are boldfaced.

Note: $C(m,\varepsilon,T) = \dfrac{2}{(n-m+1)(n-m)} \sum\limits_{s=m}^{T} \sum\limits_{t=s+1}^{T} \sum\limits_{j=0}^{m-1} I_\varepsilon(X_{s-j}, X_{t-j})$ and

$I_\varepsilon(X,Y) = \begin{cases} 1, & \text{if } |X-Y| < \varepsilon \\ 0, & |X-Y| \geq \varepsilon \end{cases}$. We use sample standard deviation as an

estimator for $\sigma(m,\varepsilon,T)$. For details, consult Kanzler (1999).

The reasons for the preference of insiders for some bonds over others could be diverse. For instance, the bankers who acquired insider information could have already invested in some bond portfolio or they may have deliberately traded in a few bonds and ignored the others to mask their insider status.[*]

We display the change in informed sentiment (Equation 27.3) of the spread in Figure 27.1. Even in the nascent Russian bond market, informed trading contributed less than 10% to the total size of the spread though this contribution is typically 40% with respect to the VAR residual from Equation 27.1 (see Table 27.5).

Between informed trading events, as identified by the regression of Section 27.4, spread volatility declines. This agrees well with the conclusions of informed trading theory (Kyle, 1985). We do not observe spread

---

[*] Obviously, once the insider status of a particular trader becomes known in the community, others can copycat his or her actions even without access to inside information to capture a share of the profits.

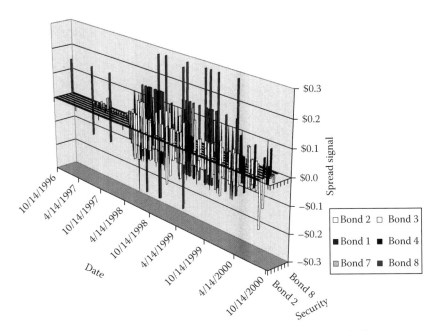

FIGURE 27.1   Informed trading component of the dollar spread of six Russian bonds in the 4 year period between 11/27/1996 and 10/25/2000. Positive signals indicate diminishing liquidity and negative signals indicate increasing liquidity. We separated informed trading events on the basis of second-stage OLS regression of the VAR(1) spread residuals. The highest density of informed trading is observed around the time of the Russian default in August 1998 and it subsides after the Russian economy stabilized.

volatility declining to zero either because of finite samples or because there are always intervening fundamental economic events. These fundamental events influence the asset price and reset the learning game for the liquidity traders to zero time. We display the realized historical volatility of spreads in the "windows" between the clusters of informed trading in Figure 27.2.

In the original Kyle theory, the volatility decays to zero as a straight line $\sigma(t) = \sigma_0(1 - \gamma(t - t_i))$, where t_i is the time of an information event and γ is an inverse characteristic time of information leakage from insiders to liquidity traders. Mendelson and Tunca (2004) proposed a different law: $\sigma(t) = \sigma_0(1 - \gamma(t - t_i))^{1/3}$[*]. The constant γ depends on the preferences of both

---

[*] This is a consequence of Mendelson and Tunca representing the prices not as a Vasiček (Ornstein–Uhlenbeck) but as a CIR-type process (Hull, 1997; Avellaneda and Laurence, 1999).

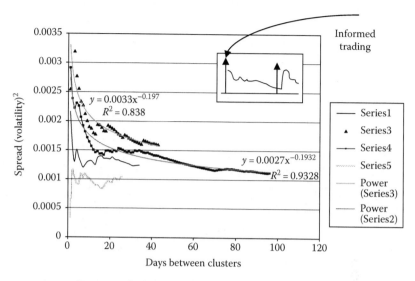

FIGURE 27.2 Decay of the spread volatility between informed trading events. The inset schematically shows the behavior of the spreads between informed events, which are indicated by vertical arrows. When informed events are rare, the decay in spread volatility can be well ($R^2 \approx 0.8$–0.9) approximated by the inverse power law with the coefficient equal to 0.2.

insider and liquidity traders and the overall frequency of trading. Most of the intervals between clusters of informed trading do not have a clear pattern and are too short for reliable statistics. For a few long intervals, we observe an approximate inverse power law: $\sigma \cong \sigma_0 \, (\sigma(t - t_i))^{-1/5}$, where the time of an informed event is determined according to the regression from Equation 27.2.

We can notice from Figure 27.1 that informed trading events become more frequent as we approach the August 17, 1998 default of the Russian Federation* but their frequency steadily declines after the first quarter of 1999. Moreover, the negative-liquidity events (i.e., growing spreads) dominated the period a few months before the default, while the positive-liquidity events dominated entire postdefault period. This agrees with the

---

view that informed agents, unlike their uninformed counterparts at the time, viewed the Russian economy as fundamentally sound. Presumably, they must have considered the default and the currency crisis of 1998 as a monetary aberration, which would subside if only the Russian Government demonstrated seriousness in tax collection.

To check our visual perception of the trends in informed trading, we use Haar and Lagrange wavelets to analyze the density of informed events (Mallat, 1999; Percival and Walden, 2000). To apply wavelet analysis, we construct two time series measuring positive (increasing liquidity) and negative (decreasing liquidity) insider sentiment:

$$p_+(t) = 1\left[\sum_{i=1}^{6} \Delta s_{it} < 0\right],$$

$$p_-(t) = 1\left[\sum_{i=1}^{i=6} \Delta s_{it} > 0\right]$$

(27.4)

In Equation 27.4, 1[.] means an indicator function and the summation is performed over all six bonds. The meaning of the variables p_+, p_- is as follows. If a combined insider sentiment on a day t favored increased liquidity or reduced the spread, then $p_+(t) = 1$, being zero otherwise. Conversely, if the insider sentiment increased the spread, signifying the perception of shrinking liquidity, then $p_-(t) = 1$, being zero otherwise. One can apply autoregressive conditional intensity [ACI] (Russell, 1999; Bauwens and Hautsch, 2006) analysis to quantify the behaviors, which we outline in the following text from the visual inspection of the wavelet plots. ACI analysis by itself does not add any new information and we omit it from this discussion.

We plot the fourth crystal of the low-pass wavelet filter (Percival and Walden 2000) to visualize the dynamics of the insider sentiment in Figure 27.3. The filter uses the information compressed for $2^4 = 16$ days to sum up the event. We used two popular wavelet filters, Haar and Lagrange, to a similar result.

Wavelet decomposition of the spread volatility is illustrated in Table 27.7 for the Haar wavelet. The similarity in the behavior of spread volatility for the positive (liquidity enhancing) and negative events bolsters our case for the division of spread components according to Equation 27.2. Positive and negative informed events arrive unpredictably and independently from one another, which is characteristic of true economic facts. We apply our analysis of informed sentiment to the discussion in the next section.

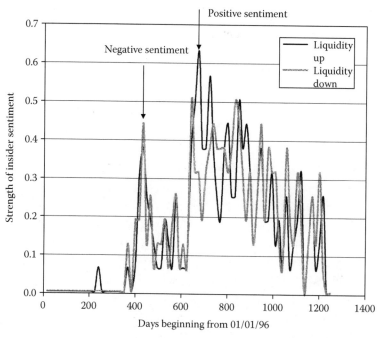

FIGURE 27.3 Fourth crystal of Haar wavelet of the time series of the positive and negative sentiments according to Equation 27.4. Time series are obtained as indicator functions of the change in the cumulative insider trading component for all six bonds (see Figure 27.1). Vertical arrows indicate large unbalanced insider signals of reduced or enhanced liquidity of the Russian Federation bonds.

TABLE 27.7   Decomposition of Spreads Volatility by the Haar Wavelets

Events	σ, Original	First Crystal	Second Crystal	Third Crystal	Fourth Crystal	Filter
Positive	0.1338	0.0675	0.0210	0.0122	0.0066	High-pass
		0.662	0.0452	0.0332	0.0268	Low-pass
Negative	0.1361	0.0671	0.0204	0.0185	0.0065	High-pass
		0.0692	0.0489	0.0306	0.0243	Low-pass

Notes: In this table, we list standard deviations of successive generations of wavelets for both positive and negative informed events according to the "insider sentiment" rule of Equation 27.4. Generations of crystals in the table are "fathered" by the high-pass filter; consequently the "low-pass" filter is their "mother" (Percival and Walden, 2000). Our main observation is that contribution of positive and negative informed events to the spreads volatility is very similar. If this were not the case, spreads would have a recognizable tendency, which could potentially be exploited by arbitrageurs.

27.6 DISCUSSION

We notice from Figure 27.3, that the positive and negative sentiments of insiders in the Russian Federation bond market in 1996–2000 were approximately balanced. Otherwise, one could suspect a deterministic trend in the market, which would be difficult to reconcile even with weak market efficiency. Large imbalances of insider sentiment are observed around September 15, 1997 (negative imbalance and reduced liquidity) and around August 27, 1998 (positive imbalance and increased liquidity). The latter date was separated from default by about 1 week of trading. If we follow our interpretation, insiders must have initiated a massive sellout of Russian Federation bonds 8–9 months before the default and a massive buyout of Russian bonds immediately after the default. We can estimate the profitability of this strategy, for instance, by using 10% Minfin (Bond 7) as the benchmark. Its price on 09/15/97 was $103.25, while on 08/27/98 it was $19.25. Our hypothetical insiders, who sold 10% Minfin in September 1997, converted the proceeds into T-bills, picked up the same bond after the default, and subsequently held it to maturity in June 2007, would receive in excess of 500% return on their initial investment. In practice, insiders would not need to wait until 2007. When the information leaks back to the market, they can realize the profits immediately by shorting requisite number of calls on their portfolio of the Russian bonds at a strike price close to par.*

Because we placed two approximate time stamps (±16 trading days, see the previous section) on the largest one-sided changes in insider sentiment, we can hypothesize the signals the insiders received. Russian default was not accompanied by widespread political turmoil unlike the more recent default of Argentina (2001). The Russian Government did not propose a forced loan restructuring program and did not threaten repudiation of its debt, again in contrast to the Argentinean case. It quickly renegotiated its debt covenants with its main foreign creditors to relieve the immediate pressure on the state budget (Gaidar, 2003). This can be considered a liquidity-enhancing event in accord with an observed positive change of sentiment.

In search of the event that triggered massive insider sellout of the Russian bonds around September 1997, we can point to the International Monetary Fund (IMF) meeting in Singapore. What information pertinent to Russia could have been discussed at the meeting, which dealt with

---

* Before that leakage occurs, these calls would be severely out-of-the-money.

the Asian financial crisis? Very likely, the insider impression of the drain of IMF resources and the reluctance of the U.S. Congress to recapitalize IMF formed the basis for the widespread pessimism toward all emerging markets (Rubin and Weisberg, 2004). We thus approach the subject of "financial contagion," which lies far outside the topic of this chapter. However, our conclusion points to an intriguing subject, namely, that financial contagion of economically unrelated markets may be driven by the activity (or lack thereof) of regulatory bodies rather than by spontaneous actions of investors.

ACKNOWLEDGMENTS

We thank participants in the finance seminar at Syracuse University, Elin Tully, Discussant, and Brian Lucey, Chair, at the Global Finance Conference (Dublin, 2005) for valuable suggestions. Our special thanks go to K. Yuan (University of Michigan) who provided us with the bond database for emerging markets. P.L. expresses gratitude to Mark Miller (Syracuse University) and Natasha Trofimenko (Kiel Institute for World Economics) for help with SAS and statistical analysis in general. This version has been reworked after suggestions by Y. Hong (Cornell). All errors are our own.

REFERENCES

Ait-Sahalia, Y., P. Mykland, and L. Zhang (2005) How often to sample a continuous-time process in the presence of market microstructure noise. *Review of Financial Studies*, 18(2): 351–415.

Amihud, Y. and H. Mendelson (1991) Liquidity, maturity and the yields on the US treasury bonds. *Journal of Finance*, 46(4): 1411–1425.

Avellaneda, M. and P. Laurence (1999) *Quantitative Modeling of Derivative Securities*. Chapman&Hall/CRC, Boca Raton, FL.

Bauwens, L. and N. Hautsch (2006) Stochastic conditional intensity process. *Journal of Financial Econometrics*, 4(3): 450–493.

Bazarkin, K. V. (1998) Future of the debt market for foreign investors. In: V. L. Kvint and J. R. Gallus (ed.), *Emerging Market of Russia*. John Wiley, New York.

Bollen, N. P. B., T. Smith, and R. Whaley (2004) Modeling the bid/ask spread: Measuring the inventory-holding premium. *Journal of Financial Economics*, 72(1): 97–141.

Borodin, A. N. and P. Salminen (2005), *Handbook of Brownian Motion—Facts and Formulae*, Birkhauser, Basel, Switzerland.

Brockman, P. and D. Chung (1999) Bid–ask spread components in an order-driven environment. *Journal of Financial Research*, 22(2): 227–246.

Broock, W. A., W. D. Dechert, J. A. Sheinkman, and B. LeBaron (1996) A test for independence based on the correlation dimension. *Econometric Reviews*, 15(3): 197–235.

Green, C. T. and E. Elton (1998) Tax and liquidity effects in pricing government bonds. *Journal of Finance*, 53(5): 1533–1562.

Dittmar, R. and K. Yuan (2005) Pricing impact of sovereign bonds. University of Michigan Working Paper, Ann Arbor, MI.

Duffie, D., L. H. Pedersen, and K. J. Singleton (2003) Modeling sovereign bond yields: A case study of Russian debt. *Journal of Finance*, 58(1): 119–160.

Fabozzi, F. and E. Pilarinu (eds.) (2002) *Investing in Emerging Fixed Income Markets.* Wiley, New York.

Gaidar, E. T. (ed.) (2003) *The Economics of Russian Transition.* MIT Press, Cambridge, MA.

George, T. J., G. Kaul, and M. Nimalendran (1991) Estimation of bid–ask spread and its components: A new approach. *Review of Financial Studies*, 4(4): 623–656.

Glosten, L. and L. Harris (1988) Estimating the components of the bid–ask spread. *Journal of Financial Economics*, 21(1): 123–142.

Glosten, L. and P. Milgrom (1985) Bid, ask and transaction prices in a specialist market with heterogeneously informed traders. *Journal of Financial Economics*, 14(1): 71–100.

Greene, W. H. (2000) *Econometric Analysis.* Prentice Hall, Upper Saddle River, NJ.

He, Y. and C. Wu (2005) The effects of decimalization on return volatility components, Serial correlation and trading costs. *Journal of Financial Research*, 28(1): 77–96.

Huang, R. D. and H. R. Stoll (1997) The components of the bid–ask spread: A general approach. *Review of Financial Studies*, 10(4): 995–1034.

Hull, J. C. (1997) *Options, Futures and Other Derivatives.* Prentice Hall, Upper Saddle River, NJ.

Kanzler, L. (1999) Very fast and correctly sized estimation of the BDS statistic. Working Paper, Oxford University, Oxon, U.K.

Karatzas, J. and S. Shreve (1997) *Brownian Motion and Stochastic Calculus.* Springer, New York.

Kyle, A. S. (1985) Continuous auctions and insider trading. *Econometrica*, 53(5): 1315–1336.

Lerner, P. (2007) Review of several hypotheses in market microstructure. Working Paper, Syracuse University, Syracuse, NY.

Mallat, S. (1999) *A Wavelet Tour of Signal Processing.* Academic Press, San Diego, CA.

Mendelson, H. and T. Tunca (2004) Strategic trading, liquidity and information acquisition. *Review of Financial Studies*, 17(2): 295–337.

Menyah, K. and K. Paudyal (2000) The components of bid–ask spreads on the London Stock Exchange. *Journal of Banking and Finance*, 24(11): 1767–1785.

O'Hara, M. (1995) *Market Microstructure.* Blackwell, Boston, MA.

Percival, D. P. and A. T. Walden (2000) *Wavelet Methods for Time Series Analysis.* Cambridge University Press, Cambridge, U.K.

Porter, D. and D. Weaver (1996) Estimating bid–ask spread components: Special versus multiple market maker system. *Review of Quantitative Finance and Accounting*, 6(2): 167–180.

Rubin R. and J. Weisberg (2004) *In an Uncertain World*. Random House, New York.

Russell, J. R. (1999) Econometric modeling of multivariate irregularly-spaced high-frequency data. Working Paper, Chicago University, Chicago, IL.

Stoll, H. (1989) Inferring the components of the bid–ask spread: Theory and empirical tests. *Journal of Finance*, 44(1): 115–134.

Stoll, H. R. (2003) Market microstructure. In G. M. Constantinides, M. Harris, and R. Stultz (eds.), *Handbook of the Economics of Finance*, Vol. 1A, Elsevier, North Holland, the Netherlands.

Strother, T. S., J. Wansley, and P. Daves (2002) The impact of electronic communications networks on the bid–ask spread. Working Paper. University of Tennessee, Knoxville, TN.

CHAPTER 28

Reaction of Fixed-Income Security Investors to Extreme Events: Evidence from Emerging Markets

Spyros Spyrou

CONTENTS

28.1 INTRODUCTION

In capital markets where investors are rational and try to maximize expected utility, asset prices will be efficient with respect to information and will incorporate information accurately and quickly; as a result price changes should not be predictable. The results of many studies, however, indicate that investors may react in a nonrational way and that this behavior leads to predictable patterns in asset returns. For instance, DeBondt and Thaler (1985) find that returns are predictable based on historical price information: prior losers (winners) become winners (losers) in the subsequent period and as a result contrarian strategies that are long in prior losers and short in prior winners consistently make long-term abnormal profits. Other studies show that these strategies are profitable for short-term horizons as well (see, among others, Jegadeesh, 1990; Lehman, 1990; and Bremer and Sweney, 1991). DeBondt and Thaler argue that this behavior is consistent with investor overreaction to information. In addition, the results of studies for the medium-term indicate that prior losers (winners) remain losers (winners) in the subsequent period, and as a result momentum strategies that are long in prior winners and short in prior losers consistently make abnormal profits (see, among others, Jegadeesh and Titman, 1993, 2001; Rouwenhorst, 1998, 1999; and Liew and Vassalou, 2000). The profits of momentum strategies are consistent with investor underreaction to information.

Attempts to explain negative serial correlation and price reversals, from a rational point of view, range from bid–ask biases (see, among others, Roll, 1984; Cox and Peterson, 1994; Jegadeesh and Titman, 1995; and Park, 1995) to multifactor asset pricing models (Fama and French, 1996) and size effects (Zarowin, 1990). Possible explanations for the momentum behavior are analyst coverage, transaction costs, book-to-market effects, size, and trading volume (see, among others, Asness, 1997; Chan et al., 2000; Hong et al., 2000; Lee and Swaminathan, 2000; Hameed and Kusnadi, 2002; Lesmond et al., 2004). Many economists attempt to explain these patterns from a behavioral point of view and discuss several channels through which investor sentiment and psychological biases may lead to inefficiencies in asset returns (see, among others, Lakonishok et al., 1994; Barberis et al., 1998; Daniel et al., 1998; Odean 1998; and Scott et al., 2003). For instance, the model suggested by Barberis et al.

(1998) indicates that the heuristics of representativeness and conservatism will lead investors to overreact to strong and salient information and underreact to information low in weight.

The overreaction–underreaction literature is voluminous and covers many aspects and implications of return behavior. This chapter contributes to the specific strand of literature that examines investor behavior following extreme events. Extreme events (or alternatively price/market shocks) act as a proxy for the arrival of unobservable information to the marketplace. More specifically, the hypothesis of efficient reaction following price shocks is investigated against the alternatives of investor overreaction and underreaction for a sample of nine emerging fixed-income security markets from different geographical regions (Argentina, Brazil, Chile, Thailand, Hong Kong, Singapore, Poland, Hungary, and Bulgaria). The vast majority of the empirical studies on asset return behavior deal with inefficiencies in equity returns; comparatively, few studies focus on bond markets despite the size and significance of fixed-income security markets. This point is even stronger for emerging bond markets: most emerging market empirical studies focus on equities rather than bonds. Thus, this chapter aims to partially fill this gap in the literature and examine investor reaction to information in emerging bond markets. Although it has been argued that fixed-income security markets may be less biased as opposed to stock markets (see Khang and King, 2004) there is evidence to suggest that return predictability exists in bond markets as well (e.g., Cutler et al., 1991). In addition, the multicountry analysis may offer important insight and help establish whether there is a cross-country pattern in securities' behavior (Fama and French, 1996). One could expect higher return predictability in emerging markets: for instance, common characteristics of emerging markets are biased investors, thin trading, and low liquidity (Dabbs et al., 1991); local information may play an important role in the determination of asset returns (Harvey, 1995); asset returns are often more predictable and have higher and not symmetric volatility, non-normal distributional characteristics along with time-varying kurtosis and skewness (see, among others, Claessens et al. 1995 and Bekaert et al., 1998).

In a previous study of investor reaction to market shocks, Schnusenberg and Madura (2001) examined the short-term reaction for six U.S. equity indexes and report one-day underreaction following positive and negative market shocks (i.e., days on which an index experiences abnormally high or low returns). They also find significant reversals over longer periods after negative shocks, a result consistent with the notion that prices react stronger to bad news rather than good news (see also, Brown et al., 1988). In a related paper, Lasfer et al. (2003) employ a similar methodology and investigate this issue for 39 international equity markets. Their results suggest that, on average, positive (negative) shocks are followed by subsequent large positive (negative) abnormal returns in both developed and emerging markets; this evidence is consistent with the short-term underreaction hypothesis. In a recent study, Kassimatis et al. (2008) examine 17 international bond markets for the 1989–2004 period and find a delayed reversal in government bond prices. This pattern is stronger after negative shocks, persists for all counties, and, furthermore, simulated trading strategies suggest that it is also economically significant.

To anticipate the results, investors in fixed-income securities of Asian markets seem to react efficiently to the arrival of unobservable information to the marketplace and incorporate all information in bond prices quickly and accurately. No predictable patterns seem to take place the following days. A similar result holds for the three European markets of the sample, with the exception of Poland where investors seem to underreact for a day for positive price shocks and underreact for at least 20 days for negative price shocks. For the Latin American markets of the sample, the situation is more complex: investors react efficiently in Chile when it comes to positive shocks but seem to overreact and subsequently reverse their behavior when it comes to negative shocks; investors in Argentina seem to react efficiently to negative shocks but underreact to positive shocks; investors in Brazil tend to underreact to both positive and negative price shocks. Furthermore, the subsequent abnormal returns are strongly related to the event day reaction; i.e., the higher the event day reaction the higher the momentum during the following days. The rest of this chapter is organized as follows: Section 28.2 discusses the data and the testing methodologies; Section 28.3 presents the results; Section 28.4 investigates whether the subsequent abnormal returns are related to the event day shock; and Section 28.5 concludes this chapter.

28.2 DATA, METHODOLOGY, AND HYPOTHESES

For the empirical analysis, daily clean prices on government bond portfolios are employed for a sample of nine emerging markets. For comparability of the results, sample markets from three different geographical regions are chosen: three Latin American markets (Argentina, Brazil, and Chile), three Asian markets (Hong Kong, Singapore, and Thailand), and three European markets (Bulgaria, Hungary, and Poland). The portfolios used in the analysis are the J.P. Morgan total return indexes for the aforementioned countries and the sample begins on 23/6/1998 and ends on 23/6/2008, covering 10 years and providing 2616 daily observations for each market. Note that the sample period includes important events that triggered to global financial volatility, such as the recent subprime loans crisis in the United States, 9/11, 2001 attacks, the Internet stock bubble incident with the collapse of the NASDAQ index in 2000, the Russian currency devaluation in 1998, etc. One important issue with bond market studies is the quality of available data. For example, fixed-income securities typically do not trade as often as other assets (stocks, exchange rates, etc.) and this may present problems due to illiquidity. That is, stale prices and illiquidity may appear in bond returns as positive serial correlation and indicate a short-term return momentum. As regards to the data set in this chapter, both the construction of the indexes and an examination for runs in the data suggest that these issues are not likely to influence the results.

In order to investigate bond investor behavior following extreme events we first have to define when an extreme event takes place. Some previous studies use absolute price changes either on a daily basis or on a weekly basis or during a certain time period. For instance, Atkins and Dyl (1990) employ the largest price change in a 300 day window, Bremer and Sweeney (1991) use a 10% daily price change, Howe (1986) employs a weekly price change of more than 50%, as a proxy for the extreme event day. Using an absolute price change of a certain percent, however, may lead to biased results since assets have different volatilities. For instance, a certain percent daily price change may be an extreme observation for an asset with low volatility but may be less significant for an asset with high volatility.

To overcome this issue, in this chapter, a different methodology is employed. More specifically, a positive (negative) price shock is assumed to occur for a market when this market's bond index return at any given day is above (below) two standard deviations the average daily return computed over the [−60 to −11] days before the given day. Note that this methodology is also employed by Lasfer et al. (2003) in a study of a large number of international equity markets. The window is chosen as to avoid unusual price lead-up prior to the price shocks. The expected return and the standard deviation for day t is also computed from the observations between day $t - 60$ and day $t - 11$. Note also that this specification accounts for time-varying risk premia, which could cause serial correlation in returns (see Ball and Kothari, 1989). Positive and negative price shocks are examined separately since previous evidence indicates that prices react stronger to bad news rather than good news (see aforementioned discussion).

Once an event day is identified, we calculate the post event abnormal return (AR) as

$$AR_{i,t} = R_{i,t} - E(R_{i,t}) \qquad (28.1)$$

where
   $R_{i,t}$ is the return of country's i bond index on day t
   $E(R_{i,t})$ is the average return of the 50 day window ending 10 trading days prior to the price shock

We then compute the abnormal return for the first day following the price shock and the cumulative abnormal returns (CARs) for each portfolio and for each event for a trading month (i.e., 20 days) following the shock ($t + 1$ until $t + 20$). Once this is done the average cumulative abnormal returns (ACARs) are estimated for each market and each shock.

As discussed in Section 28.1, if emerging bond market investors react efficiently, all information related to the extreme event should be incorporated in the price reaction on the event day; thus, no statistically significant abnormal returns (ACARs) should be observed on the following days. If, however, emerging market bond investors overreact to the information related to the extreme event and then reverse their behavior, we should observe statistically significant ACARs on the opposite direction during the following days. That is, for positive price shocks the following ACARs should be negative and statistically significant and for negative price shocks the following ACARs should be positive and statistically significant. Similarly if emerging market bond investors underreact to the information related to the extreme event (i.e., react with a drift), we should observe statistically significant ACARs on the same direction during the following days. That is, for positive price shocks, the following ACARs should be positive and statistically significant and for negative price shocks, the following ACARs should be negative and statistically significant.

28.3 RESULTS

Tables 28.1 through 28.3 present the results for the three Latin American, Asian, European markets, respectively. Panel A in each table presents results for positive shocks and Panel B presents results for negative shocks. We examine positive and negative shocks separately in order to investigate whether investors react differently to each type of shock. For each

TABLE 28.1   Market Reaction to Bond Price Shocks for Latin American Markets

	Argentina		Brazil		Chile	
	ACAR	t-Test	ACAR	t-Test	ACAR	t-Test
Panel A: Reaction to positive shocks						
N	20		19		42	
AAR-0	0.020576	2.51*	0.016948	5.78*	0.000974	4.27*
AAR-1	0.003973	1.84**	-0.00106	-0.30	0.000135	0.75
ACAR-2	0.002982	2.00*	0.000307	0.14	-0.00015	-1.18
ACAR-3	0.005758	3.28*	0.002838	1.54	-0.00025	-1.42
ACAR-4	0.002557	0.67	0.005025	1.89**	-0.00022	-1.24
ACAR-5	0.009983	3.46*	0.008805	2.31*	-0.00017	-0.84
ACAR-10	0.013997	1.70**	0.002125	0.36	-0.00024	-0.75
ACAR-15	0.023786	1.49	0.000895	0.11	-0.00024	-0.54
ACAR-20	0.022887	1.81**	-0.00231	-0.19	-0.00013	-0.23
Panel B: Reaction to negative shocks						
N	36		34		17	
AAR-0	-0.02409	-2.60*	-0.01947	-7.69*	-0.00133	-3.73*
AAR-1	-1.4E-05	-0.00	-0.00668	-3.13*	0.000593	1.68
ACAR-2	0.004997	1.11	-0.00397	-1.4	0.000888	2.49*
ACAR-3	0.007165	1.30	-0.00443	-1.04	0.000822	2.35*
ACAR-4	-0.00109	-0.42	-0.00525	-0.90	0.000697	1.71**
ACAR-5	-0.00134	-0.52	-0.00693	-1.11	0.000697	1.59
ACAR-10	0.005677	1.13	-0.00818	-1.33	0.000819	1.67**
ACAR-15	-0.00261	-0.59	-0.00224	-0.26	0.001097	2.06*
ACAR-20	0.00029	0.06	0.001556	0.14	0.001305	2.28*

Notes: AAR-0 is the average abnormal return on the extreme event day; AAR-1 is the average abnormal return on the first day after the shock; $ACAR_{t+i}$ is the average cumulative abnormal return over i days after an even ($i = 2$–20). N denotes the number of shocks observed for each time series: for example, for positive shocks, $N = 20$ suggest that 20 extreme positive events occurred during the sample period (2616 observations). The sample begins on 23/6/1998 and ends on 23/6/2008, covering 10 years.
* denotes significance at the 5%; ** denotes significance at the 10%.

market we report the average abnormal return on the event day (AAR-0), the average abnormal return on the following day (AAR-1), the ACAR from day 2 until day 20 (ACAR-2 to ACAR-20), and the corresponding t-statistics. Also, the number of shocks (N) observed for each time series is reported for both positive and negative shocks. For example, N = 20 suggests that 20 extreme events occurred during the sample period.

For the Latin American markets (Table 28.1) and positive shocks (Panel A), we observe that the average reaction to a positive extreme

TABLE 28.2   Market Reaction to Bond Price Shocks for Asian Markets

	Thailand		Singapore		Hong Kong	
	ACAR	t-Test	ACAR	t-Test	ACAR	t-Test
Panel A: Reaction to positive shocks						
N	34		36		35	
AAR-0	0.009278	10.75*	0.006186	13.43*	0.001039	6.96*
AAR-1	0.000531	0.75	−0.00066	−1.82**	−2.2E-05	−0.11
ACAR-2	0.001279	1.10	−0.00019	−0.25	−0.00019	−1.11
ACAR-3	0.002003	1.10	0.00054	0.57	−0.00031	−1.61
ACAR-4	0.001674	0.83	0.00087	0.96	−0.00044	−1.93**
ACAR-5	0.000459	0.38	0.001071	0.98	−0.00044	−1.76**
ACAR-10	0.002402	1.11	0.000561	0.28	−0.0003	−0.69
ACAR-15	0.000849	0.24	0.000254	0.08	−0.0004	−1.02
ACAR-20	0.000703	0.17	−0.00105	−0.32	−0.00034	−0.94
Panel B: Reaction to negative shocks						
N	31		32		14	
AAR-0	−0.01242	−9.48*	−0.00655	−15.52*	−0.00071	−5.71*
AAR-1	0.000335	0.26	−0.00024	−0.53	0.000101	1.04
ACAR-2	−0.00039	−0.24	0.000347	0.52	8.7E-05	0.41
ACAR-3	−0.00023	−0.11	0.000309	0.41	−0.00011	−0.46
ACAR-4	0.000536	0.24	0.000107	0.10	−0.00013	−0.52
ACAR-5	−0.00059	−0.29	0.000706	0.50	−0.00021	−0.91
ACAR-10	−0.00206	−0.57	0.00031	0.17	−0.00029	−0.77
ACAR-15	−0.00209	−0.51	0.000679	0.31	−0.0004	−0.83
ACAR-20	−0.00373	−0.73	0.000202	0.07	−0.00035	−0.63

Notes: AAR-0 is the average abnormal return on the extreme event day; AAR-1 is the average abnormal return on the first day after the shock; $ACAR_{t+i}$ is the average cumulative abnormal return over i days after an even ($i = 2$–20). N denotes the number of shocks observed for each time series: for example, for positive shocks, $N = 20$ suggest that 20 extreme positive events occurred during the sample period (2616 observations). The sample begins on 23/6/1998 and ends on 23/6/2008, covering 10 years.
* denotes significance at the 5%; ** denotes significance at the 10%.

event varies from about 2% (Argentina) and 1.69% (Brazil) to about 0.09% (Chile), and is statistically significant. Furthermore, for the Chilean market, the subsequent abnormal returns are all about 0.00% and not statistically significant for all days up to day 20, indicating an informationally efficient reaction to price shocks.

TABLE 28.3   Market Reaction to Bond Price Shocks for European Markets

	Bulgaria		Hungary		Poland	
	ACAR	t-Test	ACAR	t-Test	ACAR	t-Test
Panel A: Reaction to positive shocks						
N	36		32		31	
AAR-0	0.012966	5.22*	0.016474	4.40*	0.015126	18.68*
AAR-1	0.00291	1.00	0.002449	0.64	0.003096	2.24*
ACAR-2	0.002781	1.21	0.003517	0.90	0.002538	1.22
ACAR-3	0.001663	0.63	0.002636	0.65	0.001036	0.48
ACAR-4	0.003383	1.06	0.002214	0.53	0.001333	0.49
ACAR-5	0.002545	0.75	0.001831	0.43	0.000933	0.29
ACAR-10	0.001437	0.29	0.00545	1.15	0.003339	0.79
ACAR-15	0.002238	0.35	0.009948	1.88**	0.005027	0.99
ACAR-20	0.004452	0.62	0.009581	1.55	0.006775	1.26
Panel B: Reaction to negative shocks						
N	34		37		34	
AAR-0	−0.01084	−7.18*	−0.01723	−12.51*	−0.01594	−16.53*
AAR-1	−0.00231	−1.27	−0.00213	−1.65	−0.00385	−2.30*
ACAR-2	−0.00191	−1.70	−0.00256	−1.33	−0.00755	−4.14*
ACAR-3	−0.00086	−0.66	−0.00254	−1.08	−0.00638	−2.62*
ACAR-4	−0.001	−0.56	−0.00499	−1.83**	−0.00758	−3.00*
ACAR-5	−0.00191	−0.96	−0.0045	−1.29	−0.00769	−2.73*
ACAR-10	−0.00316	−0.96	0.00039	0.09	−0.00858	−2.43*
ACAR-15	−0.00088	−0.19	−0.00011	−0.02	−0.01062	−2.27*
ACAR-20	0.000715	0.14	−0.00014	−0.02	−0.01604	−2.92*

Notes: AAR-0 is the average abnormal return on the extreme event day; AAR-1 is the average abnormal return on the first day after the shock; $ACAR_{t+i}$ is the average cumulative abnormal return over i days after an even ($i = 2$–20). N denotes the number of shocks observed for each time series: for example, for positive shocks, $N = 20$ suggest that 20 extreme positive events occurred during the sample period (2616 observations). The sample begins on 23/6/1998 and ends on 23/6/2008, covering 10 years.
* denotes significance at the 5%; ** denotes significance at the 10%.

For Argentina and Brazil, the picture is different: the relatively high (compared to Chile) event day reaction is followed by statistically significant ACARs of the same sign for about 5 days. More specifically, for Argentina, a further 1% approximately (0.99%) is added to the initial 2.057% reaction by day 5 (see also Figure 28.1). This abnormal return is also statistically significant (t-statistic = 3.46). This grows to 2.28% by day 20 (significant

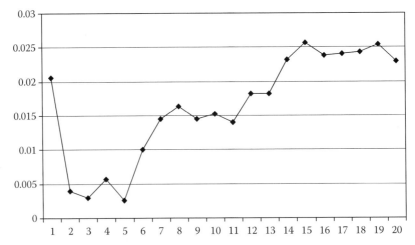

FIGURE 28.1 Market reaction to positive shocks: Argentina. AAR–0 is included in the graph.

at the 10%). For Brazil, a further 0.88% is added to the initial 1.69% reaction by day 5 and this is also statistically significant (*t*-statistic = 2.31). This momentum behavior indicates that bond investors in Argentina and Brazil tend to underreact to positive market-moving information for at least 5 days after the event.

For negative shocks (Panel B), we observe that the average reaction to a negative extreme event varies from about –2.4% (Argentina) and –1.94% (Brazil) to about –0.13% (Chile) and is statistically significant. However, the subsequent ACARs for the Argentinean and Brazilian bond portfolios are all very low and statistically indistinguishable from zero, with the exception of the first day return for Brazil (see Figure 28.2). For Chile, the subsequent ACARs are statistically significant for many days up to day 20 and of the opposite sign, indicating an initial overreaction and a subsequent reversal (see also Figure 28.3). In fact, 20 days after the initial negative shock the ACAR is 0.13% (*t*-statistic = 2.28%) indicating a complete reversal of the initial –0.13% reaction on day 0.

For the Asian markets (Table 28.2) and positive shocks (Panel A), we observe that the average reaction to a positive extreme event varies from about 0.92% (Thailand) and 0.61% (Singapore) to about 0.10% (Hong Kong) and is statistically significant. Note that the reaction to positive price shocks is much lower than the corresponding reaction for Latin American markets. The subsequent abnormal returns are all about 0.00% and not statistically significant at the 5% level of significance for all days

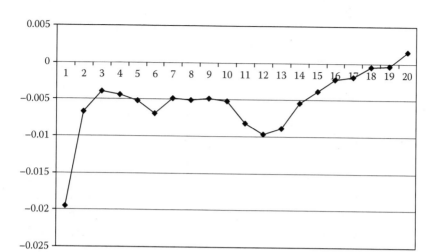

FIGURE 28.2 Market reaction to negative shocks: Brazil. AAR–0 is included in the graph.

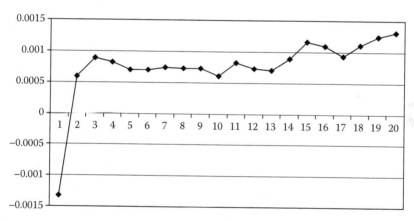

FIGURE 28.3 Market reaction to negative shocks: Chile. AAR–0 is included in the graph.

up to day 20 for all markets, indicating an informationally efficient reaction to positive price shocks. For negative shocks (Panel B) we observe that the average reaction to a negative extreme event varies from about −1.2% (Thailand) and −0.65% (Singapore) to about −0.07% (Hong Kong) and is statistically significant. As mentioned earlier, for all markets, the subsequent abnormal returns are all about 0.00% and not statistically significant for all days.

For the European markets (Table 28.3) and positive shocks (Panel A), we observe that the average reaction to a positive extreme event varies from about 1.64% (Hungary) and 1.51% (Poland) to about 1.29% (Bulgaria) and is statistically significant. With the exception of the first day abnormal return for Poland (0.30%, t-statistic = 2.24) all subsequent ACARs are small and statistically not significant at the 5% level of significance. For negative shocks (Panel B), we observe that the average reaction to a negative extreme event varies from about −1.73% (Hungary) and −1.59% (Poland) to about 1.08% (Bulgaria) and is statistically significant. However, the subsequent ACARs for the Bulgarian and Hungarian bond portfolios are all very low and statistically indistinguishable from zero at the 5% level of significance. For Poland, the subsequent ACARs are statistically significant for many days up to day 20 and of the same sign, indicating investor underreaction and momentum. In fact, 20 days after the initial negative shock the ACAR is −1.60% (t-statistic = −2.92%) nearly the same as the initial −1.59% reaction on day 0 (see also Figure 28.4).

Overall, the results seem to suggest that investors in fixed-income securities of Asian markets react efficiently to the arrival of unobservable information to the marketplace and incorporate all information in bond prices quickly and accurately. No predictable patterns seem to take place in the following days. A similar result holds for the three European markets of the sample, with the exception of Poland where investors seem to underreact for a day for positive price shocks and underreact for at least 20 days

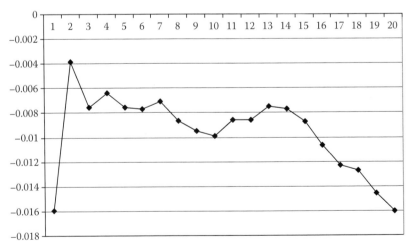

FIGURE 28.4 Market reaction to negative shocks: Poland. AAR–0 is included in the graph.

for negative price shocks. For the Latin American markets of the sample, the situation is more complex: investors react efficiently in Chile when it comes to positive shocks but seem to overreact and subsequently reverse their behavior when it comes to negative shocks; investors in Argentina seem to react efficiently to negative shocks but underreact to positive shocks; investors in Brazil tend to underreact to both positive and negative price shocks.

28.4 ARE THE SUBSEQUENT ABNORMAL RETURNS RELATED TO THE EVENT DAY SHOCK?

One important issue is whether the size of the subsequent abnormal returns is related to the event day reaction. Thus, in order to shed some light on this issue, the pooled cross-section of subsequent ACARs is regressed on the event day average abnormal return (i.e., on AAR–0):

$$ACAR_{t+i} = a + bAAR_t \qquad (28.2)$$

where

$ACAR_{t+i}$ is the average cumulative abnormal return at day $t + i$
AAR_t is the abnormal return on the event day t

Note that the $ACAR_{t+i}$ does not include AAR_t, thus, the two variables are independent. If the slope coefficient (b) is statistically significant then the size of abnormal cumulative returns is related to the event day reaction. In other words, a positive coefficient indicates that the higher the initial shock the higher the abnormal return for the days following the event and visa versa.

Table 28.4 presents the results for the pooled series (i.e., for all the markets where an inefficient reaction has been documented in the previous section) for 1, 5, 10, and 20 day ACARs. Note that for positive shocks (Panel A), all coefficients are statistically significant, indicating that the subsequent ACARs are strongly related to the event day reaction. Furthermore, the R^2 of the regressions is very high and around 0.62 to 0.64, indicating a high explanatory power. In other words, the higher the event day reaction the higher the momentum during the following days, as for positive events we have documented a momentum effect in the previous section. For negative shocks (Panel B), this is true only for the first day after the event; the R^2 is also very low with the exception of day 1 ($R^2 = 0.25$). There is also a negative coefficient for the 20 day ACAR, indicating a reversal, which is due to the documented 20 day reversal to negative shocks in Chile.

TABLE 28.4   Are CARs Related to the Event Day Abnormal Return?

Dependent Variable	α	t-Statistic	b	t-Statistic	R^2
Panel A: Positive shocks					
AAR–1	−0.0102*	−3.80	1.04*	10.49	0.6218
ACAR–5	−0.0083*	−2.99	1.14*	10.95	0.6416
ACAR–10	−0.0175*	−4.23	1.68*	10.93	0.6406
ACAR–20	−0.0252*	−4.39	2.29*	10.74	0.6329
Panel B: Negative shocks					
AAR–1	0.0021	1.38	0.42*	5.29	0.2526
ACAR–5	−0.0045	−1.06	0.07	0.34	0.0014
ACAR–10	−0.0100*	−2.25	−0.24	−1.01	0.0122
ACAR–20	−0.0295*	−4.32	−1.66*	−4.56	0.2006

Notes: The regressions are of the form: $ACAR_{t+i} = a + bAAR_t$, where $ACAR_{t+i}$ is the cumulative abnormal return at day $t+i$, and AAR_t is the abnormal return on the event day t.

* denotes significance at the 5%; ** denotes significance at the 10%.

28.5 CONCLUSION

This chapter investigated bond investor behavior following extreme events for a sample of emerging markets. The vast majority of earlier empirical studies on asset return behavior examine inefficiencies in equity returns; few studies focus on bond markets and even fewer on emerging bond markets. The null hypothesis is that of market efficiency against the alternatives of investor overreaction and/or underreaction. The findings indicate that investors in many markets, mainly Latin American markets and certain European markets, tend to underreact to the information contained in price shocks. Furthermore, the subsequent abnormal returns are strongly related to the event day reaction, i.e., the higher the event day reaction the higher the momentum during the following days.

This behavior produces momentum (or, in the case of negative shocks in Chile, reversals) that may also be economically significant. Consider, for example, investor reaction to positive shocks in Argentina: on average, a further 2.28% abnormal return is added to the initial 2.05% reaction by day 20. On an annual basis (assuming 256 trading days), this translates to a 30.7% abnormal return. In other words, a trader who follows a strategy of going long to the underlying portfolio for 20 days after a positive price shock and closes the position on day 20 could earn an annualized return of about 30% approximately. Similarly, consider investor reaction to negative shocks in Brazil: a further 0.66% decline is observed on the first day following a negative shock. Annualized this abnormal return is approximately 170%. In other words, a trader who employs a strategy of opening a short position at close on the negative shock day in Brazil and closing the position at close on the following day could earn an annualized abnormal return of about 170%. Of course, these returns are theoretical and do not take into account transaction costs and other trading restrictions (e.g., liquidity, availability of bond issues, short sale restrictions, etc.); nevertheless, they strongly indicate that bond return predictability may be to a large extent exploitable by professional investors.

The finding of short-term underreaction is also consistent with the results of previous studies on price behavior following price shocks. Recall that Schnusenberg and Madura (2001) report underreaction following positive and negative market shocks for six U.S. equity indexes and Lasfer et al. (2003) who examine 39 international equity markets also find that, on average, positive (negative) shocks are followed by subsequent large positive (negative) abnormal returns in both developed and emerging equity markets, i.e., short-term momentum. Finally, and this is another interesting finding: the multicountry comparative analysis indicated that bond return predictability with respect to extreme events is not a cross-country pattern; bond prices in Asian markets incorporate all information contained in market shocks quickly and accurately, i.e., within the event day. No predictability is observed on the following days.

REFERENCES

Asness, C. (1997). The interaction of value and momentum strategies. *Financial Analyst Journal*, 53(2): 29–36.

Atkins, A. B. and Dyl, E. (1990). Price reversals, bid–ask spreads and market efficiency. *Journal of Financial and Quantitative Analysis*, 25(4): 535–547.

Ball, R. and Kothari, S. P. (1989). Nonstationary expected returns: Implications for tests of market efficiency and serial correlations in returns. *Journal of Financial Economics*, 25(1): 51–74.

Barberis, N., Schleifer, A., and Vishny, R. (1998). A model of investor sentiment. *Journal of Financial Economics*, 49(3): 307–343.

Bekaert, G., Erb, C., Harvey, C. R., and Viskanta, T. E. (1998). Distributional characteristics of emerging market returns and asset allocation. *Journal of Portfolio Management*, 24(2): 102–116.

Bremer, M. A. and Sweeney, R. J. (1991). The reversal of large stock-price decreases. *Journal of Finance*, 46(2): 747–754.

Brown, K. C., Harlow, W. V., and Tinic, M. C. (1988). Risk aversion, uncertain information, and market efficiency. *Journal of Financial Economics*, 22(4): 355–385.

Chan, K., Hameed, A., and Tong, W. (2000). Profitability of momentum strategies in the international equity markets. *Journal of Financial and Quantitative Analysis*, 35(2): 153–174.

Claessens S., Dasgupta, S., and Glen, J. (1995). Return behaviour in emerging stock markets. *World Bank Economic Review*, 9(1): 131–51.

Cox, D. R. and Peterson, D. R. (1994). Stock returns following large one-day declines: Evidence on short-term reversals and long-term performance. *Journal of Finance*, 49(1): 255–267.

Cutler, D. P., Poterba, J., and Summers, L. (1991). Speculative dynamics. *Review of Economic Studies*, 58(3): 529–546.

Dabbs, R. E., Smith, K. L., and Brocatto, J. (1991). Test on the rationality of professional business forecasters with changing forecast horizons. *Quarterly Journal of Business and Economics*, 30(2): 28–35.

Daniel, K., Hirshleifer, D., and Subrahmanyam, A. (1998). A theory of overconfidence, self-attribution and security market under- and overreactions. *Journal of Finance*, 53(6): 1839–1886.

DeBondt, W. F. M. and Thaler, R.H. (1985). Does the stock market overreact? *Journal of Finance*, 40(3): 793–805.

Fama, E. F. and French, K. R. (1996). Multifactor explanations of asset pricing anomalies. *Journal of Finance*, 51(1) 55–84.

Hameed, A. and Kusnadi, Y. (2002). Momentum strategies: Evidence from Pacific basin stock markets. *Journal of Financial Research*, 25(3): 383–397.

Harvey, C. R. (1995). Predictable risk and returns in emerging markets. *Review of Financial Studies*, 8(3):773–816.

Hong, H., Lim, T., and Stein, J. C. (2000). Bad news travel slowly: Size, analyst coverage and the profitability of momentum strategies. *Journal of Finance*. 55(1): 265–295.

Howe, J. S. (1986). Evidence on stock market overreaction. *Financial Analysts Journal*, 42(4): 74–77.

Jegadeesh, N. (1990). Evidence of predictable behaviour of security returns. *Journal of Finance*, 45(3): 881–898.

Jegadeesh, N. and Titman, S. (1993). Returns to buying winners and selling losers: Implications for stock market efficiency. *Journal of Finance*, 48(1): 65–91.

Jegadeesh, N. and Titman, S. (1995). Short term return reversals and the bid–ask spread. *Journal of Financial Intermediation*, 4(2): 116–132.

Jegadeesh, N. and Titman, S. (2001). Profitability of momentum strategies: An evaluation of alternative explanations. *Journal of Finance*, 56(2): 699–720.

Kassimatis, K., Spyrou, S., and Galariotis, E. (2008). Short-term patterns in government bond returns following market shocks: International evidence. *International Review of Financial Analysis*, 17(5): 903–924.

Khang, K. and King, T.-H. (2004). Return reversals in bond market: Evidence and causes. *Journal of Banking and Finance*, 28(3): 569–593.

Lakonishok, J., Shleifer, A., and Vishny, R. W. (1994). Contrarian investment, Extrapolation and risk. *Journal of Finance*, 49(5): 1541–1578.

Lasfer, M. A., Melnik, A., and Thomas, D. C. (2003). Short-term reaction of stock markets in stressful circumstances. *Journal of Banking and Finance*, 27(10): 1959–1977.

Lee, C. M. C. and Swaminathan, B. (2000), Price momentum and trading volume. *Journal of Finance*, 55(5): 2017–2069.

Lehman, B. (1990). Fads, martingales and market efficiency. *Quarterly Journal of Economics*, 105(1): 1–28.

Lesmond, D. A., Schill, M. J., and Zhou, C. (2004). The illusory nature of momentum profits. *Journal of Financial Economics*, 71(2): 349–380.

Liew, J. and Vassalou, M. (2000). Can book-to-market, size and momentum be risk factors that predict economic growth? *Journal of Financial Economics*, 57(2): 221–245.

Odean, T. (1998). Volume, volatility, rice and profit when all traders are above average. *Journal of Finance*, 53(6): 1887–1934.

Park, J. (1995). A market microstructure explanation for predictable variations in stock returns following large price changes. *Journal of Financial and Quantitative Analysis*, 30(2): 241–256.

Roll, R. (1984). A simple implicit measure of the effective bid–ask spread in an efficient market. *Journal of Finance*, 39(4): 1127–1139.

Rouwenhorst, K. G. (1998). International momentum strategies. *Journal of Finance*, 53(1): 267–284.

Rouwenhorst, K. G. (1999). Local return factors and turnover in emerging stock markets. *Journal of Finance*, 54(4): 1439–1464.

Schnusenberg, O. and Madura, J. (2001). Do U.S. stock market indexes over- or underreact? *The Journal of Financial Research*, 24(2): 179–204.

Scott, J., Stumpp, M., and Xu, P. (2003). Overconfidence bias in international stock prices: Consistent across countries and trading environments. *Journal of Portfolio Management* 29(Winter): 80–89.

Zarowin, P. (1990). Size, seasonality and stock market overreaction. *Journal of Financial and Quantitative Analysis*, 25(1): 113–126.

CHAPTER 29

Market Liquidity and Investor Sentiment: Evidence from International Closed-End Funds

Paul Brockman and Gary McCormick

CONTENTS

29.1 INTRODUCTION

Investor sentiment has received increasing attention among academics and practitioners over the past decade. Many market analysts point to the sharp rise and precipitous fall of technology stocks from 1998 to 2001 as clear evidence that investor sentiment can affect market prices over a prolonged period. Although academic researchers tend to be more skeptical about the role of sentiment in asset pricing, recent theoretical models assign a central role to noise traders and uninformed investors (Barberis et al., 1998; Daniel et al., 2001). Related work by Shleifer and Vishny (1997) shows that limits to arbitrage can allow market prices to deviate from fundamental values in spite of the efforts of rational investors. Much of this previous research focuses on cross-sectional aspects of investor sentiment (i.e., variation in sentiment across firm characteristics). More recently, Baker and Stein (2002) develop a theoretical model that addresses the time-series properties of market sentiment. Their fundamental claim is that market liquidity is a sentiment indicator; that is, when liquidity is high, uninformed investors are dominant and assets are overvalued.

The purpose of this chapter is to investigate Baker and Stein's (2002) sentiment model using a unique class of assets—international closed-end funds (CEFs). Previous research has shown that CEF prices are highly susceptible to investor sentiment. Lee et al. (1991) find a significant correlation between CEF discounts and the returns on small capitalization stocks. CEF discounts narrow when small firms experience positive returns. Because CEFs and small capitalization stocks are held by individual investors, a direct implication is that uninformed investor misperceptions are positively correlated across assets (De Long et al., 1990). International CEFs are particularly prone to uninformed investor sentiment because these funds are among the smallest and least liquid of all CEF types. In addition, international CEFs tend to hold highly illiquid foreign stocks and bonds with little publicly available information. These attributes make international CEFs an ideal asset class in which to test the proposed relationships between market liquidity, investor sentiment, and asset prices.

Baker and Stein (2002) develop a theoretical model that predicts a positive relationship between market liquidity and asset prices. Although previous studies establish plausible connections between liquidity and expected returns in the cross-section (Amihud and Mendelson, 1986; Vayanos, 1998), the purpose of Baker and Stein's (2002) model is to explain the relation between liquidity and expected returns through time. Underlying their model are two fundamental assumptions: (1) irrational investors underreact to the information content of order flow and (2) short sale constraints prevent information-based arbitrage. "The short-sales constraints imply that irrational investors will only be active in the market when their valuations are higher than those of rational investors—i.e., when their sentiment is positive and when the market is, as a result, overvalued" (Baker and Stein, 2002, p. 33). Their model predicts that while positive investor sentiment leads to high share turnover and overvaluation, negative investor sentiment will have no systematic impact on prices.

We know of no asset class more susceptible to the presence of irrational investors and short sale constraints than international CEFs. In addition, CEFs allow us to measure the market price of the fund relative to the market value of its underlying assets (i.e., the net asset value [NAV]). Previous research shows that CEFs trade at discounts to their NAVs on average. We construct a time series of such discounts by comparing CEF prices with their underlying NAVs through time. If Baker and Stein's (2002) hypothesis is correct, we should find a positive relationship between market liquidity (i.e., share turnover) and changes in CEF discounts.

Our empirical results strongly support the liquidity–discount hypothesis. We examine the liquidity–discount relation using a sample of 71 international CEFs covering the 20-year period 1985 through 2004.* First, we construct an equal-weighted portfolio of all international CEFs and then regress the monthly changes in portfolio discounts against portfolio turnover and a set of control variables including Fama and French's (1992) three-factor portfolios. Our results confirm that there exists a positive and highly significant relationship between liquidity and changes in CEF discounts; that is, as turnover increases, discounts become less negative (or more positive). Following Chay and Trzcinka (1999), we divide our sample into equity and bond CEFs. We find that the positive and significant

---

* These international CEFs are predominantly emerging market CEFs. Sixty of our 71 CEFs are from emerging markets. Restricting our investigation to the 60 emerging market, CEFs provide empirical results that are consistent with those reported herein.

relation between turnover and changes in CEF discounts is driven entirely by the equity funds. These bond-versus-equity results demonstrate that the level of investor sentiment depends on the investment vehicle. Irrational noise traders make up a higher percentage of international equity fund investors than bond fund investors. We find similar results for the emerging markets' subset of CEFs.

Our second set of results is based on fund-by-fund time series regressions. For each CEF, we regress monthly discount changes against the fund's turnover and set of control variables including Fama and French's (1992) three-risk factors and a set of CEF-specific controls as suggested in the previous literature. Similar to our equal-weighted portfolio results, we find a positive and significant relation between liquidity and discount changes on a fund-by-fund basis. We again show that this positive relation is driven by the equity funds only. We find that single-country equity funds exhibit the same patterns as multicountry funds (i.e., noise traders are evenly distributed among equity fund types). Additional analysis demonstrates that liquidity and CEF prices behave as predicted by Baker and Stein (2002) in each subperiod except for 2000–2004. We offer a possible explanation as to why this period might be different.

Overall, our empirical evidence strongly supports the hypothesis that market liquidity acts as an indicator of investor sentiment. Baker and Stein's (2002) theoretical model predicts that (p. 33) "Since the irrational investors tend to make the market more liquid, measures of liquidity provide an indicator of the relative presence or absence of these investors, and hence of the level of prices relative to fundamentals." Our results confirm this prediction for international CEFs.

29.2 DATA AND METHOD OF ANALYSIS

Our sample consists of 71 international CEFs—60 of which are from emerging markets—covering the 20 year period from 1985 through 2004. The raw data include monthly prices, returns, NAVs, and turnovers (i.e., monthly trading volume divided by number of shares outstanding) for each international bond and equity fund. We obtain prices and NAVs for active funds from Lipper. All turnover data are from the Center for Research in Security Prices (CRSP) database. Consistent with prior research, we exclude data from the first 6 months after an initial public offer (IPO) because of the prevalence of underwriter price support activities (Weiss, 1989; Hanley et al., 1996). Previous empirical evidence suggests that CEF discounts stabilize 3–6 months after the IPO.

Our primary variables of interest are the CEF discount

$$\log\left(\frac{\text{CEF Price}_t}{\text{NAV}_t}\right)$$

the change in the CEF discount (ΔDiscount), and the CEF turnover. Our main hypothesis is that increases in market sentiment, as proxied by turnover, will reduce the CEF discount (or increase the CEF premium), all else being equal. Our empirical design includes various control variables that previous studies have been used to explain changes in CEF discounts.

Pontiff (1995) and others find that CEF discounts follow a mean-reverting process. We follow Dimson and Minio-Kozerski (2002) and define the following variable to capture and control for mean reversion:

$$D_{i(j),t-1} - d_{j,t-1} \tag{29.1}$$

where

$D_{i(j),t-1}$ is the equal-weighted average discount of category i, excluding fund j

$d_{j,t-1}$ is the discount for fund j

Dimson and Minio-Kozerski (2002) also show that changes in a fund's discount are related to changes in its peer funds. We define the following variable to control for peer-sector lagged effects:

$$D_{i(j),t} - D_{i(j),t-1} \tag{29.2}$$

where $D_{i(j),t-1}$ is the lagged equal-weighted average discount of category i, excluding fund j. And finally, Dimson and Minio-Kozerski (2002) find that changes in a fund's discount are related to the fund's NAV returns relative to its peer-sector NAV returns. We define the following peer performance-related measure as follows:

$$R_{\text{NAV}i(j),t-1} - R_{\text{NAV}j,t-1} \tag{29.3}$$

where

$R_{\text{NAV }i(j),t-1}$ is the lagged NAV return for the sector excluding fund j

$R_{\text{NAV}j,t-1}$ is the lagged NAV return for fund j

We refer to the CEF-related variables described in Equations 29.1 through 29.3 as Reversion, Sector, and Performance, respectively.

In addition to our CEF-related variables, we control for the January effect through the use of a dummy variable for the month of January. We also use Fama and French's (1992) three-factor model to control for risk-related changes in CEF discounts. These three factors include a market factor (Market), a size-related factor (SMB), and a market-to-book factor high minus low (HML). We expect that one or more of these three factors will control for any risk-related factor that might be correlated with our variable of interest, Turnover.

In our empirical section, we estimate two regression models; one using an aggregate index of CEFs, and a second based on individual CEF time series. In our first model, we create an equal-weighted portfolio of all CEFs and fit the following times series regression:

$$\Delta\log(\text{Discount}_t) = \beta_0 + \beta_1\text{Turnover}_t + \beta_2\text{Market}_t + \beta_3\text{SMB}_t$$
$$+ \beta_4\text{HML}_t + \varepsilon_t \tag{29.4}$$

where $\Delta\log(\text{Discount})$ is the equal-weighted change in discounts across all CEFs in month t and Turnover is the equal-weighted turnover for all CEFs during month t.* The second row of the regression consists of the three Fama and French (1992) risk factors. In this regression, we do not use the additional CEF-related control variables since all CEFs are aggregated into a single index (i.e., we cannot compare CEFs against their peers). Consistent with Baker and Stein's (2002) sentiment hypothesis, we expect to find a positive and significant coefficient for Turnover (i.e., $\beta_1 > 0$).

Our second regression model uses individual CEF time series. For each CEF in our sample, we estimate the following regression model in order to test for the hypothesized relation between changes in CEF discounts and market sentiment (i.e., Turnover):

$$\Delta\log(\text{Discount}_t) = \beta_0 + \beta_1\text{Turnover}_t + \beta_3\text{Reversion}_t + \beta_4\text{Sector}_t$$
$$+ \beta_5\text{Performance}_t + \beta_6\text{JanuaryDummy}_t + \beta_7\text{Market}_t$$
$$+ \beta_8\text{SMB}_t + \beta_9\text{HML}_t + \varepsilon_t \tag{29.5}$$

The third row of model (Equation 29.5) includes the CEF-related control variables suggested in the literature. Our statistical tests are based on

---

* In addition to equal-weighted variables, we rerun all regressions using value-weighted variables. Our conclusions are unaltered using the value-weighted variables.

time-series averages for each of the estimated coefficients. Following Fama and MacBeth (1973), the standard errors are the standard deviations of the coefficient estimates from the individual fund regressions. As in regression model (Equation 29.4), our expectation is that β_1 will be positive and statistically significant; that is, positive market sentiment induces higher market prices.

29.3 EMPIRICAL RESULTS

In Table 29.1, we present summary statistics for our 71 international CEFs. We have a total of 10,094 firm-month observations covering the 20 year period from 1985 to 2004. The mean CEF discount is −0.0668 consistent with previous studies. We find considerable variation in CEF discounts from a low of −0.7647 to a high of 0.8563. The mean change in discount, ΔDiscount, is 0.0002 and there is a relatively large range of values from a low of −0.4221 to a high of 0.8209. The average turnover is 0.7316, with a minimum of 0.0020 and a maximum of 13.7140. Our CEF-related control variables, including Reversion, Sector, and Performance, have mean values of 0.0529, 0, and −0.197, respectively.

29.3.1 Sentiment–Discount Relationship: CEF Portfolios

In Table 29.2, Panel A, we estimate two regression models: the first model includes Turnover only and the second model includes Turnover

TABLE 29.1   International Closed-End Funds—Summary Statistics

	Mean	Min	Q1	Q3	Max	Std
Discount	−0.0668	−0.7647	−0.1722	0.0163	0.8563	0.1522
ΔDiscount	0.0002	−0.4221	−0.0287	0.0277	0.8209	0.0629
Turnover	0.7316	0.0020	0.3799	0.8578	13.7140	0.6343
Mean Rev	0.0529	−1.3542	−0.0162	0.1582	0.5345	0.1609
Sector	0.0000	−0.0006	0.0000	0.0000	0.0005	0.0001
Performance	−0.0197	−1.5167	−0.0937	0.0533	1.2422	0.1628
January	0.0799	0	0	0	1	0.2712
Single entry	0.4388	0	0	1	1	0.4963

Notes: The selected statistics are from monthly observation of 61 from CEFs 1985 to 2004 inclusive (9037 observations). Discount is calculated as the log(Price/NAV). ΔDiscount is the monthly change in discount. Turnover is the monthly volume divided by shares outstanding. Mean Rev is $D_{i(j),t-1} - d_{j,t-1}$ where $D_{i(j),t-1}$ is the equal weight average discount of category i, excluding fund j and $d_{j,t-1}$ is the discount of fund j. Sector is $D_{i(j),t} - D_{i(j),t-1}$. Performance is $R_{i(j),t-1} - r_{j,t-1}$ where $R_{i(j),t-1}$ is the equal weight average NAV return of category i, excluding fund j and $r_{j,t-1}$ is the NAV return of fund j. January is a binary variable for the month of January. Single entry is a binary for single country equity funds.

602 ■ Emerging Markets: Performance, Analysis and Innovation

TABLE 29.2   All International Funds—Equal-Weighed Portfolios

Intercept	Turn	MKT	SMB	HML	Adj. R^2
Panel A: All years ($N = 120$)					
−0.0131	0.0179				0.023
(−2.15)	(2.59)				
−0.0194	0.0213	0.4132	0.0847	0.1877	0.167
(−3.34)	(3.30)	(6.34)	(1.05)	(1.90)	
Panel B: 1984–1989 ($N = 60$)					
−0.0166	0.0240				0.019
(−0.94)	(1.46)				
−0.0292	0.0331	0.5710	0.3856	−0.1168	0.214
(−1.69)	(2.12)	(2.94)	(1.05)	(−0.26)	
Panel C: 1990–1994 ($N = 60$)					
−0.0161	0.0159				0.016
(−1.44)	(1.39)				
−0.0272	0.0252	0.6778	−0.0609	0.2222	0.304
(−2.82)	(2.57)	(5.13)	(−0.33)	(1.17)	
Panel D: 1995–1999 ($N = 60$)					
−0.0442	0.0531				0.075
(−2.46)	(2.40)				
−0.0473	0.0508	0.3249	−0.0583	0.1390	0.195
(−2.79)	(2.44)	(3.20)	(−0.52)	(0.92)	
Panel E: 2000–2004 ($N = 60$)					
0.0081	−0.0102				−0.007
(1.07)	(−0.78)				
0.0106	−0.0175	0.1405	0.0738	0.0924	0.093
(1.44)	(−1.38)	(2.75)	(1.40)	(1.39)	

Notes: The dependent variable is the monthly change in fund discount. Equal-weighed portfolios are created each month. Turnover is the lagged monthly volume divided by shares outstanding. MKT, SMB, and HML are the three Fama–French factors (*t*-statistics shown in parentheses are from the parametric *t*-test).

and the Fama and French (1992) risk factors (i.e., regression model [Equation 29.4]). Our sample comprises 240 portfolio-months across 71 international bond and equity CEFs during the 20 year period 1985–2004. As described earlier, we use the total sample of 10,094 firm-months to construct an equally weighted index of 240 portfolio-months. In the first regression (without controls), our Turnover coefficient is positive (0.0179) and significant (t-value = 2.59). After adding the risk-related control variables in the second regression, the Turnover coefficient remains positive (0.0213) and significant (t-value = 3.30). In fact, the Turnover coefficient increases in magnitude and significance as more control variables are added. The adjusted R^2 also increases from 0.023 to 0.167 as we move from the first to the second regression. The control variables are all significant and consistent with expectations.* We interpret the results in Panel A as supportive of Baker and Stein's (2002) claim that market liquidity is a sentiment indicator.

In Panel B of Table 29.2, we test the relation between market liquidity and CEF discounts in each of four subperiods 1985–1989, 1990–1994, 1995–1999, and 2000–2004. We reestimate Panel A's regressions in each subperiod to examine whether the positive and significant relation between Turnover and ΔDiscount is consistent through time. In the 1984–1989 subperiod, the Turnover coefficient is positive (0.0240) and insignificant at conventional levels if we do not include our risk control variables. Using our full model (4), we find that Turnover is positive (0.0331) and statistically significant. The addition of control variables raises the adjusted R^2 from 0.019 to 0.214.

We find similar confirmatory results in the 1990–1994 and 1995–1999 subperiods. In the 1990–1994 subperiod, the Turnover coefficient is positive (0.0159) and insignificant at conventional levels without the control variables. After including our controls, we find a positive (0.0252) and significant Turnover coefficient. The addition of the control variables raises the adjusted R^2 from 0.016 to 0.304. In the 1995–1999 subperiod, the Turnover coefficients are positive (0.0531, 0.0508) and significant both with and without the control variables, respectively. As before, the addition of control variables raises the adjusted R^2 significantly from 0.016 to 0.304. In summary, our results shown in Table 29.2 are consistent with expectations and Baker and Stein's (2002) irrational investor hypothesis. We find a positive and significant relation between Turnover and CEF prices for our overall sample as well as in three of four subperiods.

In Table 29.3, we report the international equity fund results for the entire sample period in Panel A and for each of the four subperiods in Panel B. In the first regression (without controls), our Turnover coefficient

---

* We also add a January dummy variable to regression model (4) as a robustness check. Although some Turnover coefficients become less significant as a result, the overall conclusion remains unaltered.

TABLE 29.3   International Equity Funds—Equal-Weighed Portfolios

Intercept	Turn	MKT	SMB	HML	Adj. R^2
Panel A: All years ($N = 120$)					
−0.0131	0.0168				0.020
(−2.04)	(2.44)				
−0.0189	0.0193	0.4561	0.0794	0.1651	0.189
(−3.17)	(3.05)	(6.81)	(0.96)	(1.62)	
Panel B: 1984–1989 ($N = 60$)					
−0.0166	0.0240				0.019
(−0.94)	(1.46)				
−0.0292	0.0331	0.5710	0.3856	−0.1168	0.214
(−1.69)	(2.12)	(2.94)	(1.05)	(−0.26)	
Panel C: 1990–1994 ($N = 60$)					
−0.0145	0.0135				0.007
(−1.23)	(1.19)				
−0.0254	0.0219	0.7243	−0.0577	0.2551	0.297
(−2.51)	(2.26)	(5.12)	(−0.29)	(1.25)	
Panel D: 1995–1999 ($N = 60$)					
−0.0499	0.0541				0.083
(−2.55)	(2.52)				
−0.0543	0.0515	0.4058	−0.0751	0.0482	0.262
(−3.05)	(2.62)	(3.52)	(−0.59)	(0.28)	
Panel E: 2000–2004 ($N = 60$)					
0.0074	−0.0093				−0.006
(1.00)	(−0.79)				
0.0116	−0.0187	0.2081	0.0752	0.0906	0.194
(1.73)	(−1.74)	(3.80)	(1.33)	(1.27)	

Notes: The dependent variable is the monthly change in fund discount. Equal-weighed portfolios are created each month. Turnover is the lagged monthly volume divided by shares outstanding. MKT, SMB, and HML are the three Fama–French factors (t-statistics shown in parentheses are from the parametric t-test).

is positive (0.0168) and significant (t-value = 2.44). After adding the risk-related control variables in the second regression, the Turnover coefficient remains positive (0.0193) and significant (t-value = 3.05). The Turnover coefficient increases in magnitude and significance with the control variables added, and the adjusted R^2 increases from 0.020 to 0.189 as we move from the first to the second regression. The control variables are mostly significant and consistent with expectations. Overall, the international equity fund results support our hypothesis of a positive relation between market sentiment and prices.

In Panel B of Table 29.3, we test the relation between market liquidity and ΔDiscount for international equity funds in each of our four subperiods. In the 1984–1989 subperiod, the results are identical to those reported in Panel B of Table 29.2 because there are no international bond funds trading in this period. Therefore, the positive and significant results reported in Table 29.2 are attributable to international equity funds. In the next subperiod (1990–1994), the Turnover coefficient is positive (0.0135) and insignificant at conventional levels without the control variables. After including our controls, we find a positive (0.0219) and significant Turnover coefficient. The addition of the control variables raises the adjusted R^2 from 0.007 to 0.297. In the 1995–1999 subperiod, the Turnover coefficient is positive (0.0541) and significant even without control variables; it is also positive (0.0515) and significant with control variables. The addition of control variables raises the adjusted R^2 from 0.083 to 0.262. Similar to our combined sample results in Table 29.2, we do not find a significant relation between Turnover and ΔDiscount for international equity funds during 2000–2004.

In Table 29.4, we report the international bond fund results for the entire sample period in Panel A, and each of the three subperiods in Panel B. In the first regression (without controls), our Turnover coefficient is negative (−0.0065) and significant (t-value = −0.52). After adding the risk-related control variables, the Turnover coefficient remains negative (−0.0010) and insignificant (t-value = −0.08). Both adjusted R^2 figures are small (−0.005 and 0.015, respectively) compared to comparable results in Tables 29.2 and 29.3. International bond funds do not appear to be susceptible to market sentiment at least not in the overall sample. Similar to Chay and Trzcinka (1999), we find significant differences between the behavior of CEF equity and bond discounts.

We test the bond fund liquidity–discount relation for each subperiod in Panel B of Table 29.4. In the 1990–1994 subperiod, the Turnover coefficients are positive and insignificant for both regressions. The adjusted R^2 figures are negative. Both Turnover coefficients are positive and insignificant during 1995–1999, and negative and insignificant during 2000–2004. The highest adjusted R^2 during this 10 year period is only 0.039. In sharp contrast to our equity fund findings, Panel B shows that there is not a significant relation between market sentiment and ΔDiscount for bond funds in any of the subperiods.

TABLE 29.4   International Bond Funds—Equal-Weighed Portfolios

Intercept	Turn	MKT	SMB	HML	Adj. R^2
Panel A: All years ($N = 156$)					
0.0032	−0.0065				−0.005
(0.47)	(−0.52)				
−0.0004	−0.0010	−0.0072	0.0426	0.1432	0.015
(−0.06)	(−0.08)	(−0.13)	(0.74)	(2.01)	
Panel B: 1992–1994 ($N = 36$)					
−0.0064	0.0056				−0.027
(−0.66)	(0.33)				
−0.0070	0.0068	−0.1829	−0.0380	0.0615	−0.058
(−0.66)	(0.38)	(−1.16)	(−0.19)	(0.39)	
Panel C: 1995–1999 ($N = 60$)					
−0.0077	0.0097				−0.015
(−0.49)	(0.36)				
−0.0088	0.0111	0.1007	−0.0402	0.3291	0.039
(−0.56)	(0.42)	(0.94)	(−0.34)	(2.07)	
Panel D: 2000–2004 ($N = 60$)					
0.0171	−0.0291				0.006
(1.45)	(−1.17)				
0.0123	−0.0225	−0.0073	0.0670	0.0957	−0.019
(0.98)	(−0.87)	(−0.11)	(0.92)	(1.05)	

Notes: The dependent variable is the monthly change in fund discount for the period from January 1992 to December 2004. Equal-weighed portfolios are created each month. Turnover is the lagged monthly volume divided by shares outstanding. MKTt, SMB, and HML are the three Fama–French factors (*t*-statistics shown in parentheses are from the parametric *t*-test).

29.3.2 Sentiment–Discount Relationship: Individual CEFs

Beginning with Table 29.5, we examine the sentiment–discount relationship using individual CEF time series regressions. We fit regression model Equation 29.5 with additional CEF-related control variables, including Reversion, Sector, and Performance (defined earlier). We also include a dummy variable that takes the value of 1 during the month of January and 0 otherwise. Our risk-related control variables remain the same as those used in Tables 29.2 through 29.4 (i.e., Market, SMB, and HML). We follow the same sequence for Tables 29.5 through 29.7 as for Tables 29.2 through 29.4. That is, we first report regression results for the combined sample of

TABLE 29.5 All International Funds—by Fund

Intercept	Trn	Mean-Rev	Sector	Perf	Jan	MKT	SMB	HML	Adj. R^2
Panel A: All years ($N = 71$)									
-0.0060	0.0068								0.023
(-2.94)	(1.94)								
-0.0184	0.0133	0.1237	288.58	0.0264	0.0006				0.347
(-5.86)	(4.57)	(4.48)	(12.66)	(0.60)	(0.23)				
-0.0210	0.0139	0.1452	287.41	-0.0189	0.0013	0.0596	0.0107	0.0141	0.350
(-7.78)	(4.03)	(13.24)	(12.18)	(-1.79)	(0.53)	(1.97)	(0.67)	(0.62)	
Panel B: 1985–1989 ($N = 15$)									
-0.0243	0.0305								0.132
(-2.14)	(2.35)								
-0.0755	0.0319	0.3084	383.03	-0.0666	0.0298				0.374
(-3.85)	(2.86)	(4.30)	(5.81)	(-1.57)	(1.73)				
-0.0851	0.0466	0.3436	435.95	-0.0564	0.0108	0.1106	-0.5337	0.6995	0.438
(-3.26)	(3.82)	(3.93)	(3.36)	(-1.14)	(0.53)	(0.26)	(-0.97)	(1.46)	
Panel C: 1990–1994 ($N = 58$)									
-0.0207	0.0169								-0.012
(-4.69)	(2.37)								
-0.0055	-0.0061	0.3791	306.98	0.0010	-0.0013				0.378
(-0.91)	(-0.70)	(9.99)	(6.11)	(0.05)	(-0.17)				
-0.0110	0.0036	0.3423	287.37	0.0108	-0.0059	0.1592	0.0213	0.0660	0.400
(-1.86)	(0.40)	(7.93)	(7.13)	(0.69)	(-0.61)	(1.95)	(0.24)	(0.71)	

(continued)

TABLE 29.5 (continued) All International Funds—by Fund

Intercept	Trn	Mean-Rev	Sector	Perf	Jan	MKT	SMB	HML	Adj. R^2
Panel D: 1995–1999 (N = 65)									
-0.0177	0.0200								0.025
(-7.09)	(5.91)								
-0.0255	0.0054	0.2203	279.81	-0.0096	0.0050				0.398
(-5.41)	(1.25)	(11.22)	(9.88)	(-1.25)	(1.44)				
-0.0283	0.0063	0.2243	264.81	-0.0082	0.0086	0.1172	0.0131	0.1034	0.428
(-5.89)	(1.43)	(11.65)	(8.98)	(-1.05)	(2.48)	(2.66)	(0.59)	(2.73)	
Panel E: 2000–2004 (N = 71)									
0.0046	-0.0049								0.011
(2.10)	(-1.10)								
-0.0346	0.0035	0.2302	268.6369	0.0137	0.0025				0.321
(-2.99)	(0.80)	(4.48)	(10.37)	(0.31)	(0.66)				
-0.0333	0.0035	0.2357	277.4682	-0.0380	0.0005	0.0039	-0.0060	-0.0494	0.330
(-3.19)	(0.70)	(5.80)	(10.32)	(-3.14)	(0.12)	(0.12)	(-0.26)	(-1.38)	

Notes: The dependent variable is the monthly change in fund discount. Turnover is the lagged monthly volume divided by shares outstanding. Mean Rev is $D_{i(j),t-1} - d_{j,t-1}$ where $D_{i(j),t-1}$ is the equal weight average discount of category i, excluding fund j and $d_{j,t-1}$ is the discount of fund j. Sector is $D_{i(j),t} - D_{i(j),t-1}$. Performance is $R_{i(j),t-1} - r_{j,t-1}$ where $R_{i(j),t-1}$ is the equal weight average NAV return of category i, excluding fund j and $r_{j,t-1}$ is the NAV return of fund j. January is a binary variable for the month of January. Single country is a binary for single-country equity funds. MKT, SMB, and HML are the three Fama–French factors (t-statistics shown in parentheses are from the parametric t-test).

TABLE 29.6 International Equity Funds—by Fund

Intercept	Trn	Mean-Rev	Sector	Perf	Jan	MKT	SMB	HML	Adj. R^2
Panel A: All years (N = 50)									
-0.0096	0.0130								0.019
(-5.03)	(4.61)								
-0.0261	0.0177	0.1560	214.89	-0.0075	0.0000				0.319
(-9.36)	(7.34)	(12.1)	(13.74)	(-1.59)	(0.00)				
-0.0267	0.0180	0.1565	199.50	-0.0059	0.0010	0.1060	0.0007	0.0207	0.326
(-9.73)	(7.08)	(12.47)	(13.01)	(-1.25)	(0.28)	(3.55)	(0.05)	(0.72)	
Panel B: 1985–1989 (N = 15)									
-0.0243	0.0305								0.132
(-2.14)	(2.35)								
-0.0755	0.0319	0.3084	383.03	-0.0666	0.0298				0.374
(-3.85)	(2.86)	(4.30)	(5.81)	(-1.57)	(1.73)				
-0.0851	0.0466	0.3436	435.95	-0.0564	0.0108	0.1106	-0.5337	0.6995	0.438
(-3.26)	(3.82)	(3.93)	(3.36)	(-1.14)	(0.53)	(0.26)	(-0.97)	(1.46)	
Panel C: 1990–1994 (N = 41)									
-0.0188	0.0164								0.016
(-3.81)	(2.82)								
-0.0177	0.0157	0.3465	217.57	-0.0306	-0.0001				0.409
(-2.71)	(2.06)	(9.67)	(11.76)	(-1.48)	(-0.01)				
-0.0238	0.0269	0.3161	183.84	0.0038	-0.0022	0.1951	0.0317	0.0357	0.402
(-4.59)	(8.51)	(13.23)	(8.53)	(0.28)	(-0.17)	(2.08)	(0.39)	(0.39)	

(continued)

TABLE 29.6 (continued) International Equity Funds—by Fund

Intercept	Trn	Mean-Rev	Sector	Perf	Jan	MKT	SMB	HML	Adj. R^2
Panel D: 1995–1999 ($N = 50$)									
−0.0207									0.032
(−6.73)									
−0.0319	0.0127	0.2245	188.47	−0.0066	0.0070				0.383
(−5.34)	(3.23)	(9.49)	(10.31)	(−0.89)	(1.52)				
−0.0359	0.0142	0.2343	172.30	−0.0044	0.0115	0.1503	0.0117	0.1098	0.408
(−5.95)	(3.25)	(10.35)	(8.17)	(−0.56)	(2.65)	(2.58)	(0.42)	(2.13)	
Panel E: 2000–2004 ($N = 50$)									
0.0013	0.0015								−0.001
(0.61)	(0.36)								
−0.0490	0.0008	0.3089	203.22	−0.0119	0.0023				0.282
(−3.03)	(0.17)	(5.03)	(7.64)	(−1.85)	(0.44)				
−0.0443	0.0011	0.2877	199.21	−0.0179	−0.0004	0.0320	−0.0211	−0.0525	0.295
(−3.06)	(0.21)	(5.14)	(7.42)	(−2.05)	(−0.07)	(0.83)	(−0.82)	(−1.09)	

Notes: The dependent variable is the monthly change in fund discount. Turnover is the lagged monthly volume divided by shares outstanding. Mean Rev is $D_{i(j),t-1} - d_{j,t-1}$ where $D_{i(j),t-1}$ is the equal weight average discount of category i, excluding fund j and $d_{j,t-1}$ is the discount of fund j. Sector is $D_{i(j),t} - D_{i(j),t-1}$. Performance is $R_{i(j),t-1} - r_{j,t-1}$ where $R_{i(j),t-1}$ is the equal weight average NAV return of category i, excluding fund j and $r_{j,t-1}$ is the NAV return of fund j. January is a binary variable for the month of January. Single country is a binary for single-country equity funds. MKT, SMB, and HML are the three Fama–French factors (t-statistics shown in parentheses are from the parametric t-test).

TABLE 29.7 International Bond Funds—by Fund

Intercept	Trn	Mean-Rev	Sector	Perf	Jan	MKT	SMB	HML	Adj. R^2
Panel A: All years (N = 21)									
0.0022	-0.0071								0.031
(0.49)	(-0.80)								
-0.0018	0.0036	0.0532	450.0049	0.1007	0.0019				0.407
(-0.26)	(0.49)	(0.64)	(9.25)	(0.72)	(0.78)				
-0.0078	0.0044	0.1192	489.6223	-0.0487	0.0022	-0.0471	0.0336	-0.0010	0.406
(-1.48)	(0.46)	(5.63)	(11.19)	(-1.49)	(0.87)	(-0.69)	(0.87)	(-0.03)	
Panel B: 1992–1994 (N = 17)									
-0.0252	0.0181								-0.078
(-2.69)	(0.89)								
0.0261	-0.0628	0.4641	539.4197	0.0832	-0.0045				0.299
(2.67)	(-3.54)	(4.67)	(3.30)	(1.98)	(-1.17)				
0.0254	-0.0628	0.4168	582.0175	0.0309	-0.0164	0.0570	-0.0083	0.1523	0.393
(1.97)	(-2.43)	(2.71)	(5.36)	(0.64)	(-2.69)	(0.34)	(-0.03)	(0.61)	
Panel C: 1995–1999 (N = 19)									
-0.0105	0.0125								0.007
(-2.74)	(1.82)								
-0.0101	-0.0122	0.2102	500.95	-0.0169	0.0004				0.433

(continued)

TABLE 29.7 (continued)   International Bond Funds—by Fund

Intercept	Trn	Mean-Rev	Sector	Perf	Jan	MKT	SMB	HML	Adj. R^2
Panel C: 1995–1999 (N = 19)									
(−1.69)	(−1.16)	(5.83)	(8.02)	(−0.86)	(0.08)				
−0.0099	−0.0129	0.1999	488.78	−0.0174	0.0016	0.0373	0.0165	0.0880	0.475
(−1.71)	(−1.37)	(5.42)	(7.77)	(−0.92)	(0.30)	(0.74)	(0.46)	(2.33)	
Panel D: 2000–2004 (N = 21)									
0.0119	−0.0194	0.0576	411.93	0.0698	0.0028				0.038
(2.42)	(−1.84)	(0.69)	(9.02)	(0.49)	(0.72)				
−0.0030	0.0093	0.1162	457.46	−0.0843	0.0024	−0.0607	0.0287	−0.0424	0.412
(−0.41)	(1.05)	(5.00)	(10.82)	(−2.57)	(0.59)	(−0.93)	(0.64)	(−1.03)	

Notes: The dependent variable is the monthly change in fund discount. Turnover is the lagged monthly volume divided by shares outstanding. Mean-Rev is $D_{f(j),t-1} - d_{j,t-1}$ where $D_{f(j),t-1}$ is the equal weight average discount of category i, excluding fund j and $d_{j,t-1}$ is the discount of fund j. Sector is $D_{f(j),t} - D_{f(j),t-1}$. Performance is $R_{f(j),t-1} - r_{j,t-1}$ where $R_{f(j),t-1}$ is the equal weight average NAV return of category i, excluding fund j and $r_{j,t-1}$ is the NAV return of fund j. January is a binary variable for the month of January. Single country is a binary for single-country equity funds. MKT, SMB, and HML are the three Fama–French factors (t-statistics shown in parentheses are from the parametric t-test).

international CEFs, followed by the equity funds only and then the bond funds only.

We show in Panel A of Table 29.5 that sentiment and ΔDiscount are positively related in all three regressions (i.e., no controls, CEF-based controls, and CEF- and risk-based controls) for the combined international CEF sample. Our three Turnover coefficients are 0.0068 (*t*-value = 1.94), 0.0133 (*t*-value = 4.57), and 0.0139 (*t*-value = 4.03), respectively. The adjusted R^2 figures range from a low of 0.023 for the univariate regression to 0.350 for the full-model regression. These fund-by-fund time series results are consistent with Baker and Stein's (2002) assertion that market liquidity is a sentiment indicator.

Turning to the subperiods, we find that there is a positive and significant relation between Turnover and ΔDiscount during the 1985–1989 subperiod. The full-model Turnover coefficient is 0.0466 (*t*-value = 3.82) and its adjusted R^2 is 0.438. During the next three subperiods, however, the estimated Turnover coefficients tend to be insignificant. In fact, Turnover is only significant in the univariate regressions for 1990–1994 and 1995–1999. The combined bond and equity CEF results in Table 29.5 generally support our sentiment–discount hypothesis, but the subperiod findings suggest that the relation between Turnover and ΔDiscount is not stable.

Next, we examine separately the international equity (Table 29.6) and bond (Table 29.7) funds. In Panel A of Table 29.6, we observe that sentiment and ΔDiscount are positively related in all three regressions (i.e., no controls, CEF-based controls, and CEF- and risk-based controls) for the international equity CEF sample. Our three Turnover coefficients are 0.0130 (*t*-value = 4.61), 0.0177 (*t*-value = 7.34), and 0.0180 (*t*-value = 7.08), respectively. The adjusted R^2 figures range from a low of 0.019 for the univariate regression to 0.326 for the full-model regression. The international equity fund-by-fund results are stronger than their combined counterparts in Table 29.5 both in terms of coefficient magnitudes and significance levels.

In Panel B of Table 29.6, we find that there is a positive and significant relation between Turnover and ΔDiscount for each of our subperiods with the exception of 2000–2004. Although all three Turnover coefficients are positive in the 2000–2004 subperiod, none of them is statistically significant. Interestingly, this is the same subperiod that yields insignificant coefficients in Table 29.3 where we report the equally weighted CEF portfolio results for international equity funds. In the 1985–1989 subperiod, the full-model Turnover coefficient is 0.0466 (*t*-value = 3.82) and its adjusted

R^2 is 0.438. In the 1990–1994 subperiod, the full-model Turnover coefficient is 0.0269 (t-value = 8.51) and its adjusted R^2 is 0.402. The 1995–1999 subperiod yields similar results with a full-model Turnover coefficient of 0.0142 (t-value = 3.25) and an adjusted R^2 of 0.408.

In contrast to Table 29.6, our Table 29.7 results do not reveal a positive and significant relation between Turnover and ΔDiscount for international bond funds. The three Turnover coefficients are −0.0071 (t-value = −0.80), 0.0036 (t-value = 0.49), and 0.0044 (t-value = 0.46), respectively, for the overall sample in Panel A. The adjusted R^2 figures range from 0.031 for the univariate regression to 0.407 for the full-model regression. These adjusted R^2 values are comparable to those reported in Table 29.6, suggesting that our regression model fits the international bond data relatively well.

Our Panel B results also fail to find a positive and significant relation between Turnover and ΔDiscount. During the 1990–1994 subperiod, two of the regressions exhibit a negative and significant relation, although this sample includes only two years of data for 17 CEFs. Overall, the fund-by-fund findings in Table 29.7 are consistent with the CEF portfolio results in Table 29.4; that is, there does not exist a significant sentiment–discount relationship for bond funds. In addition, the combined results in Tables 29.6 and 29.7 are also consistent with Chay and Trzcinka (1999), who find significant differences between the behavior of CEF equity and bond discounts.

29.3.3 Sentiment–Discount Relationship: Single-Country Funds

We perform additional tests for the sentiment–discount relationship in Table 29.8 by examining single-country funds separately. Our sample of international equity funds consist of both single-country and multicountry CEFs. Previous research, including those of Klibanoff et al. (1998) and Hardouvelis et al. (1994), suggests that single-country fund discounts behave differently from their multicountry counterparts. For instance, fund prices are more sensitive to fundamentals in weeks when there is a front page article in the *New York Times* regarding the fund's home country (Klibanoff et al., 1998).

In Table 29.8, we report the single-country fund results for the entire sample period in Panel A, and each of the four subperiods in Panel B. In the first regression (without controls), our Turnover coefficient is positive (0.0213) and significant (t-value = 3.12). After adding the risk-related control variables in the second regression, the Turnover coefficient remains positive (0.0218) and significant (t-value = 3.45). The Turnover coefficient increases in magnitude and significance with the control variables added,

TABLE 29.8   Single-Country Funds—Equal-Weighed Portfolios

Intercept	Turn	MKT	SMB	HML	Adj. R^2
Panel A: All years ($N = 240$)					
−0.0177	0.0213				0.036
(−2.44)	(3.12)				
−0.0225	0.0218	0.5347	0.0788	0.2091	0.182
(−3.31)	(3.45)	(6.42)	(0.76)	(1.65)	
Panel B: 1984–1989 ($N = 60$)					
−0.0236	0.0296				0.035
(−1.05)	(1.76)				
−0.0298	0.0296	0.7590	0.1891	0.1140	0.170
(−1.34)	(1.83)	(2.97)	(0.38)	(0.19)	
Panel C: 1990–1994 ($N = 60$)					
−0.0184	0.0159				0.022
(−1.50)	(1.53)				
−0.0298	0.0238	0.7472	0.0149	0.3036	0.251
(−2.71)	(2.57)	(4.45)	(0.06)	(1.26)	
Panel D: 1995–1999 ($N = 60$)					
−0.0469	0.0491				0.052
(−2.07)	(2.06)				
−0.0495	0.0439	0.4518	−0.0480	0.0878	0.207
(−2.34)	(1.97)	(3.29)	(−0.32)	(0.43)	
Panel E: 2000–2004 ($N = 60$)					
0.0074	−0.0092				−0.008
(0.90)	(−0.74)				
0.0119	−0.0169	0.2228	0.0476	0.0335	0.219
(1.62)	(−1.54)	(3.82)	(0.78)	(0.44)	

Notes: The dependent variable is the monthly change in fund discount. Equal-weighed portfolios are created each month. Turnover is the lagged monthly volume divided by shares outstanding. MKT, SMB, and HML are the three Fama–French factors (t-statistics shown in parentheses are from the parametric t-test.)

and the adjusted R^2 increases from 0.036 to 0.182. In Panel B of Table 29.8, we test the relation between market liquidity and ΔDiscount for single-country funds in each of our four subperiods. Similar to our Table 29.3 results (i.e., all international equity CEFs), we find positive and mostly significant Turnover coefficients in all subperiods except 2000–2004.

In additional tests (not reported herein, but available upon request), we also examine single-country fund results for the entire sample period based on fund-by-fund regression results. Our single-country fund results confirm the same pattern that we find in our full sample of international equity funds. There is a positive and significant relation between market sentiment (Turnover) and market prices (ΔDiscount), consistent with Baker and Stein (2002). The sentiment–discount relationship remains significant after including several CEF- and risk-related control variables; it also remains significant across most time periods.

29.4 CONCLUSIONS

Investor sentiment has received considerable attention among academics and practitioners over the past decade, particularly after the sudden collapse of technology stocks during 2000–2001. Much of the research in this area focuses on the relation between liquidity and investor sentiment in the cross-section. In contrast, Baker and Stein (2002) develop a theoretical model that addresses the time-series properties of market sentiment. The fundamental claim is that market liquidity acts as sentiment indicator. When liquidity is high, uninformed investors are dominant and assets are overvalued. Baker and Stein (2002, p. 34) state that their contribution is "primarily a theoretical one, and as such do not attempt to provide a definitive empirical test of the model." The purpose of this chapter is to help fill this empirical void by testing the liquidity–sentiment relation using a unique class of assets—international CEFs.

Previous research has shown that CEF prices are highly susceptible to investor sentiment (Lee et al., 1991). Because irrational investor misperceptions are positively correlated across the assets in which such investors are prevalent, arbitrage is unable to prevent deviations between market prices and fundamental values (De Long et al., 1990). International CEFs are particularly prone to investor sentiment because these funds are among the smallest and least liquid of all CEF types. In addition, CEFs allow us to measure the market price of the fund relative to the market value of its underlying assets (i.e., the NAV). We can then track the CEF discount through time and examine its relation with liquidity.

Our empirical results strongly support the liquidity–discount hypothesis. We examine the liquidity–discount relation using a sample of 71 international CEFs funds covering the 20 year period from 1985 through 2004. We construct an equal-weighted portfolio of all international CEFs and regress the monthly changes in portfolio discounts against portfolio turnover and a set of control variables. Our results confirm that there is a positive and highly significant relationship between liquidity and CEF discounts. When turnover (liquidity) increases, CEF discounts become less negative (or more positive). We divide our sample into equity and bond CEFs following Chay and Trzcinka (1999) and show that the significant relation between turnover and discounts is driven by the equity funds. Noise trading is concentrated in the international equity CEFs.

Our second set of results is based on fund-by-fund time series regressions. Confirming our equal-weighted portfolio results, we find a positive and significant relation between liquidity and discounts using a fund-by-fund analysis. We also find that this positive relation is driven by the international equity funds. We perform various robustness tests based on subperiods and single country funds. All of these tests support the hypothesis that liquidity is an indicator of positive market sentiment and, consequently, overvaluation.

REFERENCES

Amihud, Y. and Mendelson, H. (1986) Asset pricing and the bid–ask spread. *Journal of Financial Economics*, 17(2): 223–249.

Baker, M. and Stein, J.C. (2002) Illiquidity and stock returns: Cross-section and time-series effects. *Journal of Financial Markets*, 5(1): 31–56.

Barberis, N., Shleifer, A., and Vishny, R. (1998) A model of investor sentiment. *Journal of Financial Economics*, 49(3): 307–343.

Chay, J.B. and Trzcinka, C. (1999) Managerial performance and the cross-sectional pricing of closed-end funds. *Journal of Financial Economics*, 52(3): 397–408.

Daniel, K.D., Hirshleifer, D., and Subrahmanyam, A. (2001) Overconfidence, arbitrage, and equilibrium asset pricing. *Journal of Finance*, 56: 921–965.

De Long, J.B., Shleifer, A., Summers, L., and Waldmann, R. (1990) Noise trader risk in financial markets. *Journal of Political Economy*, 98(4): 703–738.

Dimson, E. and Minio-Kozerski, C. (2002) A factor model of the closed-end fund discount. Working Paper, London Business School, London.

Fama, E. and French, K. (1992) The cross-section of expected stock returns. *Journal of Finance*, 47(2): 427–465.

Fama, E. and French, K. (1993) Common risk factors in the returns on bonds and stocks. *Journal of Financial Economics*, 33(1): 3–56.

Fama, E. and MacBeth, J.D. (1973) Risk, return, and equilibrium: Empirical tests. *Journal of Political Economy*, 81(3): 607–636.

Hanley, K.W., Lee, C.M.C., and Seguin, P.J. (1996) The marketing of closed-end fund IPOs: Evidence from transactions data. *Journal of Financial Intermediation*, 5(2): 127–159.

Hardouvelis, G., La Porta, R., and Wizman, T.A. (1994) What moves the discount on country equity funds? In: J. Frankel (Ed.), *The Internationalization of Equity Markets*. University of Chicago Press, Chicago, IL.

Klibanoff, P., Lamont, O., and Wizman, T. (1998) Investor reaction to salient news in closed-end country funds. *Journal of Finance*, 53(2): 673–700.

Lee, C., Shleifer, A., and Thaler, R.H. (1991) Investor sentiment and the closed-end fund puzzle. *Journal of Finance*, 46(1): 76–110.

Pontiff, J. (1995) Closed end fund premia and returns: Implications for financial market equilibrium. *Journal of Financial Economics*, 37(3): 341–370.

Shleifer, A. and Vishny, R. (1997) Limits of arbitrage. *Journal of Finance*, 52(1): 35–55.

Vayanos, D. (1998) Transaction costs and asset prices: A dynamic equilibrium model. *Review of Financial Studies*, 11(1): 1–58.

Weiss, K. (1989) The post-offering price performances of closed-end funds. *Financial Management*, 18(3): 57–67.

CHAPTER 30

Closed-End Funds in Emerging Markets

Michael F. Bleaney and R. Todd Smith

CONTENTS

30.1 INTRODUCTION

Emerging markets are by definition small compared with the pool of international financial wealth. They are generally characterized by poorer regulation and poorer governance than developed markets (Kaufmann et al., 2003). This makes emerging markets riskier, but their relative shortage of capital implies greater potential rewards. Small shifts in international investors' beliefs about the balance between risks and rewards in these markets can make large differences to the prices of financial assets and to the flow of funds to them. Cycles of optimism and pessimism may have sizeable effects on macroeconomic variables in the emerging markets (Tornell and Westermann, 2002). Moreover, there is evidence of contagion from one market to another—to some extent they are all tarred with the same brush (Chan-Lau et al., 2004). After the Russian crisis of 1998, investor enthusiasm for emerging markets declined sharply. As portfolio capital inflows fell, Latin American countries on floating exchange rates experienced considerable real depreciations (Brazil, Chile), while those on hard pegs suffered exchange rate pressure, culminating eventually in a full-blown crisis (Argentina, Uruguay).

Research on closed-end funds in emerging markets has raised many issues, including the effect on discounts* of restrictions on foreign investment (Bonser-Neal et al., 1990; Nishiotis, 2004), the role of such funds as vehicles for portfolio diversification for U.S. investors (Chang et al., 1995; Bekaert and Urias, 1996; Eun et al., 2002), information lags and inertia in the response of prices to net asset value (NAV) movements (Klibanoff et al., 1998; Frankel and Schmukler, 2000), and comovement with the U.S. stock market or with discounts on domestic closed-end funds (Hardouvelis et al., 1994; Bodurtha et al., 1995).

In the United States, the term "country fund" is often used to mean regional or emerging market funds as well as single-country funds. This is because single-country funds are the main type of U.S.-based closed-end funds investing in international assets. In the United Kingdom, the situation is reversed: there are relatively few single-country funds and many more multicountry funds. In both countries, emerging-market (EM) closed-end funds experienced a dramatic rise in the late 1980s and early 1990s. In that period, they traded on low discounts and there were many new issues. Since the Asian crisis of 1997, they have suffered an almost equally dramatic fall from grace. Figure 30.1 shows that the number of EM funds in the United Kingdom has fallen to less than half its 1997 level, as discounts have risen.[†] Figure 30.2 compares the average premium on domestic and EM closed-end funds in the United States. Since

---

* The discount is the difference between the net asset value (NAV) per share and the price as a percentage of the NAV. A negative discount is referred to as a premium.

† The source of these data is the Association of Investment Trust Companies.

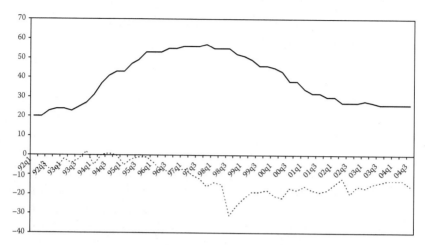

FIGURE 30.1　U.K. emerging market funds 1992–2004: number and average premium (%).

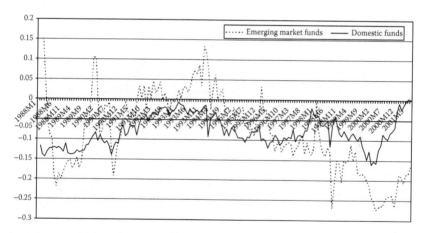

FIGURE 30.2　U.S. premia 1988–2001.

most emerging markets had liberalized their capital markets by the end of 1992, the increase in discounts since the mid-1990s cannot be attributed to a liberalization effect as documented by Bonser-Neal et al. (1990) and Nishiotis (2004).

Premia on closed-end funds are interesting for at least two reasons. First, as the supply of shares is inelastic, they may be regarded as an index of sentiment of investors. Thus, from the perspective of understanding financing flows to emerging markets, it may be useful to understand what drives premia on closed-end funds investing in emerging markets. Second, closed-end funds are a natural investment vehicle in cases where the underlying assets are illiquid, as is often true in emerging markets. Closed-end funds have a natural advantage over open-end funds in these circumstances because they do not need to trade in response to buying or selling pressure from the underlying investors. On the other hand, comovement of discounts on closed-end funds with home equity market returns reduces the potential diversification benefits. Holding EM assets in an open-end fund will reduce the variance of the portfolio return compared with holding such assets in a closed-end fund, when premia move with home stock market returns. Thus the dynamics of premia on EM closed-end funds are relevant to the underlying attractiveness of these funds as vehicles for international portfolio diversification.

In this chapter, we investigate the dynamics of discounts on EM funds in comparison with those on other international funds and on domestic funds. We do this for funds traded in the United States and the United Kingdom, the two main markets for closed-end funds.

Our first conclusion is that premia are strongly positively related to their past level—premia are persistent. Second, there is a common element of sentiment in closed-end funds that invest in domestic markets (i.e., the U.S. or U.K. markets) and in international markets. Third, we find that premia on EM funds respond positively to returns on emerging markets in general—i.e., there is a positive association between premia and returns to the EM asset class. In other words, they are indeed tarred with the same brush. Fourth, premia on EM funds generally respond inversely to the over- or underperformance of the fund (i.e., fund return minus return of the asset class). Fifth, comovement of EM fund discounts with the home stock market (i.e., the U.S. or U.K. stock market) is positive but short-lived. Finally, we also show that price inertia in international funds is if anything smaller than in domestic funds, which casts doubt on the information-lag explanation of such inertia in international funds.

The structure of this chapter is as follows. After a survey of previous research in Section 30.2, Section 30.3 discusses the design of the study, data, and empirical methodology. Section 30.4 presents the empirical results. Section 30.5 discusses the findings, while the final section concludes this chapter.

30.2 PREVIOUS RESEARCH

The following are the main findings of previous research on closed-end funds:

1. For both domestic and international closed-end equity funds, premia vary widely across funds and over time, but revert to a negative mean (Lee et al., 1991; Hardouvelis et al., 1994; Bodurtha et al., 1995; Pontiff, 1995).

2. Premia move together (Lee et al., 1991; Bodurtha et al., 1995). The presence of a common component in fund premia has been well established for domestic and international funds separately; it is unclear whether there exists a component common to domestic and country funds (or, more generally, international funds). Bodurtha et al. (1995) find little comovement of domestic fund premia and country fund premia in a fairly small sample (1986–1990). Our empirical examination includes this issue.

3. In both daily and monthly data, the premium exhibits negative short-run correlation with NAV returns for both domestic funds and country funds (Frankel and Schmukler, 2000; Bleaney and Smith, 2006); this is usually referred to as short-run price inertia. However, in the longer run, the correlation between premia and NAV returns is positive for domestic funds (Bleaney and Smith, 2003) and country funds (Frankel and Schmukler, 2000).

4. For both domestic and country funds, premia are positively related to the return on the home country's equity market (Brickley et al., 1991; Hardouvelis et al., 1994; Bodurtha et al., 1995; Chang et al., 1995; Sias, 1997; Gemmill and Thomas, 2002).

The explanation of these phenomena, and particularly of the tendency for funds to trade at a discount, is still the matter of some debate. Early explanations, based on biases in NAV (due to funds holding illiquid assets, tax liabilities, etc.), excessive management fees, and market segmentation, have generally been found wanting (Dimson and Minio-Kozerski, 1999). A more recent hypothesis is that because of comovement of premia with the stock market index, closed-end funds have higher market betas than equivalent open-end funds and must therefore yield higher returns by trading at a discount (Lee et al., 1991; Elton et al., 1998).

One possible explanation of short-run price inertia is that if investors in closed-end funds are simply less well informed than others, then closed-end fund investors may be slower to react to fundamental information (Frankel and Schmukler, 2000).* This "asymmetric information hypothesis" was largely motivated by the observation that country fund premia widen sharply during a crisis in that country, with prices only slowly catching up to the net asset value (Levy-Yeyati and Ubide, 2000). It is possible that this so-called country-fund puzzle is simply another manifestation—albeit a striking one—of the empirical fact that premia on closed-end funds generally exhibit a negative short-run correlation with NAV returns.

30.3 DESIGN OF THIS STUDY

Our main concern is the determinants of EM closed-end fund premia and whether closed-end funds in emerging markets behave exactly like other closed-end funds, or whether they display distinctive characteristics. Previous research has suggested that factors such as dividend rates, expense ratios, liquidity, and opportunities for arbitrage influence fund premia in equilibrium (Gemmill and Thomas, 2002). Since these factors are rather persistent over time, the investigator has two choices: either to model them explicitly or to use a fixed effects model that allows for fund-specific characteristics. We choose the latter option.

The basic pooled data, fixed-effects regression we estimate is

$$\text{PREM}_{j,t} = a + b\text{PREM}_{j,t-k} + c\text{DOMSR}_k + d\text{MNAVR}_k \\ + f\text{RNAVR}_{j,k} + g\Delta\text{PREMDOM}_{t,t-k} + e_j + u_{j,t} \quad (30.1)$$

where

$\text{PREM}_{j,t}$ is the log (price/NAV) of fund j at time t

e_j is a fund-specific effect

u is a random error

DOMSR_k is the k-month return on the domestic (i.e., the United States or the United Kingdom) stock market index

MNAVR_k is the k-month return to month t on the average fund in the asset class

$\text{RNAVR}_{j,k}$ is the k-month NAV return to month t on fund j minus MNAVR_k

$\text{PREMDOM}_{t,t-k}$ is the k-month change of the average premium on domestic funds (included for international funds only)

---

* In an interesting application, Klibanoff et al. (1998) find that country-fund prices overreact (underreact) to fundamental information when a related story appears (does not appear) in the *New York Times.*

This specification enables us to explore the dynamics of the premium over various horizons, while allowing for premium persistence and reversion to fund-specific mean premia. The average NAV return to all funds in the asset class (MNAVR) and the relative return (RNAVR = NAVR – MNAVR) sum to the NAV return on a particular fund (NAVR). We find that these two components of NAVR tend to have significantly different coefficients (which are not infrequently of opposite sign). We also add, for international funds, the k-month change in the average premium on domestic funds as a measure of investor sentiment in domestic markets. We estimate this equation for a short horizon ($k = 1$ month) and for a longer horizon ($k = 24$ months).

Previous research has shown that stock market liberalization has significantly reduced the premium on EM country funds, as it permits foreign equity investment by other routes (Bonser-Neal et al., 1990; Nishiotis, 2004). Most of our data are from the postliberalization period, so our results should not be influenced by the liberalization effect. Nishiotis (2004) also detects "indirect investment barriers" (the influence of other variables on the premium), but in his sample, they seem to be significant only in the preliberalization period.

The full list of funds used in this study is omitted to save space but is available on request from the authors. For U.S. funds, end-of-month price and NAV are from Bloomberg Financial Markets. Dividend yields are obtained from Datastream (where price and NAV data overlap in Bloomberg and Datastream the figures are identical). Dividend yields reported in Datastream are annualized. We therefore adjust the reported figures when the frequency of our return period is not 12 months. For U.K. funds, end-of-month data on price, NAV, and dividend yields are from Datastream.

One of our regressors is the return on the home country stock market. The market indexes are the Datastream U.S. Market Global Return Index and the Datastream U.K. Global Return Index. Another regressor is the average premium on domestic closed-end funds in the sample.

We estimate the parameters of the models by ordinary least squares (OLS). OLS standard errors are, however, questionable because both heteroscedasticity and serial correlation are likely present in the error term due to the cross-sectional nature of the data set and because previous work has found that the error term in premium regressions is serially correlated. For this reason, we estimate the standard errors of the OLS parameter estimates using a technique that is robust to both autocorrelation and heteroskedasticity (Newey and West, 1994).

This estimation method has three advantages over the commonly used procedure of estimating a cross-section regression for each date, then averaging the resulting estimates of each parameter over all dates, and calculating a standard error for this average, as in Fama and Macbeth (1973). It automatically corrects estimated standard errors for serial correlation; it corrects for arbitrary forms of heteroskedasticity and it allows for the "nonrectangular" nature of the data set, with more observations at later dates (because of the expansion in the number of funds in the market). The Fama–Macbeth method, when applied to an unbalanced panel, weights parameter estimates from dates with few observations equally with parameter estimates from dates with many observations. Implicitly, therefore, each observation from dates with few observations is exerting greater leverage over the results. This is particularly unfortunate in cases where, as here, these observations are the furthest back in time and therefore have least relevance to the future. We have, however, checked that Fama–Macbeth estimation would yield similar results.

30.4 RESULTS

We have three categories of funds traded in each of the United States and the United Kingdom: domestic funds, EM funds, and other international funds. Table 30.1 provides some basic data by type of fund. Relative to domestic funds, international funds have tended to trade on higher

TABLE 30.1   Summary Statistics for the Premium on Different Types of Funds

Fund Type	Sample Size	Mean	Standard Deviation	S.D. of Monthly Changes	Mean-Reversion Coefficient (*t*-Statistic)
U.S.-based funds					
Domestic	3967	−0.070	0.105	0.042	−0.077 (−5.53)
Emerging markets	4215	−0.112	0.167	0.067	−0.085 (−8.96)
Ind. countries	1936	−0.115	0.137	0.062	−0.108 (−7.72)
U.K.-based funds					
Domestic	5805	−0.167	0.109	0.038	−0.063 (−11.91)
Emerging markets	1659	−0.138	0.104	0.049	−0.119 (−8.93)
Ind. countries	3252	−0.115	0.094	0.043	−0.111 (−11.17)

Notes: Premium = log(share price/NAV). "Mean-reversion coefficient" is the coefficient of last month's premium in a regression of the monthly change in the premium on a constant and last month's premium. For emerging-market funds, only observations from January 1993 onward are used.

discounts, have greater month-to-month discount variability, and have faster mean-reversion. All funds show the characteristics that are well known from the closed-end fund literature and specifically reversion of the premium toward a negative mean.

Table 30.2 shows fixed effects regressions for U.S. funds with a 24 month frequency for lags and return intervals in regression (Equation 30.1). The fund premium is regressed on a 24 month lag of itself, the 24 month U.S. stock market index return, the 24 month fund NAV return, split into its relative and common components, and (for international funds only) the 24 month change in the average premium on domestic closed-end funds. For EM funds pre-1993 observations are excluded in order to avoid stock market liberalization effects (although the results are in fact very similar when these observations are included).

The premium 24 months previously is an important determinant of the current level of the premium, particularly for international funds. Also, for international funds, there is a strong positive correlation with the

TABLE 30.2   Fixed Effects Regressions for U.S.-Traded Funds

	Dependent Variable: Log Premium		
		International Funds	
Independent Variables	Domestic Funds	Emerging Market Funds	Other Funds
24 month U.S. stock market return	-0.152 (-5.05)	-0.074 (-2.10)	0.048 (1.03)
24 month fund relative NAV return	0.054 (1.01)	-0.171 (-7.48)	-0.100 (-2.77)
24 month NAV return on similar funds	0.198 (4.23)	0.123 (4.66)	-0.017 (-0.34)
24 month lagged premium	0.079 (1.11)	0.433 (8.48)	0.136 (3.54)
24 month change in domestic fund premia		0.405 (3.78)	0.705 (7.67)
R^2	0.551	0.530	0.497
Standard error	0.0711	0.117	0.0878
Sample size	3320	3582	1543

Notes: Figures in parentheses are robust t-statistics. For emerging-market funds, only observations from January 1993 onward are used.

24 month change in the average domestic premium. This implies considerable low-frequency comovement of premia on domestic and international funds, a fact that has not been documented before. Thus, since closed-end fund premia have been interpreted as investor sentiment, this suggests that there is a common element of sentiment underlying both domestic and international fund premia.

The 24 month return on the U.S. stock market has a negative coefficient where it is significant (domestic and EM funds). Thus, in the long run, good U.S. market returns are associated with lower premia on U.S.-based EM funds. One explanation for this could be trend-chasing behavior—if the domestic market performs persistently better than emerging markets, demand for EM funds falls in favor of domestic funds. This trend-chasing hypothesis is consistent with the fact that the 24 month return on similar funds has a positive coefficient where it is significant (domestic and EM funds)—i.e., demand for EM funds rises when the asset class has done well.

The 24 month *relative* NAV return—i.e., relative to the return on the asset class—has a negative and significant coefficient for international funds (but especially for EM funds). Thus a NAV return that is superior to returns on other emerging markets is associated with a *lower* premium. This result suggests that, if returns in Brazil are particularly high, U.S. investors are inclined to withdraw money from a Brazil fund (which they cannot do in the aggregate in a closed-end fund so the premium falls to a level where Brazilian funds are more attractive) in favor of other EM funds. If, however, Brazil performs only in line with other emerging markets, this effect disappears. If all EM funds do well, investors get more enthusiastic about EMs and the premia on all EM funds rises. This apparent contradiction can be explained by investors making a two-tier portfolio allocation decision: first, how much of their portfolio to allocate to EMs, and then, within that, what particular funds to hold. At the higher tier of portfolio allocation, EMs become more attractive to investors when they have yielded higher returns in the recent past. At the lower tier, a particular country fund becomes less attractive when it has outperformed EM funds as a group because investors are anxious to rebalance their EM portfolio by switching to other countries.

Table 30.3 shows the equivalent regression for U.K.-traded funds. For international funds, the main difference between EM funds and the rest is the stronger correlation with premium movements on domestic funds. Again, premia are highly persistent. As with U.S.-based EM funds, there

TABLE 30.3 Fixed Effects Regressions for U.K.-Traded Funds

	Dependent Variable: Log Premium		
		International Funds	
Independent Variables	Domestic Funds	Emerging Market Funds	Other Funds
24 month U.K. stock market return	−0.032 (−0.99)	0.033 (0.74)	−0.021 (−0.93)
24 month fund relative NAV return	0.129 (2.60)	0.023 (0.67)	0.017 (0.89)
24 month NAV return on similar funds	−0.069 (−2.15)	0.065 (2.05)	0.085 (4.40)
24 month lagged premium	0.454 (10.3)	0.355 (5.13)	0.240 (4.05)
24 month change in domestic fund premia		0.672 (5.66)	0.212 (2.44)
R^2	0.400	0.393	0.366
Standard error	0.0825	0.0824	0.0716
Sample size	5276	1419	2709

Notes: Figures in parentheses are robust t-statistics. For emerging-market funds, only observations from January 1993 onward are used.

is a positive coefficient on the 24 month return on similar funds. In contrast to U.S.-based funds, there is no relationship between the relative over- or underperformance of U.K.-based EM funds and current premia. This is likely to be related to the less geographically specialized nature of most U.K. EM funds, i.e., that they are mostly multicountry funds. In terms of the two-tier model of portfolio allocation elaborated earlier, the structure of U.K.-based EM funds means that U.K. investors are more likely to hold competing multicountry EM funds in their portfolio rather than a series of single-country funds. They are likely to react differently to the relative performance of the fund for two reasons. First, they may interpret it as a signal of quality of the manager because competing funds are likely to hold a similar portfolio of assets (unlike in the case of country funds investing in different countries). Second, because the funds are less specialized, returns will be less volatile and therefore investors will be less likely to be concerned about unbalancing their portfolio if they do not react to valuation changes. Both these reasons imply a less negative relationship between fund NAV returns and the premium for multicountry funds.

TABLE 30.4   Splitting U.S. Emerging Market Funds into Country Funds and Multicountry Funds

	Dependent Variable: Log Premium	
Independent Variables	**Single-Country Funds**	**Multicountry Funds**
24 month U.S. stock market return	−0.110 (−2.24)	−0.023 (−0.70)
24 month fund relative NAV return	−0.209 (−9.07)	−0.014 (−0.64)
24 month NAV return on similar funds	0.153 (4.43)	0.070 (2.31)
24 month lagged premium	0.422 (6.52)	0.548 (9.94)
24 month change in domestic fund premia	0.314 (2.20)	0.589 (4.59)
R^2	0.541	0.588
Standard error	0.127	0.0885
Sample size	2248	1353

Notes: Figures in parentheses are robust *t*-statistics. Only observations from January 1993 onward are used.

To test whether single-country funds are markedly different from multicountry funds, we estimate the same regression separately for these two types of U.S.-based EM funds. The results are shown in Table 30.4. The differences are highly significant ($F(6, 3549) = 29.3$, compared with a 0.01 critical value of 2.80). As suggested by the theory of portfolio allocation outlined previously, the multicountry funds look much more like the U.K. emerging market funds shown in Table 30.3 and are not characterized by a significant negative coefficient on the fund's relative NAV return. Thus it appears that highly specialized funds (such as country funds) do indeed behave differently from less specialized ones. Our interpretation, as suggested earlier, is that holders of specialized funds are more likely to rebalance their portfolios in response to valuation changes, tending to sell (buy) when a fund has markedly increased (reduced) its portfolio weight. This rebalancing appears only to apply *within* the EM component of a portfolio because good performance by *all* EMs has the opposite effect—of increasing the premium on EM funds, presumably because of anticipated momentum effects.

Table 30.5 shows short-run fixed effects regressions for the United States—i.e., with a 1 month lag of the premium and a 1 month return period. Again, the premium one month previously is highly correlated

TABLE 30.5   Short-Run Fixed Effects Regressions for U.S.-Traded Funds

| Independent Variables | Domestic Funds | Dependent Variable: Log Premium | |
| | | International Funds | |
		Emerging Market Funds	Other Funds
1 month U.S. stock market return	0.217 (7.07)	0.262 (9.09)	0.274 (6.52)
1 month fund relative NAV return	−0.394 (−4.42)	−0.277 (−13.7)	−0.180 (−6.63)
1 month NAV return on similar funds	−0.296 (−6.54)	0.076 (2.92)	−0.002 (−0.04)
1 month lagged premium	0.862 (58.3)	0.897 (130.4)	0.871 (47.2)
1 month change in domestic fund premia		0.709 (13.2)	0.672 (9.35)
R^2	0.874	0.878	0.830
Standard error	0.0369	0.0593	0.0567
Sample size	3967	4005	1866

Notes: Figures in parentheses are robust t-statistics. For emerging-market funds, only observations from January 1993 onward are used.

with the current premium. Also, for international funds, the change in the average domestic premium is highly significant, as it was for the 24 month horizon. One month return in the U.S. stock market has a significant positive coefficient for all funds. This was not evident with 24 month returns, suggesting that the finding of a positive 1 month U.S. stock market return is purely a short-run phenomenon. We also observe a significant negative relationship between the premium and the fund's own NAV return relative to the asset class. This is the familiar price inertia effect widely documented with higher frequency data. What is surprising and inconsistent with the notion that international borders delay the dissemination of information is that it is stronger for domestic funds than for international funds.

One month results for the United Kingdom are shown in Table 30.6. They are very similar to those for the United States. EM fund premia are positively related to returns on the asset class. Note that, as with the United States, price inertia—a negative coefficient on the relative NAV return—is strongest in domestic funds. We also observe, as with the United States, that the 1 month domestic stock market return is positive and significant

TABLE 30.6 Short-Run Fixed Effects Regressions for U.K.-Traded Funds

	Dependent Variable: Log Premium		
		International Funds	
Independent Variables	**Domestic Funds**	**Emerging Market Funds**	**Other Funds**
1 month U.K. stock market return	0.018 (0.53)	0.252 (5.68)	0.253 (8.63)
1 month fund relative NAV return	−0.391 (−7.58)	−0.170 (−6.31)	−0.088 (−2.38)
1 month NAV return on similar funds	0.004 (0.12)	0.087 (2.87)	−0.064 (−2.59)
1 month lagged premium	0.930 (159.0)	0.865 (72.5)	0.856 (74.1)
1 month change in domestic fund premia		0.547 (4.77)	0.323 (5.30)
R^2	0.888	0.817	0.820
Standard error	0.0367	0.0451	0.0402
Sample size	5805	1595	3192

Notes: Figures in parentheses are robust t-statistics. For emerging-market funds only observations from January 1993 onward are used.

for international funds. The correlation of monthly movements in domestic and international funds is highly significant, but not as strong as in the United States.

30.5 CONCLUSIONS

Capital flows to emerging markets are volatile. The behavior of premia on emerging-market closed-end funds is interesting because it is relevant to their attractiveness for portfolio diversification purposes and also because premia may be regarded as an index of sentiment of international investors toward emerging markets. This chapter has investigated what drives premia on closed-end funds. In general, EM funds have some features in common with domestic and other international funds, but also exhibit some significant differences. Our main findings are as follows:

1. EM fund premia are strongly positively related to the level of past premia.

2. The correlation between premium movements in domestic and international funds is strong at both short and long horizons, in both the United States and the United Kingdom.

3. At short horizons, EM premia comove with the domestic stock market, but this is a temporary effect that disappears at longer horizons.

4. Stronger returns in emerging markets raise premia on emerging-market funds, but strong returns to a particular fund that are not matched by emerging markets generally do not. Especially at short horizons, strong relative returns to a particular fund are associated with a *fall* in the premium.

5. Short-run price inertia is stronger in domestic than in international funds, in both the U.S. and the U.K. markets. This is hard to reconcile with the idea that inertia results from information differences between holders of a closed-end fund and holders of its underlying assets as one would expect that geographical distance would accentuate these differences.

REFERENCES

Bekaert, G. and Urias, M. (1996) Diversification, integration and emerging market closed-end funds. *Journal of Finance*, 51(3), 835–869.

Bleaney, M.F. and Smith, R.T. (2003) Prior performance and closed-end fund discounts. Discussion Paper no. 03/16, University of Nottingham School of Economics, Nottingham, UK.

Bleaney, M.F. and Smith, R.T. (2006) Price under-reaction to news in financial markets: Evidence from closed-end funds. *Finance Letters*, in press.

Bodurtha, J., Kim, D., and Lee, C. (1995) Closed-end country funds and U.S. market sentiment. *Review of Financial Studies*, 8(3), 879–918.

Bonser-Neal, C., Brauer, G., Neal, R., and Wheatley, S. (1990) International investment restrictions and closed-end country fund prices. *Journal of Finance* 45(2), 523–547.

Brickley, J., Manaster, S., and Schallheim, J. (1991) The tax-timing option and the discounts on closed-end investment companies. *Journal of Business*, 64(3), 287–312.

Chan-Lau, J.A., Mathieson, D.J., and Yao, J.Y. (2004) Extreme contagion in equity markets. *IMF Staff Papers*, 51(2), 386–408.

Chang, E., Eun, C., and Kolodny, R. (1995) International diversification through closed-end country funds. *Journal of Banking and Finance*, 19(7), 1237–1263.

Dimson, E. and Minio-Kozerski, C. (1999) Closed-end funds: A survey. *Financial Markets, Institutions and Instruments*, 8(2), 1–41.

Elton, E., Gruber, M., and Busse, J. (1998) Do investors care about sentiment? *Journal of Business*, 71(4), 477–500.

Eun, C., Jankiramanan, S., and Senbet, L. (2002) The pricing of emerging market country funds. *Journal of International Money and Finance*, 21(6), 833–855.

Fama, E. and MacBeth, J. (1973) Risk, return, and equilibrium: Empirical tests. *Journal of Political Economy*, 81(3), 607–636.

Frankel, J. and Schmukler, S. (2000) Country funds and asymmetric information. *International Journal of Finance and Economics*, 5(3), 177–195.

Gemmill, G. and Thomas, D. (2002) Noise trading, costly arbitrage and asset prices: Evidence from closed-end funds. *Journal of Finance*, 57(6), 2571–2594.

Hardouvelis, G., La Porta, R., and Wizman, T. (1994) What moves the discount on country equity funds. In: J. Frankel (Ed.), *The Internationalization of Equity Markets*. University of Chicago, Chicago, IL.

Kaufmann, D., Kraay, A., and Mastruzzi, M. (2003) Governance matters III: Governance indicators for 1996–2002. Working Paper, World Bank, Washington, DC.

Klibanoff, P., Lamont, O., and Wizman, T. (1998) Investor reaction to salient news in closed-end country funds. *Journal of Finance*, 53(2), 673–699.

Lee, C., Shleifer, A., and Thaler, R. (1991) Investor sentiment and the closed-end fund puzzle. *Journal of Finance*, 46(1), 75–109.

Levy-Yeyati, E. and Ubide, A. (2000) Crises, contagion, and the closed-end country fund puzzle. *IMF Staff Papers*, 47(1), 54–89.

Newey, W. and West, K. (1994) Automatic lag selection in covariance matrix estimation. *Review of Economic Studies*, 61(209), 631–653.

Nishiotis, G. (2004) Do indirect investment barriers contribute to market segmentation? *Journal of Financial and Quantitative Analysis*, 39(3), 613–630.

Pontiff, J. (1995) Closed-end fund premia and returns: Implications for financial market equilibrium. *Journal of Financial Economics*, 37(3), 341–370.

Sias, R. (1997) The sensitivity of individual and institutional investors' expectations to changing market conditions: Evidence from closed-end funds. *Review of Quantitative Finance and Accounting*, 8(3), 245–269.

Tornell, A. and Westermann, F. (2002) Boom–bust cycles in credit-constrained economies: Facts and explanation. *IMF Staff Papers*, 49 (Special Issue), 111–155.

CHAPTER 31

Financial Distress
and Emerging Markets

Stephen J. Lubben

CONTENTS

31.1 INTRODUCTION

The ability to enforce debt obligations and to distribute a debtor's assets in a rational fashion after default has important effects on the initial decision to invest in a particular jurisdiction. Seen in this light, legal rules specifying creditor rights, collateral regimes, and bankruptcy or insolvency procedures are vital parts of a developing economy* (Martin, 2005).

And when considered historically, the tools for addressing financial distress do seem to develop hand in hand with a country's economy. Thus England enacted its first between bankruptcy law for traders in 1705 (Goode, 2005) and the United States enacted a similar law in 1800 (Mann, 2002). Insolvency procedures dealing with corporate entities came in the mid- to late-nineteenth century (Lubben, 2004). China, on the other hand, enacted its first bankruptcy law in 1986 and even then the law only applied to state-owned enterprises (Zhang and Booth, 2001).

Indeed, until recently, if an emerging economy had a bankruptcy or insolvency law at all, it was typically a copy of the law of the current or former colonial parent at the time of adoption (Smart and Booth, 2001). In some jurisdictions, "the current law on corporate reorganization … still smacks of antiquated nineteenth century British experimentation gone sour" (Adeniran, 2003). Indeed, given the historical vintage of these older laws, bankruptcy or insolvency principally means liquidation (Carter, 2000). While these laws were probably sufficient when enacted, they are deficient tools for dealing with the results of increasingly globalized economies (Smart et al., 2007).

International trade has produced a much more competitive environment, especially in local markets in developing nations. In addition, emerging markets are now home to both large locally grown businesses and local assets of multinational firms. Both types of firms are apt to experience financial distress—especially during periods of global economic disruption—and often liquidation of these firms will be neither desirable nor socially efficient (Locatelli, 2008).

In recent years, several emerging economies have reacted to these realities and updated their insolvency laws (Westbrook, 2000). At the same time, many developed economies have taken great steps toward

---

* My focus throughout this chapter is the resolution of business or sovereign financial distress. I adopt the American convention of referring to such proceedings as either bankruptcy or insolvency proceedings, regardless of whether liquidation or reorganization is the intended result and with the knowledge that "bankruptcy" in some jurisdictions solely refers to the financial distress of individuals.

recognizing and coordinating insolvency procedures across borders. This chapter reviews these twin developments and also discusses the difficult resolution of the related issue of sovereign financial distress. While the world is increasingly coordinating and unifying tools for addressing corporate financial distress, sovereign defaults are still subject to competition and mistrust. In short, while the world approaches unity in one respect, much work remains to be done.

31.2 REFORM OF BANKRUPTCY AND INSOLVENCY STATUTES

The embrace of corporate rescue procedures was sparked by the U.S. adoption of chapter 11 in 1978 (Jacoby, 2006). While initially highly criticized by domestic commentators who felt the new law was not sufficiently "free market," chapter 11 eventually achieved a degree of domestic acceptance (Lubben, 2005; Westbrook, 2005). International curiosity followed, prompting a variety of new business bankruptcy reform in both developed and developing nations. The move to reorganization in place of liquidation has been further urged by international financial actors like the United Nations Commission on International Trade in its *Legislative Guide on Insolvency Law*, and the World Bank in its *Principles and Guidelines for Effective Insolvency and Creditor Rights Systems*. The latter document expressly provides that "rescue of a business should be promoted through formal and informal procedures."

Countries from Switzerland to Mozambique have thus revised their insolvency statutes to allow for a greater ability to reorganize troubled businesses. While these reforms have been global, the developments in economies that have experienced significant growth, particularly Latin America and Asia, warrant special attention.

Of course, it bears noting at the outset that the process of bankruptcy revision has not been without its problems. Argentina discovered how difficult it is to revise insolvency procedures in the midst of financial crisis when the stakes of revision are much higher. It revised its bankruptcy law at least three times in 2002 alone (Gómez Giglio, 2008). Other countries, like Thailand, adopted revised corporate rescue schemes as the result of pressure from international financial actors like the International Monetary Fund. It is unknown if such compulsion will diminish the acceptance of such systems in the long term (Miller, 2003). More generally, one has to be concerned about the effects of unthinking importation of chapter 11 into very distinct legal cultures (Martin, 2005).

And each of the bankruptcy systems discussed in the following text will only achieve its full potential if the underlying legal institutions function in a stable and efficient manner: a significant, if often neglected, footnote to all that follows in this chapter.

31.2.1 Latin America

Chile was the first nation in Latin America to revisit its bankruptcy laws in the modern era, but its amendments in the 1980s retained a liquidation-based system with little concern for business rescue (Ugarte Vial, 2007). In a recent report, the World Bank noted that these proceedings often take two or three years and result in little recovery for unsecured creditors (World Bank, 2004).

In the mid-1990s, Argentina began the real boom in bankruptcy reform by adopting a new system that included, among other things, an ostensibly chapter 11–inspired reorganization provision. The plan approved under this system does not affect secured creditors, perhaps making the more apt analogy the composition proceedings that existed under Section 12 of the 1898 Bankruptcy Act in the United States (Claps and Macdonnell, 2002). The law was drastically amended and then essentially "un-amended" during the financial crisis in 2002 as the legislature responded to domestic and international pressure resulting from the crisis.

Under the Argentinean system, a debtor must seek court approval to reorganize (Dobson, 1998). If the motion for reorganization is granted, the court appoints a provisional committee of creditors and a trustee to oversee the process. A stay is put in place and the debtor works on plan—during a rather fleeting 90 day exclusive period that can be extend by at most 30 more days (Laguna, 2003). In 2002, the law was amended to allow the debtor to seek creditor approval of a plan before filing, which might make these short periods more workable.

If the debtor does not gain approval of a plan in the exclusive period, the firm is appraised and sold. If the debtor firm does not attract any buyers, liquidation under the formal bankruptcy system results. As with many emerging market bankruptcy systems, the Argentinean approach is best suited to deal with small business default and seems ill equipped to handle the financial distress of complex, multinational firms (World Bank, 2002).

Of course, such a law is better than an antiquated law, and although Brazil's gross domestic product rivals that of Canada, and Mexico's gross domestic product rivals that of Australia, until very recently neither state had an insolvency system that worked even as well as Argentina's. Before 2000, Mexico operated under a law first enacted in 1943. It contained two basic provisions: a liquidation scheme, not unlike Chapter 7 in the United States, and a provision allowing a firm to suspend payments to its creditors. The latter provision did not facilitate any reorganization of the debtor firm and was often subject to abuse. The World Bank reported that "[t]he reality seems to be that debtors have been able to use the suspension proceedings to effectively delay creditors, while sapping virtually all value out of the estate" (Sheppard, 2001).

In 2000, Mexico replaced its business bankruptcy procedure with a new law, often said to be based on the American Bankruptcy Code, but in actuality bearing more resemblance to continental insolvency systems like Switzerland's. Under the Mexican law, all debtors file under the bankruptcy system and if they remain there they are liquidated in a Chapter 7–like proceeding. But debtors are also evaluated for possible reorganization under *concurso* proceeding (Graham-Canedo, 2007).

If the debtor is permitted to undertake the *concurso*, the court appoints a *conciliador* to monitor the debtor's operations, and the debtor has between 6 months and 1 year to formulate a plan that is acceptable to creditors. If the debtor fails to gain acceptance of their plan, the case returns to bankruptcy (Good, 2008a). Not only does this new process expedite the resolution of financial distress in Mexico, but the new law also provides for the professionalization of bankruptcy officials through the Instituto Federal de Especialistas de Concursos Mercantiles (the "Federal Institute of Business Insolvency Specialists"). Indeed, one of the few criticisms of the Mexican system is the failure to address the need for postbankruptcy financing—"DIP financing" in American terms.

Perhaps aware of this criticism, Brazil included new provisions for postbankruptcy financing when it revised its law in 2005. Brazil also made some dramatic changes in its approach to debtor–creditor law. Before 2005, secured creditors in Brazil came after labor and tax claimants. And the purchaser of a debtor's assets took subject to both types of claims. This effectively eliminated any secured creditor recovery in bankruptcy—which, not surprisingly, resulted in an observable and substantial risk premium for corporate borrowers—and also precluded any market in distressed assets (Locatelli, 2008).

As part of the 2005 revisions, secured creditors moved ahead of the taxing authorities and labor claims were capped, albeit at a still high level as compared with Anglo-Saxon bankruptcy systems. Of equal importance, the new law adopted the American concept of a bankruptcy asset sale "free and clear" of claims and interests (Felsberg et al., 2006).

The new law arrived just in time for Varig, a leading Brazilian airline: on June 9, 2005, the new statute became effective and on June 17, 2005, Varig filed for protection under it. On July 20, 2006, an auction was held in the proceeding under the auspices of the Brazilian bankruptcy court. A former subsidiary emerged as the winning bidder of most of the debtor's assets. Although the bankruptcy process itself was generally successful, the reorganized firm faltered until a European rival bought it.

Both Brazil and Mexico have put a premium on speed. This is understandable, given the early criticism of chapter 11, especially concerning cases like that of Eastern Airlines, which lingered in an American chapter 11 proceeding for 2 years before liquidating, and the longstanding criticism of the length of bankruptcy proceedings in Latin America (Sheppard, 2001). But it is also possible that rigid time limits will be ill suited to the resolution of a complex reorganization case, especially in times of market disruption.

31.2.2 Asia

In addressing the issue of corporate financial distress, Asian countries faced not only the problem of antiquated bankruptcy systems but also the added challenge of drafting systems for economies that were only beginning to accept private ownership of firms.

China is obviously the most important emerging economy in this sector, and its attempts to develop and update its insolvency system have been watched with much interest (Bufford, 2007). China's first bankruptcy law—not including Hong Kong's law, inspired by a British statute of World War II vintage—was a 1986 statute that addressed financial distress in state-owned firms (Simmons, 2004). With the incredible growth of private firms in the past decade, the need for a broader law was evident.

A key component of the new law is the ability of the debtor to continue operations and reorganize. Like most jurisdictions, other than the United States, an administrator supervises the debtor during the bankruptcy process. The administrator (sometimes referred to as a trustee) is also charged with formulating the reorganization plan within the first 6–9 months of the case. The process also anticipates the involvement of a creditors' committee, one member of which must represent labor interests (Falke, 2007).

Unlike many new bankruptcy systems, the Chinese law, like American chapter 11, allows for the adoption of a plan in the face of creditor rejection. Indeed the Chinese system directly parallels the "cramdown" provisions of chapter 11, allowing a plan to be confirmed over the objection of unsecured creditors if the plan provides they will at least as much recovery as a liquidation and the payments under the plan are consistent with the Chinese priority scheme (Qi, 2008). This gives Chinese debtors a rarely seen tool to counteract the strong creditor control seen in many insolvency systems. In addition, the new law also provides for postbankruptcy financing, a key feature of chapter 11 that allows American firms to overcome cash flow issues that might otherwise lead to liquidation.

The key challenge China will face in implementing this promising new act turns not on the law itself but rather the competence of both the courts and administrators who are key to the process (Good, 2008b). The lack of any history of judicial impendence and the need to rapidly grow a class of sophisticated insolvency professionals might delay or even thwart the law's promise (Qi, 2008).

Overall, China's new law offers the clearest example of a modern system for addressing corporate financial distress. In most other Asian jurisdictions, insolvency law is part of the overall law of business associations. And in many of the less-developed jurisdictions, like Bangladesh and Pakistan, the law still provides no opportunity for reorganization.

India is one country where there has been at least some effort to update the corporate law that governs business insolvency. The Companies (Second Amendment) Act, 2002 revised the country's 1956 corporate law to set up a National Company Law Tribunal and a related Appellate Tribunal (Chakrabarti et al., 2008). This new court system replaces the prior restructuring panel that rarely restructured any companies (Wood, 2007). While a step in the right direction, implementation of even this basic change has been painfully slow and suggestions for more substantive amendments have resulted in little actual change (Bhat, 2007).

Other key developing nations in Asia—like Indonesia and Thailand—adopted revised business bankruptcy provisions at the behest of the International Monetary Fund following the financial crisis in the late 1990s (Carter, 2002). Much like many of the newer systems previously noted in South America, these provisions, while an improvement from past regimes, seem primarily designed to address the financial distress of comparatively small firms. Limited opportunities for postbankruptcy financing and brief periods for reorganization are the rule here. While this kind of statute works for the "typical" case, the lack of a more robust procedure increases the likelihood of government intervention in the failure of a large firm.

One possible exception is South Korea, which originally updated its insolvency laws in the late 1990s in response to international pressure, but kept reviewing the laws even after the crisis had abated (Ko, 2007). This has resulted in a new statute, effective as of 2006, that includes the notion that a trustee will not be appointed in business cases. This concept of the "debtor in possession" of its own bankruptcy estate, rarely seen outside of the United States, has the potential to greatly increase the chances for reorganizations, provided that courts are capable of detecting fraud.

But as with the Chinese law, there are some doubts about whether the judiciary in Korea will be able to implement this new law. And that is the ultimate, if obvious weak link for most developing jurisdictions: the most elegantly drafted bankruptcy statute can be thwarted by ineffective implementation. Nevertheless, the increasing attention that developing nations are giving to issues of business insolvency certainly holds out the hope that future financial shocks will be handled in a more transparent and systematic manner.

31.3 CROSS-BORDER INSOLVENCY INNOVATIONS

At the same time that developing nations have been revising their internal laws on business bankruptcy, developed nations have begun to consider systems for coordinating transnational bankruptcy proceedings. These new systems have special import for developing economies, inasmuch as they offer an avenue for enforcing the newly enacted local procedures against international creditors outside of the home countries' borders. But this move toward global jurisdiction is still largely a developed nation phenomena: for example, despite the recent wave of legislative reform, with the exception of Mexico, Latin American countries have not adopted United Nations Commission on International Trade Law's Model Law on Cross-Border Insolvencies.

The leading example of this new trend is Chapter 15, a new chapter added to the United States Bankruptcy Code in 2005 (Westbrook, 2005a). It is the domestic adoption of the Model Law on Cross-Border Insolvency.

The purpose of the Model Law, and thus Chapter 15, is to provide effective mechanisms for dealing with insolvency cases involving assets and parties in more than one country. Chapter 15 allows a trustee or other representative of a foreign bankruptcy proceeding to come to the United States and have that proceeding recognized and enforced with relative ease. Chapter 15 also gives foreign creditors the right to participate in U.S. bankruptcy cases and prohibits discrimination against foreign creditors. There are no reciprocity requirements in the new chapter; essentially the United States has committed itself to respect the court systems of any country with a plausible judicial system, at least in the area of insolvency. The new chapter expressly requires the court and estate representatives to "cooperate to the maximum extent possible" with foreign courts and foreign representatives and authorizing direct communication between the court and authorized estate representatives and the foreign courts and foreign representatives.

This stands in stark contrast to the current state of the law in many developed nations and the historical posture of many American cases. For example, under the current Swiss Federal Statute on Private International Law, certain domestic creditors are expressly favored over the representative of a foreign bankruptcy court (Berti, 1997). Under Chapter 15 and the Model Law, creditors are to be treated in a global and coherent fashion regardless of where they are located (Pottow, 2007).

The Model Law has been adopted in a host of important jurisdictions—including, in addition to the United States, Japan, Canada, the United Kingdom, and Australia—which greatly increases the likelihood that an emerging markets debtor will be able to bind its important creditors to any reorganization plan formulated in it home country. The missing component is the converse situation: when a large developed economy firm enters bankruptcy in its home country, it is not clear that the proceedings will be respected with regard to assets located in the developing jurisdiction. Other than Mexico, South Africa, and a handful of Eastern European jurisdictions, the Model Law has made little headway in developing economies. This problem is further compounded by the Model Law's limited ability to deal with corporate groups (Ziegel, 2007). But clearly the possibilities for international cooperation on insolvency issues are much better than they were even 5 years ago.

31.4 SOVEREIGN FINANCIAL DISTRESS

The market for emerging sovereign debt had stabilized following the Argentinean default in the early part of this decade. But the recent financial crisis has renewed the market's concern about sovereign default: Argentina, Ecuador, Venezuela, Ukraine, and Pakistan are all presently trading in distressed territory.

While the resolution of private sector financial distress in emerging markets is a source of continued, if qualified, optimism, the issue of sovereign financial distress remains at best a muddle (Sturzenegger and Zettelmeyer, 2006). This is the result of a two-part problem. First, reorganization systems work best when there is a rough balance of power between debtors and creditors: in the United States, the debtor's power to force a plan, if certain rules are followed, and convert the case to Chapter 7, reducing value to creditors, is balanced by creditor oversight and voting rights. There is no similar balance in the sovereign context: a threat to liquidate a nation is not credible for either side and creditors lack any meaningful ability to check the sovereign debtor's behavior (Lubben, 2004a).

Second, the primary motivation for business bankruptcy systems is to solve the holdout problems that prevent an efficient solution by imposing an aggregate settlement on all parties simultaneously. In the absence of a world bankruptcy court—the International Monetary Fund proposed one through its "Sovereign Debt Restructuring Mechanism," without much success—there is no body that could impose such a solution in the case of a sovereign debtor. The courts of a key commercial jurisdiction, like the United States or the United Kingdom, could impose a plan on *most* creditors, but doing so would be predicated on the sovereign debtor submitting to jurisdiction (Buchheit and Gulati, 2002). Doing so would be fraught with political consequences for the debtor nation's leadership.

The bond market's unexpected acceptance of collective action clauses, a contractual solution to financial distress shunned until fairly recently as overly costly, and stark political reality have becalmed the International Monetary Fund's attempts at radical reform (Bratton and Gulati, 2004). But it is widely agreed that the Sovereign Debt Restructuring Mechanism, or something like it, offers at least three key advantages over contractual solutions to sovereign financial distress (Skeel, 2003). First, a codified approach to sovereign debt restructuring solves the problem of coordination among holders of different debt issues, the so-called aggregation problem. Second, the Sovereign Debt Restructuring Mechanism would facilitate postdefault lending in the sovereign context, providing needed liquidity and potentially avoiding the collateral effects of a sovereign default. Third, use of a "sovereign bankruptcy code" could provide the basis for the imposition of an automatic stay, a useful tool for addressing the problem of individual creditor action, a growing problem for most of today's approaches to sovereign debt restructuring.

For these reasons, the market may yet return to the Sovereign Debt Restructuring Mechanism model, especially if collective action clauses prove ineffective in the long run. For example, the growth of class-action litigation may undermine the utility of collective action clauses, leading to a search for stronger measures.

To be sure, further development in this regard will require continued acceptance of the globalism that supported Chapter 15 and the Model Law and avoidance of any political backlash. The Sovereign Debt Restructuring Mechanism implicates sovereignty on two fronts: it impairs the sovereignty of debtor-nations who would use the model as debtors and it threatens the sovereignty of nations whose citizens hold sovereign debt. Both have the potential to trigger political backlash. A "bankruptcy code for nations" will have to wait.

31.5 CONCLUSION

The general financial distress framework in emerging markets is still a work in progress, leaving gaps for regulators to address and investors to exploit. Many developing économies have made great progress reforming their business bankruptcy systems, although these systems are still vulnerable to weakness in local judicial institutions. At the same time, developed countries are moving to lay the groundwork for a true transnational business bankruptcy system—something that is lacking but needed in the case of sovereign financial distress.

REFERENCES

Adeniran, A. (2003) A mediation-based approach to corporate reorganizations in Nigeria. *North Carolina Journal of International Law and Commercial Regulation,* 29(2): 291–349.

Baht, V. (2007) Corporate governance in India: Past, present, and suggestions for the future. *Iowa Law Review,* 92(4): 1429–1457.

Berti, S. (1997) *Swiss Debt Enforcement and Bankruptcy Law: English Translation of the Amended Federal Statute on Debt Enforcement and Bankruptcy (SchKG),* Schulthess Polygraphischer Verlag, Zurich, Switzerland.

Bratton, W. and Gulati, G.M. (2004) Sovereign debt reform and the best interest of creditors. *Vanderbilt Law Review,* 58(1): 1–79.

Buchheit, L. and Gulati, G.M. (2002) Sovereign bonds and the collective will. *Emory Law Journal,* 51(4): 1317–1360.

Bufford, S. (2007) The new Chinese bankruptcy law: Text and limited comparative analysis. *Norton Journal of Bankruptcy Law and Practice,* 5(5): 697–749.

Carter, C. (2000) Saving face in Southeast Asia: The implementation of prepackaged plans of reorganization in Thailand, Malaysia, And Indonesia. *Bankruptcy Developments Journal,* 17(0): 295–999.

Chakrabarti, R., Megginson, M., and Yadav, P. (2008) Corporate governance in India. *Journal of Applied Corporate Finance,* 20(1): 59–72.

Claps, G. and Macdonnell, J. (2002) Secured credit and insolvency law in Argentina and the U.S.: Gaining insight from a comparative perspective. *Georgia Journal of International and Comparative Law,* 30(3): 393–442.

Dobson, J. (1998) Argentina's Bankruptcy Law of 1995. *Texas International Law Journal,* 33(1): 101–118.

Falke, M. (2007) China's new law on enterprise bankruptcy: A story with a happy end? *International Insolvency Review,* 16(1): 63–74.

Felsberg, T., Kargman, S., and Acerbi, A. (2006) Brazil overhauls restructuring regime. *International Financial Law Review,* 2006: 40–44.

Gómez Giglio, G. (2008) Argentine corporate rescue: Judicial and "out of court" (pre-packaged) reorganisation proceedings. *International Corporate Rescue,* 5(2): 77–85.

Good, M. (2008a) A "great leap forward"? or a "leap in the dark"? what happens when the new Chinese Enterprise Insolvency Law meets US courts? *International Corporate Rescue,* 5(1): 25–42.

Good, M. (2008b) More, better, faster: Gauging the effectiveness of Mexican insolvency reform. *International Corporate Rescue,* 5(5): 293–300.

Goode, R.M. (2005) *Principles of Corporate Insolvency Law,* 3rd ed. London: Sweet and Maxwell.

Graham-Canedo, J. (2007) Comparative analysis of bankruptcy legal provisions from Mexico and the United States: Which legal system is more attractive? *DePaul Business and Commercial Law Journal,* 6(1): 19–28.

Qi, L. (2008) The corporate reorganization regime under China's new Enterprise Bankruptcy Law. *International Insolvency Review,* 17(1): 13–32.

Jacoby, M. (2006) Fast, cheap, and creditor-controlled: Is corporate reorganization failing? *Buffalo Law Review,* 54(2): 401–438.

Ko, H. (2007) Korea's newly enacted Unified Bankruptcy Act: The role of the new act in facilitating (or discouraging) the transfer of corporate control. *UCLA Pacific Basin Law Journal,* 24(2): 201–224.

Laguna, F. (2003) Bankruptcy law in Argentina: Practice and procedure. *Business Credit,* 105(4): 47–54.

Locatelli, F. (2008) International Trade and Insolvency Law: Is the UNCITRAL Model Law on cross-border insolvency an answer for Brazil? (An economic analysis of its benefits on international trade). *Law and Business Review of the Americas,* 14(2): 313–345.

Lubben, S. (2004) Railroad receiverships and modern bankruptcy theory. *Cornell Law Review,* 89(6): 1420–1475.

Lubben, S. (2004a) Out of the past: Railroads and sovereign debt restructuring. *Georgetown Journal of International Law,* 35(4): 845–857.

Lubben, S. (2005) The "new and improved" Chapter 11. *Kentucky Law Journal,* 93(4): 839–866.

Mann, B. (2002) *Republic of Debtors: Bankruptcy in the Age of American Independence.* Cambridge, MA: Harvard University Press.

Martin, N. (2005) The role of history and culture in developing bankruptcy and insolvency systems: The perils of legal transplantation. *Boston College International and Comparative Law Review,* 28(1): 1–78.

Miller, J. (2003) A typology of legal transplants: Using sociology, legal history and Argentine examples to explain the transplant process. *The American Journal of Comparative Law,* 51(4): 839–885.

Pottow, J. (2007) The myth (and realities) of forum shopping in transnational insolvency. *Brooklyn Journal of International Law*, 32(3): 785–817.

Sheppard, H. (2001) The new Mexican insolvency law: Policy justifications for U.S. assistance. *UCLA Journal of International Law and Foreign Affairs*, 6(1): 45–87.

Skeel, Jr., D. (2003) Can majority voting provisions do it all? *Emory Law Journal*, 52(2): 417–425.

Smart, P. and Booth, C. (2001) Reforming corporate rescue procedures in Hong Kong. *Journal of Corporate Law Studies*, 1(2): 485–499.

Smart, P., Briscoe, S., and Booth, C. (2007) Insolvent liquidation in Hong Kong: A crisis of confidence. *International Corporate Rescue*, 4(5): 263–275.

Sturzenegger, F. and Zettelmeyer, J. (2006) *Debt Defaults and Lessons from a Decade of Crisis*. Cambridge, MA: The MIT Press.

Ugarte Vial, J. (2007) Information on bankruptcy law in Chile. *International Corporate Rescue*, 4(1): 15–21.

Westbrook, J. (2000) A global solution to multinational default. *Michigan Law Review*, 98(0): 2276–2999.

Westbrook, J. (2005a) Chapter 15 at last. *American Bankruptcy Law Journal*, 79(3): 713–729.

Westbrook, J.L. (2005) The control of wealth in bankruptcy. *Texas Law Review* 82(4): 795–862.

Wood, P. (2007) *Principles of International Insolvency*, 2nd ed. London: Sweet and Maxwell.

World Bank (2002) Report on Observance of Standards and Codes: Argentina, Insolvency and Creditor Rights Systems.

World Bank (2004) Report on Observance of Standards and Codes: Chile, Insolvency and Creditor Rights Systems.

Zhang, X. and Booth, C. (2001) Chinese bankruptcy law in an emerging market economy: The Shenzhen experience. *Columbia Journal of Asian Law*, 15(1): 1–33.

Ziegel, J. (2007) Canada–United States cross-border insolvency relations and the UNCITRAL Model Law. *Brooklyn Journal of International Law*, 32(3): 1041–1079.

CHAPTER 32

State of Corporate Governance in Ukraine

Robert W. McGee

CONTENTS

32.1 INTRODUCTION

Ukraine is one of the 15 former Soviet republics. When the Soviet Union imploded in 1991, the republics declared their independence. Ukraine is the second largest of the former Soviet republics, in terms of population, after Russia.

Culturally, Ukraine is really two countries. The eastern part of the country speaks mostly Russian. There is an active movement to speak Ukrainian in the western part. People who live in the eastern part of the country look more toward Russia whereas those in the western part look more to the West.

Russian influence has been pervasive for several hundred years. Odessa, which is within the borders of Ukraine, was founded by Catherine, the Great of Russia, and looks a lot like a Russian city, although more cosmopolitan than most Russian cities. Kiev, the capital of Ukraine, is more westernized.

Corporate governance is a relatively new concept in Ukraine and all the former Soviet republics. During the Soviet era, there were no private corporations. There was no such thing as common stock or boards of directors in the Western sense of that term. Decisions were made based on political rather than economic considerations. Fiduciary duty was a concept that was totally absent. The bureaucrats who were in charge of a particular enterprise reported to some individual, group of individuals, or commission in the capital city. They had no duty to shareholders because there were no shareholders.

All that changed when the Soviet empire collapsed. State-owned enterprises became partially or wholly privatized. Problems resulted because there was no roadmap to point the way from central planning to the market economy. The people in power did not know how to plan a market economy, although they tried, not realizing that market economies emerge spontaneously in the absence of central planning. But central planning was all the present generation of bureaucrats knew how to do and they did not do it very well. Central planning is a structurally inferior way of organizing society. Ludwig von Mises (1920, 1922) predicted the collapse of the Soviet Union several generations before it actually took place. He correctly pointed out that the lack of a price system would cause resources to be misallocated, leading to massive waste, retarded economic growth, and the eventual collapse of the system.

Ukraine and the other former Soviet republics were groping in the dark, trying to determine how to convert their inefficient state enterprises into something that looked like a Western corporation. Luckily, they received guidance from a variety of sources. The World Bank, the International

Monetary Fund (IMF), the Organisation for Economic Cooperation and Development (OECD), the United States Agency for International Development (USAID), Tacis, the European equivalent of USAID, and all four (1) KMPG, (2) Pricewaterhouse Coopers, (3) Ernst and Young and (4) Deloitte of the Big Four accounting firms pitched in to help. But the process was slow. Mistrust of the Western consultants, coupled with arrogance and incompetence on both sides, led to some false starts and suboptimum results. Some of the Western "experts" advocated various forms of central planning and those who advocated a more market-friendly approach often saw their advice ignored by the *apparatchiks* who were in control of the privatization.

32.2 LITERATURE REVIEW

The OECD has published several white papers and other documents to provide guidance to countries in transition. They published a paper on the *Principles of Corporate Governance* in 1999 (OECD, 1999) and revised it in 2004 (OECD, 2004). They also published white papers on corporate governance in South Eastern Europe (OECD, 2003a), Russia (OECD, 2002), and Latin America (OECD, 2003c) and a survey of corporate governance developments in OECD countries (OECD, 2003b). The International Finance Corporation, in conjunction with the OECD, published a study on good corporate governance practices in Latin America (IFC, 2006). McGee and Preobragenskaya (2006) devoted a chapter of a book to corporate governance in Eastern Europe.

Various studies have been made of corporate governance practices in Russia. One study looked at the Russian banking sector (IFC, 2004). Another looked at Russian industry in general (Guriev et al., 2003). The Expert Analytical Centre (2004) in Moscow published the results of interviews with Western executives working in Russia where various corporate governance issues were discussed. The Russian Institute of Directors has published several studies of corporate governance in Russia (RID, 2003; Belikov, n.d.). McGee and Preobragenskaya (2005) devoted a chapter of a book to corporate governance in Russia. Kuznetsov and Kuznetsova (2008) compared the concept and reality of corporate governance in Russia. McGee and Tarangelo (2008a,b) studied the timeliness of financial reporting in the Russian banking sector. Kryvoi (2008) examined employee ownership and corporate governance in Russia. Djatej et al. (2008) examined corporate governance in Russia from different perspectives. Lazareva et al. (2008) conducted a survey of corporate governance in Russia.

Several scholars have done studies of corporate governance in other former communist countries. Vravec and Bajus (2008) looked at corporate governance in the Slovak Republic. Kozarzewski (2008) examined the formation of corporate governance in Poland, Kyrgyzstan, Russia, and Ukraine. McGee studied corporate governance in Armenia (McGee, 2008a), Georgia (McGee, 2008b), Latvia (McGee, 2008c), the Czech Republic (McGee, 2008d), Hungary (McGee, 2008e), Poland (McGee, 2008f), and Bosnia and Herzegovina (McGee, 2008g).

32.3 CORPORATE GOVERNANCE IN UKRAINE

The World Bank has done a number of studies on corporate governance, mostly in transition and developing countries. Its study of Ukraine was conducted in 2006 (World Bank, 2006).

The methodology the World Bank used was to apply the corporate governance guidelines that the OECD used as benchmarks in its various white papers and publications on principles of corporate governance. It ranked various aspects of corporate governance into the following five categories based on how closely Ukraine corporations came to following the OECD principles:

O = Observed

LO = Largely Observed

PO = Partially Observed

MNO = Materially Not Observed

NO = Not Observed

Table 32.1 summarizes the results of the World Bank categorization.

Table 32.1 clearly shows that Ukraine has a long way to go before reaching what might be called Western standards of corporate governance. None of the items in any of the categories achieved the top rating and only one item achieved the second best rating. If we were to assign 5 points for the best rating and 1 point for the lowest rating, the scores would look as given in Table 32.2. The scores are represented graphically in Figure 32.1.

32.3.1 Disclosure and Transparency

This part of the report discusses issues of disclosure and transparency and points out that good governance includes timely and accurate disclosure

	O	LO	PO	MNO	NO
Disclosure and Transparency					
The corporate governance framework should ensure that timely and accurate disclosure is made on all material matters.				X	
Information should be prepared, audited and disclosed in accordance with high-quality standards of accounting, financial and nonfinancial disclosure, and audit.			X		
An independent audit should be conducted by an independent auditor.			X		
Channels for disseminating information should provide for fair, timely, and cost-effective access to relevant information by users.			X		
The Responsibility of the Board					
Board members should act on a fully informed basis, in good faith, with due diligence and care, and in the best interests of the company and the shareholders.					X
The board should treat all shareholders fairly.					X
The board should ensure compliance with applicable law and take into account the interests of stakeholders.				X	
The board should fulfill certain board functions.				X	
The board should be able to exercise objective judgment on corporate affairs independent from management.				X	
Board members should have access to accurate, relevant, and timely information.				X	
Shareholder Rights					
Protect shareholder rights			X		
Shareholders have the right to participate in, and to be sufficiently informed on, decisions concerning fundamental corporate changes.				X	

(continued)

TABLE 32.1 (continued) Corporate Governance in Ukraine

	O	LO	PO	MNO	NO
Shareholders should have the opportunity to participate effectively and vote in general shareholder meetings.				X	
Capital structures and arrangements that allow disproportionate control.				X	
Markets for corporate control should be allowed to function in an efficient and transparent manner.				X	
Shareholders should consider the costs and benefits of exercising their voting rights.				X	
Equitable Treatment of Shareholders					
The corporate governance framework should ensure the equitable treatment of all shareholders, including minority and foreign shareholders.				X	
Insider trading and abusive self-dealing should be prohibited.			X		
Board members and managers should be required to disclose material interests in transactions or matters affecting the corporation.					X
Role of Stakeholders in Corporate Governance					
The corporate governance framework should recognize the rights of stakeholders.			X		
Stakeholders should have the opportunity to obtain effective redress for violation of their rights.			X		
The corporate governance framework should permit performance enhancement mechanisms for stakeholder participation.		X			
Stakeholders should have access to relevant information.			X		

TABLE 32.2   Scores by Category

	Possible Points	Actual Points	Average Points
Disclosure and transparency	20	11	2.75
Responsibilities of the board	30	10	1.67
Shareholder rights	30	13	2.17
Equitable treatment of shareholders	15	6	2.00
Role of stakeholders in corporate governance	20	13	3.25
Simple average			2.37

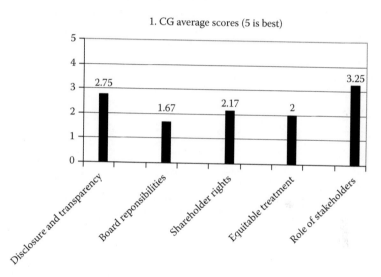

FIGURE 32.1   The scores displayed graphically.

of all material matters, including its financial situation, performance, ownership, and governance. Disclosures should include information about the company's financial and operating results, the company's objectives, major share ownership and voting rights, remuneration policy for board members and key executives, related party transactions and foreseeable risks, to name a few. The company should use high-quality accounting standards and there should be an annual independent audit.

32.3.2 Responsibilities of the Board

The board should effectively oversee the company's management and it should be held accountable to the company and its shareholders. Board members should be fully informed and work in good faith, with due diligence and care, and should always work in the best interests of the shareholders and the company. Board members should treat all shareholders equally and should take stakeholder interests into account.

32.3.3 Shareholder Rights

The report focused on several aspects of shareholder rights, including the right to secure methods of ownership registration, convey or transfer shares, obtain relevant and material company information on a timely and regular basis, participate and vote in general shareholder meetings, elect and remove board members, and share in profits of the corporation.

There was also a concern that shareholders should be able to participate in and be informed about decisions concerning fundamental corporate changes, such as amendments to statutes or articles of incorporation or other company governing documents, authorization of additional shares, and extraordinary transactions. Shareholders should also receive sufficient and timely information regarding the date, time, location, and agenda of the general meeting and should have the opportunity to ask questions.

32.3.4 Equitable Treatment of Shareholders

There was a concern that not all shareholders, especially minority and foreign shareholders, were being treated equitably. The report recommended that all shareholders should have the opportunity to obtain effective redress of grievances if their rights are violated. Minority shareholders are often abused by controlling shareholders and there are obstacles to cross-border voting. The report called for the prohibition of insider trading and abusive self-dealing and board members and key executives should be required to disclose any material interests they have with regard to particular transactions affecting the corporation.

32.3.5 Role of Shareholders in Corporate Governance

This part of the report addresses issues relating to the rights of stakeholders and their right to seek redress of grievances. Where stakeholders participate in corporate governance decisions, they should receive or have access to relevant, sufficient, and reliable information on a timely and regular basis. They should be able to freely communicate their concerns to the board.

32.4 RECOMMENDATIONS

The World Bank (2006) made several policy recommendations to improve the state of corporate governance in Ukraine. They were subcategorized into legislative, institution building, and private sector initiatives and also into immediate, medium term (1–2 years), and long term (3–5 years). Some of the main recommendations are as follows:

Legislative Changes—Immediate

- Disseminate the corporate governance reports on the observance of standards and codes (ROSC) and conduct public workshops to discuss its findings and recommendations.

- Develop an action plan to implement the agreed recommendations of the ROSC.

- Clarify the legal framework by adopting the draft law "On Acknowledgement as Invalid of Some Laws of Ukraine Due to Adoption of the Civil Code of Ukraine."

- Implement the new Law on Securities.

Legislative Changes—Medium Term

- Improve ownership disclosure.

- Reform the company law.

- Introduce the Law on Conglomerates.

Legislative Changes—Long Term

- Review the Law on Collective Investment Funds for compliance with European Union (EU) standards.

- Introduce regulation of takeovers and the market for corporate control.

- Introduce the disclosure of corporate governance policies, share voting policies, and material conflicts of interest by institutional investors.

Institutional Strengthening—Immediate

- The Securities and State Market State Commission (SSMSC) should focus its enforcement efforts.

- All the financial supervisors should be strengthened in their ability to monitor financial institutions.

- The rollout of the new business registry should be completed.

- Enforcement of existing disclosure requirements, including reporting of related party transactions, should be enhanced for publicly traded companies.
- Governance diagnostics should be prepared for the other major parts of the financial sector.

Institutional Strengthening—Medium Term

- The government, in partnership as much as possible with the private sector, should move to rapidly identify a model for future development of a central depository.
- The model for the future development of the central depository system should include higher requirements for share registries and a shareholder approval requirement on registry transfer should be imposed.
- Move more funds settlement onshore.
- A comprehensive effort should be made to standardize court practice to enable the judicial system to deal more effectively with shareholder and corporate litigation.

Institutional Strengthening—Long Term

- Strengthen auditor independence and accountability.
- Work with the private sector to develop a more streamlined process for disseminating company information.

Private Sector Initiatives—Medium Term

- The Persha Fondova Totgovelna Systema (PFTS) should enhance the listing requirements for its top tier companies, including the full adoption of international financial reporting standards (IFRS) and a requirement to "comply or explain" noncompliance with the Ukraine Corporate Governance Principles.
- An Institute of Corporate Governance should be created to provide training to board members and other company officers.

REFERENCES

Belikov, I. (n.d.) *Adoption of the Russian Code of Corporate Conduct: Accomplishments and Problems*. Russian Institute of Directors: Moscow, available at www.rid.ru/.

Djatej, A., Sarikas, R. H. S., and Senteney, D. (2008) Corporate governance in Russia: A consideration of different perspectives. In: R. McGee (Ed.), *Corporate Governance in Transition Economies*. Springer, New York.

Expert Analytical Centre (2004) *Entrepreneurial Ethics and Corporate Governance in Russia: Interviews with Western Executives Working in Russia*. Expert Analytical Centre, Moscow.

Guriev, S., Lazareva, O., Rachinsky, A., and Tsukhlo, S. (2003) *Corporate Governance in Russian Industry*. NES-CEFIR-IET, Moscow.

International Finance Corporation (2004) *A Survey of Corporate Governance Practices in the Russia Banking Sector*. International Finance Corporation, Washington, DC.

International Finance Corporation (2006) *Case Studies of Good Corporate Governance Practices: Companies Circle of the Latin American Corporate Governance Roundtable*. 2nd ed. International Finance Corporation, Washington, DC.

Kozarzewski, P. (2008) Corporate governance formation in Poland, Kyrgyzstan, Russia and Ukraine. In: R. McGee (Ed.), *Corporate Governance in Transition Economies*. Springer, New York.

Kryvoi, K. (2008) Employee ownership and corporate governance in Russia. In: R. McGee (Ed.), *Corporate Governance in Transition Economies*. Springer, New York.

Kuznetsov, A. and Kuznetsova, O. (2008) Corporate governance in Russia: Concept and reality. In: R. McGee (Ed.), *Accounting Reform in Transition and Developing Economies*. Springer, New York.

Lazareva, O., Rachinsky, A., and Stepanov, S. (2008) A survey of corporate governance in Russia. In: R. McGee (Ed.), *Corporate Governance in Transition Economies*. Springer, New York.

McGee, R.W. (2008a) An overview of corporate governance practices in Armenia. In: R. McGee (Ed.), *Corporate Governance in Transition Economies*. Springer, New York.

McGee, R.W. (2008b) An overview of corporate governance practices in Georgia. In: R. McGee (Ed.), *Corporate Governance in Transition Economies*. Springer, New York.

McGee, R.W. (2008c) An overview of corporate governance practices in Latvia. In: R. McGee (Ed.), *Corporate Governance in Transition Economies*. Springer, New York.

McGee, R.W. (2008d) An overview of corporate governance practices in the Czech Republic. In: R. McGee (Ed.), *Corporate Governance in Transition Economies*. Springer, New York.

McGee, R.W. (2008e) An overview of corporate governance practices in Hungary. In: R. McGee (Ed.), *Corporate Governance in Transition Economies.* Springer, New York.

McGee, R.W. (2008f) An overview of corporate governance practices in Poland. In: R. McGee (Ed.), *Corporate Governance in Transition Economies.* Springer, New York.

McGee, R.W. (2008g) An overview of corporate governance practices in Bosnia and Herzegovina. In: R. McGee (Ed.), *Corporate Governance in Transition Economies.* Springer, New York.

McGee, R.W. and Preobragenskaya, G.G. (2005) *Accounting and Financial System Reform in a Transition Economy: A Case Study of Russia.* Springer, New York.

McGee, R.W. and Preobragenskaya, G.G. (2006) *Accounting and Financial System Reform in Eastern Europe and Asia.* Springer, New York.

McGee, R.W. and T. Tarangelo (2008a) The timeliness of financial reporting and the Russian banking system: An empirical study. In: R. McGee (Ed.), *Accounting Reform in Transition and Developing Economies.* Springer, New York.

McGee, R.W. and Tarangelo, T. (2008b) The timeliness of financial reporting: A comparative study of Russian and non-Russian banks. In: R. McGee (Ed.), *Corporate Governance in Transition Economies.* Springer, New York.

Mises, L.v. (1920). Die Wirtschaftsrechnung im Sozialistischen Gemeinwesen. [Economic Calculation in the Socialist Commonwealth] *Archiv für Sozialwissenschaft und Sozialpolitik*, 47: 86–121.

Mises, L.v. (1922) *Die Gemeinwirtschaft: Untersuchungen über den Sozialismus.* Gustav Fischer, Jena, Germany.

Organisation for Economic Cooperation and Development (1999) *Principles of Corporate Governance.* OECD, Paris.

Organisation for Economic Cooperation and Development (2002) *White Paper on Corporate Governance in Russia.* OECD, Paris.

Organisation for Economic Cooperation and Development (2003a) *White Paper on Corporate Governance in South Eastern Europe.* OECD, Paris.

Organisation for Economic Cooperation and Development (2003b) *Survey of Corporate Governance Developments in OECD Countries.* OECD, Paris.

Organisation for Economic Cooperation and Development (2003c) *White Paper on Corporate Governance in Latin America.* OECD, Paris.

Organisation for Economic Cooperation and Development (2004) *Principles of Corporate Governance.* OECD, Paris.

Russian Institute of Directors (2003) *Disclosure of Information about Corporate Governance Practices and Compliance with the Code of Corporate Conduct Recommendations by the Russian Joint-Stock Companies.* Russian Institute of Directors, Moscow.

Vravec, J. and R. Bajus (2008) Corporate governance in the Slovak Republic. In: R. McGee (Ed.), *Accounting Reform in Transition and Developing Economies.* Springer, New York.

World Bank (2006) *Corporate Governance Country Assessment: Ukraine. Reports on the Observance of Standards and Codes (ROSC): Corporate Governance.* World Bank, Washington, DC.

CHAPTER 33

Accounting and Auditing Aspects of Corporate Governance in Emerging Economies

Robert W. McGee

CONTENTS

33.1 INTRODUCTION

Corporate governance is a topic of increasing importance in recent years. Since the collapse of the Soviet Union and the demise of central planning, corporate governance has taken on increasing importance as formerly centrally planned economies try to convert their economic systems to a market economy.

Dozens of books and hundreds, if not thousands, of articles have been written about various aspects of this transformation process. A comprehensive review of this literature is out of the question in such a limited amount of space. But a few noteworthy publications might be mentioned so that readers who might be unfamiliar with the topic will have a starting point for further research.

Two of the author's favorite books on corporate governance examined the state of corporate governance in transition economies (McGee, 2008) and developing economies (McGee, 2009). The Organisation for Economic Cooperation and Development (OECD) published a document outlining the principles of corporate governance (OECD, 1999) and revised it a few years later (OECD, 2004). It also published white papers on corporate governance in Russia (OECD, 2002), southeastern Europe (OECD, 2003a), and Latin America (OECD, 2003c) as well as a study of corporate governance in OECD countries (OECD, 2003b). The Russian Institute of Directors (2003) also published guidance on corporate governance in Russia. The International Finance Corporation published studies of corporate governance in the Russian banking center (2004) and Latin America (2006).

The World Bank has done more than 40 studies of corporate governance in various countries, most of whose economies are either developing or in transition. These studies are listed in the reference section.

The World Bank studies were structured in a way that incorporated the corporate governance principles the OECD developed as benchmarks of good corporate governance principles (OECD, 1999, 2004). The methodology of this chapter uses these benchmarks as the starting point of research as well.

33.2 METHODOLOGY

Some of the corporate governance benchmarks the OECD developed in its *Principles of Corporate Governance* (OECD, 1999, 2004) addressed accounting issues. This chapter examines these benchmarks for the 40 countries the World Bank has examined to date. The World Bank studies rated the extent of compliance with the OECD benchmarks as follows:

O = Observed

LO = Largely Observed

PO = Partially Observed

MNO = Materially Not Observed

NO = Not Observed

The tables show the relative degree of compliance with the OECD principles for each country. The author assigned numerical scores to each of the five categories so that the countries' overall scores could be compared [Observed = 5; Not Observed = 1]. Those results are also shown in the following text.

33.3 FINDINGS

The tables show how each country scored according to the World Bank reports. Separate scores are given for each benchmark that involved an accounting or auditing issue.

33.3.1 Timely and Accurate Disclosure of All Material Matters

One of the OECD principles or best practices requires disclosure and transparency. The nonexclusive list of disclosures it listed (OECD, 2004, p. 10) include

- The financial and operating results of the company
- Company objectives
- Major share ownership and voting rights
- Remuneration and other information about board members and key employees
- Related party transactions

- Material foreseeable risk factors
- Material issues regarding employees and other stakeholders
- Governance structures and policies

Table 33.1 shows the scores for all countries where the World Bank studies assigned a score.

As can be seen, none of the countries earned the top ranking and only nine countries (22.5%) made the Largely Observed category. The most frequent category was Partially Observed with 19 countries (47.5%). Moldova was the only country to earn the lowest ranking in this category.

33.3.2 Standards of Preparation, Audit, and Disclosure of Information

The OECD *Principles of Corporate Governance* (2004) state that best practices require information to be prepared, audited, and disclosed in accordance with high quality, internationally recognized accounting and auditing standards. Table 33.2 shows the extent of compliance with this benchmark.

Two countries—Jordan and Malaysia—earned the top score in this category. Twenty-three countries (57.5%) earned the Partially Observed ranking, which was the most frequent ranking. None of the countries earned the lowest ranking.

33.3.3 Independent Audit

The OECD *Principles* (2004) require an annual audit by an independent auditor. Table 33.3 shows the extent of compliance with this principle.

33.3.4 Fair, Timely, and Cost-Effective Access to Information

The OECD *Principles* (2004) state that companies should provide channels for the equal, timely. and cost-efficient dissemination of information to relevant users. Table 33.4 shows the extent of compliance with this principle.

This category had the most top scores, with eight countries (20%) earning the highest ranking. Fourteen countries (35%) earned the second highest ranking and 10 countries (25%) earned the Partially Observed designation. Two countries—Moldova and Senegal—ranked in the lowest category.

33.4 RANKINGS

Table 33.5 shows the total scores for each country. A score of 5 is assigned for Observed, 4 for Largely Observed, etc. The maximum score is 20 [5 × 4].

TABLE 33.1   Timely and Accurate Disclosure of All Material Matters

Country	O	LO	PO	MNO	NO
Armenia			X		
Azerbaijan				X	
Bhutan			X		
Bosnia and Herzegovina				X	
Brazil			X		
Bulgaria			X		
Chile			X		
Colombia			X		
Croatia				X	
Czech Republic			X		
Egypt			X		
Georgia				X	
Ghana			X		
Hungary		X			
India		X			
Indonesia			X		
Jordan		X			
Korea		X			
Latvia			X		
Lithuania			X		
Macedonia				X	
Malaysia		X			
Mauritius			X		
Mexico		X			
Moldova					X
Nepal				X	
Pakistan		X			
Panama			X		
Peru			X		
Philippines			X		
Poland		X			
Romania			X		
Senegal				X	
Slovakia				X	
Slovenia			X		
South Africa			X		
Thailand		X			
Ukraine				X	
Uruguay				X	
Vietnam				X	

TABLE 33.2 Standards of Preparation, Audit, and Disclosure of Information

Country	O	LO	PO	MNO	NO
Armenia				X	
Azerbaijan				X	
Bhutan			X		
Bosnia and Herzegovina			X		
Brazil			X		
Bulgaria			X		
Chile			X		
Colombia				X	
Croatia				X	
Czech Republic				X	
Egypt		X			
Georgia				X	
Ghana			X		
Hungary			X		
India		X			
Indonesia			X		
Jordan	X				
Korea			X		
Latvia			X		
Lithuania				X	
Macedonia			X		
Malaysia	X				
Mauritius		X			
Mexico			X		
Moldova				X	
Nepal				X	
Pakistan		X			
Panama			X		
Peru			X		
Philippines		X			
Poland		X			
Romania			X		
Senegal			X		
Slovakia			X		
Slovenia			X		
South Africa			X		
Thailand			X		
Ukraine			X		
Uruguay			X		
Vietnam			X		

TABLE 33.3   Independent Audit

Country	O	LO	PO	MNO	NO
Armenia				X	
Azerbaijan				X	
Bhutan			X		
Bosnia and Herzegovina			X		
Brazil			X		
Bulgaria			X		
Chile			X		
Colombia				X	
Croatia				X	
Czech Republic		X			
Egypt			X		
Georgia		X			
Ghana				X	
Hungary			X		
India				X	
Indonesia			X		
Jordan			X		
Korea		X			
Latvia			X		
Lithuania				X	
Macedonia			X		
Malaysia		X			
Mauritius			X		
Mexico			X		
Moldova					X
Nepal				X	
Pakistan		X			
Panama			X		
Peru				X	
Philippines			X		
Poland			X		
Romania				X	
Senegal			X		
Slovakia			X		
Slovenia			X		
South Africa			X		
Thailand		X			
Ukraine			X		
Uruguay			X		
Vietnam			X		

Notes: None of the countries earned the top score in this category. Moldova was the only country in the lowest category. Twenty-three countries (57.5%) were in the Partially Observed category, which was the most frequent ranking.

TABLE 33.4   Fair, Timely, and Cost-Effective Access to information

Country	O	LO	PO	MNO	NO
Armenia				X	
Azerbaijan				X	
Bhutan				X	
Bosnia and Herzegovina			X		
Brazil	X				
Bulgaria			X		
Chile		X			
Colombia	X				
Croatia				X	
Czech Republic		X			
Egypt		X			
Georgia				X	
Ghana		X			
Hungary	X				
India	X				
Indonesia			X		
Jordan		X			
Korea	X				
Latvia		X			
Lithuania		X			
Macedonia				X	
Malaysia		X			
Mauritius		X			
Mexico	X				
Moldova					X
Nepal			X		
Pakistan		X			
Panama	X				
Peru			X		
Philippines			X		
Poland	X				
Romania			X		
Senegal					X
Slovakia			X		
Slovenia		X			
South Africa		X			
Thailand		X			
Ukraine			X		
Uruguay		X			
Vietnam			X		

TABLE 33.5   Rankings

Rank	Country	Score
1	Malaysia	17
2	Jordan	16
3	Korea	16
4	Pakistan	16
5	Poland	16
6	Hungary	15
7	India	15
8	Mexico	15
9	Thailand	15
10	Brazil	14
11	Egypt	14
12	Mauritius	14
13	Panama	14
14	Chile	13
15	Czech Republic	13
16	Latvia	13
17	Philippines	13
18	Slovenia	13
19	South Africa	13
20	Bulgaria	12
21	Colombia	12
22	Ghana	12
23	Indonesia	12
24	Uruguay	12
25	Bhutan	11
26	Bosnia and Herzegovina	11
27	Lithuania	11
28	Peru	11
29	Romania	11
30	Slovakia	11
31	Ukraine	11
32	Vietnam	11
33	Georgia	10
34	Macedonia	10
35	Armenia	9
36	Nepal	9
37	Senegal	9
38	Azerbaijan	8
39	Croatia	8
40	Moldova	5

Malaysia had the top score—17 out of a possible 20. There was a four-way tie for second place. There is no clear pattern at the top end of the scale. Asia, the Middle East, Eastern Europe, and Latin America are all represented. But there does seem to be a pattern at the bottom end of the scale. Five of the lowest 10—Moldova, Azerbaijan, Armenia, Georgia, and Ukraine—are former Soviet republics and two more—Croatia and Macedonia—are former Yugoslav republics. Twelve of the 20 countries in the bottom half of the rankings are former or current communist countries whereas only six of the top 20 countries are former communist countries.

This result is not unexpected. Countries that are transitioning from central planning to a market economy have a longer distance to go than do countries that already have some kind of a market. It is somewhat surprising that some new European Union (EU) countries such as Slovakia, Romania, and Lithuania have such low scores because the EU has minimum standards regarding corporate governance. Perhaps their scores in other areas were sufficiently high to allow them to gain admission to the EU. If these World Bank studies were updated a few years later, the scores are likely to be better as there is a movement in transition and developing countries to upgrade their corporate governance standards.

REFERENCES

International Finance Corporation (2004) *A Survey of Corporate Governance Practices in the Russian Banking Sector.* International Finance Corporation, Washington, DC.

International Finance Corporation (2006) *Case Studies of Good Corporate Governance Practices: Companies Circle of the Latin American Corporate Governance Roundtable.* 2nd ed. International Finance Corporation, Washington, DC.

McGee. R.W. (Ed.) (2008) *Corporate Governance in Transition Economies.* Springer, New York.

McGee, R.W. (Ed.) (2009) *Corporate Governance in Developing Economies.* Springer, New York.

Organisation for Economic Cooperation and Development (1999) *Principles of Corporate Governance.* OECD, Paris.

Organisation for Economic Cooperation and Development (2002) *White Paper on Corporate Governance in Russia.* OECD, Paris.

Organisation for Economic Cooperation and Development (2003a) *White Paper on Corporate Governance in South Eastern Europe.* OECD, Paris.

Organisation for Economic Cooperation and Development (2003b) *Survey of Corporate Governance Developments in OECD Countries.* OECD, Paris.

Organisation for Economic Cooperation and Development (2003c) *White Paper on Corporate Governance in Latin America.* OECD, Paris.

Organisation for Economic Cooperation and Development (2004) *Principles of Corporate Governance*. OECD, Paris.

Russian Institute of Directors (2003) *Disclosure of Information about Corporate Governance Practices and Compliance with the Code of Corporate Conduct Recommendations by the Russian Joint-Stock Companies*. Russian Institute of Directors, Moscow.

WORLD BANK REPORTS

Armenia. World Bank (2005) Report on the Observance of Standards and Codes (ROSC), Corporate Governance Country Assessment, Armenia, April. Washington, DC: World Bank. www.worldbank.org.

Azerbaijan. World Bank (2005) Report on the Observance of Standards and Codes (ROSC), Corporate Governance Country Assessment, Azerbaijan, July. Washington, DC: World Bank. www.worldbank.org.

Bhutan. World Bank (2006) Report on the Observance of Standards and Codes (ROSC), Corporate Governance Country Assessment, Bhutan, December. Washington, DC: World Bank. www.worldbank.org.

Bosnia and Herzegovina. World Bank (2006) Report on the Observance of Standards and Codes (ROSC), Corporate Governance Country Assessment, Bosnia and Herzegovina, June. Washington, DC: World Bank. www.worldbank.org.

Brazil. World Bank (2005) Report on the Observance of Standards and Codes (ROSC), Corporate Governance Country Assessment, Brazil, May. Washington, DC: World Bank. www.worldbank.org.

Bulgaria. World Bank (2002) Report on the Observance of Standards and Codes (ROSC), Corporate Governance Country Assessment, Bulgaria, September. Washington, DC: World Bank. www.worldbank.org.

Chile. World Bank (2003) Report on the Observance of Standards and Codes (ROSC), Corporate Governance Country Assessment, Chile, May. Washington, DC: World Bank. www.worldbank.org.

Colombia. World Bank (2003) Report on the Observance of Standards and Codes (ROSC), Corporate Governance Country Assessment, Colombia, August. Washington, DC: World Bank. www.worldbank.org.

Croatia. World Bank (2001) Report on the Observance of Standards and Codes (ROSC), Corporate Governance Country Assessment, Republic of Croatia, September. Washington, DC: World Bank. www.worldbank.org.

Czech Republic. World Bank (2002) Report on the Observance of Standards and Codes (ROSC), Corporate Governance Country Assessment, Czech Republic, July. Washington, DC: World Bank. www.worldbank.org.

Egypt. World Bank (2004) Report on the Observance of Standards and Codes (ROSC), Corporate Governance Country Assessment, Egypt, March. Washington, DC: World Bank. www.worldbank.org.

Georgia. World Bank (2002) Report on the Observance of Standards and Codes (ROSC), Corporate Governance Country Assessment, Georgia, March. Washington, DC: World Bank. www.worldbank.org.

Ghana. World Bank (2005) Report on the Observance of Standards and Codes (ROSC), Corporate Governance Country Assessment, Ghana, May. Washington, DC: World Bank. www.worldbank.org.

Hungary. World Bank (2003) Report on the Observance of Standards and Codes (ROSC), Corporate Governance Country Assessment, Hungary, February. Washington, DC: World Bank. www.worldbank.org.

India. World Bank (2004) Report on the Observance of Standards and Codes (ROSC), Corporate Governance Country Assessment, India, April. Washington, DC: World Bank. www.worldbank.org.

Indonesia. World Bank (2004) Report on the Observance of Standards and Codes (ROSC), Corporate Governance Country Assessment, Republic of Indonesia, April. Washington, DC: World Bank. www.worldbank.org.

Jordan. World Bank (2004) Report on the Observance of Standards and Codes (ROSC), Corporate Governance Country Assessment, Jordan, June. Washington, DC: World Bank. www.worldbank.org.

Korea. World Bank (2003) Report on the Observance of Standards and Codes (ROSC), Corporate Governance Country Assessment, Republic of Korea, September. Washington, DC: World Bank. www.worldbank.org.

Latvia. World Bank (2002) Report on the Observance of Standards and Codes (ROSC), Corporate Governance Country Assessment, Latvia, December. Washington, DC: World Bank. www.worldbank.org.

Lithuania. World Bank (2002) Report on the Observance of Standards and Codes (ROSC), Corporate Governance Country Assessment, Republic of Lithuania, July. Washington, DC: World Bank. www.worldbank.org.

Macedonia. World Bank (2005) Report on the Observance of Standards and Codes (ROSC), Corporate Governance Country Assessment, Macedonia, June. Washington, DC: World Bank. www.worldbank.org.

Malaysia. World Bank (2005) Report on the Observance of Standards and Codes (ROSC), Corporate Governance Country Assessment, Malaysia, June. Washington, DC: World Bank. www.worldbank.org.

Mauritius. World Bank (2002) Report on the Observance of Standards and Codes (ROSC), Corporate Governance Country Assessment, Mauritius, October. Washington, DC: World Bank. www.worldbank.org.

Mexico. World Bank (2003) Report on the Observance of Standards and Codes (ROSC), Corporate Governance Country Assessment, Mexico, September. Washington, DC: World Bank. www.worldbank.org.

Moldova. World Bank (2004) Report on the Observance of Standards and Codes (ROSC), Corporate Governance Country Assessment, Moldova, May. Washington, DC: World Bank. www.worldbank.org.

Nepal. World Bank (2005) Report on the Observance of Standards and Codes (ROSC), Corporate Governance Country Assessment, Nepal, April. Washington, DC: World Bank. www.worldbank.org.

Pakistan. World Bank (2005) Report on the Observance of Standards and Codes (ROSC), Corporate Governance Country Assessment, Pakistan, June. Washington, DC: World Bank. www.worldbank.org.

Panama. World Bank (2004) Report on the Observance of Standards and Codes (ROSC), Corporate Governance Country Assessment, Panama, June. Washington, DC: World Bank. www.worldbank.org.

Peru. World Bank (2004) Report on the Observance of Standards and Codes (ROSC), Corporate Governance Country Assessment, Republic of Peru, June. Washington, DC: World Bank. www.worldbank.org.

Philippines. World Bank (2006) Report on the Observance of Standards and Codes (ROSC), Corporate Governance Country Assessment, Philippines, May. Washington, DC: World Bank. www.worldbank.org.

Poland. World Bank (2005) Report on the Observance of Standards and Codes (ROSC), Corporate Governance Country Assessment, Poland, June. Washington, DC: World Bank. www.worldbank.org.

Romania. World Bank (2004) Report on the Observance of Standards and Codes (ROSC), Corporate Governance Country Assessment, Romania, April. Washington, DC: World Bank. www.worldbank.org.

Senegal. World Bank (2006) Report on the Observance of Standards and Codes (ROSC), Corporate Governance Country Assessment, Senegal, June. Washington, DC: World Bank. www.worldbank.org.

Slovak Republic. World Bank (2003) Report on the Observance of Standards and Codes (ROSC), Corporate Governance Country Assessment, Slovak Republic, October. Washington, DC: World Bank. www.worldbank.org.

Slovenia. World Bank (2004) Report on the Observance of Standards and Codes (ROSC), Corporate Governance Country Assessment, Slovenia, May. Washington, DC: World Bank. www.worldbank.org.

South Africa. World Bank (2003) Report on the Observance of Standards and Codes (ROSC), Corporate Governance Country Assessment, Republic of South Africa, July. Washington, DC: World Bank. www.worldbank.org.

Thailand. World Bank (2005) Report on the Observance of Standards and Codes (ROSC), Corporate Governance Country Assessment, Thailand, June. Washington, DC: World Bank. www.worldbank.org.

Ukraine. World Bank (2006) Report on the Observance of Standards and Codes (ROSC), Corporate Governance Country Assessment, Ukraine, October. Washington, DC: World Bank. www.worldbank.org.

Uruguay. World Bank (2005) Report on the Observance of Standards and Codes (ROSC), Corporate Governance Country Assessment, Uruguay. September. Washington, DC: World Bank. www.worldbank.org.

Vietnam. World Bank (2006) Report on the Observance of Standards and Codes (ROSC), Corporate Governance Country Assessment, Vietnam, June. Washington, DC: World Bank. www.worldbank.org.

CHAPTER 34

Emerging Market Firms and Bonding Benefits

Eline van Niekerk, Peter Roosenboom, and Willem Schramade

CONTENTS

34.1 INTRODUCTION

According to some market observers (e.g., Faber, 2002), emerging markets will before long surpass the developed markets in importance. Not surprisingly, investors are increasingly turning their attention to these markets. However, a major concern regarding these markets is that they often lack effective protection of minority investors, which makes potential investors reluctant to invest. This could depress equity values. Moreover, it makes it considerably more difficult for a firm to raise external capital than in markets where minority protection is effective (La Porta et al., 1997; Demirguc-Kunt and Maksimovic, 1998). Coffee (1999) and Stulz (1999) argue that there is a natural response for firms to weak investor protection in that they can bond themselves to protect the interests of their minority shareholders. They can do so by agreeing to comply with the rules, regulations, and legal system of a country that does have a strong system of minority investor protection in place, as is the case in the United States. Several authors offer evidence that firms can accomplish this bonding by cross-listing their stock in the United States. For example, Doidge et al. (2004) find higher valuations for cross-listed firms than for otherwise similar firms that do not cross-list. Reese and Weisbach (2002) find that cross-listed firms issue more equity than others, which is consistent with cross-listings being more attractive to firms with a high demand for external finance: bonding is more valuable to firms with many positive net present value (NPV) projects. However, several authors cast doubt on the validity of the bonding hypothesis. For example, Licht (2000, 2003) argues that firms actually prefer to avoid bonding but are willing to accept it as a secondary effect when trying to achieve their primary objective: being listed on the dominant exchanges. Siegel (2005) finds that the Securities and Exchange Commission (SEC) has not effectively enforced the law against cross-listed foreign firms. Still, he finds that cross-listed firms that managed to uphold a clean reputation did benefit from privileged long-term access to outside finance. He therefore concludes that reputational bonding better explains cross-listings than legal bonding does.

Our chapter offers an alternative way of testing whether bonding is reputational or legal. We start out by taking the approach of Reese and Weisbach (2002) to investigate the influence of cross-listings and a firm's home country's legal origin on capital raisings following the cross-listing. However, we take their analysis one step further by also examining the influence of the method of entry, i.e., whether the firm chooses a high protection or low protection cross-listing.* This allows us to investigate whether bonding (if present at all) is of a legal or reputational nature. Following Reese and Weisbach's (2002) reasoning, legal bonding should imply that higher protection entries be more valuable to firms with high external financing needs. Thus, the high protection cross-listings should display more and larger subsequent capital raisings. To investigate this, we assembled a sample of 831 firms from 25 emerging markets (as by the Morgan Stanley capital international [MSCI]) that cross-listed in the United States for the first time between 1964 and 2004. For these firms, we collected all capital raisings and equity issues within 2 years following the cross-listing. We thus have a data set that is on one hand more limited in the number of countries covered (emerging market only), but on the other spans a longer period of time than Reese and Weisbach's (2002) analysis. This allows us to replicate their results and extend their analysis by taking into account the method of entry. We confirm their results that equity issues increase following cross-listings and that this effect is stronger for firms from countries with weaker investor protection. This suggests that firms can bond themselves through cross-listing. In addition, we indeed find a positive relation between the entry method's level of protection and subsequent equity issues, which suggests that bonding is primarily legal in nature.

The setup of this chapter is as follows. In Section 34.2, we briefly survey the literature and formulate our hypotheses, relating to whether bonding happens, and, if so, whether if differs with legal origin and/or entry method protection levels. Our sample is described in Section 34.3. Subsequently, we discuss our results and provide a conclusion in Section 34.4.

34.2 LITERATURE AND HYPOTHESES

Investors can be reluctant to invest (or apply a discount to the value of) in firms incorporated in a jurisdiction with weak protection of minority rights or poor enforcement mechanisms. However, according to the bonding hypothesis, such firms can voluntarily subject themselves to higher disclosure standards

---

* One could argue that if the SEC is ineffective versus foreign firms, the level of protection should not matter. However, we think it does as there are degrees of ineffectiveness.

and stricter enforcement in order to attract investors. These investors need to ensure that corporate insiders do not derive private benefits from the corporation beyond previously agreed levels and that insiders put the invested capital to the best available use. Corporate governance mechanisms are supposed to minimize these risks or, more technically, to minimize the adverse effects of the agency problem (see, e.g., Licht, 2003; Doidge et al., 2004). The costs of such mechanisms are called "bonding costs" (Jensen and Meckling, 1976). Reese and Weisbach (2002), King and Segal (2004), and others argue that the advantages of a U.S. cross-listing work especially well for firms with a weak legal enforcement system in their home country.

34.2.1 To Bond or Not to Bond

It is an empirical question whether bonding occurs and whether it is reputational or legal in nature. Following the line of reasoning of Reese by Weisbach (2002), we hypothesize that bonding by cross-listing is most valuable to those firms that are most in need of external capital. Therefore, the number of subsequent issues and the proceeds from those issues should be a proxy for the value of bonding. Also, firms that cross-list are expected to issue more often than comparable firms and issue more than they did before.

> *Hypothesis 1.a: Firms that cross-list will increase their frequency of issuing capital after the cross-listing.*

The same should apply to the amounts of equity these firms issue.

> *Hypothesis 1.b: Firms that cross-list will increase the amounts of capital they issue after the cross-listing.*

Moreover, if bonding is legal in nature, these proxies should be related to

1. The issuing firm's home country legal origin, and, more importantly

2. The level of protection offered by its entry method to the U.S. market

34.2.2 Legal Origin

Firm-level governance partly depends on the country's legal system. Legal systems fall into two main categories: civil law and common law. The latter laws are generally regarded as having better minority shareholder protection than the former laws. Reese and Weisbach (2002) argue that firms from civil law countries (that have weaker investor protection) should benefit more from the increased investor protection resulting from the cross-listing than firms from common law countries. Moreover, as the investors from countries with a weak protection will become more willing to invest when investor protection rises, equity issues outside the United States should increase more for civil law firms than for common law firms. Although this line of reasoning is based on legal origin, this does not necessarily make the bonding itself legal in nature.

> *Hypothesis 2.a: Cross-listed firms from civil law countries will subsequently issue more equity than firms from common law countries.*

Again, the same should apply to the amounts of equity these firms issue.

> *Hypothesis 2.b: Cross-listed firms from civil law countries will subsequently issue larger amounts of equity than firms from common law countries.*

The actual bonding, if any, is not necessarily legal in nature. Siegel (2004) states that bonding can work even without law enforcement. His Mexican case shows that market-based reputational bonding better helps firms through an economic downturn than legal bonding as the SEC is not sufficiently efficient in punishing firms that violate the rules.

34.2.3 Entry Methods and Protection Levels

To test whether bonding is legal or reputational, we distinguish between high-protection and low-protection entry methods. Level II and III American depository receipt/global depository receipt (ADR/GDR) programs are considered high-protection entry methods, which bond firms to the U.S. legal system. Level I, 144A, and Reg S ADR/GDR programs are classified as low-protection entry methods, as they carry little, if any, legal protection for minority investors. If cross-listing is legal in nature, then firms using high-protection entry methods should benefit more from cross-listing than those using low-protection entry methods, as evidenced by better subsequent market access for the former.

> *Hypothesis 3.a: Firms using high-protection entry methods will subsequently issue more equity than low-protection entrants.*

Again, the same should apply to the amounts of equity these firms issue.

> *Hypothesis 3.b: Firms using high-protection entry methods will subsequently issue larger amounts of equity than low-protection entrants.*

34.3 DATA AND SAMPLE DESCRIPTION

34.3.1 Data

The starting point for assembling our data set is a list of all ADR and GDR programs and direct listings in the United States between January 1964 and December 2004, which we obtained from the ADR department of the Bank of New York. We cross-checked this list with data from JP Morgan and the New York Stock Exchange (NYSE). In addition, the Bank of New York provided us with a list of all terminated and upgraded ADR and GDR programs. Together, these lists give an overview of firms' first ADR programs and direct listings in the specified time period. From this list, all firms incorporated in one of the 25 emerging market countries, as defined by the MSCI emerging markets index, were identified.

By limiting ourselves to the MSCI emerging markets index, we cover fewer countries than Reese and Weisbach's (2002) 48 countries, of which 20 are developed and 28 are developing countries. As stated before, we limit our research to emerging markets as we expect the possible effect of bonding through cross-listing to be strongest for emerging markets, i.e., countries with relatively weak regulations and legal systems.

As our focus is on the first entry, we select the first reported established ADR program for each entrant. The sample includes firms that have their first ADR or GDR program established in the United States between 1964 and 2004, thereby providing a 2 year time frame for firms that have listed in 2004 and have issued equity offerings in the following 2 years. 1964 is chosen as the first year because this is the year where the first reported ADR has been established. Our sampling process resulted in the identification of 831 firms that had their first entry in the U.S. market between 1964 and 2004. By contrast, Reese and Weisbach (2002) only consider the firms that were cross-listed in the United States at a certain point in time (in June 1999), plus the firms that delisted from NYSE or National Association of Securities Dealers Automated Quotation (NASDAQ) over the preceding time span (January 1985 until June 1999).

We use both ADR and GDR programs since the legal implications are essentially the same. From now on, both these two programs will be referred to as "DR." Within the DR programs, we distinguish between programs with low protection that carry little, if any, legal protection for their investors and programs with high protection that do offer such protection. As Reese and Weisbach (2002) only use the sample of NYSE and NASDAQ listings, they effectively limit themselves to high-protection methods. As a result, they focus on the investor protection offered by the country-specific characteristics and cannot distinguish between legal and reputational bonding. We are able to do so without losing the possibility to investigate the influence of the country's legal origin, antidirector, and creditor rights, and the country's political and economical rating. Thus, in many instances, we can and will run regressions similar to those of Reese and Weisbach (2002) in order to make a comparison of findings possible.

34.3.2 Descriptive Statistics

Table 34.1 shows the cross-listings in our sample, split by country and entry method. Only 176 of the 831 firms entered the U.S. market with a DR program that offers a high level of investor protection, while 655 firms, almost 80%, avoid the demanding disclosure and accounting requirements associated with exchange-listed DRs.

The table suggests that several countries have a preference for a specific type of entry method. For example, India and Poland focus mainly on bifurcated deals, whereas South Africa, Russia, Mexico, and Brazil have a clear preference for Level I DR programs. This is partly explained by timing. The SEC adopted Rule 144A and Reg S relatively late (in 1990). So, before that time, firms wishing to list through a DR program that did not require any additional disclosure had only one option: Level I ADRs. This still leaves differences in post-1990 preferences unexplained, but should not have an impact on our analyses, as we only need to make the distinction between low and high levels of protection.

Figure 34.1 shows the entries by year, split by high- and low-protection entries. Entries really took off in the 1990s (after the aforementioned adoption of Rule 144A and Reg S).

To test our hypotheses, we need to examine the number and volume of capital raisings following the cross-listing. We limit these subsequent capital raisings to the 2 year period immediately following the cross-listing, as the influence of the newly established DR program is likely to fade away over longer time periods. As not all types of DRs enable the firm to raise capital at the moment of listing (e.g., Level I and II are noncapital raising DR programs), we choose not to include these initial equity issues, but only the capital raisings subsequent to the cross-listing. As such, we can compare all DR programs and capital raisings on a similar basis. Nevertheless, we do add a control dummy for noncapital-raising DRs into the regression models. From the Securities Data Corporation (SDC),

TABLE 34.1   Cross-Listings by Type and Legal Origin

Country	Low Protection				High Protection		Total
	Reg S	144A	144A/ Reg S	Level I	Level II	Level III	
Panel A: English common law							
India	34	21	41			8	104
Israel	2		1	3	2	6	14
Malaysia				16			16
Pakistan			3				3
South Africa	1	2	3	78	16	1	101
Thailand		1	1	14			16
Subtotal	37	24	49	111	18	15	254
Panel B: German civil law							
China	1	5	3	21	2	25	57
Czech Republic	1	1	2				4
Hungary	3	1	6	2		1	13
Poland	4	1	14	1		1	21
Russia	13	2	7	41	1	4	68
South Korea	4	25	1	1	4	6	41
Taiwan	7	5	43			5	60
Subtotal	33	40	76	66	7	42	264
Panel C: French civil law							
Argentina	1	4	4	3	5	12	29
Brazil	3	4	10	43	13	8	81
Chile		1	3	1	6	18	29
Colombia	1	6	2	3		1	13
Egypt	1	1	6	1			9
Indonesia		1	3	2	1	2	9
Jordan			1	2			3
México	1	12	16	39	7	17	92
Morocco			1				1
Perú		2	3	3		2	10
Philippines		2	5	5		1	13
Turkey	2		16	5		1	24
Subtotal	9	33	70	107	32	62	313
Total	79	97	195	284	57	119	831

we obtained all equity and debt offerings by firms from the 25 emerging market countries between 1964 and 2006.

Table 34.2 shows descriptive statistics for the number and proceeds of both equity and total capital raisings after the cross-listing.

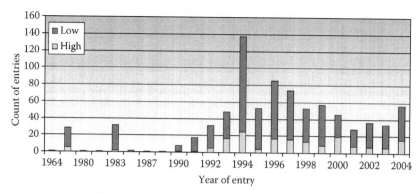

FIGURE 34.1 Method of entry by year.

TABLE 34.2 Capital Raisings

Measurement of Capital Raisings	Number of Firms	Number of Issues	Average	Min	Max	Standard Deviation
Number of capital raisings after cross-listing	339	1287	3.80	1	77	6.13
Proceeds of capital raisings after cross-listing	339	1287	468.2	0.01	9417.1	869.4
Number of equity issues after cross-listing	205	384	1.87	1	8	1.20
Proceeds of equity issues after cross-listing	205	384	264.3	0.00	5028.4	542.8

Source: Thomson's SDC new issues database.
Notes: The average, minimum, maximum, and standard deviation are conditional on the firm's decision to issue.

Forty-one percent of the firms in our sample do at least one subsequent capital raising within 2 years after the cross-listing. And those that do issue 3.8 times on average.

34.4 RESULTS

34.4.1 Subsequent Capital Raisings

One of the reasons firms are supposed to decide to cross-list in the United States is to protect minority shareholder interests. This is particularly of interest when firms plan to raise capital in the near future. Therefore, the expectation is that the number of capital raisings and the proceeds will increase after the cross-listing. To test the corresponding Hypotheses 1.a and 1.b regarding the increase in capital-raising frequency and proceeds, we compare capital raisings in the 2 years before and after the cross-listing. Table 34.3 shows the results.

As Panel A of Table 34.3 shows, the number of capital raisings actually declines by 4% in the period after the cross-listing (1287 capital raisings) versus the 2 years before (1343 raisings), which means we have to reject Hypothesis 1.a. However, and probably more meaningfully, the proceeds raised increase by 23%, confirming Hypothesis 1.b. To put some more

TABLE 34.3   Capital Raisings before and after the Cross-Listing

	English Common Law	German Civil Law	French Civil Law	Total
Panel A: Total capital raisings				
Capital raisings before listing	175	586	582	1343
Capital raisings after listing	216	574	497	1287
(% change)	23.4%	−2.0%	−14.6%	−4.2%
Proceeds of capital raisings before listing	14,860	66,271	47,588	128,719
Proceeds of capital raisings after listing	18,345	82,738	57,649	158,732
(% change)	23.5%	24.8%	21.1%	23.3%
Panel B: Equity issues				
Equity issues before listing	90	288	215	593
Equity issues after listing	61	136	187	384
(% change)	−32.2%	−52.8%	−13.0%	−35.2%
Proceeds of Equity issues before listing	7,316	22,994	33,615	63,925
Proceeds of Equity issues after listing	6,630	12,829	34,718	54,177
(% change)	−9.4%	−44.2%	3.3%	−15.2%
Panel C: Equity issues as a percentage of total capital raisings				
Number of raisings before listing	51.4%	49.1%	36.9%	44.2%
Number of raisings after listing	28.2%	23.7%	37.6%	29.8%
Proceeds before listing	49.2%	34.7%	70.6%	49.7%
Proceeds after listing	36.1%	15.5%	60.2%	34.1%

color on the issue, Panel B highlights equity issues only, which decline both in number of issues (−35%) and in proceeds (−15%). Consequently, this must be offset by a rise in debt issues to reach the overall capital raising numbers. Panel C shows that equity issues as a fraction of total capital raisings declines after the cross-listing and hence that of debt issues increases. In particular, the number of debt issues rises by 20% and their proceeds even by 60%. It seems that firms use the cross-listing as a means to access the bond market.

Table 34.4 shows cross-listings and subsequent capital raisings, split by both entry method and legal origin. For the entire sample of 831 cross-listings, there are a total of 339 firms, raising capital in the 2 years subsequent to their cross-listing on 1287 separate occasions, totaling $159 billion.

Table 34.5 makes the same split but then for subsequent equity issues only rather than for all capital raisings. Two hundred and five firms (25% of the sample firms) do a subsequent equity issue on 384 occasions, raising $54.2 billion. We will first consider legal origin, then entry method protection levels, and finally the combination of both.

34.4.2 Legal Origin and Subsequent Capital Raisings

Hypotheses 2.a and 2.b predict civil law countries to issue more often and raise more proceeds after their cross-listing than common law countries. Tables 34.4 and 34.5 seem to confirm this picture. As Table 34.4 shows, only 27% of common law firms do a subsequent capital raising, whereas 49% (German) and 45% (French) of civil law firms do so. This difference is significant at the 1% level. Average proceeds per firm are also much lower for common law firms ($266 million) than for German ($646 million) and French ($468 million) civil law firms. However, it should be noted that the civil law firms are also larger on average than those from common law countries, making the former a priori more likely do capital raisings and larger ones too. The average market capitalization of common law firms is $872 million, whereas French civil law firms have a market cap of $1180 million and German civil law firms even of $1818 million. When controlling for the firm's market capitalization, the differences remain positive, but become less significant.

The results in Table 34.5 are similar, with many more firms from German (36% of all cases) and French (22%) civil law countries raising subsequent equity than those from common law countries (only 16%). Again, the proceeds raised by civil law firms are higher as well.

TABLE 34.4   Cross-Listings and Subsequent Capital Raisings

	Reg S	144A	144A/Reg S	Level I	Subtotal "Low"	Level II	Level III	Subtotal "High"	Grand Total
Panel A: Full sample									
Number of cross-listings	79	97	195	284	655	57	119	176	831
(% of total)	9.5%	11.7%	23.5%	34.2%	78.8%	6.9%	14.3%	21.2%	100.0%
Cross-listings raising capital after listing	29	52	104	77	262	19	58	77	339
(% of cross-listings)	36.7%	53.6%	53.3%	27.1%	40.0%	33.3%	48.7%	43.8%	40.8%
Number of capital raisings after listing	75	275	415	211	976	67	244	311	1287
(Average number per capital-raising firm)	2.6	5.3	4.0	2.7	3.7	3.5	4.2	4.0	3.8
Proceeds of capital raisings after listing	11,278	20,431	53,758	24,183	109,650	14,414	34,668	49,082	158,732
(Average proceeds per capital-raising firm)	388.9	392.9	516.9	314.1	418.5	758.6	597.7	637.4	468.2
(Average proceeds per capital-raising)	150.4	74.3	129.5	114.6	112.3	215.1	142.1	157.8	123.3
Panel B: English common law									
Number of cross-listings	37	24	49	111	221	18	15	33	254
(% of total)	14.6%	9.4%	19.3%	43.7%	87.0%	7.1%	5.9%	13.0%	100.0%
Cross-listings raising capital after listing	13	10	21	17	61	1	7	8	69
(% of cross-listings)	35.1%	41.7%	42.9%	15.3%	27.6%	5.6%	46.7%	24.2%	27.2%
Number of capital raisings after listing	28	15	117	42	202	1	13	14	216
(Average number per capital-raising firm)	2.2	1.5	5.6	2.5	3.3	1.0	1.9	1.8**	3.1
Proceeds of capital raisings after listing	1,746	2,264	4,387	7,290	15,687	83	2,575	2,658	18,345
(Average proceeds per capital-raising firm)	134.3	226.4	208.9	428.8	257.2	83.4	367.9	332.3	265.9
(Average proceeds per capital-raising)	62.4	150.9	37.5	173.6	77.7	83.4	198.1	189.9	84.9

Panel C: German civil law

Number of cross-listings	33	40	76	66	215	7	42	49	264
(% of total)	12.5%	15.2%	28.8%	25.0%	81.4%	2.7%	15.9%	18.6%	100.0%
Cross-listings raising capital after listing	12	20	53	19	104	3	21	24	128
(% of cross-listings)	36.4%	50.0%	69.7%	28.8%	48.4%	42.9%	50.0%	49.0%	48.5%
Number of capital raisings after listing	41	173	179	37	430	16	128	144	574
(Average number per capital-raising firm)	3.4	8.7	3.4	1.9	4.1	5.3	6.1	6.0	4.5
Proceeds of capital raisings after listing	9,017	11,701	32,994	2,648	56,360	8,927	17,452	26,378	82,738
(Average proceeds per capital-raising firm)	751.5	585.0	622.5	139.4	541.9	2975.5	831.0	1099.1	646.4
(Average proceeds per capital-raising)	219.9	67.6	184.3	71.6	131.1	557.9	136.3	183.2	144.1

Panel D: French civil law

Number of cross-listings	9	33	70	107	219	32	62	94	313
(% of total)	2.9%	10.5%	22.4%	34.2%	70.0%	10.2%	19.8%	30.0%	100.0%
Cross-listings raising capital after listing	4	22	30	41	97	15	30	45	142
(% of cross-listings)	44.4%	66.7%	42.9%	38.3%	44.3%	46.9%	48.4%	47.9%	45.4%
Number of capital raisings after listing	6	87	119	132	344	50	103	153	497
(Average number per capital-raising firm)	1.5	4.0	4.0	3.2	3.5	3.3	3.4	3.4**	3.5
Proceeds of capital raisings after listing	515	6,466	16,378	14,244	37,603	5,404	14,641	20,046	57,649
(Average proceeds per capital-raising firm)	128.7	293.9	545.9	347.4	387.7	360.3	488.0	445.5	406.0
(Average proceeds per capital-raising)	85.8	74.3	137.6	107.9	109.3	108.1	142.2	131.0	116.0

Notes: All panels use the full sample of 831 firms that entered the U.S. market between January 1964 and December 2004. The number of capital raisings after listing is expressed as the average number of raisings per firm (conditional on its decision to raise capital). The proceeds of the capital raisings (in $ million) are expressed as an average per firm (conditional on its decision to raise capital) and per separate raising occasion.
** English Common Law and French Civil Law values are statistically different from each other at the 5% level.

TABLE 34.5   Cross-Listings and Subsequent Equity Issues

	Reg. S	144A	144A/Reg S	Level I	Subtotal "Low"	Level II	Level III	Subtotal "High"	Grand Total
Panel A: Full sample									
Number of cross-listings	79	97	195	284	655	57	119	176	831
(% of total)	9.5%	11.7%	23.5%	34.2%	78.8%	6.9%	14.3%	21.2%	100.0%
Cross-listings issuing equity after listing	18	30	68	35	151	10	44	54	205
(% of cross-listings)	22.8%	30.9%	34.9%	12.3%	23.1%	17.5%	37.0%	30.7%	24.7%
Number of equity issues after listing	31	56	126	52	265	24	95	119	384
(Average number per issuing cross-listing)	1.7	1.9	1.9	1.5	1.8	2.4	2.2	2.2	1.9
Proceeds of equity issues after listing	5,283	4,978	18,710	3,234	32,204	7,935	14,039	21,974	54,178
(Average proceeds per issuing cross-listing)	293.5	165.9	275.1	92.4	213.3	793.5	319.1	406.9	264.3
(Average proceeds per issue)	170.4	88.9	148.5	62.2	121.5	330.6	147.8	184.7	141.1
Panel B: English common law									
Number of cross-listings	37	24	49	111	221	18	15	33	254
(% of total)	14.6%	9.4%	19.3%	43.7%	87.0%	7.1%	5.9%	13.0%	100.0%
Cross-listings issuing equity after listing	9	4	13	10	36	0	5	5	41
(% of cross-listings)	24.3%	16.7%	26.5%	9.0%	16.3%	0.0%	33.3%	15.2%**	16.1%
Number of equity issues after listing	17	7	15	13	52	0	9	9	61
(Average number per issuing cross-listing)	1.9	1.8	1.2	1.3	1.4	—	1.8	1.8	1.5
Proceeds of equity issues after listing	1,457	682	1,529	785	4,454	0	2,177	2,177	6,630
(Average proceeds per issuing cross-listing)	161.9	170.6	117.6	78.5	123.7	—	435.4	435.4	161.7
(Average proceeds per issue)	85.7	97.5	101.9	60.4	85.6	—	241.9	241.9	108.7

Panel C: German civil law

Number of cross-listings	33	40	76	66	215	7	42	49	264
(% of total)	12.5%	15.2%	28.8%	25.0%	81.4%	2.7%	15.9%	18.6%	100.0%
Cross-listings issuing equity after listing	8	14	41	9	72	3	19	22	94
(% of cross-listings)	24.2%	35.0%	53.9%	13.6%	33.5%	42.9%	45.2%	44.9%**	35.6%
Number of equity issues after listing	12	27	82	10	131	8	48	56	187
(Average number per issuing cross-listing)	1.5	1.9	2.0	1.1	1.8	2.7	2.5	2.5	2.0
Proceeds of equity issues after listing	3,792	2,940	12,335	620	19,687	6,963	8,069	15,032	34,718
(Average proceeds per issuing cross-listing)	474.0	210.0	300.8	68.8	273.4	2320.9	424.7	683.3	369.3
(Average proceeds per issue)	316.0	108.9	150.4	62.0	150.3	870.3	168.1	268.4	185.7

Panel D: French civil law

Number of cross-listings	9	33	70	107	219	32	62	94	313
(% of total)	2.9%	10.5%	22.4%	34.2%	70.0%	10.2%	19.8%	30.0%	100.0%
Cross-listings issuing equity after listing	1	12	14	16	43	7	20	27	70
(% of cross-listings)	11.1%	36.4%	20.0%	15.0%	19.6%	21.9%	32.3%	28.7%	22.4%
Number of equity issues after listing	2	22	29	29	82	16	38	54	136
(Average number per issuing cross-listing)	2.0	1.8	2.1	1.8	1.9	2.3	1.9	2.0	1.9
Proceeds of equity issues after listing	34	1,355	4,846	1,829	8,064	972	3,793	4,765	12,829
(Average proceeds per issuing cross-listing)	33.9	112.9	346.1	114.3	187.5	138.9	189.7	176.5	183.3
(Average proceeds per issue)	17.0	61.6	167.1	63.1	98.3	60.8	99.8	88.2	94.3

Notes: All panels use the full sample of 831 firms that entered the U.S. market between January 1964 and December 2004. The number of equity issues after listing is expressed as the average number of equity issues per firm (conditional on its decision to issue equity). The proceeds of the equity issues (in $ million) are expressed as an average per firm (conditional on its decision to issue equity) and per separate issuing occasion.
** English Common Law and French Civil Law values are statistically different from each other at the 5% level.

34.4.3 Entry Method Protection and Subsequent Capital Raisings

Based on the legal bonding argument, Hypotheses 3.a and 3.b expect more subsequent issues and proceeds from firms that use high-protection entry methods. The columns in Tables 34.4 and 34.5 split data by entry method, showing that only 21% of cross-listings are done by means of a high-protection entry method (i.e., ADR Level II or III). These firms do in fact more often undertake subsequent capital raisings (in 44% of cases) than firms using low-protection entry methods, of which 40% do a subsequent capital raising. However, this difference is not significant (p-value of .44). When comparing proceeds, firms using high-protection entry methods do raise significantly (at the 10%) more proceeds ($637 million per firm, $158 million per raising) than low-protection firms ($419 million per firm, $112 million per raising).* A similar picture emerges from the equity issues in Table 34.5: more of the high-protection entrants subsequently issue equity (31% vs. 23%) and they also have higher average proceeds per firm ($407 million vs. $213 million) and per raising ($185 million vs. $122 million). This seems to confirm Hypotheses 3.a and 3.b and hence legal bonding.

34.4.4 Entry Method Protection and Legal Origin

If firms from countries with weaker investor protection are in higher need of bonding, they might also want to opt for higher protection entry methods. We therefore check whether legal origin and entry method are related. As both Tables 34.4 and 34.5 show, and consistent with expectations, firms from civil law countries do more often choose high-protection methods (19% for German, 30% for French) than those from common law countries (13%).

34.5 SUMMARY AND CONCLUSIONS

In this chapter, we investigate whether firms from emerging markets benefit from bonding themselves to the U.S. legal system. As in Reese and Weisbach (2002), we find evidence that cross-listings are more attractive to firms with a high demand for external finance, which is consistent with bonding. For example, cross-listings seem more valuable for firms from countries with weaker investor protection. Although the value of bonding is related to legal origin, this does not necessarily mean that the bonding itself is legal in nature. We therefore extend the analysis of previous research

---

* Conditional on the firm raising capital.

by distinguishing between high legal protection (Levels II and III ADR programs) and low legal protection entry methods (Level I ADR programs and Rule 144 ADRs). This enables us to test whether bonding is legal or reputational in nature. Consistent with legal rather than just reputational bonding, we find that firms that use high legal protection entry methods do more subsequent capital raisings than companies that use low legal protection entry methods to enter the U.S. market. Future research should shed light on how legal bonding works, for example, by studying the destination markets of subsequent issues.

REFERENCES

Coffee, J. (1999) The future as history: The prospects for global convergence in corporate governance and its implications. *Northwestern University Law Review*, 93(3): 641–708.

Demirguc-Kunt, A. and Maksimovic, V. (1998) Law, finance, and firm growth. *Journal of Finance*, 53(6): 2107–2139.

Doidge, C., Karolyi, A., and Stulz, R. (2004) Why are foreign firms listing in the U.S. worth more? *Journal of Financial Economics*, 71(2): 205–238.

Faber, M. (2002) *Tomorrow's Gold: Asia's Age of Discovery*. CLSA Books, Hong Kong.

Jensen, M.C. and Meckling, W.H. (1976) Theory of the firm: Managerial behavior, agency costs, and ownership structure. *Journal of Financial Economics*, 3(4): 305–360.

King, R. and Segal, D. (2004) International Cross-Listing and the Bonding Hypothesis. Working Paper, Bank of Canada, Ottawa.

La Porta, R., Lopez-de-Silanes, F., Shleifer, A., and Vishny, R. (1997) Legal determinants of external finance. *Journal of Finance*, 52(3): 1131–1150.

Licht, A. (2000) Genie in a bottle? Assessing managerial opportunism in international securities transactions. *Columbia Business Law Review*, 2000(1): 51–120.

Licht, A. (2003) Cross-listing and corporate governance: Bonding or avoiding? *Chicago Journal of International Law*, 4(1): 141–163.

Reese, W. and Weisbach, M. (2002) Protection of minority shareholder interests, cross-listings in the united states, and subsequent equity offerings. *Journal of Financial Economics*, 66(1): 65–104.

Siegel, J. (2005) Can foreign firms bond themselves effectively by renting U.S. securities laws? *Journal of Financial Economics*, 75(2): 319–359.

Stulz, R. (1999) Globalization of equity markets and the cost of capital. *Journal of Applied Corporate Finance*, 12(3): 8–25.

CHAPTER 35

Corruption and Public Governance: Evidence from Vietnam

Thuy Thu Nguyen and Mathijs A. van Dijk

CONTENTS

35.1 INTRODUCTION

Corruption can have an important impact on a country's socioeconomic development. Many developing countries have struggled to curb the severity of corruption and its harmful effects on the business environment and the economy. Mitigating corruption requires an understanding of the underlying economic and institutional forces that cause corruption. The literature to date investigates several channels through which corruption arises such as the availability of rents due to various government intervention mechanisms including trade restrictions, price controls, and the provision of credit (see, e.g., Ades and Di Tella, 1999; Treisman, 2000); the opportunities for public officials to initiate misconduct and the low pay for civil servants, which induces a need to collect bribes (see, e.g., Besley and McLaren, 1993).*

The focus of the literature is almost exclusively on country-level measures of corruption. We know little about how corruption varies across different companies within a country and which economic or institutional factors can explain these differences. In this chapter, we document substantial differences in the severity of corruption as experienced by companies across different provinces within a single country: Vietnam. Our data are taken from a survey among almost 900 Vietnamese firms. We specifically investigate the role of provincial public governance in determining the corruption severity in Vietnam. The public governance system is a potentially critical factor in promoting or mitigating corruption. Within-country studies can provide countries with a high level of corruption with policy advice on which local institutions matter for the prevalence and impact of corruption.

This chapter is one of a very limited number of within-country studies on the causes of corruption. This strand of the literature, to the best of our knowledge, consists of Svensson (2003) and Del Monte and Papagni (2007).† Svensson (2003) surveys 176 Ugandan firms and shows that firms' "ability to pay" (measured by their current and expected future profitability) and firms' "refusal power" (measured by the estimated alternative return on capital) can explain a large part of the variation in bribes across firms. Del Monte and Papagni (2007) investigate the determinants of corruption in southern Italian regions. They show that economic variables (e.g., government consumption and the level of development) and political and

---

* For a review of the literature on corruption, see, e.g., Bardhan (1997), Jain (2001), and Aidt (2003).

† Svensson (2003) and Fisman and Svensson (2007) show that there are significant differences in bribery payments among Ugandan firms and industries. Glaeser and Saks (2006) documents differences in the level of corruption across U.S. states.

cultural influences (e.g., party concentration, the presence of voluntary organizations, and absenteeism at national elections) have a significant effect on corruption in Italy. However, neither of these papers examines whether the design and quality of local public governance quality matter for the severity of corruption.

Corruption in Vietnam is severe. Vietnam is ranked 118 out of 163 countries in the 2007 Global Corruption Report. Vietnam has a score of 2.6 for the Corruption Perceptions Index (CPI), which ranges between 10 (highly clean) and 0 (highly corrupt). This score is similar to the scores in the 2000–2005 period, but the country's ranking has actually become worse. According to Transparency International (2007), Vietnam is one of the countries whose government's commitment to ensure adequate support for courts and their personnel has weakened, inviting corruption and undermining the rule of law. Corruption has become a serious policy issue in Vietnam, especially after a number of major corruption cases were discovered in 2006 including the case of Project Management Unit 18 (PMU18), the land corruption cases in Hai Phong, and the corruption case in purchasing equipment by 38 provincial and municipal post offices (CIEM, 2007).

Although several cross-country studies examine the role of the overall national legal effectiveness* and legal origin in affecting corruption (e.g., Treisman, 2000; Herzfeld and Weiss, 2003), these papers do not address the role of local governance structures within countries, partly due to data availability issues. We show that the severity of corruption varies substantially across 24 provinces in Vietnam, a country of French legal origin. Provinces are important administrative units under the central government in Vietnam. They have considerable ability to shape the structure and quality of local public governance. Our detailed data on Vietnam's provincial governance offer an opportunity to investigate whether and how local public governance influences the severity of corruption in the corporate sector.

We measure corruption from two perspectives: (1) the level of corruption severity in the local business environment as perceived by firms and (2) the existence of informal payments to public officials in the industry as perceived by firms. Using information from the World Bank's Productivity and Investment Climate Enterprise Survey and the Vietnam Provincial Competitiveness Index Survey in 2005, we show that the substantial differences in corruption across provinces in Vietnam can be explained by differences in the quality of local public governance (controlling for firm

---

* The legal effectiveness in Herzfeld and Weiss (2003) is defined as the citizens' willingness to accept the established institutions to make and implement laws and adjudicate disputes.

characteristics and industry effects). To the best of our knowledge, our chapter is the first to highlight the role of local public governance in shaping the severity of corruption in the business environment. In particular, we find that corruption is greater in provinces that exhibit less consistent implementation of policies, in which firms face greater costs of regulatory compliance, which have fewer policies to promote private sector development, and which have a greater level of provincial leadership's proactivity. This chapter suggests that improvements in local public governance quality would be a useful mechanism to reduce corruption and its adverse effects on a country's socioeconomic development.

35.2 DATA

We use data from two sources, both of which are based on surveys implemented in 2005. The first source concerns firm-level data obtained from the Productivity and Investment Climate Enterprise Survey conducted by the World Bank. This data set includes firm characteristics, firms' financial information, and firms' assessments of various aspects of the local business environment. The second data source consists of province-level indicators of public governance quality constructed based on the 2005 Vietnam Provincial Competitiveness Index survey. This survey was conducted by the Vietnam Competitive Initiative (VNCI) and the Vietnam Chamber of Commerce and Industry (VCCI).

We construct two measures of corruption. The first measure is the respondents' perception about the corruption level in their local business environment (CORRUPTION). This measure is the sum of scores of corruption ranks indicated by the firms in the World Bank survey. The firms are asked to rank the corruption extent of various agencies using a scale ranging from 0 (= no corruption) to 4 (= widespread corruption). The agencies to be ranked are tax department officials, officials in business registration and licensing, import/export license authorities, the customs department, construction permit authorities, the traffic police, municipal and market-control police, the land administration agency, and the district peoples' committee.* Our CORRUPTION variable thus represents

---

* We calculate Cronbach's alpha coefficient for our measure CORRUPTION. The alphas for our different samples range from 0.79 to 0.83, which means that we have a high reliability in constructing the measure. In addition, we run a factor analysis for the subscores of CORRUPTION and find that there is only one factor that has an eigenvalue greater than 1. For this factor, the factor loadings of the subscores are highly comparable (results are available upon request). These procedures indicate a reliable construction of our measure of CORRUPTION.

the general corruption severity in the local business environment as perceived by the firm. The second measure is the corruption practice in the industry in which firms are operating. PAYMENTDUM is a dummy that takes the value of 1 if the firm perceives that there are informal payments to public officials in the industry and the value of 0 otherwise.

We first consider firm-specific characteristics as determinants of corruption severity facing the firms. The variable definitions are presented in Table 35.1. We expect firm age (AGE) to have a positive relation with corruption. A firm with a longer history of operation is more likely to have working relations with authorities. In addition, the firm is more likely to be the target of corrupt public officials because a better established firm also tends to have higher ability to pay (Svensson, 2003). Similarly, a firm's size (SIZE), growth rate (GROWTH), and profitability (PROFIT) tend to increase the level of corruption that the firm faces because these factors increase the firm's ability to make informal payments. Audited firms (proxied by a dummy variable, AUDIT) are more likely to be transparent financially, which may make it more difficult for public officials to extract informal payments. We therefore expect that audited firms face lower levels of corruption. We expect corruption to occur more frequently in a bureaucratic working environment. A low speed of processing procedures and a cumbersome administrative structure can give rise to a practice of using informal payments to accelerate the operations of the system. We use the variable TIME as a proxy for bureaucracy.

Our main explanatory variables of interest concern the design and quality of local public governance. We use provincial governance indicators provided by VNCI and VCCI, which are all standardized to a 10-point scale. The indicators, whose definitions are given in Table 35.1, are (1) ENTRYCOST—the regulatory costs of firms to enter business; (2) LAND-ACCESS—firms' access to land; (3) TRANSPARENCY—transparency and access to information; (4) STATEBIAS—the bias toward state-owned enterprises; (5) TIMECOST—time costs of regulatory compliance; (6) IMPLEMENTATION—the implementation and consistency of policies; (7) PRIVSECDEV—the policies for private sector development; and (8) PROACTIVE—the proactivity of provincial leadership.

In addition to the variables described earlier, we use industry dummies as control variables, representing the following industries: food and beverage, textiles, apparel, leather products, wood and wood products, paper, chemical and chemical products, rubber and plastic products, nonmetallic mineral products, basic metals, metal products, machinery and equipment, electrical machinery, electronics, construction materials, and vehicles and other transport equipment.

TABLE 35.1 Variable Definitions

Variable	Definition
CORRUPTION	Sum of all the subscores of corruption ranks indicated by the firms (scale: from 0 = no corruption to 4 = widespread corruption). The subscores are corruption levels for tax department officials, officials in business registration and licensing, import/export license authorities, the customs department, construction permit authorities, the traffic police, municipal and other market-control police, the land administration agency, and the district peoples' committee.
PAYMENTDUM	Dummy for the presence of informal payments to public officials in the industry. Answer to the question: "We've heard that establishments are sometimes required to make gifts or informal payments to public officials to 'get things done' with regard to customs, taxes, licenses, regulations, services etc. Does it occur for establishments in your industry (not necessarily yours)?"
AGE	Firm age (in years).
SIZE	Logarithm of total sales (sales are in millions of VND).
GROWTH	Growth rate of total assets.
PROFIT	Profitability as the ratio of after-tax profits to total assets.
AUDIT	Dummy for using an external auditor to certify annual financial statements.
TIME	Fraction of total senior management's time for dealing with government regulations. The government regulations to deal with include taxes, customs, labor regulations, licensing and registration, inspections, etc.
ENTRYCOST	A provincial score that measures the time that takes firms to register, acquire land, and receive all the necessary licenses to start business.
LANDACCESS	A provincial score that measures the access of firms to land resources, i.e., whether firms possess their official land-use-right certificate, whether they have enough land for their business expansion requirements.
TRANSPARENCY	A provincial score that measures transparency and access to information.
STATEBIAS	A provincial score that measures the bias toward state-owned enterprises in terms of incentives, policy and access to capital.
TIMECOST	A provincial score that measures time costs of regulatory compliance.
IMPLEMENTATION	A provincial score that measures the implementation and consistency of policies.
PRIVSECDEV	A provincial score that measures the local policies for promoting private sector development.
PROACTIVE	A provincial score that measures the proactivity of provincial leadership.

Our initial sample consists of all firms in the World Bank survey. We exclude firms that do not have information on total assets, sales or after-tax profits. We also discard the observations with firm age equal to 0 and asset growth above 5 or below −5. Our final data sample consists of 874 Vietnamese firms based in 24 different provinces in Vietnam.

35.3 METHODOLOGY

We use regression analysis to examine the determinants of the corruption severity that firms experience in their local business environment and in their industry. The model is constructed as follows:

$$
CORRUPT_i = \beta_0 + \sum_{k=1}^{5} \beta_k FIRMVARS_{k,i} + \sum_{l=1}^{8} \beta_l GOVERNANCEVARS_{l,i}
$$
$$
+ \sum_{m=1}^{M} \beta_m CONTROLVARS_m + \varepsilon_i \qquad (35.1)
$$

where
  CORRUPT is a measure of corruption as perceived by firm i (CORRUPTION and PAYMENTDUM)
  FIRMVARS represents the firm characteristics namely, AGE, SIZE, GROWTH, PROFIT, and TIME
  GOVERNANCEVARS are the eight subindices of Vietnam's Provincial Competitiveness Index, namely, ENTRYCOST, LANDACCESS, TRANSPARENCY, TIMECOST, IMPLEMENTATION, STATEBIAS, PROACTIVE, and PRIVSECDEV
  CONTROVARS are the control variables—industry dummies

All variable definitions are included in Table 35.1. To estimate Equation 35.1, we use OLS regressions for explaining the level of corruption in local business environment (CORRUPTION) and use logit regressions for investigating the probability that a firm within a particular industry pays informal charges (PAYMENTDUM). As the corruption that individual firms perceive is unlikely to influence the structure and quality of local public governance, there is little reason to believe that endogeneity issues are important in our interpretation of the coefficients on the GOVERNANCEVARS. In all regressions, we use White standard errors to account for heteroskedasticity.

35.4 VARIATION OF CORRUPTION ACROSS FIRMS

In Table 35.2, we present summary statistics of our corruption measures as well as the other variables. The severity of corruption in Vietnam as perceived by individual firms is high. More than 60% of firms think that their industry peers pay informal charges as a common industry practice. The sampled firms estimate that the firms operating within their industry on average pay 0.7% of their revenues (sales) to corrupt public officials. In our sample, the average return on sales is 1.55%.* The informal payments that firms pay thus constitute a large fraction of their profits.

It is striking how much variation there is in the level of corruption perceived by the different firms in our sample. The score that firms assign to corruption in the local business environment ranges from 0 to 39. And

TABLE 35.2 Summary Statistics

	Mean	Median	Std. Dev.	Min	Max	Obs.
CORRUPTION	7.027	6.000	5.557	0	39	874
PAYMENTDUM	0.613	1.000	0.487	0	1	862
AGE	11.919	7.000	12.260	1	115	874
SIZE (total sales in million VND)	68,678.85	9.399	226,384.8	0	4,100,000	874
SIZE (logarithm of total sales)	9.541	9.399	1.821	3.694	15.227	874
GROWTH	0.196	0.106	0.402	−0.592	4.283	874
PROFIT	0.051	0.023	0.164	−0.409	3.399	874
AUDIT	0.296	0.000	0.457	0	1	874
TIME	0.062	0.050	0.070	0	0.80	874
ENTRYCOST	6.415	6.230	0.981	4.27	8.77	874
LANDACCESS	6.474	6.210	1.288	3.67	8.32	874
TRANSPARENCY	4.970	5.190	0.894	3.23	6.72	874
STATEBIAS	5.971	5.900	0.837	4.27	8.53	874
TIMECOST	6.584	6.560	0.699	4.64	8.35	874
IMPLEMENTATION	5.369	4.840	1.241	2.77	8.27	874
PRISECDEV	5.490	5.090	1.486	2.39	8.14	874
PROACTIVE	5.859	6.110	1.681	1.20	9.30	874

Notes: This table presents summary statistics of the measures of corruption, and corruption determinants. Definitions of variables are presented in Table 35.1.

---

* The ratio of informal payments to sales and the return on sales are computed within our sample and are taken from Nguyen and van Dijk (2008).

close to 40% of the firms indicate that they do not perceive informal payments to be part of normal industry practice.

We also observe considerable variation in firm characteristics and provincial governance scores, the latter of which we discuss in the next section. The firms in our samples are of different types of ownership from both private and state sectors. They exhibit considerable differences in terms of size, age, growth rate, profitability, audit practice, and managerial time for regulatory compliance.

We present the distribution of corruption across various subsamples in Table 35.3. The table shows that old firms tend to pay informal charges more often than young firms. Big firms pay significantly more often than small firms. Similarly, high-growth firms are also more involved with informal payments. In addition, more profitable firms face a higher level of corruption in their local business environment. All of these observations are consistent with the argument that firms' ability to pay matters for corruption (Svensson, 2003). More established, bigger firms, or firms with higher growth rates or better profitability are more likely to have the resources to pay corrupt officials and they are also more likely to become the targets that those officials approach. (We note that the variables SIZE, GROWTH, and PROFIT have correlations that are close to zero.)

In contrast to our hypothesis, auditing does not reduce corruption by increasing transparency. Auditing appears to be a service that requires, on average, more working time by the senior management and thus more bureaucracy. However, the firms with and without auditing service do not differ in terms of their perception of corruption severity.

Besides, the findings in Table 35.3 suggest that bureaucracy is a factor that stimulates the severity of corruption. Firms whose managers spend more time dealing with regulatory requirements by the government (i.e., firms with a greater value for the variable TIME) perceive higher levels of corruption and industry practice of informal payments.

35.5 VARIATION OF CORRUPTION ACROSS PROVINCES

Our sample includes data on firms from 24 different provinces in Vietnam. In Table 35.4, we present the means of our corruption and bureaucracy measures across provinces. We also perform an analysis of variance (ANOVA) to test whether the means of the variables are statistically distinguishable from each other. Table 35.4 shows that corruption and bureaucracy vary substantially across provinces in Vietnam. The highest levels of corruption in the local business environment are observed in the provinces

TABLE 35.3 Distribution of Corruption and Bureaucracy Measures by Subsample

By AGE	Young		Old		Young–Old	
	Mean	Obs.	Mean	Obs.	#	p-Value
CORRUPTION	6.792	428	7.253	446	-0.461	0.110
PAYMENTDUM	0.587	424	0.637	438	**-0.050**	0.067
TIME	0.060	428	0.065	446	-0.005	0.160

By SIZE	Small		Big		Small–Big	
	Mean	Obs.	Mean	Obs.	#	p-Value
CORRUPTION	6.811	435	7.241	439	-0.430	0.126
PAYMENTDUM	0.576	429	0.649	433	**-0.073**	0.014
TIME	0.059	435	0.065	439	-0.006	0.116

By GROWTH	Low Growth		High Growth		Low Growth–High Growth	
	Mean	Obs.	Mean	Obs.	#	p-Value
CORRUPTION	6.739	436	7.315	438	**-0.577**	0.063
PAYMENTDUM	0.575	433	0.650	429	**-0.075**	0.012
TIME	0.064	436	0.061	438	0.003	0.739

By PROFIT	Low Profitability		High Profitability		Low Profit–High Profit	
	Mean	Obs.	Mean	Obs.	#	p-Value
CORRUPTION	6.697	435	7.355	439	**-0.659**	0.040
PAYMENTDUM	0.607	428	0.618	434	-0.010	0.381
TIME	0.064	435	0.060	439	0.004	0.823

By AUDIT

	No Audit		Audit		No Audit–Audit	
	Mean	Obs.	Mean	Obs.	#	p-Value
CORRUPTION	6.911	615	7.305	259	-0.394	0.358
PAYMENTDUM	0.598	604	0.647	258	-0.050	0.167
TIME	0.058	615	0.072	259	**-0.014**	0.019

By TIME

	Low Bureaucracy		High Bureaucracy		Low Bureau.–High Bureau.	
	Mean	Obs.	Mean	Obs.	#	p-Value
CORRUPTION	6.456	406	7.524	468	**-1.068**	0.004
PAYMENTDUM	0.574	397	0.645	465	**-0.071**	0.034
TIME	0.017	406	0.102	468	-0.085	—

Notes: This table presents the means of our measures of corruption and bureaucracy across subsamples constructed on the basis of firm age, size, growth, profitability, the amount of time that managers need to deal with regulations, and audit practice. The subsamples of young and old firms, small and big firms, low-and high-growth firms, low-and high-profitability firms, and low and high regulatory-time firms consist of firms with AGE, SIZE, GROWTH, PROFIT, and TIME, respectively, below and above median. The subsamples of firms with and without audit services are based on the dummy AUDIT. The differences between subsamples' means that are significantly different from zero at the 10% level are in bold. Definitions of variables are presented in Table 35.1.

TABLE 35.4   Corruption and Bureaucracy across Vietnam's Provinces

Province	CORRUPTION Mean	CORRUPTION Obs.	PAYMENTDUM Mean	PAYMENTDUM Obs.	TIME Mean	TIME Obs.
Hanoi	9.156	109	0.583	97	0.049	109
Hai Phong	11.119	67	0.955	56	0.060	67
Ha Tay	4.231	26	0.500	24	0.064	26
Bac Ninh	7.476	21	0.714	15	0.057	21
Hai Duong	11.70	10	1.000	7	0.032	10
Nam Dinh	7.423	26	0.731	26	0.082	26
Thanh Hoa	9.434	53	0.83	49	0.063	53
Nghe An	3.222	27	0.259	22	0.103	27
Ha Tinh	2.840	25	0.458	23	0.027	25
Thua Thien Hue	7.786	14	1.000	12	0.070	14
Da Nang	4.719	32	0.531	30	0.040	32
Quang Nam	1.786	14	0.357	13	0.024	14
Quang Ngai	2.167	6	0.400	4	0.090	6
Binh Dinh	6.882	34	0.412	33	0.162	34
Khanh Hoa	6.676	34	0.588	33	0.136	34
Ho Chi Minh City	7.983	181	0.692	151	0.054	181
Binh Duong	6.552	58	0.569	47	0.056	58
Dong Nai	5.563	32	0.469	30	0.076	32
Ba Ria-Vung Tau	4.714	14	0.500	12	0.061	14
Long An	5.286	28	0.679	24	0.027	28
Dong Thap	0.000	6	0.167	6	0.093	6
An Giang	1.933	15	0.333	14	0.033	15
Tien Giang	4.063	16	0.188	13	0.032	16
Can Tho	3.615	26	0.308	24	0.043	26
ANOVA (F test) of differences in the means across provinces						
p-Value	0.000		0.029		0.000	
Correlations						
PAYMENTDUM	0.408					
TIME	0.065		0.036			

Notes: This table presents the cross-province means of the corruption and bureau-cracy measures. We follow the conventions of the Provincial Competitiveness Index report of the VNCI and VCCI and order provinces from the North to the South of Vietnam. Definitions of variables are presented in Table 35.1.

Hai Duong, Hai Phong, Thanh Hoa, Hanoi, Ho Chi Minh City, and Thua Thien Hue. For these provinces, CORRUPTION scores range from 7.8 to 11.7 and the fraction of firms that reports the existence of informal payments to public officials in the industry (PAYMENTDUM) varies from

58% to 100%. The least corrupt business environments are to be found in Dong Thap, Quang Nam, An Giang, Quang Ngai, Nghe An, and Can Tho. The provinces Binh Dinh, Khanh Hoa, and Thua Thien Hue suffer from the highest level of bureaucracy. In these provinces, the senior management of firms spends more than 10% of their working time in dealing with regulations.

The provinces with the lowest level of bureaucracy are Long An, Quang Nam, and Ha Tinh. Our ANOVA tests confirm that there are significant variations of corruption and bureaucracy measures across provinces. The correlations at the bottom of Table 35.4 are relatively low, which indicates that CORRUPTION, PAYMENTDUM, and TIME measure different aspects of the corruption and bureaucracy facing Vietnamese firms.

The statistics on corruption in Table 35.4 are confirmed by anecdotal evidence obtained from the official public media. In December 2007, for example, the vice president of the municipal people's committee in Hai Phong was brought to criminal court because of his involvement in two big corruption cases in land allocation. Hai Phong also dismissed the general secretary of the communist party in one municipal district and the director of municipal department of natural resources and environment, both of whom were convicted for land corruption. In the Thanh Hoa province, many community leaders were arrested for land corruption as well. They received sentences up to 10 years. Similarly, in July 2007, the president of the Go Vap district's people's committee, the general secretary of the communist party, and many other officials in Ho Chi Minh City were caught and sentenced for 11–25 years due to bribery and land corruption. The cases of land corruption mostly involved illegal allocation of land for private usage or for unauthorized groups of people.

From October 2006 till September 2007, over 400 corruption cases in Vietnam were brought to court involving 820 persons. Total damages in these cases were estimated to be up to VND 290 billion (or $20 million). High frequencies of corruption cases were observed in Hanoi, Ho Chi Minh City, Nghe An, Thanh Hoa, Long An, Binh Thuan (see, e.g., broadcast released on Vnexpress.net by Hoang Khue on December 17, 2007).

The provinces in our sample also exhibit a wide range of variation in public governance as measured by the subscores of the provincial competitiveness index. The governance subindices of the 24 provinces are presented in Table 35.5. ENTRYCOST is very high for Da Nang (8.77) and Binh Duong (7.65), but much lower for Ha Tay (4.27) and Hai Duong (4.50).

TABLE 35.5 Provincial Governance Subindices

Province	Entrycost	Landaccess	Transparency	Statebias	Timecost	Implementation	Privsecdev	Proactive
Hanoi	7.28	6.05	4.12	5.72	6.78	4.32	7.73	6.23
Hai Phong	7.02	5.68	5.69	5.98	6.42	4.69	5.29	5.32
Ha Tay	4.27	3.67	3.75	4.27	6.10	4.84	3.27	1.20
Bac Ninh	6.19	6.21	5.37	4.68	8.35	6.62	2.39	7.53
Hai Duong	4.50	5.26	4.18	5.39	6.18	6.62	2.93	3.39
Nam Dinh	5.82	4.23	4.19	5.85	7.41	2.77	2.56	1.60
Thanh Hoa	4.86	5.05	4.54	5.27	7.06	4.17	4.30	3.65
Nghe An	7.15	4.18	5.55	6.01	6.52	5.82	5.82	5.61
Ha Tinh	4.66	6.09	4.52	5.90	5.80	5.60	5.06	4.62
Thua Thien Hue	6.31	5.56	4.49	5.15	6.48	5.52	6.93	5.07
Da Nang	8.77	6.90	6.72	5.26	8.24	6.35	7.54	7.18
Quang Nam	6.23	6.22	4.65	5.92	5.23	8.00	7.03	7.01
Quang Ngai	5.27	5.32	3.85	5.33	5.65	5.67	3.96	4.13
Binh Dinh	5.50	6.40	6.04	5.85	5.92	7.05	5.45	7.11
Khanh Hoa	6.22	6.05	3.33	5.85	5.46	6.30	5.09	5.62
Ho Chi Minh City	6.23	8.32	5.57	6.28	6.56	4.55	4.99	6.11
Binh Duong	7.65	7.88	6.05	8.53	6.29	7.39	6.92	9.30
Dong Nai	6.52	6.42	5.19	6.30	7.88	5.30	4.58	7.74
Ba Ria-Vung Tau	5.33	7.06	4.69	5.80	6.43	6.54	5.93	6.54
Long An	7.24	6.37	3.51	6.22	6.23	5.34	5.17	5.89
Dong Thap	6.27	6.76	4.72	6.29	5.60	6.01	4.53	5.91
An Giang	6.36	7.07	4.10	4.75	4.64	7.96	4.18	5.61
Tien Giang	6.40	6.71	3.23	6.19	6.58	8.27	3.72	5.51
Can Tho	6.13	6.01	5.15	5.40	6.75	5.97	8.14	5.62

Notes: This table presents the subindices of Provincial Competitiveness Index in 2005 for the 24 Vietnamese provinces in our sample. Definitions of variables are presented in Table 35.1.

Similarly, LANDACCESS scores are very high in Ho Chi Minh City (8.32) and Binh Duong (7.88), and much lower in Ha Tay (3.67) and Nghe An (4.18). TRANSPARENCY and STATEBIAS exhibit slightly lower variability, with TRANSPARENCY ranging from 3.23 to 6.72, and STATEBIAS mostly between 5 and 6. The subindex of TIMECOST lies between 4.64 (in An Giang) and 8.35 (in Bac Ninh). The scores of IMPLEMENTATION range from 2.77 in Nam Dinh to 8.27 in Tien Giang. PRIVSECDEV also exhibits high variability among provinces, from 2.39 in Bac Ninh to 8.14 in Can Tho. The variable PROACTIVE shows the highest variation, ranging from 1.20 in Ha Tay to 9.30 in Binh Duong.

35.6 IMPACT OF PROVINCIAL PUBLIC GOVERNANCE ON CORRUPTION IN VIETNAM

In the previous section, we show that there is significant variation of corruption across Vietnam's provinces. In this section, we investigate whether provincial governance variables can explain the variation in the severity of corruption facing firms in these provinces. In Table 35.6, we present the estimation results of Equation 35.1 to identify the determinants of corruption at the firm level.

Panel A of Table 35.6 shows the results of the OLS regression to explain the corruption level in the local business environment (CORRUPTION). Several firm characteristics have a significant impact on the severity of corruption. Firm size (SIZE) and regulatory time costs for managers (TIME) are positively associated with the level of corruption as perceived by the firms. With respect to the economic significance, a one standard-deviation increase in firm size leads to an increase of 9.0% in the corruption severity based on the coefficient estimates in regression model (Equation 35.1). Similarly, a one standard-deviation increase in regulatory time cost is associated with a 6.2% increase in the corruption severity based on the coefficient estimates in regression model (8). These results are robust across all regression specifications and are consistent with the arguments that (1) bigger firms are likely to have a higher "ability to pay" and to become the target of corrupt civil servants and (2) bureaucracy stimulates corruption because informal payments are needed to accelerate the speed of processing procedures in the system.

Controlling for other firm characteristics, profitability no longer has a significant effect on CORRUPTION. Ceteris paribus, we would expect higher growth firms to face higher levels of corruption because growth should enhance their ability to pay in the view of corrupt officials. However, most regressions indicate that GROWTH is negatively associated with corruption severity, a result we find somewhat puzzling.

TABLE 35.6 The Role of Provincial Public Governance in Determining Corruption in Vietnam

Panel A: OLS Regressions—Impact of Provincial Public Governance on Corruption Severity in Local Business Environment

	(1)	(2)	(3)	(4)	(5)	(6)	(7)	(8)	(9)	(10)	(11)	(12)
AGE	−0.005	−0.007	−0.007	−0.006	−0.007	−0.006	−0.009	−0.021	−0.008	−0.009	−0.022	−0.020
	(0.773)	(0.659)	(0.665)	(0.697)	(0.656)	(0.700)	(0.570)	(0.179)	(0.637)	(0.577)	(0.161)	(0.209)
SIZE	**0.347**	**0.334**	**0.331**	**0.328**	**0.333**	**0.337**	**0.341**	**0.331**	**0.341**	**0.345**	**0.337**	**0.325**
	(0.004)	(0.003)	(0.004)	(0.004)	(0.004)	(0.003)	(0.003)	(0.004)	(0.003)	(0.003)	(0.004)	(0.005)
GROWTH	−0.541	**−0.751**	**−0.741**	**−0.732**	**−0.738**	**−0.744**	**−0.696**	**−0.673**	**−0.766**	**−0.778**	−0.434	−0.625
	(0.164)	(0.066)	(0.071)	(0.073)	(0.073)	(0.070)	(0.092)	(0.098)	(0.061)	(0.056)	(0.270)	(0.129)
PROFIT	−0.108	−0.359	−0.335	−0.358	−0.298	−0.300	−0.202	−0.642	−0.387	−0.459	−0.192	−0.292
	(0.936)	(0.755)	(0.771)	(0.750)	(0.796)	(0.792)	(0.863)	(0.570)	(0.739)	(0.700)	(0.870)	(0.785)
AUDIT	0.237	−0.345	−0.365	−0.371	−0.363	−0.393	−0.415	−0.129	−0.326	−0.273	−0.203	−0.282
	(0.601)	(0.441)	(0.424)	(0.407)	(0.417)	(0.380)	(0.357)	(0.768)	(0.470)	(0.544)	(0.653)	(0.526)
TIME	**4.973**	**5.149**	**5.223**	**5.315**	**5.123**	**5.196**	**5.755**	**6.257**	**5.034**	**5.129**	**5.355**	**5.647**
	(0.086)	(0.083)	(0.083)	(0.079)	(0.086)	(0.081)	(0.055)	(0.035)	(0.092)	(0.083)	(0.080)	(0.069)
ENTRYCOST			0.063								0.052	−0.038
			(0.758)								(0.840)	(0.888)
LANDACCESS				0.083							−0.333	−0.304
				(0.610)							(0.173)	(0.241)
TRANSPARENCY					0.162						0.277	0.152
					(0.428)						(0.272)	(0.566)
STATEBIAS						0.186					0.157	0.226
						(0.438)					(0.617)	(0.470)
TIMECOST							**0.747**				−0.288	−0.233
							(0.002)				(0.391)	(0.508)

	(13)	(14)	(15)	(16)	(17)	(18)	(19)	(20)	(21)	(22)	(23)	(24)
IMPLEMENTATION								-1.043			-1.618	-1.497
								(0.000)			(0.000)	(0.000)
PRISECDEV									-0.077		-0.236	-0.260
									(0.543)		(0.147)	(0.118)
PROACTIVE										-0.126	0.619	0.638
										(0.286)	(0.025)	(0.023)
Industry dummies	No	Yes	Yes	Yes	Yes	Yes	Yes	Yes	Yes	Yes	No	Yes
Obs.	874	874	874	874	874	874	874	874	874	874	874	874
R²	0.016	0.063	0.063	0.064	0.064	0.064	0.072	0.112	0.064	0.065	0.097	0.126
Panel B	(13)	(14)	(15)	(16)	(17)	(18)	(19)	(20)	(21)	(22)	(23)	(24)
	Logit Regressions—Impact of Provincial Public Governance on Firms' Propensity to Pay Informal Charges											
AGE	0.000	-0.002	-0.002	-0.002	-0.002	-0.002	-0.003	-0.007	-0.003	-0.004	-0.007	-0.008
	(0.945)	(0.739)	(0.558)	(0.752)	(0.737)	(0.718)	(0.684)	(0.271)	(0.605)	(0.480)	(0.262)	(0.204)
SIZE	0.141	0.173	0.178	0.172	0.173	0.172	0.176	0.178	0.188	0.185	0.165	0.188
	(0.002)	(0.000)	(0.000)	(0.000)	(0.000)	(0.000)	(0.000)	(0.000)	(0.000)	(0.000)	(0.001)	(0.000)
GROWTH	-0.004	-0.064	-0.080	-0.061	-0.059	-0.066	-0.051	-0.033	-0.095	-0.094	-0.006	-0.037
	(0.984)	(0.739)	(0.948)	(0.749)	(0.760)	(0.732)	(0.787)	(0.859)	(0.618)	(0.617)	(0.973)	(0.845)
PROFIT	-0.467	-0.609	-0.648	-0.039	-0.590	-0.620	-0.570	-0.719	-0.684	-0.724	-0.603	-0.743
	(0.400)	(0.363)	(0.379)	(0.368)	(0.379)	(0.350)	(0.386)	(0.303)	(0.278)	(0.218)	(0.316)	(0.304)
AUDIT	-0.018	-0.035	-0.003	0.083	-0.042	-0.025	-0.053	0.029	0.007	0.046	0.027	0.025
	(0.915)	(0.843)	(0.674)	(0.829)	(0.815)	(0.888)	(0.766)	(0.875)	(0.971)	(0.799)	(0.881)	(0.895)
TIME	0.891	0.713	0.588	0.732	0.700	0.706	0.852	1.112	0.462	0.719	1.003	0.919
	(0.377)	(0.491)	(0.067)	(0.481)	(0.500)	(0.495)	(0.419)	(0.291)	(0.662)	(0.486)	(0.363)	(0.402)
ENTRYCOST			-0.100								0.119	0.151
			(0.209)								(0.304)	(0.226)

(continued)

TABLE 35.6 (continued) The Role of Provincial Public Governance in Determining Corruption in Vietnam

Panel B — Logit Regressions—Impact of Provincial Public Governance on Firms' Propensity to Pay Informal Charges

Panel B	(13)	(14)	(15)	(16)	(17)	(18)	(19)	(20)	(21)	(22)	(23)	(24)
LANDACCESS				0.010							−0.035	0.021
				(0.864)							(0.720)	(0.838)
TRANSPARENCY					0.053						**0.213**	0.167
					(0.529)						(0.034)	(0.106)
STATEBIAS						−0.037					0.112	0.056
						(0.683)					(0.402)	(0.674)
TIMECOST							0.169				−0.088	−0.138
							(0.133)				(0.548)	(0.360)
IMPLEMENTATION								−0.340			**−0.388**	**−0.370**
								(0.000)			(0.000)	(0.000)
PRISECDEV									−0.164		**−0.181**	**−0.205**
									(0.001)		(0.009)	c(0.005)
PROACTIVE										−0.140	−0.038	−0.030
										(0.104)	(0.750)	(0.809)
Industry dummies	No	Yes	Yes	Yes	Yes	Yes	Yes	Yes	Yes	Yes	No	Yes
Obs.	838	838	838	838	838	838	838	838	838	838	838	838
Pseudo-R^2	0.013	0.038	0.039	0.038	0.038	0.038	0.040	0.065	0.047	0.046	0.059	0.077

Notes: This table presents the estimation results of regressions that examine the role of provincial governance variables in determining the corruption severity of local business environment (CORRUPTION, Panel A) and the probability of firms paying informal charges as part of common industry practice (PAYMENTDUM, Panel B), controlling for firm characteristics. Definitions of variables are presented in Table 35.1. Coefficients significant at the 10% level are in bold. *p*-Values (based on White standard errors) are in parentheses. Intercepts are not reported.

Regression models (3) to (12) in Panel A (Table 35.6) indicate that several provincial governance variables have a significant impact on the level of corruption severity as experienced by firms in their local business environment. Improvements in the provinces' implementation and consistency of policies (IMPLEMENTATION) significantly mitigate firms' perception about corruption in the local business environment. The effect is substantial in terms of economic significance. The magnitude of the coefficient suggests that if a Vietnamese province could improve its score of policy implementation by one standard deviation, the corruption severity score would be reduced by 18.4% (according to regression model (8)) or 28.6% (according to regression model (11)). IMPLEMENTATION is the provincial governance factor that has the most consistent and robust effect on corruption in local business environment. This variable alone significantly improves the value of R^2 of the model from 3.6% to 6.5%. Second, a higher proactivity of the provincial leadership (PROACTIVE) significantly increases corruption severity in models (11) and (12). We find this result slightly puzzling. One possible explanation is that Vietnamese firms perceive the proactivity of the provincial leaders as implying greater power to possibly induce corrupt activities. Third, in regression model (7), we find that TIMECOST is positively related to corruption. This finding indicates that high bureaucracy (measured by the costs of regulatory compliance at the province level) acts as a catalyst for corruption. A one standard deviation increase in the provincial regulatory time costs is associated with a 7.4% increase in CORRUPTION score. The impact of TIMECOST disappears when we control for all other seven aspects of local public governance, but we note that this is a rather strict test.

Overall, the results in Panel A of Table 35.6 suggests that a higher quality public governance system helps to mitigate corruption. Better governance quality can be achieved particularly by introducing more consistency in policy implementation across governmental authorities and by improving administrative procedures to save firms' time costs of regulatory compliance.

Panel B of Table 35.6 shows the results of the logit regression to explain firms' perceived propensity of payments of informal charges to public officials in the industry (PAYMENTDUM). Again, we find that firm size is positively associated with perceived corruption. In addition, the estimation results emphasize the role of provincial governance factors. Panel B provides a number of similar findings as the analysis of the determinants of the severity of corruption in the local business environment in Panel A. The results suggest that improvements in IMPLEMENTATION are the most likely to help in coping with corruption severity by reducing the likelihood that firms have to pay corrupt public officials in the industry. We again find strong and robust results with the implementation measure in all model specifications. Furthermore, we find that when provinces have more and better policies for promoting private sector development (PRIVSECDEV), firms generally experience a significantly lower likelihood of informal payments in their industry. This result is robust across our model specifications, indicating that better provincial policies for supporting the private corporate sector are effective in curbing corruption. Surprisingly, in regression (23), TRANSPARENCY shows a positive association with the likelihood of firms engaging in informal payments in their active industries. However, this effect disappears when including industry effects in the regression.

In sum, our analysis shows that provincial governance variables are important determinants of the corruption levels perceived by Vietnamese firms. Among the subindices of provincial governance quality, the implementation and consistency of policies, the promotion measures for private sector development, the costs of regulatory compliance, and the provincial leadership proactivity are the most robust and significant. Vietnam's government has a tradition to provide favorable conditions and treatments, for example, in terms of providing credit, investment and/or licenses, to state-owned enterprises due to the country's history. Better policies for private sector development involve creating a level-playing field for firms in all economic sectors. Our results suggest that improving local governance quality as well as easing the policies toward the private sector development in Vietnam is helpful for restraining corruption and thus its adverse effects on the country's development.

35.7 CONCLUSIONS

This chapter contributes to the limited literature that investigates the determinants of differences in corruption across firms within individual countries. Specifically, we analyze the impact of provincial public governance on the corruption perceived by a large number of firms in Vietnam. The large variation in corruption severity facing these firms highlights the need for in-depth research on the causes and consequences of corruption in a single-country setting.

Using a sample of nearly 900 Vietnamese firms based in 24 provinces, we present evidence that public governance factors play an important role in shaping the severity of corruption. In particular, we find that the implementation and consistency of policies, the provincial policies for private sector development, and the time costs of regulatory compliance are important determinants of corruption.

Our results underline the necessity of studying the variation in the design and quality of local public governance within countries. Within-country research is crucial for adding to our understanding of why and how corruption takes place and thus essential for enhancing central and local government policies to fight corruption. Our analysis suggests that improvements in local public governance could be important policy tools for reducing corruption and its detrimental effects on economic development and growth.

REFERENCES

Ades, A. and Di Tella, R. (1999) Rents, competition, and corruption. *American Economic Review*, 89(4): 982–993.

Aidt, T. (2003) Economic analysis of corruption: A survey. *The Economic Journal*, 113(491): F632–F652.

Bardhan, P. (1997) Corruption and development: A review of issues. *Journal of Economic Literature*, 35(September): 1320–1346.

Besley, T. and McLaren, J. (1993) Taxes and bribery: The role of wage incentives. *The Economic Journal*, 103(1): 119–141.

CIEM (Central Institute for Economic Management) (2007) *Vietnam's Economy in 2006*. Finance Publishing House, Hanoi, Vietnam.

Del Monte, A. and Papagni, E. (2007). The determinants of corruption in Italy: Regional panel data analysis. *European Journal of Political Economy*, 23(2): 379–396.

Fisman, R. and Svensson, J. (2007). Are corruption and taxation really harmful to growth? Firm level evidence. *Journal of Development Economics*, 83(1): 63–75.

Glaeser, E. and Saks, R. (2006). Corruption in America. *Journal of Public Economics*, 90(6–7): 1053–1072.

Herzfeld, T. and Weiss, C. (2003). Corruption and legal (in)effectiveness: An empirical investigation. *European Journal of Political Economy*, 19(3): 621–632.

Jain, A.K. (2001). Corruption: A review. *Journal of Economic Surveys*, 15(1): 71–121.

Nguyen, T.T. and van Dijk, M.A. (2008). Corruption and Growth: Private vs. State-Owned Firms in Vietnam. Working Paper, Rotterdam School of Management, Erasmus University, Rotterdam, the Netherlands.

Svensson, J. (2003). Who must pay bribes and how much? Evidence from a cross section of firms. *Quarterly Journal of Economics*, 118(1): 207–230.

TI (Transparency International), (2007). Global Corruption Report 2007. Cambridge University Press, Cambridge, U.K.

Treisman, D. (2000). The causes of corruption: A cross-national study. *Journal of Public Economics*, 76(3): 399–457.

CHAPTER 36

Empirical Test of New Theory of Economic Reform Using Indonesia as a Case Study (1988–2003)

Carolyn V. Currie

CONTENTS

36.1 AIMS AND OBJECTIVES

This research has one purpose: to conduct an empirical test of a new theory of economic reform. The reason for using Indonesia as a case study is due to its suitability as a country going through a reform process after a period of financial crises and regulatory failure.

Since the fall of the Berlin Wall and the opening up of Communist nations, economists have promoted new ownership structures as a solution to the maximization of welfare and a way of introducing market forces into command economies. They have advocated privatization in particular as the optimal means of achieving an increase in economic growth and lifting per capita income. Privatization is not the only means of reducing state ownership and control of enterprises. Economists have also touted public private partnerships and private finance initiatives as means of promoting economic and social development. However, regulators are still intensely debating whether privatization and other ownership structures, which are not totally dependent on state ownership and control, achieve such goals or in fact cause a deterioration in welfare levels from those existing under a command economy. For instance Stiglitz (2002)* claimed that the International Monetary Fund (IMF) advocated such policies without considering factors vital to the suitability of such reform programs to a particular economy and society. Reformists did not understand the particular history, the social capital, the political institutions, and how political forces affected political processes (Stiglitz, 1999, p. 4).† Academics and other advisers have refuted Stiglitz's trenchant criticism of development strategies, which rely largely on changing ownership structures, claiming success for such policies in China and Poland (Dabrowski et al., 2001). Dabrowski et al. (2001) point out that Stiglitz ignores the principal reasons for failure of these policies in Russia and fails to distinguish why these policies succeeded elsewhere.

The arguments of these authors (Dabrowski et al., 2001) illustrate some important flaws in contemporary thinking regarding the measurement and assessment of development yardsticks, which have implications for advocacy of an "optimal" model for reforming and restructuring an economy. They advocate policies to achieve success in reform by concentrating

---

* Stiglitz, J. (2002) *Globalization and Its Discontents*, W. Norton, New York.
† Stiglitz, J. (1999) Whither Reform: Ten Years of the Transition, Annual Bank Conference on Development Economics, Washington, DC, April 28–30. Washington, DC: World Bank.

on improving the institutional, legal, and economic conditions for rapid and sustainable growth. Hence success should be measured by the increase in output deriving from the private sector from the start of recovery, (Dabrowski et al., 2001, p. 297), and not by the overall growth of gross national product (GNP) of the whole economy, as suggested by Stiglitz, as this indicator can be affected during transition by prereform crisis conditions. Dabrowski et al. (2001) advocate a principle where loss-making state-owned enterprises (SOEs) should be shut down and not privatized, and the rapid expansion of a new private sector encouraged instead. The welfare costs associated with discontinuing and not privatizing SOEs can be regarded as an investment "needed to achieve permanent welfare gains from the better allocation of labour and other resources in the future" (Dabrowski et al., 2001, p. 298).

This research attempts to take the arguments of both Stiglitz et al. (2001) and Dabrowski et al. (2001) and test my own formulation of a new theory of economic reform. This will aid in understanding the principal factors that can constitute barriers to the success of economic reform strategies, which rely on new ownership structures to promote efficiency, and thus raise the production frontier of an economy.

36.2 NEW THEORY OF ECONOMIC REFORM

Debate regarding key factors determining the success or failure of policies of liberalization and privatization illustrates the need for a concise theoretical foundation to guide decision makers as to how, when, and where to apply policies that change underlying economic structures. In the following discussion, I outline such a framework, which is based on the perception of gradations in the process of development. It also argues that the introduction of new ownership structures, market mechanisms, and financing techniques are not necessarily solutions without providing for changes in economic, societal, and legal infrastructures.

The new theory of choice of ownership structures espoused in this chapter thus conceives of a national economy as a set of interrelating systems and subsystems. Hence the method of liberalizing an industry through changing ownership structures can be described in terms of a matrix (see Table 36.1) and a set of equations. This in turn dictates a staged approach to changing ownership structures.

In this theory, economic development (Y_1) is defined as sustainable growth. This can be measured at the level of the individual by the increase in a maintainable and stable level of income per capita and at the corporate

TABLE 36.1   Factors in the Choice of Type, Timing, and Method of Valuation
of Ownership Structures

	Ownership of Capital—proportion of direct and indirect ownership by individuals vs. institutional/elite vs. State (O)	Type of Ownership Structure, Methods of Valuation and Timing (T)
Stage of Social Development (Y_2)	Regulatory Structure (M) and Industry Structure (C)	Stage of Managerial Expertise, Organizational Structure, and Processes (E)
Stage of Economic Development (Y_1)	Compliance or stage of legal infrastructure development (L)	Political System (P) which influences Government Goals (G) and which influences the allocation of Economic Resources (R)

or institutional level by the increase in maintainable and stable accumulated earnings per capita. At the country level, improvements in the ratio of external debt and current account balance to gross domestic product (GDP) as well as increases in the level of maintainable and stable GDP per capita are appropriate measures of success. These definitions avoid the criticism of using inappropriate yardsticks of growth of GDP, as in transition economies, growth may initially be negative as prereform crisis conditions impact immediately after reform. This was one of the areas of disagreement between Dabrowski et al. (2001) and Stiglitz (1998a,b, 2001).* Also a policy may appear to be successful in the short term using growth in GDP as a yardstick, but it is the maintenance of income per capita, corporate profitability, and an increasing debt servicing ability at the national level over a long term that is the best criteria of success.

Social development (Y_2) is defined as growth in the equitable distribution of wealth, which can be measured by the dispersion and distribution of per capita income, and participation in institutions. A scale ranking the capacity to participate in the institutional framework of government may

---

* Stiglitz, J. (1998a) Must Financial Crises Be This Frequent and This Painful?, McKay Lecture, Pittsburgh, Pennsylvania, September 23; Stiglitz, J. (1998b) Towards a New Paradigm for Development: Strategies, Policies, and Processes, Prebisch Lecture, UNCTAD, Geneva, October 19; Stiglitz, J. and Ellerman, D. (2001) Not poles apart: 'Whither reform?' and 'whence reform?', *The Journal of Policy Reform*, 4(4): 325–328; Dabrowski, M., Gomulka, S., and Rostowski, J. (2001) Whence reform? A critique of the Stiglitz perspective, *The Journal of Policy Reform*, 4(4): 291–324.

be a useful adjunct to wealth distribution measures. However, active participation is in turn a function of improving *the quality of human capital,* in terms of education, knowledge, and skills, which are vital to social and economic development. This is likely to provide development strategies that adjust policies regarding ownership structures to the underlying fundamental differences between economies.

The type of government will influence government goals (G) and in turn will determine the allocation of resources. The optimum set of government goals are derived from theories of financial regulation (Sinkey, 1992). These theories emphasize the preeminence of safety of depositors in financial institutions, currency and price stability, industry structures that promote competition, convenience of users of financial services (such as access to a product or service),* and general public confidence in the financial system. Such goals are considered necessary to promote a market economy. However, how a government allocates its economic resources can determine whether such goals are achievable. An economy can expand its resource allocation through spending more than it raises as revenue, assisted by direct foreign investment, borrowings, and aid. Hence, government goals can be expressed as

$$G = f(S,S,S,C,C) = f(R) = f(D,FDI,K,A) \qquad (36.1)$$

where safety (S), stability (S), structure (S), convenience (C), and confidence (C) are a function of how economic resources (R) are allocated and may require deficit spending (D), foreign direct investment (FDI), private capital formation and borrowings (K), and foreign aid (A).†

The optimum mix of direct and indirect public ownership, where the latter is defined by ownership by pension or mutual funds or employee share ownership plans, can be seen as dependent on simultaneous design of the regulatory model and the industry structure as well as the stage of development of the legal infrastructure. The first two factors M and C will both influence and be dependent on the stage of social development more

---

* Automatic teller machines, Internet banking, or electronic commerce are examples of governments providing an environment where the financial system achieves a goal of provision of convenient products and services.
† The purpose of attempting to capture these factors and variables in mathematical form is to facilitate future research.

than L, which is closely related to the stage of economic development. This could be expressed as

The Optimum Ownership Mix or $O = M \cdot C(Y_2) \cdot L(Y_1)$ (36.2)

The type of ownership structure, whether a public–private partnership (PPP), private finance initiative (PFI), or privatization, as well as the timing and valuation method is influenced by the *E factor*—the stage of managerial expertise, organizational structure, and processes—as well as by the *P factor*. The E factor influences and is dependent on social development while the political system, the P factor, has a closer relationship with the stage of economic development. Another way of expressing this is

The Optimum Type of Ownership Structure or $T = E(Y_2) \cdot P(Y_1)$ (36.3)

It is postulated that in the aforementioned model social development, or Y_2, may decrease then increase as M, C, and E change. This is due to the initial effects of deregulation of protective measures on an industry and the time lag before learning effects of change kick in and prudential supervision increases in strength to compensate for the reduction in protective measures. The resulting increase in social development may be at a decreasing rate as decreasing returns to scale of M and C are experienced, as the regulatory model and market structure may become burdensome, clogged, and lead to obstacles. This requires input from factors affecting economic development.

The political and legal systems must allow feedback so that as social development occurs, more reliance can be made on self-regulatory market mechanisms. Hence the regulatory model and performance of the market structure will require continual monitoring. Development of the E factor will contribute to more effective prudential supervision through enhanced skill levels both within the regulator, the marketplace, and the entities subject to changed ownership.

Economic development or Y_1 may increase at a decreasing rate as efficiencies are realized or could display a Cobb Douglas pattern* due to the

---

* Cobb Douglas cycles display patterns of over then under supply, that is, overshooting then undershooting market signals. This is due to time lags in production responding to price signals. In terms of an economy, this could be evidenced by a rapid increase in economic growth followed by contractions with eventual return to a moderate state of economic growth.

political system not adapting quickly to sectoral imbalances, particularly in the supply of human capital. Alternatively, it could be due to the failure to adjust the legal infrastructure as the quality of human capital increases, which is a sequencing problem. If economic development is uneven, this could promote political instability, which in turn affects resource allocation. This would be most evident in failure to assess and adjust social development policies according to our allocated priorities of M, C, and E.

The implications of this theory, or choice matrix, are that it is vital to assess the stage of P, E, L, M, and C—the political system, human capital, legal infrastructure, and the regulatory models governing the financial system and industry—before choosing the ownership structure as well as timing and valuation methods and

- That the design of M, the regulatory model governing the financial system, is a starting point, which must take account of the type of existing, and desired market structure, i.e., C, as well as the caliber of E.

- That if all of the factors are weak at the starting point of economic reform, a careful staged approach should be used. Only a formal deregulation and program of ownership change should be considered. This will involve continual monitoring both by the government undertaking the change and by external aid agencies such as the IMF and World Bank. Otherwise an entrenched elite will take the place of the state in ownership and control without the requisite spread of the advantages of moving an economy to a full free market basis.

The interaction of the variables of P, E, L, M, and C is multidimensional, but an attempt is made in Figure 36.1 to illustrate how an economist could diagnose an economy and prescribe ownership mechanisms as well as necessary reform of factors to promote economic and social development. The purpose of Figure 36.1 is to illustrate that a staged approach to the choice of the development of ownership structures must be the overriding principle dominating an economic reform package. For instance, the left hand lower quadrant starts with state ownership and control where all factors are weak as opposed to the full market based system in the top right-hand quadrant.

We can see from Figure 36.1 that where E is rated highly together with M, C, and L that a full public float with reliance on the market mechanism should promote the optimum outcome. This is what we would expect of

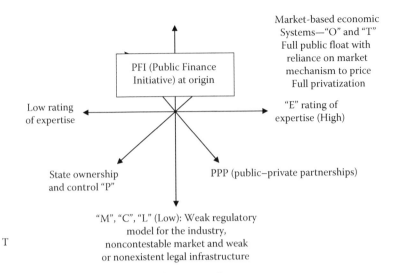

FIGURE 36.1 Interrelationships between factors.

an advanced economy with a high level of managerial expertise, organizational structure, processes, regulatory and industry structures, and legal development. Or if the industry by its nature is noncontestable, such as telecommunications, privatization can work with prudential oversight of pricing and market practices. That is the C factor must be improved.

Figure 36.1 also illustrates that a PFI is the ownership mechanism where all factors are rated at midpoint, while a PPP is appropriate where all factors receive a low rating, except for the E factor of managerial expertise, organizational structure and processes, which needs to be injected by the private sector. Indeed this is often the rationale for involvement of the private sector. PPPs can occur in an advanced economy, where due to a long history of state involvement in a particular industry such as education, the industry is one where the regulator does not have the supervisory skills or requisite powers. Also the expertise in modern market practices within the industry may be lacking and the industry may be protected from competition by legislation. Hence it appears as a pocket of state ownership and control within a free market economy. An injection of market expertise from allied industries, which have developed private sector practices, should be sought in a consortium between SOEs and private firms.

Figure 36.1 hence illustrates that where the E factor is weak, while the other factors have been improved to a moderate level, changing ownership structures is not advisable until either the E factor is improved or the other factors are strengthened.

Such a staged approach requires measurement techniques to assess the state of the input factors. The taxonomy of regulatory models described elsewhere (Currie, 2000) could be used to assess the M factor, while the C factor could be assessed by a Porter style industry analysis. The E factor could be measured by the quantitative methods to assess human capital described in Section 36.2, while the P factor assessed by stated government goals as well as the resources devoted to all the other factors, as revealed in the national accounts. The L factor could be assessed not just by compliance with the compendium of standards, but by studies of corruption and independence of the judiciary such as undertaken in Indonesia.*

36.3 NEED FOR EMPIRICAL TESTING

The theory outlined earlier has canvassed some of the essential ingredients in the resolution of the debate as to the merits of changing ownership structures on economic and social development. What is new is a theoretical framework to analyze and measure readiness of an economy for reform as well as to analyze why past attempts failed. This allows a basis to prescribe an optimum staged approach. Indonesia is a fertile database having suffered a severe financial crisis with closure and collapse of its banking system after an attempt to privatize sectors of the economy and now under IMF direction to privatize as many industries as possible. Data will relate to the measurement of factors detailed in the theory with an attempt to validate hypotheses embodied in the aforementioned three equations.

Analysis of past successes and failures in Indonesia in privatizing industry during the period 1988–2003 is to be measured by my unique indices for economic and social development (Y_1, Y_2). The ability to empirically assess the input factors of the model and then relate them to desired outcomes in terms of my economic and social development indices may lead to other approaches to economic reform through changing ownership structures. For instance, continued government involvement, by requiring a percentage of future profits or gain upon resale when economic and political opinion considers privatization a fire sale solution, has not been attempted. Neither have mechanisms to ensure employees (and customers) share in the benefits conferred by a privatized entity.

---

* http://www.worldbank.org/wbi/governance/pdf/judicial_mod_1. Refer also to World Bank, "A Diagnostic Study of Corruption in Indonesia," Partnership for Governance Reform in Indonesia, Final Report, October, 2001, and a Bookings Institutions Brief in September 2001. http://www.brook.edu/comm/policybriefs/pb89.htm.

A basis for such initial research would be a classification of regulatory models in the industries in Indonesia that did and are still to undergo ownership changes as specified by the Republic of Indonesia in conjunction with the IMF and assessment of the state of requisite inputs in terms of C, E, L, P as well as specification of G and R, government goals and resource inputs, respectively. Monitoring of the performance of the industry and entity at a micro level (which can be attempted using measures of profitability and productivity) is essential as is attempting to statistically derive functional relationships between the input of C, E, L, and P and output variables, which measure success of policies, such as increases in the proportion of GDP the private sector is contributing as well as the measures for economic and social development (Y_1, Y_2) as uniquely defined in my theory.

The key dependent variables of interest are changes in the ownership structure, the timing and valuation, and how they relate to the independent variables as specified in Equations 36.1 through 36.3. Initially multiple regression analysis will be attempted followed by use of curve fitting software to test formulations of the theory.

36.4 METHODOLOGY AND RESULTS

The first part of this research involved attempting to develop a new index of economic development known as Y_1. To do this we regressed growth in GDP per capita for the years 1988–2003 using data from the Asian Development Bank (ADB) and the World Bank against a range of variables using ANOVA techniques. After six rounds, the best result was achieved by using the variables of gross domestic capital formation as a percentage of GDP, gross domestic saving as percentage of GDP, the current account balance as a percentage of GDP, external debt as a percentage of gross national income (GNI) and Debt Service as percentage of exports of goods and services. The results are described in Table 36.2.

Hence we adopted growth in GDP per capita as a measure of sustainable economic growth. We then attempted to develop a measure of social development Y_2 by regressing two measures of social development, HDI and GINI/HDI,* against a range of selected variables including the variables

---

* World development indicators (WDI) and United Nations statistics were used to formulate the two measures of social development. HDI is a human development indicator developed by the World Bank and used to rank countries while GINI is a measure of the dispersion of economic wealth developed by an author of the same name: Gini, C. (1921) Measurement of inequality and incomes. *The Economic Journal* 31(1): 124–126.

TABLE 36.2   Summary Output (Y_1 as Growth in GDP Per Capita)

	P-Value	Significance
Intercept	0.00	
Gross domestic capital formation	0.03	S
Gross domestic saving	0.05	S
Current account balance	0.07	?
External debt as % of GNI	0.00	HS
Debt service as % of exports of goods and services	0.03	S

TABLE 36.3   Summary Output (Y_2 as GINI/HDI)

	P-Value	Significance
Intercept	0.28	
School enrollment, tertiary (% gross)—SS1	0.00	HS
School enrollment, tertiary, female (% gross)—SS3	0.00	HS
Personal computers (per 1,000 people)—SS4	0.10	
Maternal Mortality Ratio (per 100,000 live births)—SS5	0.34	
Life expectancy at birth, total (years)—SS6	0.27	

in Table 36.3 plus daily newspapers produced, immunization proxies, percentage of females in the workforce, infant mortality, and land use.

Twelve runs resulted in the percentage of the population enrolled in school and tertiary education, in particular females, having the highest correlation with measures predicting a higher index of human development adjusted for a measure of income equality. Hence this measure, GINI/HDI, was taken as a good proxy of social development or Y_2.

Given the difficulties in measuring inputs required for the first part of the matrix, that is, the determination of the optimum mix of direct and indirect ownership over the time period specified, it was decided to concentrate on the third equation:

The Optimum Type of Ownership Structure or $T = E(Y_2) \cdot P(Y_1)$ (36.3)

This was done by using Standards and Poor (S&P) ratings system as a measure of the type of political system that combines measures of democracy with political corruption. The Corruption Index provided by the World Bank only extended back to 1995.

Curve fitting was conducted in several stages.

Stage 1: E as a function of Y_2

Data for E, education levels measured by the variable of school enrollment, as function of Y_2.

The resulting equation which was the line of best fit was

$$y = a + b \ln x / x$$

where

$y = Y_2$

$x = E$

This run produced a line of best fit with the highest R^2 score where $a = 16.691747$, $b = -24.698196$. As the dependent variable was measured by GINI/HDI, the results showed that a high level of inequality was associated with a very skewed or unequal level of income and human development and was a function of very low school enrollment. As the school enrollment increased, so did our measure of social development, albeit at a decreasing rate.

Stage 2: P as a function of Y_1

Data for P, a S&P measure of the political system, was fitted as a function of Y_1.

The resulting equation which was the line of best fit was

$$y = a + b / x^2$$

where

$y = Y_1$

$x = P$

This run produced a line of best fit with the highest R^2 score where $a = 8.9270203$, $b = -506.15478$.

The results showed that as the underlying economic development indicator improved so did the political system.

Stage 3: T as a function of $E(Y_2) \cdot P(Y_1)$

T was run as 3D function of predicted-E and predicted-P.
The resulting equation was

$$\ln z = a + be^{x/wx} + c/x^2$$

where

$z = $ T
$x = $ E-predict
$y = $ P-predict

This run produced a line of best fit with 91.2% accuracy (R^2), where $a = -12.043073$, $b = -853.28273$, and $c = 2213.8574$. The results showed that z or the optimum type of ownership structure was not a function of P but of E, with increasing returns to scale up to a certain point and then decreasing returns to scale. In other words, increasing the education level could move an economy along the scale from the weakest to the strongest form of reduced government involvement in the ownership structure.

Stage 4: T as a function of E_predict and P_predict

Table 36.4 and 36.5 describe the raw data and the output for where T was run as a function of the (predicted-E × predicted-P) to see the best fit equation.

TABLE 36.4   Summary Input and Output

	T	E	Y_2	P	Y_1
1988	0.02412	8.90480	1.53847	10.60873	17.58646
1989	0.03462	9.06480	1.56611	10.02464	17.66035
1990	0.06937	9.22480	1.59375	9.44056	15.14444
1991	0.08148	9.38050	1.41783	8.85648	16.61012
1992	0.09782	9.27860	1.25449	3.00000	11.12903
1993	0.18141	10.34510	1.27470	9.60556	14.87319
1994	0.26874	11.09600	1.29004	8.73389	14.01212
1995	0.33084	11.30790	1.41636	9.00000	16.97432
1996	0.42507	11.26970	1.41600	9.00000	15.83519
1997	0.41946	11.26870	1.41668	6.50000	16.51620
1998	0.33660	12.32370	1.54931	3.80000	50.52593
1999	0.40569	13.37870	1.07131	1.33000	13.75165
2000	0.39189	14.43269	1.15571	2.00000	13.74153
2001	0.25966	15.05927	1.07839	2.25000	14.47317
2002	0.23345	15.67927	1.12279	1.00000	8.18441
2003	0.27860	16.29927	1.16719	4.50000	9.36582

Note: These are the variables for the equation $T = E(Y_2) \cdot P(Y_1)$.

TABLE 36.5 Summary Input and Output

T	E-PREDICT	P-PREDICT	T	E.P
0.02412	15.103759	1.37073	0.02412	20.70318
0.03462	14.963231	3.1568118	0.03462	47.2361
0.06937	14.144091	4.840355	0.06937	68.46242
0.08148	13.599148	6.2465353	0.08148	84.94756
0.09782	13.420381	6.2504787	0.09782	83.88381
0.18141	12.227915	6.3490631	0.18141	77.6358
0.26874	11.989031	6.5106916	0.26874	78.05688
0.33084	11.81598	6.6389212	0.33084	78.44536
0.42507	10.624672	6.7201481	0.42507	71.39937
0.41946	10.621856	6.9084821	0.41946	73.3809
0.33660	10.619251	7.0715105	0.33660	75.09414
0.40569	10.61005	7.0924358	0.40569	75.2511
0.39189	9.7760004	7.1703154	0.39189	70.09701
0.25966	9.7124232	7.2904807	0.25966	70.80823
0.23345	9.6172235	7.304147	0.23345	70.24561
0.27860	9.4687832	8.7287513	0.27860	82.65065

$$T = a + [b\ln(\text{E}\cdot\text{P})/(\text{E}\cdot\text{P})]$$

was the resulting equation where y = T and x = E·P and a = 0.42503439 and b = −2.8746624.

The results showed that as the education levels interacted with the political system, there could be decreasing returns to scale of reduced government involvement.

36.5 CONCLUSION

Despite limits to this research of one case study, lack of data, the results clearly showed a major input into the economic and social development and hence the stages by which economic reform is introduced are education levels, in particular those of the female population. This has policy implications in terms of the importance of education policies and spending on the most vital ingredient to the success of policies that aim to modernize a nation.

There is an obvious need to extend this study across a number of nations as well as test the other equations in the model. As the database for nations becomes more developed, it will make this task easier.

REFERENCES

Binhadi (1995) Financial Sector Deregulation Banking Development and Monetary Policy, The Indonesian Experience, 1983–1993, Institut Bankir Indonesia, Jakarta.

Currie, C.V. (2000) The optimum regulatory model for the next millennium—Lessons from international comparisons and the Australian–Asian experience. In: B. Gup (Ed.), *New Financial Architecture for the 21st Century*. Quorum/Greenwood Books, Westport, CT.

Dabrowski, M., Gomulka, S., and Rostowski, J. (2001) Whence reform? A critique of the Stiglitz perspective. *The Journal of Policy Reform*, 4(4): 291–324.

Sinkey Jr, J.F. (1992) *Commercial Bank Financial Management*. Maxwell MacMillan, London.

Stiglitz, J. (1998a) Must Financial Crises Be This Frequent and This Painful? McKay Lecture, Pittsburgh, Pennsylvania.

Stiglitz, J. (1999) Whither reform: Ten years of the transition, Annual Bank Conference on Development Economics, Washington, DC, April 28–30, World Bank, Washington, DC.

Stiglitz, J. (2002) *Globalization and Its Discontents*. W. Norton, New York.

CHAPTER 37

Ownership Structure and Firm Value: Evidence from the Turkish Financial Crisis

Berna Kirkulak

CONTENTS

37.1 INTRODUCTION

Most of the theories that are valid for developed markets cannot be applied to emerging markets. Emerging markets are of a different nature and are shaped by different cultural, social, political, and educational characteristics. The differences are reflected not only in economic development, but also in the nature of relationships and the business environment. This raises specific issues about corporate governance. The relationships among corporations, government bodies, and financial institutions are the core interest of corporate governance. There is growing empirical evidence regarding corporate governance in the emerging markets [see Xu and Wang (1999) for China, Yeh et al. (2001) for Taiwan, Chong and Lopez-de Silanez (2007) for Mexico, Saldana (2001) for the Philippines, Chibber and Majumdar (1999) for India, and Khanna and Rivkin (2001) for a comprehensive survey of emerging markets].

The financial crisis experienced by emerging markets provided a unique opportunity to study the change in ownership structure and the effects of the financial crisis on the firms' performance. The Asian crisis was the turning point for the issues of corporate governance. Following the Asian crisis, the consequent crisis in other emerging markets such as Russia, Mexico, Brazil, Argentina, Turkey, etc., pushed policy makers and many researchers to conduct research about the relationship between ownership structure and firm value. There is a large body of literature concerning corporate governance and the Asian crisis. Among them, Johnson et al. (2000) studied corporate governance in Asian countries and state that the Asian crisis was closely related to poor corporate governance. The severity of the crisis is attributed to the lack of transparency in the relationships among corporations, financial institutions, and governments. The informational flaws in the disclosures induced misleading assessments by the investors, and this resulted in lower values for firms. Mitton (2002) examined almost 400 firms in Asian countries and argued that firms with higher disclosure quality performed significantly better than other firms during the crisis. Claessens et al. (2000) found that Asian companies controlled by management/family groups were less likely to go bankrupt during the crisis. The findings of Lemmon and Lins (2003) supported this fact, and they showed that firm values were higher when the cash flow rights held by the controlling blockholder were higher. Furthermore, Glen and Singh (2004) examined corporate governance in the crisis-affected Asian countries. Contrary to widely accepted common beliefs about the Asian crisis, they argued that the crisis occurred not because of the flaws in corporate governance but precisely because of the financial liberalization, which a number of these crisis-ridden countries implemented prior to the crisis.

The literature has highlighted the relationship between ownership structure and firm value in Turkey. Previous evidence suggested that corporate governance had a significant impact on a firm's value in Turkey (Durukan et al., 2007; Gunduz and Tatoglu, 2003; Gursoy and Aydogan, 2002; Orbay and Yurtoglu, 2006). Yet, little is known about the role of corporate governance in determining a firm's value in particular during a financial crisis. Gonenc and Aybar (2006) examined the performance of 198 Turkish industrial firms 6 months prior and 6 months after the 2001 financial crisis. They found that ownership concentration was the main determinant of a firm's value during the crisis. Firms with high concentrated ownership were associated with low stock returns. This chapter aims to extend the literature on ownership structure and a firm's value during the 2001 financial crisis. It is a particular interest of this chapter to test whether the ownership structure of the Istanbul Stock Exchange (ISE) firms was changed during the credit crunch that characterized the crisis. Despite the importance of the topic, there have been few studies on the effect of ownership structure on Turkish firms' values, in particular with regard to group-affiliated firms between the periods before and after the 2001 financial crisis. The purpose of this chapter is to fill this gap. This chapter contributes to the previous literature in two ways: first, the ownership structure of the ISE-listed firms is examined. The subgroup analyses revealed high ownership concentration and diversified interest of holding companies in different industries. Second, this chapter concentrates on the effect of the 2001 financial crisis on the firms' values and operating performance. In order to have a better understanding of the financial crisis, the precrisis, during crisis, and postcrisis periods are examined. The empirical investigation is based on comprehensive firm-level panel data that consist of about 564 observations over the 2000–2002 period.

The rest of this chapter is organized as follows: Section 37.2 provides information about the ownership structure, legal environment, and the economy in Turkey. Section 37.3 describes the data and methodology. Section 37.4 documents the empirical findings and Section 37.5 provides conclusions.

37.2 BACKGROUND

37.2.1 Ownership Structure

Unlike corporations in the United States and the United Kingdom, which have dispersed ownership, firms in Turkey are mainly concentrated and controlled by business groups. The groups are usually in the form of holding companies that serve in diversified industries. Thus, they enjoy economies of scale, reduced transaction costs, and diversified risk by operating in different economic activities. Family-controlled firms are the predominant type of business groups, and it is common that banks and other financial institutions are part of a business group.

In order to grasp corporate governance in Turkey, a historical perspective is needed. The oldest holding companies were established after the formation of the Republic of Turkey. Yurtoglu (2003) showed that the oldest and largest Turkish holding companies, namely, Koç Holding and Çukurova Holding, started their activities in the 1920s. After the Second World War, the number of holding companies grew rapidly. They diversified their activities intensely during the 1980s. Business groups such as Sabanci Holding, Koç Holding, Alarko Holding, and Zorlu Holding dominated the Turkish business environment after the liberalization period (Yaprak et al., 2007). The liberalized economy and incentives, in particular in the textile and tourism industries, encouraged many firms to form a holding company and to operate in diversified industries. The growth of holding companies gained added impetus with the growing economy in the 1980s. However, the economy threatened to enter a recession during the 1990s. Excess public sector financing and political instability were the causes of this problem. High interest rates, increased costs of capital, and unstable economic and political environments pushed business groups to restructure their ownership and to create their own financing sources. Hence, owning a bank became fashionable among business groups during the 1990s.

With the emergence of the growing business groups in the 1980s, the ownership structure in the media was also radically changed. It became commonplace for business groups to own a TV channel or a newspaper. The result was that the business groups began financing their operations through their banks and promoted their businesses through their own media. As an outcome of the liberalization policies, some of the leading business groups such as Aydın Dogan (Doğan Group), Mehmet Emin Karamehmet (Çukurova Group), Kemal Uzan (Uzan Group), Erol Aksoy (Aksoy Group), and Turgay Çiner (Çiner Group) ended up with their own media. In Turkey, Karademir and Danisman (2007) examined the relationship between corporate governance and media. Their conclusion was that technology was a driving force behind the transformation of the ownership status of the media. Business groups not only injected capital into the media for infrastructure but also attempted to promote their own interests and to influence politicians.

The effects of concentrated ownership structure can clearly be seen in the areas of debt structure and income distribution. The corporate governance structure impeded the progress of the capital markets and promoted a heavily bank-oriented economy. Easy access to credit through holding affiliated banks played an important role in allowing business groups to have high debt ratios. However, the banks, due to their close relationships with the holdings, did not perform the monitoring and disciplinary role, which the stock market could perform. This led to a corrupted financial system and this feature spread outward, eventually culminating in the banking crisis in 2001.

As the firms have a highly concentrated ownership structure, income distribution is another main issue. In order to overcome this problem, the Capital Board of Turkey launched a mandatory dividend payment regulation. According to this regulation, listed firms on the ISE were required to distribute at least 50% of their earnings as cash dividends. This is called "the first dividend." The objective behind mandatory dividend payment can be attributed to protecting the minority shareholders' rights. In 1995, the mandatory dividend payment regulation was ceased. In 2001, second mandatory dividend payment regulation was introduced in which publicly owned companies listed on the ISE had to pay at least 20% of their earnings as dividend to their shareholders. It is important to note that the mandatory dividend payment regulation resulted in searching for alternative ways to collect the distributed dividends back. After having distributed the cash dividends, some ISE firms announced preemptive rights issues simultaneously. In his paper, Adaoglu (1999) demonstrated that rights issues were so common during the first half of the 1990s that some ISE firms ended up with negative cash dividends.

37.2.2 Legal Environment and Economy

As it is widely accepted that effective corporate governance is virtually impossible in the absence of developed capital markets, Table 37.1 explicitly presents the development of the capital market in Turkey. The characteristics of the Turkish corporate governance structure are reflected in the features of the Turkish capital market.

TABLE 37.1   ISE Indicators

Year	Firm Number	Volume of Trade (in $ Million)	Market Capitalization (in $ Million)	ISE-100 Index Closing Price	ISE-100 Index Return (%)
1986	80	113	938	1.71	—
1987	82	118	3125	6.73	293.57
1988	79	115	1,128	3.74	(44.43)
1989	76	773	6,756	22.18	493.05
1990	110	5,854	18,737	32.56	46.80
1991	134	8,502	15,564	43.69	34.18
1992	145	8,567	9,922	40.04	(8.35)
1993	160	21,770	37,824	206.83	416.56
1994	176	23,203	21,785	272.57	31.78
1995	205	52,311	20,565	400.25	46.84
1996	228	36,698	30,329	975.89	143.82
1997	258	57,178	61,348	3,451.00	253.63
1998	277	69,696	33,473	2,597.91	(24.72)
1999	285	82,931	112,276	15,208.78	485.42
2000	315	180,123	68,635	9,437.21	(37.95)
2001	310	79,945	47,189	13,782.76	46.05
2002	288	69,990	33,773	10,369.92	(24.76)
2003	285	99,406	68,624	18,625.02	79.61

Note: Compiled from ISE CD-ROMS by the author.

The roots of the ISE go back to the beginning of the 1980s. The liberalization of the financial markets was a turning point in the Turkish economy, and it provided an important infrastructure for a legal framework toward a national stock market. The ISE was established in 1986 and grew at a fast pace in terms of the number of listed firms, trading volume, market capitalization, and market return. As of 1986, 80 firms were traded with an average trading volume of $113 million and it increased up to $99,406 million in 2003.

In 1989 the Turkish government issued Decree 32, which allowed foreign investors to purchase and sell all types of securities in the ISE. Following the removal of capital controls, trading volume and market capitalization substantially increased. Another turning point in the ISE's history came in 1999. The launch of the disinflation program had a positive effect on the ISE's returns and the index showed a steep increase. The market capitalization far exceeded the largest stock markets in MENA (Middle East and North Africa).

As an emerging market, the ISE is characterized by large up and down movements in terms of returns and volatility due to the instability of the economy. During the sample period, the Turkish economy was hit by several crises. The first one occurred in 1994, when the national currency rapidly depreciated. The second crisis emerged in early 2001 due to the collapse of the disinflation program. The interest rates increased astronomically. The problems in the financial sector and the fragility of the economy became apparent. Throughout its life, the ISE also witnessed some imported crises such as the Asian crisis in 1997 and the Russian crisis in 1998.

The number of stocks listed on the ISE peaked in 2000, just before the financial turmoil in 2001. This resulted in a decline in the number of firms listed, trading volume, and market capitalization. Interestingly, the market returns did not decline in the crisis year. Furthermore, the interest of foreigners had increased substantially and the ISE index rose to near record levels. However, one year after the crash of the banking industry, a sharp decline was realized in the number of initial public offerings, trading volume, market capitalization, and market return. The ISE recovered in 2003 and has since become a magnet for foreign investors.

37.3 DATA AND METHODOLOGY

The sample includes nonfinancial firms listed on the ISE. The primary source for the sample firms is the ISE CD-ROMs, which contain income statements, balance sheets, and stock price information. The ownership data are taken from the footnotes of the annual reports for each firm. Many companies commonly use the pyramid ownership structure. In order to figure out the ultimate owners of the firms, an indirect ownership network is examined and the group affiliation is shown. The firms with missing data and reporting noncredible values (such as negative debt, negative sales, etc.) were omitted. During the sampling period, there is evidence of a wave of mergers, which created an unbalanced sample of firms. Hence, the ownership structure of some firms was changed and some firms simply disappeared. During the sampling period, 27 mergers and acquisitions (M&As) were reported.*

The financial crisis emerged in February 2001. The accumulated risk shook the economy deeply and a recovery could not be realized until 2003.

---

* The firms coded as EGBRA, GUNEY, ANBRA, TOFAS, BYRBY, and ENKA were acquired and disappeared from the ISE. The first three firms were acquired in 2000, TOFAS was acquired in 2001, and the other two firms were acquired in 2002.

In their paper, Gonenc and Aybar (2006) studied the stock performance of the nonfinancial ISE-listed firms from August 2000 to August 2001. They focused on the 12-month period around the February 2001 financial crisis. This chapter extends their study and examines firm value and operating performance of the ISE-listed firms before and after 2001.

First, an OLS regression model is run using the dummy variables to control the year effect. Since the sample is in the form of unbalanced panel data, in order to grasp the effect of the financial crisis and ownership structure on the firm's value before and after the crisis, a fixed effects regression model is also run. Tobin's Q and ROA are used as dependent variables. The equations provided in the following examine the relationship between firm value and the independent variables that are used to predict firm value:

$$Q_t = \alpha_0 + \alpha_1 Log(Sales)_t + \alpha_2 DIV/NI_t + \alpha_3 DE/TA_t + \alpha_4 Concentration_t$$
$$+ \alpha_5 Group\ Affiliation_t + \alpha_5 Crisis\ Year\ Dummy_t + \varepsilon_t \qquad (37.1)$$

$$ROA_t = \alpha_0 + \alpha_1 Log(Sales)_t + \alpha_2 DIV/NI_t + \alpha_3 DE/TA_t + \alpha_4 Concentration_t$$
$$+ \alpha_5 Group\ Affiliation_t + \alpha_5 Crisis\ Year\ Dummy_t + \varepsilon_t \qquad (37.2)$$

where

Q_t is the Tobin's Q ratio at year t

ROA_t is return on assets

ε_t is the error term in the regression models

Variables:
To test the ownership structure on the firm's performance, the primary valuation measure Tobin's Q is used. Tobin's Q is used for firm value and growth opportunity. ROA is a profitability ratio and is used as a proxy for operating performance.

Tobin's Q is the ratio of market value of equity plus book value of debt to the sum of book value of equity plus book value of debt. Tobin's Q is often interpreted as a proxy for a firm's growth opportunities and is used to assess a firm's value.

Return on Assets (ROA) is measured by the ratio of net income to total assets. ROA is used to test the company's performance and reflects the firm's efficiency in utilizing total assets. The ROA ratio gauges the operating efficiency and is used to show the accounting performance of the firm in terms of profitability.

Leverage (DE/TA) is defined as the sum of long-term and short-term debt divided by the total assets. It is expected that the higher the leverage ratio, the higher the monitoring by the lending banks, which induces higher financial performance. However, a high leverage ratio also implies a high level of debt burdens such as interest expenses.

Payout (DIV/NI) is measured as dividends paid to the shareholders divided by the net income after tax. Ownership structures may affect the willingness of the firms to pay high or low levels of dividends. For example, family-owned firms with good growth opportunities are reluctant to pay high dividends, while state-owned firms pay a higher amount of dividends.

Firm Size (Log(Sales)) is measured by the logarithm of the firm's total sales. Firm size is related to economies of scale. Large-sized firms can have easy access to funds with low costs.

Concentration is measured as the percentage of the lead share of the major shareholder. According to agency theory, a greater concentration of ownership reduces agency costs and improves the financial performance of the firm.

Group Affiliation is a dummy variable, which takes a value of 1 if a firm has a group affiliation and otherwise takes the value 0. A firm is "group affiliated" if one of the shareholders is a business group. In order to be specific about business groups, the ownership network is carefully examined. As there is no clear consensus about which firms belong to business groups, having a holding company as a shareholder is taken into consideration. Accordingly, firms are treated as group affiliated if one of the shareholders is in a form of a holding company.

Precrisis is a dummy variable, which takes a value of 1 for the year 2000 and 0 otherwise.

Crisis is a dummy variable, which takes a value of 1 for the year 2001 and 0 otherwise.

Postcrisis is a dummy variable, which takes a value of 1 for the year 2002 and 0 otherwise.

In order to capture the effect of the financial crisis, dummy variables are used for the years 2000, 2001, and 2002. A severe financial crisis emerged in 2001. The crisis was rooted in the financial institutions, namely, the banking industry, and spread to other industries. Many companies became insolvent and others suffered huge losses. As a matter of fact, there was already accumulated risk and tension in the market. The 2001 financial crisis sent its first signals in late 2000, when the Turkish Lira was devalued. In order to get a better understanding of the 2001 financial crisis, the pre-crisis period is taken into consideration. Moreover, as it is expected that the 2001 financial crisis had prolonged implications, a dummy variable for 2002 is also used as a postcrisis variable.

37.4 EMPIRICAL RESULTS

In Table 37.2, Panel A reports the number of firms based on the classification of direct major shareholders in each year. The findings show that holding companies own the largest shares in 218 observations. It is clear evidence that the majority of the ISE firms are owned by holding companies. Nonfinancial firms and families are the second and third most frequently observed major shareholders with 152 and 85 observations, respectively. In the family owned companies, the major shareholder is reported as an individual. Indeed, this type of company is owned by a family and it is the common case that the second shareholder is either a wife or a brother of the major shareholder. It is also important to note that in some of the family owned companies, the second major shareholder can be in the form of a holding company rather than a family member. The major shareholder acts as a board of director at the holding company. The state is the major shareholder in 44 observations. Some firms, where the major shareholder is Ordu Yardimlasma Kurumu (OYAK [Army Mutual Assistance Association]), are also treated as state-owned companies. The reason is that OYAK was founded as a pension program for military officers. The Turkish government sometimes uses social welfare programs such as OYAK to implement economic policy. OYAK is attached to the ministry of defense but is run by civilians (Bianchi, 1984). Foreigners and financial firms have the largest shareholdings with 41 and 18 observations, respectively. There is only one firm in which employees own the largest shares of the firm. A consortium type of major ownership is also found in only one firm. As a matter of fact, consortiums became common during privatizations.*

---

* Petrol Ofisi was owned by a consortium of Isbank and Dogan Holding after its privatization. In Turkey, some companies built a consortium in order to take ownership of state-owned enterprises when they are privatized. The logic behind this is to merge the power of the two enterprises in order to be competitive in privatization bids.

TABLE 37.2　Shareholder Structure

Major Shareholder	2000	2001	2002	Total
Panel A: Number of major shareholders across years				
Holding	74	74	70	218
Nonfinancial firms	50	53	49	152
Family	29	29	27	85
State	15	15	14	44
Foreigners	14	13	14	41
Financial firms	5	6	7	18
Employees	1	1	1	3
Consortium	1	1	0	2
Total	189	192	182	564

Panel B: Average lead share across years

Major Shareholder	2000	2001	2002	Total	2000–2002 t-Test
Holding	46.51	47.68	45.10	46.46	0.53
Nonfinancial firms	48.04	49.84	50.87	49.58	−0.73
Family	26.80	27.23	28.15	27.37	−0.37
State	62.66	62.73	63.13	62.83	−0.66
Foreigners	57.55	66.25	69.88	64.52	−1.78[c]
Financial firms	40.74	36.63	39.80	39.00	0.96
Employees	24.59	24.59	22.24	23.81	—
Consortium	51.00	51.00	—	51.00	—
Total	45.74	47.17	47.10	46.67	−0.66

Notes: Panel A reports the number of the major shareholders and Panel B documents the mean values of the lead shares for the major shareholders between the periods before and after the 2001 financial crisis. The last column of Panel B presents t-tests, which is used as a statistical significance for the difference between two sample means.

[a] Significance at 1% level.
[b] Significance at 5% level.
[c] Significance at 10% level.

Panel B documents the descriptive statistics of major shareholders. The findings show that among the major shareholders, foreigners are the most concentrated. Their average lead share is 64.52%. The state is the second most concentrated ownership structure in the sample. The results suggest that foreigners are likely to benefit the most from the 2001 financial crisis. An increase in foreign ownership concentration is clearly seen in the immediate aftermath of the crisis, with their average lead share increasing from 57.55% to 69.88%. The last column of Panel B shows that this increase is significant at the 10% level. Not only did foreigners take advantage of the financial crisis to increase their lead shares, but the state, families, and nonfinancial firms did the same. However, these increases are not statistically significant. The results further show that the firms in which the largest shareholders are employees have the least concentrated ownership. The average lead share in these firms is 23.81%.

Table 37.3 presents the industry classification for every subgroup. Each column shows the number of firm-year observations rather than the number of firms from 2000 through 2002. The findings show that most of the ISE-listed firms are operating in the metal and machinery, textile and leather, cement, chemical industries, respectively. It appears that the holding companies operate in diversified industries and are involved mainly in the metal and machinery and food and beverage industries intensively. In most of the industries, holding companies control the market. In particular, the electricity, retail and wholesale, communications and IT, and paper industries are dominated by holding companies. In the case of the mining industry, holding companies have particular interests and are the only major shareholders. Family companies have a great interest in the textile and leather industry. The state has lead shares in chemical and cement firms. It is important to note that only state-owned firms operate in the defense industry. Moreover, the foreign companies have the largest stakes in the chemical and metal and machinery industries. Financial firms are involved mainly in the food and beverage and chemical industries.

Table 37.4 presents the mean values for the financial ratios during the precrisis period (2000), crisis period (2001), and the postcrisis period (2002). The sample is divided into two groups: direct and indirect ownership structures. The direct ownership structure shows the major shareholders of the firms. The indirect ownership structure shows whether a firm is group affiliated or not. The results show that Tobin's Q increased from 2000 to 2001 and then declined in the postcrisis period. However, there is clear evidence that the financial crisis worsened the financial performances. ROA ratios turned negative during and after the financial crisis.

The average dividend payout ratios declined steadily from 2000 to 2002. State-owned firms paid the highest amount of dividends before, during, and even after the financial crisis. This result is not surprising because state-owned firms tend to pay high-dividend payouts. The reason can be attributed to populist political policies run by the elected politicians. Gugler (2003) argued that as the elected politicians have strong interest to be reelected, they may try to convince the citizens that the state-owned firms are doing well.

TABLE 37.3 Industry Classification for Major Shareholder Patterns

	Holding	Nonfinancial Firms	Family	State	Foreign	Financial Firms	Employee	Consortium	Total
Electricity	6	—	—	—	—	3	—	—	9
Metal and machinery	44	30	6	3	12	3	3	—	101
Chemical	23	10	3	17	12	4	—	2	71
Food and beverages	31	17	11	3	5	6	—	1	73
Forestry	3	3	—	—	—	—	—	—	6
Defense	—	—	—	3	—	—	—	—	3
Textile and leather	33	35	29	—	—	—	—	—	97
Transportation	3	2	—	3	3	—	—	—	22
Communications and IT	9	—	6	—	3	—	—	—	18
Mining	3	—	—	—	—	—	—	—	3
Wholesale and retail	13	6	1	—	—	2	—	—	22
Paper	25	—	—	—	5	6	—	—	36
Restaurants	7	11	3	—	—	—	—	—	21
Cement	30	12	14	15	4	3	—	—	78
Others	6	—	6	—	—	—	—	—	12
Total observation	218	152	86	44	41	18	3	2	564

TABLE 37.4 Direct and Indirect Ownership Structure

| | | Indirect Ownership | | | | | | | Direct Ownership | | |
	Holding	Nonfinancial Firms	Family	State	Foreigners	Financial Firms	Employees	Consortium	Group Affiliated	Nongroup Affiliated	Total
2000											
Tobin Q	1.618	1.495	1.532	1.801	2.261	1.574	1.060	4.473	1.643	1.652	1.647
ROA	0.006	−0.015	−0.007	0.053	0.043	−0.006	−0.255	0.210	0.013	−0.008	0.004
DIV/NI	0.174	0.182	0.116	0.408	0.186	0.019	0.000	0.861	0.182	0.190	0.185
DE/TA	0.675	0.621	0.577	0.482	0.673	0.541	0.825	0.558	0.662	0.578	0.626
Log(Sales)	7.610	7.422	7.252	8.091	7.801	7.178	8.055	9.343	7.604	7.500	7.560
2001											
Tobin Q	1.775	1.695	1.706	1.626	2.198	1.427	1.4372	5.435	1.834	1.678	1.767
ROA	−0.151	−0.165	−0.113	0.063	−0.102	−0.191	−.6094	0.294	−0.149	−0.106	−0.131
DIV/NI	0.153	0.179	0.048	0.326	0.076	0.025	0.000	0.000	0.143	0.152	0.147
DE/TA	0.827	0.799	0.706	0.513	0.821	0.694	1.258	0.535	0.813	0.721	0.774
Log(Sales)	7.533	7.605	7.449	8.323	7.925	7.552	8.208	9.620	7.773	7.651	7.722

2002

Tobin Q	1.483	1.348	1.445	1.175	2.214	1.054	1.468	—	1.462	1.454	1.458
ROA	−0.033	−0.001	−0.030	0.050	0.060	−0.092	−0.455	—	−0.025	−0.001	−0.015
DIV/NI	0.178	0.091	0.082	0.338	0.061	0.032	0.000	—	0.151	0.119	0.137
DE/TA	0.784	0.731	0.667	0.491	0.701	0.500	1.354	—	0.762	0.655	0.715
Log(Sales)	7.821	7.805	7.547	8.477	8.122	7.819	7.228	—	7.889	7.799	7.850

Notes: Tobin's Q shows the ratio of market value of equity plus book value of debt to the sum of book value of equity plus book value of debt. ROA is the return on assets. DIV/NI shows the dividend payout ratio, which is calculated by dividing the total amount of dividends by the net income. DE/TA is the leverage ratio and measured as the sum of long-term and short-term debt divided by the total assets. Log(Sales) is the natural logarithm of the firm's total sales. Firms are called group affiliated when they are attached to a business group or at least one of the shareholders is a holding company. It is important to note that there is no mean value for employee and consortium-owned firms due to the insufficient number of observations. However, the financial ratios are reported for these firms.

The firms in which foreigners are the major shareholders are the second highest dividend payers. The interpretation of the behavior of the foreign-owned firm is ambiguous. The ultimate owners of these firms can be family or state. The findings suggest that financial firms and families tend not to pay a high amount of dividends. Family members have strong cash flow incentives and have better information than outside investors. This gives flexibility in their dividend payment policy.

In general, the firms used less debt during the precrisis period of 2000 than during the period of 2001–2002. In particular, the leverage ratios peaked in the time of the crisis. With the exception of the employee-owned company, holding-owned companies had the highest leverage ratios among the others. The leverage ratio of the holding companies reached 82.70% in the year of the crisis and then fell to 78.4% in 2002, after the crisis. State-owned companies had the lowest leverage ratio during the sampling period. The leverage ratio of the companies in which the major shareholder is a financial company fell from 69.4% in 2001 to 50% in 2002. This sharp decline can be attributed to the nature of the financial crisis. Since the 2001 financial crisis emerged in the banking industry, it is inevitable that many financial institutions became insolvent and unwilling to lend.

In general, the total amount of sales increased slightly from 2000 through 2002. Among the major shareholders, the state has the highest sales records and foreigners the second highest. With the exception of the holding-owned companies, the amount of the sales for the other firms increased from 2000 to 2001. A decline in sales is noticed only in the employee-owned firm during the post-crisis period.

Table 37.5 shows the OLS and fixed effects regression results for the dependent variable Tobin's Q. The findings report differences in coefficients and significance levels, including sign reversals, between the OLS and fixed effects specifications. The OLS findings report that there is a positive and significant relationship between firm size and Tobin's Q. However, using all firm-year observations pooled, the fixed effects estimation shows a negative and significant coefficient for size variable, suggesting that small firms have growth potential. Higher firm sales are associated with lower investment before, during, and after the financial crisis.

The different signs of coefficients for size and also for dividend payout ratio variables may arise from a possible endogeneity problem. For example, in the case of the dividend payout ratio, the level of the dividends paid can be determined by the growth opportunity of a firm. The OLS regression results show a positive and significant relationship between Tobin's Q and the dividend payout ratio. Tobin's Q reflects the expectations about future

TABLE 37.5   Regression Results for Tobin's Q

	1	2	3	4
Panel A: OLS regression results				
Constant	−0.217	−0.430	−0.234	−0.390
	(−0.47)	(−0.907)	(−0.510)	(−0.85)
Log(TA)	0.148[b]	0.171[a]	0.148[b]	0.187[a]
	(2.52)	(2.86)	(2.51)	(3.16)
DIV/NI	0.442[a]	0.423[a]	0.444[a]	0.408[a]
	(3.37)	(3.22)	(3.40)	(3.14)
DE/TA	0.800[a]	0.810[a]	0.792[a]	0.801[a]
	(13.06)	(13.19)	(12.93)	(13.32)
Concentration	0.145	0.153	0.137	0.144
	(0.77)	(0.82)	(0.73)	(0.77)
Group affiliated	−0.036	−0.040	−0.037	−0.045
	(−0.47)	(−0.53)	(−0.49)	(−0.60)
Precrisis		0.145[c]		
		(1.79)		
Crisis			0.143[c]	
			(1.81)	
Postcrisis				−0.295[a]
				(−3.68)
F-statistics	34.299	29.882	29.835	32.110
Adjusted R^2	0.238	0.242	0.242	0.256
No. of observations	564	564	564	564
Panel B: Fixed effects regression results				
Constant	5.830[a]	5.670[a]	6.229[a]	2.164
	(4.08)	(2.69)	(4.37)	(1.24)
Log(Sales)	−0.603[a]	−0.584[b]	−0.647[b]	−0.118
	(−3.43)	(−2.26)	(−2.37)	(−0.54)
DIV/NI	−0.79	−0.812	−0.876	−0.135
	(−0.43)	(−0.44)	(−0.41)	(−0.74)
DE/TA	0.818[a]	0.824[a]	0.737[a]	0.881[a]
	(6.29)	(5.64)	(5.57)	(6.82)
Concentration	0.510	0.517	0.408	0.553
	(0.96)	(0.97)	(0.78)	(1.06)
Group affiliation	−0.618	−0.620	−0.625	−0.692
	(−1.33)	(−1.33)	(−1.36)	(−1.51)
Precrisis		0.009		
		(0.10)		

(*continued*)

TABLE 37.5 (continued)   Regression Results for Tobin's Q

	1	2	3	4
Crisis			0.164[a] (2.69)	
Postcrisis				−0.273[a] (−3.54)
F-statistics	12.383	10.291	11.712	12.757
R^2 within	0.1503	0.1503	0.1675	0.1798
R^2 between	0.0702	0.0734	0.0705	0.1441
R^2 overall	0.0812	0.0841	0.0705	0.1500
No. of observations	564	564	564	564

Notes: t-Statistics are given in parentheses in italics. The dependent variable is Tobin's Q. Panel A shows the OLS regression results and Panel B shows the fixed effects regression results. The independent variables are logarithm sales (Log(Sales)), dividend payout ratio (DIV/NI), leverage (DE/TA), lead share of the major shareholder (Concentration), group affiliation dummy variable, precrisis dummy variable, crisis dummy variable, and postcrisis dummy variable.

[a] Significance at 1% level.
[b] Significance at 5% level.
[c] Significance at 10% level.

earnings and market perceptions about the value of the company. When the firms have high investment opportunities, the managers have less incentive to pay high amounts of dividends. As a consequence of a high Tobin's Q, it is expected that firms are less likely to pay out dividends. To address this concern, fixed effects regressions were run and the results changed substantially. This implies that the regression coefficient in an OLS regression is biased. The results of the fixed effects regression present a negative but insignificant relationship between dividend payout and Tobin's Q.

The findings show that the coefficient on leverage is strongly positive. It implies that firms with high leverage are able to have high growth prospects. This can be related to the capital structure of a firm, suggesting that high leverage increases a firm's value due to an interest tax-shield. Furthermore, no significant relationship is found between concentration, group affiliation variables, and Tobin's Q. The market does not reward group-affiliated firms in terms of Tobin's Q. This is in line with Gunduz and Tatoglu (2003) and Gonenc and Aybar (2006). However, this finding contrasts with Orbay and Yurtoglu (2006), who found that group membership improved the market valuations of firms from 1990 through 2003. The discrepancies between their results and the results presented here may stem from either differences in the definition of group affiliation or the features of the sample used. In their paper, they used a larger sample and a longer sampling period.

It is reported that there is no significant relationship between the precrisis variable and Tobin's Q. The crisis dummy is positively and significantly associated with the firm's value. However, the effect of the crisis can clearly be seen in 2002. The negative and significant coefficient of the postcrisis variable suggests that many firms suffered after the 2001 financial crisis. Consistent with the earlier findings in Table 37.4, Tobin's Q declined substantially following the crisis year. Table 37.1 also documents a downward trend for market capitalization of the ISE-listed firms in 2002.

Table 37.6 shows the OLS and fixed effects regression results for the dependent variable ROA. The findings in show that there is a positive and significant relationship between ROA and the firm's size. The higher size, measured by the log of sales, is associated with a higher ROA. This

TABLE 37.6   Regression Results for ROA

	1	2	3	4
Panel A: OLS regression results				
Constant	−0.123	−0.179[c]	−0.117	−0.102
	(−1.41)	(−1.93)	(−1.33)	(−1.14)
Log(TA)	0.045[a]	0.050[a]	0.047[a]	0.040[a]
	(3.97)	(4.30)	(4.14)	(3.47)
DIV/NI	0.060[b]	0.056[b]	0.059[b]	0.065[b]
	(2.34)	(2.16)	(2.33)	(2.52)
DE/TA	−0.44[a]	−0.441[a]	−0.44[a] 1	−0.445[a]
	(−37.06)	(−36.80)	(−37.37)	(−37.32)
Concentration	0.042	0.044	0.045	0.042
	(1.14)	(1.20)	(1.26)	(1.13)
Group affiliated	0.023	0.022	0.023	0.024
	(1.55)	(1.49)	(1.58)	(1.63)
Precrisis		0.036[b]		
		(2.30)		
Crisis			−0.074[a]	
			(−4.89)	
Post-crisis				0.041[a]
				(2.63)
F-statistics	322.301	271.567	283.382	272.656
Adjusted R^2	0.747	0.747	0.755	0.747
No. of observations	564	564	564	564
Panel B: Fixed effects regression results				
Constant	−1.446[a]	−3.752[a]	−1.637[a]	−1.407[a]
	(−4.04)	(−7.52)	(−4.59)	(−3.18)
Log(Sales)	0.236[a]	0.515[a]	0.256[a]	0.229[a]
	(5.32)	(8.45)	(5.86)	(4.11)

(continued)

TABLE 37.6 (continued)   Regression Results for ROA

	1	2	3	4
DIV/NI	0.048	0.016	0.047	0.049
	(1.03)	*(0.37)*	*(1.03)*	*(1.05)*
DE/TA	−0.602[a]	−0.498[a]	−0.567[a]	−0.603[a]
	(−18.07)	*(−14.00)*	*(−17.01)*	*(−17.90)*
Concentration	−0.065	0.043	−0.024	−0.066
	(−0.48)	*(0.33)*	*(−0.19)*	*(−0.49)*
Group affiliation	0.058	0.025	0.061	0.059
	(0.49)	*(0.23)*	*(0.53)*	*(0.50)*
Precrisis		0.139[a]		
		(6.33)		
Crisis			−0.066[a]	
			(−4.35)	
Postcrisis				0.004
				(0.22)
F-statistics	79.645	80.346	72.862	66.201
R^2 within	0.5280	0.5759	0.5519	0.5280
R^2 between	0.8150	0.5607	0.5519	0.8195
R^2 overall	0.5759	0.5170	0.7017	0.7017
No. of observations	564	564	564	564

Notes: *t*-Statistics are given in parentheses in italics. The dependent variable is Tobin's Q. Panel A shows the OLS regression results and Panel B shows the fixed effects regression results. The independent variables are logarithm sales (Log(Sales)), dividend payout ratio (DIV/NI), leverage (DE/TA), lead share of the major shareholder (Concentration), group affiliation dummy variable, precrisis dummy variable, crisis dummy variable, and postcrisis dummy variable.

[a] Significance at 1% level.
[b] Significance at 5% level.
[c] Significance at 10% level.

is consistent with the findings of Gursoy and Aydogan (2003), who also studied the relationship between a firm's value and ROA for ISE-listed firms from 1992 to 1998. The bigger the firm is, the higher the level of profitability. Large firms are usually well diversified and can avoid risks and provide a higher level of earnings. The OLS regression results report a positive coefficient for dividend payout ratio. The higher the net income is, which is proxied by the ROA, the more dividends will be paid out to the shareholders. In order to assess the robustness of the results, the fixed effects regression models were run. The findings document positive and insignificant coefficients for the dividend payout ratio.

The ratio of debt to assets is employed as a measure of a firm's leverage. The coefficient of leverage is negative and significant, suggesting that less indebted firms are likely to have high profits. Although the cost of borrowing became high and financial firms became unwilling to lend during the financial crisis, the leverage ratio increased and the profitability of the firms decreased. The descriptive statistics shown in Table 37.4 support this.

Furthermore, the regression results show no significant relationship between ownership concentration and ROA. Concentrated ownership is not associated with better operating performance or higher firm valuation. This is consistent with Himmelberg et al. (1999) and Chen et al. (2005), who find no relationship between ownership concentration and firm value in Hong Kong. Group affiliation is found to have a negative and insignificant relationship with ROA. It implies that group affiliation is not a significant indicator in order to be efficient in using assets. The benefits associated with business groups are not associated with having a high ROA. The results presented in this chapter contradict the widely accepted belief that group-affiliated firms have superior performance in terms of profitability over nongroup-affiliated firms. The recent studies by Chu (2004) for Taiwan, Gonenc and Aybar (2006) and Gunduz and Tatoglu (2003) for Turkey support the evidence provided in this chapter.

Consistent with the descriptive results, prior to the crisis, firms used to have higher ROAs. The coefficient of the precrisis variable is positive and significant. However, during the crisis, firms became unable to benefit from their assets to generate earnings. The findings show that there is a negative and significant relationship between the crisis variable and ROA. The results of the fixed effects model suggest that although the sign of the postcrisis variables is positive, no significant relationship is found. OLS regression results present a positive and significant relationship between the postcrisis variable and ROA.

37.5 CONCLUSIONS

This chapter examines the ownership structure for ISE-listed firms from 2000 through 2002 and attempts to figure out whether the ownership structure affected firm value and operating performance. The descriptive statistics report high direct ownership concentration for the ISE firms. Among the major shareholders, foreigners are the most and employees are the least concentrated major shareholders. The second most concentrated ownership structure is the state, followed by family ownership, in which the individual(s) is dominant. One of the striking findings of this chapter suggests that foreigners increased their lead shares significantly from the precrisis to the postcrisis period. Not only did foreigners take advantage of the financial crisis to increase their lead shares, but state, families, and nonfinancial firms also did the same. However, these increases are not statistically significant.

The firms used debt intensively during the period of 2001–2002. In particular, the leverage ratios peaked in the time of the financial crisis. It is apparent that some firms were hurt by the crisis and others benefited from it. While highly indebted firms increased their values, they became unable to benefit from their assets to generate earnings. Prior to the crisis, firms had better operating performances. However, the profitability of the firms turned negative during and after the financial crisis. The prolonged effect of the crisis emerged in 2002, when the market values of the firms decreased with the economic slowdown.

Using firm-level data, the empirical findings indicate that concentrated ownership is not associated with higher firm valuation or better operating performance. Furthermore, the results show no clear evidence that the group-affiliated firms have superior performance in terms of firm value and profitability over nongroup-affiliated firms.

REFERENCES

Adaoglu, C. (1999) Regulation influence on the dividend policy of the Istanbul stock exchange (ISE) corporations. *The Istanbul Stock Exchange Review*, 3(11): 1–19.

Bianchi, R. (1984) *Interest Groups and Political Development in Turkey*. Princeton University Press, Princeton, NJ.

Chen, Z., Cheung, Y.L., Stouraitis, A., and Wong, A.W.S. (2005) Ownership concentration, firm performance, and dividend policy in Hong Kong. *Pacific-Basin Finance Journal*, 13(4): 431–449.

Chibber, P.K. and Majumdar, S.K. (1999) Foreign ownership and profitability: Property rights, control, and performance of firms in Indian industry. *The Journal of Law & Economics*, 42(1): 209–238.

Chong, A. and Lopez-de-Silanes, F. (2007) *Investor Protection and Corporate Governance: Firm-level Evidence across Latin America*. Stanford University Press, Palo Alto, CA.

Chu, W. (2004) Are group-affiliated firms really more profitable than nonaffiliated? *Small Business Economics*, 22(5): 391–405.

Claessens, S., Djankov, S., and Lang, L. (2000) The separation of ownership and control in East Asian corporations. *Journal of Financial Economics*, 58(2): 81–112.

Durukan, B., Ozkan, S., and Dalkilic, F. (2007) Measuring the Effectiveness of the Corporate Governance System in Turkey. Working Paper, Dokuz Eylul University, Izmir, Turkey.

Glen, J. and Singh, A. (2004) Comparing capital structures and rates of return in developed and emerging markets. *Emerging Markets Review*, 5(2): 161–192.

Gonenc, H. and Aybar, C.B. (2006) Financial crisis and firm performance: Empirical evidence from Turkey. *Corporate Governance*, 14(4): 297–311.

Gugler, K. (2003) Corporate governance, dividend payout policy, and the interrelation between dividends, R&D, and the capital investment. *Journal of Banking and Finance*, 27(7): 1297–1321.

Gunduz, L. and Tatoglu, E. (2003) Comparison of the financial characteristics of group affiliated and independent firms in Turkey. *European Business Review*, 15(1): 48–54.

Gursoy, G. and Aydogan, K. (2002) Equity ownership structure, risk taking and performance: An empirical investigation in Turkish companies. *Emerging Markets Finance and Trade*, 36(6): 6–25.

Himmelberg, C.P., Hubbard, R.G., and Palia, D. (1999) Understanding the determinants of managerial ownership and the link between ownership and performance. *Journal of Financial Economics*, 53(3): 353–384.

Johnson, S., Bone, P., Breach, A., and Friedman, E. (2000) Corporate governance in the Asian financial crisis. *Journal of Financial Economics*, 58(1): 141–186.

Karademir, B. and Danisman, A. (2007) Business groups and media in Turkey: A co-evolutionary approach to their interrelationships in Turkey, 1960–2005. *Problems and Perspectives in Management*, 5(3): 44–57.

Khanna, T. and Rivkin, J. (2001) Estimating the performance effects of networks in emerging markets. *Strategic Management Journal*, 22(1): 45–74.

Lemmon, M.L. and Lins, K. (2003) Ownership structure, corporate governance, and firm value: Evidence from East Asian financial crisis. *Journal of Finance*, 58(4): 1445–1468.

Mitton, T. (2002) A cross-firm analysis of the impact of corporate governance on the East Asian financial crisis. *Journal of Financial Economics*, 64(2): 215–241.

Orbay, H., and Yurtoglu, B. (2006) The impact of corporate governance structures on the investment performance in Turkey. *Corporate Governance*, 14(4): 349–363.

Saldana, C. (2001) Corporate governance environment and policy: Their impact on corporate performance and finance in the Philippines, In: OECD (Ed.) *Corporate Governance in Asia: A Comparative Perspective*, OECD Publishing, Paris.

Xu, X.N. and Wang, Y. (1999) Ownership structure and corporate governance in Chinese stock companies. *China Economic Review*, 10(1): 75–98.

Yaprak, A., Karademir, B., and Osborn, R.N. (2007) How do business groups function and evolve in emerging markets? The case of Turkish business groups. In A. Rialp and J. Rialp (Eds), *Advances in International Marketing*, Elsevier, Amsterdam, the Netherlands, pp. 275–294.

Yeh, Y.H., Lee, T.S., and Woidtke, T. (2001) Corporate governance and performance: The case of Taiwan. *International Review of Finance*, 2(1): 21–48.

Yurtoglu, B. (2003) Corporate Governance and Implications for Minority Shareholders in Turkey. Working Paper, Turkish Economic Association Working Paper, December, Ankara, Turkey.

CHAPTER 38

Nonlinear Synthesis Approach Establishing a Banking or Financial Distress Early Warning System against Corruption

Terry J. O'Neill, Jack Penm, and R.D. Terrell

CONTENTS

38.1 INTRODUCTION

Financial distress is often caused by default or credit-rating changes of the counterparty and the resulting credit risk. By measuring the credit risk technology, we can identify the default rate of the counterparty in the event of default and the impact it may have on company losses. In recent years, some publicly listed companies in Shanghai have announced restructuring, default, or delisting without warning. These financial distress events clearly challenged both the authority capacity of the Commissioner and the capacity of investors to respond in the face of confusing information and the lack of ability to measure risk, resulting in high social costs. Therefore, it is important to raise investors' credit risk awareness prior to establishing a financial distress early warning system, so advanced firms may change a deal, or if there is an indication of financial distress, investors may reduce a preventable loss. Establishment of an appropriate financial distress early warning system is the most urgent of issues.

The study of financial distress requires a suitable model in addition to selection of variables. For instance, Moody's KMV, discriminant analysis, logistic regression (LR), neural network (NN), and a synthesis model of LR–NN can be employed to observe which variables are most useful for predicting financial distress. Taking the example of LR, in statistical processing, LR transforms the original model into a probability model with resulting probability falling between 0 and 1, which allows us to compare the degree of financial distress. Besides LR, other modeling approaches may offer different benefits. Against the backdrop of a changing financial environment and the possibility of economic recession, Shanghai, like many other countries in Asia and other regions of the world, has been hit with a financial crisis (Brailsford et al., 2008) in recent years that seriously hampers the country's economic development. Thus this chapter attempts to develop a financial distress warning model based on key financial variables for publicly listed corporations in Shanghai and variables of corporate governance that can easily help to identify firms in latent crises so permit preemptive actions to be taken. This model also aims to provide a reference for the national agency to conduct risk assessment and provide advance on business directions.

The distinctive approach of this chapter is to adopt the "optimal cutoff point" approach proposed by Hosmer and Lemeshow (2000) to determine the cutoff point for financial distress in the Shanghai stock market. In addition, we examine the capacity of a relatively new technique, three two-stage synthesis models of LR–NN to predict failure in Shanghai's publicly listed corporations. These models are compared with conventional LR and NN models. The aims of this chapter are to construct a financial distress warning model in Shanghai applicable to the period of 2000–2007, which distinguish as financially sound and financially distressed firms by applying the criteria for shares (full delivery, stop trading, or delisted under the order of the regulator) during the analyzed period. In order to improve investor awareness of credit risk, prior to establishing a financial distress early warning system, advanced firms may change a deal, or if there is an indication, investors may reduce a preventable loss. The objectives of this chapter are summed up as follows: (a) through a literature review, identify important financial factors and corporate governance factors related to financial distress of publicly listed corporations in Shanghai; (b) by employing five modeling approaches, comprising LR, NN, and three two-stage models, to further explore the important factors correlated with financial distress of firms and construct a pertinent financial distress warning model for publicly listed corporations in Shanghai; and (c) by comparing the prediction accuracy of the aforementioned five models by cross validation to propose an optimal financial distress warning model for publicly listed corporations in Shanghai.

38.2 LITERATURE REVIEW

A number of studies have examined financial distress warning modeling. Johnson et al. (2000) examine the 1997 Asian financial crisis and conclude the measure of corporate governance has better explanatory power than macroeconomic measures, and in countries with weak corporate governance financial distress, problems are more likely to emerge, especially during economic recession when the agency problem exacerbates, which could lead to currency depreciation and falling stock prices. Rajan and Zingales (1998) propose that ownership concentration and inefficient corporate governance are the two major causes of the Asian financial crisis. La Porta et al. (1999) also find that many listed firms use the schemes of pyramid structures and cross-holding to gain corporate control and the financial distress of a firm is to a certain extent related to the cross-holding between a parent company and subsidiaries and poor management, and not entirely attributed to financial factors. Morck et al. (1988) propose that based on entrenchment hypothesis and convergence-of-interests hypothesis, low ownership by a board of directors limits the board's ability and willingness to supervise the management and might be positively related to the likelihood of financial distress. Their study also includes the variable of ownership by directors and supervisors to discern whether higher shareholding by directors and supervisors would increase or decrease the likelihood of financial distress. As indicated in the earlier discussions, corporate governance plays a vital role in the financial soundness of a firm. Following an overview of the relevant research, this chapter identifies the following significant financial distress variables: total debt to equity ratio, debt ratio, cash flows ratio, cash flows to long-term liabilities, equity to assets ratio, fixed assets turnover, fixed assets to long-term capital, return on equity, return on assets, return on common share's equity, profit margin, current allowance ratio, current ratio, operating income growth rate, earnings per share, quick ratio, financial leverage ratio, inventory turnover, capital interest ratio, interest coverage rate, earnings growth rate, return on owners' equity, total assets turnover, working capital turnover, accounts receivable turnover, net income before income tax, profit ratio of average capital, growth rate of real income, presence of family control, ownership by board of directors, ownership by managers, a director serving concurrently as manager, director pledge ratio, ownership by large shareholders, accountant audit opinion, and accountant replacement ratio.

38.3 RESEARCH DESIGN AND STATISTICAL METHOD

The statistical methods previously used to construct a financial distress warning model show that univariate analysis cannot establish comprehensive financial ratios and corporate governance variables to compare the unclassified and classified samples, which produce multivariate discriminant analysis (DA). Nevertheless, DA requires that the data must comply with three assumptions: the data are normally distributed; covariance matrices are equal; and the predictive variables are noncollinear.

Prior studies show that LR has better prediction power than traditional approaches. NN is a useful statistical technique that has massive computing power, powerful memory, learning ability, and fault-tolerance ability. DA takes into account multiple predictors and determines which predictor has discriminative power in a more objective sense. However, the values of independent variables in the model must have a normal distribution. As the financial data of banks used in this chapter do not conform to a normal distribution, we choose to employ LR that better fits the non-normal variables.

38.3.1 Logistic Regression

This chapter uses LR to develop the financial distress warning model and SAS for statistical analysis. LR addresses the issue of non-normal variables and the predicted probabilities of dependent variables falling between 0 and 1 depict a nonlinear phenomenon. Thus LR is adopted to compare its performance with other statistical techniques. Besides modeling dichotomous outcome, LR can be used to compute the probability of default. This chapter devotes considerable endeavor to the selection of financial ratios with explanatory power as independent variables for the financial distress warning model. In LR, the notation for conditional probability of an event is defined as $Q(Y_i = 1 | Z_i) = Q_i$ and the LR model is as follows:

$$Q_i = \frac{1}{1+e^{-(\alpha+\delta Z_i)}} = \frac{e^{\alpha+\delta Z_i}}{1+e^{\alpha+\delta Z_i}}, \tag{38.1}$$

where Q_i is the probability of event i, which is a nonlinear function of Z_i. This nonlinear function can be converted into a linear function. First, the conditional probability of a nonevent is defined as

$$1-Q_i = 1-\left(\frac{e^{\alpha+\delta Z_i}}{1+e^{\alpha+\delta Z_i}}\right) = \frac{1}{1+e^{\alpha+\delta Z_i}}.$$

Then the ratio of probability of event to probability of nonevent is

$$\frac{Q_i}{1-Q_i} = e^{\alpha+\delta Z_i},$$

which is called the odds ratio of an event. The odds ratio is definitely a positive value, for $0 < Q_i < 1$. We take the natural logarithm of odds to derive a linear function:

$$\ln\left(\frac{Q_i}{1-Q_i}\right) = \alpha + \delta Z_i. \tag{38.2}$$

In Equation 38.2, the logarithm of LR function is either called logit or logit of y, i.e., logit(y). Conceptually, Equation 38.1 expresses "LR," for it is a distribution function of LR, while Equation 38.2 expresses "logit model," for it is in logit form. Some previous research makes a distinction between the LR and logit model by determining whether the independent variable used is a continuous variable.

When there are k independent variables, Equation 38.1 can be expanded to

$$Q_i = \frac{e^{\alpha + \sum_{j=1}^{k} \delta_j Z_{ji}}}{1 + e^{\alpha + \sum_{j=1}^{k} \delta_j Z_{ji}}}.$$

Then the LR model will have the following form:

$$\ln\left(\frac{Q_i}{1-Q_i}\right) = \alpha + \sum_{j=1}^{k} \delta_j Z_{ji},$$

where

$Q_i = (Y_i = 1 | Z_{1i}, Z_{2i}, \ldots, Z_{ki})$ denotes the probability of bank i being in financial distress as estimated by the developed LR model, where $Y_i = 0, 1$ (1 denotes a financially distressed bank and 0 denotes a financially sound bank)

Z_{ji} denotes the j financial variable of bank i

α denotes a constant

δ_j denotes parameters

38.3.2 Neural Network

NN is a computing system that emulates the interconnection of neurons in organisms for complex information processing. It is an adaptive system with the ability to learn. Through different algorithms, NN can be trained to provide a desired output. NN offers the advantages of adaptive learning, inferential association, inductive judgment, and experience accumulation, the ability to memorize and forget. NN is also characterized by high-speed computing ability, high-capacity memory, powerful learning ability, and fault tolerance. The basic back-propagation network (BPN) algorithm uses the gradient steepest descent method to minimize the error function between actual output and target output of network. BPN trains the network through continuous adjustment of weights (ΔW_{ij}) in the steepest descent direction and repeats the feed forward and back propagation steps: $\Delta W_{ij} = -\eta(\partial E/\partial W)$ and $E = 0.5 \sum (T_j - Y_j)^2$, where η is the learning rate that controls the magnitude of each weight adjustment, E is the error function, T_j is the actual value, and Y_j is the output value. The learning rate is a very important parameter in the training process of NN. It affects the convergence speed of NN. A higher learning rate means faster convergence and a smaller learning rate slows down the NN convergence. However, the learning rate either being too high or too low, will adversely affect the training of NN.

38.3.3 Two-Stage Model of LR and NN

The two-stage synthesis models we propose integrates the framework of LR and NN. In stage I, influencing variables are selected using LR. In stage II, the influencing variables are taken as the input variables of BPN. It is intended that by providing the NN with a good starting point, a more precise model can be developed on the strength of its learning ability. Such a synthesis model is then compared with simple LR and NN.

Two-stage synthesis model (1):

Stage I: We use dependent variables Z_1, Z_2, ..., Z_p in the LR. We then use LR with the Wald-forward method to identify independent variables with significant influence on distress probability $Z_1^*, Z_2^*, ..., Z_K^*$.

Stage II: We use significant variables $Z_1^*, Z_2^*, ..., Z_K^*$ in the NN model as independent variables of the input layer, and Y is the dependent variable of the input layer, to obtain a set of $\hat{Q}$ based on the point of intersection of sensitivity and specificity according to Hosmer and Lemeshow (2000). We then compare the results with actual values (Y). The NN model and the cutoffs can be used for prediction based on other datasets.

Two-stage synthesis model (2):

Stage I: We use dependent variable Y and independent variables Z_1, Z_2, ..., Z_p in the LR. We then use LR with the Wald-forward method to identify independent variables with significant influence on distress probability $Z_1^*, Z_2^*, ..., Z_K^*$ and a significance model:

$$\hat{Q}_i = \frac{e^{\hat{\delta}_0 + \hat{\delta}_1 Z_1^* + \cdots + \hat{\delta}_k Z_k^*}}{1 + e^{\hat{\beta}_0 + \hat{\beta}_1 Z_1^* + \cdots + \hat{\delta}_k Z_k^*}}.$$

We obtain the predictive value for $\hat{Q}$ of financial distress probability for each data set using the aforesaid significance model. We find the financial distress cutoff point for $\hat{Q}$ based on the point of intersection of sensitivity and specificity according to Hosmer and Lemeshow (2000). We then compare the results with actual values (Y). We convert the distress probability into integrity or distress to produce a new dependent variable and a new NN model.

Stage II: We use significant variables $Z_1^*, Z_2^*, \ldots, Z_K^*$ and the new dependent variableobtained in stage I in the NN model as the independent variables and dependent variable of the input layer, to produce the predictive value $\hat{Q}^*$. As indicated in Stage 1, we find the financial distress cutoff point for $\hat{Q}^*$ and compare the results with actual values (Y). The NN model and the cutoffs can be used for prediction based on other datasets.

Two-stage synthesis model (3):

Stage I is identical to stage I of the aforementioned synthesis model (2).

Stage II: We use significant variables $Z_1^*, Z_2^*, \ldots, Z_K^*$ and $\hat{Q}$ obtained in stage I in the NN model as the independent variables and dependent variable of the input layer, to produce $\hat{Q}$. We find the financial distress cutoff point for $\hat{Q}^*$ and compare the results with Y.

38.3.4 Variables Definition

Dependent variable: The dependent variable depicts whether a company is in financial distress; 1 means a financially distressed company and 0 means a financially sound company. The definitions for financial distress are by the criteria of shares full delivery, stop trading, or delisted under the order of the regulator during the analyzed period.

Independent variables: Through overview of prior literatures and empirical analysis, this chapter derives significant variables for financial distress of corporation, including corporate governance variables and financial variables as presented in Table 38.1.

38.3.5 Research Period, Sample, and Data Source

This chapter targets 800 publicly listed corporations in Shanghai and samples data over the time period of 2000–2007. The corporations are classified

TABLE 38.1   Description of Relevant Variables

Variables	Remarks
Z_1 Earnings per share	Net income after taxes/issued and outstanding shares
Z_2 Return on assets	Net income after taxes/average total assets $\times$ 100
Z_3 Return on equity	Net income before taxes/average equity $\times$ 100
Z_4 Income after taxes ratio	Net income after taxes/net operating income $\times$ 100
Z_5 Operating margins ratio	Operating margins/net operating income $\times$ 100
Z_6 Operating income ratio	Operating income/net operating income $\times$ 100
Z_7 Operating expenses ratio	Operating expenses/net operating income
Z_8 Current ratio	Current assets/current liabilities
Z_9 Acid-test ratio	(Current assets–inventory–prepaid expenses)/ current liabilities
Z_{10} Debt ratio	Total liabilities/total assets
Z_{11} Times interest earned	Income before income taxes and interest expense/interest expense
Z_{12} Cash flow ratio	Net cash flows from operations/current liabilities
Z_{13} Equity growth ratio	(Equity of current year–equity of prior year)/ equity of prior year
Z_{14} Inventory turnover	Cost of goods sold/average inventory
Z_{15} Fixed assets turnover	Net operating income/average net fixed assets
Z_{16} Holding rate of directors and supervisors	Holding shares of directors and supervisors/ issued and outstanding shares
Z_{17} Holding rate of block shareholders	Holding shares of block shareholders/issued and outstanding shares

into financially distressed or financially sound group based on the 3 year average of sample data and the definitions for financial distress. There are 150 corporations in the financially distressed group by the criteria of shares full delivery, stop trading, or delisted under the order of the regulator during the analyzed period and 650 corporations in the financially sound group.

38.4 EMPIRICAL RESULTS

38.4.1 Statistics Summary

The descriptive statistics of empirical results in Table 38.2 reveal that the means of both financial variables and corporate governance variables of financially sound corporations are more significant than the means of financially distressed corporations. For example, in terms of earnings per share (Z_1), return on assets (Z_2), return on equity (Z_3), income after taxes ratio (Z_4), operating margins ratio (Z_5), operating income ratio (Z_6), current

TABLE 38.2   Financial and Corporate Governance Variables—Statistics

Predictive Variables	Group I—Financially Sound; Group II—Financially Distressed	Mean	Standard Deviation (S.D.)
Z_1	I	0.12	0.85
	II	−0.68	1.11
Z_2	I	0.11	0.83
	II	−0.73	1.21
Z_3	I	0.11	0.57
	II	−0.68	2.00
Z_4	I	0.05	0.50
	II	−0.30	2.21
Z_5	I	0.06	0.91
	II	−0.37	1.18
Z_6	I	0.06	0.55
	II	−0.30	2.11
Z_7	I	−0.03	0.60
	II	0.18	2.05
Z_8	I	0.01	0.92
	II	−0.17	1.36
Z_9	I	0.06	0.91
	II	−0.22	1.21
Z_{10}	I	−0.12	0.80
	II	0.53	1.52
Z_{11}	I	0.03	1.06
	II	−0.12	0.51
Z_{12}	I	0.08	0.92
	II	−0.28	1.28
Z_{13}	I	0.06	0.91
	II	−0.21	1.23
Z_{14}	I	0.03	1.08
	II	−0.31	0.52
Z_{15}	I	0.08	1.01
	II	−0.28	0.31
Z_{16}	I	0.01	0.98
	II	0.02	1.00
Z_{17}	I	0.01	0.93
	II	0.01	0.98

ratio (Z_8), acid-test ratio (Z_9), times interest earned (Z_{11}), cash flow ratio (Z_{12}), equity growth ratio (Z_{13}), inventory turnover (Z_{14}), and fixed assets turnover (Z_{15}) are positive for financially sound corporations. Alternatively, operating expenses ratio (Z_7) and debt ratio (Z_{10}) are negative for financially distressed corporations. The statistics show that as the financial risk of financially distressed corporations is greater than financially sound corporations, the probability of financial crises is also higher. On the other hand, the higher variability of the standard deviation for financially sound and distressed corporations are examined as follows: return on equity (Z_3), income after taxes ratio (Z_4), operating margins ratio (Z_5), operating expenses ratio (Z_7), and current ratio (Z_8). Unstable profitability indicates that the aforementioned variables may lead to a financial crisis (Penm, 2007) for both financially sound and distressed corporations.

38.4.2 Collinearity Diagnosis

When predictive variables exhibit collinearity, this will lead to the mean square error of parameter estimation being too high, resulting in unstable estimates. Therefore, it is necessary to eliminate variables with collinearity. Firstly, the Pearson correlation coefficient is used to measure the relevance between variables. If there is high correlation between variables, one should continue with an optional variable screening. But the Pearson correlation test is not a sufficient condition for collinearity, so this chapter further examines three indicators—conditions indicators (CI), the proportion of variance, and the variance inflation factor (VIF)—to test whether the variables show collinearity.

The Spearman's rank correlation test detects the phenomenon of correlation between variables. Table 38.3 shows that some variables with a correlation coefficient above 0.7 are highly correlated, which implies these may be a collinearity phenomenon between variables, including earnings per share (Z_1), return on assets (Z_2), current ratio (Z_8), and acid-test ratio (Z_9). Two variables are required for screening, which are incorporated into the model. "Point biserial correlation" is utilized to determine a high relationship between predictive variables and a dependent variable, so variables with a higher relationship will be included in the model. By point biserial correlation in Table 38.4, earnings per share (Z_1) and current ratio (Z_8) are discarded, after deletion remaining 15 variables. Point biserial correlation is to measure the correlation coefficient of categorical variable and continuous variable, the formula is as follows:

TABLE 38.3 Financial and Corporate Governance Variables—Correlation Statistics

	Z_1	Z_2	Z_3	Z_4	Z_5	Z_6	Z_7	Z_8	Z_9	Z_{10}	Z_{11}	Z_{12}	Z_{13}	Z_{14}	Z_{15}	Z_{16}	Z_{17}
Z_1	1																
Z_2	0.85	1															
Z_3	0.38	0.58	1														
Z_4	0.23	0.21	0.11	1													
Z_5	0.38	0.48	0.28	0.18	1												
Z_6	0.25	0.34	0.21	0.48	0.18	1											
Z_7	−0.08	−0.18	−0.10	−0.63	0.15	−0.38	1										
Z_8	0.15	0.18	0.11	−0.07	0.23	−0.16	0.14	1									
Z_9	0.18	0.23	0.12	−0.06	0.27	−0.11	0.12	0.85	1								
Z_{10}	−0.24	−0.34	−0.46	−0.12	−0.34	−0.12	0.02	−0.47	−0.46	1							
Z_{11}	0.14	0.12	0.07	0.06	0.08	0.07	−0.02	0.11	0.11	−0.10	1						
Z_{12}	0.18	0.35	0.13	0.28	0.28	0.48	−0.18	0.21	0.25	−0.28	0.02	1					
Z_{13}	0.33	0.30	0.28	0.14	0.18	0.15	−0.05	0.07	0.07	−0.11	0.03	0.02	1				
Z_{14}	0.11	0.10	0.11	0.05	0.05	0.18	−0.07	0.01	0.18	−0.06	0.02	0.08	0.05	1			
Z_{15}	0.19	0.16	0.12	0.08	−0.02	0.06	−0.07	0.05	0.06	−0.02	0.03	0.03	0.05	0.03	10		
Z_{16}	0.10	0.12	0.11	0.05	0.10	0.05	0.01	0.06	0.06	−0.08	0.04	0.03	0.02	0.02	0.03	1	
Z_{17}	0.05	0.01	0.05	0.04	−0.01	0.05	−0.04	−0.03	−0.03	0.05	0.02	0.02	−0.04	0.03	−0.02	−0.11	1

TABLE 38.4 Point Biserial Correlation (PBC)

Predictive Variables	PBC	Highly Correlated Predictive Variables	Variable Ignored
Z_1 (earnings per share)	−0.27	Z_1 and Z_2	Z (earnings per share)
Z_2 (return on assets)	−0.32		
Z_8 (current ratio)	−0.06	Z_8 and Z_9	Z (current ratio)
Z_9 (acid-test ratio)	−0.09		

$$r_{pb} = (qp)^{1/2}(\overline{Z}_1 - \overline{Z}_0 / s_z),$$

where

q is the percentage of $Y = 1$ (financial distress corporation)
p is the percentage of $Y = 0$ (financial sound corporation)
s_z^2 is the sample variance of continuous variables
$\overline{Z}_1, \overline{Z}_2$ is the mean of continuous variables of $Y = 1$ and $Y = 0$, respectively

Next, this chapter conducts collinearity test to examine the presence of collinearity and excludes variables with collinearity based on three indices, CI, tolerance (T), and VIF, as shown in Table 38.5. It is observed that Coenders and Saez (2000) point out that when VIF value is above 10, there may exist collinearity between the variables. By the threshold values of CI > 30, T < 0.2, and VIF > 10, variables with index values reaching the threshold indicate collinearity and are thereby discarded. As shown in Table 38.5, a total of 15 variables free of collinearity problem are kept.

38.4.3 Results of LR

To identify variables related to financial distress of publicity listed corporations, this chapter conducts several collinearity diagnosis procedures to exclude variables exhibiting high collinearity and use the remaining variables for LR modeling and testing. The Wald forward method is employed to select variables for the construction of financial distress warning models. The LR equation derived is as follows:

$$\ln\left(\frac{\hat{Q}}{1-\hat{Q}}\right) = -2.313 - 0.671Z_2 - 0.173Z_3 + 0.175Z_6 + 0.154Z_9 + 0.320Z_{10}$$

$$-0.287Z_{12} + 0.091Z_{13} - 1.152Z_{14} - 1.431Z_{15} + 0.244Z_{16} + 0.096Z_{17}.$$

TABLE 38.5 Results of Collinearity Tests

Dimension	1	2	3	4	5	6	7	8	9	10	11	12	13	14	15	16	VIF
CI	1.00	1.20	1.52	1.83	1.75	1.85	1.82	1.83	1.80	1.83	2.03	2.37	2.55	3.30	3.21	2.58	
Z_2	0.05	0.01	0.01	0.02	0.01	0.01	0.00	0.01	0.00	0.01	0.03	0.03	0.32	0.08	0.33	0.03	2.03
Z_3	0.02	0.01	0.05	0.02	0.02	0.02	0.00	0.01	0.02	0.01	0.12	0.07	0.03	0.01	0.38	0.03	1.55
Z_4	0.01	0.05	0.01	0.00	0.00	0.01	0.00	0.01	0.01	0.01	0.02	0.11	0.01	0.12	0.03	0.23	2.12
Z_5	0.01	0.02	0.01	0.01	0.05	0.00	0.01	0.06	0.01	0.02	0.01	0.23	0.10	0.05	0.01	0.33	1.27
Z_6	0.02	0.03	0.02	0.01	0.02	0.01	0.00	0.02	0.02	0.01	0.03	0.08	0.11	0.53	0.02	0.05	1.85
Z_7	0.02	0.07	0.00	0.02	0.031	0.00	0.01	0.03	0.02	0.02	0.01	0.02	0.01	0.02	0.01	0.38	2.18
Z_9	0.01	0.03	0.05	0.01	0.02	0.00	0.01	0.03	0.00	0.01	0.12	0.01	0.03	0.21	0.26	0.02	1.58
Z_{10}	0.02	0.03	0.02	0.00	0.01	0.00	0.02	0.08	0.03	0.02	0.02	0.03	0.35	0.01	0.31	0.01	1.58
Z_{11}	0.00	0.01	0.01	0.03	0.15	0.01	0.55	0.01	0.01	0.16	0.01	0.01	0.01	0.01	0.00	0.00	1.21
Z_{12}	0.02	0.01	0.11	0.02	0.01	0.00	0.01	0.00	0.03	0.03	0.02	0.18	0.12	0.23	0.00	0.03	1.51
Z_{13}	0.01	0.00	0.11	0.00	0.05	0.00	0.01	0.02	0.07	0.04	0.51	0.03	0.02	0.03	0.01	0.01	1.26
Z_{14}	0.00	0.02	0.00	0.15	0.08	0.00	0.18	0.10	0.03	0.23	0.01	0.01	0.02	0.00	0.00	0.01	1.11
Z_{15}	0.01	0.00	0.12	0.08	0.22	0.01	0.01	0.01	0.01	0.24	0.01	0.00	0.05	0.03	0.00	0.01	1.12
Z_{16}	0.01	0.00	0.03	0.18	0.10	0.02	0.06	0.12	0.31	0.00	0.05	0.01	0.00	0.01	0.00	0.00	1.11
Z_{17}	0.01	0.00	0.02	0.35	0.09	0.01	0.05	0.08	0.32	0.03	0.05	0.00	0.01	0.00	0.00	0.00	1.08

TABLE 38.6   LR Results Using Financial Distress Model

Variables	Estimated Coefficient	Standard Error	P-Value
Constant	−2.115	0.053	0.000**
Z_2	−0.626	0.055	0.001**
Z_3	−0.155	0.061	0.011**
Z_6	0.133	0.028	0.000**
Z_9	0.158	0.051	0.002**
Z_{10}	0.312	0.052	0.001**
Z_{12}	−0.255	0.058	0.001**
Z_{13}	0.080	0.032	0.022**
Z_{14}	−1.132	0.086	0.001**
Z_{15}	−1.521	0.152	0.001**
Z_{16}	0.255	0.031	0.000**
Z_{17}	0.083	0.033	0.011**

** Significance level 5%.

As shown in Table 38.6, significant variables for the firm's distress warning model include return on assets (Z_2), return on equity (Z_3), operating income ratio (Z_6), acid-test ratio (Z_9), debt ratio (Z_{10}), cash flow ratio (Z_{12}), equity growth ratio (Z_{13}), inventory turnover (Z_{14}), fixed assets turnover (Z_{15}), holding rate of directors and supervisors (Z_{16}), and holding rate of block shareholders (Z_{17}). The Wald method is employed to test the hypothesis (without helping the prediction). For example, the significance probability of the aforementioned variables all is indicating 5% level of significance. Thus the hypothesis is rejected. Thus we deduce that return on assets (Z_2), return on equity (Z_3), operating income ratio (Z_6), acid-test ratio (Z_9), debt ratio (Z_{10}), cash flow ratio (Z_{12}), equity growth ratio (Z_{13}), inventory turnover (Z_{14}), fixed assets turnover (Z_{15}), holding rate of directors and supervisors (Z_{16}), and holding rate of block shareholders (Z_{17}) help the prediction significantly, of which, the inventory turnover (Z_{14}) and fixed assets turnover (Z_{15}) have the biggest influence, meaning their value will significantly influence the probability of a financial distress event. The coefficient of return on assets (Z_2), return on equity (Z_3), cash flow ratio (Z_{12}), inventory turnover (Z_{14}), and fixed assets turnover (Z_{15}) carry a negative sign, meaning the higher the ratio, the lower the probability of financial distress; the coefficient of operating income ratio (Z_6), acid-test ratio (Z_9), debt ratio (Z_{10}), holding rate of directors and supervisors (Z_{16}), and holding rate of block shareholders (Z_{17}) are positive, meaning the higher the ratio, the higher the probability of financial distress.

Cross-validation is used to evaluate the prediction power of the constructed models. First, we use SAS software to randomly divide the source sample data into five groups. When testing group 1 data, the other groups (groups 2, 3, 4, and 5) are used as training data for group 1 model construction; when testing group 2 data, the data of groups 1, 3, 4, and 5 are used as training data for group 2 model construction, and so on. After five tests using the same method—that is, using training samples for model construction—the test samples are used in the model to test its prediction power. Finally, the average of the five test results is taken as the average rate of prediction accuracy.

This chapter adopts the "optimal cutoff point" approach proposed by Hosmer and Lemeshow (2000) using the point of intersection of sensitivity and specificity curves to determine the cutoff point for financial distress. It is observed that specificity indicates the percentage of accurately predicting a financially sound company when the company is in fact financially sound. Both sensitivity and specificity concern "accuracy" and have a trade-off relationship. That is, with "all other factors" staying unchanged, raising the sensitivity by adjusting the cutoff point lowers the specificity. The point at which the sensitivity and the specificity have the same probability is taken as the financial distress cutoff. Cutoff derived in such a manner is applicable to different data structures.

In summary, all remaining variables after collinearity diagnosis are subject to LR for the construction of a financial distress model, of which significant variables are identified using the Wald forward method. The financial distress probabilities are estimated, and the financial distress cutoff point is determined by taking the point of intersection of sensitivity and specificity curves according to Hosmer and Lemeshow (2000). As shown in Table 38.7, the mean financial distress cutoff value is 0.14; the

TABLE 38.7   Results of LR for Model Cross-Validation

Forecasting Rate	1	2	3	4	5
Cutoff point of financial distress	0.15	0.15	0.15	0.15	0.15
Correct pred. rate— financial distress	71.82%	67.02%	74.23%	71.15%	68.01%
Correct pred. rate—financial soundness	73.11%	75.93%	72.83%	73.22%	71.58%
Correct pred. rate of all samples	72.93%	74.35%	73.25%	72.83%	71.08%

LR model achieves a prediction accuracy of 70.48% in the financially distressed group, 73.33% in the financially sound group, and 72.89% for all remaining samples.

38.4.4 Results of NN

For the NN model, we take 15 variables out of the original 17 predictive variables following collinearity diagnosis and use them as input layer variables of the NN model. For the hidden layer, we adopt a single hidden layer and set the number of hidden layer neurons at twice the number of input variables (i.e., 30). The nonlinear transfer function used by the hidden layer is a sigmoid function. The output layer has a single variable to determine whether the firm is financially sound or distressed. The setting of BPN parameters refers to the suggestion of Cao and Tay (2001): learning rate at 0.005 and momentum term at 0.9. Similarly, the financial distress cutoff point for the predictive values obtained from the NN model is determined by the point of intersection of sensitivity and specificity curves according to Hosmer and Lemeshow (2000). As shown in Table 38.8, the mean of financial distress cutoff value for five test groups is 0.16; the NN model achieves a prediction accuracy of 72.64% in financially distressed group, 70.27% in financially sound group, and 70.61% for all samples.

The NN study is summed up as follows: Following collinearity diagnosis, variables with collinearity problem are discarded based on CI, T, and VIF. The resulting variables are used as input layer variables for the NN model. The outcome thereof better meets our expectations. The NN model achieves a prediction accuracy of 72.64% in financially distressed group, 70.27% in financially sound group, and 70.61% for all samples. It is

TABLE 38.8   Results of NN for Model Cross-Validation

Forecasting Rate	1	2	3	4	5
Cutoff point of financial distress	0.18	0.15	0.16	0.18	0.15
Correct pred. rate—financial distress	71.33%	81.60%	63.58%	69.63%	76.01%
Correct pred. rate—financial soundness	72.32%	65.33%	73.821%	68.18%	71.27%
Correct pred. rate of all samples	72.53%	67.85%	72.25%	68.35%	72.18%

clear NN model performs significantly better in the prediction of financial distress than the prediction of financial soundness. But the LR is superior to NN in overall prediction accuracy.

38.4.5 Results of Two-Stage Synthesis Model

Two-stage synthesis model (1)
First of all, LR is used to screen predictive variables significantly related to the financial distress warning model. Those significant predictive variables and the dependent variable are used in the NN model to construct a financial distress warning model.

As shown in Table 38.9, the mean financial distress cutoff value is 0.17; the two-stage synthesis model (1) achieves a prediction accuracy of 72.90% in financially distressed group, 70.01% in financially sound group, and 70.41% for all samples.

Two-stage synthesis model (2)
First of all, LR is used to screen significant independent variables, and predicted probabilities are calculated. The predicted probabilities are converted to a dichotomous variable (DV) through financial distress cutoff. The significant independent variables and the converted dependent variable are used in the NN model to construct a financial distress warning model. As shown in Table 38.10, the mean financial distress cutoff value is

TABLE 38.9   Results of Two-Stage Synthesis Model (1) for Model Cross-Validation

Forecasting Rate	1	2	3	4	5
Cutoff point of financial distress	0.18	0.18	0.18	0.15	0.18
Correct pred. rate—financial distress	70.18%	73.55%	65.11%	77.38%	78.28%
Correct pred. rate—financial soundness	72.83%	72.21%	72.78%	59.93%	72.31%
Correct pred. rate of all samples	72.43%	72.42%	71.48%	62.50%	73.21%

TABLE 38.10   Results of Two-Stage Synthesis Model (2) for Model Cross-Validation

Forecasting Rate	1	2	3	4	5
Cutoff point of financial distress	0.45	0.50	0.48	0.45	0.48
Correct pred. rate—financial distress	72.53%	74.11%	68.73%	70.23%	72.52%
Correct pred. rate—financial soundness	74.28%	73.88%	74.15%	69.73%	74.28%
Correct pred. rate of all samples	75.01%	75.01%	73.22%	68.80%	734.01%

TABLE 38.11   Results of Two-Stage Synthesis Model (3) for Model Cross-Validation

Forecasting Rate	1	2	3	4	5
Cutoff point of financial distress	0.187	0.18	0.15	0.18	0.18
Correct pred. rate—financial distress	70.18%	72.45%	66.11%	69.03%	77.12%
Correct pred. rate—financial soundness	72.41%	72.83%	72.31%	67.15%	72.31%
Correct pred. rate of all samples	72.05%	72.75%	71.28%	67.48%	73.08%

0.47; the two-stage synthesis model (2) achieves a prediction accuracy of 71.63% in financially distressed group, 73.26% in financially sound group, and 73.01% for all samples.

Two-stage synthesis model (3)

First of all, LR is used to screen significant independent variables and predicted probabilities are calculated. The significant independent variables and predicted probabilities are used in the NN model to construct a financial distress warning model.

As shown in Table 38.11, the mean financial distress cutoff value is 0.17; the two-stage synthesis model (3) achieves a prediction accuracy of 70.99% in financially distressed group, 71.41% in financially sound group, and 71.33% for all samples.

38.4.6 Comparing the Accuracy of Five Models

Table 38.12 ranks the overall prediction accuracy of different models from high to low: two-stage synthesis model (2), LR, two-stage synthesis model (3), back-propagation NN, and two-stage synthesis model (1). The five models are ranked from high to low in terms of accuracy in predicting financially sound banks: LR, two-stage synthesis model (2), two-stage synthesis model (3), back-propagation NN, and two-stage synthesis model (1).

The five models are ranked from high to low in terms of accuracy in predicting financially distressed banks: two-stage synthesis model (2), LR, two-stage synthesis model (3), back-propagation NN, and two-stage synthesis model (1).

This chapter suggests that when investors facing investment decision-making of a new target corporation, they may wish to consider using the

TABLE 38.12   Performance Assessment of Five Models

Average Pred. Rate	Financial Distress (S.D.)	Financial Soundness (S.D.)	Full Sample (S.D.)
LR	70.38%	73.33%	72.88%
	(0.025)	(0.012)	(0.011)
NN	72.63%	70.25%	70.63%
	(0.058)	(0.033)	(0.021)
Two-stage synthesis model (1)	72.58%	70.00%	70.31%
	(0.045)	(0.051)	(0.038)
Two-stage synthesis model (2)	71.61%	73.25%	73.00%
	(0.015)	(0.018)	(0.015)
Two-stage synthesis model (3)	70.98%	71.42%	71.33%
	(0.038)	(0.022)	(0.021)

highest prediction accuracy of a two-stage synthesis model (1) to model a financial distress warning system. Subject to return on asset, return on equity, operating income ratio, acid-test ratio, debt ratio, cash flow ratio, equity growth ratio, inventory turnover, fixed assets turnover, holding rate of directors and supervisors, and holding rate of block shareholders of the target corporation, the aforementioned variables can be used in a back-propagation NN in order to identify the predictive value of financial distress. Through the use of a 0.17 cutoff point of financial distress, we can determine whether the target corporation is grouping to distress or soundness, in which the prediction accuracy is about 72.90%.

38.5 CONCLUSIONS

This chapter looks at the ability of a relatively new technique, three two-stage synthesis models of LR–NN, to predict failure by Shanghai's publicly listed corporations. These models are compared with conventional LR and NN models. The aim of this chapter is to construct a financial distress warning model in Shanghai over the period of 2000–2007 and distinguishes financially sound and financially distressed corporations by the criteria for shares (full delivery, stop trading, or delisted under the order of the regulator) during the analyzed period.

The results of LR indicate that the factors of return on assets (Z_2), return on equity (Z_3), operating income ratio (Z_6), quick ratio (Z_9), debt ratio (Z_{10}), cash flow ratio (Z_{12}), equity growth ratio (Z_{13}), inventory turnover (Z_{14}), fixed assets turnover (Z_{15}), holding rate of directors and supervisors (Z_{16}), and holding rate of block shareholders (Z_{17}), are all significantly related to the financial distress status of publicly listed corporations in Shanghai. The significant variables identified in this chapter warrant close attention of the Securities and Exchange Commission and may serve as early warning signals of financial distress for banks. The NN model achieves a prediction accuracy of 72.64% in the financially distressed group, while the LR's accuracy prediction is 70.48%. In comparing the performances of single models, it is found that NN is superior to LR in the prediction of financial distress. The NN model does not hypothesize the normal distribution of variables and can recognize the existence of complex relationships between factors. Thus it is more efficient than LR that requires the absence of collinearity between independent variables and involves more complicated computing algorithms.

For the two-stage model, this chapter uses significant variables identified by LR and then uses different criteria to determine the accuracy of prediction so as to develop three different two-stage synthesis models. Results show that in the prediction accuracy of financially distressed firms, the two-stage synthesis model (1) offers the highest accuracy of 72.9%. Hence, this model demonstrates stronger prediction power than the one-stage model in identifying financially distressed publicly listed corporations in Shanghai. In the prediction related to all publicly listed corporations, all models offer an accuracy rate of more than 70%, of which, the two-stage model (2) achieves the highest overall accuracy of 73%. Consequently, we conclude that the two-stage synthesis model demonstrates better prediction power and is more suitable for the construction of a financial distress warning model for publicly listed corporation in Shanghai.

REFERENCES

Brailsford, T., O'Neill, T., and Penm, J. (2008) A new approach for estimating relationships between stock market returns: Evidence of financial integration in the Southeast Asian region. *International Finance Review*, 8(1): 17–38.

Cao, L. and Tay, F. E. (2001) Financial forecasting using support vector machines. *Neural Computing & Applications*, 11(3): 184–192.

Coenders, G. and Saez, M. (2000) Collinearity, heteroscedasticity and outlier diagnostics in regression. *Metodološki Zvezki*, 16(1): 79–94.

Hosmer, D. W. and Lemeshow, S. L. (2000) *Applied Logistic Regression*. Wiley-InterScience, New York.

Johnson, S., Boone, P., and Friedman, E. (2000) Corporate governance in the Asian financial crisis, 1997–98. *Journal of Financial Economics*, 58(2): 141–186.

La Porta, R., Lopez-de-Silanes, F., and Shleifer, A. (1999) Corporate ownership around the world. *Journal of Finance*, 54(4): 471–517.

Morck, R., Shleifer, A., and Vishny, R. (1988) Management ownership and market valuation: An empirical analysis. *Journal of Financial Economics*, 20(3): 293–316.

Penm, J. (2007) Asian rupee for a common electronic financial market of India and the ASEAN region. *International Journal of Electronic Finance*, 1(4): 473–483.

Rajan, R. and Zingales, L. (1998) Which capitalism? Lessons from the East Asian crisis. *Journal of Applied Corporate Finance*, 11(1): 40–48.

CHAPTER 39

Corporate Governance in Emerging Markets: An Overview

Serdar Özkan and A. Fatih Dalkılıç

CONTENTS

39.1 INTRODUCTION

The term "emerging markets" (EMs) refers to a broad range of countries that are rapidly integrating into the world financial system like some of East European, Asian, and Latin American countries.* These markets are called emerging since they have not fully met all the requirements of financial integration, and various EMs stand at different levels of economic development due to country-specific reasons. For example, Brazil, China, Turkey, and Russia provide better results in terms of higher market capitalization, higher foreign direct investments, and being attractive for foreign equity investments instead of debt investments, whereas many others are still suffering from their modest financial integration into the world.

The move from close, market-unfriendly, undemocratic systems to open, transparent, market oriented, and democratic systems is important but not very easy to achieve, as it requires radical changes in economic, political, and legal environment of the EMs. The fact that the institutional infrastructure of a particular country has close links with its historical background, thus with its culture and its legal heritage, makes the integration process even more complex for some of the EMs, where the institutions are heavily relationship-based instead of more effective rules-based ones.

The waves of liberalization of international trade and investment, and privatization of business in transition economies, the Asian financial crisis in the late 1990s, and the recent high-profile corporate scandals around the world have illustrated that effective corporate governance (CG) system is a special issue for both developed countries and EMs.

Yet, there is also a growing consensus among specialists of business, national, and international policy makers that the ways of making many of those radical institutional improvements are embedded in

---

* Standard and Poors' S&P/IFCG Index now classifies 32 countries as emerging: Argentina, Brazil, Chile, Colombia, Mexico, Peru, Venezuela, China, Korea, the Philippines, Taiwan China, India, Indonesia, Malaysia, Pakistan, Sri Lanka, Thailand, Bahrain, Czech Republic, Egypt, Hungary, Israel, Jordan, Morocco, Nigeria, Oman, Poland, Russia, Saudi Arabia, Slovakia, Turkey, and South Africa.

the principles of modern CG. This consensus has also motivated various regional and national development agencies (e.g., Cadbury Commission, The Organization for Economic Co-Operation and Development [OECD], Asian Development Bank, Center for International Private Enterprise [CIPE], the King Committee on Corporate Governance in South Africa) to develop and initiate governance programs in recent years.

The scope of the CG differs depending on one's view. For example, Shleifer and Vishny (1997) define CG as the deals with the ways in which suppliers of finance to firms, namely, shareholders, make sure that they get a return on their investment. Henceforth, the original need for CG mainly stems from the agency relationship, that is, the problems associated with the separation of ownership and control of the modern firm.

However, for EMs, CG is an important part of the economic and social development processes and the need for CG in EMs extends far beyond resolving problems resulting from the separation of ownership and control. Yet, in most of these countries, institutional framework is not well developed or is performing weakly, creating significant inconsistencies and uncertainties, and, thus most of the EMs are still dealing with very fundamental issues such as property rights, the abuse of minority shareholders, contract violations, asset stripping, and self-dealing (OECD, 2004a; CIPE, 2002).

In order to extend the scope and provide guidance in setting or improving CG systems for almost all countries, the OECD (2004b) approaches CG with a broader stakeholder perspective as "Corporate governance specifies the distribution of rights and responsibilities among different participants in the corporation, such as the board, managers, shareholders and other stakeholders, and spells out the rules and procedures for making decisions on corporate affairs. By doing this, it also provides the structure through which the company objectives are set, and the means of attaining those objectives and monitoring performance." In sum, CG involves a set of relationships between a company's management, its board, its shareholders, and other stakeholders. Table 39.1 shows a brief summary of the OECD principles of CG.

The complex nature of CG and institutional differences between the countries make it almost impossible to reach a consensus on the factors that determine the optimal CG structure. Thus, the convergence can be viewed as the adoption of the best practices and CG mechanisms, either by countries or individual firms (Rubach and Sebora, 1998; Yoshikawa and Phan, 2003; Davies, 2008).

TABLE 39.1   A Brief Summary of the OECD Principles of Corporate Governance

---

(a) *Implementation and enforcement* is a key issue and legal, regulatory and institutional framework is needed to ensure effective implementation and enforcement in a country,

(b) *Shareholder rights* need to be protected and the framework should facilitate the exercise of shareholders' rights,

(c) *Equitable treatment of all shareholders*, including minority and foreign shareholders, should be secured,

(d) All *stakeholders' rights* should be recognized and their cooperation with the firms should be encouraged in creating wealth, jobs, and the sustainability of financially sound enterprises,

(e) The framework should ensure *timely* and *accurate* financial reporting and additional *disclosure* about ownership, and governance of the company, and

(f) *Boards of directors have* a critical role in an effective governance system and the framework should ensure the strategic guidance of the company, the effective monitoring of management by the board, and the board's accountability to the company and the shareholders.

Source: OECD (2004b). With permission.

In the CG literature, there are different classifications of CG mechanisms (Denis, 2001; Denis and McConnel, 2003; Gillan, 2006). However, the common focus is on the board of directors, ownership structure of firms, market for corporate control, and the legal system. However, in order to asses the performance or effectiveness of a CG system in a particular market, instead of investigating CG mechanisms, the focus should be on CG outcomes (Macey 1997; Gibson, 2003), in other words, on whether the mechanisms are functioning properly in terms of reducing agency costs, hence increasing the firm value and performance. The OECD Principles and various CG indexes (i.e., S&P's CG Index) also focus on the existence of CG mechanisms in a particular market.

39.2 CORPORATE GOVERNANCE MECHANISM IN EMERGING MARKETS

This section investigates in general the CG in EMs in terms of the main CG mechanisms, namely, the board of directors, ownership structure, legal and regulatory structures, and competitive markets.

39.2.1 Board of Directors

Board of directors play advisory and monitoring roles to ensure that management of the firm is acting in the interests of the shareholders. Consistent with their primary roles, the Board can hire, fire, change, and compensate the managing people in order to ensure shareholders' wealth maximization. Today, in many countries as well as in most of the EMs, the capital market listing rules and company laws require several criteria related to the Board structure such as size (number of directors), independence (the fraction of outside directors), existence and number of subcommittees, expertise of the members, and frequency of meetings.

Board independence mainly means that the relatives of management, board members, recent employees, and officers in related companies are to be excluded. After those corporate scandals, even in the United States, the New York Stock Exchange (NYSE) listing rules were changed to require that the Board must have a majority of independent directors. In the last decade, there are similar reforms in CG imposed on EMs (i.e., following the Asian crisis, new CG reforms took place in China, Korea, Indonesia, Thailand, and Malaysia), and today legislations in many EMs require a minimum percent of independent directors on the Board.

39.2.2 Ownership Structure

The ownership structure of a firm mainly refers to the identities of a firm's shareholders and the sizes of their positions (Denis, 2001). There are two types of ownership structures that are known as dispersed and concentrated. In dispersed ownership scenario, among a large number of shareholders, each shareholder holds small numbers of shares of the firm. The shareholders have no direct control over the management, have no involvement in decision making, and mostly focus on their short-term gains. In a concentrated ownership scenario, a small number of shareholders hold the largest number of shares of the firm and also have a significant control over the firm's management. Combining ownership and control can be advantageous, as large shareholders' power to monitor management closely and their deeply involvement in decision-making may minimize the managerial expropriation (e.g., the use of transfer pricing, related-party transactions, assets stripping) on behalf of all shareholders.

In most of the EMs, the overall situation is the prevalence of ownership concentration, that is, the large shareholders are also holding the majority of controlling rights and mostly the owners are the Board members. Concentrated ownership is a key characteristic of family owned firms in the EMs like all over the world. Since the family's wealth is closely linked to the firm's performance, and families regard their firms as an asset to pass on to their descendants (Anderson and Reeb, 2003), they tend to be a major part of the controlling mechanisms of the firms, mostly via pyramidal structures, nonvoting shares, and cross shareholdings.

The main disadvantage of the concentrated ownership is that the blockholders' significant control over management may lead the Board to make decisions for blockholders' own good, but not other shareholders'. The effectiveness of concentrated ownership therefore mainly depends on the blockholders' behavior that is highly related to trade-off between common benefits of all shareholders and private benefits of blockholders. Thus, the CG mechanisms in the EMs mainly deal with the conflict between blockholders and minority shareholders, not the conflict between managers and shareholders, which is the general case in developed markets.

Moreover, state-owned enterprises and privatization are two other factors related to ownership structure debate in the EMs. The primary CG-related question of the privatization literature is whether firm performance increases when firms become privately owned (Denis and McConnel, 2003). The studies in the area provide evidence that private ownership is associated with better firm performance than state ownership and privatization is associated with greater productivity and higher productivity growth (see Denis and McConnel (2003) for a broad literature review).

39.2.3 Legal and Regulatory Structure

There is a positive correlation between a strong and effective judiciary, acting as an important formal contract enforcement institution, and economic development (Dam, 2006). Consistently, Oman et al. (2003) state that "[d]eveloping a strong, competent, politically independent and well funded judiciary therefore important for enhancing the contribution of CG to corporate performance." A country's company, bankruptcy, and capital market laws and regulations, combined with effective enforcement of these laws, determine the rights of shareholders and the performance of financial systems (Schelifer and Vishny, 1997).

Thus, a large body of research, starting with La Porta et al. (1997), studies the link between governance, law, and finance at country level. According to "law matters" school, the reason for different level of regulatory intervention across countries is their legal heritage. Civil law is less effective in protecting shareholder rights than common law. However, there are some outliers. For example, one of the subindexes that World Economic Forum (WEF) uses to calculate overall Global Competitiveness Index reflects the quality of institutions, including legal environment of the countries. The institutions index show that civil law countries such as Chile, Jordan, and Morocco have higher scores than some common law countries such as Bangladesh, Pakistan, and Uganda (WEF, 2008).* Accordingly, Cornelius (2005) suggests that some other country-specific factors like politics, cultural, and historical roots should also be considered in country-level legal environment analysis. A recent study by Graff (2008) also identifies a number of problems that cast doubt on the soundness of the empirical basis of the literature on the relationships between legal origin and shareholder protection.

On the other hand, the corporate scandals that the world witnessed in the last decade in both civil law and common law countries lead another argument that even though the quality of a country's legal environment is high; such environment may not rule out serious CG risks at the firm level. On the contrary, there exist a significant number of cases, especially in EMs, where firm-level CG practices outperform national institutions. There are several countries such as China, the United Arab Emirates, Ukraine, and Russia whose firms on average appear to follow better practices than the quality of their legal and regulatory environments (Cornelius, 2005).

Firm-level CG practices matter especially in countries with weak legal environments, potentially compensating for ineffective laws and enforcement by providing credible shareholder protection (Klapper and Love, 2003). Consistently, the coincidence of concentrated ownership (firm-level practice) with a lack of shareholder protection (country level regulation) is because shareholders who are not legally protected from controllers will seek to protect themselves by becoming controllers. In other words, concentrated ownership structure in EMs substitutes for legal protections (La Porta et al., 1998).

Legal and regulatory environment in EMs evolves through years. The most recent reforms are given in The World Bank and the International Finance Corporation (IFC)'s copublication *Doing Business* (2008). According to this report, the most protective country is Malaysia and the least protective one is Venezuela in terms of shareholder protection. Ten economies strengthened shareholder protection in 2006–2007. Georgia is the top reformer. The most popular 2006–2007 shareholder protection reforms of EMs are (a) increasing disclosure of related party transactions (Turkey, Mexico, Colombia, Poland); (b) defining duties for directors and controlling shareholders (Georgia);

---

* Cornelius (2005) provides a similar analysis which is based on WEF's World Competitiveness Report 2004.

(c) regulating approval of related party transactions (Thailand); (d) making easy to sue directors (Mexico, Korea); (e) strengthening audit committees (Indonesia, India); (f) giving shareholders access to company documents (China); and (g) increasing penalties for self-dealing (Malaysia) (World Bank-IFC, 2008).

The goal of strengthening the shareholders' rights in a country cannot be achieved just by setting the legal framework. This also requires a sound enforcement mechanism and the main driver of effective public enforcement mechanisms is the quality of courts in a particular country. La Porta et al. (1998) and Pistor et al. (2000) indicate that the quality of courts can vary independent of the quality of the law on the books. Moreover, the quality of laws does not substitute for the quality of enforcement (La Porta et al., 1998). A recent study of Safavian and Sharma (2007) finds that firms have more access to bank credit in countries with better creditor rights, but the association between creditor rights and bank credit is much weaker in countries with inefficient courts. Cross-country indicators of enforcement quality, such as the *Doing Business* measures of court quality in terms of (a) requiring less number of procedures (Czech Republic), (b) shortening the time to enforce a judgment (Uzbekistan, Korea, Lithuania), and (c) decreasing the cost of going to the courts (China, Poland, Korea) show considerable variation across EMs (World Bank-IFC, 2008).

The equity culture in a particular country also plays an important role as a complement of law and enforcement. One of the main obstacles is that shareholders in the EMs are not aware of their rights or have no willingness to exercise them. Improvement of such awareness requires collective efforts of national and international policy and rule makers. For example, the experience of Eurasian countries shows that shareholder awareness has been improved as a result of specific programs initiated by bilateral and multilateral donors together with securities commissions and market professionals (OECD, 2004).

39.2.4 Competitive Markets

Market competition is crucial in driving market efficiency and thus enhancing business productivity by ensuring that most of the efficient firms, producing goods demanded by the market, are those that thrive. Therefore, competitive markets force the management of a firm to allocate and use resources efficiently; otherwise the firm loses its market share, faces difficulties in raising funds either from potential investors or creditors, and misses investment opportunities. The performance of the management is closely linked to its capacity to run the firm successfully in a competitive environment. If the management fails to perform well, then the shareholders may sell their shares, which decrease the value of the firm. As a matter of fact, competitive markets as external mechanisms are crucially important for a strong CG structure in all countries. Consistently, there is a growing attention to make business reforms in order to establish competitive business environments among EMs. One subindex of World Competitiveness Index, Goods Efficiency Index, assesses countries' current competitive business environment position based on 15 subfactors (such as effectiveness of antimonopoly policy, prevalence of trade barriers, prevalence of foreign ownership, and number of procedures required to start a business) and ranks them. The ranks, according to overall Goods Efficiency Index and to each of those subfactors, present variations among EMs. For example Russia, Turkey, China, and Brazil are ranked at 99, 55, 51, and 101 out of 134, respectively (WEF, 2008).

39.3 CORPORATE GOVERNANCE IN TURKEY, CHINA, AND RUSSIA

This part focuses on three EMs, Turkey, Russia, and China, and aims to give overview of these different EMs. Each of these three EMs has unique characteristics; for this reason our intention is to create a brief country picture rather than providing a standardized comparison.

39.3.1 Corporate Governance in Turkey

From the establishment of Turkish Republic in 1923 until the 1980s, the Turkish economy was heavily state oriented. State-owned enterprises were established with the aim to stimulate the economy in the absence of private capital. Beginning in the 1980s, the government policy shifted from an import-led and protectionist to an export-led policy, favoring a liberal market economy and followed by privatization of the state-owed enterprises. In the course of the liberalization and orientation toward a market economy, the Capital Markets Law was enacted in 1981 followed by the establishment of the Capital Markets Board of Turkey in 1982 and the Istanbul Stock Exchange in 1986. Like many EMs, Turkey has an underdeveloped equity culture (IIF, 2005).

The studies that investigate the Turkish CG system focus mainly on the ownership structure. Since Turkey is a common law country, as stated by La Porta (1998), the emphasis is on the controlling shareholders rather than capital markets. Ararat and Ugur (2003) consider low liquidity, high volatility, high cost of capital, and limited new capital formation as characteristics of the Turkish capital market. These characteristics indicate that Turkish capital market is not perceived as a primary source of funds yet. Thus, a very small percentage of companies are listed on the Istanbul Stock Exchange. Fitch (2007) reports state that several group companies have merged subholding entities to simplify their ownership structures and shifted their stakes from financial sector to private sector in order to meet the transparency standards of international markets.

Ararat and Ugur (2003) also conclude that the shortcomings in the legal and regulatory framework increase the risk of investing in the Turkish capital market. Their argument supports La Porta et al. (1998) who rate Turkey 2 in a 6-scale assessment, 6 being highest, with respect to shareholder rights and 4 in a 10-scale assessment, 10 being highest, with respect to judicial efficiency. Moreover, World Bank issued a Report on the Observance of Standards and Codes (ROSC) on accounting and auditing in 2007, stating that the international investors and credit rating agencies assess the disclosure and transparency level of the top companies of Turkey as unsatisfactory.

Turkey is still in the process of reforming its institutional and legal structures. Capital Markets Board has built a committee including experts from ISE and Turkish Corporate Governance Forum and in June 2003 has issued the "Corporate Governance Principles of Turkey" on "comply or explain" basis and revised them in 2005. In August 2007, Istanbul Stock Exchange has launched the Corporate Governance Index, a special index where only the listed companies with minimum corporate governance score of 6 out of 10 rated by certified rating agencies of Capital Markets Board (CMB) of Turkey may participate. With the establishment of the index, foreign ownership in listed companies has increased (Ararat and Yurtoglu, 2008).

Demirag and Serter (2003) and Yurtoglu (2000), who investigate the ownership structure as an alternative CG mechanism in Turkey, state that the ownership structure is pyramidal and concentrated. Demirag and Serter (2003) further provide a thorough analysis of the ownership structures of 100 Turkish firms and conclude that family ownership is common and the firms acquire a bank in the further stages of their development. In contrast to La Porta's (1998) argument that concentrated ownership structures act as substitutes for markets for corporate control, this structure affects the corporate performance negatively and its expected role as an alternative disciplinary corporate mechanism does not translate into increased firm value as expected in Turkey (Yurtoglu, 2000;

Gonenc, 2004). It is also emphasized that CEOs in Turkey are evaluated on accounting-based measures and in relation to firm performance, indicating that Turkish CG system is not ineffective (Durukan et al., 2007). Kula (2005), on the other hand, taking a different stand, investigates small and nonlisted companies in Turkey. He concludes that the separation of chairman and general manager positions in these firms are reflected positively on the firm performance.

In Turkey, controlling shareholders play a leading role in the management and strategic decision for the group companies including the listed companies. This leads to the potential abuse of minority shareholders' rights through the imposition of commercial conditions by the controlling shareholders (Yuksel, 2008).

The present Turkish Commercial Code (TCC) was adopted in 1957 and since then only insignificant changes were made. Due to the economic and social developments, the code has been revised and the Draft TCC is still being discussed in parliament. The existing TCC does not provide the appraisal right for the minority shareholder in major corporate decisions but the draft TCC comprises provisions in response to these weaknesses in minority interest in corporate reorganizations (Hacimahmutoglu, 2007).

39.3.2 Corporate Governance in China

China has emerged as one of the largest economies in the world since introducing market-oriented reforms in the 1980s. The pace of economic growth in China has been rapid since the beginning of the economic reform. In the early 1990s, the government introduced a wide range of reforms in the state-owned enterprises with the objective of privatization. Since then, many state-owned enterprises have been transformed into corporations and listed on the Chinese and Hong Kong stock exchanges (Cheung et al., 2008). The new diversified ownership structure after privatization makes CG an important issue in China since this is a valid effect for all emerging economies (Dhardwadkar and Brandes, 2000).

Qiang (2003) documents that from the beginning of the reform to restructure Chinese state-owned enterprises into public listed companies, the concern of losing state assets and government control led to an ownership structure with three major parties. A typical listed company in China has a mixed ownership structure with three predominant groups of shareholders, which are state, legal persons (institutions), and individuals. The major form of nonstate ownership is individual shareholding; independent nonstate institutional investors are very rare (Chen, 2004).

Weifeng et al. (2008) and Ding et al. (2007) studied ownership structure in Chinese companies and reported highly concentrated ownership structure. Their results indicate that highly concentrated ownership determines the nature of the agency problem in Chinese corporations. Liu (2005) sees the reason of having highly concentrated ownership as the incompleteness of law in China. In the light of highly concentrated ownership structure of listed Chinese firms, the chairman appointed by the largest shareholder tends to be powerful and often involved in the company's daily decision-making process even without holding the position of general manager simultaneously (Kato and Long, 2006).

The two stock exchanges in mainland China were established in Shanghai and Shenzhen in 1990 and 1991, respectively, as a major initiative of economic reform (Cheung et al., 2008). The Shanghai and Shenzhen stock exchanges list more than 1500 companies with a combined market capitalization of $2658.2 billion (2008), rivaling the Hong Kong Stock Exchange ($2121.8 billion) as Asia's second-largest stock market after the Tokyo Stock Exchange ($3925.6 billion).

In September 2008, the Shanghai Stock Exchange, in cooperation with the China Securities Index, introduced the SSE 180 Corporate Governance Index, a new investment target index for investors to indicate trend of the stocks with good corporate governance in the SSE 180 Index. The Shenzhen Stock Exchange has recently signed a Memorandum of Understanding with the New York Stock Exchange to accelerate internationalization of the exchange, enhance communication, and build cordial relationship with overseas bourses. According to the Memorandum, the two parties will cooperate in many fields such as project development, information sharing, personnel communication, and others.

China's Company Law became effective in 1994 and it is an important starting point in the evolution of China's corporate governance reforms. China's Securities Law, which became effective in December 1998, regulates capital market issuance, trading activities, and related matters. The law also strictly prohibits insider trading and market manipulation (Rajagopalan and Zhang, 2008). Chinese Company Law does not allow the use of preferential shares or shares with double voting rights (Ding et al., 2007). Recent revisions to Company Law and Securities Law have focused on strengthening minority shareholder rights, increasing financial reliability, and clarifying the role of the supervisory board and chairman (IIF, 2006).

The China Securities Regulatory Commission (CSRC), the Chinese counterpart of U.S. Securities and Exchange Commission (SEC), issued new rules in the form of a "Code of CG for Listed Companies in China" in January 2002, which follows the U.S. regulatory system (Conyon and He, 2008). The code aims to establish solid CG in stock market listed companies by elevating requirements related to accounting procedures and information disclosure, introducing independent director systems, and tightening the supervision of corporate management (Shi and Weisert, 2002).

Chinese firms are characterized by having a dominant or block stockholder; in line with this the larger the percentage of shares held by the block holder, the more influence they have over the firm's management (Xu, 2004). Weifeng et al. (2008) report that the level of private benefits of control is close to countries of EMs. Conyon and He (2008) present an empirical evidence on the relation between CEO-turnover and firm performance in Chinese publicly traded firms and similar to this result, Xu (2004) provides evidence that executive turnover serves as an effective mechanism in reversing poor company performance in China.

39.3.3 Corporate Governance in Russia

Russia is one of the largest, most complex, and dynamic of the transition economies (Litwak and Sutherland, 2002). During Russia's 70 years of communism and central planning, there was no private ownership of commercial or industrial enterprises. There were no shareholders since the state was the owner of all productive assets and organizations; thus, there was no room for corporate governance. This was the reason why communism and central planning had provided little or no experience in dealing with issues of ownership and shareholder rights (McCarthy and Puffer, 2002). Putin was elected the President of the Russian Federation in 2000 and he focused on pursuing Russia's membership in the World Trade Organization as a means of achieving economic stability, increasing economic growth, enhancing attractiveness for investment, and building confidence in the country's economic future (McCarthy and Puffer, 2002).

In late 2001, the government issued a draft of a new Code of Corporate Conduct, which was expected to be the foundation of a system of CG, aimed at increasing transparency and disclosure, as well as initiating a supporting infrastructure. This system of CG has borrowed heavily from international organizations (OECD, World Economic Forum, etc.) as well as from other industrialized countries (McCarthy and Puffer, 2002).

After initially adopting a primarily Anglo-American model of governance, Engelen (2002) argues that Russia will be forced to continue with the Anglo-American model as it enters the global economy. Sprenger (2002) argues that Russia will gradually adopt the Western European model of governance. Taking a different view, Kornai (1996) states that Russia will move toward a blend of these two models. Some authors have different views on the issue; according to them, Russia will adopt its own unique form of governance (McCarthy and Puffer, 2002). Finally, some observers even argue that the model adopted will vary by industry sectors within Russia (Wright et al., 2003).

Russia has a history of a weak equity culture that tends to undervalue minority shareholders' rights; one of the reasons for this is that there has been little time for an equity culture to develop during the relatively short period, which lapsed between the end of communism and the mass privatization of the early 1990s, and another key reason relates to the continuing high level of ownership concentration (IIF, 2004).

Russia's law is based on a civil code (Buck, 2003). Law on Joint Stock Companies ("JSC Law") has been adopted over the course of the last 10 years, which considerably strengthened the protection of minority shareholder rights. The Law of Securities Market has gone through a series of refinements to fill certain empty issues. Finally, the Federal Service for Financial Market has been increasingly active in adopting a series of both mandatory and advisory measures, aimed at protecting shareholder rights, including the CG Code (Wack, 2006).

In Russia small- and medium-sized enterprises contribute less to the national economy than large businesses compared to most of the countries (Litwak and Sutherland, 2002). That is why big companies are playing a crucial role in the national economy. Many leading Russian companies raised both equity and debt capital in international financial markets, prompting them to improve their corporate governance but these companies are still owned primarily by one or a limited number of shareholders. Also most of the Russian companies remain undervalued when compared to their Western competitors (Black et al., 2006).

There are some studies about CEOs, boards, and governance scores in Russia. In some circumstances, CEOs also serve as the chairperson. This situation is named as "CEO duality," and this is problematic from an agency perspective where the CEO chairs the people who are in charge of monitoring and evaluating the CEO's performance (Judge et al., 2003). To encourage the independence of the board, a 1996 Russian law and the 2002 Code of Corporate Conduct stipulate that different individuals must serve as board chairman and CEO. Supporting such a separation of roles, a recent survey found a negative relationship between Russian firm performance and situations where the CEO and chairman roles were held by the same person (Judge et al., 2003).

39.4 CONCLUSION

The complex nature of CG and institutional differences between the countries make it almost impossible to reach a consensus on the factors that determine the optimal CG structure. The convergence can be viewed as the adoption of the best practices and CG mechanisms, either by countries or individual firms.

Since CG is a complement of economic and social development processes in EMs, there are many challenges confronting EMs. Some of them are establishing a rule-based system, dismantling pyramid ownership structures, establishing property rights systems, protecting minority shareholders' rights, improving equity culture by increasing the awareness of being shareholder, finding active owners and skilled managers, and promoting good governance within family owned and concentrated ownership structures (CIPE, 2002).

In order to realize the aforementioned goals, thus to make the institutional reforms, domestic and foreign, public and private initiatives should take place in the EMs. All of these efforts are essential elements of the integration into the world business system.

REFERENCES

Anderson, R.C. and Reeb, D.M. (2003) Founding-family ownership and firm performance: Evidence from the S&P 500. *The Journal of Finance*, 58(3): 1301–1328.

Ararat, M. and Ugur, M. (2003) Corporate governance in Turkey: An overview and some policy recommendations. *Corporate Governance*, 3(1): 58–75.

Ararat, M. and Yurtoglu, B. (2008) Rating based indexing of the ISE: Lessons from Novo Mercado's success to advance corporate governance reforms. *Focus*, 5(1): 57–77.

Black, B.S., Love, I., and Rachinsky, A. (2006) Corporate governance indices and firm's market values: Time series evidence from Russia. *Emerging Markets Review*, 7(4): 361–379.

Buck, T. (2003) Modern Russian corporate governance: Convergent forces or product of Russia's history? *Journal of World Business*, 38(4): 299–313.

Chen, J.J. (2004) Determinants of capital structure of Chinese-listed firms. *Journal of Business Research*, 57(12): 1341–1351.

Cheung, Y.L., Jiang, P., Limpaphayom, P., and Lu, T. (2008) Does corporate governance matter in China? *China Economic Review*, 19(3): 460–479.

CIPE (2002) *Instituting Corporate Governance in Developing, Emerging and Transitional Economies A Handbook*. The Center for International Private Enterprise, Washington, DC.

Conyon, M. and He, L. (2008) Firm performance and corporate governance reforms in China. Available at SSRN. http://ssrn.com/abstract=1084396.

Cornelius, P. (2005) Good corporate practices in poor corporate governance systems. Some evidence from the Global Competitiveness Report. *Corporate Governance*, 5(3): 12–23.

Dam, K.W. (2006) The Judiciary and Economic Development. Working Paper, University of Chicago, Chicago, IL.

Davies, M. (2008) The impracticality of an international "one size fits all" corporate governance code of best practice. *Managerial Auditing Journal*, 23(6): 532–544.

Demirag, I. and Serter, M. (2003) Ownership patterns and control in Turkish listed companies. *Corporate Governance*, 11(1): 40–51.

Denis, K.D. (2001) Twenty-five years of corporate governance research … and counting. *Review of Financial Economics*, 10(3): 191–212.

Denis, K.D. and McConnel, J.J. (2003) International corporate governance. *Journal of Financial and Quantitative Analysis*, 38(1): 1–36.

Ding, Y., Zhang, H., and Zhang, J. (2007) Private vs state ownership and earnings management: Evidence from Chinese listed companies. *Corporate Governance*, 15(2): 223–238.

Durukan, B., Özkan, S., and Dalkiliç, A.F. (2007) Effectiveness of the Turkish Corporate Governance System: CEO Changes and Performance Measures. Corporate Governance in Emerging Markets Conference Proceedings, Sabanci University, Istanbul, Turkey.

Engelen, E. (2002) Corporate governance, property and democracy: A conceptual critique of shareholder ideology. *Economy and Society*, 31(3): 391–414.

Fitch Ratings (2007) Corporate Governance: The Turkish Perspective Special Report. Washington, DC.

Gibson, M.S. (2003) Is corporate governance ineffective in emerging markets. *Journal of Financial and Quantitative Analysis*, 38(1): 231–250.

Gonenc, H. (2004) Sermaye Sahipligi Yapısı, Yatırımcıların Korunması ve Firma Degeri: Türkiye, A.B.D. ve Japonya Karşılaştırması. *8. Ulusal Finans Sempozyumu*, pp. 157–167.

Graff, M. (2008) Law and finance: Common law and civil law countries compared—an empirical critique. *Economica*, 75(297): 60–83.

Hacimahmutoglu, S. (2007) The problems of minority protection and their solutions within the legal framework in Turkish corporate governance. *Journal of Banking Regulation*, 8(2): 131–158.

IIF (2004) Corporate Governance in Russia—An Investor Perspective, Task Force Report, Institute of International Finance, Washington, DC.

IIF (2005) Corporate Governance in Turkey—An Investor Perspective, Task Force Report. Institute of International Finance, Washington, DC.

IIF (2006) Corporate Governance in China—An Investor Perspective, Task Force Report. Institute of International Finance, Washington, DC.

Judge, W.Q., Naumova, I., and Koutzevol, N. (2003) Corporate governance and firm performance in Russia: An empirical study. *Journal of World Business*, 38(4): 385–396.

Pistor, K., Raiser, M., and Stanislaw, G. (2000) Law and finance in transition economies. *Economics of Transition*, 8(2): 325–368.

Kato, T. and Long, C. (2006) CEO turnover, firm performance, and enterprise reform in China: Evidence from micro data. *Journal of Comparative Economics*, 34(4): 796–817.

Klapper, L.F. and Love, I. (2003) Corporate Governance, Investor Protection, and Performance In Emerging Markets. Policy Research Working Paper No. 2818, The World Bank, Washington, DC.

Kornai, J. (1996) New Approaches to Russian Privatization. Presented at the Fletcher School of Law and Diplomacy, Harvard University, Boston.

Kula, V. (2005) The impact of the roles, structure and process of boards on firm performance: Evidence from Turkey. *Corporate Governance*, 13(2): 265–277.

La Porta, R., Lopez De-Silanes, F., Shleifer, A., and Vishny, R. (1997) Legal determinants of external finance. *Journal of Finance*, 52(3): 1131–1150.

La Porta, R., Lopez De-Silanes, F., Shleifer, A., and Vishny, R. (1998) Law and Finance. *The Journal of Political Economy*, 106(6): 1113–1155.

Litwak, J. and Sutherland, D. (2002) *OECD Economic Surveys, 2001–2002: Russian Federation*. New York, OECD.

Liu, S.G. (2005) Comparative corporate governance: The experience between China and UK. *Corporate Governance*, 13(1): 1–4.

Macey, J.R. (1997) Institutional investors and corporate monitoring: A demand-side perspective. *Managerial and Decision Economics*, 18(7) and (8): 601–610.

McCarthy, D. and Puffer, S. (2002) Corporate governance in Russia: Towards a European, US, or Russian Model? *European Management Journal*, 20(6): 630–640.

Oman, C., Fries, S., and Buiter, W. (2003) Corporate Governance in Developing, Transition and Emerging Market Economies. Policy Brief No 23, OECD, New York.

OECD (2004a) Corporate Governance in Eurasia: A Comparative Overview. Organization for Co-Operation and Development, New York.

OECD (2004b) OECD Principles of Corporate Governance. Organization for Co-operation and Development, New York.

Qiang, Q. (2003) Corporate governance and state-owned shares in China listed companies. *Journal of Asian Economics*, 14(5): 771–783.

Rajagopalan, N. and Zhang, Y. (2008) Corporate governance reforms in China and India: Challenges and opportunities. *Business Horizons,* 51(1): 55–64.

Rubach, M. and Sebora, T.C. (1998) Comparative corporate governance: Competitive implications of an emerging convergence. *Journal of World Business*, 33(2): 167–184.

Safavian, M. and Sharma, S. (2007) When do creditor rights work. Working Paper, The World Bank Financial and Private Sector Vice Presidency Enterprise Analysis Unit.

Shi, S. and Weisert, D. (2002) Corporate governance with Chinese characteristics. *The China Business Review*, 29(5): 40–44.

Shleifer, A. and Vishny, R. (1997) A survey of corporate governance. *The Journal of Finance*, 52(2): 737–783.

Sprenger, C. (2002) Ownership and corporate governance in Russian industry: A survey. Working Paper, European Bank for Reconstruction and Development.

Wack, D. (2006) Minority shareholder rights under Russian law. *International Journal of Disclosure and Governance*, 3(14): 317–326.

WEF (2008) *The Global Competitiveness Report 2008–2009*. World Economic Forum.

Weifeng, H., Zhaoguo, Z., and Shasha, Z. (2008) Ownership structure and the private benefits of control: An analysis of Chinese firms. *Corporate Governance*, 8(3): 286–298.

World Bank (2007) *Turkey- Report on The Observance of Standards and Codes (ROSC) Accounting and Auditing*. World Bank.

World Bank and IFC (2008) *Doing Business 2008*. World Bank and International Finance Corporation.

Wright, M., Filototchev, I., and Bishop, K. (2003) Is stakeholder governance appropriate in Russia? *Journal of Management and Governance*, 7(3): 263–290.

Xu, L. (2004) Types of large shareholders, corporate governance, and firm performance: Evidence from China's listed firms. Working Paper, Zhongshan University.

Yoshikawa, T. and Phan, P.H. (2003) The performance implications of ownership-driven governance reform. *European Management Journal*, 21(6): 698–706.

Yuksel, C. (2008) Recent developments of corporate governance in the global economy and the new Turkish commercial draft law reforms. *Journal of International Commercial Law and Technology*, 3(2): 101–111.

Yurtoglu, B.B. (2000) Ownership, control and performance of Turkish listed firms. *Empirica*, 27(2): 193–222.

CHAPTER 40

Government Corruption and Transactional Impediments in Emerging Markets

Mark D. Griffiths and Jill R. Kickul

CONTENTS

40.1 INTRODUCTION

By 2015, it is estimated that the combined gross domestic product (GDP) of emerging-market nations will surpass that of the top 20 developed economies. The current World Bank definition of a developing country is one in which the gross national income (GNI) per capita is $10,725 or less. Over 141 countries, representing 84% of the world's population, meet this criterion.

Vital Wave Consulting (VWC), a company specializing in the information technology and telecommunications industries, performs considerable research in emerging-market business growth, including market research. VWC (2008a) reports that while developing countries have always had a larger population than developed countries, emerging-market incomes and purchasing power have increased significantly since the 1990s. Further, advances in international transportation, finance, and communications have turned developing countries into viable markets. The company estimates that from 1998 to 2012 inflation-adjusted GDP is expected to grow by 5.7% in developing countries versus 2.5% in developed countries. Within emerging markets, an estimated 69% of the workforce employed in the formal sector work in micro, small, or medium-sized companies.

That entrepreneurial growth is crucial to national economic growth is often overlooked. According to the World Bank Group, while the share of small- and medium-sized enterprises (SMEs) in the total economic activity varies across countries and even, for any given country over time, they play a substantial role in virtually every economy and represent a large segment of the private sector. It is reported that:

> [I]n low-income countries with GNP per capita between $100 and $500, SMEs account for over 60 percent of GDP and 70 percent of total employment; in middle-income countries they produce close to 70 percent of GDP and 95 percent of total employment. Even within OECD countries, SMEs comprise the majority of firms and contribute over 55 percent of GDP and 65 percent of total employment.*

The competitiveness of entrepreneurial firms, as well as regions and countries, is closely linked to the innovation competency that drives their

---

* Available at: http://www1.worldbank.org/devoutreach/mar05/textonly.asp?id=286

growth (Dunning, 1993; Hay and Kamshad, 1994). However, when it comes to linking entrepreneurship to growth at the national level, there is a relative void despite recent efforts of the Global Entrepreneurship Monitor (GEM) research program (Reynolds et al., 2001).

Anokhin and Schultze (2008) find in a longitudinal study of 64 nations that efforts to control corruption increase levels of trust in the ability of the state and market institutions to reliably and impartially enforce law and the rules of trade. The authors argue that better control of corruption is associated with increased levels of innovation and entrepreneurship.

In this chapter, we examine the governmental and transactional factors as well as an emerging market's level of competitiveness and economic growth using specific categories of emerging markets. We examine a sample of 72 emerging market countries for which we have complete data. The sample comprises nations from East Asia and the Pacific ($n = 7$), Middle East and North Africa ($n = 5$), Eastern Europe and Central Asia ($n = 23$), Latin America and the Caribbean ($n = 19$), Southern Asia ($n = 5$), and Sub-Saharan Africa ($n = 13$). Please refer to Table 40.1. To distinguish between categories of emerging markets, we employ the VWC (2008b) definitions of strategic opportunity markets ($n = 16$), niche opportunity markets ($n = 46$), and long-term opportunity markets ($n = 10$) to examine the relationships between levels of government corruption (as defined by Transparency International), costs of doing business (as defined by the World Bank), level of national competitiveness, and growth in GDP per capita. Please refer to Table 40.1, Panels A and B.

The chapter proceeds as follows. In Section 40.2, we discuss the sources of our data and the methods used in this chapter. Section 40.3 presents our preliminary descriptive results while Section 40.4 presents the results from our more in-depth analysis. Section 40.5 concludes with a discussion of our results.

40.2 DATA SOURCES AND METHODS

We obtained our data from a number of data sources available online. The single-largest source of data was the CIA World Fact book (available at: https://www.cia.gov/library/publications/the-world-factbook/). From this source, we obtained the real growth rate in GDP, the purchasing power parity (PPP) GDP per capita in U.S. dollars, the unemployment rate, the percentage of the population living below the poverty line, and the inflation rate; the majority of these variables are calculated as of 2007. It should be noted that, in general, emerging markets tend to be quite fragmented

TABLE 40.1   Panel A: Emerging Market Sample by VWC Category and Location

	Country	VWC Category	Location
1	Cambodia	Niche opportunity	Eastern Asia and Pacific
2	Malaysia	Niche opportunity	Eastern Asia and Pacific
3	China	Strategic opportunity	Eastern Asia and Pacific
4	Indonesia	Strategic opportunity	Eastern Asia and Pacific
5	Philippines	Strategic opportunity	Eastern Asia and Pacific
6	Thailand	Strategic opportunity	Eastern Asia and Pacific
7	Vietnam	Strategic opportunity	Eastern Asia and Pacific
8	Algeria	Niche opportunity	Middle East and North Africa
9	Jordan	Niche opportunity	Middle East and North Africa
10	Morocco	Niche opportunity	Middle East and North Africa
11	Tunisia	Niche opportunity	Middle East and North Africa
12	Egypt	Strategic opportunity	Middle East and North Africa
13	Kyrgyzstan	Long-term opportunity	Eastern Europe and Central Asia
14	Tajikistan	Long-term opportunity	Eastern Europe and Central Asia
15	Albania	Niche opportunity	Eastern Europe and Central Asia
16	Armenia	Niche opportunity	Eastern Europe and Central Asia
17	Azerbaijan	Niche opportunity	Eastern Europe and Central Asia
18	Bosnia and Herzegovina	Niche opportunity	Eastern Europe and Central Asia
19	Bulgaria	Niche opportunity	Eastern Europe and Central Asia
20	Croatia	Niche opportunity	Eastern Europe and Central Asia

21	Estonia	Niche opportunity	Eastern Europe and Central Asia
22	Georgia	Niche opportunity	Eastern Europe and Central Asia
23	Hungary	Niche opportunity	Eastern Europe and Central Asia
24	Kazakhstan	Niche opportunity	Eastern Europe and Central Asia
25	Latvia	Niche opportunity	Eastern Europe and Central Asia
26	Lithuania	Niche opportunity	Eastern Europe and Central Asia
27	Macedonia, FYR	Niche opportunity	Eastern Europe and Central Asia
28	Moldova	Niche opportunity	Eastern Europe and Central Asia
29	Mongolia	Niche opportunity	Eastern Europe and Central Asia
30	Poland	Niche opportunity	Eastern Europe and Central Asia
31	Romania	Niche opportunity	Eastern Europe and Central Asia
32	Slovakia	Niche opportunity	Eastern Europe and Central Asia
33	Russia	Strategic opportunity	Eastern Europe and Central Asia
34	Turkey	Strategic opportunity	Eastern Europe and Central Asia
35	Ukraine	Strategic opportunity	Eastern Europe and Central Asia
36	Argentina	Niche opportunity	Latin America and Caribbean
37	Chile	Niche opportunity	Latin America and Caribbean
38	Costa Rica	Niche opportunity	Latin America and Caribbean
39	Dominican Republic	Niche opportunity	Latin America and Caribbean
40	Ecuador	Niche opportunity	Latin America and Caribbean
41	El Salvador	Niche opportunity	Latin America and Caribbean
42	Guatemala	Niche opportunity	Latin America and Caribbean

(continued)

TABLE 40.1 (continued) Panel A: Emerging Market Sample by VWC Category and Location

	Country	VWC Category	Location
43	Guyana	Niche opportunity	Latin America and Caribbean
44	Honduras	Niche opportunity	Latin America and Caribbean
45	Jamaica	Niche opportunity	Latin America and Caribbean
46	Nicaragua	Niche opportunity	Latin America and Caribbean
47	Panama	Niche opportunity	Latin America and Caribbean
48	Paraguay	Niche opportunity	Latin America and Caribbean
49	Peru	Niche opportunity	Latin America and Caribbean
50	Uruguay	Niche opportunity	Latin America and Caribbean
51	Venezuela	Niche opportunity	Latin America and Caribbean
52	Brazil	Strategic opportunity	Latin America and Caribbean
53	Colombia	Strategic opportunity	Latin America and Caribbean
54	Mexico	Strategic opportunity	Latin America and Caribbean
55	Nepal	Long-term opportunity	Southern Asia
56	Sri Lanka	Niche opportunity	Southern Asia
57	Bangladesh	Strategic opportunity	Southern Asia
58	India	Strategic opportunity	Southern Asia
59	Pakistan	Strategic opportunity	Southern Asia
60	Burkina Faso	Long-term opportunity	Sub-Saharan Africa
61	Ethiopia	Long-term opportunity	Sub-Saharan Africa
62	Kenya	Long-term opportunity	Sub-Saharan Africa
63	Mozambique	Long-term opportunity	Sub-Saharan Africa

64	Nigeria	Long-term opportunity	Sub-Saharan Africa
65	Tanzania	Long-term opportunity	Sub-Saharan Africa
66	Uganda	Long-term opportunity	Sub-Saharan Africa
67	Angola	Niche opportunity	Sub-Saharan Africa
68	Cameroon	Niche opportunity	Sub-Saharan Africa
69	Lesotho	Niche opportunity	Sub-Saharan Africa
70	Mauritius	Niche opportunity	Sub-Saharan Africa
71	Namibia	Niche opportunity	Sub-Saharan Africa
72	South Africa	Strategic opportunity	Sub-Saharan Africa

Panel B: Emerging Market Summary Statistics

Location	Number of Countries	VWC Category	Number of Countries
Eastern Asia and Pacific	7	Strategic opportunity	16
Middle East and North Africa	5	Niche opportunity	46
Eastern Europe and Central Asia	23	Long-term opportunity	10
Latin America and Caribbean	19		
Southern Asia	5		
Sub-Saharan Africa	13		
Total	72		72

and accordingly, it is often difficult to obtain accurate statistics on such variables as unemployment, income, and percentage of the population living below the poverty line. Some of the data are estimates from as early as 2001 and 2004. Unemployment numbers are official government statistics, which may or may not include population segments that are underemployed. Certain nations' statistics carry warnings such as the following for Burkina Faso: "a large part of the male labor force migrates annually to neighboring countries for seasonal employment;" Angola: "extensive unemployment and underemployment affecting more than half the population (2001 est.);" Jordan: "13.5% official rate; unofficial rate is approximately 30% (2007 est.);" and El Salvador: "6.2% official rate; but the economy has much underemployment (2007 est.)."

According to VWC (available at: www.vitalwaveconsulting.com), strategic opportunity markets are the largest and most economically attractive for a multinational corporation that is looking to grow its customer base. These markets have a population of over 40 million and strong real GDP growth. GNI per capita for these nations is over $2000 per year in terms of PPP. Niche opportunity markets are similar countries but with a population under 40 million and also have average incomes over $2000 in PPP terms and strong real GDP growth. These markets are often seen by multinational companies as opportunities to grow their markets on a smaller scale, or they may be "gateways" to larger nearby markets. Long-term opportunity markets are the least attractive markets to a multinational corporation. These markets exhibit a low standard of living with a GNI per capita under $2000 per year. VWC states that in these countries, persistent poverty, corruption, and political instability may be hampering economic growth. These countries may become viable markets in the long term with the benefit of consistent political and economic reform.

We supplemented our data with information from Transparency International (TI). TI, founded in 1993, is an organization leading the fight against corruption. It brings together relevant participants from government, civil society, business, and the media to promote transparency in elections, public administration, procurement, and business. TI's global network of chapters and contacts also use advocacy campaigns to lobby governments to implement anticorruption reforms. The organization does not undertake investigations of alleged corruption or expose individual cases, but at times will work in conjunction with other organizations that do undertake such activities.

Specifically, we extracted from their 2008 Corruption Perception Index the relevant evaluation for each country for which we had complete data. This index focuses on corruption in the public sector and defines corruption as the abuse of public office for private gain. The surveys used in compiling the index ask questions that relate to the misuse of public power for private benefit, for example, bribery of public officials, kickbacks in public procurement, embezzlement of public funds, or questions that probe the strength of anticorruption policies, thereby encompassing both administrative and political corruption. This broader definition of corruption includes nepotism, cronyism, insider trading, and issues involving government/administrative discretionary actions such as the granting (or not) of licenses and permits.

The 2008 index, which evaluates 180 nations, draws on 13 different polls and surveys from 11 independent institutions taken in the current and previous year. To qualify for inclusion, the data must be well documented and sufficient to permit a judgment of its reliability. All sources must provide a ranking of nations and must measure the overall extent of corruption. The expertise reflected in the index draws upon the understanding of corrupt practices held by those based in both the industrialized and developing world. The surveys also use two types of samples both nonresident and resident.

TI notes that resident experts' viewpoints correlate well with those of nonresident experts. The Corruption Perception Index is scaled from 0 (high levels of perceived corruption) to 10 (low levels of perceived corruption). Our purpose in including these data was to highlight the costs of corruption as put forth by Sullivan and Shkolnikov (2004) who contend that, among other things, corruption leads to misallocation of resources, a lack of competitiveness and efficiency, lower public revenues for essential goods and services, lower productivity and lower levels of innovation, and lower growth and private sector employment rates.

Many researchers (Donckels and Courtmans, 1990; Butler and Hansen, 1991; Drori and Lerner, 2002; Reynolds et al., 2004; Reynolds, 2007) have noted that transactional impediments, in other words, the ease of doing business within a region or nation can affect the growth rates and success of new ventures. The World Bank also ranks economies based upon the perceived ease of doing business, from 1 to 178, with first place being the best. A high ranking on the ease of doing business index means the regulatory environment is conducive to new business operations. These data were obtained from *Doing Business—The World Bank Group* (available at: www.doingbusiness.org/MethodologySurveys/methodolog-ynote.aspx). On the World Bank Web site, it states

> The *Doing Business* methodology offers several advantages. It is transparent, using factual information about what laws and regulations say and allowing multiple interactions with local respondents to clarify potential misinterpretations of questions. Having representative samples of respondents is not an issue, as the texts of the relevant laws and regulations are collected and answers checked for accuracy. The methodology is inexpensive and easily replicable, so data can be collected in a large sample of economies. Because standard assumptions are used in the data collection, comparisons and benchmarks are valid across countries.

The index is the ranking of the simple average of country percentile rankings on each of the 10 topics (starting a firm, licenses, employing workers, registering property, getting credit, protecting investors, paying taxes, trading across borders, enforcing contracts, and closing a firm) each of which, in turn, is the simple average of the percentile rankings of its component indicators. If a nation has no laws or regulations covering a specific area, e.g., bankruptcy, it receives a "no practice" or "not possible" mark. Similarly, an economy receives a similar mark if regulation exists but is never used in practice or if competing regulations prohibit such practice. Either way, such marks puts the country at the bottom of the rankings on the relevant indicator.

To proxy for national competitiveness, we downloaded each nation's Global Competitiveness Scores and Index rank from the World Economic Forum (http://www.weforum.org/). Produced in collaboration with leading academics and a global network of research institutes, this report provides a comprehensive data set on a broad array of competitiveness indicators for a large number of industrialized and developing economies. The 2007 edition provided data on 131 economies, accounting for more than 98% of the world's GDP.

In addition to quantitative data from leading international sources, these Index includes the annual results of the Executive Opinion Survey carried out by the World Economic Forum. This Survey is designed to capture the perceptions of thousands of business leaders across the countries covered on topics related to national competitiveness. The Index ranks nations from 1 (highly competitive) to 131 (least competitive).

TABLE 40.2 Correlation Matrix of Principal Variables of Interest

Correlations	GDP per Capita	Government Corruption	Ease of Doing Business	Global Competitiveness Index
GDP per capita	1.000	−0.637***	0.540***	0.685***
Government corruption		1.000	0.606***	−0.653***
Ease of doing business			1.000	0.600***
Global Competitiveness Index				1.000

Note: Indices have been reversed scored to assist in interpretation of the table.
*** Significant at the 1% level (two-tailed).

According to the World Economic Forum, nations can be categorized in terms of their development on the basis of GDP per capita. Stage 1: *Factor-driven economies* realize GDP per capita levels of less than $2000. Firms making the transition from Stage 1 to Stage 2: *Efficiency-driven economies* experience GDP levels between $2000 and $3000. Nations in Stage 2 generally realize $3000–$9000 in GDP per capita. Nations in transition from Stage 2 to Stage 3: *Innovation-driven economies* experience GDP levels between $9000 and $17,000 per capita (see, e.g., Schwab and Porter, 2008). In our sample, 5 nations are in Stage 3, 18 nations including Brazil and Russia are making the transition from Stage 2 to Stage 3, 30 countries including China are at Stage 2, 9 countries including India are making the transition to Stage 2, and remaining 10 nations are at the Stage 1 level of development.

We commence with a brief comparison of the relations between our major variables of interest: GDP per capita, the TI corruption index, the ease of doing business index, and the World Economic Forum (WEF) Competitiveness Index. Please refer to Table 40.2.

Consistent with prior expectations, higher levels of GDP per capita are associated with lower levels of government corruption, greater ease of doing business, and higher scores of global competitiveness. These results are consistent with earlier international studies (see, e.g., Broadman and Recanatini, 2002; Tanzi and Davoodi, 1998, 2001, 2002) that find that corruption is correlated with low GDP, reduced investment, low growth, low levels of education, and increased fractionalization.

40.3 PRELIMINARY RESULTS

We commence by categorizing the emerging markets in our sample according to the VWC groupings mentioned earlier. We then present descriptive statistics (mean and standard deviation) for the major variables of interest, namely, government corruption, an estimate of national competitiveness on a global scale, the ease of doing business, unemployment levels, the percentage of the population living below the poverty line and an estimate of GDP per capita. Please refer to Table 40.3.

In general, we find that the average GDP per capita is greater (but not significantly so) in the niche opportunity markets than in the strategic opportunity markets although this may be a function of the difference in the size of the population (less than 40 million versus greater than 40 million). Both markets have approximately the same percentage size of the population living below the poverty line although unemployment levels tend to be higher (but not statistically significantly so) in the niche opportunity markets. Interestingly, we find that niche opportunity markets have higher corruption scores, suggesting that the level of corruption is lower in these nations. Although according to Transparency International, any nation with a corruption score below 5.0 indicates serious corruption issues in the public sector. Only 53 of the 180 countries evaluated scored greater than 5.0. In our sample, nine nations scored above this level: Chile, Uruguay, Costa Rica in Latin America and the Caribbean, Estonia,

TABLE 40.3   Mean and Standard Deviation Values for Corruption, Competitiveness, Ease of Doing Business, Unemployment, Percentage of Population Living below the Poverty Line, and GDP per Capita by VWC Category

	Strategic Opportunities	Niche Opportunities	Long-Term Opportunities	Test of Equality of Group Means (F-Statistic)
Corruption	3.15	3.72	2.49	4.099**
	(0.84)	(1.30)	(0.58)	
Competitiveness	4.14	3.96	3.41	9.872***
	(0.28)	(0.40)	(0.22)	
Doing business	90.19	82.53	119.00	2.668*
	(38.11)	(40.08)	(32.05)	
Unemployment	7.73	11.22	29.33	8.487***
	(5.33)	(10.31)	(26.03)	
Poverty level	27.39	26.93	52.47	9.848***
	(13.59)	(14.50)	(14.92)	
GDP per capita	$6,781	$8,693	$1,542	7.183***
	($4,177)	($5,189)	($454)	

Note: *, **, *** Significant at the 10%, 5%, 1% level (respectively).

Hungary, Latvia, Slovakia in Eastern Europe, Mauritius in Sub-Saharan Africa and Jordan in the Middle East.

The nations categorized as long-term opportunities are statistically significantly worse off than countries in the other two categories by all measures. The government is more corrupt, the level of competitiveness is lower, the cost of doing business is higher, poverty is rampant, and GDP per capita is the smallest.

Thus, it would appear that both the strategic opportunity countries and the countries in the niche opportunity category do represent the best emerging markets most conducive to new firm creation and ease of entry into the market. Similarly, it seems clear that the long-term opportunity countries are indeed only viable markets if consistent political and economic reform takes place.

40.4 DETAILED ANALYSIS

We next performed a discriminant function analysis, with the purpose to determine how well our governmental, economic, transactional variables along with the proxy for corruption predicted the three specific categories of emerging markets. We found that the combination of variables were indeed "discriminating" factors that had various levels depending on the nature of the emerging market (as confirmed earlier in our descriptive analyses).

Our analysis also indicates that there are two discriminate functions at work. The first comprises the economic factors and the transactional impediments to business (Wilks' Lambda = 0.46, $p < 0.01$), while the second reflects the role of corruption (Wilks' Lambda = 0.72, $p < 0.01$). Indicated below in Table 40.4 is the Structure Matrix table of the correlations of each variable with each discriminant function. These correlations serve in a fashion similar to loadings in factor analysis. By identifying the largest absolute correlations associated with each discriminant function, the researcher gains insight into how to name each function.

While the structural matrix is useful in determining the overall factor structure, the individual discriminant coefficients capture the relative importance of each of the variables in predicting our two functions, including

$$\text{DisFn}_1 = 0.07\text{Corrsc} - 0.42\text{Compsc} - 0.16\text{EaseofBiz} + 0.59\text{Unemp} + 0.47\text{Poverty} - 0.7\text{GDPpc} + \varepsilon$$

$$\text{DisFn}_2 = 0.84\text{Corrsc} + 1.54\text{Compsc} - 0.22\text{EaseofBiz} + 0.17\text{Unemp} + 0.51\text{Poverty} - 0.59\text{GDPpc} + \varepsilon$$

TABLE 40.4   Structure Matrix for Discriminant Analysis
Functions

Structure Matrix		
	Discriminant Function	
	1	2
Compsc	−0.727*	0.140
Poverty	0.712*	0.224
Unemp	0.682*	−0.016
GDPpc	−0.518*	−0.436
EaseofBiz	0.332*	0.234
Corrsc	−0.315	−0.434*

Notes: Pooled within-groups correlations between discriminating
        variables and standardized canonical discriminant func-
        tions. Variables ordered by absolute size of correlation
        within function.
* Largest absolute correlation between each variable and any
  discriminant function.

where

Corrsc = the nation's corruption score as defined by Transparency
International

Compsc = the nation's competitiveness score as defined by the World
Economic Forum

EaseofBiz = the nation's Ease of Doing Business rank as defined by the
World Bank Group

Unemp = unemployment rate as reported in the CIA World Fact Book

Poverty = the proportion of the nation's population living below the
poverty line rate as reported in the CIA World Fact Book

GDPpc = the nation's GDP per capita

Based on the above analyses, the two most significant factors contributing to our first discriminant function were our baseline economic factors of GDP and unemployment and not necessarily the transactional impediments to business. In the second discriminant function and as indicated by its label, corruption overwhelmingly comprised this second factor.

A benefit in conducting these analyses is that the captured discriminant scores can be employed in a fashion similar to using predicted values in regression. Since we also have data on the 2007 Innovation Index published by the World Economic Forum (available at: www.weforum.org), we can then determine the role and relationship each of our discriminating factors has on a country's innovation profile. To illustrate, we regressed innovation on both of the calculated discriminating scores. We found that both significantly predicted innovation ($\beta = 0.69$, $p < 0.01$ for function 1, economic/transactional impediments; $\beta = 0.20$, $p < 0.01$ for function 2, corruption). However, the role of corruption predicted innovation *above and beyond* the role of our economic and transactional impediments ($\Delta R^2 = 0.04$; $p = 0.025$).

40.5 DISCUSSION

A major concern in any study examining corruption, growth, and innovation is the role of reverse causality. Corruption has been shown (Tanzi and Davoodi, 1998, 2002) to reduce the productivity of public investment and dampen growth. Such effects would be seen in the differing GDP per capita in countries ranked differently by Transparency International.

A strong relationship of higher per capita incomes (Treisman, 2000) has been found to be associated with reduced corruption: "Rich countries are perceived to be less corrupt than poor ones." However, Kaufmann and Kraay (2002) find evidence of negative feedback from rising per capita incomes toward better governance outcomes. They explain this somewhat peculiar result by arguing that higher incomes do not necessarily lead to demands for better institutions, but may be accompanied, initially by "crony capitalism," elite influence, regulatory capture, or "state capture"; these phenomena have been observed in varying degrees in East Asia, Latin America, and the transition economies of Central and Eastern Europe, even during upswings in output.

This begs the question: "Does a poor inefficient economy generate the necessary conditions for corruption?" The answer is "possibly." Suppose, for example, that bureaucratic efficiency increases with wages. This could occur in situations where government workers are paid higher wages to have them refuse bribes and inducements. Poorer countries in general may have difficulty keeping government wages high relative to corrupt alternatives. Thus, as a country gets richer, it can pay higher wages and becomes less corrupt but it is the better economic conditions that create higher wages *not the reverse*. See, for example, Klick and Tabarrok (2005) and Glaeser and Saks (2006).

Sullivan and Shkolnikov (2004) contend that, among other things, corruption leads to misallocation of resources, and a lack of competitiveness and efficiency, lower public revenues for essential goods and services, lower productivity and lower levels of innovation, and lower growth and private sector employment rates. Further, Sullivan (2000) asserts that

> "Firms that refuse to participate in corrupt transactions may find themselves forced out of certain markets ... (D)omestic firms, especially small businesses are much more vulnerable ... they (too) can leave the market ... they can emigrate into the informal or underground economy." (p. 1)

As an example, studies in Slovakia, one of the niche opportunity markets in our sample and ranking 52nd out of the 180 countries studied in terms of the greatest level of corruption—i.e., one of the nine least corrupt nations in our sample—was found to have an underground economy estimated between 20% and 40% of GDP (Sicakova, 1999). This implicitly suggests that the existence of corruption does not necessarily dampen growth totally. Because corrupt transactions are by definition illegal, or at the very least, part of the "informal economy," they are not recorded and thus impossible to measure directly.

Thus, one should be careful in drawing conclusions about the relation between corruption and growth. Previous studies have found some causal effect particularly from higher income and better education to less corruption but the effect of corruption on growth, especially given the measurement difficulties involved, is relatively weak.

40.6 CONCLUSION

The institutional context and environment that includes a confluence of the governmental, economic, and business transactional factors influences an emerging market's characteristics and overall profile. While the purpose of our chapter was to reveal the initial role that many of these factors had on defining each of our emerging markets, we were also able to uncover the factors that contributed to comprising the profile of these emerging markets. Although the descriptive analyses revealed the differences between our groups, our follow-up discriminant function analyses revealed the most salient were at the economic and governmental levels. Lastly, we were also able to discern how our discriminating factors contributed to a nation's innovation level. While much work remains, it is our hope that additional research incorporates a diverse set of governments and economies to better understand what changes and initiatives that may need to be proposed and implemented to foster a nation's innovative prowess and future economic growth.

REFERENCES

Anokhin, S.M. and Schulze, W.S. (2008) Entrepreneurship, innovation, and corruption. *Journal of Business Venturing*, in press.

Broadman, H.G. and Recanatini, F. (2002) Corruption and policy: Back to the roots. *Journal of Policy Reform*, 5(1): 37–49.

Butler, J. and Hansen, G. (1991) Network evolution, entrepreneurial success, and regional development. *Entrepreneurship & Regional Development*, 3(1): 1–16.

Donckels, R. and Courtmans, A. (1990) Big brother is watching over you: The counseling of growing SMEs in Belgium. *Entrepreneurship & Regional Development*, 2(3): 211–224.

Drori, I. and Lerner, M. (2002) The dynamics of limited breaking out: The case of the Arab manufacturing businesses in Israel. *Entrepreneurship & Regional Development*, 14(2): 135–154.

Dunning, J.H. (1993) *The Globalization of Business*. Routledge, London.

Glaeser, E.L. and Saks, R.E. (2006) Corruption in America. *Journal of Public Economics*, 90(6–7): 1053–1072.

Hay, M. and Kamshad, K. (1994) Small firm growth: Intentions, implementation and impediments. *Business Strategy Review*, 5(3): 49–68.

Kaufmann, D. and Kraay, A. (2002) Growth Without Governance. Economia, Fall, 2002. World Bank, Washington, DC.

Klick, J. and Tabarrok, A. (2005) Using terror alert levels to estimate the effect of police on crime. *The Journal of Law & Economics*, 48(1): 267–279.

Reynolds, P. (2007) *Entrepreneurship in the United States*. Springer, New York.

Reynolds, P.D., Camp, S.M., Bygrave, W.D., Autio, E., and Hay, M. (2001) Global Entrepreneurship Monitor—2001 Executive Report. Babson College, IBM, Kauffman Center for Entrepreneurial Leadership and London Business School, available through http://www.gemconsortium.org.

Reynolds, P., Bygrave, W., and Autio, E. (2004) Global Entrepreneurship Monitor: 2003 Summary Report. Babson College, Babson Park, MA.

Schwab, K. and Porter, M.E. (2008) *The Global Competitiveness Report 2008–2009*. World Economic Forum, Geneva, Switzerland.

Sicakova, E. (1999) Transparency and Hidden Economy—Mutually Contradicting Phenomena. Center for Economic Development, Slovak Republic.

Sullivan, J. (2000) Development as a Two-Way Street: Merging Social Progress with Financial Profits. Center for International Private Enterprise, U.S. Chamber of Commerce, Sixth Annual Harvard International Development Conference, Washington, DC.

Sullivan, J. and Shkolnikov, A. (2004) Combating Corruption: Private Sector Prespectives and Solutions. *Economic Reform* (September).

Tanzi, V. and Davoodi, H. (1998) Corruption, Public Investment, and Growth. IMF Working Paper 97/139. International Monetary Fund, Washington, DC.

Tanzi, V. and Davoodi, H. (2001) Corruption, growth, and public finances. In: Arvind K. Jain (Ed.), *The Political Economy of Corruption*. Taylor & Francis, London.

Tanzi, V. and Davoodi, H. (2002) Corruption, Public Investment and Growth. In: G.T. Abed and S. Gupta (Eds.), *Governance, Corruption and Economic Performance*. International Monetary Fund, Washington, DC.

Treisman, D. (2000) The causes of corruption: A cross-national study. *Journal of Public Economics*, 76(3): 399–457.

Vital Wave Consulting. (2008a) 10 Facts about Emerging Markets, Mimeo available at: www.vitalwaveconsulting.com.

Vital Wave Consulting. (2008b) Emerging Markets Definition and World Market Groups, Mimeo available at: www.vitalwaveconsulting.com.

Index

A